DAMAGE AND FAILURE OF INTERFACES

PROCEEDINGS OF THE FIRST INTERNATIONAL CONFERENCE ON DAMAGE AND FAILURE OF INTERFACES – DFI -1 / VIENNA / AUSTRIA / 22-24 SEPTEMBER 1997

Damage and Failure of Interfaces

Edited by

HANS-PETER ROSSMANITH

Institute of Mechanics, Vienna University of Technology, Austria

A.A. BALKEMA / ROTTERDAM / BROOKFIELD / 1997

The texts of the various papers in this volume were set individually by typists under the supervision of each of the authors concerned.

Published by
A.A. Balkema, P.O. Box 1675, 3000 BR Rotterdam, Netherlands (Fax: +31.10.413.5947)
A.A. Balkema Publishers, Old Post Road, Brookfield, VT 05036-9704, USA (Fax: 802.276.3837)

ISBN 90 5410 899 1

Printed in the Netherlands

Damage and Failure of Interfaces, Rossmanith (ed.) © 1997 Balkema, Rotterdam, ISBN 90 5410 899 1

Table of contents

Failure of fibrous composites

Dynamics and impact

Damage and Failure of Interfaces, Rossmanith (ed.) © *1997 Balkema, Rotterdam, ISBN 90 5410 899 1*

Preface

This volume contains selected papers presented at the International Conference on *Damage and Failure of Interfaces* which was held in Vienna, Austria, at the Vienna University of Technology, during September 22-24, 1997.

Interface problems arise in almost all fields of engineering, and the papers included in the present volume demonstrate not only the diversity of interface problems, but also the wide range of applicability, covering mining engineering problems as well as polymer composites and micro-electronic devices.

Research regarding interface phenomena, particularly modelling of static and dynamic behaviour has undergone explosive growth during the last three decades, primarily due to the development and availability of powerful high resolution techniques and advanced numerical modelling methods based on ever faster computer hardware and software.

As interfaces constitute surfaces, and these surfaces are in contact either directly or via an intermediate layer, practical and theoretical problems quickly arise which are associated with interface strength, reliability and integrity. The solution of these problems requires an interdisciplinary approach based on the expertise of chemists, physicists, surface technologists, materials scientists, etc.

Interface problems arise in connection with all engineering materials spanning several orders of magnitude in scale, from nanostructures to macro-size engineering structures. Intensive studies have revealed that the properties of polycrystalline materials and composite structures are determined to a large extent by those of their internal interfaces. In fact, the study of the structural integrity of polycrystalline materials and fibrous and laminated structures is often reduced to the study of the behaviour of their interfaces.

In polymer science and engineering, bonding of polymers and metals has become common in many applications because of the improved performance and ease of manufacture associated with polymer/metal structures. In structural engineering, the classical fastening techniques such as welding and jointing are gradually being replaced by polymer-based adhesive bonded sheet-metal joints, which often show improved fatigue resistance and superior dielectric properties.

The fracture resistance of interfaces is important as it controls the integrity of the structure. Today, thermal fatigue of solder joints leading to the development of fatigue cracks or progressive growth of defects is one of the most insidious failures threatening the reliability and durability of many micro-electronic components.

From a geometrical point of view interfaces come in many varieties among which the mathematically sharp straight perfectly bonded elastic interface is the simplest case. More complex interfaces are characterised by non-straight, possibly undulating surfaces, where certain parts of the two media are in direct contact and the remainder of the interface gap is filled with a filler material, which displays a different material behaviour. In the theory of elasticity, despite having a well established history spanning several decades, the singular character of many interface problems of a more general type is not known. This pertains particularly to coupled problems, where elasticity and thermal, electrical, electro-magnetical, hygro- and rheological conditions apply at the interface.

Subinterface cracks, i.e. cracks which develop near an interface between dissimilar materials, have long been of interest in fracture mechanics. Also, cracks which develop from a boundary discontinuity, such as at the end of surface stiffening elements, pose considerable threat to the integrity of a structure.

Ductile-brittle interfaces occur in a variety of advanced engineering material combinations, such as composites, cements, polycrystalline intermetallic alloys, etc. Interface failure in such materials is a common occurrence caused by the propagation and coalescence of pre-existing micro-cracks along and off the interface. Although numerous contributions may be found in the literature for the elastic-elastic interface, non-elastic problems are at a preliminary stage of development at this time. With elastic-plastic material interfaces, ultimate failure is most often preceded by slow stable crack extension. In fibrous composites, bridging of macro-cracks by only partially pulled out fibres is a significant source of toughness and this corresponds to the increase in toughness due to stretching of molecular chains in certain polymers.

From a wave dynamics point of view fully or partially bonded interfaces between two solids or a solid and a liquid can support a range of interface waves with a number of surprising characteristic features. Such interface waves, resembling in many ways generalised Rayleigh-waves, may induce separation of the interface. The resulting dynamic wave field is highly complicated and experimental and numerical solutions are still missing even for the most simple problems. As faults and joints in geology represent interfaces of a general kind, the interaction of such structural geological features with bulk and surface waves is of greatest importance in seismology and in mining in order to understand and hopefully prevent large scale destruction on the surface or in mining environments.

This volume contains 57 contributions by 147 authors from 17 countries. Three plenary lectures introduce into *Fracture mechanics of interfaces* (F. Erdogan, USA), *Numerical simulation methodologies for dynamic crack propagation in homogeneous and non-homogeneous materials* (T. Nishioka, Japan), and *Dynamic properties of interfaces* (L. R. Myer, USA).

The following seven chapters comprise:

– Basics: Contributions consider stress singularities associated with terminating cracks, interface corners, axisymmetric deformation, free edge effects, time-dependent singular stresses, analogy between interface crack problems and punch problems and the effect of micro-voids on the stress intensity factor at interface cracks, as well as moving discontinuities.

– Interface degradation and delamination: Seven papers deal with matrix cracking and delamination, identification and prediction of interface damage, cyclic loading, and three papers are devoted to delamination and cracking due to global buckling, blistering and plate bending.

– Thermal problems: Five papers study thermal problems associated with fibre reinforced and polycrystalline materials and laminates, inclusions, barrier coatings and composite nozzles with interfaces.

– Fibrous composites: Four papers focus on single and multiple fibre breaks, creep, finite element modelling, stress analysis.

– Dynamics and impact: Ten papers concentrate on wave interaction and impact problems associated with interfaces featuring elastic-plastic interfaces, intersonic interface fracture, fault dynamics, interaction of stress waves with mining discontinuities, flexural vibrations, impact damage at single and repeated impacts, gas-gun impact welding and ordnance impact.

– Debonding and pull-out: Eight contributions cover the phenomenon of fibre-pull-out, analysis of micromechanical tests, acoustic emission, trapping mechanisms and dowel action.

– Miscellaneous: The seven concluding papers pertain to biomechanics, bone-prosthesis-interface, interface fracture in simulations of sequential mining, modelling of damage in finger joints of wood, welded steel joints and mixed-mode fatigue cracking.

The range of papers contributed gives testimony to the wide variety of interface problems in engineering. It is hoped that this collection of papers will stimulate further work and progress in this exciting and extremely up-front field of research.

The Editor would like to acknowledge the effect of all authors in providing the manuscripts in time and making the conference a great success. The financial support of the Austrian National Science Foundation (FWF) under project number P 10326 GEO is kindly appreciated. The Editor is particularly grateful to Dr Natalia Kouzniak who has been instrumental in the organisation of the DFI-1 Conference.

H. P. Rossmanith
Vienna, September 1997

Plenary lectures

Damage and Failure of Interfaces, Rossmanith (ed.) © *1997 Balkema, Rotterdam, ISBN 90 5410 899 1*

Fracture mechanics of interfaces

F. Erdogan
Department of Mechanical Engineering and Mechanics, Lehigh University, Bethlehem, Pa., USA

ABSTRACT: The processing of composite materials and structures is a very old concept. In designing new materials it is generally used in order to take advantage of favorable properties of different homogeneous constituents. The distinguishing feature of these materials is that they generally consist of two bonded dissimilar phases having a particulate, layered or fiber or filament-reinforced structure. In applications the constituents invariably are metals, polymers or ceramics. In many cases the structural performance of these materials depends very heavily on the mechanical behavior of the interfaces or interfacial regions between the two phases. In particular, the fracture-related failures in most cases seem to initiate at the interfaces. Consequently, the interface fracture is a rather common mode of failure that needs to be considered in designing with composite materials. In this article we review the progress made in recent past on the fracture mechanics of interfaces. Among the topics covered are the classification, description and analysis of various models used to represent the interfacial regions and the adherents and the techniques used in each case to calculate the crack driving force.

1. INTRODUCTION

In recent past, concerns with mechanical failures initiating largely at interfacial regions in such engineering materials as modern composites, thermal barrier coatings and a wide variety of other bonded materials have led to rather extensive studies for the purpose of understanding the interaction between the flaws that may exist in these regions and the applied loads and the environmental conditions. In applications the optimal design of interfacial regions generally involves tradeoffs between strength and toughness. For example, the untreated carbon is known to have very poor adhesion to epoxy and good adhesion can be achieved by using proper coupling agents, e.g., a variety of silanes (Manson 1985). However, joints of such high interfacial strength usually also have poor toughness. Similarly, in fiber-reinforced composites considerations regarding fiber integrity require a relatively weak or compliant interface. Thus, in studying the failure of interfacial regions, mechanical and strength parameters are expected to play a major role. In particular, in fracture related failures the crack driving force would heavily depend on the thermomechanical properties of the adherents, the thickness, structure and properties of the interfacial zone and the size, location and the orientation of the crack as well as on the nature of the applied loads and the geometry of the medium (Hutchinson and Suo 1992, Rice et al. 1989, Erdogan 1972).

In studying the fracture mechanics of bonded materials it is generally assumed that the medium is piecewise homogeneous. Thus, the interfacial zone is replaced by an ideal interface across which the physical properties of the medium exhibit a jump discontinuity. The consequences of such a model are the anomalous behavior of crack tip stress oscillations (Erdogan 1963, Erdogan 1965, Rice and Sih 1965, England 1965) and nonsquare-root stress singularities (Cook and Erdogan 1972, Erdogan and Biricikoglu 1973). Despite this, the model is very useful and for many applications is quite adequate (Hutchinson and Suo 1992). On the other hand, a close examination of bonded materials would indicate that in nearly all cases the interfacial region has a structure which is generally different from that of the adherents and, for physical or analytical reasons, may require different modeling. In some cases the interfacial region may simply consist of the reaction zone with very small thickness and generally inhomogeneous properties due to interdiffusion and compound formation. In other bonded materials interfacial regions with distinct structure and thickness may develop as a result of surface preparation and the technique used for processing. In still other materials, such as adhesive joints and certain coatings, separately introduced interfacial region is part of the material design. In some cases these regions may have a columnar or a lamellar structure (Manson 1985, Ishida 1984) and in some others may exhibit an inhomogeneous behavior due to material property grading in thickness direction (Kurihara et al. 1990, Lee et al. 1996).

In this article after some brief remarks on the fracture criteria that may be used in studying the failure of bonded materials, the existing models for the interfacial zones and adherents will be reviewed and the salient features of each model will be briefly discussed. For a somewhat more thorough examination of the field the reader may refer to Hutchinson and Suo (1992) and the review articles in Rühle et al. (1990).

2. ON THE FRACTURE CRITERIA

Despite quite considerable advances made during the past four decades on the mechanics of fracture, the application of the technique, particularly to bonded materials, is still far from being satisfactory. This is largely due to the lack of sufficiently realistic models for the underlying fracture processes and the analytical difficulties involved in solving the related mechanics problems. In the absence of large-scale inelastic deformations, the concept of energy balance as proposed by Griffith (1921, 1924) and expanded by Irwin (1957) remains to be valid and applicable to fracture involving interfaces. This is mostly due to the fact that deformations and stresses near the tip of a propagating interface crack are self-similar and are controlled by the stress intensity factors. Consequently, the energy balance theory would lead to a single parameter fracture criterion provided the interface is the weak fracture plane and the shear component of the external load is sufficiently small so that the effect of crack surface contact may be neglected.

The fracture of interfaces as any other fracture has two aspects. One is the fracture *initiation* which is a local phenomenon and takes place at the tip of a crack or flaw when certain necessary critical conditions are met. These conditions may be based on maximum (tensile) stress or strain as in rupture, maximum resolved shear stress along a slip plane as in dislocation emissions or on energy considerations as in decohesion at the crack tip. Along the crack front Rice defines the (critical) strain energy release rates $\mathcal{G}_{c\ell}$ and $\mathcal{G}_{di}$ necessary for the inception of cleavage and dislocation emission, respectively (Rice et al. 1990). If $\mathcal{G}_{c\ell} > \mathcal{G}_{di}$, then the dislocation emission is more likely to occur, crack tip would be blunted and the ensuing fracture would be by some ductile mechanism. On the other hand if $\mathcal{G}_{di} > \mathcal{G}_{c\ell}$ the crack tip would remain atomically sharp leading to brittle interfacial fracture. One would argue that the former is a stress and the latter an energy-based criterion. In the case of cleavage decohesion $\mathcal{G}_{c\ell}$ is essentially the work of adhesion W_a which may be expressed as (Rice et al. 1990, Derby 1990)

$$W_a = \gamma_{s1} + \gamma_{s2} - \gamma_{12} \quad (1)$$

where γ_{s1} and γ_{s2} are surface free energies of the adherents prior to bonding and γ_{12} is the free energy of the interface before separation.

The second aspect of the interfacial fracture is the further growth of the interface crack which always requires a global energy balance. In order to propagate the crack the rate of externally added or internally released energy must exceed the rate of all dissipative energies associated with the crack growth. The latter constitutes the resistance of the solid to fracture, the major component of which is plastic or viscoplastic work. Other contributions to the interface toughness come from the work of adhesion, resistance induced by contacting and locking of crack surface asperities, (Hutchinson and Suo 1992) and impurity and solute segregation in the interfacial region. In interface fracture the toughness appears to be significantly influenced by the in-plane shear component of the stress state at the crack tip (Hutchinson and Suo 1992, Evans et al. 1990). The generally accepted measure of this influence is the so-called phase angle ψ defined by

$$\psi = \arctan\frac{k_2}{k_1} \quad (2)$$

where k_1 and k_2 are the modes I and II stress intensity factors (or the real and imaginary parts of the complex stress intensity factor) at the interface crack tip. The calculated plastic zone size (as well as the measured toughness) increases with increasing ψ.

The global energy balance or fracture criterion consists of the comparison of $\mathcal{G}$, the crack driving force representing the rate of input energy and $\mathcal{G}_c$, the "toughness" representing the fracture resistance. In most cases of small scale yielding, $\mathcal{G}$ can be calculated with reasonable accuracy by using linear elasticity and, for self-similarly growing cracks and constant ψ , $\mathcal{G}_c$ may be considered as being a material constant. If ψ varies with the crack length then the dependence of $\mathcal{G}_c$ on the crack length must also be taken into account in applying the fracture criterion. In the presence of large-scale inelastic deformations, an R-curve type approach along with a full field numerical solution may thus be necessary (Shih et al. 1990, Varias et al. 1990).

3. MODELING OF INTERFACIAL REGIONS AND ADHERENTS

The mechanical strength of interfacial zones are heavily influenced by their microstructure which, in turn, is determined by factors relating largely to processing techniques such as, for example, cleanliness and preparation of surfaces, selection of coupling agents, the degree of vacuum, time-temperature profile and bonding pressure. The nominal "thickness" of the interfacial zone may be of the order of a few lattice distances in diffusion bonded materials, submicrons in grain boundaries, and tens of microns in structural adhesive joints and other intentionally introduced interlayers. The modeling of various interfacial zones, therefore, depends partly on one's purpose for analyzing the problem and partly on the structure, thickness an stiffness of the adherents as well as of the interfacial zone. Let the bonded medium consist of two adherents of thicknesses h_1 and h_2 and the interfacial zone of thickness h_0. Most of the existing interfacial zone models may then be classified as follows:

(a) Ideal interface $(h_0 = 0)$
(b) Shear spring $(h_0 \neq 0)$
(c) Shear/tension spring $(h_0 \neq 0)$
(d) Homogeneous continuum $(h_0 \neq 0)$
(e) Inhomogeneous continuum $(h_0 \neq 0)$.

Similarly, in analyzing the problem one may model the adherents as

(i) Membrane
(ii) Plate or shell
(iii) Homogeneous or inhomogeneous continuum

Analytically, the interface models are defined by certain idealized continuity conditions. For example, in two-dimensional problems the interface conditions corresponding to the models (a)-(c), respectively, are

$$u_1^- = u_2^+, \; v_1^- = v_2^+, \; \sigma_{1yy}^- = \sigma_{2yy}^+, \; \sigma_{1xy}^- = \sigma_{2xy}^+, \tag{3}$$

$$u_2^+ - u_1^- = \frac{h_0}{\mu_0}\sigma_{0xy}, \; \sigma_{0yy} = 0, \tag{4}$$

$$u_2^+ - u_1^- = \frac{h_0}{\mu_0}\sigma_{0xy}, \; v_2^+ - v_1^- = \frac{h_0}{E_0^*}\sigma_{0yy}, \tag{5}$$

whereas in models d and e the interfacial layer itself is a continuum and, therefore, is subject to the set of continuity conditions of the form (3) along its boundaries. In (4) and (5) μ_0 and E_0^* are the measure of shear and tensile stiffness of the interfacial region and u, v, σ_{ij}, (i,j = x, y) refer to the displacement and stress components in the usual notation. The superscripts - and + refer to the limiting values from negative and positive y directions, respectively. If at least one of the adherents is approximated by a membrane (e.g., a thin film), then the interface would carry only shear stress and consequently has to be modeled as either an ideal interface or a shear spring. Similarly, in bonded materials consisting of any combination of plate and continuum adherents, the interface must be modeled by (a) or (c). The primary factors influencing the choice of a particular model for the interface and the adherents are generally various ratios of the characteristic joint dimensions and adequacy of the available analytical tools needed to solve the problem.

4. IDEAL INTERFACES - CONTINUUM ADHERENTS

In this section we discuss some salient features of the fracture of bonded materials involving ideal and continuum interfaces and continuum adherents. Needless to say, the great bulk for the past and current research on the fracture mechanics of interfaces falls into this category. The primary objective of the research in the field has been the evaluation of the strain energy release rate $\mathcal{G}$ and the stress intensity factors k_1 and k_2. In Section 5 we will discuss certain approximate solutions by introducing spring models for the interfacial zones and, along with elastic continuum, membrane or plate models for the adherents.

4.1 *Crack Parallel to an Interface*

To shed some light on the interface crack problem we first consider the plane elasticity problem for

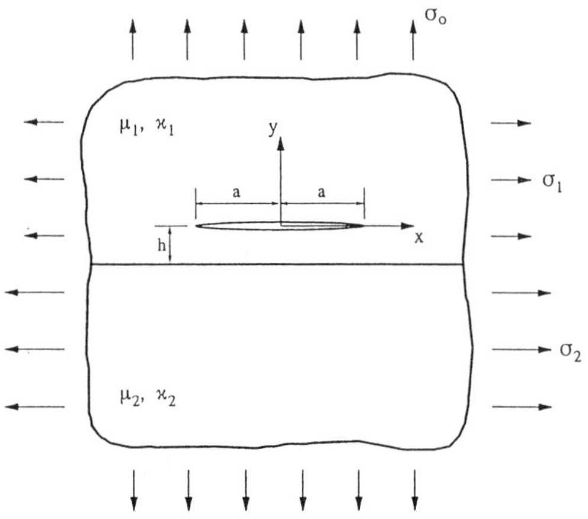

Figure 1. The geometry of a crack parallel to an interface subjected to mode I loading,

two bonded dissimilar materials containing a finite crack parallel to the interface and subjected to mode I loading, $\sigma_{yy}(x,\pm\infty) = \sigma_0$ (Figure 1). We also assume that one or both planes are under appropriate uniform tension in x direction such that $\varepsilon^1_{xx}(\pm\infty,+0) = \varepsilon^2_{xx}(\pm\infty,-0)$. By defining

$$f_1(x) = \frac{\partial}{\partial x}[v_1(x,h+0) - v_1(x,h-0)],$$
$$f_2(x) = \frac{\partial}{\partial x}[u_1(x,h+0) - u_1(x,h+0)], \tag{6}$$

the integral equations of the problem may be expressed as

$$\frac{1}{\pi}\int_{-a}^{a}\sum_{1}^{2}\left[\frac{\delta_{ij}}{t-x} + k_{ij}(x,t)\right]f_j(t)dt = \frac{1+\kappa_1}{2\mu_1}g_i(x), \quad -a<x<a, \ (i=1,2) \tag{7}$$

where κ_i and $\mu_i, (i=1,2)$ are the elastic constants ($\kappa = 3-4\nu$ for plane strain and $\kappa = (3-\nu)/(1+\nu)$ for plane stress) and

$$k_{11} = -\frac{4c_1h^2(t-x)[12h^2-(t-x)^2]}{[4h^2+(t-x^2)]^3} - \frac{8c_1h^2(t-x)}{[4h^2(t-x)^2]^2} - \left(\frac{c_1}{2}-c_2\right)\frac{t-x}{4h^2+(t-x)^2},$$

$$k_{22} = -\frac{4c_1h^2(t-x)[12h^2-(t-x)^2]}{[4h^2+(t-x)^2]^3} + \frac{8c_1h^2(t-x)}{[4h^2+(t-x)^2]^2} - \left(\frac{c_1}{2}-c_2\right)\frac{t-x}{4h^2+(t-x)^2},$$

$$k_{12} = -k_{21} = \frac{8c_1h^3[4h^2-3(t-x)^2]}{[4h^2+(t-x)^2]^3} - \left(\frac{c_1}{2}+c_2\right)\frac{2h}{4h^2+(t-x)^2}; \tag{8a-c}$$

$$c_1 = \frac{\mu_1-\mu_2}{\mu_1+\kappa_1\mu_2}, c_2 = \frac{1}{2}\frac{\kappa_1\mu_2-\kappa_2\mu_1}{\mu_2+\kappa_2\mu_1}; \tag{9}$$

$$g_1(x) = \sigma_{1yy}(x,h), \ g_2(x) = \sigma_{1xy}(x,h), \quad -a<x<a. \tag{10}$$

The crack surface tractions g_1 and g_2 are known functions and represent the applied loads. In the example shown in Fig. 1, $g_1 = -\sigma_o, g_2 = 0$. The solution of (7) is of the form (Muskhelishvili 1953)

$$f_i(x) = F_i(x)/\sqrt{a^2-x^2}, \ (i=1,2) \tag{11}$$

where F_1 and F_2 are bounded with $F_i(\pm a) \neq 0$. Upon solving (7) the modes I and II stress intensity factors, for example, at x = a may be obtained from

$$k_1(a) = \lim_{x\to a}\sqrt{2(x-a)}\sigma_{yy}(x,0) = -\frac{2\mu_1}{1+\kappa_1}\frac{F_1(a)}{\sqrt{a}},$$
$$k_2(a) = \lim_{x\to a}\sqrt{2(x-a)}\sigma_{xy}(x,0) = -\frac{2\mu_1}{1+\kappa_1}\frac{F_2(a)}{\sqrt{a}}. \tag{12a,b}$$

From (11) and (12) the "energy release rate" at the crack tip x = a may be expressed as

$$\mathcal{G} = \frac{\pi(1+\kappa_1)}{8\mu_1}(k_1^2+k_2^2). \tag{13}$$

4.1.1 *The Interface Crack*

For h > 0 the solution of (7) may be obtained numerically in a straightforward manner. For $h \to 0$ we have

$$\frac{t-x}{h^2+(t-x)^2} \to \frac{1}{t-x},$$

$$\frac{1}{\pi}\int_{-a}^{a}\frac{2hf_i(t)dt}{4h^2+(t-x)^2} \to f_i(x),\ i=1,2, \tag{14}$$

and contributions from all other terms in k_{ij} disappear. Thus integral equations (7) become

$$\frac{1}{\pi}\int_{-a}^{a}\frac{f_1(t)}{t-x}dt - \beta f_2(x) = \frac{1+\kappa_1}{c\mu_1}g_1(x),$$

$$\frac{1}{\pi}\int_{-a}^{a}\frac{f_2(t)}{t-x}dt + \beta f_1(x) = \frac{1+\kappa_1}{c\mu_1}g_2(x),\ -a<x<a, \tag{15}$$

$$c = 2\frac{1+\alpha}{1-\beta^2},\ \beta = \frac{(\kappa_1-1)\mu_2-(\kappa_2-1)\mu_1}{(\kappa_1+1)\mu_2+(\kappa_2+1)\mu_1},$$

$$\alpha = \frac{(\kappa_1+1)\mu_2-(\kappa_2+1)\mu_1}{(\kappa_1+1)\mu_2+(\kappa_2+1)\mu_1}, \tag{16}$$

where α and β are Dundurs constants and $(1+\kappa_1)/\mu_1$ is a scaling factor. The solution of (15) may be obtained by using either the complex function theory (Muskhelishvili 1953) or properties of the Jacobi polynomials (Erdogan and Gupta 1971). In the latter case the solution may be written down simply by inspection. For example, under mode I loading $g_1(x) = -\sigma_0$, $g_2(x) = 0$, the solution of (15) may be expressed as (Erdogan and Gupta 1971)

$$f_1 + if_2 = -\frac{2\sigma_0(1+\kappa_1)}{c\mu_1\sqrt{1-\beta^2}}w(\xi)(\frac{\xi}{2}-i\omega),\ \xi = \frac{x}{a}, \tag{17}$$

$$w(\xi) = (1-\xi)^{\gamma}(1+\xi)^{\delta},\gamma = -\frac{1}{2}-i\omega,$$

$$\delta = -\frac{1}{2}+i\omega,\omega = \frac{1}{2\pi}\log\left(\frac{1-\beta}{1+\beta}\right), \tag{18}$$

$$\sigma_{xy}(x,0) - i\sigma_{yy}(x,0) = i\sigma_0$$

$$-\sigma_0(x-a)^{\gamma}(x+a)^{\delta}(2\omega a+ix),\ x>a, \tag{19}$$

$$\frac{c\mu_1}{(1+\kappa_1)\sigma_0}[v(x)+iu(x)]$$

$$= \frac{\sqrt{a^2-x^2}}{\sqrt{1-\beta^2}}[\cos(\omega\log\frac{a+x}{a-x},$$

$$+ i\sin(\omega\log\frac{a+x}{a-x})], \tag{20}$$

$$v(x) = v_1(x,+0) - v_2(x,-0),$$
$$u(x) = u_1(x,+0) - u_2(x,-0). \tag{21}$$

The complex stress intensity factor may then be defined by and obtained from (Erdogan and Gupta 1971b)

$$k_1 + ik_2 = \lim_{x\to a}\sqrt{2(x-a)}\left(\frac{x-a}{2a}\right)^{i\omega}[\sigma_{yy}(x,0)$$
$$+ i\sigma_{xy}(x,0)] = \sigma_0\sqrt{a}(1-2i\omega). \tag{22}$$

Also, by using the concept of crack closure energy, form (20) and (22) the energy release rate for interfacial crack growth is found to be (Malyshev and Salganik 1965, Erdogan and Wu 1993b)

$$\mathcal{G}_{12} = \frac{1+\kappa_1}{c\mu_1}\frac{\pi}{4}(k_1^2+k_2^2) = \frac{1+\kappa_1}{c\mu_1}\frac{\pi}{4}\sigma_0^2 a(1+4\omega^2). \tag{23a,b}$$

Before discussing the results for the crack near the interface one should observe that, because of the crack surface interference near the crack tips implied by (20), the results given by the linear elasticity solution (17)-(23) are physically inadmissible. One may attempt to remove this objectionable feature of the solution within the confines of continuum solid mechanics in various ways. One way would be in an ad-hoc fashion to assume that, because of the tendency toward interpenetration, near the crack tips the crack surfaces would come in smooth contact and form a cusp, and the resulting contact region would consist of a single uninterrupted zone rather than the sum of a series of discrete zones as implied by the oscillatory nature of the elastic solution (Comninou 1977, Atkinson 1982). Another way is to assume that near the crack tip the linear theory is not valid and to use a large deformation non-linear theory. An asymptotic analysis using such a theory was provided by Knowles and Sternberg (1983) for the plane stress interface crack problem in two bonded dissimilar incompressible Neo-Hookean materials which shows no oscillatory behavior for stresses or displacements near the crack tips. However, the exact solution obtained by Varley (1989) for another finite elasticity problem of a half space (in which the strain energy function is assumed to be quadratic in the nonlinear strain measure) bonded to a rigid rectangular stamp under plane strain conditions indicates that, as in the linear case, the oscillations disappear only if the elastic medium is incompressible. Thus, the theoretical question regarding the oscillatory behavior of stresses and displacements near the tip of interface cracks does not seem to have been completely resolved yet. A third approach to deal with the question involves the introduction of an inhomogeneous interfacial zone which will be discussed briefly later in this section.

4.1.2 *Crack Closure Model*

In the case of certain mixed boundary value problems, most notably the crack and contact problems, for some material combinations and boundary conditions, complex eigenvalues are unavoidable. A physically acceptable solution would then require the elimination of the implied stress and displacement oscillations and, hence, an alternate formulation of the problem by modifying the mixed boundary conditions. In the interface crack problem under consideration, the formulation given by (15) actually assumes that the crack surfaces remain open near and up to the end points x = a - 0 and x = -a + 0. The solution should be physically admissible if this condition is relaxed by allowing crack closure near the ends and by satisfying all other conditions of the boundary value problem. This is the question addressed by Comninou (1977) and Atkinson (1982). Because of the exceedingly small size of the related contact zones, the numerical solutions attempted by Comninou (1977) had rather severe convergence problems. Allowing crack tip closures, the integral equations (15) may be modified as follows (Figure 1, h = 0):

$$\frac{1}{\pi}\int_{-d}^{b}\frac{f_1(t)}{t-x}dt-\beta H(a^2-x^2)f_2(x)$$
$$=\frac{1+\kappa_1}{c\mu_1}\sigma_{yy}(x,0),-d<x<b,$$

$$\frac{1}{\pi}\int_{-a}^{a}\frac{f_2(t)}{t-x}dt+\beta H(b-x)H(d+x)f_1(x)$$
$$=\frac{1+\kappa_1}{c\mu_1}\sigma_{xy}(x,0),\ -a<x<a, \tag{24}$$

where -a and a are the actual physical crack tips, the crack is assumed to be closed (that is, $v_1^+ - v_2^- = 0$, $u_1^+ - u_2^- \neq 0$) along $-a<x<-d$ and $b<x<a$, and H(x) is the Heaviside function. The single-valuedness conditions under which the integral equations must be solved would thus be modified as

$$\int_{-d}^{b} f_1(t)dt = 0,\quad \int_{-a}^{a} f_2(t)dt = 0. \tag{25}$$

In this problem the constants b and d defining the size of the contact zones are unknown and are to be determined from the smooth closure at x = b and x = -d which may be expressed as

$$f_1(b)=0,\ f_1(-d)=0 \tag{26}$$

together with the condition that $v_1^+ - v_2^- = 0$ for $-a<x<-d$ and $b<x<a$. These conditions imply that the mode I stress intensity factors are zero, namely $k_1(b)=0$, $k_1(-d)=0$, whereas $f_2(x)$ is expected to have the standard square-root singularity giving nonvanishing mode II stress intensity factors at the crack tips $x=\mp a$. It is assumed that the coefficient of friction along the contact zones is zero. The nonlinear problem formulated by (24)-(26) was treated analytically by Gautesen and Dundurs (1987, 1988). Because of the exceedingly difficult nature of the exact solution given by Gautesen and Dundurs (1987, 1988), Gautesen (1992,1993) reconsidered the problem, reduced it to an eigenvalue problem for a second order ordinary differential equation and, by using the method of matched asymptotic expansions, was able to obtain a highly accurate approximate analytical solution for constant tractions

$$\sigma_{yy}(x,0)=-\sigma_0,\ \sigma_{xy}(x,0)=-\tau_0. \tag{27}$$

For mode I loading $\sigma_0>0$, $\tau_0=0$, Gautesen (1992) obtained the following results*

* Note that material 1 occupies y < 0 half plane in Gautesen's notation, whereas it occupies y > 0 half plane in the notation used in Figure 1 leading to the definition of β in (16).

$$\frac{a-b}{a}=8\exp\left[-\frac{\pi}{2t(\beta)}\left(\pi+2\arctan(\frac{2}{\pi}t(\beta))\right)\right],\ d=b, \tag{28}$$

$$t(\beta)=\text{arctanh}(-\beta),\ \lambda=\frac{2}{\pi}t(\beta), \tag{29}$$

$$k_{2c}(\mp a)=\mp\sqrt{1+\lambda^2}\sigma_0\sqrt{a} \tag{30}$$

where k_{2c} refer to the mode II stress intensity factor calculated by assuming crack closure. For example, for $\beta=-0.1,-0.2,-0.3,-0.4,-0.5$ $(\mu_2<\mu_1)$, from (27) it can be shown that (a - b)/a = 4.740×10^{-22}, 2.937×10^{-11}, 1.322×10^{-7}, 9.905×10^{-6}, 1.465×10^{-4}, respectively (Gautesen 1992). Also, for the pair material 1 aluminum and material 2 epoxy ($\nu_1=0.3$, $\nu_2=0.35$, $E_2/E_1=0.045$) we find $\beta=-0.1612$, $(a-b)/a=7.2199\times10^{-14}$ and $k_{2c}(a)$ $=1.0053\,\sigma_0\sqrt{a}$.

Along the interface y = 0, since $\sigma_{yy}(x,0)$ is bounded, $k_1(\mp a)=0$ and the strain energy release rate given by (23a) becomes

$$\mathcal{G}_{12}=\frac{1+\kappa_1}{c\mu_1}\frac{\pi}{4}k_{2c}^2. \tag{31}$$

From (23b), (30) and (31) it follows that

$$k_{2c}(a)=\sigma_0\sqrt{a}\sqrt{1+4\omega^2}. \tag{32}$$

Comparing (30) and (32) it may be seen that

$$\lambda=2\omega \tag{33}$$

which can be verified by using (29), the following definition of ω:

$$\omega=\frac{1}{2\pi}\log\left(\frac{1-\beta}{1+\beta}\right), \tag{34}$$

and the identity

$$\text{arctan}\,\text{h}x=\tfrac{1}{2}\log\left(\frac{1+x}{1-x}\right),\ 0\le x^2<1. \tag{35}$$

For the mixed mode loading condition $\sigma_0\neq0$, $\tau_0\neq0$, the relevant results found by Gautesen (1993) may be summarized as

$$\lambda=\frac{2}{\pi}\text{arctan}\,\text{h}(-\beta), \tag{36}$$

$$\lambda\,\text{arctan}\,\text{h}(\gamma^{-1})-\arctan(\lambda\gamma)-\arctan(\sigma_0/\tau_0)=0, \tag{37}$$

$$b/a=2\gamma^{-2}-1, \tag{38}$$

$$\frac{a-d}{a}=32\left(\frac{a+b}{a-b}\right)\exp\left[-\frac{2(\pi+2\arctan(\lambda))}{\lambda}\right], \tag{39}$$

$$k_{2c}(a)\cong-k_{2c}(-a)\cong\left[\frac{(1+\frac{b}{a})(\sigma_0^2+\tau_0^2)}{1+\frac{b}{a}+2\lambda^2}\right]^{1/2}(1+\lambda^2) \tag{40}$$

where γ must be obtained numerically from (37). For example, for β = -0.25 Table 1 shows some calculated results (Gautesen 1993). In Table 1 $\tau_0>0$ and $-0.4\tau_0\le\sigma_0\le0.4\tau_0$; that is, the last two columns give the contact zones for shear and compressive loading. From a theoretical viewpoint the importance of (33) cannot be overemphasized. It shows that one obtains identical strain energy release rates for the crack problem by using two entirely different formulations. In the expression of $\mathcal{G}$ given by (23b) the dominant stress intensity factor is k_1, whereas in (31), which is obtained from the crack closure model, k_1 = 0 and $\mathcal{G}$ is dependent on k_2 only. Most likely this agreement is due to the fact that in the mode I loading problem under consideration the contact zones are very small and the asymptotic analysis performed by Gautesen (1992) was based on $\varepsilon=(a-b)/a$ being a "small" parameter. However, there is no reason to expect that a similar agreement would be valid if the medium is under mixed mode loading. In fact from (40) it may be seen that even under pure mode II loading (i.e., for $\sigma_0=0$) $k_{2c}(-a)=-k_{2c}(a)$. This is a result one would expect only under pure mode I loading in conventional interface crack solutions. The reason for this seemingly paradoxical result must be found in the relative crack closures. As shown by Table 1, for $\sigma_0=0$ the crack tip at x = -a is nearly open, whereas near x = a almost a third of the crack is closed.

4.1.3 *Some Results for a Crack Parallel to the Interface*

In two bonded half planes under mode I loading described by Figure 1, if h/a is greater than a characteristic distance h*/a, then the crack opening would have a parabolic behavior and k_I would be positive. However, as qualitatively described in Figure 2, if $h \le h_1^*$ for the crack in the less stiff material and $|h| \le h_2^*$ for the crack in the stiffer material, a contact region of size Δ_1 or Δ_2 is developed near the crack tip, the crack forms a cusp, mode I stress intensity factor becomes zero and $\Delta_1 \to \Delta_0$ and $\Delta_2 \to \Delta_0$ as $|h| \to 0$. Even though for $|h| > 0$ the crack is fully embedded in a homogeneous medium, its correct solution cannot be obtained without taking into account the crack closure near the ends. The constants h_1^* and h_2^* may be obtained by imposing the closure condition $F_1(a) = 0$ in solving (7) (see (6) and (11)) (Erdogan and Joseph 1988).

In a hypothetical material pair described by $\alpha = 0.96$, $\beta = 0.49$ (an exaggerated material system selected to give a relatively large Δ_0), Figure 3 shows the normalized crack opening displacement defined by $v(x) = v_i^+ - v_i^-$, $(i = 1,2)$, for $h \ne 0$ and $v(x) = v_1^+ - v_2^-$ for $h = 0$, where again material 1 is the less stiff medium. The lower part of Figure 3 shows the magnified version of $v(x) = v_1^+ - v_1^-$ around the crack tip, essentially verifying the qualitative results given in Figure 2. A different presentation of the normal crack opening is shown in Figure 4. Note that for $0 < h < \infty$ the normal crack opening may be expressed as

$$v(x) = v_1^+ - v_1^- = \frac{\sigma_0(\kappa_1 + 1)}{2\mu_1} \bar{v}(x)\sqrt{a^2 - x^2} \qquad (41)$$

where

$$\bar{v}(a) = k_1(a) / \sigma_0\sqrt{a}, \ \lim_{h\to\infty} \bar{v}(a) = 1. \qquad (42)$$

The figure shows $\bar{v}(x)$. It is seen that $\bar{v}(a)$ decreases with decreasing h/a and becomes zero for $h = h_1^*$. The figure also shows the interface crack opening (the dashed lines) normalized with respect to the same factor given in (41).

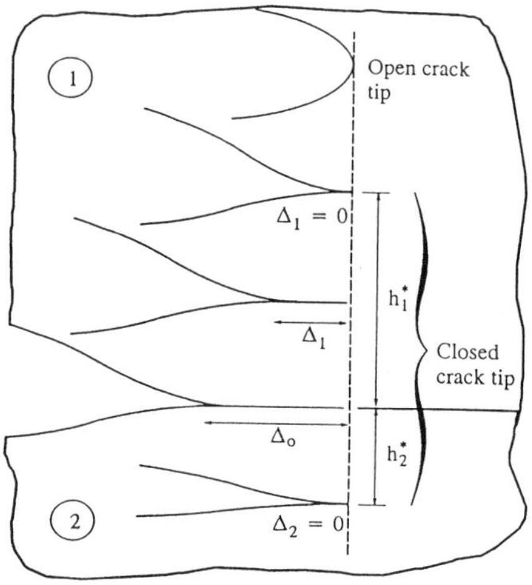

Figure 2. Crack surface contact for small |h|/a.

Somewhat more interesting results for a crack parallel to the interface in bonded half planes under mode I loading are shown in Figures 5 and 6. Figure 5 shows the modes I and II stress intensity factors as functions of the normalized crack distance h/a. Note that for $h \to \mp\infty$, as expected, $k_1 \to \sigma_0\sqrt{a}$ and $k_2 \to 0$. For relatively large values of $|h/a|$ k_1 is the dominant stress intensity factor. However, as $|h/a| \to 0$ $k_1 \to 0$ and $k_2 \to k_{2c}$ which is calculated from (30) by using the interface crack closure model (Gautesen 1992). These limits may better be seen in the lower part of Figure 5 where the scale is magnified. Figure 6 shows the normalized strain energy release rate $\mathcal{G}/\mathcal{G}_1$ as a function of h/a. Note that the normalizing strain energy release rate $\mathcal{G}_1$ corresponds to that in material 1 at $h = \infty$. $\mathcal{G} \to \mathcal{G}_1$ for $h \to \infty$ and $\mathcal{G} \to \mathcal{G}_2$ as $h \to -\infty$ where

$$\mathcal{G}_1 = \frac{\pi(\kappa_1 + 1)}{8\mu_1}\sigma_0^2 a, \ \mathcal{G}_2 = \frac{\pi(\kappa_2 + 1)}{8\mu_2}\sigma_0^2 a. \qquad (43)$$

Thus $\mathcal{G}/\mathcal{G}_1 \to 1$ for $h \to \infty$ and $\mathcal{G}/\mathcal{G}_1 \to (\kappa_2 + 1)\mu_1 / (\kappa_1 + 1)\mu_2 = (1 - \alpha)(1 + \alpha)$ for $h \to -\infty$. $(\mathcal{G}/\mathcal{G}_1)_I$ and $(\mathcal{G}/\mathcal{G}_1)_{II}$ shown in the lower part of the figure with magnified scale are, respectively, mode I and mode II contributions and are given by

$$(\mathcal{G}/\mathcal{G}_1)_I = k_1^2 / \sigma_0^2 a, \ (\mathcal{G}/\mathcal{G}_1)_{II} = k_2^2 / \sigma_0^2 a. \qquad (44)$$

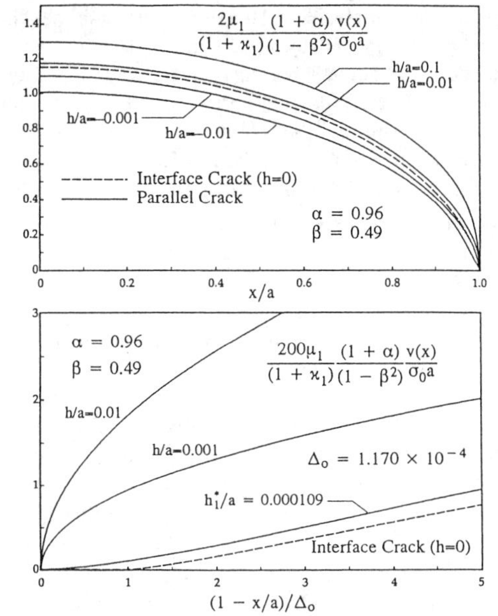

Figure 3 Normalized crack opening displacement. The crack is in the less stiff material for $h/a > 0$ and in the stiffer material for $h/a < 0$. The dashed lines correspond to the interface crack, $h = 0$.

Note that unlike the stress intensity factors which are discontinuous at $h = 0$, the strain energy release rate $\mathcal{G}$ and its components $\mathcal{G}_I$ and $\mathcal{G}_{II}$ are continuous functions of h in $-\infty < h < \infty$. Furthermore, the lower part of the figure clearly shows that as $|h| \to 0$ the strain energy release rate $\mathcal{G}$ approaches $\mathcal{G}_{12}$ calculated independently from (31) by using interface crack closure model (Gautesen 1992). From (43), (31) and (32) the calculated limit may be expressed as

$$\lim_{|h|\to 0} \mathcal{G}/\mathcal{G}_1 = \mathcal{G}_{12}/\mathcal{G}_1 = \frac{1-\beta^2}{1+\alpha}(1+4\omega^2). \qquad (45)$$

The conclusion which may be drawn from Figures 1-6 is that Gautesen's model for the interface crack is a natural limit of an embedded crack problem which is well-understood and which can be accurately treated.

In bonded half planes under mode I loading shown in Figure 1, generally k_2 decreases monotonically as $|h/a|$ increases. This however, is not always the case (Hutchinson et al. 1987, Erdogan and Joseph 1988). Table 2 shows the results of an example considered by Hutchinson et al. (1987) and Erdogan and Joseph (1988) for a semi-infinite crack parallel to the interface and the result for a finite crack of length 2a. The table clearly indicates that the sign of k_2 indeed changes near the interface and the asymptotic results given by Hutchinson et al. (1987) are surprisingly accurate. The table also shows the smooth convergence of $\mathcal{G}$ calculated from (13) for the embedded cracks to $\mathcal{G}_{12}$ calculated from (23) for the interface crack as $|h/a| \to 0$. In the other pair of materials Cu/Si for which k_2 becomes zero near the interface, the corresponding values of h/a are (0.0847, -0.07235) and (0.08293, -0.07077) as obtained from the asymptotic and "exact" solutions, respectively.

Based on physical considerations to the effect that in homogeneous solids the crack would tend to grow toward free surfaces or less stiff materials, it should be pointed out that in the crack problem described by Figure 1 the sign change in k_2 must occur at least twice. This may be once in each half plane as shown in Table 2 or, in very rare cases, possibly twice in the same material. At the tip of a crack under mixed-mode loading, a good estimate of the crack growth (or kinking) angle θ_0 may be obtained from (Erdogan and Sih 1963)

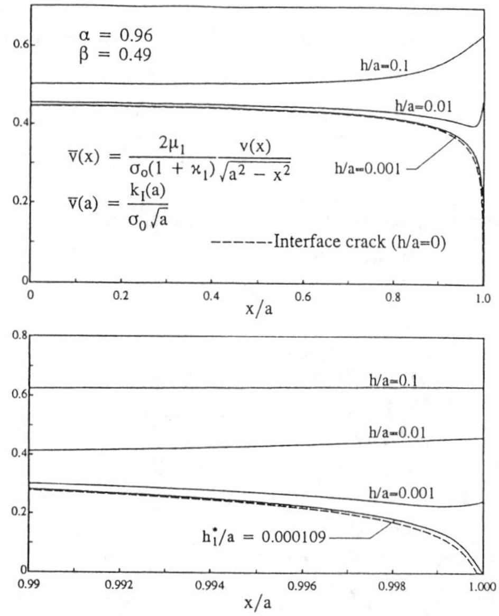

Figure 4. Normalized crack opening displacement for various values of h/a.

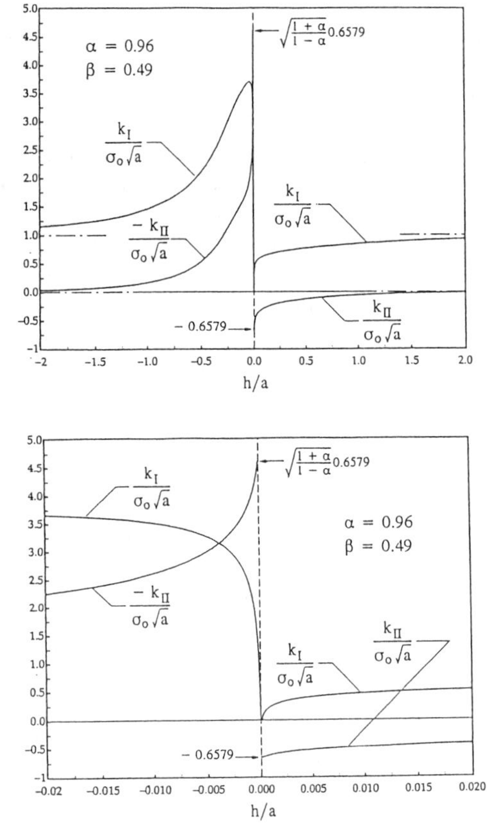

Figure 5. Normalized mode I and mode II stress intensity factors for a crack parallel to the interface.

$$\tan^2\frac{\theta_0}{2} - \frac{k_1}{2k_2}\tan\frac{\theta_0}{2} - \frac{1}{2} = 0, \quad \sigma_{\theta\theta}(r,\theta_0) > 0 \qquad (46)$$

where θ_0 is measured from the plane of the crack. In the problem considered, for sufficiently large $|h|$, $k_2 < 0$ and $\theta_0 > 0$ in both half planes, hence the even number of sign reversals in k_2.

Thus far the discussion given in this section is concerned with a crack parallel to or along the interface in bonded dissimilar materials under mode I loading. The vast literature on the subject indicates that for realistic material pairs the oscillation zone size is much smaller than the microstructural length parameter of the materials (Erdogan 1965) and the strain energy release rate $\mathcal{G}$ and the stress intensity ratio k_2/k_1 (k_1 and k_2 being the real and imaginary part of the complex k) calculated by ignoring the local crack tip contact shown in Figures 2 and 3 are perfectly adequate for the elastic fracture mechanics analysis. However, in practice the loading is not always pure mode I and, consequently, the contact region is not always local and small. Therefore, the correct treatment of the problem requires the solution of the nonlinear "global" crack/contact problem. Returning again to Figure 1, for $A\ell_2O_3/Cu$ pair ($\alpha = 0.5, \beta = 0.089$) the problem was considered by Hutchinson et al. (1987) by ignoring the crack surface contact and by Yang and Kim (1993) by taking it into account. For mode I loading the calculated stress intensity factors in the two studies seem to be in reasonably good agreement. Under pure shear, however, k_1 found by Yang and Kim (1993) at the open end of the crack is roughly 46% greater than that found by Hutchinson et al. (1987). Similar discrepancies may be found for interface cracks under mixed mode loading conditions (Yuuki et al. 1994) implying that ignoring the global crack surface contact, if it exists, may lead to erroneous results.

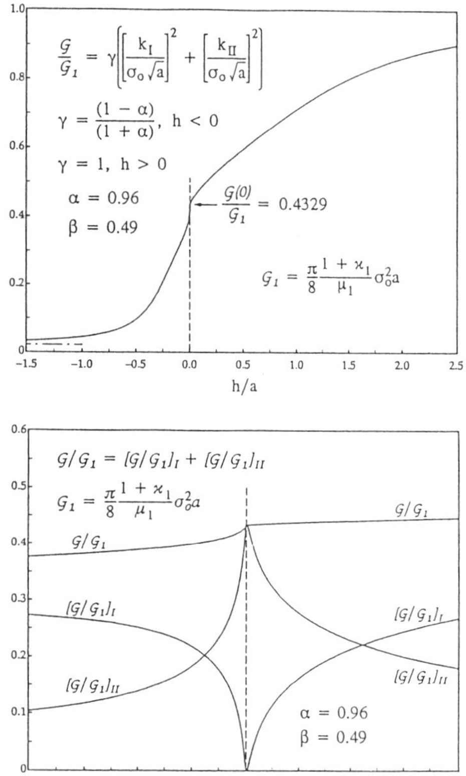

Figure 6. The normalized strain energy release rate as a function of h/a.

4.2 *Homogeneous Continuum Interfacial Zone*

In many cases during the bonding process a thermodynamically stable phase such as an intermetallic may form along the interface having mechanical properties different from that of the adherents. In such materials a simple way to model the interfacial zone would be to assume that the intermediate zone is a thin layer of homogeneous continuum bonded to the adherents through ideal interfaces. In studying fracture mechanics of the medium, one may again assume that the interfacial zone contains a flaw which may be modeled as a planar crack. The crack may be either fully embedded in the interfacial zone (Figure 7) or lie along one of the (ideal) interfaces. Let y = 0 be the plane of the crack. By defining the unknown functions

$$f_1(x) = \frac{\partial}{\partial x}(v^+ - v^-),$$
$$f_2(x) = \frac{\partial}{\partial x}(u^+ - u^-), \; -a < x < a, \tag{47}$$

the integral equations for the plane elasticity problem may be expressed as (Erdogan and Gupta, 1971a, 1971b)

$$\frac{1}{\pi}\int_{-a}^{a}\frac{f_1(t)}{t-x}dt - \beta f_2(x) + \int_{-a}^{a}\sum_{1}^{2}k_{1j}(x,t)f_j(t)dt = \frac{1+\kappa^+}{c\mu^+}g_1(x)$$
$$\frac{1}{\pi}\int_{-a}^{a}\frac{f_2(t)}{t-x}dt + \beta f_1(x) + \int_{-a}^{a}\sum_{1}^{2}k_{2j}(x,t)f_j(t)dt = \frac{1+\kappa^+}{c\mu^+}g_2(x), \; |x|< a, \tag{48}$$

$$\beta = \frac{\mu^-(\kappa^+ - 1) - \mu^+(\kappa^- - 1)}{\mu^-(\kappa^+ + 1) + \mu^+(\kappa^+ - 1)}, \tag{49}$$

where k_{ij} are known bounded functions, superscripts + and - refer to quantities on the positive and the negative side of the crack and

$$g_1(x) = \sigma_{yy}(x,0), \; g_2(x) = \sigma_{xy}(x,0), \; |x|< a \tag{50}$$

are the crack surface tractions. Note that $\beta = 0$ for an embedded crack and $\beta \neq 0$ for interface cracks.

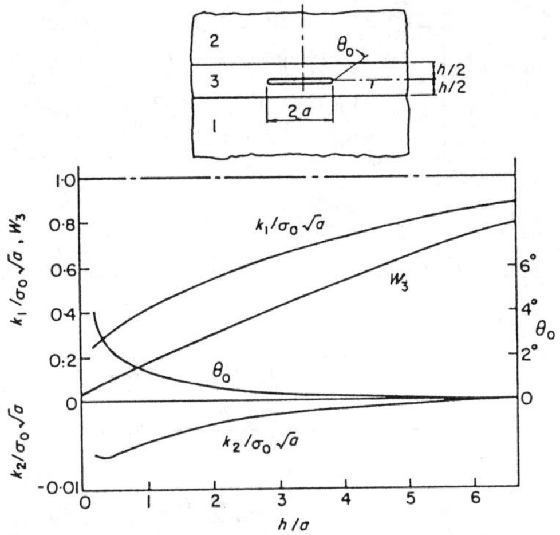

Figure 7. Normalized stress intensity factors, strain energy release rate and the probable crack growth angle for a crack fully embedded in a continuum interfacial zone.

The general solution of (48) is of the form

$$f_1(x) + if_2(x) = \sum_{0}^{\infty}A_n P_n^{(\gamma,\delta)}(x/a)w(x/a),$$
$$-a < x < a \tag{51}$$

where 2a is the crack length, A_n are unknown constants determined from (48) and the single-valuedness conditions, $P_n^{(\gamma,\delta)}(x/a)$ is the Jacobi polynomial and w is the fundamental solution of (45) and is given by (18). After solving (45) the stress intensity factors and the strain energy release rate may be obtained from (22a) and (23a), respectively. Note that for embedded cracks $\beta = 0$, the associated orthogonal polynomials $P_n^{(\gamma,\delta)}$ become T_n, the Chebyshev polynomials of the first kind, and f_i, k_i, (i = 1,2) and $\mathcal{G}$ would be given by (11), (12) and (13) respectively. As an example, Figure 7 shows the normalized stress intensity factors, normalized strain energy release rate and the probable direction of the crack growth θ_0 obtained from (46) at the crack tip x = a for the materials $E_1/E_2 = 3$, $E_3/E_2 = 0.045$, $\nu_1 = \nu_2 = 0.3$, $\nu_3 = 0.35$ (Steel/Epoxy/Aluminum) and the external load $\sigma_{yy}(x,\mp\infty) = \sigma_0$. The normalized energy release rate is defined by

Table 1. Relative contact zone sizes in an interface crack under mixed-mode loading

σ_0/τ_0:	0.4	0.2	0	-0.2	-0.4
(a-d)/a:	1.967×10^{-15}	2.110×10^{-16}	2.110×10^{-17}	2.226×10^{-18}	1.364×10^{-18}
(a-b)/a:	0.01011	0.0907	0.621	1.426	1.760

Table 2: Stress intensity factors and the energy release rate for a crack parallel to the interface. Mat. 1: Ni($\mu_1 = 8.08\times10^{10}\,N/m^2$, $\nu_1 = 0.324$); Mat. 2: MgO($\mu_2 = 12.83\times10^{10}\,N/m^2$, $\nu_2 = 0.175$).

h/a	Hutchinson et al. 1987		Erdogan and Joseph 1988		
	$k_1/\sigma_0\sqrt{a}$	$k_2/\sigma_0\sqrt{a}$	$k_1/\sigma_0\sqrt{a}$	$k_2/\sigma_0\sqrt{a}$	$\mathcal{G}/\mathcal{G}_1$
∞			1.0	0	1.0
1.0	0.9370	-0.0244	0.9572	-0.0146	0.9165
0.5	0.9371	-0.0212	0.9436	-0.0211	0.8909
0.2	0.9372	-0.0170	0.9386	-0.0181	0.8812
0.1	0.9372	-0.0138	0.9376	-0.0142	0.8793
0.01	0.9373	-0.0031	0.9372	-0.0031	0.8784
0.00510	0.9373	0.0000			
0.005091			0.9373	0.0000	0.8785
.001	0.9373	0.0075	0.9373	0.0075	0.8785
0					0.8786
-0.001	1.077	0.0053	1.077	0.0053	0.8786
-0.002705			1.077	0.0000	0.8786
-0.00270	1.077	0.0000			
-0.01	1.077	-0.0070	1.077	-0.0070	0.8786
-0.1	1.077	-0.0192	1.076	-0.0198	0.8770
-0.2	1.077	-0.0229	1.074	-0.0241	0.8743
-0.5	1.077	-0.0277	1.066	-0.0262	0.8619
-1.0	1.077	-0.0314	1.047	-0.0164	0.8307
$-\infty$			1.0	0	0.7574

$$W_3 = \frac{\mathcal{G}_h}{\mathcal{G}_0},\ \mathcal{G}_0 = \mathcal{G}(\infty) = \frac{\pi(1+\kappa_3)}{8\mu_3}\sigma_0^2 a,$$
$$W_3 = \frac{k_1^2+k_2^2}{\sigma_0^2 a}. \tag{52}$$

As $h\to\infty$ the limits are $k_1\to\sigma_0\sqrt{a}$, $k_2\to 0$, $\theta_0\to 0$, and $\mathcal{G}\to\mathcal{G}_0$, which may also be observed in Figure 7. On the other hand, for h = 0 the problem becomes one of two bonded half planes and because of the complex singularity the limits of k_1,k_2 and θ_0 for $h\to 0$ are not defined. The strain energy release rate, however, is

$$\mathcal{G}_{12} = \frac{\kappa_2+1}{4c_{12}\mu_2}\pi\left(k_1^2+k_2^2\right),\ c_{12} = \frac{2(1-\alpha)}{1-\beta^2} \tag{53}$$

where α and β are given by (16), k_1 and k_2 by (22b) and $\mathcal{G}_{12}$ by (23b). Physically it is clear that

$$\lim_{h\to0}\mathcal{G} = \mathcal{G}_{12}, \tag{54}$$

or, for the materials considered

$$\lim_{h\to0} W_3 = \frac{\mathcal{G}_{12}}{\mathcal{G}_0} = \frac{2\mu_3(\kappa_2+1)}{c_{12}\mu_2(\kappa_3+1)}\frac{(k_1^2+k_2^2)_{12}}{\sigma_0^2 a} = 0.0303. \tag{55}$$

As another example we consider the same material system as in Figure 7 ($E_1/E_2 = 3$, $E_3/E_2 = 0.045$, $\nu_1 = \nu_2 = 0.3$, $\nu_3 = 0.35$) with a crack of length 2a along the interface of materials 1 and 3 and under remote tension $\sigma_{yy} = \sigma_0$. The normal-

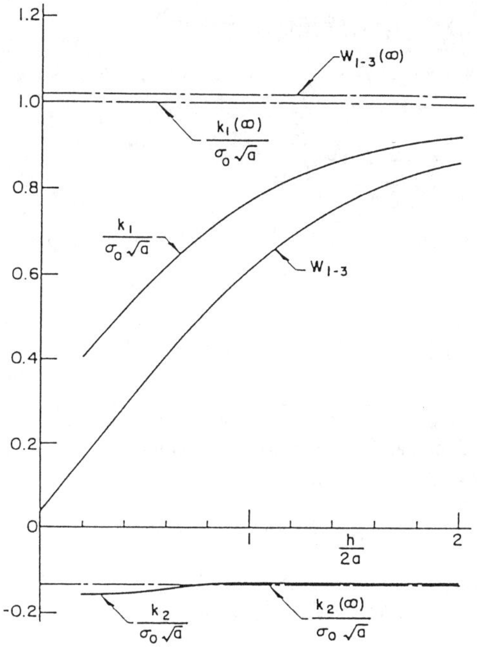

Figure 8. Normalized stress intensity factors and strain energy release rate for a crack along the interface between materials 1 and 3.

ized results are shown in Figure 8. In this case the calculated limits for $h \to \infty$ are

$$(k_1 + ik_2)/\sigma_0\sqrt{a} = 1 - 0.1443i. \tag{56}$$

The normalized strain energy release rate is defined by

$$W_{13} = \frac{\mathcal{G}_{13}(h)}{\mathcal{G}_{13}(\infty)},\ \mathcal{G}_{13}(h) = \frac{\kappa_3+1}{4c_{13}\mu_3}\pi\left(k_1^2+k_2^2\right)_{13},$$
$$\mathcal{G}_{13}(\infty) = \frac{\kappa_3+1}{4c_{13}\mu_3}\pi\sigma_0^2 a(1+4\omega_{13}^2), \tag{57}$$

where $\omega_{13} = 0.07215$ (see (18)). Thus

$$\lim_{h\to\infty} W_{13} = W_{13}(\infty) = 1.0208. \tag{58}$$

For $h \to 0$, again the limits of k_1 and k_2 are not defined and for the materials under consideration the limit of W_{13} may be determined from

$$\lim_{h\to 0}\mathcal{G}_{13}(h) = \mathcal{G}_{12},$$

$$\lim_{h\to 0} W_{13} = \frac{\mathcal{G}_{12}}{\mathcal{G}_{13}(\infty)}$$

$$= \frac{\kappa_2+1}{\kappa_3+1}\frac{(k_1^2+k_2^2)_{12}}{\sigma_0^2 a}\frac{\mu_3}{\mu_2}\frac{c_{13}}{c_{12}}\frac{1}{1+4\omega_{13}^2} = 0.0541. \tag{59}$$

Similar results for a penny-shaped crack may be found in Arin and Erdogan 1971, Erdogan and Arin 1972). A thorough discussion of wide ranging aspects of mixed mode and interface fracture is given by Hutchinson and Suo (1992).

4.3. *Effect of Material Orthotropy*

In some bonded materials the interfacial region may be identified as a thin layer of oriented medium with a columnar or a lamellar structure. This is generally the consequence of the processing technique. Furthermore, the adherents themselves may be naturally anisotropic. The solution of the interface crack problem for two bonded semi-infinite materials having various forms of anisotropies was considered, among others, by Gotoh (1967), Clements (1971), Willis (1971), Ting (1986), Bassani and Qu (1989) and Suo (1990). The general plane elasticity problem of two orthotropic adherents bonded through an orthotropic interfacial region and containing a series of collinear cracks along one of the interfaces is considered by Erdogan and Wu (1993a) (Figure 9). The formulation is somewhat simplified by introducing the parameters

$$E_0 = \sqrt{E_{11}E_{22}},\ \nu_0 = \sqrt{\nu_{12}\nu_{21}},$$

$$c^4 = \frac{E_{11}}{E_{22}} = \frac{\nu_{12}}{\nu_{21}},\ \kappa_0 = \frac{E_0}{2\mathcal{G}_{12}} - \nu_0 \tag{60}$$

to replace the engineering constants E_{11}, E_{22}, $\mathcal{G}_{12}$, ν_{12}. The definitions (57) are for plane stress case. For plane strain E_{11}, E_{22}, ν_{12} and ν_{21} are respectively replaced by

$$\frac{E_{11}}{1-\nu_{13}\nu_{31}},\ \frac{E_{22}}{1-\nu_{23}\nu_{32}},$$

$$\frac{\nu_{12}+\nu_{13}\nu_{32}}{1-\nu_{13}\nu_{31}},\ \frac{\nu_{21}+\nu_{23}\nu_{31}}{1-\nu_{23}\nu_{32}}. \tag{61}$$

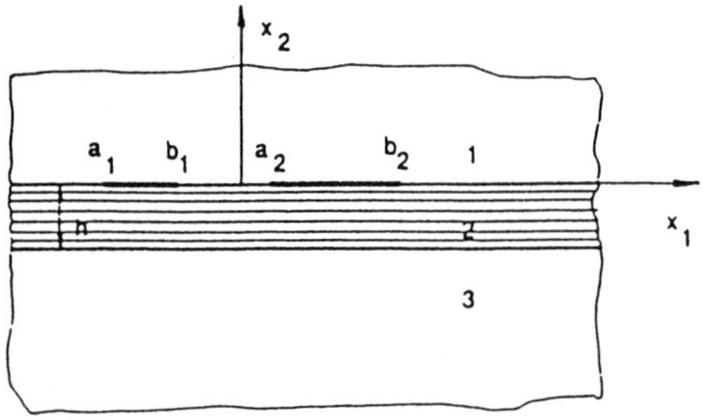

Figure 9. Geometry of bonded orthotropic materials with interface cracks.

Referring to Figure 9 and defining the unknown functions and the crack surface tractions by

$$f_1(x_1) = \frac{\partial}{\partial x_1}\left[u_{11}(x_1,0) - u_{21}(x_1,0)\right],$$

$$f_2(x_1) = \frac{\partial}{\partial x_1}\left[u_{12}(x_1,0) - u_{22}(x_1,0)\right]; \; x_1 \varepsilon L \tag{62}$$

$$\sigma_{122}(x_1,0) = p_1(x_1),$$

$$\sigma_{112}(x_1,0) = p_2(x_1), \; x_1 \varepsilon L, \tag{63}$$

the integral equations of the problem may be expressed as

$$\gamma_{11} f_1(x_1) + \gamma_{12}\frac{1}{\pi}\int_L \frac{f_2(t)}{t-x_1}dt + \int_L \sum_1^2 k_{1j}(x_1,t) f_j(t)dt = p_1(x_1),$$

$$\gamma_{21}\frac{1}{\pi}\int_L \frac{f_1(t)}{t-x_1} + \gamma_{22} f_2(x) + \int_L \sum_1^2 k_{2j}(x_1,t) f_j(t)dt = p_2(x_1), \; x_1 \varepsilon L \tag{64}$$

where u_{ki} is the ith component of the displacement and σ_{kij} are the stresses in material k, γ_{ij} are known bimaterial constants, k_{ij} are known functions and L is the sum of cracks $L = \sum_1^n L_j$, $L_j = (a_j, b_j)$ (Erdogan and Wu, 1993a, 1993b). If we further define

$$f(x_1) = \frac{1}{\eta} f_2(x_1) + i f_1(x_1),$$

$$p(x_1) = p_2(x_1) - i\eta p_2(x_1), \tag{65}$$

$$\gamma = \frac{\gamma_{11}}{\sqrt{\gamma_{12}\gamma_{21}}}, \quad \eta = \sqrt{\frac{\gamma_{21}}{\gamma_{12}}}, \tag{66}$$

The principal part of (64) (which also corresponds to $h = \infty$) may be written as

$$\frac{1}{\pi_i}\int_L \frac{f(t)}{t-x_1}dt - \gamma f(x_1) = \frac{1}{\gamma_{21}} p(x_1), \; x_1 \varepsilon L. \tag{67}$$

Equation (67) can be solved in closed form (Erdogan and Wu 1993a). The solution of (67) is of the form

$$f(t) = F(t)\prod_1^n (b_k - t)^{\alpha}(t - a_k)^{\beta}, \tag{68}$$

$$\alpha = -\tfrac{1}{2} - i\omega, \quad \beta = -\tfrac{1}{2} + i\omega,$$

$$\omega = \frac{1}{2\pi}\log\left(\frac{1+\gamma}{1-\gamma}\right) \tag{69}$$

where F(t) is a bounded function (Erdogan and Wu 1993a). for the interface cracks the complex stress intensity factors and the strain energy release rate at the crack tips b_k and a_k may be obtained as

$$k(b_k) = \eta k_1 + i k_2 = \lim_{x_1 \to b_k} \sqrt{2}\,(x_1 - b_k)^{-\alpha}[\eta\sigma_{22}(x_1,0) + i\sigma_{12}(x_1,0)] \tag{70}$$

$$k(a_k) = \eta k_1 + i k_2 = \lim_{x_1 \to a_k} \sqrt{2}\,(a_k - x_1)^{-\beta}[\eta\sigma_{22}(x_1,0) + i\sigma_{12}(x_1,0)], \tag{71}$$

$$\mathcal{G}(b_k) = \frac{\pi k(b_k)\bar{k}(b_k)}{4\gamma_{21}}, \quad \mathcal{G}(a_k) = \frac{\pi k(a_k)\bar{k}(a_k)}{4\gamma_{21}}. \tag{72}$$

For example, for a single interface crack ($n = 1$, $a_1 = -a$, $b_1 = a$) in two bonded orthotropic half planes subjected to remote loading $\sigma_{22} = \sigma_0$, $\sigma_{12} = \tau_0$ (Figure 9, $h = \infty$) it was shown that (Erdogan and Wu, 1993, Erdogan and Kadioglu 1994, Kadioglu and Erdogan 1995)

$$k(a) = (2a)^{i\omega}(1-2i\omega)(\eta\sigma_0 + i\tau_0)\sqrt{a},$$

$$k(-a) = (2a)^{-i\omega}(1+2i\omega)(\eta\sigma_0 + i\tau_0)\sqrt{a}, \qquad (73)$$

$$\mathcal{G}(-a) = \mathcal{G}(a) = \frac{\pi}{4}\frac{1+4\omega^2}{\gamma_{21}}(\eta^2\sigma_0^2 + \tau_0^2)a. \qquad (74)$$

If the two planes are isotropic γ_{ij} and $\mathcal{G}$ would reduce to the following known results (Erdogan and Gupta 1971a, Erdogan and Wu 1993a)

$$\gamma_{12} = \gamma_{21} = \frac{\mu_1\mu_2[\mu_1(\kappa_2+1)+\mu_2(\kappa_1+1)]}{(\mu_1+\mu_2\kappa_1)(\mu_2+\mu_1\kappa_2)},$$

$$\gamma_{11} = \frac{\mu_1\mu_2[\mu_1(\kappa_2-1)-\mu_2(\kappa_1-1)]}{(\mu_1+\mu_2\kappa_1)(\mu_2+\mu_1\kappa_2)},$$

$$\gamma = \frac{\gamma_{11}}{\sqrt{\gamma_{12}\gamma_{21}}}\frac{\mu_1(\kappa_2-1)-\mu_2(\kappa_1-1)}{\mu_1(\kappa_2+1)+\mu_2(\kappa_1+1)}, \quad \eta = 1, \qquad (75)$$

$$\mathcal{G} = \frac{\pi k\bar{k}}{4}(1-\gamma^2)\left(\frac{\kappa_2+1}{4\mu_2} + \frac{\kappa_1+1}{4\mu_1}\right). \qquad (76)$$

A typical example showing the strain energy release rate is given by Figure 10, where medium 1 is a high modulus carbon (E = 380 GPa, $\nu = 0.32$), medium 3 is epoxy (E = 3.1 GPa, $\nu = 0.35$) and medium 2 is an orthotropic material with a lamellar (e = 0.25) or a columnar (e = 4) structure (e = E_{222}/E_3, $G_{212} = E_{222}/2(1+\nu_3)$) (Erdogan and Wu 1993b). Here the structure of medium 2 is assumed to be due to crystallization or copolymerization of epoxy near the interface. The results given in Figure 10 are obtained under remote tension σ_0 and plane strain conditions. The normalizing strain energy release rate G_0 shown in Figure 10 corresponds to h = 0; that is, to the material pair 1 and 3 (carbon/epoxy). Note that, as physically expected, $\mathcal{G} > \mathcal{G}_0$ if the stiffness of the interfacial region is smaller than that of medium 3 (e < 1) and $\mathcal{G} < \mathcal{G}_0$ if e > 1. In either case $\mathcal{G}$ is bounded by $\mathcal{G}_0$ and $\mathcal{G}_\infty$, corresponding to h = 0 and h = ∞, respectively. For the material system considered, these limits are

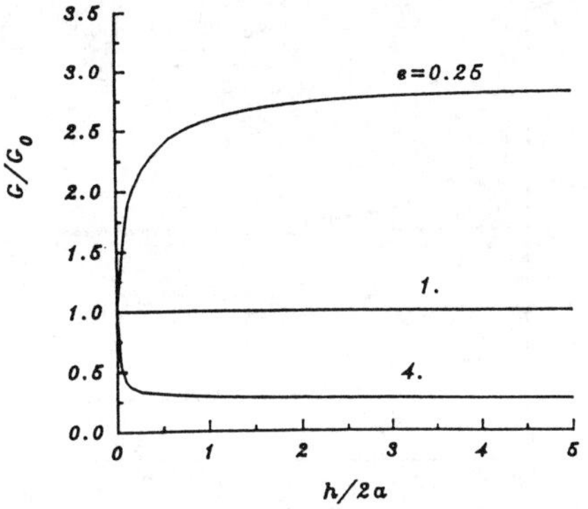

Figure 10. Normalized strain energy release rate for an interface crack in dissimilar isotropic materials bonded through an orthotropic interfacial zone (see Figure 9 for geometry).

e	$\mathcal{G}_0/\sigma_0^2 a$ $(GPa)^{-1}$	$\mathcal{G}_\infty/\sigma_0^2 a$ $(GPa)^{-1}$	$\mathcal{G}_\infty/\mathcal{G}_0$
0.25	0.4344	1.2349	2.8428
4	0.4344	0.1167	0.2686

Figure 11 shows the results for two collinear cracks of lengths 2a and 2c, c = 0.2a, in an isotropic medium (full lines) and two interface cracks in bonded isotropic materials (dashed lines) (Figure 9, E_2/E_1 = 15, $\nu_1 = \nu_2 = 0.1$, $\omega = -0.1306$). In each case $\mathcal{G}^*$ is obtained for a single crack of length 2a for the particular material or material combination. Thus, the limiting values are

$$\frac{\mathcal{G}_A}{\mathcal{G}^*} = \frac{\mathcal{G}_B}{\mathcal{G}^*} = 1, \; \frac{\mathcal{G}_C}{\mathcal{G}^*} = \frac{\mathcal{G}_D}{\mathcal{G}^*} = c/a \quad \text{for } d/a \to \infty, \qquad (77)$$

$$\mathcal{G}_A \to \infty, \; \mathcal{G}_D \to \infty, \; \frac{\mathcal{G}_B}{\mathcal{G}^*} = \frac{\mathcal{G}_C}{\mathcal{G}^*} = \frac{(c+a)}{a} \quad \text{for } \frac{d}{a} = 0 \qquad (78)$$

where d is the distance between the inner crack tips A and D. These and more extensive results given by Erdogan and Wu (1993b) show that the normalized strain energy release rate for collinear cracks in bonded materials may be approximated by that obtained for the homogeneous isotropic medium having the same crack geometry. Further results obtained from a wide range of material combina-

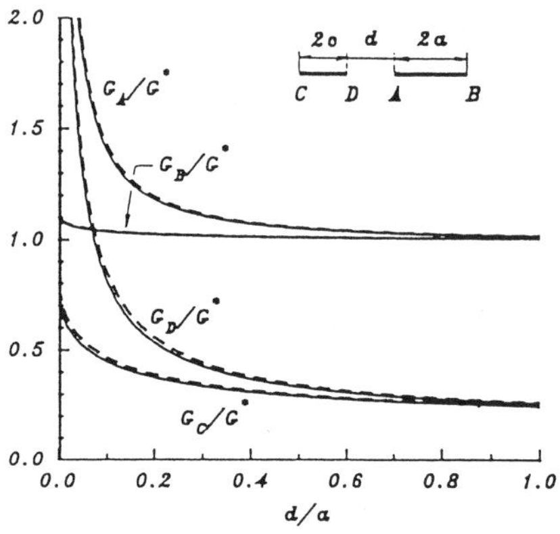

Figure 11. Normalized strain energy release rates at various crack tips in a homogeneous isotropic medium (full lines) and in two bonded dissimilar isotropic half planes (dashed lines, $E_2 / E_1 = 15$, $\nu_1 = \nu_2 = 0.1$, $\omega = -0.1306$).

tions and crack geometries were given by Erdogan and Wu (1993a, 1993b).

The interface crack problems involving orthotropic coatings bonded to orthotropic substrates are considered by Erdogan and Kadioglu (1994) and Kadioglu and Erdogan (1995). Erdogan and Kadioglu (1994) considered the T-shaped crack, and Kadioglu and Erdogan (1995) studied an interface crack starting from a stress-free end under thermal as well as mechanical loading.

4.4 *Effect of Material Inhomogeneity-Graded Interfacial Zones*

Recent studies seem to indicate that, largely as a consequence of material processing, in some bonded materials the thermomechanical properties of the interfacial region vary continuously in thickness direction, resulting in a highly inhomogeneous layer between the adherents. Electron microbe line scans and scanning Auger depth profiles indeed show that in some diffusion bonded materials atomic composition of the two materials varies continuously across the nominal interface (Shiau et al. 1988, Brennan 1991). Similarly, during some deposition processes, as a consequence of sputtering there is a certain amount of mixing of the two species, giving again an inhomogeneous interfacial region (Batakis and Vogan 1985, Houck 1987). There is also the new class of materials called *functionally graded materials* (FGMs) used mostly as coatings and interfacial zones in which the material properties are intentionally graded for the purpose of enhancing the bonding strength (Kurihara 1990) and reducing the residual and thermal stresses (Lee et al. 1996, Lee and Erdogan 1994). (See Yamanouchi et al. 1990, Holt et al. 1992 and Ilschner and Cherradi 1994 for extensive review and references.)

In replacing the "ideal" interface by a thin graded layer, on one hand the problem is made much more difficult due to the fact that the material parameters are functions of the space variables, and consequently the governing equations are now a system of partial differential equations with variable coefficients. On the other hand, by assuming the material parameters to be continuous functions, the anomalies regarding the crack tip stress and displacement oscillations and non-square-root stress singularities associated with the "ideal interfaces" or with the material property discontinuities are completely eliminated. Comparison of the two typical crack problems for piecewise homogeneous materials and for inhomogeneous materials with continuous properties and piecewise continuous derivatives is shown in Figure 12. Defining f_i and g_i, $(i = 1,2)$ as in (6) and (10), the integral equations of the crack problems (a) and (b) may be expressed as

$$\frac{1}{\pi}\int_{-a}^{a}\sum_{1}^{2}\left[\frac{\delta_{ij}}{t-x}+k_{ij}^{s}(x,t)+k_{ij}^{f}(x,t)\right]f_j(t)dt = \frac{1+\kappa_1}{2\mu_1(0)}g_i(x), |x|<a \tag{79}$$

where the kernels k_{ij} are known functions which depend on the geometry of the medium and properties of the materials, k_{ij}^{s} are generally associated with the infinite medium and k_{ij}^{f} represent the details of the geometry and are bounded for all values of h. For $h > 0$ in both cases the problem is one of embedded crack, k_{ij}^{s} as well as k_{ij}^{f} are bounded and the solution has the standard square-root singularity (Konda and Erdogan 1994). For $h = 0$ in problem (a), k_{11}^{s} and k_{22}^{s} would become a Cauchy kernel $1/(t-x)$ and k_{12}^{s} and k_{21}^{s} would degenerate into a delta function $\delta(t-x)$, leading to the singular integral equations of the second kind for an ideal interface crack.

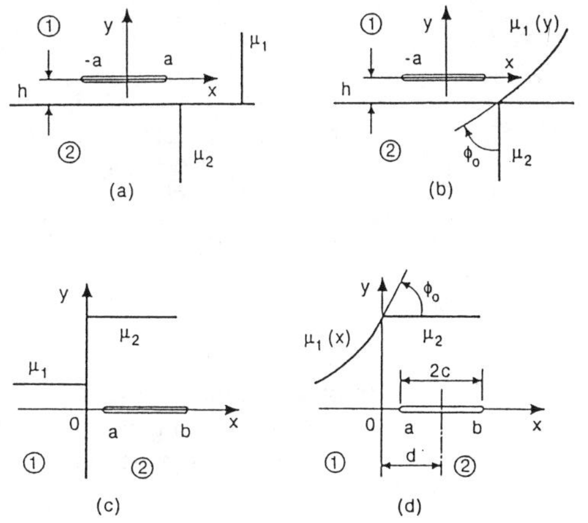

Figure 12. Typical crack geometries in piecewise homogeneous and in inhomogeneous materials.

On the other hand in problem (b) (Figure 12) it can be shown that for $h = 0$, k_{ij}^s become (Delale and Erdogan 1988b, Chen and Erdogan 1996, Ozturk and Erdogan 1996)

$$k_{11}^s = k_{22}^s = -\frac{\pi\gamma}{8}\frac{|t-x|}{t-x},$$
$$k_{12}^s = k_{21}^s = \frac{\gamma}{4}\log|t-x|, \quad \gamma = \tan\phi_0. \tag{80}$$

The important consequence of (80) is that, aside from the Cauchy kernels, since all kernels in (79) are square integrable, the solution would have standard square-root singularity. It can also be shown that the stress state around the crack tip has the asymptotic behavior

$$\sigma_{ij}(r,\theta) = \exp[r\varphi_1(\theta)]\frac{1}{\sqrt{2r}}\left[k_1 f_{1ij}(\theta) + k_2 f_{2ij}(\theta)\right], \quad (i,j = r,\theta), \tag{81}$$

$$\sigma_{iz}(r,\theta) = \exp[r\varphi_2(\theta)]\frac{k_3}{\sqrt{2r}} f_{3i\theta}, \ (i = r,\theta) \tag{82}$$

where φ_1 and φ_2 are known functions, f_{1ij}, f_{2ij}, f_{3i} are the standard functions corresponding to modes I, II and III crack tip stress distributions in homogeneous materials (Delale and Erdogan 1988b) and k_1, k_2, and k_3 are the stress intensity factors.

Similarly, in mode I crack problems shown in Figures 12(c) and (d), it can be shown that by defining

$$f(x) = \frac{\partial}{\partial x}(v^+ - v^-), \ p(x) = \sigma_{yy}(x,0), \tag{83}$$

the integral equation may be expressed as

$$\frac{1}{\pi}\int_a^b \left[\frac{1}{t-x} + k_s(x,t) + k_f(x,t)\right] f(t)dt = \frac{1+\kappa_2}{2\mu_2} p(x), \ |x| < a. \tag{84}$$

Again, for $a > 0$, k_s as well as k_f is bounded and the solution has the standard square-root singularity. In problem (c) for $a = 0$, k_s has the form (Cook and Erdogan 1972, Erdogan and Biricikoglu 1973)

$$k_s(x,t) = \frac{c_1}{t+x} + \frac{c_2 x}{(t+x)^2} + \frac{c_3 x^2}{(t+x)^3}, \quad 0 < (t,x) < b \tag{85}$$

where c_1, c_2 and c_3 are known bimaterial constants (Cook and Erdogan 1972). Note that as x and t go to zero, k_s becomes unbounded and would contribute to the singular behavior of the solution, giving

$$\sigma_{ij}(r,\theta) = \frac{k_1}{r^\alpha} g_{ij}(\theta), \ 0 \le \theta \le \pi, \ (i,j = x,y), \quad 0 < \alpha < 1 \tag{86}$$

where k_1 is a "stress intensity factor," g_{ij} are known functions, α is real and is obtained from the material constants. Generally, $\alpha \ge \frac{1}{2}$ for $\mu_2 \ge \mu_1$ and $\alpha \le \frac{1}{2}$ for $\mu_2 \le \mu_1$.

If we now "smooth" the material property distribution by eliminating the discontinuity (problem d) it can be shown that for $a = 0$ the leading terms in k_s become (Erdogan et al. 1991)

$$k_s(x,t) = \frac{d_1 t}{t+x} + \frac{d_2 x}{t+x} + \frac{d_3 tx}{(t+x)^2} + d_4 \log(t+x), \tag{87}$$

where $d_1,\ldots,d_4$ are known bimaterial constants (Erdogan et al. 1991). Note that k_s and k_f are square integrable and, therefore, would make no contribution to the singular behavior of the stress state at the crack tip; that is, the square-root power singularity and the angular distribution of the stresses would remain identical to that found for homogeneous materials. Similar results are found for a crack crossing the interface (Kasmalkar 1997). In piece-wise homogeneous materials the stresses are singular (with a power always less than 1/2) (Erdogan and Biricikoglu 1973) and in inhomogeneous materials (with $\phi_0 \neq 0$) the stress state is always bounded (Kasmalkar 1997). Needless to say, because of the standard square-root singularity, the strain energy release rate for a planar crack in an inhomogeneous medium under general loading conditions may be expressed as

$$\mathcal{G} = \frac{\pi(1+\kappa_a)}{8\mu_a}(k_1^2 + k_2^2) + \frac{\pi k_3^2}{2\mu_a} \tag{88}$$

where k_1, k_2 and k_3 are the standard modes I, II and II stress intensity factors, μ_a and κ_a are elastic constants calculated at the crack tip x = a and $\mathcal{G}$ is for unit length of the crack front.

In bonded materials with an inhomogeneous intefacial zone, the weak fracture plane may be either along one of the nominal interfaces (e.g., z = 0 or z = h in Figure 13) or in the interfacial zone itself (insert in Figure 14). Some sample results showing the stress intensity factors and the strain energy release rate are given in Figures 14-17. Figures 14 and 15 show the results for in-plane loading $\sigma_{yy}(x,0) = -p_0$, $\sigma_{xy}(x,0) = 0$, $-a < x < a$ of two dissimilar homogeneous materials bonded through an inhomogeneous medium containing a crack parallel to the nominal interfaces (Delale and Erdogan 1988a). The inhomogeneity is assumed to be of the form

$$E(y) = E_0 e^{\beta y}, \nu(y) = (A_0 + B_0 y)e^{\beta y}, \; -h_2 < y < h_1 \tag{89}$$

where from the insert in Figure 15 it follows that

$$E_0 = E_1 e^{-\beta h_1}, \; B = \frac{1}{h_0}\log(E_1/E_2),$$

$$A_0 = \nu_2 e^{\beta h} + B_0 h_2, \; B_0 = \frac{1}{h_0}(\nu_1 e^{-\beta h_1} - \nu_2 e^{-\beta h_2}). \tag{90}$$

Figure 14 shows the strain energy release rate obtained from (88) by using the values of E and ν

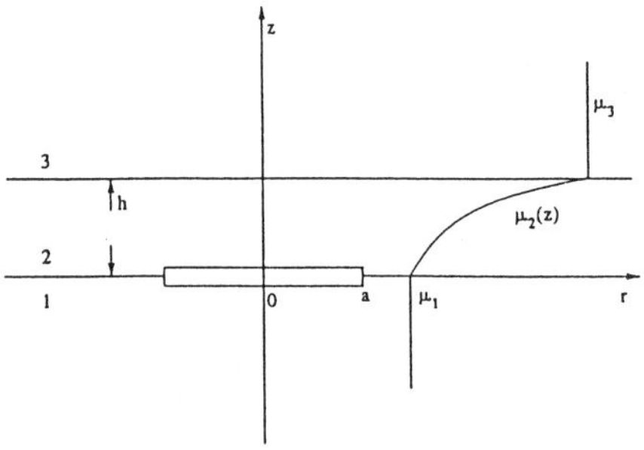

Figure 13. Geometry and notation for a penny-shaped interface crack in dissimilar materials bonded through an inhomogeneous interfacial zone.

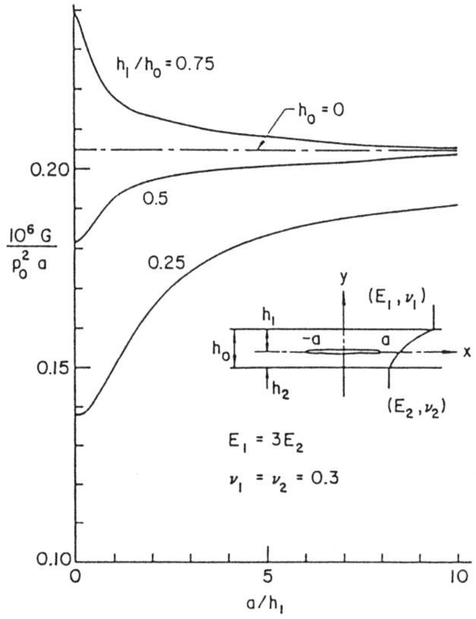

Figure 14. Normalized strain energy release rate for a pressurized crack embedded into an inhomogeneous interfacial region.

at (k = a, y = 0). In this example the material constants are $E_1 = 20.7\times10^{10}$Pa, $\nu_1 = 0.3$, $E_2 = 6.9\times10^{10}$Pa, $\nu_2 = 0.3$. The figure also shows $\mathcal{G} = \mathcal{G}_{12}$ corresponding to $h_0 = 0$ which is obtained from (23) for the ideal interface crack problem. It is seen that as a/h_0 tends to infinity, $\mathcal{G}(h)$ approaches $\mathcal{G}_{12}$.

For the material pair $E_1 = 3.1\times10^9$Pa, $\nu_1 = 0.35$, $E_2 = 12.96\times10^{10}$Pa, $\nu_2 = 0.25$, Figure 15 shows the stress intensity factors and the probable direction θ_0 of further crack growth obtained from (24) (Erdogan and Sih 1963). Further results for the interfacial zone model described by (89) are given by Delale and Erdogan (1988a).

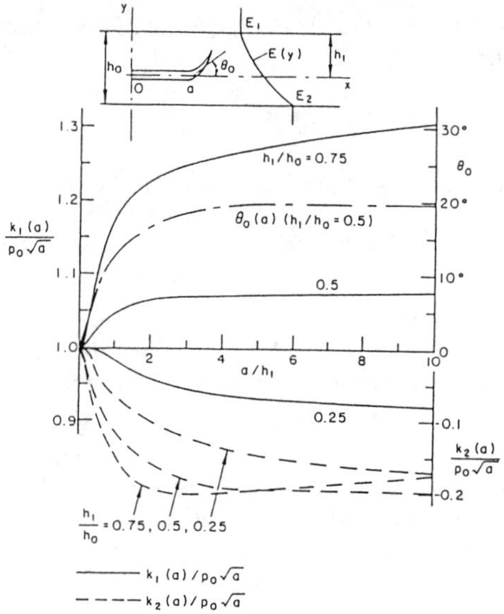

Figure 15. Normalized stress intensity factors and the probable kinking angle θ_0 for a crack in an inhomogeneous interfacial zone.

Some results for the axisymetric interface crack problem for homogeneous dissimilar materials bonded through an inhomogeneous interfacial zone are given in Figures 16 and 17 (Ozturk and Erdogan 1996). In this example κ is assumed to be constant throughout the medium and the shear modulus is given by

$$\mu_2(z) = \mu_1 e^{\alpha z}, \ \alpha = \frac{1}{h}\log(\mu_3/\mu_1). \tag{91}$$

Figure 16 shows the normalized strain energy release rate $\mathcal{G}/\mathcal{G}_0$ for various values of μ_3/μ_1, where $\mathcal{G}$ is obtained from (88) by using $\mu_a = \mu_1$ and $\mathcal{G}_0 = \pi(1+\kappa_1)p_0^2 a/8\mu_1$ is the corresponding plane strain result in a homogeneous medium (μ_1, κ_1), both for unit crack length along the crack front. The external load is a remote tension $\sigma_{zz} = p_0$. $\mathcal{G}$ is a function of h/a, the dimensionless length parameter of the medium. For each given μ_3/μ_1 and κ, the limiting value $\mathcal{G}_{13}$ of $\mathcal{G}(h)$ as $h \to 0$ is obtained independently from the corresponding interface crack problem for which the complex stress intensity factor is given by Kassir and Bergman (1972) and Erdogan and Arin (1972) as follows:

$$k_1 + ik_2 = 2p_0\left(\frac{a}{\pi}\right)^{1/2}\frac{\Gamma(2+i\omega)}{\Gamma(\frac{1}{2}+i\omega)},$$

$$\omega = \frac{1}{2\pi}\log\left(\frac{\mu_3+\mu_1\kappa}{\mu_1+\mu_3\kappa}\right), \tag{92}$$

where $\Gamma(z)$ is the Gamma function. With k_1 and k_2 known, $\mathcal{G}_{13}$ is obtained from (23a). Note that for $h \to \infty$ the problem reduces to that of a homogeneous medium (μ_1, κ) containing a penny-shaped crack of radius a for which we have

$$\lim_{h\to\infty}\mathcal{G}(h) = \frac{4}{\pi^2}\mathcal{G}_0. \tag{93}$$

Thus, in Figure 16 for $(h/a) \to \infty$, $\mathcal{G}/\mathcal{G}_0$ would be expected to approach $(4/\pi^2) = 0.4053$. This trend, too, may be observed in Figure 16 for all values of μ_3/μ_1. Note that as physically expected $\mathcal{G}/\mathcal{G}_\infty > 1$ if the global stiffness of the medium is smaller than μ_1 or $\mu_3 < \mu_1$ and $\mathcal{G}/\mathcal{G}(\infty) < 1$ if $\mu_3 > \mu_1$.

Some sample results showing the dependence of $\mathcal{G}$, k_1 and k_2 on the stiffness ratio μ_3/μ_1 is given by Figures 17a and 17b for the crack surface fractions $\sigma_{zz}(r,0) = -p_0$, $\sigma_{rz}(r,0) = 0$ and $\sigma_{zz}(r,0) = 0$, $\sigma_{rz}(r,0) = -q_0$, respectively. For $\mu_3/\mu_1 = 1$ the medium is homogeneous and the results are

$$k_1/p_0\sqrt{a} = \frac{2}{\pi}, \ k_2/p_0\sqrt{a} = 0,$$

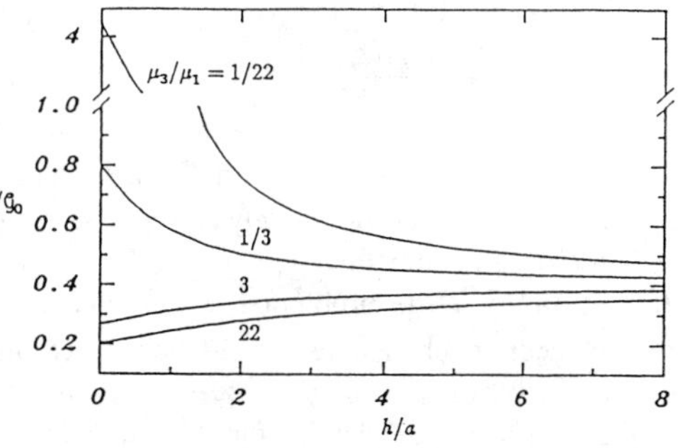

Figure 16. Normalized strain energy release rate for a penny-shaped interface crack (Figure 13) in a medium under remote tension.

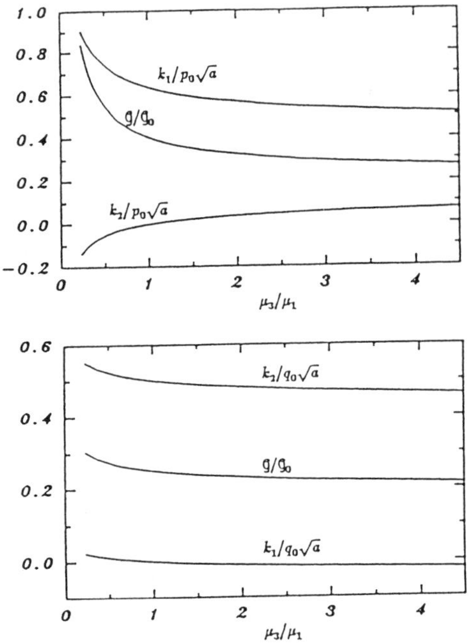

Figure 17. Normalized stress intensity factors and strain energy release rate for a penny-shaped interface crack in dissimilar materials bonded through an inhomogeneous interfacial zone (Figure 13); h/a = 0.5, $\nu = 0.3, \sigma_{zz}(r,0) = -p_0, \sigma_{rz}(r,0) = 0$ (top figure), $\sigma_{rz}(r,0) = -q_0$ (bottom figure).

$$\mathcal{G}/\mathcal{G}_0 = \frac{4}{\pi^2}, \mathcal{G}_0 = \frac{\pi(1+\kappa)}{8\mu_1} p_0^2 a, \tag{94}$$

$$k_1/q_0\sqrt{a} = 0, \; k_2/q_0\sqrt{a} = \tfrac{1}{2},$$
$$\mathcal{G}/\mathcal{G}_0 = \frac{1}{4}, \mathcal{G}_0 = \frac{\pi(1+\kappa)}{8\mu_1} q_0^2 a. \tag{95}$$

These results may easily be verified in Figure 17. As μ_3/μ_1 approaches zero, the stiffness of the half space z > 0 decreases indefinitely and, as a result k_1, k_2 and $\mathcal{G}$ become unbounded. On the other hand, as $(\mu_3/\mu_1) \to \infty$, the problem becomes one of an elastic medium bonded to a rigid half space and, consequently, k_1, k_2 and $\mathcal{G}$ approach certain finite limits. For analytical details, further results and for the discussion of probable crack growth direction, we refer to Ozturk and Erdogan (1996). Additional results for crack problems involving inhomogeneous interfaces and coatings may be found in (Delale and Erdogan 1988b) and (Chen and Erdogan 1996) for plane strain and in (Ozturk and Erdogan 1992 and 1993) for antiplane shear loading.

4.5 *Crack Kinking*

In homogeneous isotropic brittle solids the direction of crack kinking may be obtained from (24) (Erdogan and Sih, 1963) which maximizes the cleavage stress $\sigma_{\theta\theta}(r,\theta)$. It may be observed that if $\sigma_{\theta\theta}(r,\theta_0) = \max.$ then $\sigma_{r\theta}(r,\theta_0) = 0$. Physically, a somewhat more acceptable criterion for crack kinking would be to assume that θ_0 corresponds to $\mathcal{G}^t(\theta) = \max$, $\mathcal{G}^t(\theta)$ being the strain energy releae rate for a "small" crack kinked at an angle θ measured from the direction of the parent crack. If the stress intensity factors for the parent crack prior to kinking are k_1 and k_2, then the stress intensity factors at the tip of a kinked crack may be expressed as

$$k_i^t = \sum_1^2 c_{ij} k_j + b_i T\sqrt{\Delta a}, \; (i = 1,2) \tag{96}$$

where T is the uniform remote stress acting parallel to the crack, the θ − dependence of the coefficients c_{ij} are given by Hayashi and Nemat-Nasser (1981) and by He and Hutchinson (1989) and b_i are given by He et al. (1991). With k_i^t known, the strain energy release rate for the kinked crack may be obtained from

$$\mathcal{G}^t(\theta) = \pi \frac{1+\kappa}{8\mu}\left(k_1^{t^2} + k_2^{t^2}\right). \tag{97}$$

Observing that k_1^t and k_2^t are functions of θ, the kink angle θ_0 is obtained by maximizing $\mathcal{G}^t(\theta)$. The actual crack kinking would take place at the level of the external loads corresponding to $\mathcal{G}^t_{max} = \mathcal{G}^t(\theta_0) = \mathcal{G}_{IC}$, where $\mathcal{G}_{IC}$ is the fracture toughness of the brittle solid. It was shown that θ_0 obtained from $\mathcal{G}^t(\theta)$ = maximum is very nearly identical to the kinking angle given by $k_2^t(\theta) = 0$ (Hutchinson and Suo, 1992). The kinking angle θ_0 is a function of k_2/k_1 explicitly or as expressed through the angle $\psi = \arctan(k_2/k_1)$.

Similar to the kinked crack in a homogeneous medium, one may define the stress intensity factors k_1^t and k_2^t at the tip of a "small" crack of length Δa kinking from the parent crack which now lies along the interface. k_1^t and k_2^t may thus be expressed as

$$k_1^t + ik_2^t = ck(\Delta a)^{i\omega} + d\bar{k}(\Delta a)^{i\omega} + bT\sqrt{\Delta a} \quad (98)$$

where k is the complex stress intensity factor for the parent interface crack, T is the remote stress parallel to the interface, the constants c and d are given by He and Hutchinson (1989) and b is given by He et al. (1991). Thus, once k_1^t and k_2^t are known, the strain energy release rate for the kinked crack may be obtained from (97). If $\mathcal{G}_{IC}$ for the homogeneous medium and the fracture toughness $\mathcal{G}_C^i$ of the interface are known, then kinking would be favored over coplanar growth of the interface crack provided

$$\mathcal{G}^i / \mathcal{G}_{max}^t < \mathcal{G}_C^i / \mathcal{G}_{IC} \quad (99)$$

where $\mathcal{G}^i = \mathcal{G}_{12}$ is the strain energy release rate for the interface crack given by (23a), $\mathcal{G}_C^i$ may be heavily dependent on ψ and again $\mathcal{G}_{max}^t$ is obtained by maximizing $\mathcal{G}^t(\theta)$. For further discussion of the crack kinking we refer to He and Hutchinson (1989), He et al. (1991) and Hutchinson and Suo (1992). The related problem of a crack terminating at the interface and either growing into the adjacent medium or kinking along the interface was considered by Lu and Erdogan (1983) for isotropic materials and by Martinez and Gupta (1994) and Kadioglu and Erdogan (1994) for orthotropic materials.

It should be pointed out that the analytical continuum elasticity solutions described in Section 4 of this article invariably involve semi-infinite spaces, simple crack geometries and simple loading conditions. They, nevertheless, provide very valuable general information and, in many cases, very useful benchmark solutions. However, for more realistic part/crack geometries, material systems and loading conditions, a fully numerical approach such as the application of a finite element, finite difference or boundary element method is unavoidable. Since all fracture problems involve stress singularities, during the past few decades quite considerable effort has been devoted to improve the efficiency and accuracy of these numerical techniques by embedding the correct asymptotic behavior of the solution near the crack tips into the numerical analysis. Needless to say, the literature on the subject is quite extensive. As examples we may mention the work of Kaya and Nied (1993) describing the development of enriched finite elements and transition elements at the tip of an interface crack between two dissimilar orthotropic materials. Similar enriched and transition elements for interface cracks in bonded dissimilar materials and in graded materials was developed by Lee and Erdogan (1997).

4.6 *Effect of Plasticity*

As mentioned previously, if the fracture process zone around the crack tip is "sufficiently small" for a given phase angle ψ, the coplanar growth of the crack may be represented by a single parameter fracture criterion by, for example, comparing the strain energy release rate $\mathcal{G}$ with the interfacial fracture toughness $\mathcal{G}_C^i$. In this simplified fracture model $\mathcal{G}_C^i$ is assumed to be independent of the crack size. If these conditions are not satisfied, then the plastic deformations in the medium associated with the interface crack need to be taken into account, which invariably lead to a multiparameter or an R-curve type fracture characterization.

There are generally two approaches to account for plasticity effects in ductile fracture. In the first and the simpler of the two approaches, it is assumed that the plastic deformations are confined to a "thin" elastic-plastic strip along the crack plane which, in turn, may be represented by a nonlinear spring (Hutchinson and Suo, 1992). This is known as the Dugdale or Dugdale-Barenblatt model. In applying the model the size of the plastic zone or the cohesive zone is determined by requiring the stress singularity to vanish. As for the fracture criterion, one either compares the maximum value of the crack opening displacement (usually the opening at the actual crack tip) with a critical crack opening stretch or expresses the condition of global energy balance based on J-integral type considerations.

The second approach in dealing with plasticity effects in interface fracture is based on full-field elastic-plastic analysis and a global energy balance criterion. The problem is studied generally either under the assumption of small scale yielding (Shih and Asaro, 1988, 1989, 1990) or finite deformation (Shih and Asaro, 1991). In the small scale yielding studies the interface crack is assumed to be semi-infinite and the loading and geometry are characterized by a complex stress intensity factor describing the elastic far-field. A good deal of the existing solutions in this area are obtained by Asaro, Shih and their colleagues using a finite element method (see Shih et al. 1990 for review and references). The elastic-plastic boundary is shown to be given by

$$r_p(\theta) = \frac{k\bar{k}}{\sigma_0^2}\pi R(\theta, \psi) \tag{100}$$

where (r,θ) are the polar coordinates at the crack tip, k is the far-field complex stress intensity factor, σ_0 is the smaller of the two yield strengths of bonded elastic-plastic solids, and R is a dimensionless function (Shih et al. 1990). Each material is characterized by the Ramberg-Osgood relation or by a uniaxial stress-strain relation of the form

$$\frac{\sigma}{\sigma_0} = \frac{\varepsilon}{\varepsilon_0} + \alpha\left(\frac{\sigma}{\sigma_0}\right)^n, \quad \varepsilon_0 = \frac{\sigma_0}{E} \tag{101}$$

where $\sigma_0 = \min(\sigma_{01}, \sigma_{02})$, $n = \max(n_1, n_2)$, E, ν and α are the parameters of material 1. Thus the crack tip stress field may be expressed as (Shih et al. 1990)

$$\sigma_{ij} = \sigma_0\left(\frac{J}{\alpha\sigma_0\varepsilon_0 r}\right)^{\frac{1}{n+1}} h_{ij}(\theta, \bar{r}, \psi, n), \tag{102}$$

$$\bar{r} = \frac{r\sigma_0^2}{k\bar{k}\pi}, \tag{103}$$

where h_{ij} are a bounded functions. Note that the solution for stationary interface cracks exhibits strong similarities to mixed-mode fields for homogeneous materials, which have the form (Shih and Asaro, 1988)

$$\sigma_{ij} = \sigma_0\left(\frac{J}{\alpha\sigma_0\varepsilon_0 r}\right)^{\frac{1}{n+1}} \bar{\sigma}_{ij}(\theta, \psi_p, n), \tag{104}$$

where ψ_p is the measure of mode mixity based on the relative magnitudes of $\sigma_{\theta\theta}(r,0)$ and $\sigma_{r\theta}(r,0)$ as $r \to 0$ and the functions $\bar{\sigma}_{ij}$ are given in tabular form.

An exhaustive analysis of elastic-plastic interface crack problem under plane strain conditions was given by Sharma and Aravas (1993). In this study one of the adherents was assumed to be rigid, and again J_2 deformation theory with power law hardening was used to describe the other. It was shown that the two term asymptotic solution developed may provide a suitable basis for a two parameter fracture criterion.

The problem of steadily growing interface crack between a brittle and a ductile material under small scale yielding conditions was recently considered by Bose and Ponte Castañeda (1995). Limited solutions are available for finite deformation studies in bonded materials. For sample results obtained by a finite strain viscoplastic J_2 flow theory see Shih et al. (1990).

5. INTERFACIAL ZONES APPROXIMATED BY SPRINGS

In many components such as, for example, adhesively bonded joints used in aerospace structures, because of the complexity of the geometry of the medium, the exact analytical treatment of the problem may not be tractable and, by introducing certain judicious approximations, it may be possible to obtain physically acceptable and yet analytically manageable solutions. Generally, the primary factors influencing such approximations are the adhesive-to-adherent and adherent-to-adherent thickness ratios, the ratio of various thicknesses to the characteristic in-plane dimension of the joint and the degree of accuracy required. The models described in Section 4 of this article dealt with ideal interfaces and continuum adherents and interfacial zones. That is, the problem was treated by using continuum solid mechanics and standard material models and no approximation was made in formulating the related crack problems. In this section we will briefly discuss such approximate models as membranes and plates for adherents and shear and shear/tension springs for the interfacial zone along with the ideal interfaces and continuum adherents whenever appropriate.

5.1 *Shear Spring-Membrane and Continuum Adherents*

As indicated in Section 3, if at least one of the adherents is approximated by a membrane (e.g., thin films, cover plates and various other relatively thin "reinforcements"), then the interfacial zone would have to be modeled either as an ideal interface ($h_0 = 0$) or as a shear spring ($h_0 > 0$), h_0 being the interfacial zone thickness. The generalized plane stress analysis of lap joints shows that if the thickness variation of stresses is neglected and if the Poisson's ratios of the two bonded elastic layers are equal, then the shear stresses as well as the normal stress at the interface are zero and the load transfer would take place along the edges of the bonded region only (Muki and Sternberg 1968). Even when the Poisson's ratios are different, the magnitude of the contact shear at the interface away from the edges are found to be extremely small, implying very high stress concentrations near the edges.

First we consider the two-dimensional elasticity problem of a thin film of thickness h(x), length 2a and elastic constants μ_1 and $\kappa_1, (\kappa_1 = 4 - 3\nu_1)$ bonded to a substrate of elastic constants μ and κ, $(\kappa = 4 - 3\nu)$. The thickness of the substrate is very large compared to that of the film. Hence the substrate and the film are modeled as a semi-infinite continuum $(-\infty < x < \infty, -\infty < y < 0)$ and a membrane, respectively. Initially it is assumed that the bonding is through an ideal interface. The substrate is subjected to a uniform strain $\varepsilon_{xx}(\mp\infty, y) = \varepsilon_0$ at infinity and the bonded medium may undergo a uniform temperature change T. By using the elasticity solution for the half space and the continuity of the displacements at the interface, the problem may be reduced to the following integral equation for the unknown function $\sigma_{xy}(x,0) = \tau(x)$ (Erdogan and Joseph 1990 Part I, Erdogan and Civelek 1974)

$$\frac{1}{\pi}\int_{-a}^{a}\frac{\tau(t)}{t-x}dt - \frac{\lambda}{h(x)}\int_{-a}^{x}\tau(t)dt = -p, \quad -a < x < a, \qquad (105)$$

where

$$\lambda = \frac{\mu(\kappa_1 + 1)}{2\mu_1(\kappa + 1)}, \qquad (106)$$

$$p = \frac{4\mu}{1+\kappa}\left[\varepsilon_0(1 - \nu\nu_1) + (\alpha^* - \alpha_1^*)(1+\nu)T\right], \qquad (107)$$

and α^* and α_1^* are the thermal expansion coefficients of the substrate and the film, respectively. If there are no external forces acting on the film, then (105) must be solved under the following equilibrium condition

$$\int_{-a}^{a}\tau(x)dx = 0. \qquad (108)$$

If we now assume that the interfacial zone has a thickness h_0 and shear modulus μ_0, and $h_0 << h$, then it may be modeled by a shear spring and the interface condition may be expressed as

$$\frac{u_1(x) - u(x,0)}{h_0} = \frac{\tau(x)}{\mu_0}, \quad -a < x < a \qquad (109)$$

where u_1 and u are the x-component of the displacement in the film and the substrate, respectively. By using the membrane and the half plane solutions, the interface condition (109) would give (Erdogan and Joseph, 1990)

$$\frac{1}{\pi}\int_{-a}^{a}\tau(t)\log|t-x|\,dt + \lambda\int_{-a}^{x}K(x,t)\tau(t)dt - \omega\tau(t) = px + C, \quad -a < x < a \qquad (110)$$

where

$$K(x,t) = \int_{t}^{x}\frac{dr}{h(r)}, \quad \omega = \frac{4\mu h_0}{(\kappa+1)\mu_0}, \qquad (111)$$

and C is an unknown constant to be determined from the equilibrium condition (108).

After solving (105) or (110), the stress in the film may be obtained from

$$\sigma(x) = \frac{1}{h(x)}\int_{-a}^{x}\tau(t)dt. \qquad (112)$$

In the case of an ideal interface and constant film thickness h(x) = h, the solution of (105) is of the form

$$\tau(x) = \frac{g(x)}{\sqrt{a^2 - x^2}}, \qquad (113)$$

where the bounded unknown function g(x) has no closed form expression but may be obtained numerically in a simple and efficient way (Erdogan and Joseph 1990). If the membrane thickness has the form

$$h(x) = f(x)\sqrt{1 - (x^2/a^2)}, \qquad (114)$$

the problem can be reduced to a system of linear algebraic equations. In particular if f(x) is a polynomial, the interface shear $\tau(x)$ can be expressed in closed form (Erdogan and Joseph 1990). In the special case of f(x) = A_0 = constant, it may be shown that (Erdogan and Joseph 1990)

$$\tau(x) = -\frac{pA_0}{A_0 + \lambda a}\frac{x}{\sqrt{a^2 - x^2}}, \quad -a < x < a. \qquad (115)$$

Since the solution has a standard square-root singularity, the mode II stress intensity factor may be defined by and evaluated from

$$k_2 = \lim_{x\to a}\sqrt{2(a-x)}\,\tau(x) = \frac{g(a)}{\sqrt{a}}. \qquad (116)$$

For example, for a membrane having an elliptic thickness, from (115) and (116) it follows that

$$k_2 = -\frac{A_0 p\sqrt{a}}{A_0 + \lambda a}. \tag{117}$$

Also note that if the membrane is inextensible, $\mu_1 = \infty$, $\lambda = 0$ and solution of (105) becomes

$$\tau(x) = -\frac{px}{\sqrt{a^2 - x^2}}, \quad -a < x < a. \tag{118}$$

As one might expect, the singular behavior of the shear stress at the end points $\mp a$ would depend on the contact angle or on $\partial h / \partial x$ at $x = \mp a$. Near the crack tip x = -a. This may be seen, for example, expressing the film thickness and the solution by

$$h(x) = F(x)(x+a)^\gamma, \ \gamma \geq 0, \ 0 < F(-a) < \infty \tag{119}$$

$$\tau(x) = \frac{g(x)}{(a-x)^\beta (a+x)^\alpha},$$

$$0 < (\alpha, \beta) < 1, \ g(x) = \sum_0^\infty c_n (x+a)^n. \tag{120}$$

A simple asymptotic analysis would show that (Erdogan and Joseph 1990)

$$\alpha = 1/2 \ \text{for} \ 0 \leq \gamma < 1, \tag{121}$$

$$\cot \pi\alpha - \frac{\lambda}{F(a)} \frac{1}{1-\alpha} = 0, \ 0 < \alpha < 1/2 \ \text{for} \ \gamma = 1. \tag{122}$$

These are the two interesting cases corresponding to $h'(-a) = \infty$ and $h'(-a) =$ a positive constant, respectively. It can also be shown that for $\gamma > 1$ or for $h'(-a) = 0$, $\tau(-a) = 0$ (Erdogan and Joseph 1990).

If the interfacial zone is a shear spring, the corresponding integral equation (110) is a Fredholm equation of the second kind and its solution would be bounded in the closed interval $-a \leq x \leq a$. By differentiation it can also be shown that near x = a we have

$$\tau'(x) \sim \frac{1}{\pi\omega} \int_{-a}^{a} \frac{\tau(t)}{t-x} dt \sim \frac{\tau(a)}{\pi\omega} \log(a-x). \tag{123}$$

From (123) it may be seen that, even though $\tau(x)$ is bounded at the ends $\mp a$, since its derivative is unbounded, $\tau(x)$ is highly ill-defined as $x \to \mp a$.

The membrane adherent/ideal interface model described in this section and the resulting singularity α are clearly approximate. The exact value of α may be obtained from the asymptotic analysis of bonded elastic wedges (Hein and Erdogan 1971). For example, for a contact angle $\theta_c = \pi/2$ ($h'(a) = \infty$), the membrane theory gives $\alpha = 1/2$ regardless of the material properties, whereas in bonded elastic wedges, depending on θ_c and the material properties, α may not even be real and the $\text{Re}(\alpha)$ is always less than ½. The comparison of the powers of singularity obtained by using the two methods is shown in Figures 18 and 19.

In Figure 18 the membrane is assumed to be of the same material as the substrate. Thus, in the membrane model $\lambda = 1/2$, $F(a) = \tan\theta_c$, and α is obtained from (122) as a function of θ_c. In the elasticity model α is simply obtained from a wedge of angle $\pi + \theta_c$ (Hein and Erdogan 1971). The results are shown in Figure 18. The maximum error is at $\theta_c = \frac{\pi}{2}$ and is approximately 8.9%.

The comparison of singularities for a Si_3N_4 film ($\mu_1 = 2.8 \times 10^6 \, N/m^2$, $\nu_1 = 0.23$) bonded to a Si substrate ($\mu = 6.9 \times 10^6 \, N/m^2$, $\nu_1 = 0.22$) is shown in Figure 19. Note that in the elasticity solution for $0 < \theta_c < 85^\circ$ there is only one root, for

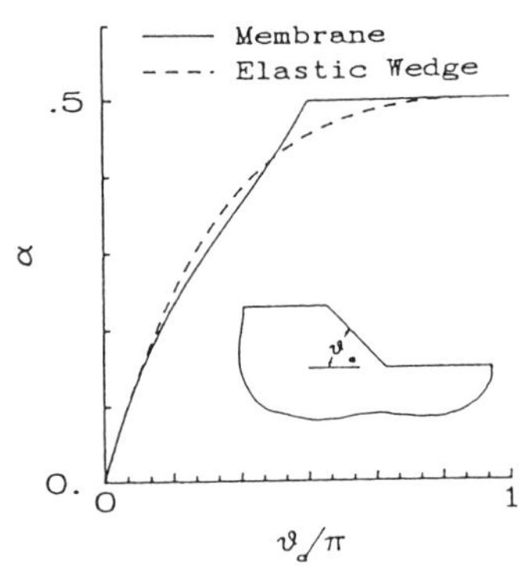

Figure 18. Power of singularity α obtained from the membrane model and from the elastic wedge of angle $\theta_c + \pi$ for identical film and substrate materials.

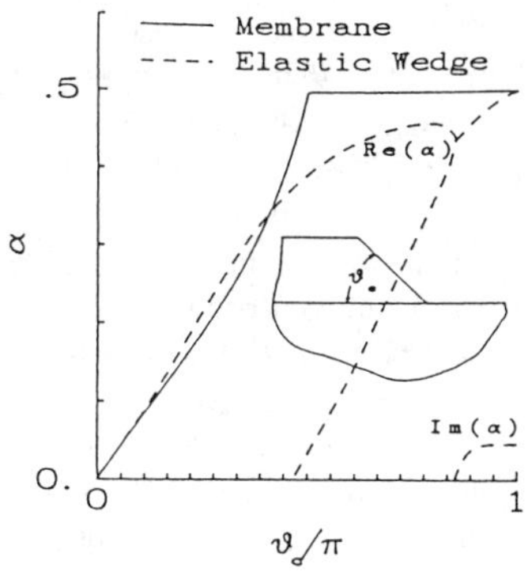

Figure 19. Power of singularity α obtained from membrane and elasticity solutions for dissimilar materials (substrate: silicon, membrane: silicon nitride).

$85^\circ < \theta_c < 153^\circ$ there are two roots which are both real and for $153^\circ < \theta_c < 180^\circ$ the root is complex (for $\theta_c = 180$, $\alpha = 1/2 + 0.046133i$). In the range of $0 < \theta_c < \pi/2$ the maximum difference between the two singularities is approximately 20% and is, again, at $\theta_c = \pi/2$.

For $h_0 = 0$ (the ideal interface) some sample results showing the interface shear τ and the film stress σ are given in Figure 20. In this example it is assumed that the film thickness is

$$h(x) = h_1\left(1 - \frac{x^2}{a^2}\right)^{\gamma}, \quad 0 \le \gamma \le 1, \tag{124}$$

and $\lambda a / h_1 = 5$. By examining the asymptotic behavior of the solution near the end points it can be shown that (Erdogan and Joseph 1990) as $x \to -a$, for example, $\sigma \to 0$ if $0 \le \gamma \le 1/2$, $\sigma =$constant if $\gamma = 1/2$, and $\sigma \to \infty$ if $1/2 < \gamma \le 1$. These trends may also be observed in Figure 20. Note that $\alpha = 1/2$ for $0 \le \gamma < 1$ and $0 < \alpha < 1/2$ for $\gamma = 1$. Also note that for $0 \le \gamma < 1/2$ the maximum film stress is at x = 0, meaning that if the bond is adequately strong, then the failure may initiate as film cracking.

Sample results for interface shear τ and film stress σ for a Si_3N_4 film of constant thickness h bonded to a Si substrate through a SiO_2 interfacial zone ($\mu_0 = 2.9 \times 10^{10}\,N/m^2$, $\nu_0 = 0.2$) of thickness h_0 are shown in Figure 21. The dashed lines shown in

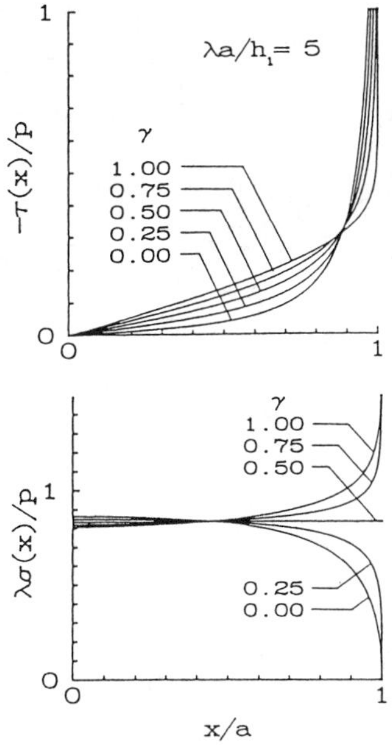

Figure 20. Interface shear $\tau(x)$ and film stress $\sigma(x)$ for a film of thickness $h(x) = h_1(1\text{-}x^2/a^2)$ bonded to an elastic substrate.

the figure corresponds to the perfect bonding case ($h_0 = 0$). For $h_0 > 0$, τ is bounded as $x \to \mp a$ and its values are marked in the figure for various values of h_0/h. On the other hand, for $h_0 = 0$ $\tau(\mp a)$ is unbounded.

5.2 *Strain Energy Release Rate*

In any fracture propagation process, if the stress state near the fracture front remains self-similar, then one can define a "strain energy release rate." In the two interface models described in Section 5.1 this would be the case for ideal interface if $h'(a) = \infty$ and for shear spring interfacial zone if $h'(a) > 0$. In the former a mode II condition with a square-root singularity and in the latter a bounded stress state at the end point is maintained as the fracture front propagates. For the purposes of analysis it is sufficient to consider "fixed grip" conditions and compute the energy available for fracture as the sum of energies internally released from the adherents and the interfacial zone. In the direct adhesion case ($h_0 = 0$) the energy released by the substrate $\mathcal{G}$ may be expressed as

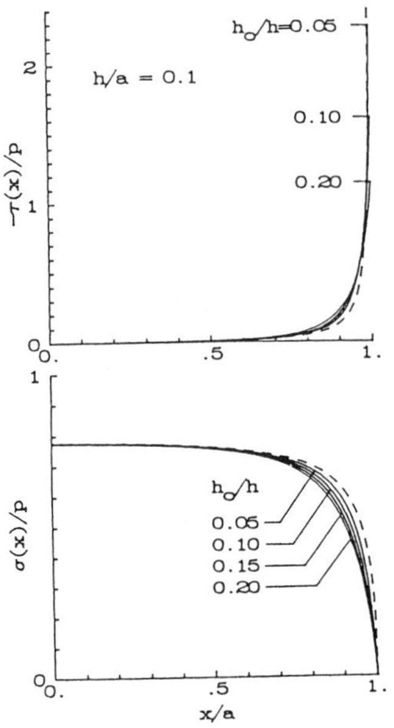

Figure 21. The interfacial shear $\tau(x)$ and film stress $\sigma(x)$ for a Si_3N_4 film of constant thickness h bonded to Si substrate through a SiO_2 layer of thickness h_0. The dashed lines correspond to the solution obtained from the ideal interface $h_0 = 0$.

$$\mathcal{G}_s = \frac{\pi(\kappa+1)}{16\mu} k_2^2 \tag{125}$$

by observing that a pure mode II stress state is maintained at the end point x = a. To calculate the rate of energy released by the film $\mathcal{G}_f$ we note that near the end point x = -a we have

$$\sigma(x) = \frac{1}{h}\int_{-a}^{x} \tau(t)dt \cong \frac{k_2}{h}\sqrt{2(x+a)} \tag{126}$$

$$\begin{aligned} dU_f &= \mathcal{G}_f da = \frac{1+\kappa_1}{8\mu_1}\int_{-a}^{-a+da}\frac{1}{2}\sigma^2 h dx \\ &= \frac{(1+\kappa_1)k_2^2}{4\mu_1 h} k_2^2 (da)^2, \end{aligned} \tag{127}$$

$$\mathcal{G}_f = \frac{1+\kappa_1}{4\mu_1 h} k_2^2 da \cong 0. \tag{128}$$

Thus, the total strain energy release rate would consist of $\mathcal{G}_s$ only. Note that if the contact angle is less than $\pi/2$, since $\alpha < 1/2$ initially the stress state near the end is not self-similar. However, upon initiation of a small interface crack, for example at x = -a, the membrane model dictates that the film thickness at the end becomes nonzero, θ_c becomes $\pi/2$ and (125) and (128) can be used to determine the strain energy release rate.

By using the asymptotic results developed one may easily show that for $0 \le \alpha < 1/2$; that is, for the contact angle $0 \le \theta_c < \pi/2$ in direct film/substrate bonding and for $h_0 > 0$ in bonding through an interfacial zone, $\mathcal{G}_s$ would be zero. Also $\mathcal{G}_f$ is found to be zero regardless of the interface and the end conditions (Erdogan and Joseph 1990). Thus, in the case of an interfacial zone with $h_0 > 0$ the total released energy can only come from the homogeneous interfacial region. If $\tau(-a)$ is the shear stress at the end x = -a, then the energy released by the interface for a "crack" of length da becomes

$$dU_i = \mathcal{G}_i da = \int_{-a}^{-a+da} \frac{h_0}{2\mu_0}[\tau(x)]^2 dx \cong \frac{h_0}{2\mu_0}[\tau(-a)]^2 da,$$

$$\mathcal{G}_i = \frac{h_0}{2\mu_0}[\tau(-a)]^2. \tag{129}$$

The strain energy release would then be $\mathcal{G} = \mathcal{G}_s$ for $h_0 = 0$ and $\mathcal{G} = \mathcal{G}_i$ for $h_0 > 0$.

For a single film of length 2a and thickness h bonded to a substrate through an interfacial zone of thickness h_0 some sample results showing the dependence of the strain energy release rate on the dimensionless stiffness parameters $h/\lambda a$ and $\bar{\beta}$ are given in Figure 22, where

$$\bar{\beta} = \frac{\lambda\omega}{h} = \frac{2h_0(1+\kappa_1)\mu^2}{h(1+\kappa)^2\mu_1\mu_0},$$

$$\lambda = \frac{(1+\kappa_1)\mu}{2(1+\kappa)\mu_1}, \tag{130}$$

and the normalizing strain energy release rate is defined by

$$\mathcal{G}_0 = \frac{1+\kappa}{16\mu}\pi p^2 a \tag{131}$$

which corresponds to the inextensible membrane/ideal interface case. Also note that $\overline{\beta}=0$ corresponds to the ideal interface. The dashed line shown in the figure is obtained from the asymptotic solution for "small" values of $\varepsilon = h/\lambda a$ (Erdogan and Joseph 1990). The asymptotic solution also shows that $\lim_{\varepsilon\to 0} \mathcal{G}/(\varepsilon\mathcal{G}_0) = 2/\pi$.

The results of an example for film cracking and subsequent debonding are shown in Figure 23. Initial film length is 2b and the length of the debond crack is 2a. The problem has a symmetry and $\mathcal{G}(a)$ and $\mathcal{G}(b)$ refer to the strain energy release rates at $x=\mp a$ and $x=\mp b$, respectively. The results are obtained for direct bonding ($h_0 = 0$) and for homogeneous interfacial shear layer ($h_0 = 0.1h$) from (125) and (129), respectively. Note that the difference between $\mathcal{G}(a)$ and $\mathcal{G}(b)$ is insignificant except near $a/b=0$. For $h_0 = 0$, $\mathcal{G}(a)$ becomes unbounded as $a/b \to 0$. This is due to a "strong" singularity caused by the cracked film (Erdogan 1971). A somewhat surprising result is that the strain energy release rates for $h_0 = 0$ and $h_0 > 0$ are not significantly different.

5.3 *Shear/Tension Springs - Plate or Continuum Adherents*

Consider now two elastic half planes bonded through an interfacial zone with distinct structure and nonzero thickness (see the insert in Figure 24). By using the interface conditions (Figure 24)

$$v_2(x,+0)-v_1(x,-0)=\frac{h_3}{E_3^*}\sigma_{yy}(x,0),\ -a<x<a,$$
$$u_2(x,+0)-u_1(x,-0)=\frac{h_3}{\mu_3}\sigma_{xy}(x,0),\ -a<x<a, \tag{132}$$

and the elasticity solution for the half planes, the integral equations for the unknown functions

$$p_1(x)=\sigma_{yy}(x,0),\ p_2(x)=\sigma_{xy}(x,0),\ -a<x<a$$

may be expressed as

$$\int_{-a}^{a} p_2(t)\log\left|\frac{t-x}{t}\right|dt+\beta\pi\int_0^x p_1(t)dt-bp_2(x)+e=0,\quad |x|<a,$$

$$\int_{-a}^{a} p_1(t)\log\left|\frac{t-x}{t}\right|dt-\beta\pi\int_0^x p_2(t)dt-cp_1(x)+d=0,\quad |x|<a, \tag{133}$$

where $E_3^*=E_3$ for plane stress, $E_3^*=E_3/(1-\nu_3^2)$ for plane strain, and

$$\beta=\frac{(\kappa_2-1)\mu_1-(\kappa_1-1)\mu_2}{(\kappa_2+1)\mu_1+(\kappa_1+1)\mu_2}, \tag{134}$$

$$b=\frac{h_3}{\mu_3\gamma_2},\quad c=\frac{h_3}{E_3^*\gamma_2}, \tag{135}$$

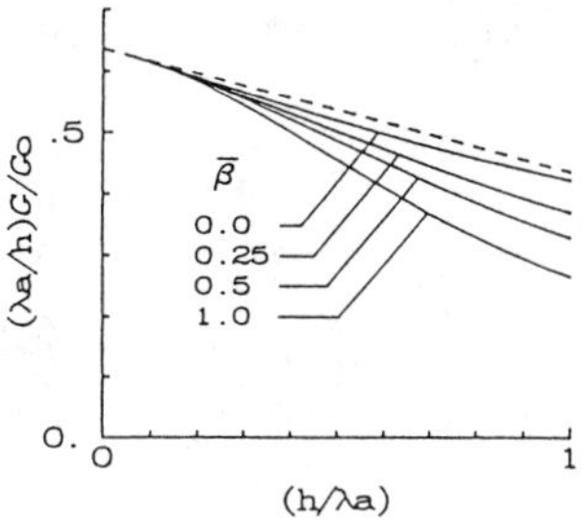

Figure 22. Normalized strain energy release rate for a membrane bonded to an elastic substrate through an interfacial zone of thickness h_0, $\overline{\beta}=0$ for $h_0 = 0$.

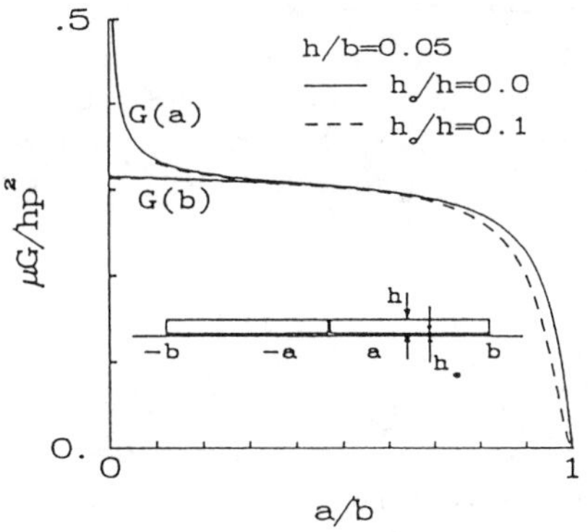

Figure 23. Strain energy release rates for a film of constant thickness cracked in its midsection.

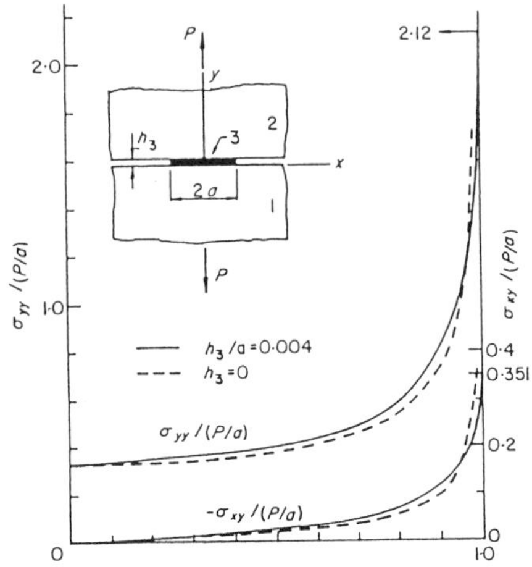

Figure 24. Contact stresses in two dissimilar elastic half planes bonded through a thin interfacial zone ($E_1/E_2 = 3$, $E_3/E_2 = 0.045$, $\nu_1 = \nu_2 = 0.3$, $\nu_3 = 0.35$).

$$\gamma_2 = \frac{1}{4\pi}\left(\frac{\kappa_1+1}{\mu_1}+\frac{\kappa_2+1}{\mu_2}\right). \tag{136}$$

The unknown constants d and e are determined from the following equilibrium conditions:

$$\int_{-a}^{a} p_1(t)dt = P, \quad \int_{-a}^{a} p_2(t)dt = Q. \tag{137}$$

The integral equations (133) are of the second kind and have square-integrable kernels. Consequently, for $h_3 > 0$, the solution is bounded in the closed interval $-a \le x \le a$. In this case the asymptotic analysis of (133) shows that at $x = \mp a$, even though p_1 and p_2 are bounded, their derivatives have a logarithmic singularity; that is, for small values of, for example, a - x we have

$$\frac{dp_i}{dx} \cong C_i \log(a - x), \quad (i = 1,2) \tag{138}$$

where C_i is a constant.

For the symmetric loading Q = 0, the normalized contact stresses $\sigma_{yy}(x,0)$ and $\sigma_{xy}(x,0)$ are shown in Figure 24. The figure indeed shows that, because of (138), at x = a σ_{yy} and σ_{xy} are rather ill-defined. The figure also shows the contact stresses for the limiting case of direct adhesion, $h_3 = 0$ which is given by (Erdogan 1965)

$$p_1(x) = \frac{P}{2\pi}\frac{\alpha+1}{\sqrt{\alpha}}\frac{1}{\sqrt{a^2-x^2}}\cos\left(\omega\log\left(\frac{a-x}{a+x}\right)\right),$$

$$p_2(x) = \frac{P}{2\pi}\frac{\alpha+1}{\sqrt{\alpha}}\frac{1}{\sqrt{a^2-x^2}}\sin\left(\omega\log\left(\frac{a-x}{a+x}\right)\right), \tag{139}$$

where

$$\alpha = (1-\beta)/(1+\beta),$$
$$\omega = \frac{1}{2\pi}\log\alpha. \tag{140}$$

Note that for $h_3 = 0$, the stresses are singular at $x = \mp a$ with the oscillating behavior discussed in Section 4 of this article (see(139). One may also note that for $h_3 > 0$ the stresses are bounded everywhere and the "stress concentration" is expected to decrease with increasing thickness of the interfacial zone. For the material system considered in Figure 24 this may indeed be observed from Table 3.

Table 3. Stress concentration factor for two elastic half planes bonded through an adhesive layer of thickness h_3.

h_3/a	0.004	0.010	0.040
$p_1(a)/(P/a)$	2.120	1.450	0.836
$p_2(a)/(P/a)$	-0.351	-0.166	-0.057

In the approximate analysis of bonded materials, if the adherents are relatively "thick" they may have to be considered as elastic continua. If they are sufficiently "thin" so that their bending stiffness may be neglected, then they could be treated as thin films or membranes. There are, however, structural components that fall between these two models. In this third category of components the thickness is usually small compared to the in-plane dimensions of the layered medium, but not small enough to have negligible bending stiffness. The obvious choice for such components is a plate or shell model, generally taking into account the transverse shear deformations. There is a rather extensive literature on the subject investigating a wide variety of plate theories and linear and nonlinear adhesive models. Invariably the problem is reduced to a system of differential equations in contact stresses by using the interface conditions (see, for example, Delale et al. 1981 where the Reissner's plate theory and a modification of tension/shear spring are used to obtain closed form solutions for various joint geometries and loading conditions). In the absence of finite deflections, the plate approximation can, very often, give reasonably accurate results. Figure 25 shows the shear and normal contact stresses τ and σ in an aluminum panel reinforced by a boron/epoxy composite laminate under tension. The results were obtained by using a transverse shear plate theory and from a full-field finite element method (Delale et al. 1981). For the simple example considered, the agreement is rather good.

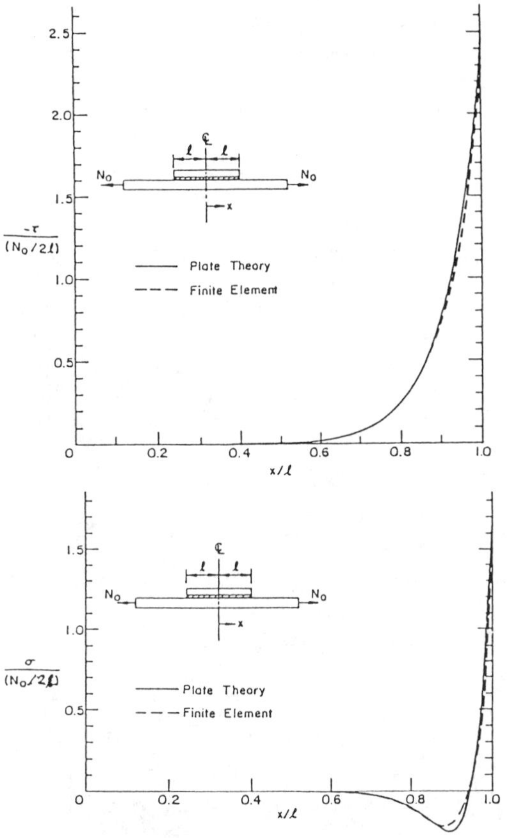

Figure 25. Comparison of contact stresses in a reinforced panel obtained from a plate theory and from the finite element method (Base plate: aluminum h_1 = 2.3 mm., cover plate: boron/epoxy laminate h_2 = 0.76 mm., adhesive: epoxy h_3 = 0.1 mm.

5.4 *Time/Temperature Effects*

In bonded materials usually the operating temperatures are sufficiently low so that no rheological effects need to be taken into account in analyzing the composite medium. However, particularly in adhesively bonded joints, even though the adherents may not be affected, the viscoelastic effects in the adhesive, usually an epoxy, may have to be taken into account. One of the advantages of modeling the interfacial zone as a homogeneous shear layer or tension/shear layer is that such time/temperature effects can be introduced into the analysis without difficulty. Using, for example, hereditary integrals and observing that, for the practical range of stress and temperature levels in adhesive joints under hydrostatic stress most viscoelastic materials behave elastically, the constitutive equations of the adhesive layer may be expressed as (Pipkin 1972, Findley et al. 1976)

$$s_{ij} = 2\int_{-\infty}^{t} G(T, t-\xi)\frac{\partial e_{ij}}{\partial \xi} d\xi, \quad (i, j = 1,2,3), \tag{141}$$

$$e = \frac{s}{3K(T)} + \alpha_3(T)(T - T_o)H(t - t_1) \tag{142}$$

where s_{ij} and e_{ij} are the deviatoric components and s and e the hydrostatic components of respectively the stress and the strain tensor, G is the relaxation modulus, K is the bulk modulus, α_3 is the thermal expansion coefficient of the adhesive and H is the Heaviside function. By using (141), (142) and the plate equations, the interface conditions may be reduced to a system of integro-differential equations which can be solved generally by applying Laplace transforms (Delale and Erdogan 1981a). Similarly, the problem may be formulated by expressing the constitutive equations of the viscoelastic adhesive layer in terms of differential operators assuming that (Pipkin 1972, Flügge 1975)

$$P_1(s_{ij}) = Q_1(e_{ij}), \quad (i, j = 1,2,3), \tag{143}$$

$$P_2(s) = Q_2(e) \tag{144}$$

where P_1, P_2, Q_1 and Q_2 are differential operators of the form $\sum_{0}^{n} a_k(t)\partial^k / \partial t^k$ and a_k is generally temperature-dependent. The application of these techniques to viscoelastic analysis of lap joints under tension, bending, or transverse shear is described by Delale and Erdogan (1981a, 1981b).

Some simple results showing the time dependence of the contact stresses are given in Figures 26 and 27. The single lap-joint problem is solved under the assumptions that the adherents are identical (aluminum plates), the operating temperature T is constant, the external load is a membrane force $N_{xx} = N_0 H(t)$ applied at t = 0 and the relaxation and bulk moduli of the adhesive layer are

$$G(t,T) = \left[\left(\mu_0(T) - \mu_\infty(T)\right)e^{-t/\varepsilon(T)} + \mu_\infty(T)\right]H(t), \tag{145}$$

$$K(T) = \frac{E_0(T)\mu_0(T)}{3\left[3\mu_0(T) - E_0(T)\right]}, \tag{146}$$

$$\varepsilon(T) = \frac{\mu_\infty(T)}{\mu_0(T)} t_0(T) \tag{147}$$

where $\mu_0(T)$ is the shear modulus at t = 0, $\mu_\infty(T)$ at $t = \infty$, $E_0(T)$ is the Young's modulus at t = 0 and $t_0(T)$ corresponds to the retardation time. E_0, μ_0, μ_∞ and t_0 are known functions of temperature. The dependence of interfacial shear stress $\tau(x,t)$ on temperature and time is shown in Figure 26. Note that as time increases, there is a tendency to more uniform distribution of $\tau(x,t)$ which must satisfy the following equilibrium condition

$$\int_{-\ell}^{\ell} \tau(x,t)dx = N_0. \tag{148}$$

At any given time the contact stresses τ and σ are at their peak for $x = \mp\ell$, the ends of the bonding region. These peak values are dependent on temperature and relax in time. Figure 27 shows the time and temperature dependence of these peaks $\tau(\ell,t)$ and $\sigma(\ell,t)$ in the single aluminum/epoxy lap joint described by Figure 26. Further results on the time/temperature effects in bonded joints may be found in Delale and Erdogan (1981a, 1981b).

6. FRACTURE RESISTANCE CHARACTERIZATION

Clearly, any discussion of the fracture mechanics of interfaces would be incomplete without some remarks on the experimental characterization of the resistance of bonded materials to interfacial fracture. As mentioned in Section 2 of this article, the underlying physical principle that provides the necessary guidance to both analyses and experiments is simple enough: the macroscopic fracture process is governed by an energy balance criterion. The difficulty is in the details concerning (i) the determination of the appropriate mode or alternate modes of fracture and the associated fracture criteria (e.g., kinking vs. coplanar crack growth, local cleavage vs. crack blunting, quasi-brittle vs. high energy fracture), (ii) sufficiently accurate modeling and analysis of the mechanics problem corresponding to the particular mode of fracture and calculation of the representative crack driving force for each alternate fracture mode (e.g., the energy release rate, stress intensity factors, J-integral, crack opening stretch, etc.) and (iii) the experimental determination of the fracture resistance parameter(s) of the particular material system for each conjectured mode of fracture.

Generally, the base level fracture resistance is the work of adhesion. However, particularly in interface fracture, the actual fracture resistance of the medium is quite considerably greater than the work of adhesion. In real bonded materials two major sources of contributions to fracture resistance appear to be mechanical energies due to plastic deformations and that associated with contacting crack surface asperities which shield the crack tip from the full effect of shear loading represented by mode II stress intensity factor k_2 (Evans et al. 1990). Because of the absence of symmetry, the interface fracture is always one of mixed mode. Thus, depending on the external loads, the component/crack geometry and material parameters, the

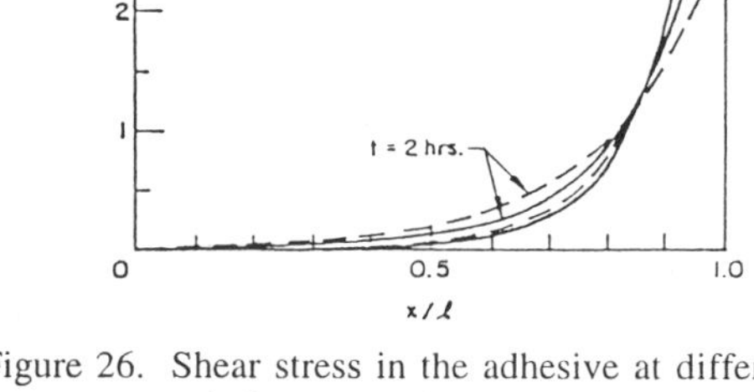

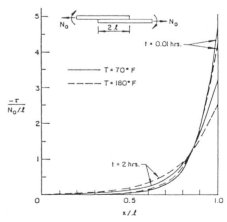

Figure 26. Shear stress in the adhesive at different temperatures (adherents: aluminum, h = 2.3 mm., adhesive: epoxy, $h_0 = 0.1$ mm., $2\ell = 25$mm.)

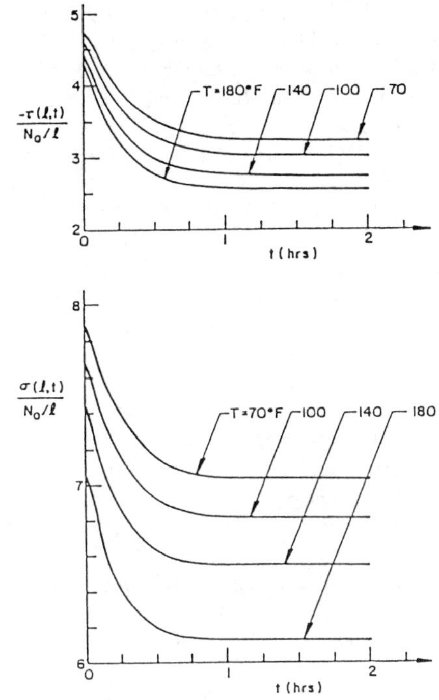

Figure 27. Relaxation of peak values of $\tau(x,t)$ and $\sigma(x,t)$, the shear and normal components of adhesive stresses at various temperatures in a single lap joint.

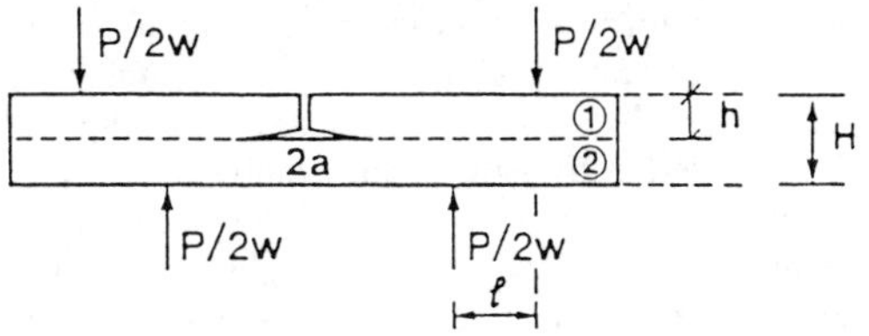

Figure 28. Four point bending test specimen used in interfacial fracture toughness measurements.

nominal value of k_2/k_1 may vary between $-\infty$ and $+\infty$. Therefore, k_2 is expected to and does play a major role in developing relevant fracture criteria and experimental methods for fracture characterization. First, the interface fracture toughness $\mathcal{G}_c$ is known to be heavily dependent on k_2/k_1 or the phase angle $\psi = \arctan(k_2 / k_1)$. This is an experimental observation (Charalambides et al. 1990, Cao and Evans 1989) and there are also some mechanical models that attempt to justify it (Evans and Hutchinson 1989). Secondly, k_2 seems to control the crack kinking. Therefore, in properly designed experiments for measuring the interface fracture toughness, generally a great deal of thought must be given to assessing the relative influence of k_2.

Among the test methods that have been successful in measuring the interface fracture toughness, one may mention the four point bending test, the peel test, double cantilever beam test, composite cylinder test, edge notched specimens and Brazil-nut specimens (Hutchinson and Suo 1992, Charalambides et al. 1990, Evans et al. 1989). The geometry of the four point bending specimen is shown in Figure 28. This is one of the more versatile and more accurately controlled test methods for measuring the interface fracture resistance. For more detailed description of the underlying analysis and the experimental technique, particularly for the discussion of the effect of residual strains and the friction at load transfer points, we refer to Charalambides et al. (1990), Cao and Evans (1989) and Evans et al. (1990).

Another test specimen with highly promising potential for measuring the interface fracture toughness is the Brazil-nut specimen. The original disk-shaped specimen was developed for testing under compression and the phase angle ψ was varied by changing the direction of loading relative to the plane of the interface crack (Wang and Suo 1990). Yuuki et al. (1994) modified the test procedure by loading the specimen under tension as well as compression, thereby covering a broader range of stress intensity ratio k_2/k_1.

CONCLUDING REMARKS

Regarding the fracture mechanics of interfaces, many of the theoretical aspects of the phenomenon have been clarified in recent years largely due to the efforts of Hutchinson and his colleagues. However, in light of the preliminary results presented in Section 4.1 of this article, perhaps it may not be very wise to keep ignoring the effect of crack surface contact and its consequences. Particularly important could be its influence on crack kinking. Also important could be the influence of very high compressive stress acting on a sizable part of the crack surface on the fracture resistance of the interface, especially if the friction is not negligible.

In most of the recent studies of interface fracture it is assumed that the medium consists of two bonded semi-infinite half planes, the interface crack is semi-infinite, and the external load is the elastic asymptotic far field controlled by a complex stress intensity factor. This appears to be the case even in most small-scale yielding and some full-field plasticity studies. After eliminating the physical length scale (such as the crack length), one can no longer claim that the crack tip contact zone is infinitesimally small and can be neglected, because the size of the contact zone would now be the only length parameter left that can be used for scaling. To study and assess the effect of the crack surface contact properly, it appears that at some point one has to consider simple but realistic part/crack geometries and loading conditions.

An additional theoretical difficulty would arise if the friction is taken into account along the contact region. As shown by Comninou (1977), in this case the singularity for in-plane shear stress at the crack tip is different from $1/\sqrt{r}$ and is given by

$$\tau_{r\theta}(r,0) \sim r^{-\lambda}, \quad \cot\pi\lambda = f\beta, \tag{149}$$

where the crack is on $\theta = \mp\pi$, f is the coefficient of friction and β is the standard Dundurs constant. It was further shown that $f\beta$ is always positive and hence λ is always less than ½; that is, in the presence of friction the singularity of the stress state around the crack tip is always weaker than square-root. This may further complicate the question regarding the crack kinking.

REFERENCES

Arin, K. & F. Erdogan 1971. *Int. J. Engrg. Sci.* 9:213-232.

Atkinson, C., R. E. Smelser & J. Sanchez 1982. *Int. J. Fracture* 18:279-291.

Bassani, J.L. & J. Qu, 1989. *J. Mech. Phys. Solids* 37:435-452.

Batakis, A.P. & J.W. Vogan, 1985. *Rocket Trust Chamber Thermal Barrier Coatings* NASA CR-175022.

Bose, K. & P. Ponte-Castañeda 1995. In F. Erdogan (ed). *Fracture Mechanics* 25th Volume: 106-124.

Brennan, J.J. 1991. Interfacial Studies of Refractory Glass-Ceramic Matrix/Advanced SiC Fiber Reinforced Composites. Annual Report, ONR Contract N00014-87-C-0699, United Technologies Research Center.

Cao, H.C. & A.G. Evans 1989. *Mechanics of Materials* 7:295-305.

Charalambides, P.G., H.C. Cao, J. Lund & A.G. Evans 1990. *Mechanics of Materials* 8:269-283.

Chen, Y.F., F. Erdogan 1996. *J. Mech. Phys. Solids.* 44:771-787.

Clements, D.L. 1971. *Int. J. Engng. Sci.* 9:257-268.

Comninou, M. 1977. *ASME J. Appl. Mech.* 44:631-636.

Cook, T.S. & F. Erdogan 1972. *Int. J. Engrg. Sci.* 10:667-697.

Derby, B. 1990. In M. Rühle, A.G. Evans, A.G. Ashby & J.P. Hirth (eds), *Metal-Ceramic Interfaces* 161-167, New York: Pergamon Press.

Delale, F., F. Erdogan & M. N. Aydinoglu 1981. *Journal of Composite Materials* 15:249-271.

Delale, F. & F. Erdogan 1981a. *Journal of Composite Materials* 5:561-581.

Delale, F. & F. Erdogan 1981b. *ASME J. Appl. Mech.* 331-338.

Delale, F. & F. Erdogan 1988a. *ASME J. Appl. Mech.* 55:317-324.

Delale, F. & F. Erdogan 1988b. *Int. J. Engng. Sci.* 26:559-568.

England, A.H. 1965. *ASME J. Appl. Mech.* 32:400-402.

Erdogan, F. 1963. *ASME J. Appl. Mech.* 30: 232-236.

Erdogan, F. & Sih, G.C. 1963. *J. Basic Eng., Trans. ASME,* 85, Series D:519-526.

Erdogan, F. 1965. *ASME J. Appl. Mech.* 32:403-410.

Erdogan, F. & G.D. Gupta 1971a. *Int. J. Solids Structures 7:*39-61.

Erdogan, F. and G.D. Gupta 1971b. *Int. J. Solids Structures* 7:1089-1107.

Erdogan, F. 1971. In *Developments in Mechanics*, Vol. 6, Proc. 12th Midwestern Mechanics Conference: 817-829.

Erdogan, F. & K. Arin 1972. *Int. J. Engrg. Sci.* 10: 115-125.

Erdogan, F. 1972. *J. Engrg. Fracture Mechanics* 4:811-840.

Erdogan, F. & V. Biricikoglu 1973. *Int. J. Engrg. Sci.* 11:745-766.

Erdogan, F. & M.B. Civelek, 1974 *ASME J. Appl. Mech.* 41:1014-1018.

Erdogan, F. & P.F. Joseph 1988. In S.L. Koh & C.G. Speciale (eds) *Recent Advances in Engineering Science*, Proc. The Eringen Symposium, Berkeley, California.

Erdogan, F. 1995. *J. Composites Engineering* 5:753-770.

Erdogan, F. & P.F. Joseph 1990. *ASME J. Electronic Packaging* 112:309-326.

Erdogan, F., A.C. Kaya & P.F. Joseph 1991. *ASME J. of Appl. Mech.* 58:410-418.

Erdogan, F. & M. Ozturk 1992. *Int. J. Engng. Sci.* 30:1507-1523.

Erdogan, F. & B-H. Wu 1993a. *Materials Science and Engineering* A 162:199-214.

Erdogan, F. & M. Ozturk 1993a. *Int. J. Engng. Sci.* 31:1641-1657.

Erdogan, F. & B-H. Wu 1993b. *J. Mech. Phys. Solids* 41:889-917.

Erdogan, F. & S. Kadioglu 1994. *Int. J. of Fracture* 67:273-300.

Erdogan, F. 1995. *J. Composites Engineering,* 5:753-770.

Evans, A.G. & J. W. Hutchinson 1989. *Acta Metall. Mater.* 37:909-916.

Evans, A.G., M. Rühle, B.J. Dalgleish & P.G. Charalambides 1990. In M. Rühle, A.G. Evans, A.G. Ashby & J.P. Hirth (eds), *Metal-Ceramic Interfaces* 345-364. New York: Pergamon Press.

Findley, W.N., J.S. Lai & K. Onaran 1976. *Creep and Relaxation of Nonlinear Viscoelastic Materials.* North Holland Publ. Co.

Flügge, W. 1975. *Viscoelasticity.* Springer-Verlag.

Gautesen, A.K. & J. Dundurs, 1987. *ASME J. Appl. Mech.* 54:93-98.

Gautesen, A.K. & J. Dundurs, 1988. *ASME J. Appl. Mech.* 55:580-586.

Gautesen, A.K. 1992. *Int. J. of Fracture*, 55:261-271.

Gautesen, A.K. 1993. *Int. J. of Fracture*, 60:349-361.

Gotoh, M. 1967. *Int. J. Fracture Mechanics* 3:253-263.

Griffith, A.A. 1921. *Phil. Trans. Roy. Soc. London*, A221: 163-198.

Griffith, A.A. 1924. *Proc. 1st Int. Congr. Appl. Mech.* 55:55-93.

Hayashi, K. & S. Nemat-Nasser 1981. *ASME J. Appl. Mech.* 48:520-524.

He, M.-Y., A. Bartlett, A.G. Evans & J.W. Hutchinson 1991. *J. Am. Ceram. Soc* . 74:767-771.

He, M.-Y. & J.W. Hutchinson 1989a. *.ASME J. Appl. Mech.* 56:270-278.

Hein, V.L. & F. Erdogan 1971. *Int. J. Fracture Mechanics* 7:317-330.

Holt, J.B., M. Koizumi, T. Hirai & Z.A. Munir (eds) 1993. *Ceramic Transactions - FGM* 94. Westerville OH: American Ceramic Society.

Houck, D.L. (ed) 1987. *Thermal Spray: Advances in Coatings Technology, Proc. of the National Thermal Spray Conference*, Orlando, FL: ASM International.

Hutchinson, J.W., M.E. Mear & J.R. Rice 1987. *J. Appl. Mech.* 54:828-832.

Hutchinson, J.W. & Z. Suo 1992. In J. W. Hutchinson & T.Y. Wu (eds) *Advances in Applied Mechanics* 63-191. New York: Academic Press, Inc.

Ilschner, B. & N. Cherradi (eds) 1995. *Proc. 3rd Int. Symp. on Structural and Functional Gradient Materials*, Lausanne, Switzerland: Presses Polytechniques et Universitaires Romandes.

Irwin, G.R. 1957. *ASME J. Appl Mech.* 24:361-364.

Ishida, H. 1984. *Polymer Composites* 5:101-106.

Kadioglu, S. & F. Erdogan 1995. *Int. J. of Fracture* 33:1105-1120.

Kasmalkar, M. 1997. *The Surface Crack Problem for a Functionally Graded Coating Bonded to a Homogeneous Layer.* Ph.D. Dissertation, Lehigh University, Bethlehem, PA.

Kassir, M.K. & A.M. Bergman 1972. *ASME J. Appl. Mech.* 39: 308-310.

Kaya, A.C. & H.F. Nied 1993. In. K. Kokini (ed) *Ceramic Coatings* ASME, MD 44:47-71.

Knowles, J.K. & E. Sternberg 1983. *J. Elasticity* 13:257-275.

Konda, N. & F. Erdogan 1994. *Engineering Fracture Mechanics* 47:533-545.

Kurihara, K., K. Sasaki & M. Kawarada 1990. In M. Yamanouchi, M. Koizumi, T. Hirai & I. Shiota (eds). FGM-90 *Proc. 1st Int. Symp. on Functionally Graded Materials*, Sendai, Japan: FGM Forum.

Lee, W.Y., D.P. Stinton, C.C. Berndt, F. Erdogan, Y.D. Lee & Z. Mutasim 1996. *J. Amer. Ceram. Soc.* 79:3003-3012.

Lee, Y-D. & F. Erdogan 1994. *Int. J. of Fracture* 69:145-165.

Lee, Y-D. & F. Erdogan 1997. *Int. J. of Fracture* (in press).

Lu, Ming-Che & F. Erdogan 1983. *Engineering Fracture Mechanics* 18:491-528.

Malyshev, B.M. & R.L. Salganik 1965. *Int. J. Fracture Mech.* 5:114-128.

Manson, J.J. 1985. *Pure and Applied Chemistry* 57:1667-1671.

Martinez, D. & V. Gupta 1994. *J. Mech. Phys. Solids* 42:1247-1271.

Muki, R. & E. Steraberg 1968. *Int. J. Solids Structures* 4:75-94.

Muskhelishvili, N.I. 1953. *Singular Integral Equations.* Groningen-Holland: P. Noordhoff Ltd.

Ozturk, M. & F. Erdogan 1996. *Int. J. Solids Structures* 33:193-217.

Pipkin, A.C. 1972. Lectures on Viscoelasticity Theory. Springer-Verlag.

Rice, J.R. & G.C. Sih 1965. *J. Appl. Mech.* 32:418-423.

Rice, J.R., Z. Suo & J.W. Wang 1990. In *Metal-Ceramic Interfaces*, M. Rühle, A.G. Evans, M.F. Ashby & J.P. Hirth (eds). Pergamon Press, New York 269-294.

Rühle, M., A.G. Evans, M.F. Ashby & J.P. Hirth (eds) 1990. *Metal-Ceramic Interfaces*, New York: Pergamon Press.

Sharma, S.M. & N. Aravas 1993. *Int. J. Solids Structures* 30:695-723.

Shiau, F.Y., Y. Zuo, X.Y. Zeng, J.C. Lin & Y.A. Chang, 1988. In D.M. Mattox, J.E.E. Baglin, R. J. Gottshall & C.D. Batich (eds), *MRS 119* Pittsburgh.

Shih, C.F. & R.J. Asaro 1988. *ASME J. Appl. Mech.* 55:299-316.

Shih, C.F. & R.J. Asaro 1989. *ASME J. Appl. Mech.* 56:763-779.

Shih, C.F. & R.J. Asaro 1990. *Int. J. of Fracture* 42:101-116.

Shih, C.F., R.J. Asaro & N.P. O'Dowd 1990. In M. Rühle, A.G. Evans, A.G. Ashby & J.P. Hirth (eds), *Metal-Ceramic Inerfaces* 313-325. New York: Pergamon Press.

Shih, C.F. & R.J. Asaro 1991. *ASME J. Appl. Mech.* 58:450-463.

Suo, Z. 1990. *Proc. R. Soc.* London A 427:331-358.

Ting, T.C.T. 1986. *Int. J. Solids Structures* 22:965-983.

Varias, A.G., N.P. O'Dowd, R.J. Asaro & C.F. Shih 1990. In M. Rühle, A.G. Evans, A.G. Ashby & J.P. Hirth (eds), *Metal-Ceramic Interfaces* 375-382. New York: Pergamon Press.

Varley, E. 1989. Private Communication.

Wang, J.S. & Z. Suo 1990. *Acta Met.* 38:1279-1290.

Willis, R.J. 1971. *J. Mech. Phys. Solids* 19:353-373.

Yamanouchi, M., M. Koizumi, T. Hirai & I. Shiota (eds) 1990. *FGM-90 Proc. 1st. Int. Symp. on Functionally Graded Materials*, Sendai, Japan: FGM Forum.

Yang, Mingfa & K-S. Kim 1993. *Engng. Fracture Mechanics* 44:155-165.

Yuuki, R., J-Q. Lin, T-Q. Xu, T. Ohira & T. Ono 1994. *Engng. Fracture Mechanics* 47:367-377.

Damage and Failure of Interfaces, Rossmanith (ed.) © *1997 Balkema, Rotterdam, ISBN 90 5410 899 1*

Numerical simulation methodologies for dynamic crack propagation in homogeneous and nonhomogeneous materials

T. Nishioka
Department of Ocean Mechanical Engineering, Kobe University of Mercantile Marine, Japan

ABSTRACT: Numerical simulations of dynamic fracture phenomena involve many inherent difficulties. To overcome such difficulties, the author and coworkers have developed various numerical simulation methodologies for dynamic crack propagation in homogeneous as well as in nonhomogeneous materials. Some of these developments are summarized in this paper. The contents are related to (i) several expressions for the dynamic J integral in homogeneous and nonhomogeneous materials, (ii) the concept of mixed-phase simulation for non-self-similar crack propagation, (iii) moving finite element methods for self-similar and for non-self-similar dynamic crack propagation, and (iv) numerical simulation results for dynamic fracture phenomena in homogenous and nonhomogeneous materials.

1 INTRODUCTION

Numerical simulations of dynamic fracture phenomena involve many inherent difficulties. Main reasons of these difficulties may be listed as follows:

(I) Moving singularities at the tips of dynamically propagating cracks should be treated accurately. The ordinary numerical methods cannot treat the moving singularities.

(II) When a propagating crack tip passes a material point, the material point instantaneously separates into at least two parts. In the ordinary numerical models with nodal release techniques, this sudden unloading process often produces spurious oscillations.

(III) Cracks may curve, kink or bifurcate. Automatic mesh generations for these non-self-similar fast fracture phenomena are extremely difficult.

(IV) The crack propagation velocity along a bimaterial interface can exceed the shear wave speed of the compliant material (Lambros & Rosakis 1995).

To overcome such difficulties, the author and coworkers have developed various numerical simulation methodologies for dynamic crack propagation in homogeneous as well as in nonhomogeneous materials. Some of these developments are summarized in this paper. The contents are related to (i) several expressions for the dynamic J integral in homogeneous and nonhomogeneous materials, (ii) the concept of mixed-phase simulation for non-self-similar crack propagation, (iii) moving finite element methods for self-similar and for non-self-similar dynamic crack propagation, and (iv) numerical simulation results for dynamic fracture phenomena in homogenous and nonhomogeneous materials.

2 DYNAMIC J INTEGRAL IN HOMOGENEOUS AND NONHOMOGENEOUS MATERIALS

2.1 *Path Independent Dynamic J Integral for Homogeneous Materials*

The static J integral (Rice 1968) has played an important role in *static fracture mechanics*. From the theoretical and computational points of view, the static J integral has the following salient features: (i) it physically represents the energy release rate; (ii) it has the property of the path-independent integral, which gives a unique value for an arbitrary integral path surrounding the crack tip; (iii) it can be related to the stress intensity factors by arbitrarily shrinking the integral path to the crack tip.

In the case of *dynamic fracture mechanics*, Nishioka & Atluri (1983) have derived the dynamic J integral (J') which has the aforementioned three features as:

$$J' = J'_1 \cos\theta_0 + J'_2 \sin\theta_0 \tag{1}$$

$$J'_k = \lim_{\varepsilon\to 0}\int_{\Gamma_\varepsilon}\left[(W+K)n_k - t_i u_{i,k}\right]dS \tag{2.a}$$

$$= \lim_{\varepsilon\to 0}\left\{\int_{\Gamma+\Gamma_c}\left[(W+K)n_k - t_i u_{i,k}\right]dS + \int_{V_\Gamma - V_\varepsilon}\left[\rho\ddot{u}_i u_{i,k} - \rho\dot{u}_i\dot{u}_{i,k}\right]dV\right\} \tag{2.b}$$

where W and K are the strain and kinetic energy densities, respectively, and $(\ )_{,k} = \partial(\ )/\partial X_k$. Integral paths are defined in Figure 1. Physically, the near-tip region V_ε can be considered as the process zone in which micro-processes associated with fracture occur.

The crack-axis components of the dynamic J integral can be evaluated by the coordinate transformation. Thus the dynamic energy release rate G can be expressed by

$$G = J' = J'^{0}_{1} \tag{3}$$

where J' is the magnitude of the vectorial dynamic J integral and J'^{0}_{1} is the tangential component of the dynamic J integral along the crack axis. The crack-axis components J'^{0}_{k} can be related to the dynamic stress intensity factors (Nishioka & Atluri 1983).

In most numerical analyses, the far-field integrals are usually used to evaluate the values of the dynamic J integral. In this case it is convenient to consider the following expression:

$$J'_k = \int_{\Gamma+\Gamma_c} \left[(W+K)n_k - t_i u_{i,k}\right] dS + \int_{V_\Gamma} \left[\rho \ddot{u}_i u_{i,k} - \rho \dot{u}_i \dot{u}_{i,k}\right] dV \tag{4}$$

2.2 *Extension of the Dynamic J Integral for Nonhomogeneous Materials*

The path independence of the dynamic J integral is an important feature for evaluating the dynamic J integral values by a numerical analysis such as the finite element method. Since the dynamic J integral was originally defined for homogeneous materials as stated in the previous section, the original version of the dynamic J integral is not path independent if the path intercepts a material-interface in a nonhomogeneous material. This is due to the discontinuities arising in the strains and stresses at the interface. When a crack-tip locates in the vicinity of a material-interface, it is difficult to set up a path without intercepting the interface.

In this paper, two path independent forms of the dynamic J integral for nonhomogeneous materials are explained.

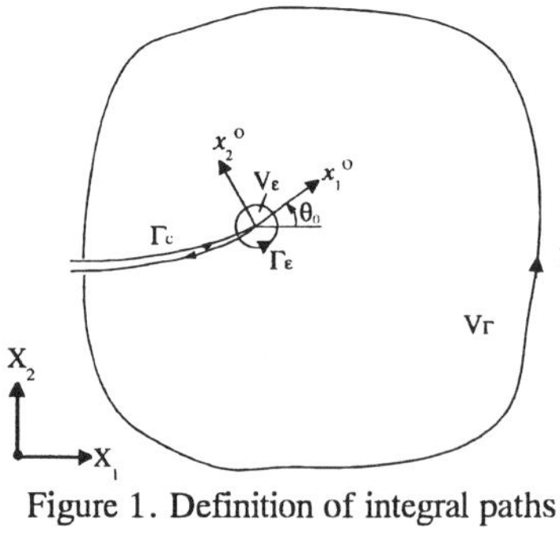

Figure 1. Definition of integral paths

2.2.1 Contour integral expression

Consider a crack in a multilayer-material of which different materials were perfectly bonded, as shown in Figure 2. The integral path Γ intercepts the interfaces 1 to m. The detailed derivation of the path independent dynamic J integral for such multilayered materials will be presented elsewhere (Nishioka 1997). In this article, this will be explained briefly.

From the original definition of the dynamic J integral, the far-field dynamic J along a path Γ for the multilayer-material may be expressed by

$$J'_k = \lim_{\varepsilon \to 0} \int_{\Gamma_\varepsilon} \left[(W+K)n_k - t_i u_{i,k}\right] dS \tag{5.a}$$

$$= \lim_{\varepsilon \to 0} \left\{ \int_{\Gamma+\Gamma_c} \left[(W+K)n_k - t_i u_{i,k}\right] dS + \int_{V_\Gamma - V_\varepsilon} \left[\rho \ddot{u}_i u_{i,k} - \rho \dot{u}_i \dot{u}_{i,k}\right] dV + R \right\} \tag{5.b}$$

where R is the additional term for the nonhomogeneous material. The additional term R can be expressed as

$$R = \int_{-\Sigma_{i=1}^{m} \Gamma_{Ii}} \left[(W+K)n_k - t_i u_{i,k}\right] dS \tag{6}$$

of which its physical meaning can be interpreted as the clockwise integral along the all interface paths within the far-field path Γ.

Consequently, the path independent dynamic J integral for the nonhomogeneous multilayer-material can be expressed as

$$J'_k = \int_{\Gamma+\Gamma_c-\sum_{i=1}^{m} \Gamma_{Ii}} \left[(W+K)n_k - t_i u_{i,k}\right] dS + \int_{V_\Gamma} \left[\rho \ddot{u}_i u_{i,k} - \rho \dot{u}_i \dot{u}_{i,k}\right] dV \tag{7}$$

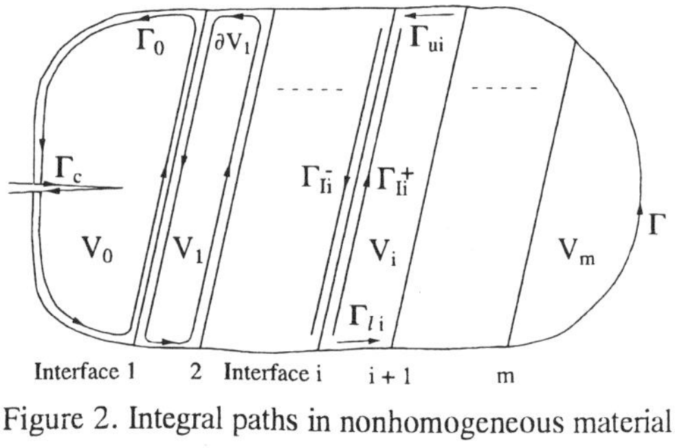

Figure 2. Integral paths in nonhomogeneous material

It is noted that a similar modification to the static J integral for a bimaterial with one-interface has been done by Kikuchi & Miyamoto (1986). However, in actual nonhomogeneous materials, it is not usually possible to bond the materials with one interface, since the material properties vary gradually near the joint. Furthermore, in functionally gradient materials, which are recent focus of considerable attention, multilayer models are often used with varying the material properties for element by element. The path independent dynamic J integral expressed by Eq.(7) overcomes the above difficulties in the single-interface model.

2.2.2 Equivalent domain integral expression

In this section, an extended form of the dynamic J integral for a multilayer-material using the equivalent domain integral expression (Nikishkov & Atluri 1987) is briefly explained. The detailed formulations will be presented in a separate paper (Nishioka 1997) together with the contour integral expression explained in the previous section.

For homogeneous materials, the dynamic J integral is defined by Eq.(2). Introducing a continuous function s such that s=1 at Γ_ε and s=0 at $\Gamma+\Gamma_C$, Eq.(2) can be rewritten by

$$J'_k = \lim_{\varepsilon\to 0}\int_{\Gamma_\varepsilon-(\Gamma+\Gamma_c)} \left[(W+K)n_k - t_i u_{i,k}\right] s\, dS \tag{8}$$

Using the divergence theorem, the equivalent domain integral of dynamic J integral for homogeneous materials can be derived as

$$J'_k=\int_{V_\Gamma} [\sigma_{ij}u_{i,k}s_{,j}-(W+K)s_{,k}+\rho\ddot{u}_i u_{i,k}s-\rho\dot{u}_i\dot{u}_{i,k}s]dV \tag{9.a}$$

$$\text{and}\begin{cases} s=1 & \text{at the crack tip}\\ s=0 & \text{at } \Gamma \text{ and } \Gamma_C\end{cases} \tag{9.b}$$

The above integral is independent from the size of V_Γ. Thus it is path independent integral in this sense.

For multilayer-materials, using a similar technique in the previous section, the equivalent domain integral expression of dynamic J can be summarized as follow:

$$J'_k=\int_{V_\Gamma} [\sigma_{ij}u_{i,k}s_{,j}-(W+K)s_{,k}+\rho\ddot{u}_i u_{i,k}s-\rho\dot{u}_i\dot{u}_{i,k}s]dV \tag{10.a}$$

$$\text{and}\begin{cases} s=1 & \text{at the crack tip}\\ s=0 & \text{at } \Gamma, \Gamma_c \text{ and } \sum_{i=1}^{m}\Gamma_{Ii}\end{cases} \tag{10.b}$$

The value of the s function in each element can be interpolated by the same shape function used for the finite element displacement field. For L-noded elements, the s function is expressed by

$$s = \sum_{q=1}^{L} N_q s_q \tag{11}$$

in which N_q denotes the shape function and s_q denotes a nodal value of the s function. Using Eq.(11) in Eq.(10), the dynamic J can be simplified as

$$J'_k=\sum_{p=1}^{M}\left(\sum_{q=1}^{L} R_{kpq}s_q\right) \tag{12}$$

and

$$R_{kpq}=\int_{V_p} [\sigma_{ij}u_{i,k}N_{q,j}-(W+K)N_{q,k}+\rho(\ddot{u}_i u_{i,k}-\dot{u}_i\dot{u}_{i,k})N_q]dV \tag{13}$$

in which V_p denotes the volume of an element inside the path V_Γ. Since R_{kpq} can be calculated in advance for each element, the equivalent domain integral expression (10) or Eq.(12) can easily treat nonhomogeneous material conditions.

3 MIXED-PHASE SIMULATION FOR NON-SELF-SIMILAR CRACK PROPAGATION

In a nonhomogeneous material, dynamic crack propagation usually occurs along an interface of bonded materials. However as the crack velocity becomes higher, the crack propagation tends to deviate from the interface, and to continue along a curved path.

For non-self-similar fracture such as curving crack growth, three types of numerical simulation can be considered, as proposed by Nishioka et al. (1996). First, the generation phase simulation can be conducted similarly with the generation phase simulation (Kanninen 1978) for self-similar dynamic fracture, except additionally using experimental data on the curved fracture-path history (see Figure 3(a)).

On the other hand, in the application phase simulation for curving crack growth, two criteria must be postulated or predetermined as shown in Figure 3(b). One is the crack-propagation criterion that is almost the same with the one used in the application phase simulation (Kanninen 1978) for self-similar fracture. However, the crack-propagation criterion for the curving crack growth may involve mixed-mode fracture parameters. Thus, it should be described by a fracture parameter taking into account mixed-mode conditions, such as the dynamic J integral (Nishioka & Atluri 1983). The other one is a criterion for predicting the direction of crack propagation (propagation-direction criterion or growth-direction criterion).

However, the application phase simulations of curving crack growth have not been fully established, due to several critical difficulties in those simulations.

For instance, in dynamic brittle fracture, the crack-propagation criterion described by fracture-toughness versus crack-velocity relation itself has several unsolved problems. Furthermore, the crack-propagation criteria may also be influenced by the geometry of fracture specimen.

To verify only the propagation-direction criterion such as the maximum energy release rate criterion, Nishioka (1996) has proposed *"mixed-phase simulation"* as depicted in Figure 3(c). Regarding the crack-propagation history, the same experimental data for the a-t relation used in the generation phase simulation can be used in the mixed-phase simulation. Thus, the increment of crack propagation is prescribed for the given time-step sizes in the numerical simulation. Then the propagation-direction criterion predicts the direction of fracture path in each time step. Simulated final fracture path will be compared with the actual one obtained by the experiment. This mode of the mixed-phase simulation may be called *"fracture-path prediction mode"*(see Figure 3(c)-(i)).

Another mode of the mixed-phase simulation can be considered as depicted in Figure 3(c)-(ii), i.e., *"crack-growth prediction mode"*. In this mode, the experimental data for the fracture-path history and the crack-propagation criterion are used simultaneously. In this case, the crack is forced to propagate along the actual fracture path during the numerical simulation. Simulated crack-propagation history should agree with the experimentally obtained actual one if the postulated crack-propagation criterion is valid.

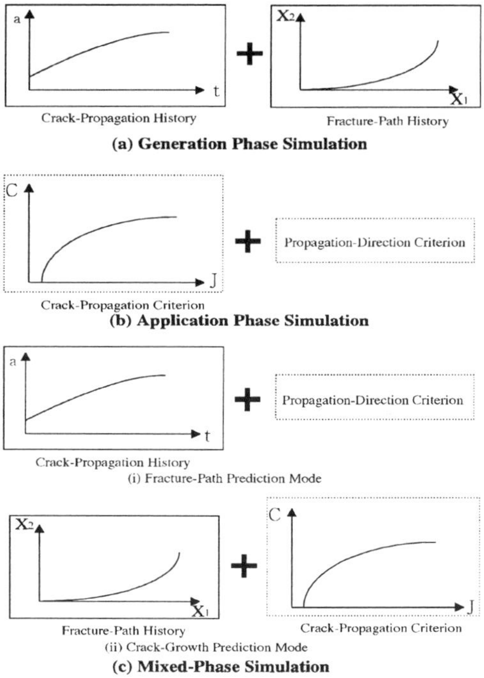

Figure 3. Mixed-phase simulation for non-self-similar dynamic fracture

4 MOVING FINITE ELEMENT METHODS FOR SELF-SIMILAR AND NON-SELF-SIMILAR DYNAMIC CRACK PROPAGATION

4.1 *Moving Element Procedures for Self-Similar Dynamic Crack Propagation*

In numerical simulation of dynamic crack propagation, one must treat a moving-singularity boundary-value problem. Since the moving singularity induces various errors in numerical models, it is critical to develop accurate solution procedures for the advancement of dynamic fracture mechanics.

To simulate the crack propagation in solids, two different concepts of computational modeling can be considered, i.e., (i) the stationary (fixed) element procedure, and (ii) the moving (distorting) element procedure, as reviewed in (Nishioka & Atluri 1986, and Nishioka 1994). In this section, only moving finite element procedures are explained.

Nishioka & Atluri (1980 & 1984) have developed various types of moving finite element procedures, as shown in Figure 4. In these procedures, the mesh pattern for the element(s) near the crack-tip translate(s) in each time step for which crack growth occurs. Thus, the crack-tip always remains at the center of the moving element(s) throughout the analysis. The regular isoparametric elements surrounding the moving element(s) are continuously distorted. To simulate a large amount of crack propagation, the

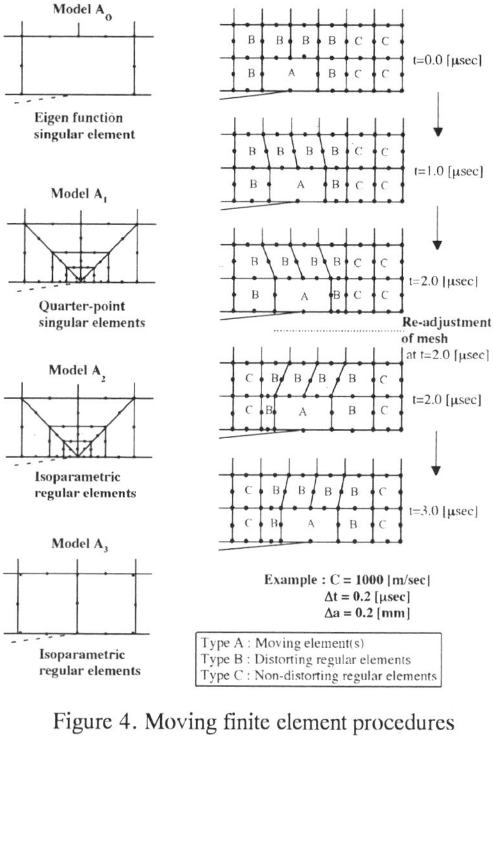

Figure 4. Moving finite element procedures

mesh pattern around the moving element(s) is periodically readjusted as also shown in Figure 4.

In the models A_0~A_2, a relatively elaborate "bookkeeping" system of nodal numbering to readjust the mesh pattern periodically is required while this is not the case when a simplified moving element procedure (Nishioka, Fujihara & Yagami 1987) with the model A_3 is used. However, the simplified moving element procedure requires more refined mesh to accurately model the crack-tip region. Therefore, in the present study, we employ the model A_2 in most cases of recent applications.

An important feature that distinguishes the moving element procedures from the fixed element procedures with various nodal release techniques often used in literature is that the displacement boundary conditions ahead of the crack-tip can be satisfied exactly in the moving element procedures.

4.2 *Moving Element Procedures for Non-Self-Similar Dynamic Crack Propagation*

In order to simulate non-self-similar dynamic crack propagation such as dynamic crack curving and dynamic crack branching, the curved crack path(s) should be modeled accurately by the finite element mesh pattern. Moreover, it is difficult to move a group of isoparametric elements around the crack-tip, in accordance with the curved fracture path. To overcome these difficulties, Nishioka et al. (1990) have developed an automatic element-control method using a mapping technique. The concept of this technique is illustrated in Figure 5.

The procedures of the mesh movement and readjustment are controlled in the element-controlling plane. The mesh pattern in the real plane is created by the mapping from the element-controlling plane through a mapping function. The mapping function for the entire region consists of several Lagrangian elements. The shape functions of Lagrangian element (see Figure 6) can be constructed by

$$N_{kj}=L_k^m(\zeta)\cdot L_j^n(\lambda);\quad (k=1,2,\cdot\cdot,m;\ \text{and}\ j=1,2,\cdot\cdot,n) \tag{14}$$

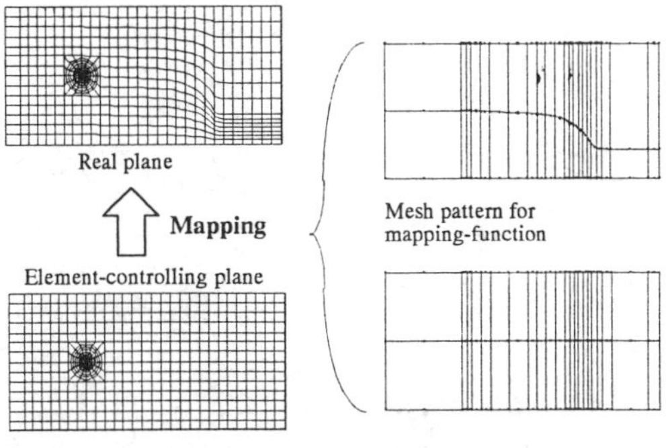

Figure 5. Mapping between the real plane and the element-controlling plane

where $L_k^m(\zeta)$ and $L_j^n(\lambda)$ are Lagrange's polynomials, and can be written as

$$L_k^m(\zeta)=\frac{(\zeta-\zeta_1)\cdot\cdot(\zeta-\zeta_{k-1})(\zeta-\zeta_{k+1})\cdot\cdot(\zeta-\zeta_m)}{(\zeta_k-\zeta_1)\cdot\cdot(\zeta_k-\zeta_{k-1})(\zeta_k-\zeta_{k+1})\cdot\cdot(\zeta_k-\zeta_m)} \tag{15.a}$$

$$L_j^n(\lambda)=\frac{(\lambda-\lambda_1)\cdot\cdot(\lambda-\lambda_{j-1})(\lambda-\lambda_{j+1})\cdot\cdot(\lambda-\lambda_n)}{(\lambda_j-\lambda_1)\cdot\cdot(\lambda_j-\lambda_{j-1})(\lambda_j-\lambda_{j+1})\cdot\cdot(\lambda_j-\lambda_n)} \tag{15.b}$$

in which ζ_k denotes the coordinate value of the k-th node in the ζ direction.

Any point $P_v(x_v,y_v)$ on the virtual element-controlling plane is mapped onto a point $P_r(x_r,y_r)$ on the real plane according to the following relation:

$$\begin{Bmatrix} x_r \\ y_r \end{Bmatrix}=\sum_{k=1}^{m}\sum_{j=1}^{n} N_{kj}(\zeta_v,\lambda_v)\begin{Bmatrix} x_{rkj} \\ y_{rkj} \end{Bmatrix} \tag{16}$$

where ζ_v, λ_v are the natural coordinates of the point $P_v(x_v,y_v)$ in the Lagrangian element, and (x_{rkj}, y_{rkj}) are the nodal coordinates of the Lagrangian element. As shown in Figure 5, the Lagrangian-element

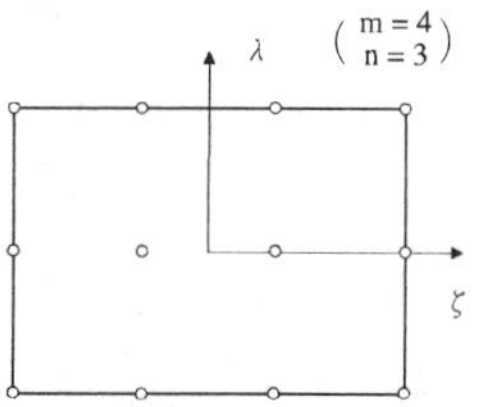

Figure 6. Lagrangian element

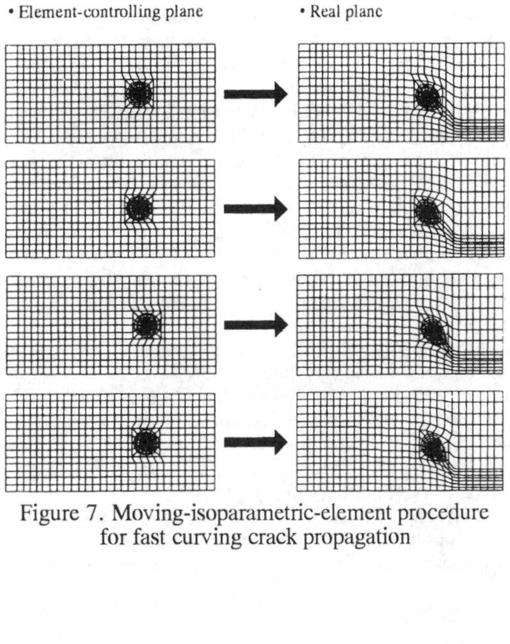

Figure 7. Moving-isoparametric-element procedure for fast curving crack propagation

mapping function can be obtained by using only the coordinates of the crack path and the external boundaries.

In the moving element procedure, the mesh pattern for the elements near the crack-tip translates along the curved crack path in each time step for which crack growth occurs as illustrated in Figure 7. Thus, the crack-tip always remains at the center of the moving elements throughout the analysis. The regular isoparametric elements surrounding the moving elements are continuously distorted. To simulate a large amount of crack propagation, the mesh pattern around the moving elements is periodically readjusted, as is also shown in Figure 7.

5 NUMERICAL SIMULATION RESULTS FOR DYNAMIC FRACTURE PHENOMENA IN HOMOGENOUS AND NONHOMOGENEOUS MATERIALS

5.1 *Generation-Phase Simulation of Dynamic Crack Propagation with Crack Acceleration Effects*

The effects of crack acceleration on the dynamic stress intensity factor were experimentally studied by Arakawa & Takahashi (1987). Figure 8 shows the geometry of a typical specimen used in the dynamic fracture experiment. The material of the specimen was Homalite 100.

The crack-propagation history measured by the experiment is shown in Figure 9. The circular points in the figure are the experimental data, while the solid lines are least-square fitted curves and are used as the input data for the generation-phase simulations.

The generation-phase simulation was carried out by Nishioka & Okizuka (1997), using the moving finite element method explained in the earlier section. Excellent path independence of the dynamic J integral was observed throughout the simulation. At each time step, the average value of the dynamic J integral values was converted to the dynamic stress intensity factor.

The relations between the dynamic stress intensity factor and crack velocity (or the dynamic fracture toughness K_{ID} vs C curves) are shown in Figure 10. As seen in the figure, the K_{ID} vs C curves are not unique, and depend on the crack acceleration. For a same crack velocity, the K_{ID} value at the acceleration stage ($\ddot{a}>0$) is lower than that at the deceleration stage ($\ddot{a}<0$). This indicates that the crack can propagate with the lower stress intensity factors at the acceleration stage while the higher stress intensity factors are needed to continue the crack propagation at the deceleration stage.

Furthermore, in Nishioka & Okizuka (1997), using the analytical expressions for the asymptotic eigen stress field derived by Nishioka & Kondo (1995), the near-tip stress fields are visualized at the acceleration and deceleration stages. The mechanism of the crack acceleration and deceleration effects on the dynamic fracture toughness versus crack velocity relation are successfully explained by the unsteady higher-order parts in the asymptotic stress field ahead of the propagating crack-tip.

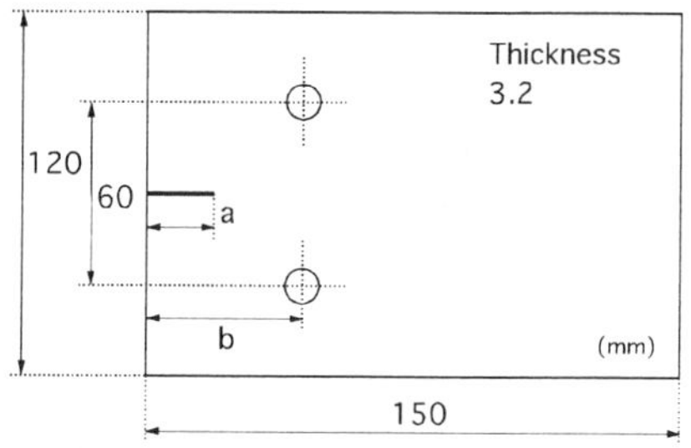

Figure 8. Geometry of dynamic fracture specimen

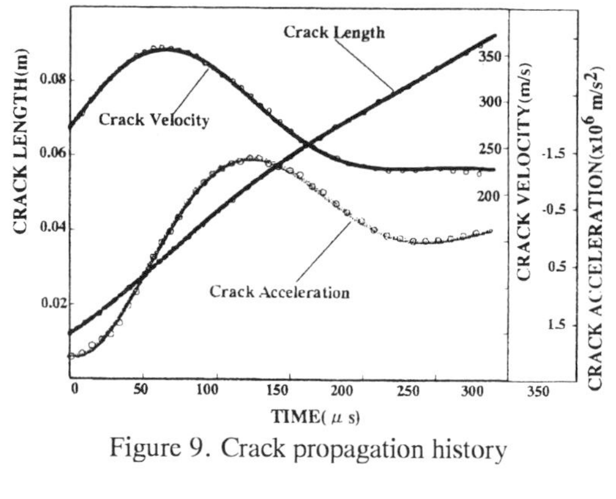

Figure 9. Crack propagation history

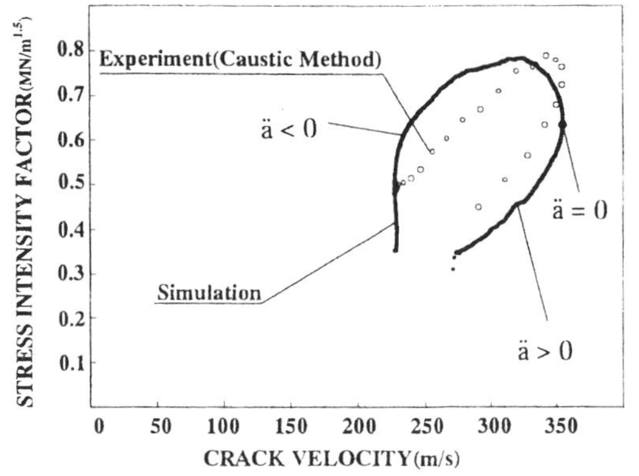

Figure 10. Hysteresis loop in K_{ID} vs C relation

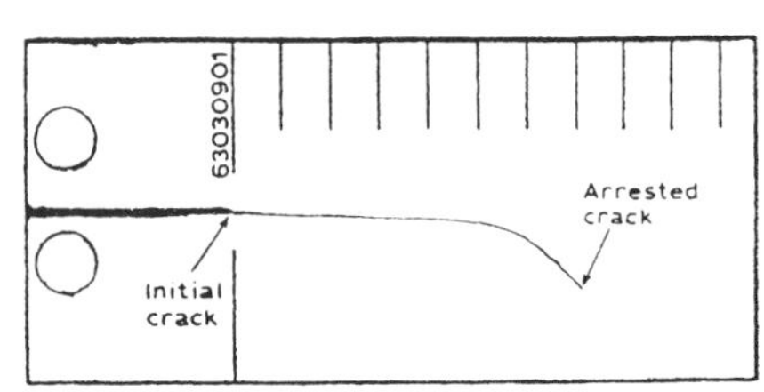

Figure 11. Dynamically curved fracture specimen

5.2 *Mixed-Phase Simulation of Dynamic Curving Fracture*

Figure 11 shows a typical fractured specimen obtained in fast curving fracture experiments (Nishioka et al. 1990b). The history of fast curving crack propagation was measured by a high-speed camera in the experiment. The maximum crack velocity was approximately 300 m/sec. The crack attested at t=1264 μsec.

Figure 12 schematically explains the numerical procedures for the path-prediction mode of the mixed-phase simulation. In each time step, the crack is advanced by the small increment according with the experimental history (crack-length versus time curve). The fracture path is predicted as follows: At a generic time step n, as the first trial, the crack is advanced in the tangential direction at the crack tip of the step n-1. If an employed propagation-direction criterion, for example K_{II}=0 criterion, is satisfied at the attempted crack-tip location, the crack is advanced in this direction. If not, K_{II} values are evaluated at the other two crack tips in attempted directions. Then the exact direction θ_c for K_{II}=0 is sought.

If the maximum K_I criterion is employed, the K_I values at the crack tips of three attempted directions are calculated (see Figure 12). Then the fracture direction θ_c that exactly satisfies the employed propagation-direction criterion is determined by the K_I versus θ curve.

The global components of the dynamic J integral J'_1 and J'_2 were evaluated along the circular paths in the web-like mesh (see Figure 7). Excellent path independence of the dynamic J integral values was obtained throughout the simulations. Then these dynamic J integral values were converted to the dynamic stress intensity factors by using the component separation method (Nishioka et al. 1990a).

All simulated fracture paths are compared with the experimental one in Figure 13. As long as the criteria tested here, the local symmetry criterion (K_{II}=0 criterion) predicts the closest fracture path to the experimental one.

The numerical results of the K_I and K_{II} values obtained by the mixed-phase simulations were summarized in (Nishioka, Okada & Nakatani 1997).

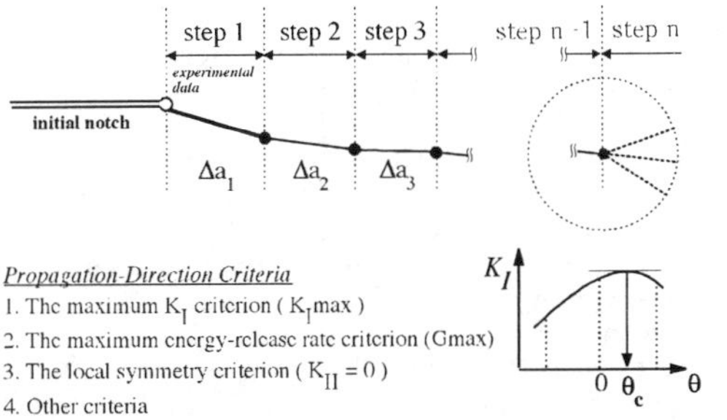

Figure 12. Procedures for fracture-path prediction

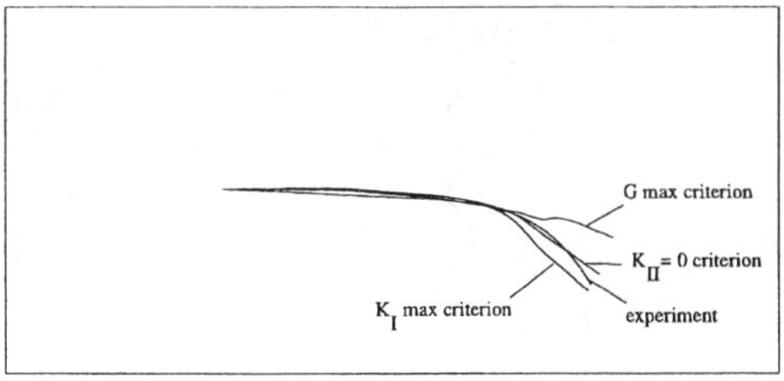

Figure 13. Simulated fracture paths

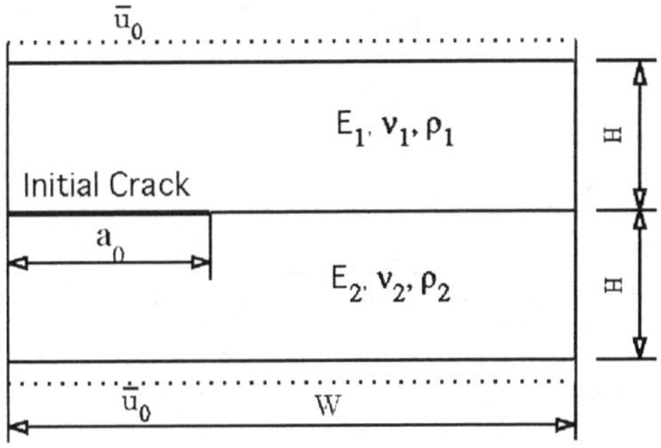

Figure 14. A cracked bimaterial plate

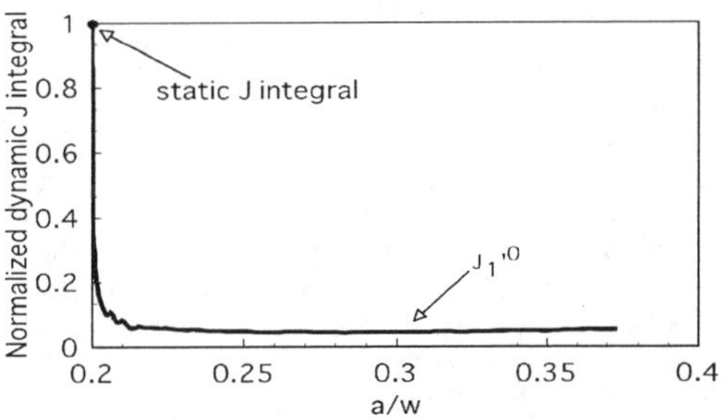

Figure 15. Energy release rate during dynamic interfacial crack propagation

5.3 *Interfacial Transonic Crack Propagation*

In this section, dynamic crack propagation along the interface of a bimaterial plate (see Figure 14) is considered. The compliant material is placed in the upper part of the plate, and denoted by the material 1, while the stiff one is denoted by the material 2.

The plate is subject to constrained displacements $\bar{u}_0$ in the X_2 direction at the upper and lower ends of the plate. Then at the time t=0, the crack is assumed to propagate with a constant velocity C.

Detailed numerical results for a parametric study varying the mismatch parameter $E^{(2)}/E^{(1)}$ and normalized crack velocity $C/C_S^{(1)}$ will be presented elsewhere (Nishioka & Yasin 1997). In this article, only the numerical results for the case of $E^{(2)}/E^{(1)}$=3.0

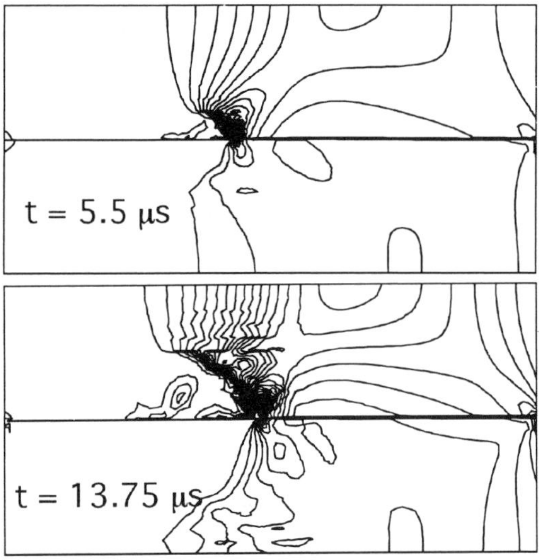

Figure 16. Strain energy density distribution

and $C/C_s^{(1)}$=1.200 ($C/C_d^{(1)}$=0.663) are explained. In this case, the crack velocity exceeds the shear wave velocity of the compliant material $C_s^{(1)}$. Thus the crack velocity is in the transonic range. For simplicity, the Poisson's ratios and the mass densities of both materials are assumed to be the same. The numerical simulations were carried out using the moving finite element method explained in Section 4.

The dynamic J integral values calculated by Eq.(7) indicated excellent path independence. The variation of the dynamic energy release rate calculated by the dynamic J integral J'^{0}_{1} (see Eq.(3)) is shown in Figure 15. The energy release rate drastically drops immediately after the onset of dynamic crack propagation, and then remains almost constant during transonic crack propagation.

Figure 16(a) and (b) show the distributions of the strain energy density at t=5.5 μs and 13.75 μs, respectively. The shock waves can be observed along the 3π/4 direction in the compliant material. The crack velocity faster than the shear wave velocity of the compliant material created the shock waves.

ACKNOWLEDGMENTS:

This study was supported by the Grant-in-Aid for Scientific Research (No.08455063) from the Ministry of Education, Science and Culture in Japan. The author acknowledges also Mr. Anwer Yasin who made the finite element calculation of the interfacial dynamic crack propagation.

REFERENCES

Arakawa, K. & Takahashi, K. 1987. Effects of crack velocity and acceleration on dynamic fracture toughness of Homalite-100. *Trans. JSME.* 53(485): Ser. A: 128-134.

Kanninen, M.F. 1978. A critical appraisal of solution techniques in dynamic fracture mechanics. In D.R.J. Owen & A.R. Luxmoore (eds) *Numerical Methods in Fracture Mechanics*, Pineridge Press: 612-634.

Kikuchi, M., Miyamoto, H. & Sugawara, S. 1986. Evaluation of the J integral of a crack in a pressure vessel under thermal transient loadings. *J. Press. Vess. Tech.* 108: 312-319.

Lambros, J. & Rosakis, A.J. 1995. Dynamic decohesion of bimaterials: experimental observations and failure criteria. *Int. J. Solids Struct.* 32(17-18): 2677-702.

Nikishkov, G.P. & Atluri, S.N. 1987. Calculation of fracture mechanics parameters for an arbitrary three-dimensional crack, by the 'equivalent domain integral' method. *Int. J. Num. Meth. in Eng.* 24: 1801-1821.

Nishioka, T. & Atluri, S.N. 1980. Numerical modeling of dynamic crack propagation in finite bodies by moving singular elements, part 1: formulation. *J. Appl. Mech.* 47(3): 570-576.

Nishioka, T. & Atluri, S. N. 1983. Path independent integrals, energy release rates and general solutions of near-tip fields in mixed-mode dynamic fracture mechanics. *Eng. Fract. Mech.* 18(1): 1-22.

Nishioka, T. & Atluri, S. N. 1984. A path-independent integral and moving isoparametric elements for dynamic crack propagation. *AIAA J.* 22(3): 409-414.

Nishioka, T. & Atluri, S.N. 1986. Computational methods in dynamic fracture. In S.N. Atluri (ed) *Computational Methods in the Mechanics of Fracture*, Elsevier Science Publ.: 336-383.

Nishioka, T., Fujihara, H. & Yagami, H. 1986. Finite element analyses of stress intensity factors in dynamic crack propagation using path independent J' integral. *Role of Fracture Mechanics in Modern Technology*, North-Holland: 561-573.

Nishioka, T. Murakami, R. & Takemoto, Y. 1990a. The use of the dynamic J integral (J') in finite element simulation of mode I and mixed-mode dynamic crack propagation. *Int. J. Press. Vess. Pip.* 44: 329-352.

Nishioka, T. et al. 1990b. A laser caustic method for the measurement of mixed-mode dynamic stress intensity factors in fast curving fracture tests. *Int. J. Press. Vess. Pip.* 44: 17-33.

Nishioka, T. 1994. The state of the art in computational dynamic fracture mechanics. *JSME Int. J.* Ser. A, 37: 313-333.

Nishioka, T. & Kondo, K. 1995. A unified derivation of explicit expressions for transient asymptotic solutions of dynamically propagating cracks under the mode I, II and III unsteady state conditions. In R.C. Batra (ed) *Contemporary Research in Engineering Science*, Springer-Verlag: 393-417.

Nishioka, T et al. 1996. Numerical simulations of self-similar and non-self-similar dynamic fracture phenomena. Proc. *Asian Pacific Conf. for Fract. & Strength '96*, Kyongju, Korea: 699-704.

Nishioka, T., Okada, T. & Nakatani, M. 1997. Mixed-phase simulation with fracture-path prediction mode for dynamically curving fracture. in B.L. Kalihaloo et al. (eds) *Advances in Fracture Research*, Vol. 4, Pergamon: 2063-2070.
Nishioka, T. & Okizuka, S. 1997. Generation-phase simulation of brittle fast fracture with crack acceleration and deceleration using moving finite element method. In S.N. Atluri & G. Yagawa, (eds) *Advances in Computational Engineering Science*, Tech Science Press: 165-170.
Nishioka, T. & Yasin, A. 1997. *Trans. JSME* Ser. A (in Preparation).
Nishioka, T. 1997. *Eng. Fract. Mech.* (in preparation).
Rice, J.R. 1968. A path independent integral and approximate analysis of strain concentration by notches and cracks. *J. Appl. Mech.* 35: 379-386.

Damage and Failure of Interfaces, Rossmanith (ed.) © *1997 Balkema, Rotterdam, ISBN 90 5410 899 1*

Dynamic properties of interfaces

L. R. Myer, K. T. Nihei & S. Nakagawa
Ernest Orlando Lawrence Berkeley National Laboratory, Calif., USA

ABSTRACT: Rocks are discontinuous at all scales, from the grain to tectonic scale. Many discontinuities may be characterized as non-welded interfaces, that is part solid and part void. If the dimensions of the voids are small compared to the wavelength of an elastic wave, the interface can be well represented as a displacement discontinuity, which may be elastic or rheologic. Using these models, a number of dynamic properties of compliant interfaces, which are useful for characterization, have been explored.

1 INTRODUCTION

Interfaces are present at almost all scales in geologic media. At the macroscopic scale, interfaces are present at lithologic boundaries where rock type changes, and joints or fractures represent interfaces across which the rock type may not change. At the mesoscale faults are interfaces between tectonic plates. One characteristic that many of these interfaces have in common is that, at the scale of observation, there will be portions of the interface where the material is in intimate contact and portions where voids are present. Observations indicate that these voids often have large aspect ratios. A joint or fracture can therefore be idealized as a collection of co-planar flat cracks in an otherwise homogenous elastic material. Water and/or alteration products such as clay may be present in the void space.

When an elastic wave is incident upon a coplanar collection of cracks, part of the energy will be scattered. Angel and Achenbach (1985) showed that if the wavelength is long compared to the crack size and spacing, the collection of cracks can be replaced by a zero thickness compliant layer for purposes of calculating the far field response. When a wave crosses such an interface, there is a jump in the displacement field which is proportional to the compliance of the interface, so that the model is sometimes referred to as a displacement discontinuity model. Modeling an imperfect interface as a displacement discontinuity greatly simplifies calculations while yielding many interesting wave propagation effects which can be used to characterize the interface. In addition, details of the geometry of the void space in geologic interfaces are rarely, if ever, known, whereas compliance (or stiffness) can be measured and is a parameter used by engineers in many applications.

In seismic geophysics, it has been long recognized that joints and fractures reduce the stiffness of the rock. Most often the approach has been to derive effective media properties (e.g. O'Connell and Budiansky 1974, Crampin 1981, Hudson 1981, Thomsen 1986 and others) thereby "smearing" the effects of individual fractures. In addition, it has usually been assumed that a joint or fracture at the macroscale is represented by a single flat crack, as opposed to a collection of coplanar cracks. The alternative approach, which is the subject of this paper, is to model joints and fractures as discrete, imperfect interfaces. The result is a number of effects which have not previously been recognized in geophysics. Outside of geophysics, use of the displacement discontinuity model for an imperfect interface has been widely used in ultrasonic nondestructive testing of welds and adhesive bonds (Nagy et al. 1990, Rokhlin and Wang 1991).

2 NORMAL INCIDENCE TRANSMISSION AND REFLECTION

To understand how an imperfect interface affects a propagating wave, it is instructive to begin with the simple case of a plane wave normally incident upon the interface. Transmission, reflection and group time delay will be discussed for different interface constitutive properties.

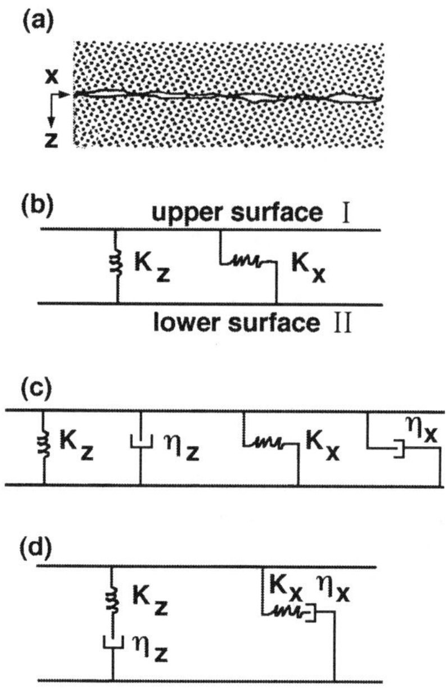

Figure 1. (a) Imperfect interface, (b) Elastic constitutive representation, (c) Kelvin rheologic model, (d) Maxwell rheologic model.

The simplest constitutive model is elastic, in which the imperfect interface is characterized by stiffness values (Figure 1a,b). To obtain a plane wave solution it is assumed that the interface is a boundary in the x-y plane between two elastic half-spaces.

The boundary conditions for an incident compressional (P-) wave or shear wave polarized in the x-z plane (S_v-wave) are:

$$u_z^I - u_z^{II} = \tau_{zz}/\kappa_z, \ \tau_{zz}^I = \tau_{zz}^{II}, \tag{1}$$

$$u_x^I - u_x^{II} = \tau_{zx}/\kappa_x, \ \tau_{zx}^I = \tau_{zx}^{II}$$

where

u = particle displacement

τ = stress

κ = stiffness of the interface in the x or z directions

I,II = superscripts referring to media above and below interface

For an incident shear wave with polarization in the x-y plane (S_h-wave), the boundary conditions are:

$$u_y^I - u_y^{II} = \tau_{zy}/\kappa_y, \ \tau_{zy}^I = \tau_{zy}^{II}, \tag{2}$$

For seismic waves normal to the interface with the same material properties in both half-spaces, the reflection ($R(\omega)$) and transmission ($T(\omega)$) coefficients for P, S_v and S_h waves are (Schoenberg 1980, Pyrak-Nolte et al. 1990 and others):

$$R_p(\omega) = \frac{i\omega}{-i\omega + 2(\kappa_z/z_p)}, \ T_p(\omega) = \frac{2(\kappa_z/z_p)}{-i\omega + 2(\kappa_z/z_p)}$$

$$R_{sv}(\omega) = \frac{-i\omega}{-i\omega + 2(\kappa_x/z_s)}, \ T_{sv}(\omega) = \frac{2(\kappa_x/z_s)}{-i\omega + 2(\kappa_x/z_s)} \tag{3}$$

$$R_{sh}(\omega) = \frac{-i\omega}{-i\omega + 2(\kappa_y/z_s)}, \ T_{sh}(\omega) = \frac{2(\kappa_y/z_s)}{-i\omega + 2(\kappa_y/z_s)}$$

where

ω = angular frequency

$z = \rho c_P$ for P-waves, or ρc_S for S-waves

$c_P = \sqrt{\lambda + 2\mu/p}$

$c_S = \sqrt{\mu/\rho}$

ρ = density

λ,μ = Lame's constants for media on either side of interface

From the phase of $T(\omega)$ a group time delay, t_g, can be found. For normally incident waves and the same materials in both half-spaces, the group time delay for the transmitted wave, t_{gT}, is given by:

$$t_{gT} = \frac{2(\kappa/z)}{4(\kappa/z)^2 + \omega^2} \tag{4}$$

Equations (3) and (4) show that both the reflection and transmission coefficients, as well as the group time delay are dependent upon the stiffness of the interface and the frequency of the propagating wave. In addition,

$$|R(\omega)|^2 + |T(\omega)|^2 = 1 \tag{5}$$

Thus energy is conserved at an elastic interface.

Figure 2 presents a plot of $|T|$, $|R|$, and t_{gT} as a function of the dimensionless parameter ($\omega z/\kappa$). For high stiffness $|T| \to 1$, and since energy is conserved, $|R| \to 0$. Thus, the case of infinite stiffness corresponds to the "welded" boundary condition assumed in classic seismologic analyses of multilayer systems. For zero stiffness $|T| \to 0$

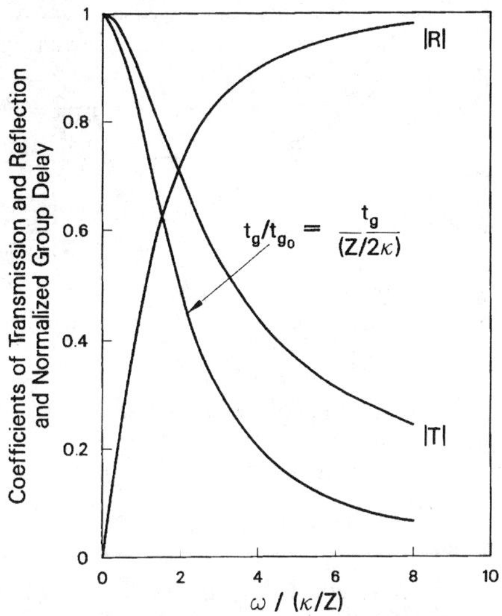

Figure 2. Magnitudes of the reflection and transmission coefficients and normalized group delay for a seismic wave normally incident upon an imperfect elastic interface.

and $|R| \rightarrow 1$, so in this limit the interface acts as a free surface. At intermediate values of stiffness high frequencies are preferentially reflected, so the fracture acts as a low pass filter. The elastic interface has been used successfully to model results of laboratory and small scale field tests of transmission across imperfect interfaces (Pyrak-Nolte et al. 1990, Myer 1991, Myer et al. 1995).

The simple elastic constitutive model does not account for dissipation of energy at the interface. Energy dissipation would be expected under conditions of partial liquid saturation and the presence of infilling materials such as clay. The constitutive model can be modified accordingly by incorporating a specific viscosity (units of viscosity per length) property of the interfaces.

The boundary conditions for the wave equation have been formulated in two ways, which, because of their similarity to well-known rheological models are referred to as the Kelvin and Maxwell models (Myer et al. 1990).

Assuming the Kelvin model, for an incident P- and S_v-wave, the boundary conditions are (Figure 1c):

$$\kappa_z(u_z^I - u_z^{II}) + \eta_z(\dot{u}_z^I - \dot{u}_z^{II}) = \tau_{zz}$$

$$\kappa_x(u_x^I - u_x^{II}) + \eta_x(\dot{u}_x^I - \dot{u}_x^{II}) = \tau_{zx} \qquad (6)$$

$$\tau_{zz}^I = \tau_{zz}^{II}$$
$$\tau_{zx}^I = \tau_{zx}^{II} \qquad (7)$$

where

η = specific viscosity

and the dot refers to time derivative. For an incident S_h-wave the boundary conditions are:

$$\kappa_y(u_y^I - u_y^{II}) + \eta_y(\dot{u}_y^I - \dot{u}_y^{II}) = \tau_{zy} \qquad (8)$$

$$\tau_{zy}^I = \tau_{zy}^{II} \qquad (9)$$

Assuming the Maxwell model (Figure 1d) boundary conditions for an incident P- and S_v-wave become:

$$(\dot{u}_z^I - \dot{u}_z^{II}) = \dot{\tau}_{zz}/\kappa_z + \tau_{zz}/\eta_z$$
$$(\dot{u}_x^I - \dot{u}_x^{II}) = \dot{\tau}_{zx}/\kappa_x + \tau_{zx}/\eta_x \qquad (10)$$

and for an incident S_h-wave become:

$$(\dot{u}_y^I - \dot{u}_y^{II}) = \dot{\tau}_{zy}/\kappa_y + \tau_{zy}/\eta_y \qquad (11)$$

For waves normally incident upon an imperfect interface where the two half-spaces have the same elastic properties, the reflection and transmission coefficients for P-, S_v-, and S_h-waves for the Kelvin model are:

$$R_p(\omega) = \frac{i\omega z_p}{2\kappa_z - i\omega(2\eta_z + z_p)},$$

$$T_p(\omega) = \frac{2(\kappa_z - i\omega\eta_z)}{2\kappa_z - i\omega(2\eta_z + z_s)}$$

$$R_{sv}(\omega) = \frac{-i\omega z_s}{2\kappa_x - i\omega(2\eta_x + z_s)},$$

$$T_{sv}(\omega) = \frac{2(\kappa_x - i\omega\eta_x)}{2\kappa_x - i\omega(2\eta_x + z_s)} \qquad (12)$$

$$R_{sh}(\omega) = \frac{-i\omega z_s}{2\kappa_y - i\omega(2\eta_y + z_s)},$$

$$T_{sh}(\omega) = \frac{2(\kappa_y - i\omega\eta_x)}{2\kappa_y - i\omega(2\eta_y + z_s)}$$

For the Maxwell model, the reflection and transmission coefficients are:

$$R_p(\omega)=\frac{-\frac{z_p}{2\eta_z}\left(1+\frac{z_p}{2\eta_z}\right)+\frac{\omega z_p}{2\kappa_z}\left(i-\frac{\omega z_p}{2\kappa_z}\right)}{\left(1+\frac{z_p}{2\eta_z}\right)^2+\left(\frac{\omega z_p}{2\kappa_z}\right)^2}$$

$$T_p(\omega)=\frac{\left(1+\frac{z_p}{2\eta_z}\right)+\frac{i\omega z_p}{2\kappa_z}}{\left(1+\frac{z_p}{2\eta_z}\right)^2+\left(\frac{\omega z_p}{2\kappa_z}\right)^2},$$

$$R_{sv}(\omega)=\frac{\frac{z_s}{2\eta_x}\left(1+\frac{z_s}{2\eta_x}\right)+\frac{\omega z_s}{2\kappa_x}\left(\frac{\omega z_s}{2\kappa_x}-i\right)}{\left(1+\frac{z_s}{2\eta_x}\right)^2+\left(\frac{\omega z_s}{2\kappa_x}\right)^2}, \tag{13}$$

$$T_{sv}(\omega)=\frac{\left(1+\frac{z_s}{2\eta_z}\right)+\frac{i\omega z_s}{2\kappa_x}}{\left(1+\frac{z_s}{2\eta_x}\right)^2+\left(\frac{\omega z_s}{2\kappa_x}\right)^2},$$

$$R_{sh}(\omega)=\frac{\frac{z_s}{2\eta_y}\left(1+\frac{z_s}{2\eta_y}\right)+\frac{\omega z_s}{2\kappa_y}\left(\frac{\omega z_s}{2\eta_y}-i\right)}{\left(1+\frac{z_s}{2\eta_y}\right)^2+\left(\frac{\omega z_s}{2\kappa_y}\right)^2},$$

$$T_{sh}(\omega)=\frac{\left(1+\frac{z_s}{2\eta_y}\right)+\frac{i\omega z_s}{2\kappa_y}}{\left(1+\frac{z_s}{2\eta_y}\right)^2+\left(\frac{\omega z_s}{2\kappa_y}\right)^2}$$

It should be noted that for both the Maxwell and Kelvin models, $|R(\omega)|^2+|T(\omega)|^2\neq 1$ because of viscous losses at the interface.

The magnitude of the transmission coefficient (of similar form for P-, S_v- and S_h-waves) for the Kelvin model is plotted in Figure 3. The curve for $\eta/z=0$ (specific viscosity of zero) is equivalent to that which would be obtained for the elastic model. Compared to the solution for the elastic model, the effect of finite values of specific viscosity is to reduce energy transmitted at low frequencies and increase the energy transmitted at high frequencies.

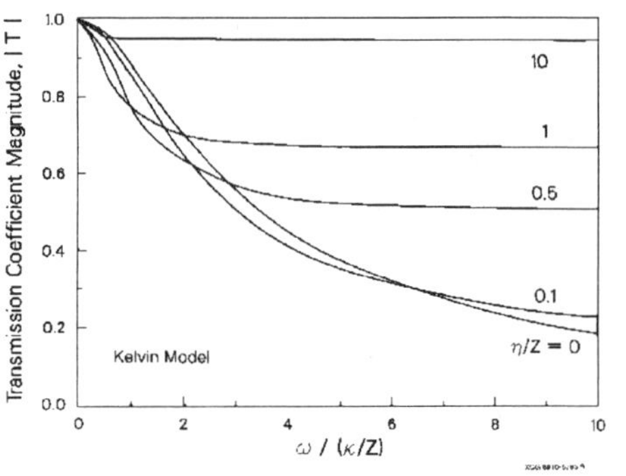

Figure 3. Magnitude of the transmission coefficient as predicted by the Kelvin model as a function of normalized frequency for a range of normalized specific viscosities.

For a given value of specific stiffness and at a particular frequency, increasing the specific viscosity results in larger values of $|T|$.

For comparison with the Kelvin model the magnitude of the transmission coefficient as predicted by the Maxwell model is plotted in Figure 4. In this case it is seen that very high ($\eta/z\rightarrow\infty$) specific viscosity values in combination with finite specific stiffness values yield a curve similar to that for the elastic model. Reducing the value of specific viscosity (while holding specific stiffness constant) has the effect of reducing energy transmitted at low frequencies relative to that transmitted at high frequencies. When terms involving η dominate the solution $|T|$ becomes frequency independent.

Pyrak-Nolte et al. 1990 found that the Kelvin model sucessfully replicated results of laboratory S-

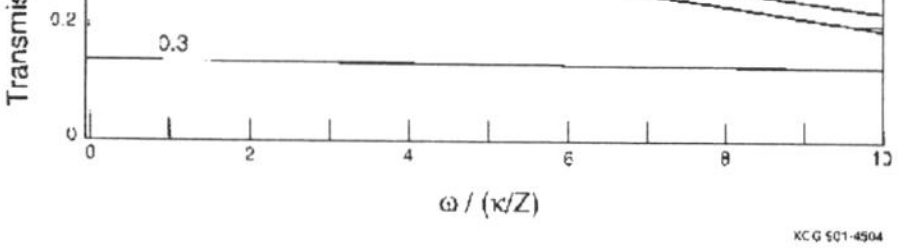

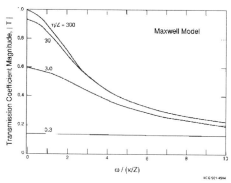

Figure 4. Magnitude of the transmission coefficient as predicted by the Maxwell model as a function of normalized frequency for a range of normalized specific viscosities.

wave transmission experiments on a fluid-filled single fracture in granite. Suárez-Rivera et al. 1992 used the Maxwell model to successfully simulate both the reflection and transmission of S-waves normally incident upon thin clay layers. These laboratory tests included different pore fluid chemistries as well as clay mineral compositions.

A final consideration is the effect of cross-coupling terms in the stiffness matrix describing the constitutive behavior of the interface. In Eqs. (1) and (2) the elastic interface is completely described by three stiffness values, so that the force displacement relationships in the normal and shear directions are independent. A situation in which coupling could be important is if the two opposing surfaces of an interface are very rough and loaded in shear. This would result in nonhomogeneous loading on the surfaces.

The boundary conditions for an incident P- or S_v-wave become:

$$\begin{aligned}(u_z^I - u_z^{II})\kappa_{zz} + (u_x^I - u_x^{II})\kappa_{zx} &= \tau_{zz}\\ (u_z^I - u_z^{II})\kappa_{zx} + (u_x^I - u_x^{II})\kappa_{xx} &= \tau_{zx}\end{aligned} \tag{14}$$

and

$$\begin{aligned}\tau_{zx}^I &= \tau_{zx}^{II}\\ \tau_{zz}^I &= \tau_{zz}^{II}\end{aligned} \tag{15}$$

For a normally incident P-wave the effect of the cross coupling stiffness term κ_{zx} is to generate a transmitted and reflected S_v-wave in addition to the transmitted and reflected P-wave. The transmission and reflection coefficients therefore become for the P-wave:

$$\begin{aligned}T_P(\omega) &= \frac{i\beta_{zx}^P}{(1+i\beta_{zz}^P)(1+i\beta_{xx}^S)+\beta_{zx}^P\beta_{xz}^S}\\[2ex]
T_{SV}(\omega) &= \frac{i\beta_{xx}^S(1+i\beta_{zz}^P)+\beta_{zx}^P\beta_{xz}^S}{(1+i\beta_{zz}^P)(1+i\beta_{xx}^S)+\beta_{zx}^P\beta_{xz}^S}\\[2ex]
R_P(\omega) &= \frac{i\beta_{zx}^P}{(1+i\beta_{zz}^P)(1+i\beta_{xx}^S)+\beta_{zx}^P\beta_{xz}^S}\\[2ex]
R_{SV}(\omega) &= \frac{1+i\beta_{zz}^P}{(1+i\beta_{zz}^P)(1+i\beta_{xx}^S)+\beta_{zx}^P\beta_{xz}^S}\end{aligned} \tag{16}$$

and for the normally incident S_v-wave:

$$\begin{aligned}T_p(\omega) &= \frac{i\beta_{zz}^P(1+i\beta_{xx}^S)+\beta_{zx}^P\beta_{xz}^S}{(1+i\beta_{zz}^P)(1+i\beta_{xx}^S)+\beta_{zx}^P\beta_{xz}^S}\\[2ex]
T_{SV}(\omega) &= \frac{i\beta_{xz}^S}{(1+i\beta_{zz}^P)(1+i\beta_{xx}^S)+\beta_{zx}^P\beta_{xz}^S}\\[2ex]
R_p(\omega) &= \frac{1+i\beta_{xx}^S}{(1+i\beta_{zz}^P)(1+i\beta_{xx}^S)+\beta_{zx}^P\beta_{xz}^S}\\[2ex]
R_{SV}(\omega) &= \frac{-ib_{xz}^S}{(1+i\beta_{zz}^P)(1+i\beta_{xx}^S)+\beta_{zx}^P\beta_{xz}^S}\end{aligned} \tag{17}$$

where

$$\begin{bmatrix}\beta_{xx}^P\beta_{xz}^P\\ \beta_{zx}^P\beta_{zz}^P\end{bmatrix} = \begin{bmatrix}\dfrac{2\kappa_{xx}/\omega}{z_P} & \dfrac{2\kappa_{zx}/\omega}{z_P}\\[2ex] \dfrac{2\kappa_{zx}/\omega}{z_P} & \dfrac{2\kappa_{zz}/\omega}{z_P}\end{bmatrix},$$

$$\begin{bmatrix}\beta_{xx}^P\beta_{xz}^P\\ \beta_{zx}^P\beta_{zz}^P\end{bmatrix} = \begin{bmatrix}\dfrac{2\kappa_{xx}/\omega}{z_S} & \dfrac{2\kappa_{zx}/\omega}{z_S}\\[2ex] \dfrac{2\kappa_{zx}/\omega}{z_S} & \dfrac{2\kappa_{zz}/\omega}{z_S}\end{bmatrix} \tag{18}$$

Since the interface is elastic $|T|^2 + |R|^2 = 1$.

The amplitudes of the transmission and reflection coefficients for a normally incident P-wave are shown in Figure 5. To illustrate the effect of the cross coupling terms, plots are shown for a range of values of the ratio $R = \kappa_{zx}/\kappa_{zz}$. For $R = 0$ the plot is identical to Figure 2. For nonzero values of κ_{zx}, energy is transferred from the transmitted and reflected P-wave to transmitted and reflected S-waves which have equal amplitudes. The amplitude of the converted waves are frequency dependent and stiffness dependent. In the limit, when the interface approaches either a welded boundary or a free surface, amplitudes of the converted waves approach zero. As shown in the plots, the maximum amplitudes occur at intermediate values of frequency and stiffness.

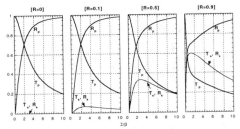

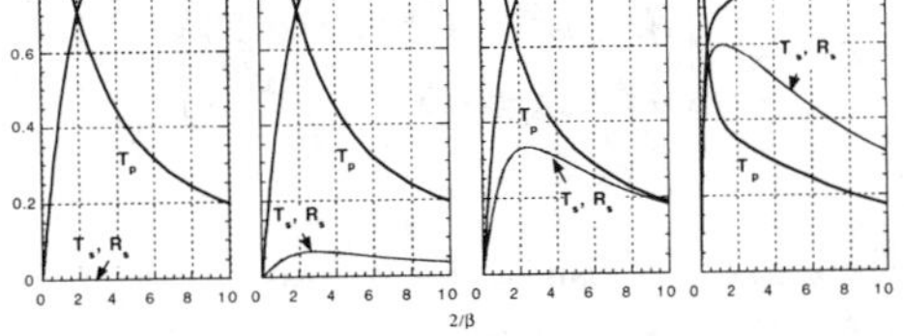

Figure 5. Magnitude of transmission and reflection coefficient for P-wave normally incident upon an elastic interface with cross coupled stiffness.

3 OBLIQUE ANGLES OF INCIDENCE

Transmission and reflection coefficients are calculated from solution of the wave equation for a plane wave incident upon an elastic interface. Snell's law defines the relationship between the angles of incidence, reflection and refraction. If the material properties on either side of the interface are equal, then transmission and reflection angles must equal incidence angles. Matrix equations for determining transmission and reflection coefficients have been presented by a number of authors (Schoenberg 1980, Pyrak-Nolte et al. 1990). Closed-form expressions, which provide somewhat better physical insight, were obtained by Gu et al. 1996. Thus, for an incident P-wave:

$$R_{P\to P} = \frac{r_{P\to P}}{fa\cdot fs},\ R_{P\to S} = \frac{r_{P\to S}}{fa\cdot fs}$$
$$T_{P\to P} = \frac{t_{P\to P}}{fa\cdot fs},\ T_{P\to S} = \frac{t_{P\to S}}{fa\cdot fs} \tag{19}$$

and for an incident S_v-wave:

$$R_{S\to P} = \frac{r_{S\to P}}{fa\cdot fs},\ R_{S\to S} = \frac{r_{S\to S}}{fa\cdot fs}$$
$$T_{S\to P} = \frac{t_{S\to P}}{fa\cdot fs},\ T_{S\to S} = \frac{t_{S\to S}}{fa\cdot fs} \tag{20}$$

where

$$\begin{aligned}
r_{P\to P} &= -\cos^4 2\phi + \xi^4 \sin^2 2\theta \sin^2 2\phi \\
&\quad - i2\bar{\kappa}_x \cos^2 2\phi \cos\phi + i2\xi^4 \bar{\kappa}_z \sin^2 2\theta \cos\phi, \\
r_{P\to S} &= 2\xi \sin 2\theta \cos 2\phi \begin{pmatrix} \cos^2 2\phi + \xi^2 \sin 2\theta \sin 2\phi \\ + i\bar{\kappa}_x \cos\phi + ic\bar{\kappa}_z \cos\theta \end{pmatrix}, \\
t_{P\to S} &= 2\xi \sin 2\theta \cos 2\phi \left(-i\bar{\kappa}_x \cos\phi + ic\bar{\kappa}_z \cos\theta\right), \\
t_{P\to P} &= 2\xi \cos\theta \begin{pmatrix} i4\bar{\kappa}_x \sin^2\phi \cos 2\phi \\ + i\bar{\kappa}_z \cos^2 2\phi - 2\bar{\kappa}_x \bar{\kappa}_z \cos\phi \end{pmatrix}, \\
r_{S\to P} &= -\xi \sin 4\phi \begin{pmatrix} \cos^2 2\phi + \xi^2 \sin 2\theta \sin 2\phi \\ + i\bar{\kappa}_x \cos\phi + i\xi\bar{\kappa}_z \cos\theta \end{pmatrix}, \\
r_{S\to S} &= -\cos^4 2\phi + \xi^4 \sin^2 2\theta \sin^2 2\phi \\
&\quad + i2\xi\bar{\kappa}_x \cos\theta \sin^2 2\phi - i2\xi\bar{\kappa}_z \cos\theta \cos^2 2\phi, \\
t_{S\to P} &= \xi \sin 4\phi \left(-i\bar{\kappa}_x \cos\phi + i\xi\bar{\kappa}_z \cos\theta\right), \\
t_{S\to S} &= 2\cos\phi \begin{pmatrix} i\bar{\kappa}_x \cos^2 2\phi + i\xi^4 \bar{\kappa}_z \sin^2 2\theta \\ -2\xi\bar{\kappa}_x \bar{\kappa}_z \cos\phi \end{pmatrix}.
\end{aligned} \tag{21}$$

and

$$fs = \cos^2 2\phi + \xi^2 \sin 2\theta \sin 2\phi + i2\xi\,\bar{\kappa}_z \cos\theta \tag{22}$$
$$fa = \cos^2 2\phi + \xi^2 \sin 2\theta \sin 2\phi + i2\,\bar{\kappa}_x \cos\phi$$

$\xi = c_s/c_p$, the ratio of S- to P-wave velocities
$\bar{\kappa}_z = \kappa_z/(\omega z_P)$
$\bar{\kappa}_x = \kappa_x/(\omega z_S)$

and θ and ϕ are defined in Figure 6. It is interesting to note that the terms fs and fa are identical to the dispersion relations for generalized Rayleigh interface waves which can exist on an interface. These will be discussed further in the next section.

The amplitude of the transmitted and reflected waves as a function of the S_v wave incidence angle are shown in Figure 7, where $\phi = 0$ is normal incidence. For S_v incident waves it is seen that a critical angle exists (at 37.8° in this case) above which the transmitted and reflected waves are no longer real-valued. At high values of interface stiffness, the transmitted S-wave coefficient ($|T_{s\to s}|$) is nearly unity and consequently the amplitudes of all other waves are nearly zero. $|T_{s\to s}|$ shows little variation with angle of incidence. $|R_{s\to s}|$ is largest for low values of stiffness, and near unity at normal incidence. However at angles of incidence between zero and the critical angle, energy is transferred to the converted P-waves.

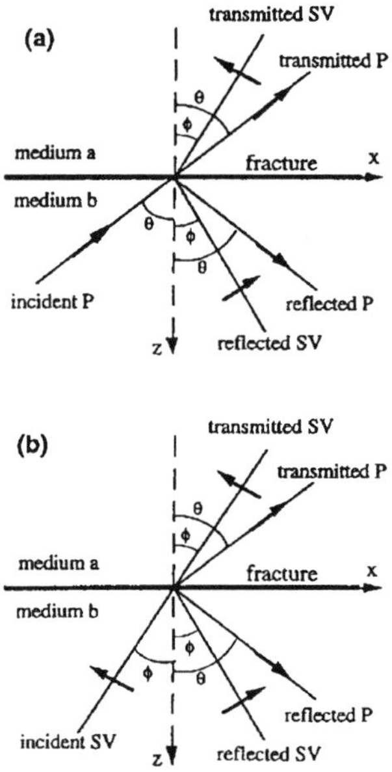

Figure 6. Coordinate system used in the analysis of reflection and transmission coefficients for (a) a P-wave incidence and (b) an S_v wave incidence. The arrows indicate the positive direction of the particle motion of the waves.

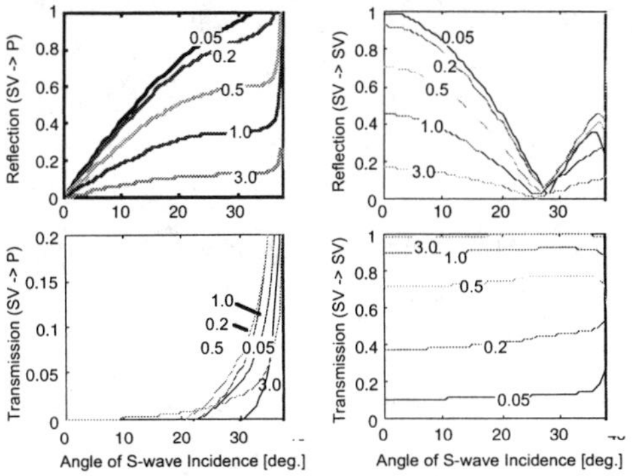

Figure 7. Magnitude of transmission and reflection coefficients as a function of S_v wave angle of incidence on an interface in a homogeneous medium with Poisson's ratio of 0.2. Numbers on curves are values of normalized stiffness. Critical angle at 37.8°.

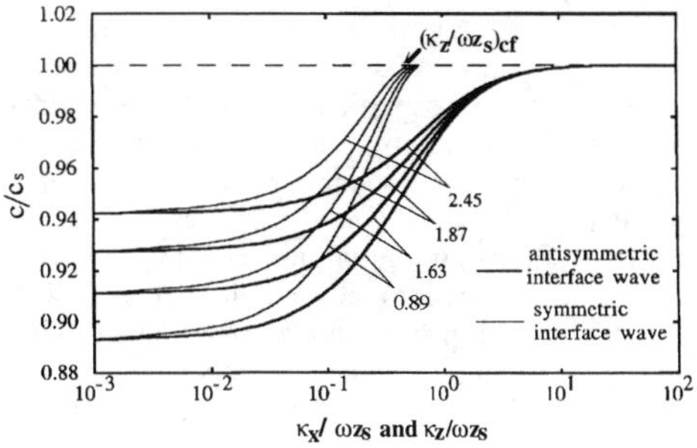

Figure 8. Normalized phase valocities of the symmetric and antisymmetric interface wave. Numbers on curves are values of c_P/c_S ratio.

P-waves are well behaved at all angles of incidence up to 90°, though the form of the variation with angle of incidence for the converted waves is different from S_v waves. Results for P-waves can be found in Gu et al. 1996.

4 INTERFACE WAVES

As noted above, the solution for a plane wave obliquely incident upon an imperfect interface suggests that such an interface should support generalized Rayleigh waves. To explore the properties of these waves, solutions for an inhomogeneous plane wave propagating along an elastic imperfect interface with an amplitude that decays exponentially with distance from the interface have been obtained (Pyrak-Nolte and Cook 1987, Gu et al. 1996). Nihei et al. 1995 showed that the solution can be decomposed into two dispersion equation, one representing a symmetric interface wave:

$$4\vartheta^2(\vartheta^2-\xi)^{1/2}\left[\frac{\kappa_z}{\omega z_S}+(\vartheta^2-1)^{1/2}\right] -(1-2\vartheta^2)\left[(1-2\vartheta^2)-2(\vartheta^2-\xi^2)^{1/2}\frac{\kappa_z}{\omega z_S}\right]=0 \qquad (23)$$

and an antisymmetric interface wave:

$$4\vartheta^2(\vartheta^2-1)^{1/2}\left[\frac{\kappa_x}{\omega z_S}+(\vartheta^2-\xi^2)^{1/2}\right] -(1-2\vartheta^2)\left[(1-2\vartheta^2)-2(\vartheta^2-1)^{1/2}\frac{\kappa_x}{\omega z_S}\right]=0 \qquad (24)$$

where $\vartheta = c_s/c$.

Equations (23) and (24) are obtained from Eqs. (22) using the identities:

$$(c^{-2}-c_P^{-2})^{1/2} = \cos\theta/c_P$$
$$(c^{-2}-c_S^{-2})^{1/2} = \cos\phi/c_S$$

When the stiffness of the interface is set to zero, Eqs. (23) and (24) degenerate to the nondispersive Rayleigh equation for surface waves on a free surface. Since Eq. (23) depends only on κ_z and Eq. (24) only on κ_x, the symmetric interface wave depends only on coupling in a direction normal to the interface, while the antisymmetric wave depends only on tangential or shear coupling.

Phase velocities of the interface waves are shown in Figure 8 for various values of the ratio c_P/c_S. The phase velocities lie between the Rayleigh surface wave velocity and the shear wave velocity of the medium, and the symmetric wave propagates faster than the antisymmetric wave. The symmetric interface wave ceases to exist for values greater than the dimensionless ratio $(\kappa_z/\omega z_S)_{cf}$.

Boundary element simulations have been carried out (Gu et al. 1996) to explore the wavefield generated by point sources placed on or near an elastic imperfect interface. These simulations not only confirmed the results of the analyses discussed above, but also indicated that another interface wave not predicted by the inhomogeneous plane wave theory should exist. This wave has elliptic particle motion like that of a Rayleigh wave, and has an amplitude sensitive to interface stiffness, but travels at a velocity close to the P-wave velocity of the medium.

5 INTERFACE CHANNEL WAVES

Having demonstrated that energy can be trapped by a single imperfect interface, it is natural to ask if energy might be trapped between two parallel interfaces in an otherwise homogeneous medium. The existence and properties of interface channel waves are explored by developing dispersion relations for an inhomogeneous plane wave that propagates between two fractures with a sinusoidally varying amplitude while decaying exponentially in the surrounding half spaces. Nihei et al. 1997 developed a modal determinant with roots that are the frequency dependent phase velocities of the modes that can exist in the waveguide defined by the interfaces.

For a waveguide defined by two elastic imperfect interfaces in a homogeneous elastic medium, Nihei et al. 1997 found that the solution for the trapped waves could be separated into a part with x-direction particle motions symmetric with respect to the center of the layer and a part with antisymmetric x-direction motion. The determinant for the symmetric part is given by:

$$A_S = \begin{pmatrix} a_{11} & a_{12} & 2a_{14} & 2a_{15} \\ a_{21} & a_{22} & 2a_{24} & 2a_{25} \\ a_{31} & a_{32} & 2a_{34} & 2a_{35} \\ a_{41} & a_{42} & 2a_{44} & 2a_{45} \end{pmatrix} \quad (25)$$

and for the antisymetric part is:

$$A_a = \begin{pmatrix} a_{11} & a_{12} & 2a_{13} & 2a_{16} \\ a_{21} & a_{22} & 2a_{23} & 2a_{26} \\ a_{31} & a_{32} & 2a_{33} & 2a_{36} \\ a_{41} & a_{42} & 2a_{43} & 2a_{46} \end{pmatrix} \quad (26)$$

where:

$$a_{11} = ik\left(1+\frac{2\mu p}{\kappa_x}\right)e^{-ph},\ a_{12} = \left[q+\frac{\mu(k^2+q^2)}{\kappa_x}\right]e^{-qh},$$

$$a_{13} = -ik\sin(ph),$$

$$a_{14} = -ik\cos(ph),\ a_{15} = q\cos(qh),\ a_{16} = -q\sin(qh)$$

$$a_{21} = \left[-p+\left(\frac{\lambda k^2 - \rho c_p^2 p^2}{\kappa_z}\right)e^{-ph},\right]$$

$$a_{22} = ik\left(1+\frac{2\mu q}{\kappa_z}\right)e^{-qh},\ a_{23} = -p\cos(ph),$$

$$a_{24} = p\sin(ph),\ a_{25} = -ik\sin(qh),\ a_{26} = -ik\cos(qh)$$

(27)

$$a_{31} = -2i\mu kpe^{-ph},\ a_{32} = -\mu e^{-qh}(k^2+q^2),$$

$$a_{33} = -2i\mu kp\cos(ph),$$

$$a_{34} = 2i\mu kp\sin(ph),\ a_{35} = \mu(k^2-q^2)\sin(qh),$$

$$a_{36} = \mu(k^2-q^2)\cos(qh)$$

$$a_{41} = (-\lambda k^2+\rho c_p^2 p^2)e^{-ph},\ a_{42} = -2i\mu kqe^{-qh},$$

$$a_{43} = (\lambda k^2+\rho c_p^2 p^2)\sin(ph),$$

$$a_{44} = (\lambda k^2+\rho c_p^2 p^2)\cos(ph),\ a_{45} = -2i\mu kq\cos(qh),$$

$$a_{46} = 2i\mu kq\sin(qh)$$

and:

$$p^2 = k^2 - k_p^2$$

$$q^2 = k^2 - k_S^2$$

$$k = \omega/c,\ k_p = \omega/c_P,\ k_s = \omega/c_S$$

$2h$ = distance between the interfaces

The nontrivial solutions of $\det A_S = 0$ and $\det A_a = 0$ yield the dispersion curves for the symmetric and antisymmetric trapped modes.

Dispersion curves are shown in Figure 9 for a particular case of $c_P = 3118$m/s, $c_S = 1800$m/s, $\rho = 2100$kg/m^3 and h = 1m. A family of trapped wave modes is predicted for each value of interface stiffness. Only the fundamental and first symmetric (S) and antisymmetric (A) modes are plotted in each case. When the stiffness is very low (10^7Pa/m), the layer between the interfaces is nearly decoupled from the surrounding medium and the solution degenerates to the classic Rayleigh-Lamb solution for a plate. At the other extreme, if interface stiffness is infinite (not shown), the interfaces become welded and the solution degenerates to the S-wave velocity of the medium.

In general, the phase velocities lie between the Rayleigh wave velocity and shear wave velocity of the medium. The exception is at low stiffness where the fundamental antisymmetric mode can have velocities less than the Rayleigh wave velocity. The effect of increasing stiffness is to shift the cutoff frequencies of the modes to higher values.

6 CONCLUSIONS

Joints and fractures in rock are characterized as imperfect interfaces. If there is no energy disipation mechanism in the interface, it is elastic and is completely described by a value of stiffness which

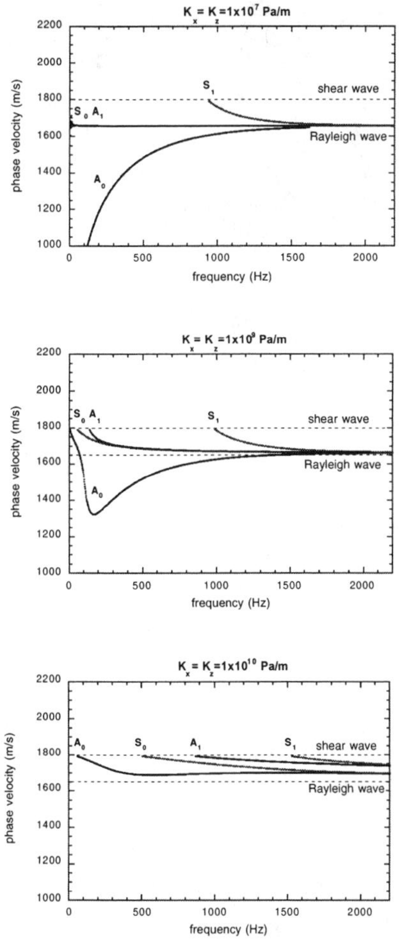

Figure 9. Dispersion curves for symmetries (S) and antisymmetries (A) modes trapped between two interfaces with different values of stiffness in a homogeneous medium.

may vary with direction. For an elastic interface, solutions have been found for plane homogeneous waves at arbitrary angles of incidence and inhomogeneous plane waves propagating along and between interfaces. Solutions are well behaved in that they degenerate to classical solutions for limiting values of stiffness. That is, for zero stiffness the interface becomes a free surface and for infinite stiffness the effect of the interface vanishes. If energy is lost at the interface, as modeled by Kelvin or Maxwell rheologies, properties of transmitted and reflected plane waves are significantly affected. Effects of rheology on interface and trapped waves have yet to be investigated but are presumed to be significant.

Application of these results should lead to greatly enhanced capabilities for characterization of fractured rock masses. Transmitted and reflected seismic data can be better interpreted. New approaches to characterization could involve using trapped waves on or between fractures.

7 ACKNOWLEDGMENTS

The insight of Neville Cook has significantly influenced this work. The work was carried out under U.S. Department of Energy Contract No. DE-AC03-76SF00098. Support was provided by the Director, DOE Office of Energy Research, Office of Basic Energy Science, and the Gas Research Institute.

REFERENCES

Angel, Y.C. & J.D. Achenbach (1985). Reflection and transmission of elastic waves by a periodic array of cracks, *J. Appl. Mech.*, 52, 33-41.

Crampin, S. (1981). A review of wave motion in anisotropic and cracked elastic media, *Wave Motion*, 3, 343-391.

Gu, B., R. Suarez-Rivera, K. Nihei, and L.R. Myer (1996). Incidence of plane waves upon a fracture, *J. Geophys. Res.*, 101 (B11), 25337-25346.

Gu, B., K. Nihei, L. Myer, L. Pyrak-Nolte (1996). Fracture interface waves, *J. Geophys. Res.*, 101 (B1), 827-835.

Hudson, J. A. (1981). Wave speeds and attenuation of elastic waves in material containing cracks, *Geophys. J.R. Astron. Soc.*, 64(1), 133-150.

Myer, L.R., Pyrak-Nolte, L.J. and Cook, N.G.W. (1990). Effects of single fracture on seismic wave propagation, in *Rock Joints*, Barton N. and Stephansson, O., eds., Proceedings of the International Symposium on Rock Joints, Loen, Norway, pp. 467-474.

Myer, L.R., Hopkins, D., Peterson, J., and Cook, N. (1995). Seismic wave propagation across multiple fractures in *Fractured and Jointed Rock Masses*, Myer, Cook, Goodman and Tsang eds. Balkema Publishers, Rotterdam, pp. 105-110.

Myer, L.R. (1991). Hydromechanical and seismic properties of fractures, *Proceedings of the 7th International Congress on Rock Mechanics*, vol. 1, pp. 397-404.

Nagy, P.B., D.V. Rypien, & L. Adler (1990). Dispersive properties of leaky interface waves in adhesive layers, in *Review of Progress in Quantitative Nondestructive Evaluation*, D.O. Thompson and D.E. Chimenti (eds), 9, 1247-1254.

Nihei, K.T., B. Gu, L.R. Myer, L.J. Pyrak-Nolte, and N.G.W. Cook (1995). Elastic interface wave propagation along a fracture, *Proc. 8th Int. Cong.*

Rock Mech. Tokyo, Sept. 25-29, Balkema Publishers, Rotterdam.
Nihei, K., W. Yi, L. Myer, and N. Cook (1997). Fracture channel waves, in preparation.
O'Connell, R.J., and B. Budiansky (1974). Seismic velocities in dry and saturated cracked solids, *J. Geophys. Res.*, 79, 5412-5426.
Pyrak-Nolte, L.J. & N.G.W. Cook (1987). Elastic interface waves along a fracture, *Geophys. Res. Lett.*, 14, 1107-1110.
Pyrak-Nolte, L.J., L.R. Myer, & N.G.W. Cook (1990). Transmission of seismic waves across single natural fractures, *J. Geophys. Res.*, 95, 8617-8638.
Rokhlin, S.I. & Y.J. Wang (1991). Analysis of boundary conditions for elastic wave interaction with an interface between two solids, *J. Acoust. Soc. Am.*, 89, 503-515.
Schoenberg, M. (1980). Elastic wave behavior across linear slip interfaces, *J. Acoust. Soc. Am.*, 68(5), 1516-1521.
Suárez-Rivera, R., Cook, N.G.W., and Myer, L.R. (1992). Study on the transmissions of shear waves across thin liquid films and thin clay layers, *Proceedings of 33rd U.S. Rock Mechanics Symposium*, J.R. Tillerson and W. Wawersik, eds., Balkema, pp. 937-946.
Thomsen, L. (1986). Weak elastic anisotropy, *Geophysics*, 51(10), 1954-1966.

Basics

Damage and Failure of Interfaces, Rossmanith (ed.) © 1997 Balkema, Rotterdam, ISBN 90 5410 899 1

Stress singularity near a crack tip terminating at a nonideal interface

G. S. Mishuris
Department of Mathematics, Rzeszów University of Technology, Poland

ABSTRACT: Modelling problems for a semi-infinite crack terminating at a bimaterial interface are considered. Instead of the "ideal contact" interfacial conditions, usually done, we have introduced a thin intermediate zone between the materials, and we model it as a thin elastic region the width of which changes according to an exponential law. Depending on parameters defining the thickness of the intermediate zone, we discuss the displacement distribution and stress field at the neighbourhood of the crack tip. It is shown that the geometry of the thin region influences essentially the state of strain and stress not only qualitatively (the character of the stress singularity near the crack tip), but also quantitatively (the increase of a number of singular terms in the asymptotics).

1 INTRODUCTION

Problems of nonhomogeneous solids with cracks terminating at bimaterial interfaces, as a rule, are discussed in the case of the so-called "ideal contact" conditions. They consist of the continuity of displacements and the traction vectors along the interface. It is known (Williams 1959, Zak & Williams 1963) that in such problems the exponent of the stress singularity $\omega - 1$, as a rule, is not equal to -0.5. Moreover, in the case of an interfacial crack, the model of the "ideal contact" leads to such inconsistencies as oscillating character of the stress and strain components and overlapping of the crack surfaces in the vicinity of crack tips; both unacceptable from the standpoint of physics. These facts have been the subject of many controversies. For extensive literature on this topic see Comninou (1979) and Rice (1988). Let us note in addition that in all these cases, the usually applied Griffith-Irwin's criterion cannot be directly employed.

To eliminate such inconsistencies, in the case of an interfacial crack *a modification of the contact conditions* along the crack surfaces near the crack tip has been proposed (Comninou 1979). Another way to investigate the problem when the crack tip is situated on the bimaterial interface is to use so-called "kinked crack approach" (He & Hutchinson 1989_1, 1989_2). A third way is to assume that there exists a thin *interfacial zone*, mechanical properties of which vary between the different materials (Atkinson 1977, the author 1985, Erdogan, et al. 1991). Then the fracture mechanics analysis can be done in terms of the stress intensity factor (SIF) with Griffith-Irwin's criterion.

In contradiction to the other models, the "intermediate region" allows us to take into account the influence of the real adhesive region not only at the stage of strength analysis by the theory of adhesion (Cherepanov 1983), but also at the first stage of finding the displacements and stress fields near the crack tip.

However, when the thin interlayer is of infinitesimal thickness, the problem has to be investigated another way. In this case, the adhesive zone could be considered as a thin elastic inclusion of characteristic thickness h_*. Usually, we do not have any additional information on the exact form of the adhesive layer near the crack tip. In order to take into account an arbitrary form of the thin interlayer we assume that its thickness $h(r)$ varies according to the law

$$h(r) = h_* r^\alpha, \quad 0 \le h_* \ll D_*, \quad 0 \le \alpha < \infty, \quad (1)$$

where r is a distance from the crack tip to a point on the interface, while D_* is a characteristic size of the body under consideration. The corresponding geometry of the adhesive zone in the neighbourhood of the crack tip is presented in Figures 1b – 1e depending on the value of parameter α. Here μ_0, μ_1 and ν_0, ν_1 are the shear moduli and Poisson's ratios of the elastic materials being in contact.

In the Cartesian coordinate system x, y we con-

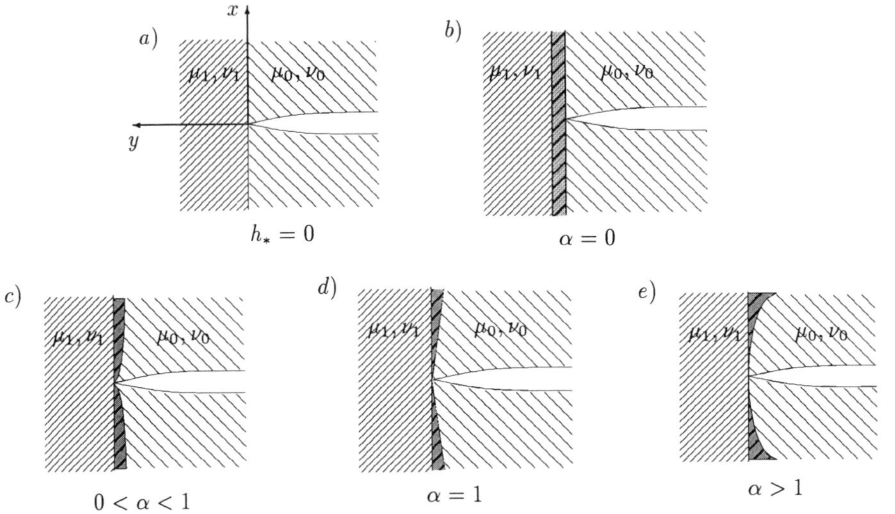

Figure 1. Modelling geometry of the intermediate zone near the crack tip.

sider a two-dimensional state of the bimaterial body with the thin adhesive interlayer. Applying a standard technique for thin inclusion, we can integrate respective equilibrium equations of the intermediate zone by the parameter determining normal direction to surfaces of the thin region. Then the following interfacial conditions appear:

$$[\boldsymbol{\sigma}]_{|y=0} = \mathbf{o}, \quad ([\mathbf{u}] - r^{\alpha}\boldsymbol{\tau}\boldsymbol{\sigma})_{|y=0} = \mathbf{o}. \qquad (2)$$

Here $[\mathbf{u}]$, $[\boldsymbol{\sigma}]$ are jumps of the vectors of displacements and tractions, respectively, along the bimaterial interface ($y = 0$), while $\boldsymbol{\tau}$ is a diagonal matrix of the components: (see Cherepanov 1974):

$$\tau_1 = h_*/\mu^*_{xy}, \quad \tau_2 = h_*/E_*, \quad \tau_3 = h_*/\mu^*_{yz},$$

where E_* and μ^*_{xy}, μ^*_{yz} are the Young's and the shear moduli of the elastic inclusion. Conditions (2) can be considered independently of an assumed model of the thin interlayer. Then parameters τ_j, α have to be determined experimentally.

Such an approach allows us to investigate in one scheme different forms of the intermediate zone near the crack tip. Thus, if $\tau = 0$, we have the usual "ideal contact" (Figure 1a). When $\alpha = 0$, there is a thin interlayer of constant thickness between the materials (Figure 1b). The case where $0 < \alpha < 1$ can be interpreted as a thin intermediate zone with a damage near the crack tip (Figure 1c). When $\alpha = 1$, the materials are contracted by the thin wedges (Figure 1d). Finally, $1 < \alpha < \infty$ (Figure 1e) can represent an "almost ideal contact" between materials near the crack tip.

In the paper the Mode I, II, III, problems for infinite bimaterial plane with semi-infinite crack terminating at the thin adhesive zone between the materials are considered. Along the bimaterial interface, conditions (2) are assumed to be true. We show that the stress singularity at the crack tip depends essentially on parameter α and is not equal to -0.5, in general. In works by Cherepenov (1984), Erdogan and coauthors (1991), the reader can find arguments underlining why such character of the stress singularity plays an important role in the fracture mechanics. In fact, if an additional parameter of length dimension (length of the contact zone, thickness of the intermediate layer or length of a kinked crack) tends to zero in the respective models mentioned above, then the behaviour of the values of SIF cannot be satisfactory.

2 THE MODE III PROBLEM

Let us consider a modelling problem for a bimaterial plane with a semi-infinity crack terminating perpendicularly at the interface. We shall seek for a nonzero component of the displacement $u = u_z$, which is a harmonic function in domains $y > 0$, $y < 0$, and satisfy exterior boundary conditions along the crack surfaces:

$$\sigma_{\theta r}|_{\theta=-\pi/2\pm 0} = -g(r). \qquad (3)$$

We assume that the function $g(r)$ is sufficiently smooth and vanishes at zero and infinity points (for example, $g \in C_0^{\infty}(\mathbb{R}_+)$), then any singularities of the solutions are connected with the interior properties of the problems only. In view of symmetry of the problem geometry and the conditions (3), we can conclude that

$$u_{|\theta=\pi/2} = 0, \tag{4}$$

on the crack line ahead ($x = 0, y > 0$).

Along the bimaterial interface ($y = 0, x > 0$), the interfacial conditions follow from (2)

$$\left([u] - \bar{\tau}_3\mu_0^{-1}r^\alpha\sigma_{\theta z}\right)_{|\theta=0}, \quad [\sigma_{\theta z}]_{|\theta=0} = 0, \tag{5}$$

with certain known dimensionless parameters $\bar{\tau}_3 = \mu_0\tau_3$, $\alpha \geq 0$.

We shall seek for the solution $u(x,y)$ of problem (3) - (5) which meets with the conditions at the singular points:

$$u = O(r^{\vartheta_0}), \ \sigma = O(r^{\gamma_0 - 1}), \quad r \to 0,$$
$$u = O(r^{-\vartheta_\infty}), \ \sigma = O(r^{-\gamma_\infty - 1}), \ r \to \infty, \tag{6}$$

where σ is the corresponding stress tensor. Unknown constants $\vartheta_0, \vartheta_\infty \geq 0$, $\gamma_0, \gamma_\infty > 0$ ($\vartheta_0 + \vartheta_\infty > 0$) depend on the values of mechanical parameters $\mu_0, \mu_1, \bar{\tau}_3, \alpha$, and shall be calculated while the problem is solved.

2.1 *Reduction to a functional equation*

Applying the Mellin transform technique we obtain the following functional equation:

$$\bar{\tau}_3 p(s + \alpha - 1) + F(s)p(s) = G(s), \tag{7}$$

with the additional condition $p(0) = -\tilde{g}(0)$. Here an unknown function $p(s)$ is the Mellin transform of the traction along the interface:

$$p(s) = \int_0^\infty \sigma_{\theta z}(r,\theta)_{|\theta=0} r^s ds,$$

which is analytic in the strip $-\gamma_0 <$ Re $< \gamma_\infty$ in view of *a priori* estimates (6). Besides, the following notations are introduced in (7):

$$F(s) = \frac{2(\kappa - \cos \pi s)}{(1-\kappa)s \sin \pi s}, \quad \kappa = \frac{\mu_0 - \mu_1}{\mu_0 + \mu_1},$$

$$G(s) = \frac{\tilde{g}(s)}{s \sin(\pi s/2)}, \quad \tilde{g}(s) = \int_0^\infty g(r) r^s dr.$$

In the case of the "ideal contact" ($\bar{\tau}_3 = 0$), solution of equation (7) is found in a closed form, and is well known. Thus, function $p(s)$ is analytic in the strip $-\omega_3 < Re\ s < \omega_3$, and has simple poles at points $s = \pm\omega_3$, where $\omega_3 \in (0,1)$ is the first zero of the function $F(s)$ which is the nearest to the imaginary axis. Hence, we can conclude that $\gamma_0 = \gamma_\infty = \vartheta_\infty = \vartheta_0 = \omega_3$.

2.2 *Solution to the problem*

First of all, let us note that in case $\alpha = 1$, equation (7) is solved in a closed form. Function $p(s)$ is analytic in the strip $-\omega_*(\bar{\tau}_3) < Re\ s < \omega_*(\bar{\tau}_3)$, and has simple poles at points $s = \pm\omega_*(\bar{\tau}_3)$. Here $\omega_*(\bar{\tau}_3) \in (0,1)$ is the first zero of function $F(s) + \bar{\tau}_3$. The corresponding graphs of the values of $\omega_*(\bar{\tau}_3)$ are presented in Figure 2.

In case $\alpha \in [0,1)$, equation (7) is valid in the strip $0 < Re\ s < \omega_3$, at least. Hence, the function $p(s)$ should be analytic in the strip $\alpha - 1 < Re\ s < \omega_3$, and has simple poles at points $s = \alpha - 1$ ($\alpha \neq 0$) and $s = \omega_3$, but in case $\alpha = 0$ there exists a double pole at point $s = -1$.

In case $\alpha > 1$, we can conclude that equation (7) holds true in the strip $-\omega_3 < Re\ s < 0$. Hence, the function $p(s)$ has simple poles at points $s = -\omega_3$ and $s = \alpha - 1$ and is analytic in the strip $-\omega_3 < Re\ s < \alpha - 1$.

It can be proved that the functional equation (7) has a unique solution in both cases. They can be found from some singular integral equations with fixed point singularities. In case $\alpha = 0$, the corresponding results have been presented earlier (Mishuris 1997_1).

2.3 *Analysis of the solutions*

Now we investigate asymptotics of displacement $u(r,\theta)$ near the crack tip ($u_0(r,\theta)$ and $u_1(r,\theta)$ in the domain $-\pi/2 < \theta < 0$ and $0 < \theta < \pi/2$, respectively). Thus, when $\alpha = 0$, we can obtain

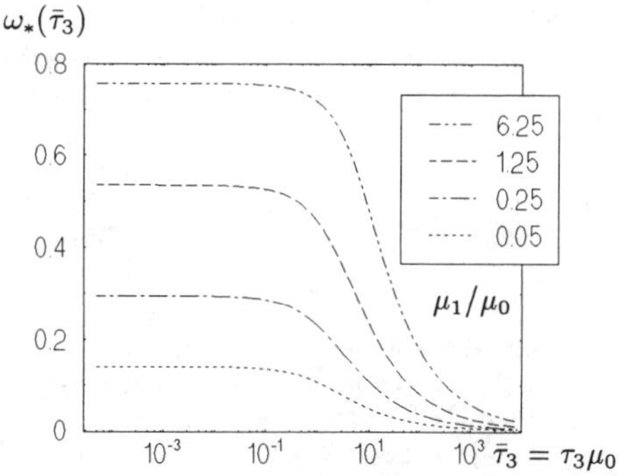

Figure 2: Graph of parameter $\omega_*(\bar{\tau}_3)$ determining the exponent of stress singularity $\gamma_0 - 1 = \omega_*(\bar{\tau}_3) - 1$ for case $\alpha = 1$.

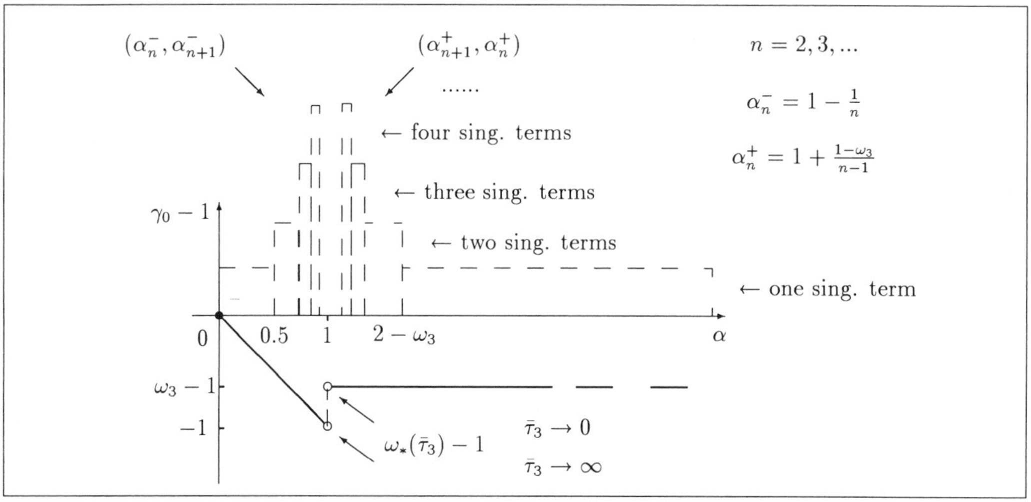

Figure 3. Graph of the stress singularity, and a scheme of distributions of the number of singular terms in the asymptotics of stress near the crack tip for the Mode III problem.

$$
\begin{aligned}
&u_0(r,\theta) = C_1 - C_1\bar{\tau}_3^{-1} r \sin\theta + O(r^{1+\omega_3}), \ r \to 0, \\
&u_1(r,\theta) = -\frac{C_1\mu_0 r}{\pi\mu_1\bar{\tau}_3}\big[(\pi - 2\theta)\sin\theta + \\
&\quad +2(C_2 + \ln r)\cos\theta\big] + O(r^{1+\omega_3}), \ r \to 0, \\
&u_j(r,\theta) = O(r^{-\omega_3}), \quad r \to \infty,
\end{aligned} \tag{8}
$$

where C_1, C_2 are constants (Mishuris 1997_1).

For all remaining cases ($\alpha > 0$), the corresponding relations can be rewritten in a common form as follows ($\phi = \pi/2 - \theta$):

$$
\begin{aligned}
&u_0 = C_0 + \frac{K_3}{\mu_0\gamma_0} r^{\gamma_0} \operatorname{ctg}\frac{\gamma_0\pi}{2}\cos\gamma_0(\pi - \phi) + O(r^{\gamma_1}), \\
&u_1 = K_3(\mu_1\gamma_0)^{-1} r^{\gamma_0}\sin\gamma_0\phi + O(r^{\gamma_1}), \qquad r \to 0, \\
&\left.\begin{aligned} u_0(r,\theta) &= C_\infty + O(r^{-\gamma_\infty}), \\ u_1(r,\theta) &= O(r^{-\gamma_\infty}), \end{aligned}\right\} \ r \to \infty,
\end{aligned} \tag{9}
$$

where the constants from *a priori* estimations (6) are calculated taking into account the behaviour of function $p(s)$:

$$
\begin{aligned}
\gamma_0 = \gamma_\infty = \vartheta_\infty = \vartheta_0 = \omega_*(\bar{\tau}_3), \quad & \alpha = 1, \qquad (10) \\
\gamma_0 = 1 - \alpha, \gamma_\infty = \vartheta_\infty = \omega_3, \vartheta_0 = 0, \quad & 0 \le \alpha < 1, \\
\gamma_\infty = 1 - \alpha, \gamma_0 = \vartheta_0 = \omega_3, \vartheta_\infty = 0, \quad & 1 < \alpha < \infty.
\end{aligned}
$$

In equations (9), constant K_3 in the main term of stress singularity is so-called generalized SIF. Its value is calculated by the corresponding constant of the pole of function $p(s)$.

Corresponding nonzero components of the tensor of stress can be calculated using (8), (9). Let us note that situations can appear where there is a number of singular terms of stress near the crack tip ($\gamma_1 < 1$).

In Figure 3, a graph of the main exponent of stress singularity $\gamma_0 - 1$ at the crack terminating at the bimaterial interface is presented with respect to the value of parameter α. Besides, a scheme demonstrating the distribution of the number of singular terms in the stress asymptotics in the neighbourhood of the crack tip is shown.

- For $\alpha = 0$, there is not an exponential singularity of stress near the crack tip for any values of the mechanical parameters μ_0, μ_1, $\bar{\tau}_3$. In this case, stress concentration appears only in the domain $y > 0$ (on the crack line ahead), and it has a logarithmic character (see Mishuris $(1997)_1$). In regards to the displacement field, there is displacement discontinuity near the crack tip along the bimaterial interfacial contact ($C_1 \neq 0$ in (8)).
- If $\alpha \in (0, 0.5]$, only one singular term in asymptotics of stress in the neighbourhood of the crack tip occurs ($\gamma_1 \geq 1$). Corresponding exponent in the interval $(-1, 0)$ is $\gamma_0 - 1 = -\alpha$, and does not depend on the values of μ_0, μ_1, $\bar{\tau}_3$.
- For case $\alpha \in (0.5, 1)$, or more precisely $\alpha \in (\alpha_n^-, \alpha_{n+1}^-)$, $(n = 2, 3, ...)$, where $\alpha_n^- = 1 - 1/n$, there are exactly n singular terms in the asymptotics of stress with the exponents: $\gamma_0 - 1 = -\alpha$, $\gamma_j^- - 1 = (j+1)(1-\alpha) - 1 \in (-1, 0)$, $j = 1, ..., n$ (see the diagram in Figure 3). Moreover, if $\alpha \to 1$, then the number of singular terms tends to infinity $n \to \infty$! In the last two cases ($\alpha \in (0, 1)$), the displacement discontinuity near the crack tip also appears ($C_0 \neq 0$ in (9)).
- If $\alpha = 1$ there is one singular term of asymptotics ($\gamma_1 \geq 1$) with the exponent $\gamma_0 - 1 = \omega_*(\bar{\tau}_3) - 1 \in (-1, \omega_3 - 1)$ depending essentially on the val-

ues of the mechanical parameters μ_0, μ_1, $\bar{\tau}_3$ (see Figure 2). Thus, if $\bar{\tau}_3 \to 0$ then $\gamma_0 - 1 \to \omega_3 - 1$, which coincides with the result for the "ideal contact". In this case ($\alpha = 1$) and in the next one ($\alpha > 1$) the displacement field is continuous near the crack tip ($C_0 = 0$). (However, it is discontinuous on any distance from the crack tip along the bimaterial contact in view of the conditions $(5)_1$).

- For case $\alpha \in (1, 2 - \omega_3)$, or more precisely $\alpha \in (\alpha^+_{n+1}, \alpha^+_n)$, $(n = 2, ...)$, where $\alpha^+_n = 1 + (1 - \omega_3)/(n-1)$, there are accurately n singular terms in asymptotics of stress with the exponents: $\gamma_0 - 1 = \omega_3 - 1$, $\gamma^+_j - 1 = -j(\alpha - 1) + \omega_3 - 1 \in (-1, 0)$, $j = 1, ..., n$ (see the diagram in Figure 3). As above, $n \to \infty$ when $\alpha \to 1$.
- Finally, in case $\alpha \in [2 - \omega_3, \infty)$, one singular term of stress asymptotics appears ($\gamma_1 \geq 1$). The corresponding exponent $\gamma_0 - 1 = \omega_3 - 1$ is similar to that for the "ideal contact" model, and does not depend on the remaining problem parameters.

Numerical results for SIF and the displacement discontinuity near the crack tip are presented in the papers (Mishuris 1997_1, 1997_2) for different values of the mechanical parameters μ_0, μ_1, $\bar{\tau}_3$. In particular, as it has been shown for $0 < \alpha < 1$, estimations are true:

$$C_0 \sim \bar{\tau}_3^{\omega_3 + \alpha}, \quad K_3 \sim \bar{\tau}_3^{\omega_3 + \alpha - 1}, \quad \bar{\tau}_3 \to 0. \qquad (11)$$

In the opposite case ($1 < \alpha$), the following relation is obtained ($C_0 = 0$):

$$K_3 = K_3^{id} + O(\bar{\tau}_3^{\beta}), \quad \bar{\tau}_3 \to 0, \qquad (12)$$

where K_3^{id} is generalized SIF in the case of the "ideal contact", but $\beta = \beta(\alpha) > 0$ is some constant. In particular, $\beta(1 + \omega_3) = 1$. Hence, in all of the cases, asymptotics (9) is rebuilt to those for the "ideal bimaterial contact" when $\bar{\tau}_3 \to 0$.

3 THE MODE I, II PROBLEMS

Now we seek for vector of displacements $\mathbf{u}(r, \theta)$, and tensor of stress $\boldsymbol{\sigma}(r, \theta)$, which components satisfy the equilibrium equations in each of the domain ($y < 0$, $y > 0$). Boundary conditions along the crack surfaces are described in the form (the Mode I and Mode II, respectively):

$$\begin{array}{ll} \mathrm{I} : & \sigma_\theta|_{\theta=-\pi/2} = -g_1(r), \quad \sigma_{r\theta}|_{\theta=-\pi/2} = 0, \\ \mathrm{II} : & \sigma_\theta|_{\theta=-\pi/2} = 0, \quad \sigma_{r\theta}|_{\theta=-\pi/2} = -g_2(r), \end{array} \qquad (13)$$

Besides, we can conclude, that the following relations hold true on the crack line ahead

$$\begin{array}{ll} \mathrm{I} : & u_\theta|_{\theta=\pi/2} = 0, \quad \sigma_{r\theta}|_{\theta=\pi/2} = 0, \\ \mathrm{II} : & u_r|_{\theta=\pi/2} = 0, \quad \sigma_\theta|_{\theta=\pi/2} = 0. \end{array} \qquad (14)$$

Finally, along the interface between the materials ($y = 0$), interfacial conditions (2) are assumed to be true.

As above, functions g_1, g_2 are sufficiently smooth and have compact supports. We shall seek for solutions of the problems meeting conditions (6), where constants $\vartheta_0, \vartheta_\infty \geq 0$, $\gamma_0, \gamma_\infty > 0$, ($\vartheta_0 + \vartheta_\infty > 0$) depend on the values of mechanical parameters μ_0, μ_1, ν_0, ν_1, α, τ_1, τ_2 and calculating below.

3.1 *Reduction to systems of functional equations*

Applying the Mellin transform we obtain the following systems of functional equations:

$$\mathbf{\Lambda p}(s + \alpha - 1) + \mathbf{\Phi}_k(s)\mathbf{p}(s) = \mathbf{G}_k(s), \qquad (15)$$

where $k = 1, 2$ for the Mode I and Mode II, respectively. Here, unknown vector-functions $\mathbf{p}(s)$ are the Mellin transforms of the tractions along the interface

$$\mathbf{p}(s) = \left(\tilde{\sigma}_\theta(s, 0), \tilde{\sigma}_{r\theta}(s, 0)\right)^{\mathsf{T}}, \qquad (16)$$

are analytic in the strip $-\gamma_0 < Re\ s < \gamma_\infty$, and should satisfy additional conditions

$$\mathbf{p}(0) = \begin{cases} \left(0, -\tilde{g}_1(0)\right)^{\mathsf{T}}, & \text{Mode I}, \\ \left(-\tilde{g}_2(0), 0\right)^{\mathsf{T}}, & \text{Mode II}. \end{cases}$$

Matrices $\mathbf{\Lambda}$, $\mathbf{\Phi}_k(s)$ and the vectors $\mathbf{G}_k(s)$ are defined in the following manner:

$$\mathbf{\Lambda} = \begin{pmatrix} \bar{\tau}_\theta & 0 \\ 0 & \bar{\tau}_r \end{pmatrix}, \quad \mathbf{\Phi}_2 = \mathbf{E\Phi}_1\mathbf{E}, \quad \mathbf{E} = \begin{pmatrix} 0 & 1 \\ 1 & 0 \end{pmatrix},$$

$$\mathbf{G}_1(s) = \frac{-\tilde{g}_1(s)}{s[s^2 - \sin^2(\pi s/2)]} \begin{pmatrix} s\cos(\pi s/2) \\ (s+1)\sin(\pi s/2) \end{pmatrix},$$

$$\mathbf{G}_2(s) = \frac{\tilde{g}_2(s)}{s[s^2 - \sin^2(\pi s/2)]} \begin{pmatrix} (1-s)\sin(\pi s/2) \\ s\cos(\pi s/2) \end{pmatrix}.$$

The corresponding components $\phi_{ij}(s)$ of matrix-function $\mathbf{\Phi}_1(s)$ are calculated by the relations

$$\phi_{11}(s) = \frac{\sin \pi s}{s[2s^2 - 1 + \cos \pi s]} - \frac{b}{s}\mathrm{ctg}\frac{\pi s}{2},$$

$$\phi_{22}(s) = \frac{\sin \pi s}{s[2s^2 - 1 + \cos \pi s]} + \frac{b}{s}\mathrm{tg}\frac{\pi s}{2},$$

$$\phi_{12}(s) = \frac{2s - 1 + \cos \pi s}{s[2s^2 - 1 + \cos \pi s]} - \frac{a + b}{s},$$

$$\phi_{21}(s) = \frac{2s + 1 - \cos \pi s}{s[2s^2 - 1 + \cos \pi s]} + \frac{a + b}{s}.$$

Here dimensionless parameters a, b, $\bar{\tau}_{r(\theta)}$ are calculated as follows:

$$a = \frac{\mu_1 - \mu_0}{2\mu_1(1 - \nu_0^*)}, \quad b = \frac{\mu_0(1 - \nu_1^*)}{\mu_1(1 - \nu_0^*)}, \quad \bar{\tau}_{r(\theta)} = \frac{\mu_0}{1 - \nu_0^*}\tau_{1(2)},$$

where $\nu_j^* = \nu_j$ under plane strain deformations, but $\nu_j^* = \nu_j/(1 - \nu_j)$ under plane stress conditions.

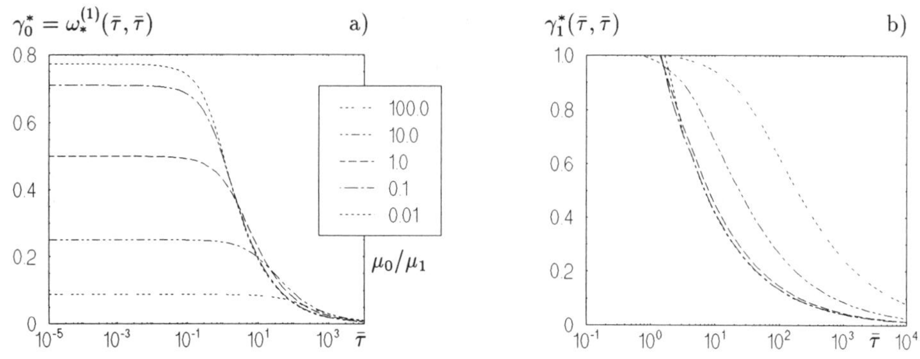

Figure 4. Graphs of parameters γ_0^*, γ_1^* determing the exponents of two first terms in asymptotics of stress ($\gamma_0^* - 1$, $\gamma_1^* - 1$) for the case $\alpha = 1$. Here Poisson's ratios are $\nu_0 = \nu_1 = 0.2$, $\bar\tau_\theta = \bar\tau_r = \bar\tau$, but μ_0/μ_1 are of different values.

3.2 *Solutions to the systems*

In the case of the "ideal contact" ($\mathbf{\Lambda} = \mathbf{0}$), solutions of the Mode I and Mode II are found in closed forms (Zak & Williams 1963). The corresponding vector-functions $\mathbf{p}(s)$ in (15) are analytic in the strip $-\omega_1 < Re\ s < \omega_1$, and have simple poles at points $s = \pm\omega_1$. Here, $\omega_1 \in (0, 1)$ is the first zero of the determinant $\det \mathbf{\Phi}_k(s)$ ($\det \mathbf{\Phi}_1(s) = \det \mathbf{\Phi}_2(s)$) which is the nearest to the imaginary axis. So, in this case we can conclude that $\gamma_0 = \gamma_\infty = \vartheta_\infty = \vartheta_0 = \omega_1$.

When $\alpha = 1$, systems (15) are solved in closed forms also. Vector-functions $\mathbf{p}(s)$ are analytic in the strip $-\omega_*^{(k)}(\bar\tau_\theta, \bar\tau_r) < Re\ s < \omega_*^{(k)}(\bar\tau_\theta, \bar\tau_r)$, and have simple poles at points $s = \pm\omega_*^{(k)}$. Here $\omega_*^{(k)} \in (0, 1)$ are the first zeros of functions $\det\mathbf{\Psi}_k(s, \bar\tau_\theta, \bar\tau_r)$ ($\mathbf{\Psi}_k = \mathbf{\Phi}_k + \mathbf{\Lambda}$). Let us note that

$$\det\mathbf{\Psi}_2(s, \bar\tau_\theta, \bar\tau_r) = \det\mathbf{\Psi}_1(s, \bar\tau_r, \bar\tau_\theta). \qquad (17)$$

Hence, exponents $\omega_*^{(1)} - 1$, $\omega_*^{(2)} - 1$ of the stress singularity are different for the Mode I and Mode II problems! Moreover, situations can appear when two singular terms of stress asymptotics exist in contradiction to the Mode III problem in the case of the thin adhesive wedge ($\alpha = 1$).

In Figure 4 a,b graphs of the values of two first real zeros γ_0^*, γ_1^* of function $\det\mathbf{\Psi}_1(s, \bar\tau_\theta, \bar\tau_r)$ for various values of $\bar\tau_\theta = \bar\tau_r = \bar\tau$, a different ratio μ_0/μ_1, when Poisson's ratios of the materials are equal to $\nu_0 = \nu_1 = 0.2$. As it can be seen, the second singular term of stress appears ($\gamma_1^* < 1$) when $\bar\tau > 1$.

In case $\bar\tau_\theta = 0$, only one real zero $\omega_*^{(1)}(0, \bar\tau_r)$ of function $\det\mathbf{\Psi}_1(s, 0, \bar\tau_r)$ appears, which belongs in interval (0,1). In Figure 5 graphs of the corresponding values of $\omega_*^{(1)}(0, \bar\tau_r)$ are presented for different magnitudes of ratio μ_0/μ_1. As in Figures 4,

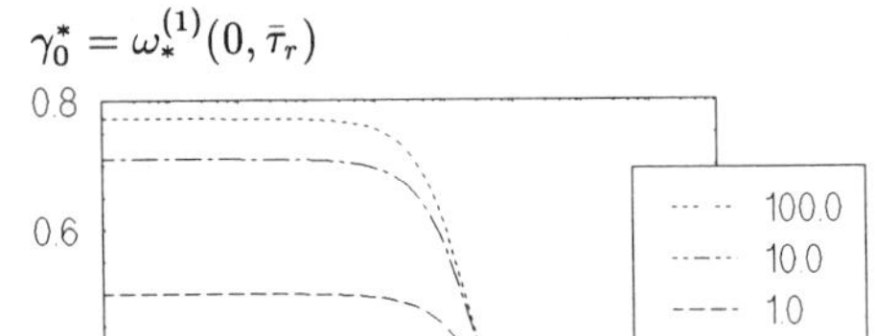

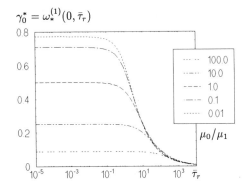

Figure 5: Graphs of parameter γ_0^* determining the stress singularity in the case $\alpha = 1$, $\nu_0 = \nu_1 = 0.2$, for different values of ratio μ_0/μ_1, and $\bar\tau_\theta = 0$.

Poisson's ratios of the materials are $\nu_0 = \nu_1 = 0.2$.

When $\bar\tau_r = 0$, two real zeros γ_0^*, γ_1^* of function $\det\mathbf{\Psi}_1(s, \bar\tau_\theta, 0)$ also exist. In Figures 6 the corresponding graphs are presented for the same Poisson's ratios of the materials for various values of $\bar\tau_\theta$.

Let us note, that exponents $\omega_*^{(2)}(\bar\tau_\theta, \bar\tau_r) - 1$ of the stress singularity for the Mode II problem can be calculated taking into account relation (17).

In the both remaining cases ($0 \le \alpha < 1$ and $\alpha > 1$), we can conclude that vector-functions $\mathbf{p}(s)$ should be analytic in the strips $\alpha - 1 < Re\ s < \omega_1$, and $-\omega_1 < Re\ s < \alpha - 1$, respectively. They have simple poles at points $s = \alpha - 1$, $s = \omega_1$ ($\alpha \neq 0$) and $s = -\omega_1$, $s = \alpha - 1$, but in case $\alpha = 0$ there exist a double pole at point $s = -1$.

As for the Mode III problem, it can be proved that systems of functional equations (15) have also unique solutions for the Mode I and Mode II

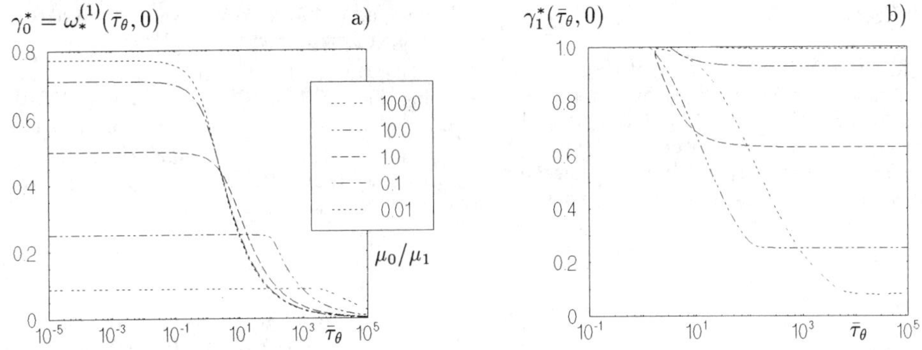

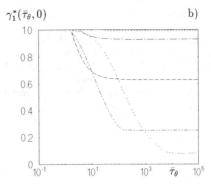

Figure 6. Graphs of parameters γ_0^*, γ_1^* determing the exponents of two first terms in asymptotics of stress ($\gamma_0^* - 1$, $\gamma_1^* - 1$) for the case $\alpha = 1$, for different values of μ_0/μ_1. Here Poisson's ratios are $\nu_0 = \nu_1 = 0.2$, and $\bar{\tau}_r = 0$.

problems for all values of parameter α. They can be found from respective systems of singular integral equations with fixed point singularities.

3.3 *Analysis of the solutions*

Now we discuss asymptotics of displacements and stress fields near the crack tip. Corresponding relations in all cases ($0 < \alpha < \infty$) can be written in the following forms ($r \to 0$):

$$\sigma_\theta^{(j)} = K_l f(\theta) r^{\gamma_0 - 1} + O(r^{\gamma_1 - 1}),$$

$$\sigma_{r\theta}^{(j)} = -K_l(\gamma_0 + 1)^{-1} f' r^{\gamma_0 - 1} + O(r^{\gamma_1 - 1}), \qquad (18)$$

$$\sigma_r^{(j)} = \frac{K_l}{\gamma_0}\left[\frac{1}{\gamma_0 + 1} f'' + f\right] r^{\gamma_0 - 1} + O(r^{\gamma_1 - 1}),$$

$$u_r^{(j)} = \frac{K_l r^{\gamma_0}}{2\gamma_0^2 \mu_j}\left[[1 - \nu_j(1 + \gamma_0)] f + \frac{1 - \nu_j}{\gamma_0 + 1} f''\right] +$$

$$C_{r0}^{(j)} + O(r^{\gamma_1}), \quad u_\theta^{(j)} = C_{\theta 0}^{(j)} - \frac{K_l (1 - \nu_j) r^{\gamma_0}}{2\gamma_0^2 \mu_j (\gamma_0^2 - 1)} \cdot$$

$$\left[\left(\frac{\gamma_0(\gamma_0 - 1)}{1 - \nu_j} + (1 + \gamma_0)^2\right) f' + f'''\right] + O(r^{\gamma_1}),$$

where $C_{r0}^{(1)} = C_{\theta 0}^{(1)} \equiv 0$, but constants K_l ($l = 1, 2$) are generalized SIF (for the Mode I and Mode II, respectively). In relations (18), $f' = f'_\theta(\theta)$ is the derivative of function $f(\theta) = f_l^{(j)}(\gamma_0, \theta)$ which has different form in each of the domain ($j = 0; 1$ for $-\pi/2 < \theta < 0$; $0 < \theta < \pi/2$, respectively). Thus, for $j = 1$ ($\phi = \pi/2 - \theta \in [-\pi/2, \pi/2]$) we obtain:

$$f_l^{(1)} = \begin{Bmatrix} \cos \\ \sin \end{Bmatrix}[(1 - \gamma_0)\phi] + B_l \begin{Bmatrix} \cos \\ \sin \end{Bmatrix}[(1 + \gamma_0)\phi], \quad \begin{matrix} l = 1 \\ l = 2 \end{matrix}$$

$$B_l = (-1)^l \frac{\delta_0 p_1 + t_l p_2}{p_1 - t_l p_2}, \quad t_l = \mathrm{tg}\left[\frac{\pi}{2}(l + \gamma_0 - 1)\right],$$

where $\delta_0 = (1 - \gamma_0)/(1 + \gamma_0)$, but the values of p_1, p_2 are calculated by relation $(s + \gamma_0)p(s) \sim [p_1, p_2]^\top$, $s \to -\gamma_0$. We do not present here the forms of functions $f_l^{(0)}(\gamma_0, \theta)$ determining the asymptotics in domain $-\pi/2 < \theta < 0$, because they are cumbersome, and the volume of the paper is limited.

In equations (18), constants from *a priori* estimations (6) are calculated taking into account the behaviour of vector-functions $p(s)$ by relations (10) with the respective values of parameters: ω_1, $\omega_*^{(k)}(\bar{\tau}_\theta, \bar{\tau}_r)$ instead of ω_3, $\omega_*(\bar{\tau}_3)$.

For the Mode I and Mode II problems, all conclusions drawn in subsection 2.2 for the Mode III problems are also true. Therefore, parameter $\bar{\tau}_3$ should be replaced by parameters $\bar{\tau}_\theta$ and $\bar{\tau}_r$ in the diagram of Figure 3. Besides, the value of the first zero ω_3 of function $F(s)$ is different from the value ω_1 ($= \omega_2$) of the first zero of function $\det \mathbf{\Phi}_k(s)$. Differences in contradiction to the Mode III problem only appear in the case of the thin adhesive wedge ($\alpha = 1$), when two real singular terms in asymptotics of stress near the crack tip can appear for the Mode I and II problems (see the previous subsection 3.2). Besides, parameter B_l in the relation for function $f_l^{(j)}$ depends on tractions along the crack surfaces, in general. And only for the following two cases: a) $\mathbf{\Lambda} = \mathbf{0}$; d) $\alpha = 1$, $\mathbf{\Lambda} \neq \mathbf{0}$; its values do not connect with the loading, and are defined by the mechanical parameters of the problems. The last fact means that two parameters arise in the asymptotics of stress for the "nonideal contact" conditions (K_l, B_l), what contradicts to the situation for the "ideal contact", when only one parameter K_1 (K_2) (generalized SIF) exists for the Mode I (the Mode II) problem.

We do not write here unwieldy expressions for the asymptotics of displacements and stress for the Mode I and Mode II problems in case $\alpha = 0$, when the stress singularity has a logarithmic character.

4 CONCLUSIONS

As it could be expected, geometry of the thin intermediate zone between the different materials essentially influences the stress singularity near the crack tip terminating at the interface. Moreover, this influence has not only qualitative character (different values of the stress singularity near the crack tip), but also a quantitative one (the increase of the number of singular terms in the asymptotics of stress). Besides, in the case $\alpha = 1$ (when thin adhesive zone is represented by thin wedge) different values of stress singularity appear for the Mode I and Mode II problems.

Of course, these theoretical results should be experimentally verified. Let us note in this connection that for a real adhesive intermediate zone the value of the parameter τ_j should be very small, as a rule. Hence, on the distance from the crack tip $r \sim \tau_j$, the asymptotics is rebuilt to that for the "ideal bimaterial contact". This fact should be taken into account when the corresponding experimental results are interpreted.

Although in most of the considered cases the stress singularity is not equal to -0.5, in general, such fracture mechanics criteria as the critical crack opening criteria (Leonov & Panasyuk 1959, Dugdale 1960, Wells (1969), or the effective stress criteria (Novozhylov 1969, Seweryn & Mróz 1995) allow us to consider arbitrary displacements and stress fields taking into account the corresponding (elastic, plastic) properties of the materials.

In fact, the main singular terms of displacements and stresses near a defect tip are usually represented in the following form:

$$u \sim K_j \gamma_0^{-1} f_u(\theta) r^{\gamma_0}, \quad \sigma \sim K_j f_\sigma(\theta) r^{\gamma_0 - 1}, \quad r \to 0,$$

Consequently, due to any of the mentioned criteria, the value of $K_j d^{\gamma_0}/\gamma_0$ has to be taken into account in fracture mechanics analysis, instead of the value of the generalized SIF K_j. Here the small parameter d of length dimension has different physical interpretation in frame of each of the criterion. Corresponding investigation is not a goal of this paper.

REFERENCES

Atkinson, C. 1977. On stress singularities and interfaces in linear elastic fracture mechanics, *Int. J. of Fract.*, 13(6): 807-820.

Cherepanov, G.P. 1974. *Mechanics of Brittle Fracture*, Nauka, Moscow, (in Russian).

Cherepanov, G.P. 1983. *Fracture Mechanics of Composite Materials*, Nauka, Moscow, (in Russian).

Comninou, M. 1979. An overview of interface crack, Engng Fract. Mech. 37: 197-208.

Dugdale, D.S. 1960. Yielding of steel sheets containing slits, *J. Mech. Phys. Solids*, 8(2): 100-108.

Erdogan, F., A.C. Kaya & P.F. Joseph 1991. The crack problem in bonded nonhomogeneous materials, *J. Appl. Mech.*, 58: 410-418.

Erdogan, F., A.C. Kaya & P.F. Joseph 1991. The mode III crack problem in bonded materials with a nonhomogeneous interfacial zone, *J. Appl. Mech.*, 58: 419-427.

He, M.Y. & J.W. Hutchinson 1989. Kinking of a crack out of an interface, *J. Appl. Mech.*, 57: 270-278.

He, M.Y. & J.W. Hutchinson 1989. Crack deflection at an interface between dissimilar elastic materials. *Int. J. Solids Structures*, 25: 1053-1067.

Leonov, M.Ya. & V.V. Panasyuk 1959. Evolution of small cracks in solid bodies. *Appl. Mech.*, 5(4): 391-401, (In Ukrainian).

Mishuris, G.S. 1985. On models of the interface between two elastic media one of which weakened by an angular cut, *Vestnik Leningradskogo Un-ta*, 22: 62-66, (in Russian).

Mishuris, G.S. 1997. Influence of interfacial models on a stress field near a crack terminating at a bimaterial interface, *Int. J. Solids Structures*, 34(1): 31-46.

Mishuris, G.S. 1997. On interaction between mode III crack and a nonideal bimaterial interface, (XXXI Polish SolMec Conference), Mierki, 1996, *Engng Trans.*, 45(1): 103-118.

Novozhilov, V.V. 1969. On nessecary and sufficient criterion of brittle fracture. *Appl. Math. Mech.*, 33(5): 797-812. (in Russian).

Rice, J.R. 1988. Elastic fracture mechanics concepts for interfacial cracks. *J. Appl. Mech.*, 55: 98-103.

Seweryn, A. 1994. Brittle fracture criterion for structures with sharp notes. *Engng Fract. Mech.* 47: 673-681.

Seweryn, A. & Z. Mróz 1995. A non-local stress failure condition for structural elements under multiaxial loading. *Engng Fract. Mech.* 51: 955-973.

Wells, A.A. 1961. Critical tip opening as fracture criterion, *In: Proc. Crack Propagation Symp. Grangield*, Cranfield, 1961, 1: 210–221.

Williams, M.L. 1959. The stress around a fault or crack in dissimilar media. *Bull. Seismological Soc. Am*, 49: 199–204.

Zak, A.R. & M.L. Williams 1963. Crack point stress singularities at a bimaterial interface. *J. Appl. Mech.*, 30(1): 142–143.

Damage and Failure of Interfaces, Rossmanith (ed.) © *1997 Balkema, Rotterdam, ISBN 90 5410 899 1*

Stress fields at interface-corners and cracks for non-linear deformations

M. Scherzer
Department of Mathematics, Technical University Chemnitz-Zwickau, Germany

ABSTRACT: This paper briefly presents the findings of the author referring to asymptotic equations, eigenfunction expansions and their coupling to usual finite elements and the conclusions for fracture characterizations of isotropic flow and deformation theory statements at interface corners in the geometrically linear deformation range. So-called power-law approaches are discussed critically.

1 ASYMPTOTIC EQUATIONS

1.1 Deformation theory and flow theory

The first step to introduce non-linear material behaviour in isotropic solid mechanics can be taken by the classical deformation theory. Its constitutive equations have the form:

$$e_{ij} = \frac{\gamma(\tau)}{2\tau} S_{ij} + \frac{(1-2\nu)\varsigma}{E} \delta_{ij}. \qquad (1)$$

e_{ij}, S_{ij}, E, ν, and δ_{ij} denote the components of strain tensor, the components of stress deviator, Young's modulus, Poisson's coefficient, and Kronecker's symbol, respectively. Futher the following notations are introduced: $-\varsigma = -\frac{\sigma_{11}+\sigma_{22}+\sigma_{33}}{3}$ -hydrostatic pressure, $\gamma = \sqrt{\frac{4}{3}\sum_{ij} \epsilon_{ij}\epsilon_{ij}}$ -octahedron shear, $\tau = \sqrt{\frac{1}{3}\sum_{ij} S_{ij}S_{ij}}$-octahedron stress. ϵ_{ij} mark the components of strain deviator and σ_{ij} stand for the stress tensor components. From equation (1) it can be seen that γ is related to τ by a material function $\gamma(\tau)$. For $\tau < \tau_t$ (τ_t-yield stress) $\frac{\gamma}{2\tau} = \frac{1}{2\mu} = \frac{1+\nu}{E}$ is valid. Here μ denotes the shear modulus and equation (1) coincides with isotropic Hooke's law. The specification of the further material behaviour ($\gamma(\tau)$ for $\tau > \tau_t$) is not necessary. Introducing a polar co-ordinate system (r, θ) which is chosen in such way that the singularity (critical point) occurs at the origin ($r = 0$) the final equations for two-dimensional plane strain solid statements for the deformation theory (1) end up in (Scherzer (1994)):

$$\frac{\tau}{\gamma(\tau)} L^{1mv}_{r\theta}(U_{ij}) = -\frac{E}{2(1-2\nu)}(U_{rr} + U_{\theta\theta})_{,r},$$

$$\frac{\tau}{\gamma(\tau)} L^{2mv}_{r\theta}(U_{ij}) = -\frac{E}{2(1-2\nu)r}(U_{rr} + U_{\theta\theta})_{,\theta},$$

$$L_{r\theta}(U_{ij}) + r\,\triangle\,(U_{rr} + U_{\theta\theta}) = 0, \quad m = \frac{\gamma'(\tau)\tau}{\gamma(\tau)} \qquad (2)$$

In (2) the new variables $U_{ij} = \frac{\gamma(\tau)}{\tau} S_{ij}$ are introduced and $\gamma'(\tau)$ denote the derivative of $\gamma(\tau)$. $\triangle$ marks the Laplace Operator and $L_{r\theta}(U_{ij})$ which follows from the compatibility equation is completely linear in relation to U_{ij}. It can be shown (Scherzer (1994)) that the operators $L_{r\theta}(U_{ij})$, $(U_{rr} + U_{\theta\theta})_{,r}$, $\frac{1}{r}(U_{rr} + U_{\theta\theta})_{,\theta}$, $r\,\triangle\,(U_{rr} + U_{\theta\theta})$, $L^{1mv}_{r\theta}(U_{ij})$ and $L^{2mv}_{r\theta}(U_{ij})$ all have the same order in the singularity neighbourhood. The function m, which occurs in the coefficients of the differential operators indicating m, can be used as a boundary layer parameter, having a concrete value at the singularity point. For ordinary materials ($m \geq 1$) two different possibilities exist:

$$\lim_{\substack{r\to 0\\ \tau\to\infty}} \frac{\gamma(\tau)}{\tau} = \infty \text{ (a)} \quad \text{and} \quad \lim_{\substack{r\to 0\\ \tau\to\infty}} \frac{\gamma(\tau)}{\tau} < \infty, \text{ (b)}$$

It can be seen (Scherzer (1994)) that in the case of (a), which is equivalent to $m > 1$, a bad conditioned non-linear asymptotic system of equations follows for which, in principle, it is impossible to construct a mathematically complete eigenfunction system. On the contrary, for (b) because of $m = 1$ a linear system results and the corresponding eigenfunction expansion can be given in an easy manner (Scherzer and Meyer (1996)). It is clear that the function $\gamma(\tau)$ has to be determined at the tip neighbourhood in inverse statements. For this reason the application of $\gamma(\tau)$ with the quality (a) is not necessary while the use of material laws which belong to case (b) is very effective.

The deformation theory introduced above is able to describe the non-linear elasticity effects. The real plasticity behaviour i.e. local load trajectory dependent material properties can only be characterized by the flow theory. Let us introduce the yield surface function f:

$$f = g(\tau) - \chi = 0, \qquad \chi = \int \sqrt{\tfrac{4}{3} d\epsilon_{ij}^{(pl)} d\epsilon_{ij}^{(pl)}} \qquad (3)$$

In (3) g, χ, and $d\epsilon_{ij}^{(pl)}$ represent a material function, the Odkvist-parameter, and the increments (differentials, rates) of the plastic strain tensor (deviator), respectively. On the basis of the usual conditions of the plastic flow theory with isotropic hardening the material equations for active plastic loading read:

$$de_{ij} = \tfrac{1+\nu}{E} dS_{ij} + \tfrac{1-2\nu}{E} d\varsigma \delta_{ij} + \tfrac{S_{ij}}{\tau}\left(\tfrac{\gamma'(\tau)}{2} - \tfrac{1+\nu}{E}\right) d\tau \qquad (4)$$

Naturally, the function $\gamma(\tau)$ in (4) is the same as introduced above. In general equations (4) cannot be integrated to get deformation theory like equations between e_{ij} and σ_{ij}. But for the condition $t_{ij} = \frac{S_{ij}}{\tau} = c_{ij}(r,\theta)$ where $c_{ij}(r,\theta)$ are constants during the whole loading process solely depending on the space coordinates the classical deformation theory (1) results from (4). This is an expression of the well known fact that the deformation theory (1) is only a very simple special case of the flow theory (4). On the other hand, to obtain practicable flow theory relations for the investigation of the asymptotic behaviour in singular points it is useful to introduce appropriate suppositions for the components t_{ij} (Scherzer and Meyer (1996)):

$$\tfrac{S_{ij}}{\tau} = C_{ij}(r,\theta), \; C_{ij}(r,\theta) = \text{ constant for each loading increment, or } dt_{ij} = \tfrac{dS_{ij}}{\tau} - \tfrac{S_{ij}}{\tau^2} d\tau = 0 \qquad (5)$$

$C_{ij}(r,\theta)$ may be different in different loading increments. The condition (5) reduces the constitutive equations (4) to a very simple form and finally the asymptotic equations for $\gamma''(\tau = \infty) = 0$ (which in principle coincides with (b)) and plane strain conditions end up in (Scherzer (1994)):

$$L_{r\theta}(\widetilde{dU}) + r \,\triangle\, (\widetilde{dU}_{rr} + \widetilde{dU}_{\theta\theta}) = 0$$

$$\tfrac{1}{\gamma'}\{\widetilde{dU}_{rr,r} + \tfrac{1}{r}\widetilde{dU}_{\theta r,\theta} + \tfrac{1}{r}(\widetilde{dU}_{rr} - \widetilde{dU}_{\theta\theta})\} = = -\tfrac{E}{2(1-2\nu)}(\widetilde{dU}_{rr} + \widetilde{dU}_{\theta\theta})_{,r}$$

$$\tfrac{1}{\gamma'}(\widetilde{dU}_{\theta r,r} + \tfrac{1}{r}\widetilde{dU}_{\theta\theta,\theta} + \tfrac{2}{r}\widetilde{dU}_{\theta r}) = = -\tfrac{E}{2(1-2\nu)r}(\widetilde{dU}_{rr} + \widetilde{dU}_{\theta\theta})_{,\theta} \qquad (6)$$

with the notations

$$\widetilde{dU}_{rr} = \gamma'(\tau) dS_{rr}, \; \widetilde{dU}_{\theta\theta} = \gamma'(\tau) dS_{\theta\theta}, \; \widetilde{dU}_{\theta r} = \gamma'(\tau) dS_{\theta r}.$$

It can be shown (Scherzer and Meyer (1996)) that for this system the eigenfunction expansions at interface corners can also be given in the same way as for the deformation theory (case (b)). On the other hand it is interesting to note that these eigenfunction expansions are also ascertainable in the very general situation when condition (5) fails incrementally (Scherzer (1994)).

1.2 About the usefulness of the so-called power-law material in asymptotic expansions

It was shown above that in general the real plastic material behaviour at critical points can only be characterized by the asymptotic equations of the flow theory. It is clear that in this context the concrete form of the material function $\gamma(\tau)$ (for $\tau > \tau_t$) is essential for the construction of the eigenfunction expansions not in general but only in numerical sense. Hence the use of $\gamma(\tau)$ with quality (a) is effectless. Nevertheless for the special case of deformation theory applications there are attempts to construct eigenfunction expansions of power-law material which is prolongated for $\tau \to \infty$ to the tip and this way belongs to the case (a). Beside the classical work of Cherepanov (1967), Rice and Rosengren (1968) and Hutchinson (1968) the articles of Aravas and Sharma (1991), Yang et. al. (1993), Nikishkov (1995) and others could be mentioned. The fact that these approaches are not very effective also results from the following reasons:

1. Provided that $\gamma(\tau)$ is known from experiments the region $\tau_t < \tau < \infty$ can be divided into intervals of equal lenght l_1. $\gamma(\tau)$ may be approximated by splines of degree m_1 on each interval. To achieve a given accuracy l_1 has to be chosen sufficiently small. This way a power-law hardening material with the power exponent m_1 is established. Provided that it is possible to construct a mathematically complete asymptotic eigenfunction expansion for power-law hardening material the solution of the boundary value problem at the interface tip can be given except the unknown coefficients of the eigenfunctions. On the other hand the same

procedure can be done by splines of degree m_2 ($m_2 \neq m_1$) and the corresponding $l_2 \neq l_1$ with equal accuracy in order to get another eigenfunction expansion with other coefficients of the eigenfunctions and which is valid for the same material function $\gamma(\tau)$ and represents the same solution of the boundary value problem. So, whether or not to choose power-law material has no physical importance and the concrete choice is only an approximation. In this sense, the approximation of $\gamma(\tau)$ in consideration of (*b*) will be the most effective one in comparison to all power-law prolongations to the tip. The asymptotic behaviour is connected with $\gamma(\tau)$ in the vicinity of the tip only, and not, for instance, around τ_t.

2. The expansions for power-law hardening material of the articles mentioned above will not have the mathematical quality of completeness which was provided in the first point and saves convergent solutions. This can be seen on the simple example of a crack or of an interface crack. For a crack in homogenous isotropic non-linear material it is obvious that double roots (or maybe complex roots together with their corresponding conjugate complex ones) in the solvability condition and therefore at least two independent eigenfunctions for the same power exponent of r must exist. As a consequence of the fact that symmetrical in relation to the crack loading at infinity cannot be represented by anti-symmetrical eigenfunctions at the tip and (vice versa) that anti-symmetrical in relation to the crack loading at infinity cannot be represented by symmetrical eigenfunctions at the tip this circumstance is self-evident. In the mentioned articles the eigenfunction systems are constructed from a non-linear boundary value problem numerically (shooting technique) and the symmetry (or anti-symmetry) is introduced as a given condition and does not follow automatically from either the solvability condition or from the equation system. This way the authors can only give expansions of the symmetrical (or anti-symmetrical) case alone. On the other hand the superposition principle is not applicable because of the non-linear statements and so, the very crucial conclusion that there is no way to get any expansions for non-symmetrical and non-anti-symmetrical statements follows. And therefore as it will be seen at the end of this paper there is no way to characterize and predict fracture situations for these statements. It is clear that mathematically complete eigenfunction expansions cannot depend on the concrete form of loads at infinity.

3. In general practicable extensions to more realistic material behaviour (such as for instance anisotropy and geometrical non-linearities) are impossible in these very restricted and mathematically incorrect approaches.

2 REPLACEMENT OF ASYMPTOTIC SOLUTION BY STIFFNESS ACTIONS

Let us consider two-dimensional statements in the neighbourhood of an interface corner consisting of two material ranges ($0 \leq \theta \leq \theta_o$ and $0 \geq \theta \geq -\theta_u$). At a distance of $\xi = \xi_0$ from the corner the finite element nodes of a regular net are established in the polar co-ordinate system of $\xi = \frac{r}{b}$ (b is a length scale) and θ together with the displacement degrees of freedom $u_k(\xi_0, \theta_j)$ (see Figure 1).
The following boundary conditions are given for formulations of geometrically linear and physically non-linear isotropic statements:
- Vanishing normal and tangent stresses ($\sigma_{\theta\theta}, \sigma_{\xi\theta}$) at $\theta = \theta_0, -\theta_u$
- Continuity of normal and tangent stresses and displacements (u_ξ, u_θ) at $\theta = 0$.
The main idea of the singular and non-singular stress and deformation field calculation at interface corners presented here characterizes a replacement of the corner neighbourhood ($\xi < \xi_0$) effect to the surrounding body ($\xi > \xi_0$) by introducing stiffness actions at $\xi = \xi_0$ which can be

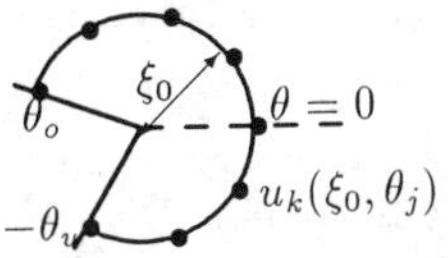

Figure 1. Neighbourhood of an interface corner together with the finite element nodes

assembled in a conventional way together with the other element stiffness matrices to the global stiffness matrix of the body. For $\xi < \xi_0$ the following relations are valid:

$$\begin{aligned} \sigma &= \sum_i C_i f_i^{(\sigma)}(\xi,\theta), \quad \sigma = \{\sigma_{\xi\xi}, \ \sigma_{\xi\theta}\}; \\ u &= \sum_i C_i f_i^{(u)}(\xi,\theta), \quad u = \{u_\xi, u_\theta\}. \end{aligned} \tag{7}$$

In (7) the marks $f_i^{(u)}(\xi,\theta) = \{f_{i\xi}^{(u)}(\xi,\theta), f_{i\theta}^{(u)}(\xi,\theta)\}$ and $f_i^{(\sigma)}(\xi,\theta) = \{f_{i\xi\xi}^{(\sigma)}(\xi,\theta), f_{i\xi\theta}^{(\sigma)}(\xi,\theta)\}$ denote the corresponding vectors of the displacement and stress eigenfunctions whose concrete forms are given by Scherzer and Meyer (1996). In the case of flow theory applications the corresponding eigenfunctions for the increments of stresses and displacements have to be used in the way shown by Scherzer (1994) and Scherzer and Meyer (1996). Because of the boundary conditions at the circle $\xi = \xi_0$ it is not necessary to include the stress component $\sigma_{\theta\theta}$ in σ. The constants C_i can be related to $u_k(\xi_0, \theta_j)$ $(k = \xi, \theta)$ by

$$\begin{aligned} u_k(\xi_0,\theta_j) &= \sum_i C_i f_{ik}^{(u)}(\xi_0,\theta_j), \\ &\quad (k = \xi, \theta; j = 1,2,3,...) \end{aligned} \tag{8}$$

and solving (8) one gets

$$\begin{aligned} C_i &= \sum_j b_{ij}(\xi_0,\theta_1,...)v_j(\xi_0), \\ &\quad \theta_1 = -\theta_u, ...\theta_N = \theta_o, \\ &\quad v_1(\xi_0) = u_\xi(\xi_0,\theta_1), v_2(\xi_0) = u_\theta(\xi_0,\theta_1), \\ &\quad v_3(\xi_0) = u_\xi(\xi_0,\theta_2), v_4(\xi_0) = u_\theta(\xi_0,\theta_2), \\ &\quad v_5(\xi_0) = u_\xi(\xi_0,\theta_3), ... \end{aligned} \tag{9}$$

To obtain stiffness actions it is necessary to calculate the virtual work δA of stresses on the circle $\xi = \xi_0$:

$$\begin{aligned} \delta A &= \xi_0 \int_{-\theta_u}^{\theta_o} \sigma \star \delta u \, d\theta \\ &= \sum_{k,l} q_{kl} v_k(\xi_0) \delta v_l(\xi_0), \end{aligned}$$

$$q_{kl} = \xi_0 \sum_{i,j} b_{ik} b_{jl} \int_{-\theta_u}^{\theta_o} f_i^{(\sigma)}(\xi_0,\theta) \star f_j^{(u)}(\xi_0,\theta)\, d\theta \tag{10}$$

The symbol "$\star$" marks scalar products of corresponding vectors. The kl (k-th column, l-th row) element of the wanted stiffness action matrix is determined in (10) as the factor of $v_k(\xi_0)\delta v_l(\xi_0)$ i.e. q_{kl}. Computations of stiffness actions by (8), (9) and (10) lead to an introduction of n eigenfunctions if n degrees of freedom exist at $\xi = \xi_0$. Avoiding this non-effective procedure it is possible to orthogonalize the eigenfunctions.

3 ORTHOGONALIZATION OF EIGENFUNCTIONS

A more detailed description of the orthogonalization procedure can be found in the publication of Scherzer and Meyer (1996). In this paper only the final formulas will be given. After orthogonalization the virtual work δA, the stresses und displacements at $\xi = \xi_0$ result in:

$$\begin{aligned} u(\xi,\theta)_{|\xi=\xi_0} &= \sqrt{\xi_0} \sum_i \overline{C_i} \overline{f_i^{(u)}}(\theta), \\ \sigma(\xi,\theta)_{|\xi=\xi_0} &= \frac{1}{\sqrt{\xi_0}} \sum_i \overline{C_j} \overline{f_i^{(\sigma)}}(\theta), \end{aligned} \tag{11}$$

$$\begin{aligned} \delta A &= \xi_0 \sum_i \overline{C_i} \delta \overline{C_i}, \\ \overline{C_i} &= \int_{-\theta_u}^{\theta_o} \overline{f_i^{(\sigma)}}(\theta) \star \frac{u(\xi_0,\theta)}{\sqrt{\xi_0}} d\theta. \end{aligned} \tag{12}$$

The vector functions $\overline{f_i^{(\sigma)}}(\theta)$ and $\overline{f_j^{(u)}}(\theta)$ fulfil the condition

$$\int_{-\theta_u}^{\theta_o} \overline{f_i^{(\sigma)}}(\theta) \star \overline{f_j^{(u)}}(\theta)\, d\theta = \delta_{ij}. \tag{13}$$

Note that the coefficients $\overline{C_i}$ depend on ξ_0 ($\overline{C_i} = \overline{C_i}(\xi_0)$) while C_i are constants. The relations (11) and (12) allow an excellent determination of the wanted stiffness actions in δA after a possible choice of the $\theta-$ finite element approximation for the displacements $u(\xi_0, \theta)$:

$$u(\xi_0,\theta) = \sum_k N_k(\theta) v_k(\xi_0)$$

with $N_k(\theta)$ as one-dimensional vector shape functions at the circle $\xi = \xi_0$. Then δA and $\overline{C_i}$ get the representations:

$$\delta A = \sum_{k,l,i} \hat{q}_{ik} \hat{q}_{il} v_k(\xi_0) \delta(v_l(\xi_0)) \tag{14}$$

$$\overline{C_i}(\xi_0) = \frac{1}{\sqrt{\xi_0}} \sum_k \hat{q}_{ik} v_k(\xi_0), \tag{15}$$

whereby $\hat{q}_{ik}$ follow from

$$\hat{q}_{ik} = \int_{-\theta_u}^{\theta_o} \left(\overline{f_i^{(\sigma)}}(\theta) \star N_k(\theta) \right) d\theta. \tag{16}$$

Thus the wanted stiffness actions q_{kl} can be calculated by:

$$q_{kl} = \sum_i \hat{q}_{ik}\hat{q}_{il}, \qquad (17)$$

The interesting feature of these stiffness actions is their independence of ξ_0 which was found by Scherzer and Meyer (1996). This circumstance is very important in flow theory applications where invariant parameters for fracture predictions have to be introduced.

The calculation of the constants C_i from $\overline{C_i} = \overline{C_i}(\xi_0)$ is possible by means of (12) and $u(\xi,\theta) = \sum_j C_j f_j^{(u)}(\xi,\theta)$. For the roots of the solvability condition $\lambda_1 = \alpha_1$, $\lambda_2 = \alpha_2$, $\lambda_3 = \alpha_3 + \imath\beta_3$, $(\imath = \sqrt{-1})$, $\lambda_4 = \alpha_3 - \beta_3\imath, \ldots, \alpha_1 < \alpha_2 < \alpha_3 \ldots$ the following system of equations is valid (Scherzer and Meyer (1996)):

$$\begin{aligned}
\overline{C_1} &= \xi_0^{\alpha_1+\frac{1}{2}} K_{11}C_1 + \xi_0^{\alpha_2+\frac{1}{2}} K_{12}C_2 + \\
&\qquad + \xi_0^{\alpha_3+\frac{1}{2}}(K_{13}C_3 + K_{14}C_4) + \ldots \\
\overline{C_2} &= \xi_0^{\alpha_2+\frac{1}{2}} K_{22}C_2 + \xi_0^{\alpha_3+\frac{1}{2}}(K_{23}C_3 + K_{24}C_4) + \ldots \\
\overline{C_3} &= \xi_0^{\alpha_3+\frac{1}{2}}(K_{33}C_3 + K_{34}C_4) + \ldots \\
\overline{C_4} &= \xi_0^{\alpha_3+\frac{1}{2}} K_{44}C_4 + \ldots \\
\vdots &= \qquad \vdots \qquad\qquad (18)
\end{aligned}$$

In (18) the quantities $K_{ij} = K_{ij}(g_{kl})$ in general depend on the integrals g_{kl} over the scalar product of the functions $g_k^{(\sigma)}(\theta)$ and $g_l^{(u)}(\theta)$ which are the θ-dependent parts of the initial eigenfunctions $f_l^{(u)}(\xi,\theta)$ and $f_k^{(\sigma)}(\xi,\theta)$ by:

$$g_{kl} = \int_{-\theta_u}^{\theta_o} g_k^{(\sigma)}(\theta) \star g_l^{(u)}(\theta)d\theta.$$

The fact that $K_{21} = K_{31} = K_{32} = K_{41} = \ldots = 0$ is a consequence of the orthogonalized eigenfunction system construction. If $u(\xi_0,\theta)$ is known from the global solution of the solid $\overline{C_i}$ can be determined by (15), and C_i follow from (18).

4 EXAMPLE OF AN INTERFACE CRACK

The theoretical results described above were implemented in a computer program on heavy parallel computers. Using the Domain Decomposition (DD) as parallelizing concept the effective parallel preconditioned conjugate gradient method developed by Meyer (1990) for high order degree of freedom finite element systems is applied. Results of test calculations will be explained. An interface crack specimen (750*1500 dimensionless extension, crack in the middle of the specimen with a length of 375) is strained homogenously at the elastic softer specimen end and clamped right opposite at the elastic harder specimen end. For pure elastic calculations the following material parameters are used:

$$\boxed{\eta = 0.0505,\ \kappa^o = 4.5,\ \kappa^u = 3.25}$$

η is the ratio of the applied shear moduli (μ), κ^o and κ^u are the ratios of bulk and shear modulus multiplied by 3/2 of the elastic softer and the elastic harder specimen half. Then the roots of the solvability condition have the values:

$$\boxed{\begin{array}{c}\lambda_i = -0.5 \pm \imath 0.06591194,\ 0.0, 0.5 \pm \imath 0.06591194, \\ 1,\ 1.5 \pm \imath 0.06591194,\ 2,\ \ldots\end{array}}$$

Figures 2 to 4 show the stress fields σ_{yy} (the $y-$axis is perpendicular to crack and interface) related to μ of the elastic softer material for different ξ_0 and equal zoom radii $\xi_z = 1.0$. The crack tip lies in the centre of the Figures and the interface on the horizontal straight line (x-axis) on the right side as prolongation of the crack. For all further Figures the same conventions are used. The zipper-like domains on both crack flanks result from postprocessing approximation errors. It should be mentioned that first stresses are calculated in Gauss points (best approximation) and then they are extrapolated to node points. If in Figures 3 and 4 one changes ξ_z from 1.0 to 0.1 and 0.01, respectively, for a corresponding stress scale in these two cases the same Figure 2 will be

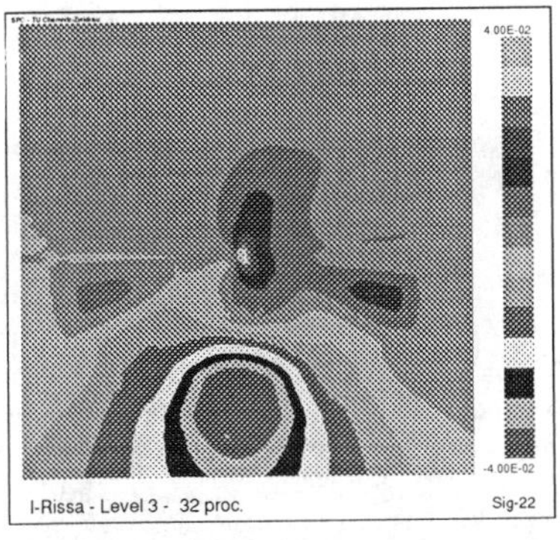

Figure 2. $\xi_0 = 0.75$

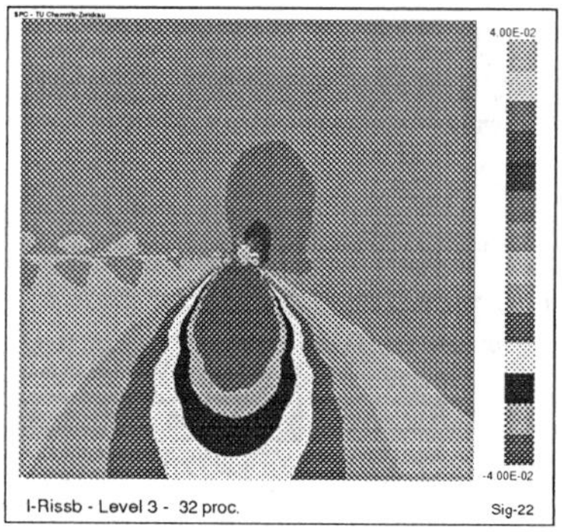

Figure 3. $\xi_0 = 0.075$

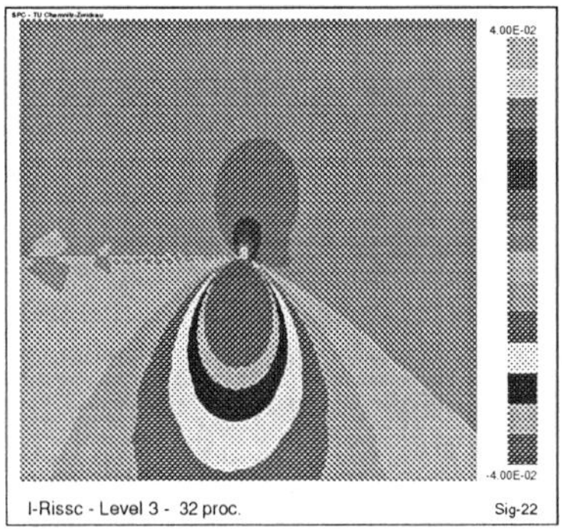

Figure 4. $\xi_0 = 0.0075$

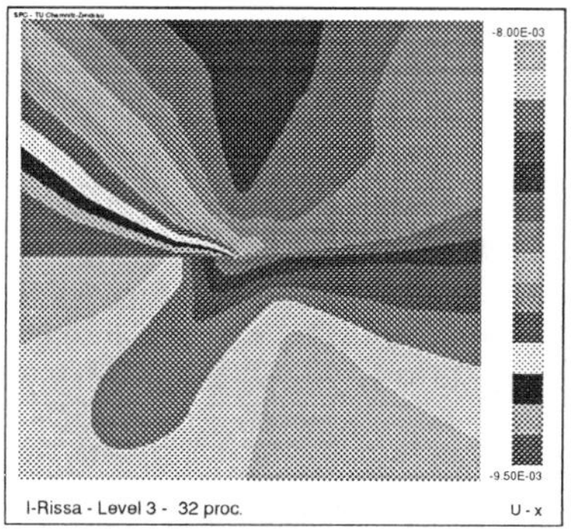

Figure 5. u_x with asymptotics, $\xi_z = 1.0$

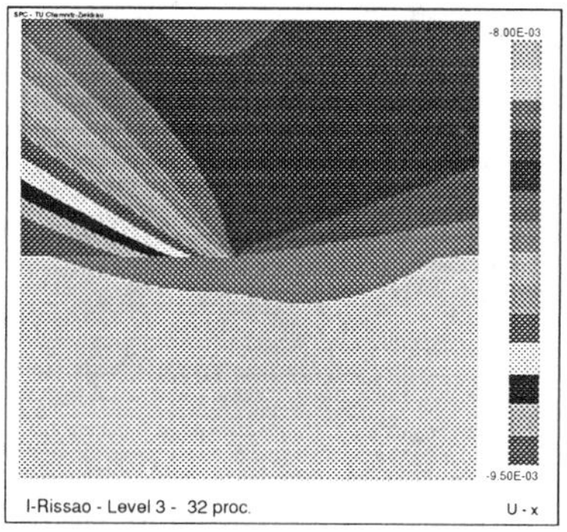

Figure 6. u_x without asymptotics, $\xi_z = 1.0$

produced illustrating the stiffness independence of ξ_0 mentioned above. On the other hand the Figures 2 to 4 express convergent stress fields at the interface crack tip.

Further, the pressure stresses ($\sigma_{yy} < 0$) under the crack tip are remarkable. They result from the different Poisson effects in both material domains. Both the soft and the hard material are exposed to tension in $y-$direction which causes different contractions in $x-$direction (u_x). At the interface crack tip the hard material impedes the contractions of the soft one and gets the pressure stresses as a reaction to this impediment. These contractions are demonstrated in Figures 5 to 8 for two different zoom radii ξ_z and in comparison to usual results of finite element calculations in which the special asymptotic behaviour is not considered ("without asymptotics"). Figures 7 and 8 illustrate the impediment of the displacements u_x at the crack flank of the upper softer material region which looks like a "sausage" and causes pressure stresses under the interface crack tip. Figures 5 and 6 show the differences between the solution of usual finite element computations ("without asymptotics") and the solution following from the technique introduced above at the interface tip directly. It is clear that these differences will grow if the stress fields are considered. It can be shown that for instance the $\sigma_{yy}-$field of the solution "without asymptotics" gives pressure stresses in a small region under the crack flank only (Scherzer and Meyer (1996)). This way it is seen that usual finite element methods without special asymptotics cannot give correct solutions at interface cracks. The pure finite element approach is too coarse to feel the described effects. Also mesh refinements cannot save the situation of usual finite elements at interface cracks because they produce wrong asymptotic behaviour at the tip. In connection with the chosen material parameters this behaviour affects the surrounding body more or less. It is self-evident that the circumstances described above have corresponding consequences for the deformation field around

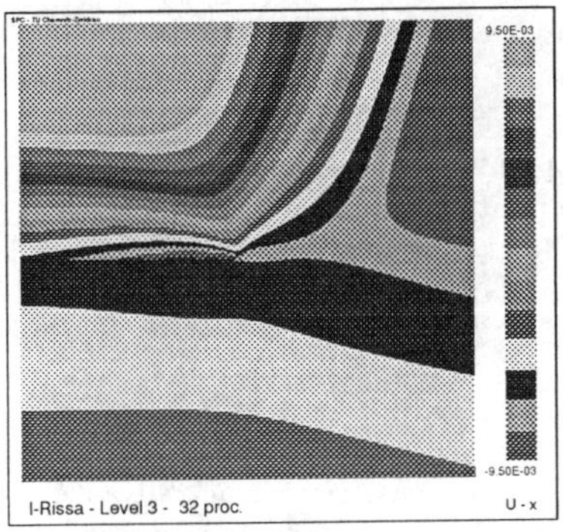

Figure 7. u_x with asymptotics, $\xi_z = 250$

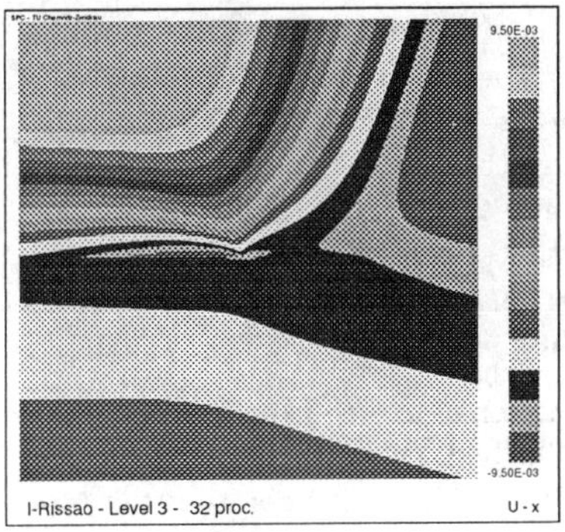

Figure 8. u_x without asymptotics, $\xi_z = 250$

the interface crack tip allowing non-linear constitutive behaviour on the basis of flow theory. This fact cannot be discussed here. More details are presented by Scherzer and Meyer (1996).

5 CHARACTERIZATION OF DEFORMATION AND STRESS STATES

As it is known the application of solid mechanics in modern microelectronics leads to the calculation of very complicated stress and deformation fields depending on geometry, material laws, boundary conditions, loads and so on. On the other hand it must be pointed out that sole the calculation of stresses, strains, displacements etc does not give the answer to the main question:

Does the structure fail or not?

To estimate or predict fracture it is necessary:

1. to know the material laws which can only be identified by experimental realizations together with the solutions of solid boundary value problems in inverse statements at the considered critical points,
2. to know experimantally and to characterize critical failure situations by parameters following from the solid solutions

It must be said that these solid solutions have to be comparable with other solutions of corresponding statements. This results from the fact that fracture occurs in strong connection to local stress and deformation states. In other words it is necessary to distinguish the possible local stress and deformation states for different fracture situations. For example using the flow theory where material behaviour depends on local load trajectories the stress conditions change during the loading process and can result in different possible local stress states of fracture.

To describe these circumstances let us introduce main stresses. The main stresses σ_1, σ_2 and σ_3 related to the von Mieses stress ($\sigma_v = \frac{3}{\sqrt{2}}\tau$) can be expressed by the constraint factor ($h = \frac{\sqrt{2}\varsigma}{6\tau}$) and the similarity angle of the stress deviator ($\omega = \frac{1}{3}\arccos\frac{\sqrt{2}\sum_{ijk}(S_{ij}S_{ik}S_{kj})}{3\tau^3}$) which are independent of concrete material laws (Scherzer and Meyer (1996)). Main stresses are the roots of the corresponding eigenvalue equation of stress tensor. For two-dimensional plane strain conditions the constraint factor and the similarity angle are interdependent. Thus it can be shown that in this case the deformation components e_{xx}, e_{yy} and e_{xy} ($e_{zz} = 0$) for isotropic deformation theory materials following from the main stress representations have the form:

$$\begin{aligned} e_{xx} &= \frac{6\gamma(\tau)}{8}[\sin(2\alpha - \omega + \tfrac{\pi}{3}) + \cos(2\alpha + \omega + \tfrac{\pi}{3}) \\ &\quad + 2\sqrt{3}\sin(\omega + \tfrac{\pi}{6})] \\ e_{yy} &= \frac{6\gamma(\tau)}{8}[\sin(2\alpha - \omega + \tfrac{\pi}{3}) + \cos(2\alpha + \omega + \tfrac{\pi}{3}) \\ &\quad - 2\sqrt{3}\sin(\omega + \tfrac{\pi}{6})] \\ e_{xy} &= \frac{6\gamma(\tau)}{8}[\cos(2\alpha - \omega + \tfrac{\pi}{3}) \\ &\quad - \sin(2\alpha + \omega + \tfrac{\pi}{6})]. \end{aligned} \tag{19}$$

For the flow theory (19) is analogical using the corresponding increments. The angle α which depends on stress components ($\tan(2\alpha) = \frac{2\sigma_{xy}}{\sigma_{xx} - \sigma_{yy}}$) characterizes the location of main stress axes referring to the arbitrary $xy-$ co-ordinate system. From these relations it can be seen that the deformation components which qualify the solid straining can be calculated if τ, κ and α are known. τ acts as an amplitude characterizing the magnitude of strains and κ and α, which are finite in singular points, specify the kind of deformation. If one puts the asymptotic expansions into τ it can be concluded that the magnitude of deformation is only dominated by the coefficients of the first eigenfunctions because of the discrete solvability condition roots λ_i. These facts suggest the following fracture analysis for solids:

1. Determination of $\mathbf{u}(\xi, \theta)$ from the whole solution of the body by finite elements or other methods taking into consideration the correct asymptotic behaviour using the stiffness actions described above which do not depend on the distance to the considered point.
2. Computation of $\overline{C_i}$ by (15).
3. Calculation of C_i by the triangle system (18) $\overline{C_i} = K_{ij}C_j$ with $K_{ij} = K_{ij}(g_{ij})$.
   - g_{ij} can be interpreted as "inner deformation metric" of the local point.
   - $C_i \approx \frac{\overline{C_i}}{K_{ii}} \xi_0^{-(\alpha_i + \frac{1}{2})}$ is valid for small ξ_0.
4. C_i are the fracture parameters which get critical values in failure situations. Because of (15), (18) and g_{ij} they express integral magnitudes of the stress and deformation field around the considered point. C_1 acts as the main factor (and if the root λ_1 is double then C_1 and C_2 both do) and the other ones operate as higher order moments.
5. κ and α characterize critical stress states for the critical C_i−values. If such critical C_i−values are determined in an experimental way together with the numerical technique described above then fracture predictions are possible in connection with the concrete realized κ and α.

It can be shown that κ and α at crack tip neighbourhoods remain the same for arbitrary load values in the linear material law case, provided the "form" of boundary conditions does not change (Scherzer and Meyer (1996)). If non-linear material behaviour (flow theory) has been applied they change during loading for the same "form" of boundary conditions characterizing different stress conditions (local load trajectory). This way they distinguish critical fracture states if the realized boundary value problem for any load level corresponds to failure at the interface crack tip.

Summarizing it can be concluded that macroscopic fracture estimation of solids leads to the question of describing deformations by solid models at critical points. If this question is solved, meaning the constitutive behaviour is known, the corresponding fracture characteristics result from the asymptotic analysis presented above.

REFERENCES

Cherepanov, G.P. (1967). About crack advance in solids (in Russian). *Prikladna'a matematika i mekhanika*, vol. 31, pp. 476–488.

Hutchinson, J.W. (1968). Singular behaviour at the end of a tensile crack in a hardening material. *J. Mech. Phys. Solids*, vol. 16, pp. 13–31.

Meyer, A (1990). A parallel preconditioned conjugate gradient method using domain decomposition and inexact solvers on each subdomain. *Computing*, vol. 45, pp. 217–234.

Nikishkov, G.P. (1995). An algorithm and a computer program for the three-term asymptotic expansion of elastic-plastic crack tip stress and displacement fields. *Engineering Fracture Mechanics*, vol. 50, 1, pp. 65–83.

Rice, J. R. and Rosengren, G. F. (1968). Plane strain deformation near a crack tip in a power-law hardening material, *J. Mech. Phys. Solids*, vol. 16, pp. 1–12.

Scherzer, M. (1994). Asymptotic Equations for Bimaterial Interface Corners in the Non-Linear Deformation Field, in: *Mis-Matching of Welds*, ESIS (Edited by K.-H. Schwalbe and M. Kocak), Mechanical Engineering Publications, London, pp. 161–175.

Scherzer, M. and Meyer, A. (1996). Zur Berechnung von Spannungs- und Deformationsfeldern an Interface-Ecken im nichtlinearen Deformationsbereich auf Parallelrechnern. *TU Chemnitz-Zwickau, Preprint-Reihe des Sonderforschungsbereiches 393 „Numerische Simulation auf massiv parallelen Rechnern"*, SFB393/96-03, Chemnitz.

Sharma, S.M. and Aravas, N. (1991). Determination of higher-order terms in asymptotic elastoplastic crack tip solutions. *J. Mech. Phys. Solids*, vol. 39, 8, pp. 1043–1072.

Yang, S., Chao, Y.J. and Sutton, M.A. (1993). Higher oder asymptotic crack tip fields in a power-law hardening material. *Engineering Fracture Mechanics*, vol. 45, 1, pp. 1–20.

Damage and Failure of Interfaces, Rossmanith (ed.) © 1997 Balkema, Rotterdam, ISBN 90 5410 899 1

Stress singularities at the tip of interfaces in polycrystals

T.C.T.Ting
Department of Civil and Materials Engineering, University of Illinois at Chicago, Ill., USA

ABSTRACT: A polycrystal may have n interfaces converging at one point. If we take that point as the origin of a coordinate system we have a composite space that consists of n wedges of different homogeneous anisotropic elastic materials. We will study the stress singularities at the origin for a composite wedge and a composite space. A general result is obtained for which the wedge angle and the material property in each wedge can be arbitrary. We then specialize the result to the case when the whole wedge consists of periodic wedges in which each period consists of s wedges. A particular case is when s=1. This means that each wedge is of the same material and same wedge angle ω.

1 INTRODUCTION

The problem of finding stress singularities in an isotropic elastic wedge was first considered by Knein (1926) and Williams (1952). The technique has been applied by many researchers to wedges that consist of more than one material. In particular, Bogy (1970, 1971a,b) and Bogy and Wang (1971) have studied extensively isotropic bimaterial wedges, and Dempsey and Sinclair (1979, 1981) gave an analysis for a composite wedge that consists of an arbitrary number of isotropic elastic wedges. Extension to anisotropic elastic bimaterial wedges was investigated by Bogy (1972, 1974) and Kuo and Bogy (1974). They considered materials to be orthotropic materials. A composite wedge or a composite space that consists of an arbitrary number of general anisotropic elastic wedges has not been studied.

We will consider a wedge that consists of n different homogeneous anisotropic elastic materials of wedge angles ω_1, ω_2,..., ω_n, respectively. The deformation is assumed to be two-dimensional so that the displacement depends on x_1 and x_2 only. For a general linear anisotropic elastic material the displacement component u_3 is coupled with the u_1, u_2 components. In the (x_1,x_2)-plane, let r be the radial distance from the origin and θ be the polar angle. The kth wedge occupies the region (Fig. 1)

$$\theta_{k-1} \le \theta \le \theta_k, \quad \omega_k = \theta_k - \theta_{k-1}. \tag{1.1}$$

The total wedge angle ψ is

$$\psi = \theta_n - \theta_0 = \omega_1 + \omega_2 + \ldots + \omega_n. \tag{1.2}$$

It is not necessarily advantageous to choose θ_0=0 for an anisotropic composite wedge. If there is a symmetry plane of the material in one of the wedges, we may choose the

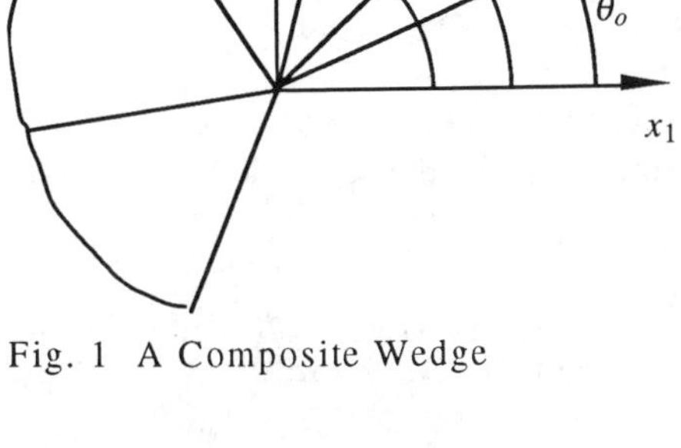

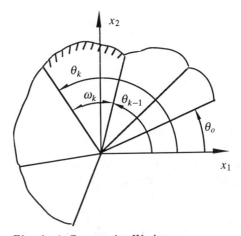

Fig. 1 A Composite Wedge

coordinate system so that one of the coordinate planes coincides with the symmetry plane. In any case all wedges refer to the same coordinate system. A *composite wedge* has $\psi \le 2\pi$, with the sides $\theta = \theta_0$ and $\theta = \theta_n$ either fixed or traction free. An interface crack is a special case of composite wedge with $\psi = 2\pi$ and with the sides $\theta = \theta_0$ and $\theta = \theta_n = \theta_0 + 2\pi$ being traction free. A *composite space* has $\psi = 2\pi$, and the surfaces $\theta = \theta_0$ and $\theta = \theta_n = \theta_0 + 2\pi$ are the same surface. We will first derive the conditions for the stress singularities δ at the wedge apex for a single wedge. We then consider the general case in which the composite wedge consists of an arbitrary number of wedges. The special case of a periodic wedge is then discussed. In particular, we consider the case when each wedge is of the same monoclinic material with a wedge angle ω but the orientation of the material from one wedge to the next wedge differs by the wedge angle ω.

2 BASIC EQUATIONS

In a fixed rectangular coordinate system x_i (i=1,2,3) let u_i and σ_{ij} be, respectively, the displacement and stress in an anisotropic elastic material. The stress-strain laws and the equations of equilibrium are

$$\sigma_{ij} = C_{ijks} u_{k,s}, \tag{2.1}$$

$$C_{ijks} u_{k,sj} = 0, \tag{2.2}$$

in which a comma denotes differentiation, repeated indices imply summation and C_{ijks} are the elastic stiffnesses which are assumed to possess the full symmetry

$$C_{ijks} = C_{jiks} = C_{ksij} = C_{ijsk}. \tag{2.3}$$

A general solution to (2.2) is (Eshelby et al., 1953)

$$u_i = a_i f(z), \quad \text{or} \quad \mathbf{u} = \mathbf{a} f(z), \tag{2.4}$$

where

$$z = x_1 + p x_2. \tag{2.5}$$

In the above f is an arbitrary function of z, and p and $\mathbf{a}$ satisfy the eigenrelation

$$\left\{\mathbf{Q} + p(\mathbf{R} + \mathbf{R}^T) + p^2 \mathbf{T}\right\}\mathbf{a} = \mathbf{0}, \tag{2.6}$$

in which the superscript T denotes the transpose and $\mathbf{Q}$, $\mathbf{R}$, $\mathbf{T}$ are 3×3 matrices whose components are

$$Q_{ik} = C_{i1k1}, \quad R_{ik} = C_{i1k2}, \quad T_{ik} = C_{i2k2}. \tag{2.7}$$

If the strain energy is positive, $\mathbf{Q}$ and $\mathbf{T}$ are symmetric and positive definite, and the six eigenvalues p_α (α=1,2,...,6) obtained from (2.6) must be complex. If $\mathbf{a}_\alpha$ (α=1,2,...,6) are the associated eigenvectors, we let

$$\operatorname{Im} p_\alpha > 0, \quad p_{\alpha+3} = \bar{p}_\alpha, \quad \mathbf{a}_{\alpha+3} = \bar{\mathbf{a}}_\alpha, \tag{2.8}$$

(α = 1,2,3), where Im stands for the imaginary part and the overbar denotes the complex conjugate. Assuming that p are distinct, the general solution obtained by superposing six solutions of $(2.4)_2$ is

$$\mathbf{u} = \sum_{\alpha=1}^{3}\left\{\mathbf{a}_\alpha f_\alpha(z_\alpha) + \bar{\mathbf{a}}_\alpha f_{\alpha+3}(\bar{z}_\alpha)\right\}, \tag{2.9}$$

where f_α (α=1,2,...,6) are arbitrary functions of their arguments and

$$z_\alpha = x_1 + p_\alpha x_2. \tag{2.10}$$

Introducing the vector $\mathbf{b}$ by

$$\mathbf{b} = (\mathbf{R}^T + p\mathbf{T})\mathbf{a} = -\frac{1}{p}(\mathbf{Q} + p\mathbf{R})\mathbf{a}, \tag{2.11}$$

in which the second equality follows from (2.6), the stress obtained by inserting (2.9) into (2.1) can be written as (Stroh, 1958)

$$\sigma_{i1} = -\phi_{i,2}, \quad \sigma_{i2} = \phi_{i,1}, \tag{2.12}$$

where ϕ_i (i=1,2,3) are the components of the stress function vector

$$\boldsymbol{\phi} = \sum_{\alpha=1}^{3}\left\{\mathbf{b}_\alpha f_\alpha(z_\alpha) + \bar{\mathbf{b}}_\alpha f_{\alpha+3}(\bar{z}_\alpha)\right\}. \tag{2.13}$$

For the analysis of stress singularities we choose

$$f_\alpha(z_\alpha) = q_\alpha z_\alpha^{\delta+1}, \quad f_{\alpha+3}(\bar{z}_\alpha) = \tilde{q}_\alpha \bar{z}_\alpha^{\delta+1},$$

where δ is the order of stress singularity and q_α, $\tilde{q}_\alpha$ (α=1,2,3) are arbitrary constants. Equations (2.9) and (2.13) can be written as

$$\mathbf{u} = \mathbf{A}\left\langle z_*^{\delta+1}\right\rangle\mathbf{q} + \bar{\mathbf{A}}\left\langle \bar{z}_*^{\delta+1}\right\rangle\tilde{\mathbf{q}},$$
$$\boldsymbol{\phi} = \mathbf{B}\left\langle z_*^{\delta+1}\right\rangle\mathbf{q} + \bar{\mathbf{B}}\left\langle \bar{z}_*^{\delta+1}\right\rangle\tilde{\mathbf{q}}, \tag{2.14}$$

in which the 3×3 matrices $\mathbf{A}$ and $\mathbf{B}$ are

$$\mathbf{A} = [\mathbf{a}_1, \mathbf{a}_2, \mathbf{a}_3], \quad \mathbf{B} = [\mathbf{b}_1, \mathbf{b}_2, \mathbf{b}_3], \tag{2.15}$$

and $\langle z_*^{\delta+1}\rangle$ is the diagonal matrix given by

$$\langle z_*^{\delta+1}\rangle=\text{diag}(z_1^{\delta+1},\ z_2^{\delta+1},\ z_3^{\delta+1}). \tag{2.16}$$

Since $x_1=r\cos\theta$ and $x_2=r\sin\theta$,

$$z_\alpha = x_1 + p_\alpha x_2 = r\zeta_\alpha(\theta), \tag{2.17}$$

$$\zeta_\alpha(\theta) = \cos\theta + p_\alpha \sin\theta, \tag{2.18}$$

and (2.14) can be written as

$$\begin{aligned}\mathbf{u}(\theta) &= r^{\delta+1}\left\{\mathbf{A}\langle \zeta_*^{\delta+1}(\theta)\rangle \mathbf{q} + \overline{\mathbf{A}}\langle \overline{\zeta}_*^{\delta+1}(\theta)\rangle \tilde{\mathbf{q}}\right\},\\ \boldsymbol{\phi}(\theta) &= r^{\delta+1}\left\{\mathbf{B}\langle \zeta_*^{\delta+1}(\theta)\rangle \mathbf{q} + \overline{\mathbf{B}}\langle \overline{\zeta}_*^{\delta+1}(\theta)\rangle \tilde{\mathbf{q}}\right\}.\end{aligned} \tag{2.19}$$

The dependence of $\mathbf{u}$ and $\boldsymbol{\phi}$ on θ is now given explicitly, while the dependence on r is understood.

If δ is real, $\tilde{\mathbf{q}}$ is the complex conjugate of $\mathbf{q}$, and $\mathbf{u}$ and $\boldsymbol{\phi}$ are real. Insertion of $(2.19)_2$ into (2.12) tells us that the stresses are singular at $r=0$ when $\delta<0$. If δ is complex, $\tilde{\mathbf{q}}$ is not necessarily the complex conjugate of $\mathbf{q}$, and $\mathbf{u}$ and $\boldsymbol{\phi}$ may not be real. It can be shown that if a complex δ is a solution, so is its complex conjugate $\overline{\delta}$. One can then superimpose two solutions associated with δ and $\overline{\delta}$ to obtain real values for $\mathbf{u}$ and $\boldsymbol{\phi}$. Again, insertion of $(2.19)_2$ into (2.12) tells us that the stresses are singular at $r=0$ when the real part of δ is negative. Thus the δ that has a negative real part represents the order of stress singularity.

3 SINGLE WEDGE–TRANSFER MATRIX

The two equations in (2.19) can be combined into one and written as

$$\mathbf{w}(\theta) = r^{\delta+1}\mathbf{X}\mathbf{Z}^{\delta+1}(\theta)\mathbf{t}, \tag{3.1}$$

in which

$$\mathbf{w}(\theta)=\begin{bmatrix}\mathbf{u}(\theta)\\ \boldsymbol{\phi}(\theta)\end{bmatrix},\quad \mathbf{X}=\begin{bmatrix}\mathbf{A} & \overline{\mathbf{A}}\\ \mathbf{B} & \overline{\mathbf{B}}\end{bmatrix},\quad \mathbf{t}=\begin{bmatrix}\mathbf{q}\\ \tilde{\mathbf{q}}\end{bmatrix}, \tag{3.2}$$

$$\mathbf{Z}(\theta)=\begin{bmatrix}\langle \zeta_*(\theta)\rangle & \mathbf{0}\\ \mathbf{0} & \langle \overline{\zeta}_*(\theta)\rangle\end{bmatrix}. \tag{3.3}$$

When the vectors $\mathbf{a}$ and $\mathbf{b}$ are normalized according to

$$2\mathbf{a}\cdot\mathbf{b}=1, \tag{3.4}$$

it can be shown that (Barnett & Lothe, 1973; Chadwick & Smith, 1977)

$$\mathbf{X}^{-1}=\begin{bmatrix}\mathbf{B}^T & \mathbf{A}^T\\ \overline{\mathbf{B}}^T & \overline{\mathbf{A}}^T\end{bmatrix}. \tag{3.5}$$

Consider a single wedge that occupies the region

$$\theta^- \le \theta \le \theta^+.$$

Equation (3.1) for $\theta=\theta^-$ and θ^+ are

$$\begin{aligned}\mathbf{w}(\theta^-) &= r^{\delta+1}\mathbf{X}\mathbf{Z}^{\delta+1}(\theta^-)\mathbf{t},\\ \mathbf{w}(\theta^+) &= r^{\delta+1}\mathbf{X}\mathbf{Z}^{\delta+1}(\theta^+)\mathbf{t}.\end{aligned} \tag{3.6}$$

Insertion of $\mathbf{t}$ obtained from $(3.6)_1$ into $(3.6)_2$ leads to

$$\mathbf{w}(\theta^+)=\mathbf{E}\mathbf{w}(\theta^-), \tag{3.7}$$

where

$$\mathbf{E}=\mathbf{X}\mathbf{Z}^{\delta+1}(\theta^+,\theta^-)\mathbf{X}^{-1}. \tag{3.8}$$

In the above (Hwu & Ting, 1989)

$$\mathbf{Z}(\theta^+,\theta^-)=\begin{bmatrix}\langle \zeta_*(\theta^+,\theta^-)\rangle & \mathbf{0}\\ \mathbf{0} & \langle \overline{\zeta}_*(\theta^+,\theta^-)\rangle\end{bmatrix}, \tag{3.9}$$

$$\zeta_\alpha(\theta^+,\theta^-) = \cos\omega + p_\alpha(\theta^-)\sin\omega, \tag{3.10}$$

where $\omega=\theta^+-\theta^-$ is the wedge angle and

$$p_\alpha(\theta^-)=\frac{p_\alpha\cos\theta^- - \sin\theta^-}{p_\alpha\sin\theta^- + \cos\theta^-}. \tag{3.11}$$

In the special case $\theta^-=0$, $p_\alpha(\theta^-)=p_\alpha$ and (3.10) reduces to (2.18).

The 6×6 matrix $\mathbf{E}$ is the *transfer matrix* in the sense of Bufler (1971) (see also Spencer et al., 1993), not in the sense of Dempsey and Sinclair (1979). It transfers the value of $\mathbf{w}$ at θ^- to $\mathbf{w}$ at θ^+. If we divide the 6×6 matrix $\mathbf{E}$ as

$$\mathbf{E}=\begin{bmatrix}\mathbf{E}^{(1)} & \mathbf{E}^{(2)}\\ \mathbf{E}^{(3)} & \mathbf{E}^{(4)}\end{bmatrix},\quad \mathbf{E}^{(4)}=\left(\mathbf{E}^{(1)}\right)^T, \tag{3.12}$$

in which $\mathbf{E}^{(i)}$ $(i=1,2,3,4)$ are 3×3 matrices, we obtain from (3.8) with the use of $(3.2)_2$, (3.5) and (3.9),

$$\mathbf{E}^{(1)} = \mathbf{A}\langle \zeta_*^{\delta+1}(\theta^+,\theta^-)\rangle \mathbf{B}^T + \overline{\mathbf{A}}\langle \overline{\zeta}_*^{\delta+1}(\theta^+,\theta^-)\rangle \overline{\mathbf{B}}^T,$$

$$\mathbf{E}^{(2)} = \mathbf{A}\langle \zeta_*^{\delta+1}(\theta^+,\theta^-)\rangle \mathbf{A}^T + \overline{\mathbf{A}}\langle \overline{\zeta}_*^{\delta+1}(\theta^+,\theta^-)\rangle \overline{\mathbf{A}}^T,$$

$$\mathbf{E}^{(3)} = \mathbf{B}\langle \zeta_*^{\delta+1}(\theta^+,\theta^-)\rangle \mathbf{B}^T + \overline{\mathbf{B}}\langle \overline{\zeta}_*^{\delta+1}(\theta^+,\theta^-)\rangle \overline{\mathbf{B}}^T. \quad (3.13)$$

The matrix $\mathbf{E}^{(4)}$ is the transpose of $\mathbf{E}^{(1)}$ as stated in (3.12).

For a single wedge for which the surfaces at $\theta = \theta^-$ and θ^+ are either traction free or rigidly clamped, we write (3.7) as

$$\mathbf{u}(\theta^+) = \mathbf{E}^{(1)}\mathbf{u}(\theta^-) + \mathbf{E}^{(2)}\phi(\theta^-),$$
$$\phi(\theta^+) = \mathbf{E}^{(3)}\mathbf{u}(\theta^-) + \mathbf{E}^{(4)}\phi(\theta^-). \quad (3.14)$$

Case I: The free-free wedge. When the surfaces at $\theta = \theta^-$ and θ^+ are both traction free, $\phi(\theta^-) = \phi(\theta^+) = \mathbf{0}$ so that $(3.14)_2$ gives

$$\mathbf{E}^{(3)}\mathbf{u}(\theta^-) = \mathbf{0}.$$

This has a non-trivial solution for $\mathbf{u}(\theta^-)$ if

$$|\mathbf{E}^{(3)}| = 0. \quad (3.15)$$

The roots of the determinant provide the stress singularities δ. It is clear from $(3.13)_3$ that, if δ is a root, so is its complex conjugate $\overline{\delta}$.

Case II: The fixed-fixed wedge. If the surfaces at $\theta = \theta^-$ and θ^+ are both fixed, we have $\mathbf{u}(\theta^-) = \mathbf{u}(\theta^+) = \mathbf{0}$. A similar argument using $(3.14)_1$ leads to

$$|\mathbf{E}^{(2)}| = 0. \quad (3.16)$$

Case III: The free-fixed wedge. When the surface at $\theta = \theta^-$ is traction free while the surface at $\theta = \theta^+$ is fixed, the result is

$$|\mathbf{E}^{(1)}| = 0. \quad (3.17)$$

When the surface at $\theta = \theta^-$ is fixed while the surface at $\theta = \theta^+$ is traction free, the condition is

$$|\mathbf{E}^{(4)}| = 0. \quad (3.18)$$

This is identical to (3.17) because $\mathbf{E}^{(4)}$ is the transpose of $\mathbf{E}^{(1)}$.

Equations (3.15)-(3.17) recover the results obtained by Ting (1996, pp. 333-337) who also presented alternative expressions that may be more convenient for numerical computation, and for the degenerate cases when $p_1 = p_2$ or $p_1 = p_2 = p_3$.

4 COMPOSITE WEDGE AND SPACE

We now apply the solutions (3.1) and (3.7) to each wedge in a composite wedge or composite space. The kth wedge has the wedge angle ω_k and occupies the region

$$\theta_{k-1} \le \theta \le \theta_k, \qquad \omega_k = \theta_k - \theta_{k-1}.$$

Denoting by a subscript k the quantity for the kth wedge, (3.1) and (3.7) are

$$\mathbf{w}_k(\theta) = r^{\delta+1}\mathbf{X}_k\mathbf{Z}_k^{\delta+1}(\theta)\mathbf{t}_k. \quad (4.1)$$

$$\mathbf{w}_k(\theta_k) = \mathbf{E}_k\mathbf{w}_k(\theta_{k-1}), \quad (4.2)$$

where, by (3.8),

$$\mathbf{E}_k = \mathbf{X}_k\mathbf{Z}_k^{\delta+1}(\theta_k,\theta_{k-1})\mathbf{X}_k^{-1}. \quad (4.3)$$

Continuity of the displacement and surface traction across the interface boundary θ_{k-1} demands that

$$\mathbf{w}_k(\theta_{k-1}) = \mathbf{w}_{k-1}(\theta_{k-1}). \quad (4.4)$$

Repeated applications of (4.2) and (4.4) leads to

$$\mathbf{w}_n(\theta_n) = \mathbf{K}_n\mathbf{w}_1(\theta_0) \quad (4.5)$$

where

$$\mathbf{K}_n = \mathbf{E}_n\mathbf{E}_{n-1}\cdots\mathbf{E}_1. \quad (4.6)$$

For a composite space, $\theta = \theta_n$ and θ_0 are the same surface with $\theta_n - \theta_0 = 2\pi$. The continuity of displacement and surface traction across this surface demands that

$$\mathbf{w}_n(\theta_n) = \mathbf{w}_1(\theta_0). \quad (4.7)$$

We then obtain from (4.5)

$$(\mathbf{K}_n - \mathbf{I})\mathbf{w}_1(\theta_0) = \mathbf{0}, \quad (4.8)$$

where $\mathbf{I}$ is the identity matrix. A nontrivial solution for $\mathbf{w}_1(\theta_0)$ exists if

$$|\mathbf{K}_n - \mathbf{I}| = 0. \quad (4.9)$$

The roots of the determinant provide the stress singularities δ.

For a composite wedge for which the surfaces at $\theta = \theta_0$ and θ_n can be traction free or fixed, the analysis is similar to the one presented in Section 3 for a single wedge. If we write the 6×6 matrix $\mathbf{K}_n$ as

$$\mathbf{K}_n = \begin{bmatrix} \mathbf{K}_n^{(1)} & \mathbf{K}_n^{(2)} \\ \mathbf{K}_n^{(3)} & \mathbf{K}_n^{(4)} \end{bmatrix} \tag{4.10}$$

where $\mathbf{K}_n^{(i)}$ (i=1,2,3,4) are 3×3 matrices, the results can be stated as follows.

Case I: The free-free wedge. When the surfaces at $\theta = \theta_0$ and θ_n are both traction free, we have

$$|\mathbf{K}_n^{(3)}| = 0. \tag{4.11}$$

Case II: The fixed-fixed wedge. If the surfaces at $\theta = \theta_0$ and θ_n are both fixed, we obtain

$$|\mathbf{K}_n^{(2)}| = 0. \tag{4.12}$$

Case III: The free-fixed wedge. When the surface at $\theta = \theta_0$ is traction free while the surface at $\theta = \theta_n$ is fixed, the result is

$$|\mathbf{K}_n^{(1)}| = 0. \tag{4.13}$$

Case IV: The fixed-free wedge. When the surface at $\theta = \theta_0$ is fixed while the surface at $\theta = \theta_n$ is traction free, we have

$$|\mathbf{K}_n^{(4)}| = 0. \tag{4.14}$$

Unlike $\mathbf{E}^{(4)}$, which is the transpose of $\mathbf{E}^{(1)}$, $\mathbf{K}_n^{(4)}$ is not the transpose of $\mathbf{K}_n^{(1)}$. Therefore (4.13) and (4.14) may yield different stress singularities.

5 PERIODIC WEDGE

We now consider a special case in which the n wedges consist of m identical units of composite wedges, each unit comprising s wedges. Thus the composite wedge has the periodicity of

$$\omega = \theta_s - \theta_0 = \omega_1 + \omega_2 + \ldots + \omega_s .$$

The total wedge angle is $m\omega$. The vectors **a**, **b** and the matrices **A**, **B** can be shown to be tensors of rank one when the transformation is a rotation about the x_3-axis (Ting, 1982). Hence

$$\mathbf{A}_{k-s} = \Omega(\omega)\mathbf{A}_k, \quad \mathbf{B}_{k-s} = \Omega(\omega)\mathbf{B}_k,$$

$$\Omega(\omega) = \begin{bmatrix} \cos\omega & \sin\omega & 0 \\ -\sin\omega & \cos\omega & 0 \\ 0 & 0 & 1 \end{bmatrix}, \quad \Omega^{-1}(\omega) = \Omega^T(\omega). \tag{5.1}$$

From $(3.2)_2$ we have

$$\mathbf{X}_{k-s} = \mathbf{Y}(\omega)\mathbf{X}_k, \tag{5.2}$$

$$\mathbf{Y}(\omega) = \begin{bmatrix} \Omega(\omega) & \mathbf{0} \\ \mathbf{0} & \Omega(\omega) \end{bmatrix}, \ldots \mathbf{Y}^{-1}(\omega) = \mathbf{Y}^T(\omega). \tag{5.3}$$

The $p_\alpha(\theta^-)$ defined in (3.11) represents the Stroh eigenvalues p_α referred to a new coordinate system obtained by rotating the original coordinate system about the x_3-axis until the x_1-axis coincides with $\theta = \theta^-$. Thus $p_\alpha(\theta^-)$ is independent of the choice of the original coordinate system. When the materials in the kth wedge and in the $k-s$ wedge are identical, we have

$$p_\alpha(\theta_k) = p_\alpha(\theta_{k-s}).$$

This implies that

$$\mathbf{Z}_k^{\delta+1}(\theta_k, \theta_{k-1}) = \mathbf{Z}_{k-s}^{\delta+1}(\theta_{k-s}, \theta_{k-s-1}). \tag{5.4}$$

It follows from (5.2) and (4.3) that

$$\mathbf{E}_{k-s} = \mathbf{Y}(\omega)\mathbf{E}_k\mathbf{Y}^T(\omega) \quad \text{or}$$

$$\mathbf{E}_k = \mathbf{Y}^T(\omega)\mathbf{E}_{k-s}\mathbf{Y}(\omega). \tag{5.5}$$

Repeated application of (5.4) and (5.5) leads to

$$\mathbf{Z}_k^{\delta+1}(\theta_k, \theta_{k-1}) = \mathbf{Z}_q^{\delta+1}(\theta_q, \theta_{q-1}),$$

$$\mathbf{E}_k = [\mathbf{Y}^T(\omega)]^{j-1}\mathbf{E}_q[\mathbf{Y}(\omega)]^{j-1},$$

where $q=k-js$, j being a positive integer. If we define

$$\mathbf{K}_s = \mathbf{E}_s\mathbf{E}_{s-1}\cdots\mathbf{E}_1,$$

we have

$$\mathbf{E}_{2s}\mathbf{E}_{2s-1}\cdots\mathbf{E}_{s+1} = \mathbf{Y}^T(\omega)\mathbf{K}_s\mathbf{Y}(\omega),$$

$$\mathbf{E}_{3s}\mathbf{E}_{3s-1}\cdots\mathbf{E}_{2s+1} = [\mathbf{Y}^T(\omega)]^2\mathbf{K}_s\mathbf{Y}^2(\omega),$$

....

$$\mathbf{E}_{ms}\mathbf{E}_{ms-1}\cdots\mathbf{E}_{(m-1)s+1} = [\mathbf{Y}^T(\omega)]^{m-1}\mathbf{K}_s\mathbf{Y}^{m-1}(\omega).$$

The $\mathbf{K}_n$ of (4.6) is, with $n=ms$,

$$\mathbf{K}_n = [\mathbf{Y}^T(\omega)]^m[\mathbf{Y}(\omega)\mathbf{K}_s]^m = \mathbf{Y}^T(m\omega)[\mathbf{Y}(\omega)\mathbf{K}_s]^m, \tag{5.6}$$

in which we have employed the identity

$$\Omega^j(\omega) = \Omega(j\omega). \tag{5.7}$$

For a composite space, $m\omega=2\pi$ so that $\mathbf{Y}(m\omega)=\mathbf{I}$. From (5.6) and (4.9) we have

$$\left|[\mathbf{Y}(\omega)\mathbf{K}_s]^m - \mathbf{I}\right| = 0. \tag{5.8a}$$

This is equivalent to

$$\left|\mathbf{Y}(\omega)\mathbf{K}_s - e^{2\pi ik/m}\mathbf{I}\right| = 0, \tag{5.8b}$$

where k is any integer.

For a composite wedge, we again divide the $\mathbf{K}_n$ of (5.6) into four 3×3 matrices according to (4.10). The stress singularities δ are computed from (4.11)-(4.14) depending on the boundary conditions at the wedge surfaces θ_0 and θ_n.

6 THE SPECIAL CASE OF $s=1$

A special case of periodic wedge is when $s=1$. This means that $m=n$ and (5.6) simplifies to

$$\mathbf{K}_n = \mathbf{Y}^T(n\omega)[\mathbf{Y}(\omega)\mathbf{E}_1]^n \tag{6.1}$$

because $\mathbf{K}_1=\mathbf{E}_1$. This is the case when the wedge consists of n identical wedges, each has the same wedge angle ω. The orientation of the material in the kth wedge is obtained by rotating the material in the $(k-1)$th wedge about the x_3-axis counter-clockwise an angle ω, Fig. 2.

For a composite space, $n\omega=2\pi$ so that (4.9) reduces to

$$\left|[\mathbf{Y}(\omega)\mathbf{E}_1]^n - \mathbf{I}\right| = 0, \text{ or} \tag{6.2a}$$

$$\left|\mathbf{Y}(\omega)\mathbf{E}_1 - e^{2\pi ik/n}\mathbf{I}\right| = 0, \tag{6.2b}$$

where k is any integer.

When the materials in the wedges are monoclinic materials with the symmetry plane at $x_3=0$, the plane strain deformation and the anti-plane deformation are uncoupled. For the plane strain deformation the stress singularity δ depends on p_1, p_2 of

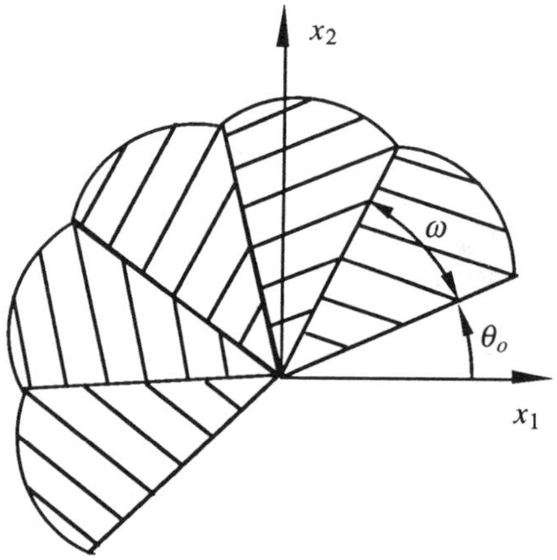

Fig. 2 Periodic wedge with $s=1$.

the materials in each wedge and the generalized Dundurs constants for every pair of materials that have a common boundary (Ting, 1995). The Dundurs constants are invariant with the rotation of the material about the x_3-axis. If the two materials with a common boundary are of the same material with different orientations, the generalized Dundurs constants vanish. For the periodic wedge considered here, this means that the stress singularity δ depends on p_1, p_2 of the material in wedge 1 and the mismatch orientation ω. The same statement applies to a composite wedge if the boundaries at $\theta=\theta_0$ and θ_n are traction free.

To show that this is indeed the case, we rewrite the transfer matrix $\mathbf{E}$ in (3.8) in terms of the inverse of the impedance matrix $\mathbf{M}$ defined by (Ingebrigtsen and Tonning, 1969)

$$\mathbf{M}^{-1} = i\mathbf{A}\mathbf{B}^{-1} = \mathbf{L}^{-1} - i\mathbf{S}\mathbf{L}^{-1}, \tag{6.3}$$

in which $\mathbf{S}$ and $\mathbf{L}$ are two of the three Barnett-Lothe tensors (Barnett and Lothe, 1973; Chadwick and Ting, 1987). They are real. $\mathbf{L}$ is symmetric and positive definite while $\mathbf{S}\mathbf{L}^{-1}$ is skew-symmetric. In computing $\mathbf{M}^{-1}$ from (6.3), the vectors $\mathbf{a}$, $\mathbf{b}$ that form the columns of the matrices $\mathbf{A}$, $\mathbf{B}$ need not be normalized according to (3.4) (Ting, 1992). Using (6.3), let

$$\hat{\mathbf{X}} = \mathbf{X}\begin{bmatrix} \mathbf{B}^{-1} & \mathbf{0} \\ \mathbf{0} & \overline{\mathbf{B}}^{-1} \end{bmatrix} = \begin{bmatrix} -i\mathbf{M}^{-1} & i\overline{\mathbf{M}}^{-1} \\ \mathbf{I} & \mathbf{I} \end{bmatrix}, \tag{6.4}$$

$$\hat{\mathbf{Z}}(\theta^+,\theta^-) = \begin{bmatrix} \mathbf{B} & \mathbf{0} \\ \mathbf{0} & \overline{\mathbf{B}} \end{bmatrix} \mathbf{Z}(\theta^+,\theta^-) \begin{bmatrix} \mathbf{B}^{-1} & \mathbf{0} \\ \mathbf{0} & \overline{\mathbf{B}}^{-1} \end{bmatrix}$$

$$= \begin{bmatrix} \mathbf{F}(\theta^+,\theta^-) & \mathbf{0} \\ \mathbf{0} & \overline{\mathbf{F}}(\theta^+,\theta^-) \end{bmatrix}, \tag{6.5}$$

$$\mathbf{F}(\theta^+,\theta^-) = \mathbf{B}\langle \zeta_*(\theta^+,\theta^-)\rangle \mathbf{B}^{-1}.$$

Again, the columns of $\mathbf{B}$ in (6.5) need not be normalized. The transfer matrix $\mathbf{E}$ of (3.8) has the expression

$$\mathbf{E} = \hat{\mathbf{X}}\hat{\mathbf{Z}}^{\delta+1}(\theta^+,\theta^-)\hat{\mathbf{X}}^{-1}.$$

The inverse of $\hat{\mathbf{X}}$ obtained from (6.4) can be shown to be

$$\hat{\mathbf{X}}^{-1} = \frac{1}{2}\begin{bmatrix} i\mathbf{L} & \mathbf{L}\overline{\mathbf{M}}^{-1} \\ -i\mathbf{L} & \mathbf{L}\mathbf{M}^{-1} \end{bmatrix}. \tag{6.6}$$

It is easily verified that $\hat{\mathbf{X}}^{-1}\hat{\mathbf{X}} = \mathbf{I}$. The $\mathbf{K}_n$ of (6.1) is now written as

$$\begin{aligned} \mathbf{K}_n &= \mathbf{Y}^T(n\omega)[\mathbf{Y}(\omega)\hat{\mathbf{X}}\hat{\mathbf{Z}}^{\delta+1}(\theta_1,\theta_0)\hat{\mathbf{X}}^{-1}]^n \\ &= \mathbf{Y}^T(n\omega)\hat{\mathbf{X}}\mathbf{V}^n\hat{\mathbf{X}}^{-1}, \end{aligned} \tag{6.7}$$

$$\mathbf{V} = \hat{\mathbf{X}}^{-1}\mathbf{Y}(\omega)\hat{\mathbf{X}}\hat{\mathbf{Z}}^{\delta+1}(\theta_1,\theta_0). \tag{6.8}$$

We have dropped the subscript 1 for $\hat{\mathbf{X}}$ and $\hat{\mathbf{Z}}$. A direct calculation employing (5.3), (6.3), (6.4) and (6.6) shows that

$$\hat{\mathbf{X}}^{-1}\mathbf{Y}(\omega)\hat{\mathbf{X}} = \frac{1}{2}\begin{bmatrix} \mathbf{C}^+ & \mathbf{C}^- \\ \mathbf{C}^- & \mathbf{C}^+ \end{bmatrix} + \frac{i}{2}\begin{bmatrix} -\mathbf{LG} & -\mathbf{LG} \\ \mathbf{LG} & \mathbf{LG} \end{bmatrix}, \tag{6.9}$$

$$\begin{aligned} \mathbf{C}^\pm &= \Omega(\omega) \pm \mathbf{L}\Omega(\omega)\mathbf{L}^{-1}, \\ \mathbf{G} &= \Omega(\omega)\mathbf{SL}^{-1} - \mathbf{SL}^{-1}\Omega(\omega). \end{aligned} \tag{6.10}$$

For a monoclinic material with the symmetry plane at $x_3=0$ we have (Stroh, 1958)

$$\mathbf{B} = \begin{bmatrix} p_1 & p_2 \\ -1 & -1 \end{bmatrix}. \tag{6.11}$$

The 3×3 matrix $\mathbf{B}$ has been reduced to a 2×2 matrix because we are considering two-dimensional plane strain deformations. With (6.11), it is easily seen from (6.5) that $\hat{\mathbf{Z}}(\theta_1,\theta_0)$ in (6.8) depends on p_1, p_2, ω and θ_0 only. Let

$$p_1 + p_2 = a + ib, \quad p_1p_2 = c + id, \tag{6.12}$$

where a, b, c, d are real. It is proved in (Ting, 1992) that

$$\mathbf{L}^{-1} = s'_{11}\begin{bmatrix} b & d \\ d & e \end{bmatrix}, \quad \mathbf{SL}^{-1} = \kappa\begin{bmatrix} 0 & -1 \\ 1 & 0 \end{bmatrix}, \tag{6.13}$$

where $e=ad-bc$, and s'_{11} and κ are material constants. One can then show that $\mathbf{G}$ of (6.10) vanishes while

$$\mathbf{L}\Omega(\omega)\mathbf{L}^{-1} = \frac{1}{be-d^2}\begin{bmatrix} e & -d \\ -d & b \end{bmatrix}\Omega(\omega)\begin{bmatrix} b & d \\ d & e \end{bmatrix}. \tag{6.14}$$

It depends on p_1, p_2 and ω only. Thus $\mathbf{V}$ of (6.8) depends on p_1, p_2, ω and θ_0 only.

For a composite space for which $n\omega=2\pi$ and $\mathbf{Y}(n\pi)=\mathbf{I}$, insertion of (6.7) into (4.9) leads to

$$|\mathbf{V}^n - \mathbf{I}| = 0. \tag{6.15a}$$

This is equivalent to

$$|\mathbf{V} - e^{2\pi ik/n}\mathbf{I}| = 0, \tag{6.15b}$$

where k is any integer.

For a composite wedge with the surfaces θ_0 and θ_n being traction free, the condition is the vanishing of the determinant of $\mathbf{K}^{(3)}$ as shown in (4.11). If we divide the 6×6 matrix $\mathbf{V}^n$ as

$$\mathbf{V}^n = \begin{bmatrix} \mathbf{V}^{(1)} & \mathbf{V}^{(2)} \\ \mathbf{V}^{(3)} & \mathbf{V}^{(4)} \end{bmatrix} \tag{6.16}$$

in which $\mathbf{V}^{(i)}$ (i=1,2,3,4) are 3×3 matrices, it is easily shown that $\mathbf{K}^{(3)}$ obtained from (6.7) is

$$\mathbf{K}^{(3)} = \frac{i}{2}\Omega^T(n\omega)(\mathbf{V}^{(1)} - \mathbf{V}^{(2)} + \mathbf{V}^{(3)} - \mathbf{V}^{(4)})\mathbf{L}.$$

Therefore the condition is

$$|\mathbf{V}^{(1)} - \mathbf{V}^{(2)} + \mathbf{V}^{(3)} - \mathbf{V}^{(4)}| = 0. \tag{6.17}$$

The stress singularity δ computed from (6.15) or (6.17) depends on p_1, p_2, ω and θ_0 only.

REFERENCES

Barnett, D. M. & J. Lothe 1973. Synthesis of the sextic and the integral formalism for dislocations, Greens function and surface waves in anisotropic elastic solids. *Phys. Norv.* **7**: 13–19.

Bogy, D. B. 1970. On the problem of edge-bonded elastic quarter plane loaded at the boundary. *Int. J. Solids Structures* **6**: 1287–1313.

Bogy, D. B. 1971a. Two edge-bonded elastic wedges of different materials and wedge angles under surface tractions. *J. Appl. Mech.* **38**: 377–386.

Bogy, D. B. 1971b. On the plane elastostatics problem of a loaded crack terminating at a material interface. *J. Appl. Mech.* **38**: 911–918.

Bogy, D. B. 1972. The plane solution for anisotropic elastic wedges under normal and shear loading. *J. Appl. Mech.* **39**: 1103–1109.

Bogy, D. B. 1974. Plane solutions for traction problems on orthotropic unsymmetrical wedges and symmetrically twinned wedges. *J. Appl. Mech.* **41**: 203–208.

Bogy, D. B. & K. C. Wang 1971. Stress singularities at interface corners in bonded dissimilar isotropic elastic materials. *Int. J. Solids Structures* **7**: 993–1005.

Bufler, H. 1971. Theory of elasticity of a multilayered medium. *J. Elasticity* **1**: 125-143.

Chadwick, P. & G. D. Smith 1977. Foundations of the theory of surface waves in anisotropic elastic materials. *Adv. Appl. Mech.* **17**: 303–376.

Chadwick, P. & T. C. T. Ting 1987. On the structure and invariance of the Barnett–Lothe tensors. *Q. Appl. Math.* **45**: 419–427.

Dempsey J. P. & G. B. Sinclair 1979. On the stress singularities in the plane elasticity of the composite wedge. *J. Elasticity* **9**: 373–391.

Dempsey J. P. & G. B. Sinclair 1981. On the singular behavior at the vertex of a bi-material wedge. *J. Elasticity* **11**: 317–327.

Eshelby, J. D., W. T. Read & W. Shockley 1953. Anisotropic elasticity with applications to dislocation theory. *Acta Metall.* **1**: 251-259.

Hwu, Chyanbin & T. C. T. Ting 1989. Two-dimensional problems of the anisotropic elastic solid with an elliptic inclusion. *Q. J. Mech. Appl. Math.* **42**: 553–572.

Ingebrigtsen, K. A. & A. Tonning 1969. Elastic surface waves in crystal. *Phys. Rev.* **184**: 942–951.

Knein, M. 1926. Zur theorie des druckversuchs. *Zeit. Ang. Math. Mech.* **6**: 414–416.

Kuo, M. C. & D. B. Bogy 1974. Plane solutions for the displacement and traction-displacement problems for anisotropic elastic wedges. *J. Appl. Mech.* **41**: 197–203.

Spencer, A. J. M., T. G. Rogers & P. Watson 1993. Transfer matrix methods applied to anisotropic inhomogeneous, non-circular elastic cylinders. *ASME AMD-Vol.* **158**: 7-11.

Stroh, A. N. 1958. Dislocations and cracks in anisotropic elasticity. *Phil. Mag.* **3**: 625–646.

Ting, T. C. T. 1982. Effects of change of reference coordinates on the stress analyses of anisotropic elastic materials. *Int. J. Solids Structures* **18**: 139–152.

Ting, T. C. T. 1992. Barnett–Lothe tensors and their associated tensors for monoclinic materials with the symmetry plane at $x_3 = 0$. *J. Elasticity* **27**: 143–165.

Ting, T. C. T. 1995. Generalized Dundurs constants for aniso-tropic bimaterials. *Int. J. Solids Structures* **32**: 483-500.

Ting, T. C. T. 1996. *Anisotropic Elasticity: Theory and Applications*. New York: Oxford University Press.

Williams, M. L. 1952. Stress singularities resulting from various boundary conditions in angular corners of plates in extension. *J. Appl. Mech.* **19**: 526–528.

Damage and Failure of Interfaces, Rossmanith (ed.) © 1997 Balkema, Rotterdam, ISBN 90 5410 899 1

Stress singularities in a bimaterial joint under axisymmetric deformation

Y.L.Li & S.Y.Hu
Forschungszentrum Karlsruhe, IMF II, Germany (On leave from: Lanzhou University, People's Republic of China)

D.Munz
Forschungszentrum Karlsruhe, IMF II & University of Karlsruhe, IZSM, Germany

Y.Y.Yang
University of Karlsruhe, IZSM, Germany

ABSTRACT: The stress singularities near the free edge of the interface in a bimaterial joint under axisymmetric deformation are studied. The characteristic equation for determining the singular exponents is derived, which is found to be the same as that of the plane joint under plane strain deformations. The associated stress angular functions and displacement angular functions are obtained analytically. In an axisymmetric deformation problem, the singular stress term only can't describe the stresses in the vicinity of the singular point well. The effect of regular stress terms should be considered. In this work, the regular stress term which is independent of the distance from the singular point, is presented. An asymptotic description is developed for the stress field near the singular point.

1 INTRODUCTION

For many special applications dissimilar materials have to be joined. Due to the differences in the mechanical and thermal properties of the joined materials, however, high stresses are induced near the free edge of the interface during mechanical or thermal loading. In such cases the high stresses may exceed the strength of the joined materials and cause failure.

A lot of investigations for the stress analysis near the edge of the interface in joints have been published for the plane problems. In this work we consider the case in which the specimen shown in Figure 1 is an axisymmetric joint with two different materials under an axisymmetric deformation. Both materials are assumed to be linear elastic, isotropic and homogeneous; and the bonding between the two materiäls is perfect. The stress singularities at a notch in a homogenous material under axisymmetric deformation were investigated by Zak[1964] for isotropic material and by Ting etc.[1985] for transversely isotropic material. In this work, the method used by Zak is extended to study the stress singularity at the intersection of the interface and the free edge of the axisymmetric joint, denoted as a singular point S in Figure 1. Besides the singular stress term, the regular stress term which is independent of the distance from the singular point, is presented, and an asymptotic description for the stress field near the singular point is developed.

2 BASIC FORMULATIONS

Let $\{u_r, u_z\}$ and $\{\sigma_r, \sigma_t, \sigma_z, \tau_{rz}\}$ be the non-zero displacement and stress components in the

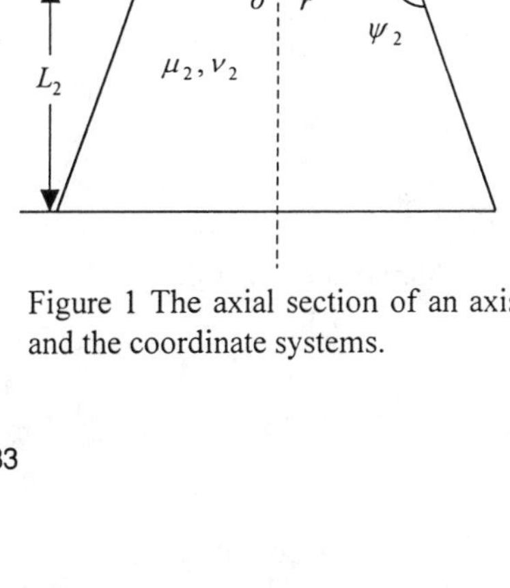

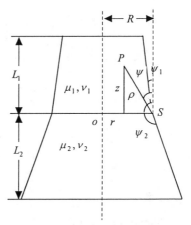

Figure 1 The axial section of an axisymmetric joint and the coordinate systems.

cylindrical coordinate system (r,z) defined in Figure 1. The equilibrium equations without body force and the relation between stresses and displacements of each material can be expressed as

$$\frac{\partial \sigma_r}{\partial r}+\frac{\partial \tau_{rz}}{\partial z}+\frac{\sigma_r-\sigma_t}{r}=0 , \tag{1a}$$

$$\frac{\partial \tau_{rz}}{\partial r}+\frac{\partial \sigma_z}{\partial z}+\frac{\tau_{rz}}{r}=0 , \tag{1b}$$

$$\sigma_r = 2\mu\left(\frac{\nu}{1-2\nu}e+\frac{\partial u_r}{\partial r}\right) , \tag{2a}$$

$$\sigma_t = 2\mu\left(\frac{\nu}{1-2\nu}e+\frac{u_r}{r}\right) , \tag{2b}$$

$$\sigma_z = 2\mu\left(\frac{\nu}{1-2\nu}e+\frac{\partial u_z}{\partial z}\right) , \tag{2c}$$

$$\tau_{rz} = \mu\left(\frac{\partial u_r}{\partial z}+\frac{\partial u_z}{\partial r}\right) , \tag{2d}$$

where $e=\frac{\partial u_r}{\partial r}+\frac{u_r}{r}+\frac{\partial u_z}{\partial z}$, μ is the shear modulus, ν is the Poisson's ratio.

Introducing stress function $\Phi=\Phi(r,z)$, the stresses and displacements are given by

$$\sigma_r = \frac{\partial}{\partial z}\left(\nu\nabla^2\Phi-\frac{\partial^2\Phi}{\partial r^2}\right) , \tag{3a}$$

$$\sigma_t = \frac{\partial}{\partial z}\left(\nu\nabla^2\Phi-\frac{1}{r}\frac{\partial\Phi}{\partial r}\right) , \tag{3b}$$

$$\sigma_z = \frac{\partial}{\partial z}\left((2-\nu)\nabla^2\Phi-\frac{\partial^2\Phi}{\partial z^2}\right) , \tag{3c}$$

$$\tau_{rz} = \frac{\partial}{\partial r}\left((1-\nu)\nabla^2\Phi-\frac{\partial^2\Phi}{\partial z^2}\right) , \tag{3d}$$

$$2\mu u_r = -\frac{\partial^2\Phi}{\partial r\partial z} , \tag{4a}$$

$$2\mu u_z = 2(1-\nu)\nabla^2\Phi-\frac{\partial^2\Phi}{\partial z^2} . \tag{4b}$$

The equations of equilibrium are satisfied, if

$$\nabla^4\Phi=\nabla^2\nabla^2\Phi=\left(\frac{\partial^2}{\partial r^2}+\frac{1}{r}\frac{\partial}{\partial r}+\frac{\partial^2}{\partial z^2}\right)^2\Phi=0 . \tag{5}$$

3 STRESS SINGULARITY

To describe the stress field near the singular point, a new coordinate system (ρ,ψ) is introduced, which originated at the singular point S. Any point P inside the body which has the cylindrical coordinates r and z can be defined in the new coordinate system by ρ and ψ. The transformation relations between the two coordinate systems are

$$z=\rho\cos\psi , \quad r=R-\rho\sin\psi , \tag{6}$$

where R is the radius of the circular interface. Now Eq.(5) is transformed from r, z to ρ, ψ to give

$$\left[\frac{\partial^2}{\partial\rho^2}+\frac{1}{\rho^2}\frac{\partial^2}{\partial\psi^2}+\left(\frac{1}{\rho}-\frac{\sin\psi}{R-\rho\sin\psi}\right)\frac{\partial}{\partial\rho} -\frac{\cos\psi}{\rho(R-\rho\sin\psi)}\frac{\partial}{\partial\psi}\right]^2\Phi=0 . \tag{7}$$

It is very difficult to find an analytical solution for Eq.(7). In order to analyse the stress singularities, only the neighbourhood of $\rho=0$ is interesting, where $\frac{\rho}{R}\ll 1$. So, we neglect the term of $\frac{1}{R-\rho\sin\psi}$ with respect to the term of $\frac{1}{\rho}$ and Eq.(7) is simplified as

$$\left[\frac{\partial^2}{\partial\rho^2}+\frac{1}{\rho}\frac{\partial}{\partial\rho}+\frac{1}{\rho^2}\frac{\partial^2}{\partial\psi^2}\right]^2\Phi=0 . \tag{8}$$

The solution of Eq.(8) can be taken in the form

$$\Phi(\rho,\psi)=\rho^{3-\omega}\left[\frac{A}{-3+\omega}\sin(3-\omega)\psi +\frac{B}{-3+\omega}\cos(3-\omega)\psi+C\sin(1-\omega)\psi+D\cos(1-\omega)\right] , \tag{9}$$

where A,B,C,D and ω are unknown parameters.

Carrying out the variable transformation in Eq.(6) and making the same approximation as above to Eqs.(3-4), the stresses and displacements can be expressed in terms of the solution in Eq.(9) as follows

$$\sigma_r = (1-\omega)(2-\omega)\rho^{-\omega}\{ A\sin\omega\psi - B\cos\omega\psi + C[(1-4\nu)\sin\omega\psi + \omega\sin(2+\omega)\psi] + D[-(1-4\nu)\cos\omega\psi - \omega\cos(2+\omega)\psi]\} \quad (10a)$$

$$\sigma_t = 4\nu(1-\omega)(2-\omega)\rho^{-\omega}[-C\sin\omega\psi + D\cos\omega\psi], \quad (10b)$$

$$\sigma_z = (1-\omega)(2-\omega)\rho^{-\omega}\{ -A\sin\omega\psi + B\cos\omega\psi + C[-(5-4\nu)\sin\omega\psi - \omega\sin(2+\omega)\psi] + D[(5-4\nu)\cos\omega\psi + \omega\cos(2+\omega)\psi]\} \quad (10c)$$

$$\tau_{rz} = (1-\omega)(2-\omega)\rho^{-\omega}\{ -A\cos\omega\psi - B\sin\omega\psi + C[-(3-4\nu)\cos\omega\psi - \omega\cos(2+\omega)\psi] + D[-(3-4\nu)\sin\omega\psi - \omega\sin(2+\omega)\psi]\} \quad (10d)$$

$$u_r = \frac{1}{2\mu}(2-\omega)\rho^{-\omega+1}\{-A\cos(\omega-1)\psi - B\sin(\omega-1)\psi + C(1-\omega)\cos(\omega+1)\psi + D(1-\omega)\sin(\omega+1)\psi\}, \quad (11a)$$

$$u_z = \frac{1}{2\mu}(2-\omega)\rho^{-\omega+1}\{-A\sin(\omega-1)\psi + B\cos(\omega-1)\psi + C[-(6-8\nu)\sin(\omega-1)\psi + (1-\omega)\sin\omega\psi] + D[(6-8\nu)\cos(\omega-1)\psi - (1-\omega)\cos\omega\psi]\}. \quad (11b)$$

The problem is reduced to the determination of the unknown parameters A,B,C,D and ω. To obtain them, the stress-free boundary conditions at $\psi=\psi_1$ and $\psi=\psi_2$, and the interfacial continuous conditions at $\psi=\frac{\pi}{2}$ have to be used. They are

$$\sigma_{\psi 1}(\rho,\psi_1)=0 , \quad (12a)$$

$$\tau_{\rho\psi 1}(\rho,\psi_1)=0 , \quad (12b)$$

$$\sigma_{\psi 2}(\rho,\psi_2)=0 , \quad (12c)$$

$$\tau_{\rho\psi 2}(\rho,\psi_2)=0 , \quad (12d)$$

$$\sigma_{\psi 1}\left(\rho,\frac{\pi}{2}\right)=\sigma_{\psi 2}\left(\rho,\frac{\pi}{2}\right) , \quad (12e)$$

$$\tau_{\rho\psi 1}\left(\rho,\frac{\pi}{2}\right)=\tau_{\rho\psi 2}\left(\rho,\frac{\pi}{2}\right) , \quad (12f)$$

$$u_{\rho 1}\left(\rho,\frac{\pi}{2}\right)=u_{\rho 2}\left(\rho,\frac{\pi}{2}\right) , \quad (12g)$$

$$u_{\psi 1}\left(\rho,\frac{\pi}{2}\right)=u_{\psi 2}\left(\rho,\frac{\pi}{2}\right) , \quad (12h)$$

in which $\{\sigma_\psi,\tau_{\rho\psi}\}$, $\{u_\rho,u_\psi\}$ are the stress and displacement components in the polar coordinates (ρ,ψ), and the subscripts 1 and 2 refer to the materials 1 and 2 in the joint. Making the coordinate transform on stresses and displacements and substituting them in Eq.(12), eight equations for A_j, B_j, C_j and D_j $(j=1,2)$ are yielded by considering the arbitrariness of ρ, which are expressed in the matrix form as:

$$[M]\{X\}=\{0\} , \quad (13)$$

where $[M]$ is the coefficient matrix, and $\{X\}=\{A_1,B_1,C_1,D_1,A_2,B_2,C_2,D_2\}^T$. In order to obtain a non-trivial solution from the homogeneous linear algebraic equations in Eq.(13), the determinant of its coefficient matrix has to be zero, which leads to a characteristic equation for ω:

$$|M| = \Pi(\omega,\alpha,\beta)=0 , \quad (14)$$

where α, β are the Dundurs parameters which are $\alpha=\frac{(1-\nu_2)-\kappa(1-\nu_1)}{(1-\nu_2)+\kappa(1-\nu_1)}$, $\beta=\frac{\left(\frac{1}{2}-\nu_2\right)-\kappa\left(\frac{1}{2}-\nu_1\right)}{(1-\nu_2)+\kappa(1-\nu_1)}$ and $\kappa=\frac{\mu_2}{\mu_1}$. Eq.(14) is a transcendental equation for ω. When $\psi_1=0,\psi_2=180^0$, i.e., the joint becomes a cylindrical joint, the characteristic equation (14) is verified to be identical to that of the quarter plane joint under plane strain deformation [see Li etc., 1997]. For an arbitrary ψ_1 and ψ_2, the roots of Eq.(14) in the range of $-1<\omega<1$ are examined numerically to be the same as those in the plane joint under the plane strain deformation.

For a given ω, the coefficients $\{X\}$ in Eq.(13) can be determined, but one of them is arbitrary. Substituting them in Eqs.(10-11), the associated stresses and displacements denoted with the superscript s are expressed as

$$\sigma_r^s(\rho,\psi)=K\left(\frac{\rho}{R}\right)^{-\omega}\zeta_r(\psi), \tag{15a}$$

$$\sigma_t^s(\rho,\psi)=K\left(\frac{\rho}{R}\right)^{-\omega}\zeta_t(\psi), \tag{15b}$$

$$\sigma_z^s(\rho,\psi)=K\left(\frac{\rho}{R}\right)^{-\omega}\zeta_z(\psi), \tag{15c}$$

$$\tau_{rz}^s(\rho,\psi)=K\left(\frac{\rho}{R}\right)^{-\omega}\zeta_{rz}(\psi), \tag{15d}$$

$$u_r^s(\rho,\psi)=R\frac{K}{2\mu}\left(\frac{\rho}{R}\right)^{-\omega+1}\chi_r(\psi), \tag{16a}$$

$$u_z^s(\rho,\psi)=R\frac{K}{2\mu}\left(\frac{\rho}{R}\right)^{-\omega+1}\chi_z(\psi), \tag{16b}$$

where $\zeta_{kl}(\psi)$ and $\chi_k(\psi)$ are the stress angular functions and displacement angular functions, respectively. They can be calculated analytically with $\zeta_z\left(\frac{\pi}{2}\right)=1$. The only unknown parameter K which is the stress intensity factor depends on the material combination, the geometry of the joint and the loading.

4 REGULAR STRESS TERM

It was pointed out [see Li etc., 1997] that the singular term in Eq.(15) only can't describe correctly the stress field in the vicinity of the singular point S. The regular stress terms should be considered. Here the regular term which is independent of the distance ρ from the singular point S is considered. There are two parts in this regular stress term. One results from the r-direction displacement of the singular point, denoted as u_0. From the relation between stresses and displacements in Eq.(2), it can be seen that not only the differential of the displacement u_r but also the displacement u_r itself produces non-zero stresses. For this, a particular solution is introduced, which satisfies the equilibrium equation and the relation between the stresses and displacements, i.e.

$$u_r=u_0\frac{r}{R}=u_0-u_0\frac{\rho}{R}\sin\psi, \tag{17a}$$

$$u_z=0, \tag{17a}$$

$$\sigma_r=\frac{2\mu}{1-2\nu}\frac{u_0}{R}, \tag{18a}$$

$$\sigma_t=\frac{2\mu}{1-2\nu}\frac{u_0}{R}, \tag{18b}$$

$$\sigma_z=\frac{4\nu\mu}{1-2\nu}\frac{u_0}{R}, \tag{18c}$$

$$\tau_{rz}=0\ . \tag{18d}$$

The second part results from the following stress function $\Phi_0(\rho,\psi)$ which satisfies Eq.(5),

$$\begin{aligned}\Phi_0(\rho,\psi)&=E\rho^3\ln\rho\left[\left(1-\frac{4}{3}\nu\right)\sin3\psi+\sin\psi\right]\\&+\rho^3\left\{E\psi\left[\left(1-\frac{4}{3}\nu\right)\cos3\psi+\cos\psi\right]\right.\\&+F\cos3\psi+G\left[\left(1-\frac{4}{3}\nu\right)\sin3\psi+\sin\psi\right]\\&\left.+H\cos\psi+Q\frac{1}{3}\sin3\psi\right\}\end{aligned}, \tag{19}$$

where E,F,G,H,Q are unknown coefficients. By superposing Eqs. (17) and (18) on those displacements and stresses resulting from $\Phi_0(\rho,\psi)$ in Eq.(19), the regular stresses and the associated displacements, denoted by the superscript 0, have the expressions

$$\begin{aligned}\sigma_r^0&=E(4\psi-2\sin2\psi)+6F+(-2+8\nu)H\\&+\frac{2\mu}{1-2\nu}\frac{u_0}{R}\end{aligned}, \tag{20a}$$

$$\sigma_t^0=E8\nu\psi+8\nu H+\frac{2\mu}{1-2\nu}\frac{u_0}{R}, \tag{20b}$$

$$\begin{aligned}\sigma_z^0&=E(4\psi+2\sin2\psi)-6F+(10-8\nu)H\\&+\frac{4\nu\mu}{1-2\nu}\frac{u_0}{R}\end{aligned}, \tag{20c}$$

$$\tau_{rz}^0 = E\left(2-\frac{8}{3}\nu+2\cos 2\psi\right)+2Q \ , \tag{20d}$$

$$\begin{aligned} u_r^0 = u_0\left(1-\frac{\rho}{R}\sin\psi\right)+\frac{\rho}{\mu}\{&E(4-4\nu)\ln\rho\cos\psi \\ &+E\left[(-2+4\nu)\psi\sin\psi+\left(4-\frac{10}{3}\nu\right)\cos\psi\right] \\ &-3F\sin\psi+G(4-4\nu)\cos\psi \\ &+H\sin\psi+Q\cos\psi\} \end{aligned} , \tag{21a}$$

$$\begin{aligned} u_z^0 = \frac{\rho}{\mu}\{&E(4-4\nu)\ln\rho\sin\psi \\ &+E\left[(2-4\nu)\psi\cos\psi+\left(2-\frac{2}{3}\nu\right)\sin\psi\right] \\ &-3F\cos\psi+G(4-4\nu)\sin\psi \\ &+H(5-8\nu)\cos\psi-Q\sin\psi\} \end{aligned} . \tag{21b}$$

After using the stress free boundary conditions and the interface continuous conditions given in Eq.(12), we can obtained nine equations for ten unknown coefficients $\{E_j, F_j, G_j, H_j, Q_j\}$, $(j=1,2)$, by considering the arbitrary of ρ in the displacement components. With the exception of the rigid rotation constant G_1 or G_2, all the unknown coefficients in Eqs.(20-21) can be determined uniquely. Not losing generality, let $G_2 = 0$. Then all the coefficients, the regular stresses and the associated displacements are proportional to the value of u_0. For the special case of $\psi_2 - \psi_1 = \pi$, the regular stresses are constants and can be expressed as:

$$\sigma_r^0 = \frac{2\mu_2(\nu_1-\nu_2)\sin^2\psi_1}{\nu_2-\kappa\nu_1+(\kappa-1)\sin^2\psi_1}\frac{u_0}{R} \ , \tag{22a}$$

$$\sigma_{t1}^0 = 2\mu_1\left[1+\frac{\nu_1(1-\kappa)(\nu_2-\sin^2\psi_1)}{\nu_2-\kappa\nu_1+(\kappa-1)\sin^2\psi_1}\right]\frac{u_0}{R} \ , \tag{22b}$$

$$\sigma_{t2}^0 = 2\mu_2\left[1+\frac{\nu_2(1-\kappa)(\nu_1-\sin^2\psi_1)}{\nu_2-\kappa\nu_1+(\kappa-1)\sin^2\psi_1}\right]\frac{u_0}{R} \ , \tag{22c}$$

$$\sigma_z^0 = \frac{2\mu_2(\nu_1-\nu_2)\cos^2\psi_1}{\nu_2-\kappa\nu_1+(\kappa-1)\sin^2\psi_1}\frac{u_0}{R} \ , \tag{22d}$$

$$\tau_{rz}^0 = -\frac{2\mu_2(\nu_1-\nu_2)\cos\psi_1\sin\psi_1}{\nu_2-\kappa\nu_1+(\kappa-1)\sin^2\psi_1}\frac{u_0}{R} \ , \tag{22e}$$

in which the subscript 1 or 2 refers to the bonded material 1 or 2.

5 ASYMPTOTIC DESCRIPTION OF THE SINGULAR STRESS FIELD

By summing Eq.(15) and Eq.(20), an asymptotic description for the singular stress field near the singular point S is developed as

$$\begin{aligned}\sigma_r(\rho,\psi) &= \sigma_r^s(\rho,\psi)+\sigma_r^0(\psi) \\ &= K\left(\frac{\rho}{R}\right)^{-\omega}\zeta_r(\psi)+\sigma_r^0(\psi)\end{aligned} , \tag{23a}$$

$$\begin{aligned}\sigma_t(\rho,\psi) &= \sigma_t^s(\rho,\psi)+\sigma_t^0(\psi) \\ &= K\left(\frac{\rho}{R}\right)^{-\omega}\zeta_t(\psi)+\sigma_t^0(\psi)\end{aligned} , \tag{23b}$$

$$\begin{aligned}\sigma_z(\rho,\psi) &= \sigma_z^s(\rho,\psi)+\sigma_z^0(\psi) \\ &= K\left(\frac{\rho}{R}\right)^{-\omega}\zeta_z(\psi)+\sigma_z^0(\psi)\end{aligned} , \tag{23c}$$

$$\begin{aligned}\tau_{rz}(\rho,\psi) &= \tau_{rz}^s(\rho,\psi)+\tau_{rz}^0(\psi) \\ &= K\left(\frac{\rho}{R}\right)^{-\omega}\zeta_{rz}(\psi)+\tau_{rz}^0(\psi)\end{aligned} . \tag{23d}$$

In Eq.(23), there are two unknown parameters. One is the stress intensity factor K, the other is the r-direction displacement u_0 of the singular point S included in the regular stress term. Both can be obtained by matching the solution of the present analysis with the results obtained by a numerical method, for example, the Boundary Element Method (BEM), which is used in our work.

6 EXAMPLES

As an example, the axisymmetric joint with Al_2O_3 and $Al-alloy$ under uniform tensile is studied. Their shear modulus and Poisson's ratio are

Al_2O_3: $\mu_1 = 147.6GPa$; $\nu_1 = 0.27$;
$Al-alloy$: $\mu_2 = 26.7GPa$, $\nu_2 = 0.33$.

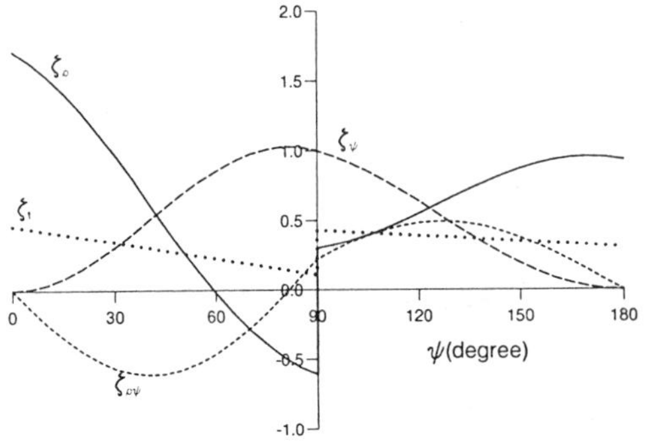

Figure 2 The stress angular functions in geometry A.

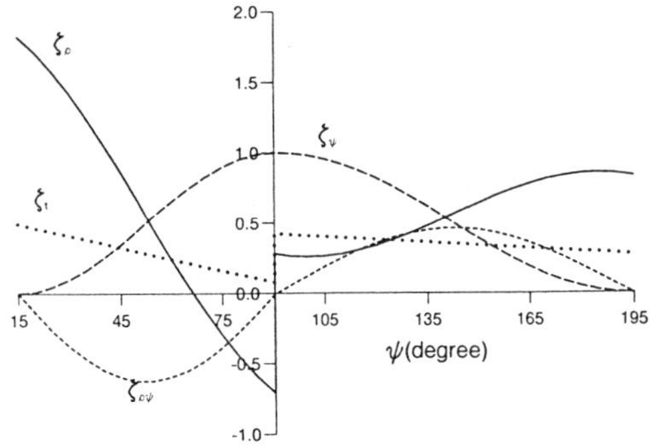

Figure 3 The stress angular functions in geometry B.

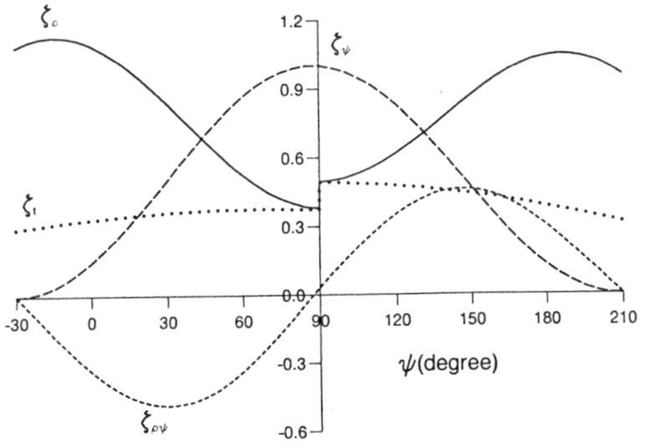

Figure 4 The stress angular functions in geometry C.

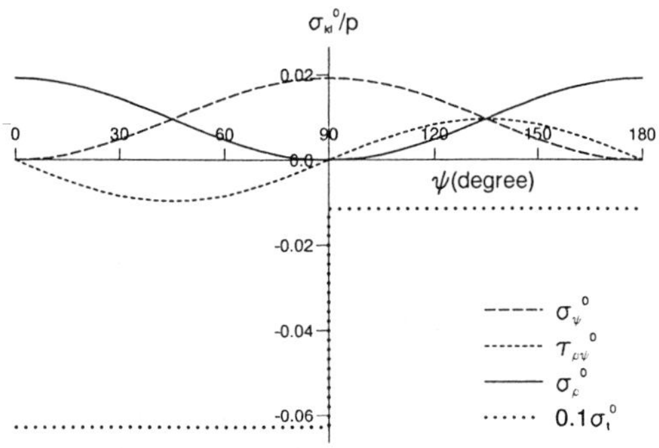

Figure 5 The regular stress term in geometry A.

The height of the joint is provided to be $L_1 = L_2 = R$ and the following three geometries are considered:

A: $\psi_1 = 0, \psi_2 = 180^0$;

B: $\psi_1 = 15^0, \psi_2 = 195^0$;

C: $\psi_1 = -30^0, \psi_2 = 210^0$.

6.1 *The singular exponent ω and the stress angular functions*

For the three joint geometries, the singular exponents can be obtained by solving Eq.(14). It is found that in the range of $0 < \omega < 1$ there is only one root for each of the above three geometries. They are $\omega_A = 0.1571$, $\omega_B = 0.1071$ and $\omega_C = 0.3933$, respectively. For the given ω the stress angular functions $\zeta_{kl}(\psi)$ can be calculated analytically and they are drawn in Figures 2-4.

6.2 *The regular stress term*

The regular stress term in Eqs. (20) or (22) depends on the r-direction displacement u_0. u_0 can be obtained by solving the problem numerically. In this work, the BEASY program based on BEM is employed for solving the problem. Let p_1 and p_2 be the tensile traction on the upper and lower ends of the joint. For equilibrium, p_1 and p_2 are given as

$$p_1 = p\frac{R^2}{(R - L_1 \tan\psi_1)^2}, \qquad p_2 = p\frac{R^2}{(R - L_2 \tan\psi_2)^2},$$

where p is a constant. For the three geometries, the obtained u_0 are $\frac{u_{0A}}{R} = -0.2493\frac{p}{\mu_1}$, $\frac{u_{0B}}{R} = -0.3089\frac{p}{\mu_1}$, $\frac{u_{0C}}{R} = 0.08795\frac{p}{\mu_1}$. The regular stress term can then be calculated analytically. The figures 5-7 show the regular stress term versus the angle ψ.

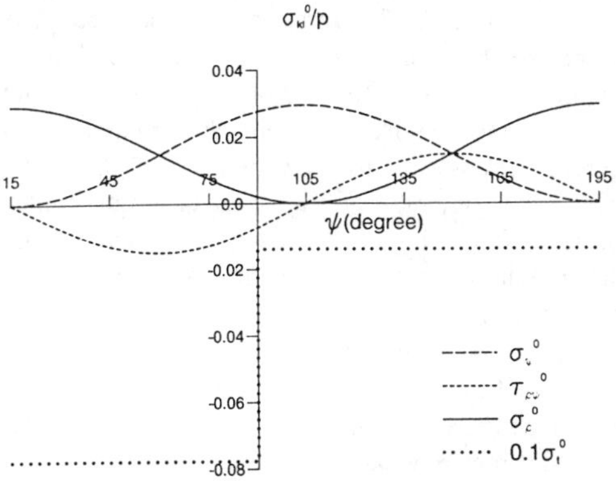

Figure 6 The regular stress term in geometry B.

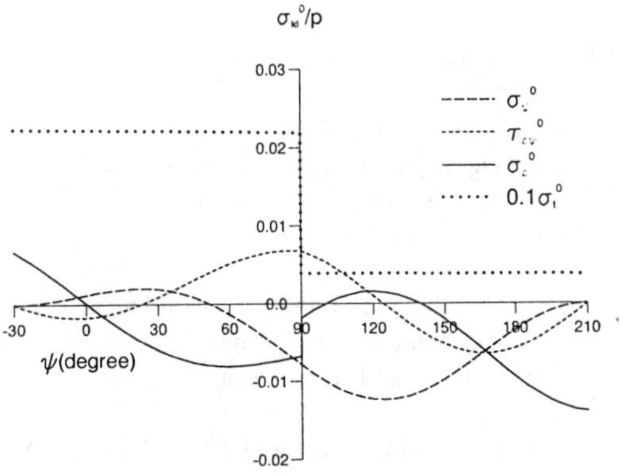

Figure 7 The regular stress term in geometry C.

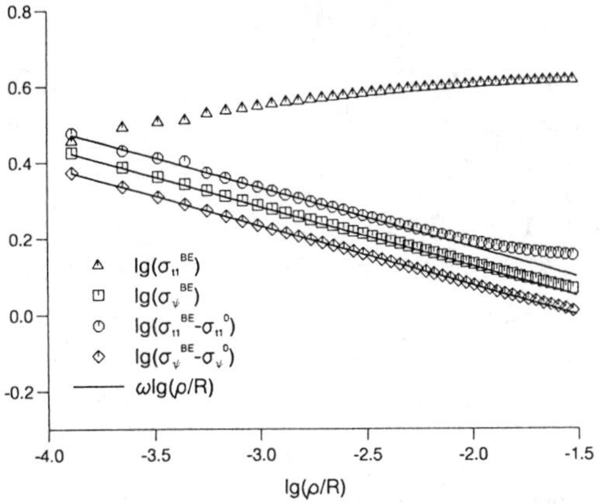

Figure 8 The $lg(\sigma_{kl}) \sim lg(\rho / R)$ plots along the interface in geometry A.

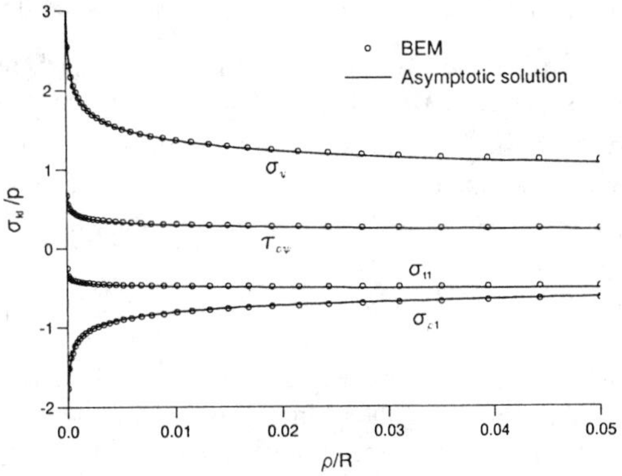

Figure 9 The stresses near the singular point along the interface in geometry A.

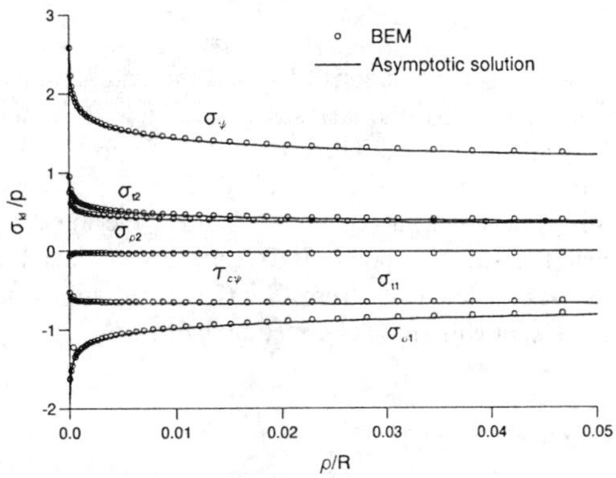

Figure 10 The stresses near the singular point along the interface in geometry B.

It is found that the regular stress σ_t^0 is much larger than the other stress components. This means that the effect of σ_t^0 on σ_t is larger than the effect of other σ_{kl}^0 on its corresponding σ_{kl}. This will be seen in the next section.

6.3 The stress intensity factor and the stresses in the vicinity of the singular point S

It is known that the singular stress term can describe the stress field in the vicinity of the singular point well for a joint under mechanical loading [Munz and Yang, 1992]. But in the axisymmetric deformation problem, this conclusion is not valid. This can be seen easily from the plot in Figure 8. In this figure,

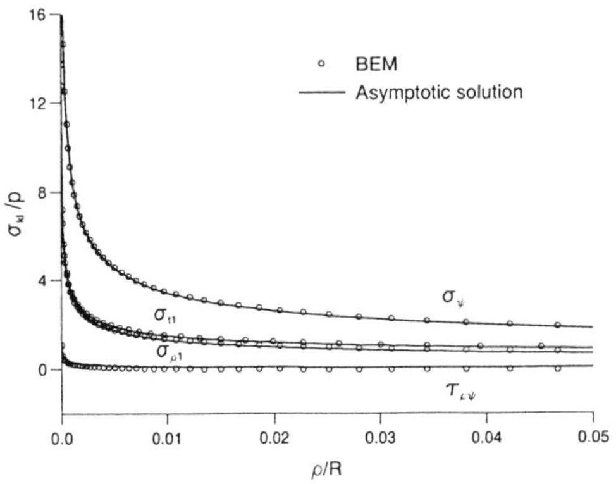

Figure 11 The stresses near the singular point along the interface in geometry C.

the plot of $\lg(\sigma_{t1}^{BE}) \sim \lg\left(\frac{\rho}{R}\right)$ is not parallel to the standard straight line with the slope of $-\omega$, where the superscript BE refers to the results calculated with BEM. Only the subtracted stresses $\sigma_{kl}^{BE} - \sigma_{kl}^{0}$ are of the singularity of ω. It also can be seen that the effect of σ_t^0 on σ_t is larger than the effect of σ_ψ^0 on σ_ψ.

Now, only the stress intensity factor K is unknown in the stress expression Eq.(23). It can be obtained by fitting the results of BEM with those of Eq.(23) using the least square method. The fitting points should be chosen in the range of straight line shown in Figure 8. For the geometries A, B and C, the stress intensity factor K is evaluated $\frac{K_A}{p} = 0.6487$, $\frac{K_B}{p} = 0.8539$ and $\frac{K_C}{p} = 0.5644$.

With the substitutions of K and the regular stress term in Eq.(23), the stresses in the vicinity of the singular point S can be calculated analytically. For comparison, the stresses calculated from Eq.(23) and BEM are drawn in Figures 9-11 along the interface near the singular point. It is easy to see that the stresses calculated from Eq.(23) are in good agreement with those obtained by BEM.

7 CONCLUSION

In this paper, the stress singularities near the free edge of the interface in an axisymmetric joint under 3-D axisymmetric deformation are analysed. The characteristic equation for determining the singular exponent ω and the associated stress and displacement angular functions are obtained. An asymptotic description is developed for the stress field in the vicinity of the singular point, which consists of one singular term and one regular term. The regular stress term is independent of the distance ρ from the singular point and depends on the r-direction displacement of the singular point. It is very different from the plane deformation problem, where the displacements at the singular point do not produce any stress. The numerical results show that the asymptotic description (23) can describe the singular stress field well.

REFERENCES

Li, Y. L., Hu, S. Y., Munz, D. and Yang, Y. Y., 1997, The asymptotic description of the singular stress field around the bond edge of a cylinder joint, to be published.

Munz, D. and Yang, Y. Y., 1992, Stress singularities at the interface in bonded dissimilar materials under mechanical and thermal loading, ASME Journal of Applied Mechanics, Vol. 59, pp. 857-861.

Ting, T. C. T., Jin Yijian and Chou, S. C., 1985, Eigenfunctions at a singular point in transversely isotropic materials under axisymmetric deformations, ASME Journal of Applied Mechanics, Vol. 52, pp. 565-570.

Zak, A. R., 1964, Stresses in the vicinity of boundary discontinuities in bodies of revolution, ASME Journal of Applied Mechanics, Vol. 31, pp. 150-152.

ACKNOWLEDGEMENT

The author Y. L. Li gratefully acknowledges the financial support provided by the Alexander von Humboldt Foundation.

Damage and Failure of Interfaces, Rossmanith (ed.) © 1997 Balkema, Rotterdam, ISBN 90 5410 899 1

Free edge effects at interfaces in selectively reinforced structures

C.M.Chimani, H.J.Böhm & F.G.Rammerstorfer
Institute of Light Weight Structures and Aerospace Engineering, Vienna University of Technology, Austria

C.Hausmann
Swiss Federal Laboratories for Material Testing and Research, Thun, Switzerland

ABSTRACT: At intersections between interface and free edges of multi material structures complex tri-axial stress states occur which often are critical with respect to damage. In the following an axisymmetric component selectively reinforced with a metal matrix composite is analyzed under thermal loading, special consideration being given to the free edge effect. When the stress fields are studied in terms of a bimaterial junction problem using homogenized material descriptions singular stress fields are typically predicted. An analytical method was utilized to improve the interface design with respect to the free edge effect. It employs the Airy stress function to set up the corresponding boundary value problem, which is solved *via* the Mellin transform.
Using a micromechanical approach it is then shown, that under certain conditions these theoretically derived stress singularities disappear when the inhomogeneous micro structure of a selectively reinforced component is accounted for explicitly.

1 INTRODUCTION

The common tendency in the automotive and aerospace industries to aim at reducing the weight of their vehicles has enhanced the use of light metal alloys. Especially magnesium alloys are increasingly used for load bearing components. Major restrictions to these applications are due to the relatively low stiffness of magnesium alloys and their tendency to creep at moderate temperatures. To overcome these weaknesses such structures can be designed and produced with particle or fiber reinforced regions that are able to withstand higher loads. Thus, locally the material properties are adjusted to the service conditions, and material costs are kept at a low level since reinforcing material is only applied to relatively small zones.
Numerical and experimental investigations of selectively reinforced axisymmetric components have indicated that the interfaces between the reinforced and the unreinforced material are their most critical regions with respect to failure. There are several reasons for this. On the one hand, at the interface the material strength is reduced because of the formation of oxide layers and intermetallic phases. On the other hand, as a consequence of the mismatch in the material properties thermal loading gives rise to eigenstresses of first and second order, e.g. stresses on the macro scale due the hybrid structure and stresses on the micro scale within the reinforced material. These stresses reach maximum values close to the material interface. In addition, the intersection points between a material interface and the free surface are locations of complicated tri-axial stress states which are the consequence of the step-like, i.e. discontinuous, variation of the mechanical material parameters.
The first part of the present contribution studies the possibility of reducing the stress concentrations by a proper geometrical design in the vicinity of the intersection between interface and free surface. The applied method is based on the analytical solution of the bimaterial wedge problem, special attention being given to the influence of the angle of inclination between the material interface and the free surface. The second part discusses the validity of the singular solution for the stress distributions close to such intersection points in hybrid components if the micro structure is accounted for explicitly.

2 REDUCTION OF THE FREE EDGE EFFECT BY PROPER INTERFACE DESIGN

In the following section possibilities are discussed for reducing stress concentrations at the material interface of two bonded axisymmetric rings. Figure 1 shows a cross-section of the structure considered in the following, which is a generic model of a selectively reinforced axisymmetric cast component. The inner ring, i.e. the insert, is made of light metal alloys reinforced by circumferentially aligned fibers. The dimensions of the insert are 50mm for the inner diameter and 60mm for the outer one, its height being 12mm. The matrix material of the insert is the magnesium alloy AZ91 and the fibers are T300 carbon fibers with a volume fraction of $\zeta = 0.5$. The material parameters are taken from (Aune & Westengen, 1995) and (Weeton et al., 1987), respectively. The outer part of the structure is made of a homogeneous casting alloy, AZ91 magnesium, without any reinforcement. Within section 2 of the paper additional material combinations are considered, e.g. an Altex/Al metal matrix composite (MMC) surrounded by Al and an Altex/Al MMC surrounded by AZ91. In section 3 the only material combination studied is T300/AZ91–AZ91.

The homogenized material parameters of the composite material are obtained *via* the Mori–Tanaka method, see (Mori & Tanaka, 1973). Therefore, within the structural analysis both parts, i.e. the MMC insert and the monolithic ring, are treated as having homogenized material properties. Assuming two dissimilar homogeneous materials we have to expect singular stress fields at the intersection points A and B marked in Figure 1 (possible singularities in the vicinity of the lower right corner of the insert are not discussed here). Accordingly, arbitrary loading conditions lead to a stress state close to the intersection points which can be described by the equation

$$\sigma_{ij}(r,\theta) = K r^{-\lambda} F_{ij}(\theta). \quad (1)$$

Here, polar coordinates (r, θ) are used and the origin of the coordinate system is taken to be situated at the intersection of the interface with the free surface, i.e. at point S in Figure 10. K is the stress intensity factor, F_{ij} is a function of θ only, and the order of the singularity is determined by the exponent λ. Parts of the solution for this singular stress field can be found using the analytical techniques developed by (Hein & Erdogan, 1971) and (Bogy, 1971). An analytical method for determining stress singularities accounting for anisotropic materials was presented in (Mikhailov, 1979). The full solution for finite sized components can be obtained by using a combined analytical–numerical approach, see (Yang, 1992). Comprehensive reviews of numerical techniques are given in (Withcomb et al., 1982) and (Gu & Reddy, 1992).

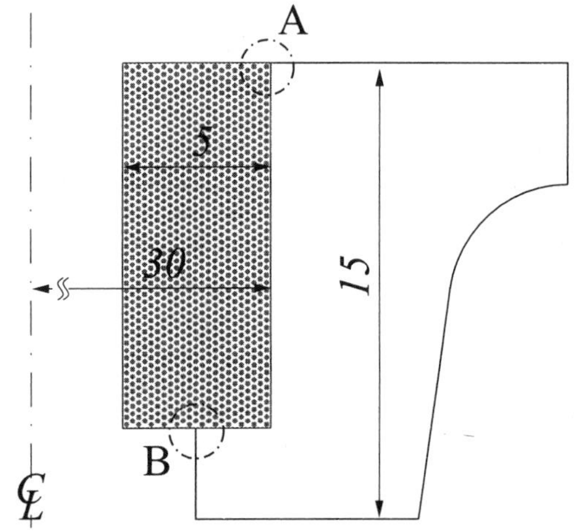

Figure 1: Cross section of a selectively reinforced ring. Regions where singular stress fields may occur are marked as A and B

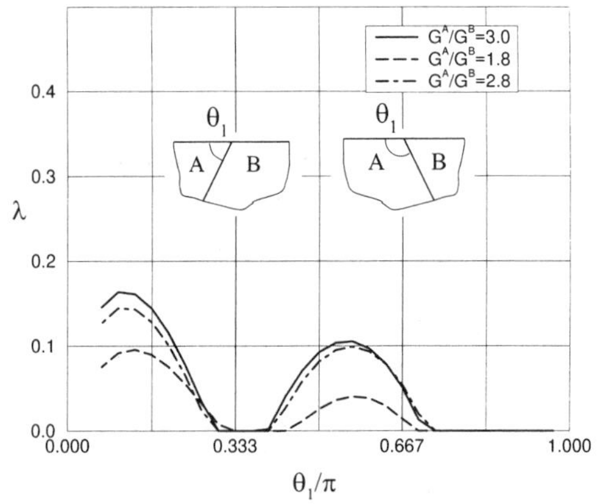

Figure 2: Variation of the order of singularity λ for different inclination angles θ_1 at intersection point A, keeping $\theta_1 + \theta_2 = \pi$

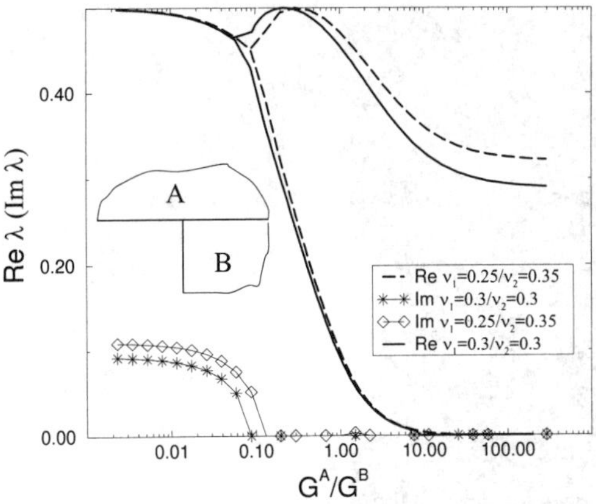

Figure 3: The order of singularity λ for a bimaterial wedge with $\theta_1 = \pi$, $\theta_2 = \pi/2$, as a function of the elastic contrast G^A/G^B

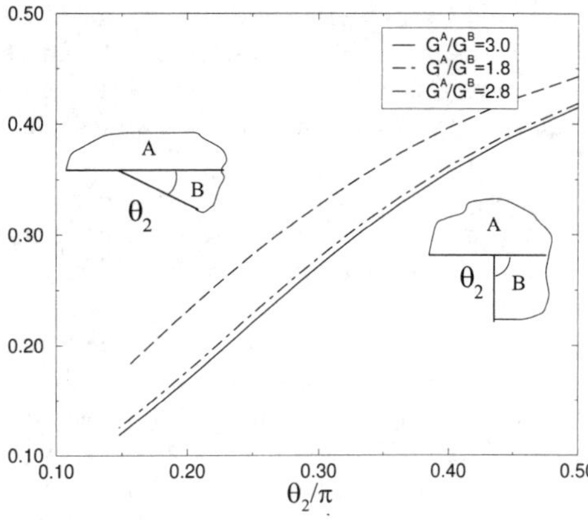

Figure 4: Variation of the order of singularity λ for different inclination angles θ_2 ($\theta_1 = \pi$, and θ_2 varies from 0 to $\pi/2$)

Details of the analytical treatment employed within the present study are given in the Appendix. Using the simplifications that the materials are homogeneous and isotropic on both sides of the interface and that a plane strain stress state is present, the order of the singularity λ can be found by determining the eigenvalues of the matrix $\mathbf{K}$ in eqn. (11) for each individual material combination and interface geometry. The resulting solutions are used to arrive at a proper interface design which minimizes the order of the stress singularity at the intersection points A and B defined in Figure 1. A complementary numerical treatment of the singular stress fields in selectively reinforced axisymmetric structures is given in (Chimani et al., 1997b). Within that study the above simplifications are not used and the overall anisotropy of the composite material is accounted for explicitly.

Structural analyses of hybrid rings under thermal and mechanical loading indicated that intersection point A is the more critical one, the stress levels in its vicinity being much higher than those near intersection point B. Accordingly, we first discuss intersection point A, for which the the order of the singularity is plotted in Figure 2 as a function of the angle θ_1 defined in Figure 10. The angle of inclination of the interface is varied while keeping $\theta_1 + \theta_2 = \pi$ and three different elastic contrasts are considered to account for different material combinations, $G^A/G^B = 3$ corresponding to the material combination T300/AZ91–AZ91, $G^A/G^B = 1.8$ to Altex/Al–Al, and $G^A/G^B = 2.8$ to Altex/Al–AZ91, respectively.

The qualitative behavior of the order of singularity is predicted to be equal for all material combinations. The curves show two maxima for λ, one at $\theta_1 \approx 0.1\pi$ and another at $\theta_1 \approx 0.55\pi$. For $\theta_1 \approx \pi/3$ and for $\theta_1 \geq 0.7\pi$ one obtains $\lambda = 0$, i.e. no singular stress field is found. This demonstrates the possibility of reducing or avoiding the singular free edge effect by a proper interface design.

For the hybrid structure structure sketched in Figure 1 this means that an inclination of the interface of more than 40° from the axial direction should avoid stress singularities in the vicinity of point A. For material combinations producing a weak interface between the inner ring and the outer casting such an inclined interface has the additional advantage of acting as geometrical support for the MMC insert.

A global inspection of the results of structural analyses of hybrid rings gives the impression that the stress concentration at point B is rather weak. However, a more detailed consideration of bimaterial wedge configurations with $\theta_1 = \pi$ and $\theta_2 = \pi/2$ indicates that this point may be critical, too, since the order of the singularity is high and approaches the values reached at crack tips. This is evident from Figure 3, where λ is close to 0.5 over a wide range of elastic contrasts. For stiffness ratios smaller than 0.1, λ becomes complex and the value of the real part is about 0.5. Close to

$G^A/G^B = 0.1$ a bifurcation occurs and for higher elastic contrasts two eigenvalues are found, so that the stress field has to be described using two singular terms. Analogous solutions were given in (Hein & Erdogan, 1971).

In Figure 4 the order of the singularity λ is plotted as a function of the wedge angle θ_2, with θ_1 kept constant at a value of $\theta_1 = \pi$. Only the first eigenvalue is displayed which is the more important one for the singular stress field close to the intersection point. Again, we find that by decreasing the angle θ_2 the order of the singularity can be reduced. Thus, a smooth transition, which could be realized by a radius instead of the sharp corner, would avoid most of free edge effect in this intersection region. It is also interesting to note that the order of singularity λ is higher for smaller elastic contrasts, an analogous behavior being evident in Figure 3. For elastic contrasts larger than 1 the slope of the first eigenvalue is negative, i.e. the order of singularity decreases with increasing elastic contrast.

3 MICROMECHANICAL STUDY OF THE SINGULAR STRESS FIELD UNDER THERMAL LOADING

The analytical investigations presented in the previous section point out ways for reducing the free edge effect in the given axisymmetric hybrid structure (Figure 1). In earlier studies, e.g. (Chimani et al., 1997a), it was demonstrated that the singular solution of the stress field close to the intersection point of the free surface and the material interface calculated on the basis of homogenized material properties is not always valid, since in the case of micro structured materials the gradients of the stress fields and the micro structural gradients are much too large to allow the use of a homogenized material description. This effect is even more pronounced if the material transition is smeared out at the micro scale and a distinct interface no longer exists, as is the case for the material combinations covered here.

In Figures 5 and 6 metallographic sections of the transition zone from carbon fiber reinforced magnesium to pure magnesium alloy are shown. These REM pictures represent cross-sectional views perpendicular to the fiber direction. The "dotted" dark zone on the left side corresponds to the car-

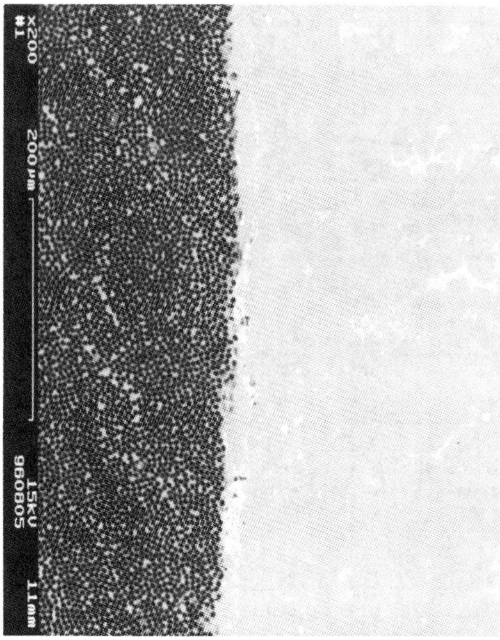

Figure 5: Cross-section of the macro interface perpendicular to the fiber direction obtained by REM with a magnification of 200

bon/magnesium composite, whereas at the right pure magnesium alloy is visible. The very light areas are intermetallic $Mg_{17}Al_{12}$ precipitates. Although on the structural level a sharp interface would be expected, it is obvious that there is no distinct interface at the micro scale.

From the analytical considerations we know that the order of the singularity at the intersection point A in Figure 1 is rather low. Thus, the region affected by the singular stress field has a length scale which is comparable to or even smaller than the scale of the micro structure as given e.g. by the mean fiber distance.

A combined macro–micromechanical finite element based approach as introduced in (Chimani et al., 1997a) is used for an assessment of the stresses close to the intersection region. An axisymmetric micro scale submodel is centered around intersection point A. It has the dimensions of 0.2mm in height, 0.4mm in width, the outer diameter of the insert being 60mm, see Figure 7. These dimensions are chosen to be sufficiently large to cover the region affected by the singular stress field derived from the macro model.

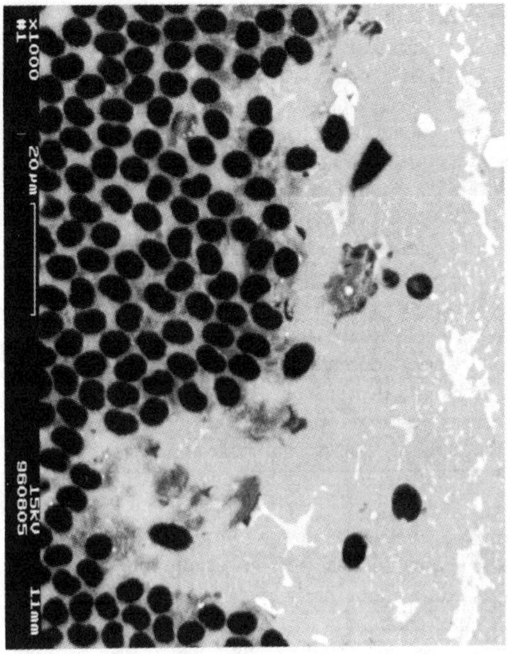

Figure 6: Detail cross-section of the macro interface perpendicular to the fiber direction obtained by REM with a magnification of 1000

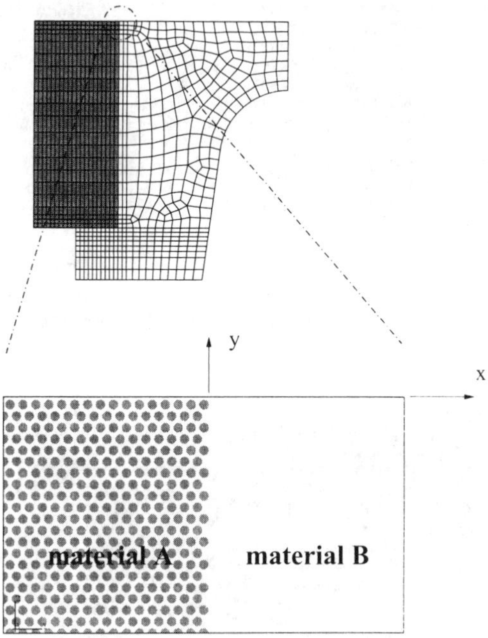

Figure 7: Micromechanically based submodel for local investigations of the region around intersection point A

Within the submodel the heterogeneous microstructure of the MMC is modeled explicitly. The diameter of the fibers is chosen as 10μm, which is comparable to their actual size, see Figure 6, and for simplicity a periodic hexagonal arrangement of fibers is used within the insert. The model micro geometry is generated by a hexagonal cell tiling approach (i.e. the computational domain is split into regular hexagons that are assigned to either fibers or matrix and subsequently modified to obtain the required fiber volume fraction and cross section) and discretized with 6-noded triangular elements, compare (Chimani et al., 1997a). The global structure was loaded by a homogeneous temperature change of $\Delta T = +10$K and the boundary conditions for the submodel are derived from this global analysis, i.e. displacements obtained from the macroscopic model are prescribed at its left, right, and bottom boundaries, and the same thermal loads are applied.

The results obtained with the submodel display no marked stress concentrations at the intersection between the macro interface and the free surface, as can be seen in Figure 8, a fringe plot of the radial stress component, σ_{xx}, and in Figure 9, which displays the axial stress component, σ_{zz}. Both stress components increase as the free surface is approached from the interior of the hybrid ring, but they actually decrease near the surface close to the macro interface. These predictions can be seen as evidence that in the immediate vicinity of the interface the micro stresses within the constituents of the MMC are relieved due to the presence of the monolithic material in the homogeneous region. In addition, these stress distributions strongly indicate that in the vicinity of the junction between MMC and homogeneous metal the conditions for using homogenized material properties, which lead to the prediction of a stress singularity, are not met. Support for the above predictions comes from a series of experiments, in which special test specimens consisting of a cylindrical MMC insert embedded in an axisymmetric monolithic casting were produced for a structural push out test in order to characterize the strength of the macro interface. Investigations of the fracture surfaces after push out showed composite material on both

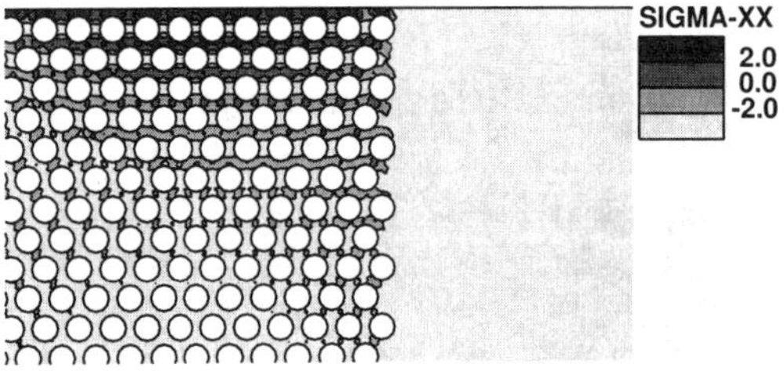

Figure 8: Detail fringe plot of the predicted distribution of the radial stresses σ_{xx} in the vicinity of the intersection between the macro interface and the free surface.

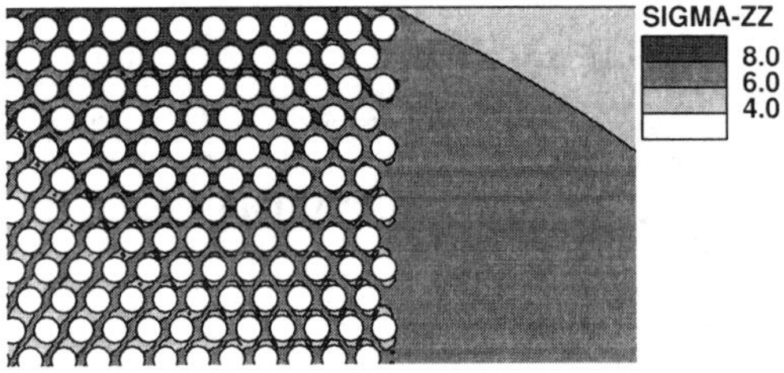

Figure 9: Detail fringe plot of the predicted distribution of the axial stresses σ_{zz} in the vicinity of the intersection between the macro interface and the free surface.

sides of the fracture surface, which indicates that damage initiated within the MMC a few fiber diameters away from the interface.

The prediction that the maximum stresses occur at a distance of a few fiber diameters away from the interface also is in good agreement with results given in (Buryachenko & Rammerstorfer, 1997). In that contribution the effective elastic properties within and around a cluster of inclusions embedded in an infinite matrix are studied using the nonlocal version of the multiparticle effective field method. It is found that the effective elastic tensor varies continuously from the cluster to the matrix within a boundary layer having a width of a few inclusion diameters. This is a further theoretical proof that no singularity occurs in configurations of the type discussed here.

4 CONCLUSION

In the first part of the contribution an analytical treatment of the bimaterial wedge problem was utilized for improving the geometrical interface design with respect to free edge effects. Assuming an effective homogeneous material for both components meeting at the interface, singular stress and strain fields were predicted at the intersection points of the material interface and the free surface. It was demonstrated that the singularity can be reduced or even avoided by a proper design of the interface, thus mitigating free edge effects.

In the second part a combined macro- and micromechanical study of a ring shaped hybrid structure under thermal loading was presented. The results demonstrate that the free edge singularities predicted at interfaces between dissimilar homogeneous elastic materials may not be evident if one of them shows a matrix–inclusion micro structure. At a size scale comparable to that of the micro structure the simple homogenized material description for the inhomogeneous material fails, since major assumptions of the mean field theory are not fulfilled. Stresses predicted using homogenized material models tend to overestimate the free edge effect in hybrid components, and a proper strength assessment can only be performed by explicitly including the micro structure into the models.

ACKNOWLEDGMENT: Parts of the present work were performed within the BRITE EURAM Project BE'95–1183.

REFERENCES

Aune, T. K. & H. Westengen (1995). Magnesium die casting properties. *Automotive Engineering* 87–92.

Bogy, D.B. (1971). Two edge-bonded elastic wedges of different materials and wedge angles under surface tractions. *J. Appl. Mech.* 38:377–386.

Buryachenko, V. & F.G. Rammerstorfer (1997). Micromechanics and nonlocal effects in graded random structure matrix composites. To appear in *Proceedings of the IUTAM Symposium on Transformation Problems in Composite and Active Materials.*

Chimani, C. M., Böhm, H. J. & F.G. Rammerstorfer (1997a). On the stress singularity at free edges of bimaterial junctions — a micromechanical study. *Scr. Mat.*, 36(8):943–947.

Chimani, C., Böhm, H. J. & F.G. Rammerstorfer (1997b). A micromechanical investigation of thermally introduced stress singularities in heterogeneous hybrid structures. To appear in *Proceedings of the International Conference on Residual Stresses 5.*

Gu, Q. & J.N. Reddy (1992). Non-linear analysis of free edge effects in composite laminates subjected to axial loads. *Int. J. Non-Linear Mech.*, 27:27–41.

Hein, V. L. & F. Erdogan (1971). Stress singularities in a two-material wedge. *J. Fract. Mech.*, 7:317–330.

Ledermann, W. & S. Vajda (1982). *Handbook of Applicable Mathematics*, Vol. IV: Analysis. Chichester: John Wiley & Sons.

Mikhailov, S.E. (1979). Stress singularity in a compound arbitrarily anisotropic body and applications to composites. *Izv. AN SSSR. Mekhanika Tverdogo Tela (Mechanics of Solids)*, 14:33–42.

Mori, T. & K. Tanaka (1973). Average stress in the matrix and average elastic energy of materials with misfitting inclusions. *Acta Metall.*, 21:571–574.

Weeton, J.W., Peters, D.M. & K.L. Thomas (1987). *Engineer's guide to composite materials.* Metals Park: American Society for Metals.

Withcomb, J.D., Raju, I.S. & J.G. Goree (1982). Reliability of the finite element method for calculating free edge stresses in composite laminates. *Comput. Struct.*, 15:23–37.

Yang, Y.Y. (1992). *Spannungssingularitäten in Zweistoffverbunden bei mechanischer und thermischer Belastung.* PhD thesis, Universität Karlsruhe.

APPENDIX

In the following a brief description of an analytical solution method is presented which is capable of solving the 2D boundary value problem of two bonded dissimilar homogeneous wedges. Using the assumption of isotropic linear elastic material behavior the problem of two bonded dissimilar wedges for configurations of the type given in Figure 10 can be treated in the following way, compare (Bogy, 1971) and (Hein & Erdogan, 1971). The stress and displacement fields in polar coordinates are obtained by solving for the Airy stress function $\phi(r,\theta)$ which satisfies the equation

$$\nabla^4\phi = 0 \qquad (2)$$

$$\text{with} \quad \nabla^2 = \frac{\partial^2}{\partial r^2} + \frac{1}{r}\frac{\partial}{\partial r} + \frac{1}{r^2}\frac{\partial^2}{\partial \theta^2}.$$

The stresses and displacements in polar coordinates are related to ϕ by

$$\sigma_{rr} = \frac{1}{r}\frac{\partial\phi}{\partial r} + \frac{1}{r^2}\frac{\partial^2\phi}{\partial\theta^2}, \qquad \sigma_{\theta\theta} = \frac{\partial^2\phi}{\partial r^2},$$

$$\sigma_{r\theta} = \frac{\partial^2\phi}{\partial r\partial\theta} + \frac{1}{r^2}\frac{\partial\phi}{\partial\theta}, \qquad (3)$$

and

$$\frac{\partial u_r}{\partial r} = \frac{1}{2G}\left[\frac{1}{r}\frac{\partial\phi}{\partial r} + \frac{1}{r^2}\frac{\partial^2\phi}{\partial\theta^2} - (1-m/4)\nabla^2\phi\right],$$

$$\frac{\partial u_\theta}{\partial r} - \frac{u_\theta}{r} + \frac{1}{r}\frac{\partial u_r}{\partial\theta} = \frac{1}{G}\left(-\frac{1}{r}\frac{\partial^2\phi}{\partial r\partial\theta} + \frac{1}{r^2}\frac{\partial\phi}{\partial\theta}\right), \qquad (4)$$

respectively, where G is the shear modulus corresponding to domains A or B (compare Figure 10), and m takes the values of $m = 4(1-\nu)$ for plane strain or $m = 4(1+\nu)^{-1}$ for plane stress analyses. The stress and displacement components have to satisfy the boundary conditions

$$\sigma^A_{\theta\theta}(r,-\theta_1) = N^A(r), \quad \sigma^A_{r\theta}(r,-\theta_1) = T^A(r),$$

$$\sigma^B_{\theta\theta}(r,\theta_2) = N^B(r), \quad \sigma^B_{r\theta}(r,\theta_2) = T^B(r), \qquad (5)$$

as well as the bonding conditions at the interface (continuity of tractions and displacements)

$$\sigma^A_{\theta\theta}(r,0) = \sigma^B_{\theta\theta}(r,0), \quad \sigma^A_{r\theta}(r,0) = \sigma^B_{r\theta}(r,0),$$

$$u^A_r(r,0) = u^B_r(r,0), \quad u^A_\theta(r,0) = u^B_\theta(r,0). \qquad (6)$$

Here $N^A(r), T^A(r), N^B(r)$, and $T^B(r)$ stand for the normal and tangential components of the tractions at the surfaces of materials A and B, respectively. The above singular boundary value problems are solved *via* the Mellin transform.

The Mellin transform of a function $\phi(r,\theta)$ has the following definition (Ledermann & Vajda, 1982):

$$\hat{\phi}(s,\theta) = \int_0^\infty \phi(r,\theta) r^{(s-1)} dr, \qquad (7)$$

where s is the complex transform parameter, and

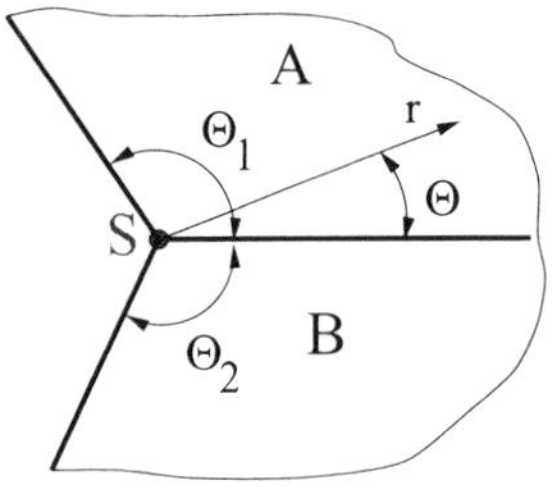

Figure 10: Sketch of the generic bimaterial wedge problem

the inverse Mellin transform is defined as

$$\phi(r,\theta) = \frac{1}{2\pi i}\int_{y-i\infty}^{y+i\infty} \hat{\phi}(s,\theta)r^{-s}ds \quad , \tag{8}$$

where y determines the path of the integration in the complex line integral. Of course, y must be chosen in such a way that the integral exists.
The application of the Mellin transform to eqn. (2) leads to an ordinary differential equation for $\hat{\phi}(s,\theta)$ of the form

$$(D^2 + s^2)[D^2 + (s+2)^2]\hat{\phi}(s,\theta) = 0, \tag{9}$$

with $\quad D = \dfrac{\partial}{\partial\theta},$

which holds for both wedges A and B. A general solution for this equation in the transformed notation takes the form

$$\hat{\phi}(s,\theta) = a(s)\sin(s\theta) + b(s)\cos(s\theta) + c(s)\sin(s\theta+2\theta) + d(s)\cos(s\theta+2\theta) \tag{10}$$

within materials A and B. Here $a(s)^A$, ... $d(s)^A$ for $\hat{\phi}(s,\theta)^A$ and $a(s)^B$, ... $d(s)^B$ for $\hat{\phi}(s,\theta)^B$ have to be determined through the transformation of eqn. (3) and from the transformed boundary and bonding conditions, eqns. (5) and (6). Now we end up with a system of eight equations and eight unknown coefficients $a(s)^A$, ... $d(s)^A$ and $a(s)^B$, ... $d(s)^B$, that depend on the transform parameter s. In matrix notation this system can be written as

$$\mathbf{K}\,\mathbf{x} = \mathbf{d}. \tag{11}$$

When the Mellin transform is applied to eqn. (3), back transforming of the solution $\hat{\sigma}_{ij}$ *via* eqn. (8) gives rise to expressions of the type

$$\sigma_{ij} = \frac{1}{2\pi i}\int_{y-i\infty}^{y+i\infty} \hat{\sigma}_{ij}(s,\theta)r^{-(s+2)}ds \tag{12}$$

for the stress components. Comparison with eqn. (1) shows that the order of singularity λ must be related to the transform parameter s by $\lambda = s+2$. Furthermore, it can be shown that the solution of eqn. (8) is analytic in the open strip $-2 \le \mathrm{Re}(s) \le -1$ except for poles that may occur at the zeros of the determinant of the matrix $\mathbf{K}$. Thus λ, the order of the singularity can be determined from the eigenvalues of the matrix $\mathbf{K}$ defined in eqn. (11).
The solutions for $s > -1$ are unphysical, since the displacements at $r = 0$ would be infinite. From eqn. (7) it is obvious that one has to distinguish between two types of zeros, first zeros lying within the interval $-2 \le Re(s_1) \le -1$ and second zeros $s_2 = -2$. The latter ones correspond to $\lambda = 0$, and thus to a constant stress term σ_0, see (Yang, 1992). For the first case we come up with a singular stress field of the type

$$\sigma_{ij}(r,\theta) = Kr^{-\lambda}F_{ij}(\theta) \tag{13}$$

if $\quad \lambda \in \mathbb{R}, \quad$ or

$$\sigma_{ij}(r,\theta) = Kr^{-\zeta}\Big(\cos\big(\eta\log r F^c_{ij}(\theta)\big) + \sin\big(\eta\log r F^s_{ij}(\theta)\big)\Big) \tag{14}$$

if $\quad \lambda = \zeta + i\eta \in \mathbb{C}.$

The global solution for the stress and displacement fields can be found by applying the inverse Mellin transform to the corresponding equation in the transformed notation. This can be done either numerically or analytically by using the residual principle (Ledermann & Vajda, 1982). For the present study the primary interest was to find the geometrical influence of the angles (θ_1, θ_2) on the singular behavior. Therefore, the discussion is restricted to the determination of λ.

Damage and Failure of Interfaces, Rossmanith (ed.) © *1997 Balkema, Rotterdam, ISBN 90 5410 899 1*

Time dependent boundary element analysis of singular stresses at the interface corner of viscoelastic adhesive layer

Sang Soon Lee
Reactor Mechanical Engineering Group, Korea Power Engineering Company, Taejon, South Korea

ABSTRACT: The purpose of this paper is to investigate the stress singularity at the interface corner between the adhesive layer and the rigid adherend subjected to a uniform transverse tensile strain. The adhesive is assumed to be a linear viscoelastic material. The standard Laplace transform technique is employed to get the characteristic equation and the order of the singularity is obtained numerically for given viscoelastic models. The time-domain boundary element method is used to investigate the behavior of stresses for the whole interface. For the viscoelastic models considered, it is shown that the free-edge stress intensity factors are relaxed with time while the order of the singularity increases with time .

1. INTRODUCTION

The problem of bonded quarter planes consisting of two isotropic and elastic materials has received much attention [Bogy 1968, Reedy 1990, Tsai and Morton 1991]. It was shown by Bogy(1968) that a stress singularity of type r^{δ} exists at the interface corner between bonded elastic quarter planes and δ is the solution of a characteristic equation. The order of the singularity at the free edge of the interface is dependent on the elastic constants of the two materials.

The interface of adhesively bonded materials would suffer from a stress system in the vicinity of the free surface under a transverse tensile loading. In such a region two interacting free surface effects occur, and very large interface stresses can be produced. A stress singularity which exists at the interface corner between the adherend and the adhesive layer might lead to adherend-adhesive debonding[Lee[a] 1996, Lee[b] 1996]. In this study, the stress singularity at the interface corner between the rigid adherend and the viscoelastic adhesive subjected to a uniform transverse tensile strain is investigated. At room temperature the adhesive remains in its initial glassy stage through the entire loading period and hence it is not necessary to consider the time-dependent behavior of the stress-strain relationships in performing the stress analysis of bonded materials.

In certain application, however, the temperature and time dependence of the loading may be such that the rheological behavior of the adhesive materials may no longer be negligible. Hence, provision is made for the adhesive assumed to be a linear viscoelastic material.

The interface stresses in viscoelastic adhesive layers have been studied by several investigators. Weitsman(1979) considered a pair of interfaces in which the adhesive material is viscoelastic and the adherend is rigid. Delale and Erdogan(1981) analyzed an adhesively bonded lap joint by assuming that one material is elastic and the other is viscoelastic. The results exhibited a redistribution of the very large stresses near the edge of the interface, but no singularities were encountered because of the simplifying assumptions with regard to the modeling of joining structural members.

In this study, the transformed characteristic equation for perfectly bonded rigid adherend and viscoelastic adhesive materials is first derived, following Williams(1952), with the use of the Laplace transform with respect to time *t*. This equation is inverted analytically for given viscoelastic models into the time-dependent viscoelastic equation which is readily solved using standard numerical procedure. The time-domain boundary element method(BEM) is then employed to investigate the behavior of stresses at the interface of viscoelastic adhesive layers subjected to a uniform transverse tensile strain.

2. ORDER OF THE SINGULARITY AT THE INTERFACE CORNER

The region near the interface corner between perfectly bonded viscoelastic and rigid quarter planes is shown in Figure 1. In the following, a condition of plane strain is considered.

A solution of

$$\nabla^4 \Phi(r,\theta;t) = 0 \tag{1}$$

is to be found such that the normal stress, $\sigma_{\theta\theta}$, and shear stress, $\tau_{r\theta}$, vanish along $\theta = -\frac{\pi}{2}$, further that the displacements are zero across the common interface line $\theta = 0$. The solution of this problem is facilitated by the Laplace transform, defined as

$$\Phi^*(r,\theta;s) = \int_0^\infty \Phi(r,\theta;t)\, e^{-st} dt \tag{2}$$

where Φ^* denotes the Laplace transform of Φ and s is the transform parameter. Then eqn(1) can be rewritten using eqn(2) as follows:

$$\nabla^4 \Phi^*(r,\theta;s) = 0 \tag{3}$$

By definition, the viscoelastic stresses in the Laplace transformed space are found from the stress function Φ^* in the following manner:

$$\begin{aligned} \sigma^*_{rr} &= \frac{1}{r}\Phi^*_{,r} + \frac{1}{r^2}\Phi^*_{,\theta\theta} \\ \sigma^*_{\theta\theta} &= \Phi^*_{,rr} \\ \tau^*_{r\theta} &= \frac{1}{r^2}\Phi^*_{,\theta} - \frac{1}{r}\Phi^*_{,r\theta} \end{aligned} \tag{4}$$

and the strains can be shown to be given by

$$\begin{aligned} u^*_{r,r} &= \frac{1}{2s\mu^*(s)}\left[\frac{1}{r}\Phi^*_{,r} + \frac{1}{r^2}\Phi^*_{,\theta\theta} - s\nu^*(s)\nabla^2\Phi^*\right] \\ u^*_{\theta,r} - \frac{u^*_\theta}{r} + \frac{u^*_{r,\theta}}{r} &= \frac{1}{s\mu^*(s)}\left[\frac{1}{r^2}\Phi^*_{,\theta} - \frac{1}{r}\Phi^*_{,r\theta}\right] \end{aligned} \tag{5}$$

where σ_{ij}^* and u_i^* are the Laplace transformed stresses and displacements, respectively, and μ^* and ν^* are Laplace transforms of the shear relaxation modulus $\mu(t)$ and the viscoelastic Poisson's ratio $\nu(t)$. Combining these equations with the traction-free boundary conditions (at $\theta = -\frac{\pi}{2}$)

$$\sigma^*_{\theta\theta} = \tau^*_{r\theta} = 0 \tag{6}$$

and the interface conditions (at $\theta = 0$)

$$u^*_r = u^*_\theta = 0 \tag{7}$$

one can solve the problem.

Using a method similar to that described in Williams(1952), a stress function of the form

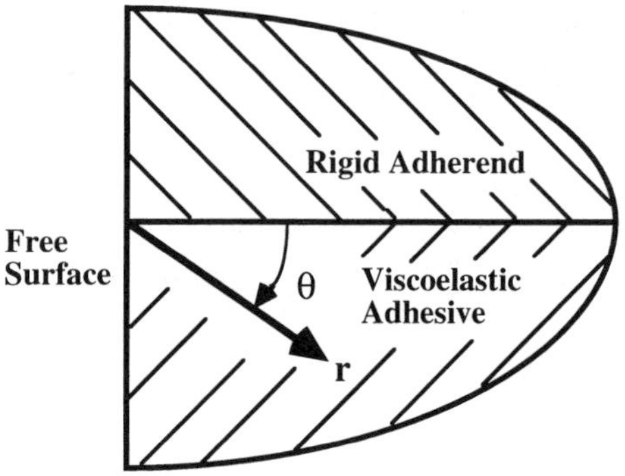

Figure 1. Region near interface corner between the viscoelastic adhesive layer and the rigid adherend.

$$\Phi^*(r,\theta;s) = r^{\lambda+1} f(\theta;s),\ -\pi/2 \le \theta \le 0,\ r > 0 \tag{8}$$

is assumed, where r and θ are defined in Figure 1. Typical solution for $f(\theta;s)$ is chosen of the form

$$\begin{aligned} f(\theta;s) = {} & c_1(s)\sin(\lambda+1)\theta + c_2(s)\cos(\lambda+1)\theta \\ & + c_3(s)\sin(\lambda-1)\theta + c_4(s)\cos(\lambda-1)\theta \end{aligned} \tag{9}$$

where c_i are arbitrary constants. Then, one can get the homogeneous system of four equations. A nontrivial solution to the equation exists only if the determinant of the coefficient matrix vanishes. This occurs when λ satisfies the following equation

$$\begin{aligned} & \frac{2\lambda^2}{s} - 8s[\nu^*(s)]^2 + 12\nu^*(s) - \frac{5}{s} \\ & - \left[\frac{3}{s} - 4\nu^*(s)\right]\cos(\lambda\pi) = 0 \end{aligned} \tag{10}$$

The time-dependent behavior of the problem is recovered by inverting eqn(10) into the real time space.

In order to examine the viscoelastic behavior at the interface corner of a viscoelastic adhesive layer bonded between rigid adherends, the viscoelastic model characterized by a standard solid shear relaxation modulus and a constant bulk modulus is taken as follows:

$$\begin{aligned} \mu(t) &= g_o + g_1 \exp\left(-\frac{t}{t^*}\right) \\ k(t) &= k_o \end{aligned} \tag{11}$$

where $\mu(t)$ is a shear relaxation modulus, $k(t)$ is a bulk modulus, g_o, g_1 and k_o are positive constants, and t^* is the relaxation time. Clearly,

$$\begin{aligned} \mu(0) &= g_o + g_1 \\ \mu(\infty) &= g_o < \mu(0) \end{aligned} \tag{12}$$

Introducing eqn(11) into (10) and rearranging the resulting equation, we have

$$\frac{2\lambda^2}{s} - 8\,A_1{}^*(s) + 12\,A_2{}^*(s) - \frac{5}{s} - \left[\frac{3}{s} - 4A_2{}^*(s)\right]\cos(\lambda\pi) = 0 \qquad (13)$$

where

$$A_1{}^*(s) = \frac{1}{4}\left[\frac{3k_o - 2\mu(0)}{3k_o + \mu(0)}\right]^2 \frac{\left[s + \dfrac{3k_o - 2\mu(\infty)}{3k_o - 2\mu(0)}\,\dfrac{1}{t^*}\right]^2}{s\left[s + \dfrac{3k_o + \mu(\infty)}{3k\;\; + \mu(0)}\,\dfrac{1}{t^*}\right]^2}$$

$$A_2{}^*(s) = \frac{[3k_o - 2\mu(0)]}{2[3k_o + \mu(0)]}\,\frac{\left[s + \dfrac{3k_o - 2\mu(\infty)}{3k_o - 2\mu(0)}\,\dfrac{1}{t^*}\right]}{s\left[s + \dfrac{3k_o + \mu(\infty)}{3k\;\; + \mu(0)}\,\dfrac{1}{t^*}\right]} \qquad (14)$$

Eqn(13) can be inverted analytically as follows:

$$2\lambda^2 - 8\,A_1(t) + 12\,A_2(t) - 5 - [3 - 4A_2(t)]\cos(\lambda\pi) = 0 \qquad (15)$$

where

$$A_1(t) = \frac{1}{4}\left[\frac{3k_o - 2\mu(0)}{3k_o + \mu(0)}\right]^2 \left[\beta_1{}^2 + \left(1 - \beta_1{}^2 + \beta_2\frac{t}{*}\right)\exp\left(-\gamma\frac{t}{*}\right)\right]$$

$$A_2(t) = \frac{[3k_o - 2\mu(0)]}{2\,[3k_o + \mu(0)]}\left[\beta_1 + (1 - \beta_1)\;\exp\left(-\gamma\frac{t}{t^*}\right)\right] \qquad (16)$$

and

$$\beta_1 = \frac{[3k_o + \mu(0)]\,[3k_o - 2\mu(\infty)]}{[3k_o + \mu(\infty)]\,[3k_o - 2\mu(0)]}$$

$$\beta_2 = 2\,\frac{3k_o - 2\mu(\infty)}{3k_o - 2\mu(0)} - \frac{3k_o + \mu(\infty)}{3k_o + \mu(0)} - \frac{3k_o + \mu(0)}{3k\;\; + \mu(\infty)}\left[\frac{3k_o - 2\mu(\infty)}{3k\;\; - 2\mu(0)}\right]^2$$

$$\gamma = \frac{3k_o + \mu(\infty)}{3k_o + \mu(0)} \qquad (17)$$

It can be easily verified that eqn(15) for $t = 0$ and $t \rightarrow \infty$ is written as follows;

for $t = 0$,

$$2\lambda^2 - 8\,[\nu(0)]^2 + 12\,\nu(0) - 5 - [3 - 4\nu(0)]\cos(\lambda\pi) = 0 \qquad (18)$$

for $t \rightarrow \infty$,

$$2\lambda^2 - 8\,[\nu(\infty)]^2 + 12\,\nu(\infty) - 5 - [3 - 4\nu(\infty)]\cos(\lambda\pi) = 0 \qquad (19)$$

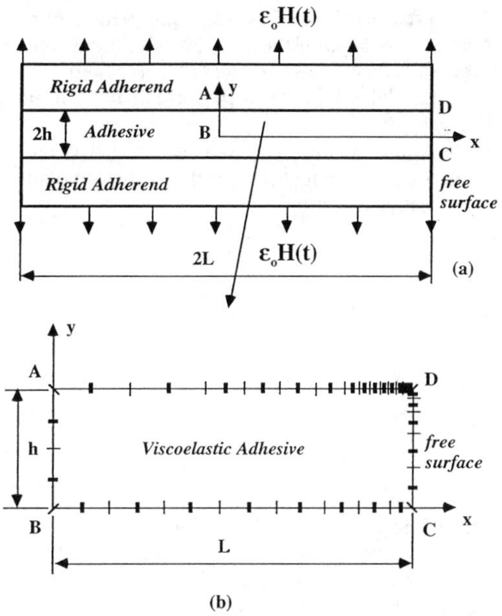

Figure 2. Analysis Model for Determination of Interface Stresses Developed in the Viscoelastic Adhesive Layer

Eqn(18) and (19) have a form identical with that of an elastic adhesive layer bonded between rigid adherends and are equivalent to that reported by Bogy(1968). The singularity at the interface corner has a form of $r^{1-\lambda}$. Roots of eqn(15) with $0 < \mathrm{Re}(\lambda) < 1$ are of main interest. The calculation of the zeros of eqn(15) must be carried out numerically for given values of material properties. For $0 < \nu(t) < 0.5$, there is at most one root λ_1 with $0 < \mathrm{Re}(\lambda) < 1$, and that root is real. A more detailed discussion of the root of eqn(15) is presented in Bogy(1968).

3. BOUNDARY ELEMENT SOLUTION FOR THE INTENSITY OF STRESS SINGULARITY

Figure 2-(a) shows an idealized configuration that models a viscoelastic adhesive layer bonded between rigid adherends , subjected to a uniform transverse tensile strain $\varepsilon_o H(t)$. Here $H(t)$ represents Heaviside unit step function. The adhesive and the adherend are considered to be perfectly bonded, with no defects or cracks. The layer has thickness $2h$, and length $2L$. Due to the symmetry, only one quarter of the layer needs to be modeled. Figure

2-(b) represents the two-dimensional plane strain model for analysis of the stresses which develop at the interface between the adhesive and the adherend. Calculations are performed for $L/h = 25$.

Assuming that no body forces exist, the boundary integral equations for the analysis model can be written as follows:

$$c_{ij}(\mathbf{y})\, u_j(\mathbf{y}, t)$$

$$+ \int_S \left[u_j(\mathbf{y}',t) T_{ij}(\mathbf{y},\mathbf{y}';0+) + \int_{0+}^{t} u_j(\mathbf{y}',t-t') \frac{\partial T_{ij}(\mathbf{y},\mathbf{y}';t')}{\partial t'} dt' \right] dS(\mathbf{y}')$$

$$= \int_S \left[t_j(\mathbf{y}', t) U_{ij}(\mathbf{y},\mathbf{y}';0+) + \int_{0+}^{t} t_j(\mathbf{y}', t-t') \frac{\partial U_{ij}(\mathbf{y},\mathbf{y}';t')}{\partial t'} dt' \right] dS(\mathbf{y}') \tag{20}$$

where u_j and t_j denote displacements and tractions, and S is the boundary of the given domain. The arguments *(y,t)* imply that the variables are dependent upon both the position $\mathbf{y}$ and the time t. $c_{ij}(\mathbf{y})$ is dependent only upon the local geometry of the boundary. For $\mathbf{y}$ on a smooth surface, the free-term $c_{ij}(\mathbf{y})$ is simply a diagonal matrix $0.5\ \delta_{ij}$. The viscoelastic fundamental solutions U_{ij} and T_{ij} can be obtained by applying the elastic-viscoelastic correspondence principle to the elastic fundamental solutions.

Closed-form integrations of eqn(20) are not, in general, possible and therefore numerical quadrature must be used. Approximations are required in both time and space. In this study, eqn(20) is solved in a step-by-step fashion in time by using the modified Simpson's rule for the time integrals and employing the standard BEM for the surface integrals. The detailed calculation procedure for eqn(20) is provided in Lee and Westmann(1995). The resulting system of equations is obtained in the matrix form as follows:

$$[\mathbf{H}]\{\mathbf{u}\} = [\mathbf{G}]\{\mathbf{t}\} + \{\mathbf{R}\} \tag{21}$$

In eqn(21), **H** and **G** are influence matrices and **R** is the hereditary effect due to the viscoelastic history. The above equation(21) can be solved by taking account of the external boundary conditions. The resulting boundary conditions for the analysis model are given as follows:

$$\begin{aligned}
&\tau_{xy} = 0, \quad u_x = 0 && \text{along A-B} \\
&\tau_{xy} = 0, \quad u_y = 0 && \text{along B-C} \\
&\tau_{xy} = 0, \quad \sigma_{xx} = 0 && \text{along C-D} \\
&u_x = 0, \quad u_y = h\,\varepsilon_o && \text{along D-A}
\end{aligned} \tag{22}$$

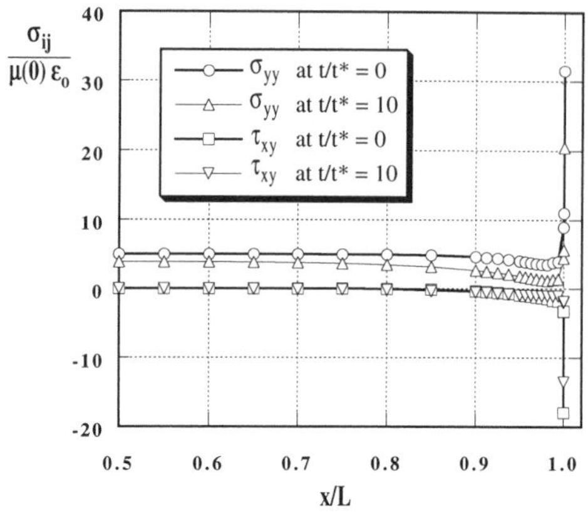

Figure 3. Distribution of interface normal and shear stresses at time t/t* = 0 and 10.

Applying the above boundary conditions to eqn(21) and solving the final system of equations at each time step lead to determination of all boundary displacements and tractions.

In order to examine the viscoelastic behavior along the interface line of the analysis model subjected to a transverse tensile strain $\varepsilon_o H(t)$, the viscoelastic model characterized by eqn(11) is employed. The numerical values used in this example are as follows:

$$\begin{aligned}
&\mu(0) = 0.55 \times 10^3 \quad MPa \\
&\mu(\infty) = 0.11 \times 10^3 \quad MPa \\
&k_o = 2.0 \times 10^3 \quad MPa \\
&t^* = 10 \quad \text{min} \\
&\varepsilon_o = 0.01
\end{aligned} \tag{23}$$

A suitable mesh density was determined for the analysis based upon a convergence study for mesh refinement. The refined mesh was used near the interface corner. The boundary element discretization consisting of 29 line elements was employed. In this study, quadratic shape functions were used to describe both the geometry and functional variations. Viscoelastic stress profiles were plotted along interface to investigate the nature of stresses. Figure 3 shows the distribution of normal stress σ_{yy} and shear stress τ_{xy} on the interface at nondimensional times $t/t^* = 0$ and 10. The numerical results exhibit the relaxation of interface stresses and large gradients are observed in the vicinity of the free surface.

The singular stress levels near the free-edge can be characterized by two parameters: the order of the singularity and the free-edge stress intensity

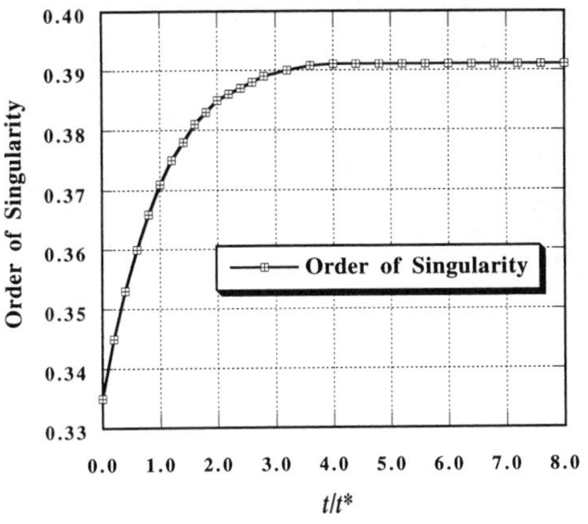

Figure 4. Variation of the order of the singularity

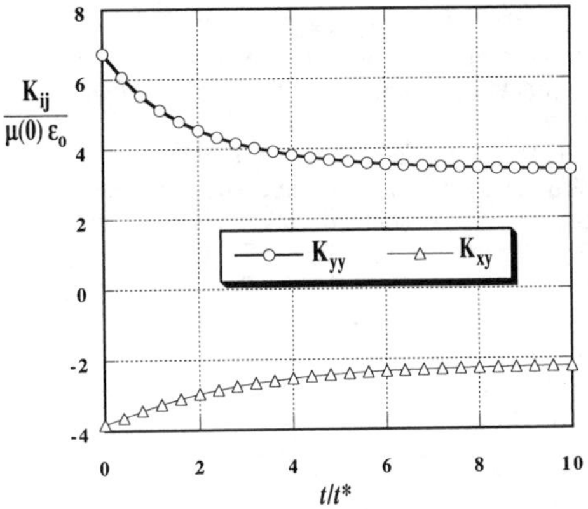

Figure 5. Variation of the free-edge stress intensity factors

factor. The order of the singularity must be determined from the roots of the characteristic equation(15). The free-edge stress intensity factor was defined first by Wang and Choi(1982). In this study, the free-edge intensity factor is normalized by the quantity $h^{1-\lambda}$, giving it stress units, as follows [Lee[c] 1997]:

$$K_{ij} = \lim_{r \to 0} \left(\frac{r}{h}\right)^{1-\lambda} \sigma_{ij}(r,0;t) \qquad (24)$$

Figure 4 shows the variation of the order of the singularity with time for the material properties given by eqn(23). Since the value of Poisson's ratio of the viscoelastic adhesive becomes greater with time, the order of the singularity increases with time. To check the accuracy of the results, it is interesting to consider an elastic case for the viscoelastic adhesive material with shear modulus $\mu(\infty)$; i.e., the viscoelastic adhesive layer of Figure 2-(b) is replaced by an elastic material with $\mu(\infty)$ and the rigid adherend remains unchanged. At greater times, the order of the singularity in Figure 4 approaches that for the analysis model consisting of elastic adhesive with $\mu(\infty)$ and rigid adherend. Figure 5 shows the variation of the free-edge stress intensity factor. It is shown that the free-edge stress intensity factor is relaxed with time while the order of the singularity increases with time. It is , however, unclear how these competing effects will effect failure or adhesive-adherend debonding.

4. CONCLUSIONS

The singular stresses at the interface corner between the viscoelastic adhesive layer and the rigid adherend subjected to a uniform transverse tensile strain have been investigated by using the time-domain boundary element method. Numerical results show that very large stress gradients are present at the interface corner and such stress singularity dominates a very small region relative to layer thickness. It is also shown that the free-edge stress intensity factor is relaxed with time while the order of the singularity increases with time for viscoelastic models considered here. Since the exceedingly large stresses at the interface corner can not be borne by adhesive layer, local yielding or adhesive-adherend debonding can occur in the vicinity of free surface.

REFERENCES

Bogy, D.B. 1968. Edge-bonded dissimilar orthogonal elastic wedges under normal and shear loading. *ASME Journal of Applied Mechanics*. 35: 460-466.

Delale, F. and Erdogan, F. 1981. Viscoelastic analysis of adhesively bonded joints. *ASME Journal of Applied Mechanics*. 48: 331-338.

Lee, S.S. and Westmann, R.A. 1995. Application of high-order quadrature rules to time-domain boundary element analysis of viscoelasticity. *Int. J. Numerical Methods in Engineering*. 38: 607-629.

Lee, S.S.[a] 1996. Boundary element evaluation of stress intensity factors for interface edge cracks in a unidirectional composite. *Engineering Fracture Mechanics*. 55: 1-6.

Lee, S.S.[b] 1996. Time-dependent boundary element analysis for an interface crack in a two-dimensional unidirectional viscoelastic model composite. *Int. Journal of Fracture*. 77: 15-28.

Lee, S.S.[c] 1997. Free-edge stress singularity in a two-dimensional unidirectional viscoelastic laminate model. *ASME Journal of Applied Mechanics*. 64: (in press).

Reedy, E.D., Jr. 1990. Intensity of the stress singularity at the interface corner between a bonded elastic and rigid layer. *Engineering Fracture Mechanics*. 36: 575-583.

Tsai, M.Y. and Morton, J. 1991. The stresses in a thermally loaded bimaterial interface. *Int. J. Solids and Structures*. 28: 1053-1075 .

Wang, S.S. and Choi, I. 1982. Boundary layer effects in composite laminates : Part 2- Free-edge stress solutions and basic characteristics. *ASME Journal of Applied Mechanics*. 49: 549-560 .

Weitsman, Y. 1979. Interfacial stresses in viscoelastic adhesive-layers due to moisture sorption. *Int. J. Solids and Structures*. 15:701-713.

Williams, M.L. 1952. Stress singularities resulting from various boundary conditions in angular corners of plates in extension. *ASME Journal of Applied Mechanics*. 74: 526-528.

Damage and Failure of Interfaces, Rossmanith (ed.) © *1997 Balkema, Rotterdam, ISBN 90 5410 899 1*

The analogy between interface crack problems and punch problems

Chyanbin Hwu & C.W. Fan
Institute of Aeronautics and Astronautics, National Cheng Kung University, Tainan, Taiwan

ABSTRACT: The interface crack problems and the punch problems are usually solved independently without considering their possible connections. In this paper, we try to re-express the boundary conditions for these two problems and rewrite their corresponding solutions into an analogous form. From the reorganized expressions, we see that if the material above the interface is rigid the solution *forms* of these two problems can be made to be equivalent by only interchanging the material eigenvectors $\underset{\sim}{A}$ and $\underset{\sim}{B}$. To get a complete analogous solution, all the boundary conditions including the conditions of the outer boundary should be analogous to each other. A traction-free interface crack problem and a flat-ended punch problem are solved completely in a similar way to illustrate the analogy. From the stress distributions of these two problems, we also observe that the stresses near the interface crack tips and the punch ends possess the same singular order.

1 INTRODUCTION

Both of the interface crack problems and the punch problems belong to the mixed type boundary value problems. The outlook of the boundary conditions of these two problems are very different. The former states the displacement and traction continuity across the uncracked portions, and the traction-prescribed conditions along the cracked portions. The latter (if the punch is considered to be rigid) states the displacement-prescribed condition along the contact regions, and the traction-free condition along the uncontact regions. Because of this difference, these two problems are usually solved independently (Muskhelishvili, 1954; Hwu, 1993a; Fan and Hwu, 1996). However, their solutions show some similarities such as the stress oscillatory singularity characteristics near the interface crack tips or the punch corners. This similarity stimulates us to find the connection between these two problems. By carefully reviewing these two different boundary conditions, we find that the punch problem is just a counterpart of the interface crack problem with one of the materials to be rigid. Hence, similar to the analogy between forces and dislocations, cracks and rigid line inclusions, or holes and rigid inclusion, we may now solve the punch problems by analogy with the interface crack problems, or *vice versa.* This finding is useful not only in analysis but also in experiment. Because one may understand the physical behavior of the interface crack by doing the experiment of punch problems, or *vice versa.*

2 GOVERNING EQUATION

The basic equations for linear anisotropic elasticity are the strain-displacement equations, the stress-strain laws and the equations of equilibrium, which can be expressed in a fixed rectangular coordinate system x_i, $i = 1,2,3$ as (the symbols x_1 and x_2 will be replaced by x and y for the convenience of presentation)

$$\begin{aligned} \varepsilon_{ij} &= \frac{1}{2}(u_{i,j} + u_{j,i}) \, , \\ \sigma_{ij} &= C_{ijks}\varepsilon_{ks} \, , \\ \sigma_{ij,j} &= C_{ijks}u_{k,sj} = 0 \, , \end{aligned} \tag{2.1}$$

where u_i, σ_{ij} and ε_{ij} are respectively the displacement, stress and strain; the repeated indices imply summation; a comma stands for differentiation and C_{ijks} are the elastic constants which are assumed to be fully symmetric and positive definite.

For two-dimensional problems in which x_3 does not appear in the basic equations or the boundary conditions, the general solution to equations (2.1) may be expressed in terms of three holomorphic functions of complex variables (Stroh, 1958; Lekhnitskii, 1963). This enables us to apply many of the powerful results of complex function theory to the two-dimensional elasticity. For the later use of derivation, we now list a compact matrix form solution (Stroh, 1958; Ting, 1986) which satisfies all the basic equations given in (2.1), i.e.,

$$\underset{\sim}{u} = 2Re\{\underset{\sim}{A}\underset{\sim}{f}(z)\} = \underset{\sim}{A}\underset{\sim}{f}(z) + \overline{\underset{\sim}{A}\underset{\sim}{f}(z)},$$
$$\underset{\sim}{\phi} = 2Re\{\underset{\sim}{B}\underset{\sim}{f}(z)\} = \underset{\sim}{B}\underset{\sim}{f}(z) + \overline{\underset{\sim}{B}\underset{\sim}{f}(z)}, \qquad (2.2a)$$

where

$$\underset{\sim}{A} = [\underset{\sim}{a}_1 \quad \underset{\sim}{a}_2 \quad \underset{\sim}{a}_3], \qquad \underset{\sim}{B} = [\underset{\sim}{b}_1 \quad \underset{\sim}{b}_2 \quad \underset{\sim}{b}_3],$$
$$\underset{\sim}{f}(z) = [f_1(z_1) \quad f_2(z_2) \quad f_3(z_3)]^T,$$
$$z_\alpha = x + p_\alpha y, \quad \alpha = 1, 2, 3. \qquad (2.2b)$$

In the above equations, $\underset{\sim}{u} = (u_1, u_2, u_3)$ is the vector form of displacement; $\underset{\sim}{\phi} = (\phi_1, \phi_2, \phi_3)$ stands for the stress function vector which is related to the stresses σ_{ij} and surface traction $\underset{\sim}{t}$ by

$$\sigma_{i1} = -\phi_{i,2}, \qquad \sigma_{i2} = \phi_{i,1}, \qquad (2.2c)$$

and

$$\underset{\sim}{t} = \frac{\partial \underset{\sim}{\phi}}{\partial s}, \qquad (2.2d)$$

where s is the arc length measured along the curved boundary. p_α, α=1,2,3, are the material eigenvalues whose imaginary parts have been arranged to be positive; $(\underset{\sim}{a}_\alpha, \underset{\sim}{b}_\alpha)$, α=1,2,3, are their associated eigenvectors; $f_\alpha(z_\alpha)$, $\alpha = 1, 2, 3$, are three holomorphic complex functions to be determined by satisfying the boundary conditions of the problems considered. The superscript T denotes the transpose and the overbar represents the conjugate of a complex number.

3 BOUNDARY CONDITIONS

3.1 *Interface crack problems*

Consider a set of cracks L lying along the interface of two dissimilar anisotropic materials. The materials are assumed to be perfectly bonded at all points of the interface except those lying in the region of cracks. To describe the boundary conditions of this kind of interface crack problems, we need to consider the displacement and traction continuity across the uncracked portions, and the traction-prescribed conditions along the cracked portions. Thus,

$$\underset{\sim}{\phi}'_1(x) = \underset{\sim}{\phi}'_2(x) = \hat{\underset{\sim}{t}}, \qquad x \in L,$$
$$\underset{\sim}{u}_1(x) = \underset{\sim}{u}_2(x), \qquad \underset{\sim}{\phi}_1(x) = \underset{\sim}{\phi}_2(x), \qquad x \notin L, \qquad (3.1)$$

where $\hat{\underset{\sim}{t}}$ is the prescribed traction along the crack surface; prime (′) denotes differentiation with respect to its argument. The symbols marked with the subscripts 1 and 2 represents, respectively, the quauntities pertaining to the materials located upper and lower the interface. If we consider the material above the interface to be rigid, the boundary conditions (3.1) will then be specialized to

$$\underset{\sim}{\phi}'(x) = \hat{\underset{\sim}{t}}(x), \qquad x \in L,$$
$$\underset{\sim}{u}(x) = \underset{\sim}{0}, \qquad x \notin L. \qquad (3.2)$$

Note that in eqn.(3.2) and the following derivation the subscript 2 is dropped for the convenience of presentation, and the subscript 1 will not enter into the boundary conditions since material 1 is assumed to be rigid.

Substituting (2.2a) into (3.2), we have

$$2Re\{\underset{\sim}{B}\underset{\sim}{f}'(x)\} = \hat{\underset{\sim}{t}}(x), \qquad x \in L,$$
$$2Re\{\underset{\sim}{A}\underset{\sim}{f}(x)\} = \underset{\sim}{0}, \qquad x \notin L. \qquad (3.3)$$

3.2 *Punch problems*

Consider the case that a set of rigid punches L of given profiles are brought into contact with the surface of the half-plane and are allowed to indent the surface in such a way that the punches completely adhere to the half-plane on initial contact and during the subsequent indentation no slip occurs and the contact region does not change. The boundary conditions of this kind of punch problems may be expressed by the displacement-prescribed condition along the contact regions, and the traction-free condition along the uncontact regions. Hence,

$$\underset{\sim}{u}(x) = \hat{\underset{\sim}{u}}(x), \qquad x \in L,$$
$$\underset{\sim}{\phi}'(x) = \underset{\sim}{0}, \qquad x \notin L. \qquad (3.4)$$

Substituting (2.2a) into (3.4), we have

$$2Re\{\underset{\sim}{A}\underset{\sim}{f}(x)\} = \hat{\underset{\sim}{u}}(x), \qquad x \in L,$$
$$2Re\{\underset{\sim}{B}\underset{\sim}{f}'(x)\} = \underset{\sim}{0}, \qquad x \notin L. \qquad (3.5)$$

By comparison between eqns.(3.3) and (3.5), we see that they are counterpart of each other. Therefore, we may deal with any one of the problems by analogy with the other problem.

4 THE ANALOGY

One of the special features of the Stroh's formalism is that the solution form, eqn.(2.2), is neat and elegant. Due to its elegancy, many important characteristics can be found at the first glance of the solution form. For example, the displacements and stress functions shown in eqn.(2.2a) are distinguished only by the material eigenvector matrices $\underset{\sim}{A}$ and $\underset{\sim}{B}$. Thus, the relevant boundary conditions of the displacement prescribed problems differ from those of the traction prescribed problems only in the appearance of the symbols $\underset{\sim}{A}$ and $\underset{\sim}{B}$. Since the mathematical formulations for the displacement prescribed problems and the traction prescribed problems are identical, their solutions should also be identical with $\underset{\sim}{A}$ and $\underset{\sim}{B}$ interchanged.

Both of the interface crack problems and the punch problems belong to the mixed type boundary value problems. Although the outlook of the boundary conditions of these two problems shown in (3.1) and (3.4) are very different, eqn.(3.2) (which is a special case of (3.1)) and eqn.(3.4) are almost identical. In order to see more clearly about their equivalency, we now differentiate the second equation of (3.3) and the first equation of (3.5) with respect to x. After differentiating, eqns.(3.3) and (3.5) may be rewritten as

interface crack problem: (the material above the interface is rigid)

$$\begin{aligned} 2Re\{\underset{\sim}{B}\underset{\sim}{f}'(x)\} &= \underset{\sim}{\widehat{t}}(x), \quad x \in L, \\ 2Re\{\underset{\sim}{A}\underset{\sim}{f}'(x)\} &= \underset{\sim}{0}, \quad x \notin L; \end{aligned} \tag{4.1}$$

punch problem: (the punch is rigid)

$$\begin{aligned} 2Re\{\underset{\sim}{A}\underset{\sim}{f}'(x)\} &= \underset{\sim}{\widehat{u}}'(x), \quad x \in L, \\ 2Re\{\underset{\sim}{B}\underset{\sim}{f}'(x)\} &= \underset{\sim}{0}, \quad x \notin L. \end{aligned} \tag{4.2}$$

Because $\underset{\sim}{\widehat{t}}(x)$ and $\underset{\sim}{\widehat{u}}(x)$ are given function values in our problem formulation, (4.1) and (4.2) are identical with $\underset{\sim}{A}$ and $\underset{\sim}{B}$ interchanged. Therefore, the solutions $\underset{\sim}{f}(z)$ to these two problems should also be identical with $\underset{\sim}{A}$ and $\underset{\sim}{B}$ interchanged. In order to testify our observation, we now list the solutions $\underset{\sim}{f}(z)$ found in the literature.

Interface crack problem: (the material above the interface is rigid) (Hwu, 1993a)

$$\begin{aligned} \underset{\sim}{f}(z) =& \underset{\sim}{B}^{-1}\underset{\sim}{M}\overline{\underset{\sim}{M}}^{-1}\underset{\sim}{\psi}(z), \\ \underset{\sim}{\psi}'(z) =& \frac{1}{2\pi i}\underset{\sim}{X}_o(z)\int_L \frac{1}{s-z}[\underset{\sim}{X}_o^+(s)]^{-1}\underset{\sim}{\widehat{t}}(s)ds \\ &+ \underset{\sim}{X}_o(z)\underset{\sim}{p}_n(z), \end{aligned} \tag{4.3a}$$

where

$$\begin{aligned} &\underset{\sim}{X}_o^+(x) = \underset{\sim}{X}_o^-(x), \qquad x \notin L, \\ &\underset{\sim}{X}_o^+(x) + \underset{\sim}{M}\overline{\underset{\sim}{M}}^{-1}\underset{\sim}{X}_o^-(x) = \underset{\sim}{0}, \qquad x \in L. \end{aligned} \tag{4.3b}$$

Punch problem: (the punch is rigid) (Fan and Hwu, 1996)

$$\begin{aligned} \underset{\sim}{f}(z) =& \underset{\sim}{A}^{-1}\underset{\sim}{M}^{-1}\overline{\underset{\sim}{M}}\underset{\sim}{\psi}(z), \\ \underset{\sim}{\psi}'(z) =& \frac{1}{2\pi i}\underset{\sim}{X}_o^*(z)\int_L \frac{1}{s-z}[\underset{\sim}{X}_o^{*+}(s)]^{-1}\underset{\sim}{\widehat{u}}'(s)ds \\ &+ \underset{\sim}{X}_o^*(z)\underset{\sim}{p}_n(z), \end{aligned} \tag{4.4a}$$

where

$$\begin{aligned} &\underset{\sim}{X}_o^{*+}(x) = \underset{\sim}{X}_o^{*-}(x), \qquad x \notin L, \\ &\underset{\sim}{X}_o^{*+}(x) + \underset{\sim}{M}^{-1}\overline{\underset{\sim}{M}}\underset{\sim}{X}_o^{*-}(x) = \underset{\sim}{0}, \qquad x \in L. \end{aligned} \tag{4.4b}$$

In the above equations, $\underset{\sim}{M}$ is the impedance matrix defined as

$$\underset{\sim}{M} = -i\underset{\sim}{B}\underset{\sim}{A}^{-1}. \tag{4.5}$$

By this definition, it can easily be proved that the interchange of $\underset{\sim}{A}$ and $\underset{\sim}{B}$ will lead to the interchange of $\underset{\sim}{M}$ and $-\underset{\sim}{M}^{-1}$. $\underset{\sim}{p}_n(z)$ is an arbitrary polynomial vector with the degree not higher than the number of cracks (or punches) n, which may be determined by the infinity conditions and the single-valuedness requirement of displacements (or equilibrium conditions of each punch); $\underset{\sim}{X}_o(z)$ or $\underset{\sim}{X}_o^*(z)$ satisfying eqn.(4.3b) or (4.4b) is called the basic Plemelj function matrix whose solution can be found in the Appendix.

Note that the solution form listed in eqn.(4.3) and (4.4) are not exactly the same as that presented in (Hwu, 1993a; Fan and Hwu, 1996). The equivalency between the present solution form and that presented in the literature has been proved in (Hwu and Fan, 1997).

In addition to the analogy shown in (4.1)-(4.4), we also like to present some simple results for the analogy between the surface traction and deformation. It is known that the stress function vector $\underset{\sim}{\phi}$ and the displacement vector $\underset{\sim}{u}$ have the following relation (Yeh, *et.al.*, 1993a,b)

$$\underset{\sim}{A}^T\underset{\sim}{\phi} + \underset{\sim}{B}^T\underset{\sim}{u} = \underset{\sim}{f}(z). \tag{4.6}$$

By this relation, many physical quantities can be obtained easily . For the interface crack problems, crack opening displacement $\underset{\sim}{u}$ along the crack surface ($x \in L$) can be found by substituting $(4.1)_1$ into (4.6), and the stress distribution $\underset{\sim}{t}$ along the interface ($x \notin L$) can be found by substituting $(4.1)_2$ into (4.6). For the punch problems, the contact pressure $\underset{\sim}{t}$ under the punch ($x \in L$) can be found by substituting $(4.2)_1$ into (4.6), and the surface deformation $\underset{\sim}{u}$ outside the punch ($x \notin L$) can be found by substituting $(4.2)_2$ into (4.6). The results are

interface crack problem: (the material above the interface is rigid)

$$\begin{aligned} &\underset{\sim}{u}'(x) = (\underset{\sim}{B}^T)^{-1}[\underset{\sim}{f}'(x) - \underset{\sim}{A}^T\underset{\sim}{\hat{t}}(x)], \quad x \in L, \\ &\underset{\sim}{t}(x) = (\underset{\sim}{A}^T)^{-1}\underset{\sim}{f}'(x), \quad x \notin L; \end{aligned} \tag{4.7}$$

punch problem: (the punch is rigid)

$$\begin{aligned} &\underset{\sim}{t}(x) = (\underset{\sim}{A}^T)^{-1}[\underset{\sim}{f}'(x) - \underset{\sim}{B}^T\underset{\sim}{\hat{u}}'(x)], \quad x \in L, \\ &\underset{\sim}{u}'(x) = (\underset{\sim}{B}^T)^{-1}\underset{\sim}{f}'(x), \quad x \notin L. \end{aligned} \tag{4.8}$$

The results obtained in (4.7) and (4.8) show that the surface traction and the displacement gradient are analogous to each other for these two different problems.

5 COMPLETENESS

For a real problem, the boundary conditions given in (4.1) and (4.2) are not complete since they only state the conditions along the interface or the half-plane surface. For a finite body, there should be a condition describing the outer boundary. For an infinity body, the outer boundary condition is the so called infinity condition. Therefore, without knowing the conditions for the outer boundary (or infinity), the solutions given in (4.3) and (4.4) are incomplete, and the polynomial function vector $\underset{\sim}{p}_n(z)$ remains undetermined. If the outer boundary condition (or infinity condition) does not possess the analogous characteristics like those shown in (4.1) and (4.2), the solutions of $\underset{\sim}{p}_n(z)$ may not have the analogous form for the interface crack problems and the punch problems. In order to have a better understanding about the analogy of $\underset{\sim}{p}_n(z)$, we now choose the examples whose $\underset{\sim}{\hat{t}}(x)$ (or $\underset{\sim}{\hat{u}}(x)$) is zero along the crack surface (or the contact region).

5.1 *A traction-free interface crack*: (the material above the interface is rigid)

Consider a finite interface crack located on $(-a, a)$ subjected to a uniform loading σ_{ij}^{∞} at infinity If the crack surface is traction free, $\underset{\sim}{t}(x) = \underset{\sim}{0}$ and $\underset{\sim}{\psi}'(z)$ given in $(4.3\text{a})_2$ can be simplified to

$$\underset{\sim}{\psi}'(z) = \underset{\sim}{X}_o(z)\underset{\sim}{p}_1(z), \tag{5.1a}$$

where

$$\begin{aligned} &\underset{\sim}{X}_o(z) = \underset{\sim}{\Lambda} \ll \chi_\alpha(z) \gg, \\ &\underset{\sim}{p}_1(z) = \underset{\sim}{c}_1 z + \underset{\sim}{c}_o, \end{aligned} \tag{5.1b}$$

and

$$\chi_\alpha(z) = \frac{1}{\sqrt{z^2 - a^2}}\left(\frac{z-a}{z+a}\right)^{i\epsilon_\alpha}. \tag{5.1c}$$

$\underset{\sim}{\Lambda}$ is a matrix satisfing the following eigenrelation

$$\overline{\underset{\sim}{M}}^{-1}\underset{\sim}{\Lambda} = \underset{\sim}{M}^{-1}\underset{\sim}{\Lambda} \ll e^{-2\pi\epsilon_\alpha} \gg. \tag{5.1d}$$

ϵ_α is the oscillatory index determined by the material elastic properties (Appendix). $\underset{\sim}{c}_o$ and $\underset{\sim}{c}_1$ are coefficient vectors of the polynomial $\underset{\sim}{p}_1(z)$, which will be determined by the infinity condition and the single-valuedness requirement. These two conditions may be expressed as

$$\begin{aligned} &\underset{\sim}{t} = \underset{\sim}{t}^{\infty}, \qquad \text{when} \quad |z| \to \infty, \\ &\int_{-a}^{a} [\underset{\sim}{u}'(x, 0^+) - \underset{\sim}{u}'(x, 0^-)]dx \\ &= -\int_{-a}^{a} \underset{\sim}{u}'(x, 0^-)dx = \underset{\sim}{0}, \qquad \text{when} \quad |x| \le a, \end{aligned} \tag{5.2a}$$

where

$$\underset{\sim}{t}^{\infty} = \{\sigma_{12}^{\infty}\ \sigma_{22}^{\infty}\ \sigma_{32}^{\infty}\}^T. \tag{5.2b}$$

Applying the results given in (4.7), eqn.(5.2) can be rewritten in terms of $\underset{\sim}{f}(x)$ as

$$\begin{aligned} &(\underset{\sim}{A}^T)^{-1}\underset{\sim}{f}'(x) = \underset{\sim}{t}^{\infty}, \quad \text{when} \quad |x| \to \infty, \\ &(\underset{\sim}{B}^T)^{-1}\int_{-a}^{a} \underset{\sim}{f}'(x^-)dx = \underset{\sim}{0}, \quad \text{when} \quad |x| \le a. \end{aligned} \tag{5.3}$$

Substituting $(4.3\text{a})_1$ and (5.1) into (5.3) we have

$$\begin{aligned} &(\underset{\sim}{A}^T)^{-1}\underset{\sim}{B}^{-1}\underset{\sim}{M}\overline{\underset{\sim}{M}}^{-1}\underset{\sim}{\Lambda} \ll \chi_\alpha(x) \gg (\underset{\sim}{c}_1 x + \underset{\sim}{c}_o) \\ &= \underset{\sim}{t}^{\infty}, \quad \text{when} \quad |x| \to \infty, \\ &\int_{-a}^{a} \chi_\alpha(x^-)(\underset{\sim}{c}_1 x + \underset{\sim}{c}_o)dx = \underset{\sim}{0}, \quad |x| \le a. \end{aligned} \tag{5.4}$$

With the aid of the following integrals (Hwu, 1992)

$$\begin{aligned} &\int_{-a}^{a} \frac{1}{\sqrt{a^2 - t^2}}\left(\frac{a-t}{a+t}\right)^{i\epsilon_\alpha} dt = \frac{\pi}{\cosh \pi\epsilon_\alpha}, \\ &\int_{-a}^{a} \frac{t}{\sqrt{a^2 - t^2}}\left(\frac{a-t}{a+t}\right)^{i\epsilon_\alpha} dt = \frac{-2i\pi a\epsilon_\alpha}{\cosh \pi\epsilon_\alpha}, \end{aligned} \tag{5.5}$$

eqn.(5.4) now leads to

$$\mathbf{c}_0 = \ll 2ia\epsilon_\alpha \gg \mathbf{c}_1, \quad \mathbf{c}_1 = \Lambda^{-1}\overline{\mathbf{B}}\mathbf{A}^T\mathbf{t}^\infty. \tag{5.6}$$

Combining the results obtained in (4.3a)$_1$, (5.1) and (5.6), the complete solution of $\mathbf{f}(z)$ can be written as

$$\mathbf{f}'(z) = \mathbf{B}^{-1}\mathbf{M}\overline{\mathbf{M}}^{-1}\Lambda \ll (z + 2ia\epsilon_\alpha)\chi_\alpha(z) \gg \Lambda^{-1}\overline{\mathbf{B}}\mathbf{A}^T\mathbf{t}^\infty. \tag{5.7}$$

Substituting (5.7) into (4.7)$_2$, and using the identities (A7) and

$$\overline{\mathbf{B}}\mathbf{A}^T = (\mathbf{I} + \mathbf{M}\overline{\mathbf{M}}^{-1})^{-1}, \tag{5.8}$$

the stress distribution along the interface can be obtained as

$$\mathbf{t}(x) = \Lambda \ll (z + 2ia\epsilon_\alpha)\chi_\alpha(x) \gg \Lambda^{-1}\mathbf{t}^\infty, \quad \text{when} \quad |x| > a. \tag{5.9}$$

5.2 *A flat-ended punch*: (the punch is rigid)

Consider the indentation by a single flat-ended punch which makes contact with the half-plane over the region $|x| \leq a$, and the force $\mathbf{q}$ applied on the punch is given. Since the punch end profile is flat, $\widehat{\mathbf{u}}'(x) = \mathbf{0}$ and $\boldsymbol{\psi}'(z)$ given in (4.4a)$_2$ can be simplified to

$$\boldsymbol{\psi}'(z) = \Lambda^* \ll \chi_\alpha(z) \gg (\mathbf{c}_1 z + \mathbf{c}_0), \tag{5.10}$$

where $\Lambda^*(= \overline{\mathbf{M}}^{-1}\overline{\Lambda})$ is an eigenvector matrix associated with $\mathbf{X}_o^*(z)$ (Appendix), and the coefficient vectors $\mathbf{c}_0$ and $\mathbf{c}_1$ will be determined by the infinity condition and the force equilibrium condition of each punch. These two conditions may be expressed as

$$\mathbf{t} = \mathbf{0}, \quad \text{when} \quad |z| \to \infty,$$
$$\int_{-a}^{a} \mathbf{t}(x, 0^-)dx = \mathbf{q}, \quad \text{when} \quad |x| \leq a. \tag{5.11}$$

Applying the results given in (4.6) and (4.8), eqn.(5.11) can be rewritten in terms of $\mathbf{f}(x)$ as

$$\mathbf{f}'(x) = \mathbf{0}, \quad \text{when} \quad |x| \to \infty,$$
$$(\mathbf{A}^T)^{-1}\int_{-a}^{a} \mathbf{f}'(x^-)dx = \mathbf{q}, \quad \text{when} \quad |x| \leq a. \tag{5.12}$$

Substituting (4.4a)$_1$ and (5.10) into (5.12), and using the identity (A13), we obtain

$$\mathbf{c}_0 = \frac{-1}{2\pi}\Lambda^{*^{-1}}\overline{\mathbf{M}}^{-1}\mathbf{q}, \quad \mathbf{c}_1 = \mathbf{0}. \tag{5.13}$$

Combining the results obtained in (4.4a)$_1$, (5.10) and (5.13), the complete solution of $\mathbf{f}(z)$ can be written as

$$\mathbf{f}'(z) = \frac{-1}{2\pi}\mathbf{A}^{-1}\mathbf{M}^{-1}\overline{\mathbf{M}}\Lambda^* \ll \chi_\alpha(x) \gg \Lambda^{*^{-1}}\overline{\mathbf{M}}^{-1}\mathbf{q}, \tag{5.14}$$

which can be proved to be identical to that presented in (Fan and Hwu, 1996). Substituting (5.13) into (4.8)$_1$, and using the identities (A7),(A11) and (5.8), the contact pressure under the punch can be obtained as

$$\mathbf{t}(x) = \frac{1}{\pi i}\Lambda \ll e^{-\pi\epsilon_\alpha}\cosh\epsilon_\alpha\chi_\alpha(x) \gg \Lambda^{-1}\mathbf{q}, \quad \text{when} \quad |x| < a. \tag{5.15}$$

6 CONCLUDING REMARKS

In this paper, the analogy between the interface crack problems and the punch problems is presented under the consideration that the materials above the interface and the punch are rigid. From the boundary conditions shown in (4.1) and (4.2) along the interface and the half-plane surface, we observe that the solutions to the interface crack problems and the punch problems should have the same forms with $\mathbf{A}$ and $\mathbf{B}$ interchanged. This observation is verified in (4.3) and (4.4). In addition to the analogy of the general solutions, the solution forms of the surface traction and the displacement gradient are also analogous to each other, which are shown in (4.7) and (4.8).

It should be emphasized that we state the analogy only by the solution *form* not the solution itself. Because for a real problem, the boundary conditions stated in (4.1) or (4.2) are not complete enough. For a finite body, there should be a condition describing the outer boundary. For an infinity body, the outer boundary condition is the so called infinity condition. Therefore, if we want to get an exactly analogous solution, all the boundary conditions should be analogous to each other. The examples presented in Section 5 show that the additional boundary conditions (5.2) or (5.3) for a traction-free interface crack problem, and (5.11) or (5.12) for a flat-ended punch, are not exactly analogous to each other (although (5.3) and (5.12) look alike). Hence, solutions of $\mathbf{f}'(z)$

found in (5.7) for a traction-free interface crack problem and in (5.14) for a flat-ended punch problem cannot be communicated by only interchanging $\underset{\sim}{A}$ and $\underset{\sim}{B}$.

Although (5.14) cannot be obtained by (5.7) with $\underset{\sim}{A}$ and $\underset{\sim}{B}$ interchanged, or *vice versa*, they really reveal the same oscillatory singularity behavior. From the stress distribution shown in (5.9) and (5.15), we observe that the oscillatory singularity characteristics is dominated by the function $\chi_\alpha(x)$ which is exactly the same for a traction-free interface crack and a flat-ended punch.

The analogy given in this paper is for a rigid punch and a rigid material above the interface. It is hoped that the analogy concept may be extended to the interface crack problems and the elastic contact problems between two dissimilar anisotropic media.

ACKNOWLEDGEMENTS

The authors would like to thank the support by National Science Council, Republic of China, through Grant No. NSC 83-0424-E006-010, and thank Professor K.C. Wu of the Institute of Applied Mechanics in National Taiwan University for his suggestion of the present subject.

REFERENCES

Fan, C.W. and C. Hwu 1996. Punch problems for an anisotropic elastic half-plane. *J. Applied Mechanics*,63(1):69-76.

Hwu, C. 1992. Thermoelastic interface crack problems in dissimilar anisotropic media. *International Journal of Solids and Structures*, 29: 2077-2090.

Hwu, C. 1993a. Explicit solutions for the collinear interface crack problems. *International Journal of Solids and Structures*, 30:301-312.

Hwu, C. 1993b. Fracture parameters for orthotropic bimaterial interface cracks. *Engineering Fracture Mechanics*, 45(1):89-97.

Hwu, C. and C.W. Fan 1997. Solving the punch problems by analogy with the interface crack problems. (submitted for publication)

Lekhnitskii, S.G. 1963. *Theory of elasticity of an anisotropic body*, MIR, Moscow.

Muskhelishvili, N.I. 1954. *Some basic problems of the mathematical theory of elasticity*, Noordhoff Pub., Groningen.

Stroh, A.N. 1958. Dislocations and cracks in anisotropic elasticity," *Philosophical Magazine*, 7: 625-646.

Ting, T.C.T. 1986. Explicit solution and invariance of the singularities at an interface crack in anisotropic composites. *International Journal of Solids and Structures*, 22:965-983.

Yeh, C.S., Y.C. Shu and K.C. Wu 1993a. Conservation laws in anisotropic elasticity I. basic frame work. *Proc. Roy. Soc. Lon. A* A443:139-151.

Yeh, C.S., Y.C. Shu and K.C. Wu 1993b. Conservation laws in anisotropic elasticity II. extension and application to thermoelasticity. *Proc. Roy. Soc. Lon. A* A443:153-161.

APPENDIX:
Plemelj function matrix $\underset{\sim}{X}_o(z)$ or $\underset{\sim}{X}_o^*(z)$

(i) $\underset{\sim}{X}_o(z)$

The plemelj function matrix $\underset{\sim}{X}_o(z)$ is a sectionally holomorphic function matrix satisfing the following relations

$$\begin{aligned} &\underset{\sim}{X}_o^+(x) = \underset{\sim}{X}_o^-(x), \qquad x \notin L, \\ &\underset{\sim}{X}_o^+(x) + \underset{\sim}{M}\overline{\underset{\sim}{M}}^{-1}\underset{\sim}{X}_o^-(x) = \underset{\sim}{0}, \qquad x \in L. \end{aligned} \tag{A1}$$

The solution to (A1) is (Hwu, 1992)

$$\underset{\sim}{X}_o(z) = \underset{\sim}{\Lambda}\underset{\sim}{\Gamma}(z), \tag{A2a}$$

where

$$\underset{\sim}{\Lambda} = [\underset{\sim}{\lambda}_1, \underset{\sim}{\lambda}_2, \underset{\sim}{\lambda}_3],$$

$$\underset{\sim}{\Gamma}(z) = \ll \prod_{j=1}^{n}(z-a_j)^{-(1+\delta_\alpha)}(z-b_j)^{\delta_\alpha} \gg . \tag{A2b}$$

δ_α and $\underset{\sim}{\lambda}_\alpha$, $\alpha = 1,2,3$ of (A2b) are the eigenvalues and eigenvectors of

$$(e^{2\pi i\delta}\underset{\sim}{I} + \underset{\sim}{M}\overline{\underset{\sim}{M}}^{-1})\underset{\sim}{\lambda} = \underset{\sim}{0}, \tag{A3a}$$

which can also be expressed as

$$(e^{2\pi i\delta}\underset{\sim}{M}^{-1} + \overline{\underset{\sim}{M}}^{-1})\underset{\sim}{\lambda} = \underset{\sim}{0}. \tag{A3b}$$

The explicit solutions for the eigenvalues δ are (Ting, 1986)

$$\delta_\alpha = -\frac{1}{2} + i\epsilon_\alpha, \qquad \alpha = 1,2,3, \tag{A4a}$$

where

$$\begin{aligned} &\epsilon_1 = \epsilon = \frac{1}{2\pi}\ln\frac{1+\beta}{1-\beta}, \qquad \epsilon_2 = -\epsilon, \qquad \epsilon_3 = 0, \\ &\beta = [-\frac{1}{2}tr(\underset{\sim}{S}^2)]^{\frac{1}{2}}, \qquad \underset{\sim}{S} = i(2\underset{\sim}{A}\underset{\sim}{B}^T - \underset{\sim}{I}). \end{aligned} \tag{A4b}$$

tr stands for the trace of matrix. Substituting (A4) into (A3b), we get

$$\overline{\underset{\sim}{M}}^{-1}\underset{\sim}{\Lambda} = \underset{\sim}{M}^{-1}\underset{\sim}{\Lambda} \ll e^{-2\pi\epsilon_\alpha} \gg . \qquad (A5)$$

To have a unique eigenvector matrix, $\underset{\sim}{\Lambda}$ is normalized by

$$\frac{1}{2}\overline{\underset{\sim}{\Lambda}}^T(\underset{\sim}{M}^{-1} + \overline{\underset{\sim}{M}}^{-1})\underset{\sim}{\Lambda} = \underset{\sim}{I} . \qquad (A6)$$

Combining the normalization equation (A6) with the eigenrelation (A5), we may obtain many useful identities like those presented in (Hwu, 1993b). Shown below is an example which has been used when simplifying the stress distribution vector $\underset{\sim}{t}(x)$ in (5.9).

$$(\underset{\sim}{I} + \underset{\sim}{M}\overline{\underset{\sim}{M}}^{-1})\underset{\sim}{\Lambda} = 2\underset{\sim}{\Lambda} \ll e^{-\pi\epsilon_\alpha}\cosh(\pi\epsilon_\alpha) \gg . \qquad (A7)$$

(ii) $\underset{\sim}{X}_o^*(z)$

The plemelj function matrix $\underset{\sim}{X}_o^*(z)$ is a sectionally holomorphic function matrix satisfing the following relations

$$\underset{\sim}{X}_o^{*+}(x) = \underset{\sim}{X}_o^{*-}(x), \qquad x \notin L,$$

$$\underset{\sim}{X}_o^{*+}(x) + \underset{\sim}{M}^{-1}\overline{\underset{\sim}{M}}\underset{\sim}{X}_o^{*-}(x) = \underset{\sim}{0}, \qquad x \in L. \qquad (A8)$$

The solution to (A8) is (Hwu, 1992)

$$\underset{\sim}{X}_o^*(z) = \underset{\sim}{\Lambda}^*\underset{\sim}{\Gamma}(z), \qquad (A9)$$

where $\underset{\sim}{\Lambda}^*$ satisfies the following eigenrelation

$$\overline{\underset{\sim}{M}}\underset{\sim}{\Lambda}^* = \underset{\sim}{M}\underset{\sim}{\Lambda}^* \ll e^{-2\pi\epsilon_\alpha} \gg . \qquad (A10)$$

From the sectionally holomorphic relations shown in (A1) and (A8), it can be proved that the oscillatory index $\epsilon_\alpha, \alpha = 1,2,3$, of (A10) are identical to those given in (A4), and

$$\overline{\underset{\sim}{M}}\underset{\sim}{\Lambda}^* = \overline{\underset{\sim}{\Lambda}}. \qquad (A11)$$

To have a unique eigenvector matrix, $\underset{\sim}{\Lambda}^*$ is normalized by

$$\frac{1}{2}\overline{\underset{\sim}{\Lambda}^*}^T(\underset{\sim}{M} + \overline{\underset{\sim}{M}})\underset{\sim}{\Lambda}^* = \underset{\sim}{I} . \qquad (A12)$$

Combining the normalization equation (A12) with the eigenrelation (A10), we may also obtain many useful identities like those presented in (Hwu, 1993b). Shown below is an example which has been used when deriving the coefficient vector $\underset{\sim}{c}_0$ in (5.13).

$$\underset{\sim}{\Lambda}^* \ll e^{\pi\epsilon_\alpha}\cosh^{-1}(\pi\epsilon_\alpha) \gg = \underset{\sim}{H}\underset{\sim}{M}\underset{\sim}{\Lambda}^*, \qquad (A13)$$

where $\underset{\sim}{H}$ is a real matrix defined as $\underset{\sim}{H} = 2i\underset{\sim}{A}\underset{\sim}{A}^T$ whose inverse equals to $\frac{1}{2}(\underset{\sim}{M} + \overline{\underset{\sim}{M}})$.

Damage and Failure of Interfaces, Rossmanith (ed.) © *1997 Balkema, Rotterdam, ISBN 90 5410 899 1*

Influence of micro-voids on toughness of interfaces

D. Leguillon
Laboratoire de Modélisation en Mécanique, CNRS URA, Université P. et M. Curie, Paris, France

ABSTRACT: In order to take into account the microstructure or the thickness of interfaces, the matched asymptotic expansions method can be an option to structural computations by finite elements based on refined meshes in the regions under consideration. It gives a description of the solution by means of two expansions, the outer one is meaningful at a distance of the interface under consideration whereas the inner one is valid in the vicinity of the interface. This approach allows to express the influence of a porous interface, which microstructure is made of microvoids regularly distributed, on the stiffness and toughness of an assembly. Moreover, it is proved that even if the macroscopic crack growth along this interface is stable, it is made of spontaneous microscopic crack increments.

1 INTRODUCTION

As usual in structural computations by finite elements, the various components of a body are generally assumed to be perfectly bonded together. The interfaces are considered as ideal lines or surfaces with perfect transmission conditions. Displacements are assumed to be continuous across the lines or surfaces as well as the normal stresses. However, interfaces between resin and fibers for instance are far from perfect, they are made of bridges of resin separated by voids. Interfaces between disoriented layers in a laminate are made of pure resin, without fibers, and in that case the assumption of a negligible thickness is questionable. Thus, the simplified usual model needs to be revisited, especially if the aim of the analysis is a fracture process along these interfaces (Ryvkin *et al.* 1995, Ryvkin 1996, Atkinson & Chen 1996).
Unfortunately, taking into account the internal structure of the joints would lead to refine drastically the meshes in the vicinity of these areas. Thus, a two-scale model seems to be a good option to describe these situations. The boundary layer method matching outer and inner asymptotic expansions provides a useful tool (Van Dyke 1964, Il'in 1992). The first term of the outer expansion is solution to the classical unperturbed problem and is solved by f.e.. The second outer term enjoys a particular feature, it involves jumps through the lines or surfaces characterizing (at the macro-scale) the interfaces (Leguillon, 1994, 1995, 1996, Berrehili, 1997). However, this approach requires smoothness properties which fail in general at the ends of the ideal interfaces, at a stress free edge (Wang & Choi 1982 and many other authors, see for instance the bibliography in Leguillon & Sanchez-Palencia 1987) or at a crack tip for instance . These points undergo singular elastic solutions and as a consequence, the jumps define strongly singular unusual transmission conditions. The problem can be solved using a superimposition principle and the theory of singularities.
The above mentioned jumps derive from the matching conditions. The inner expansions are settled on a stretched domain which permit a precise drawing of the internal structure of the joint (the micro-scale). If it is a simple isotropic homogeneous layer, computations depend on a single space variable and can be performed explicitely. Another interesting and tractable situation concerns an inperfect (porous) interface containing regularly distributed voids. Under the assumption that this distribution is periodic the inner terms are numerically computed once for all (they are independent of the external applied loads) on an infinite strip orthogonal to the joint and which

width corresponds to a single period. The matching conditions are defined by the behaviour at infinity of these solutions.
Indeed, the periodicity assumption leads to some simplifications but is frequently under discussion. It allows to reduce the inner domain to a strip instead of a complete unbounded space. Moreover, it can be considered as a good approximation which excludes only the random voids distributions (in size and location). The interaction between the successive cavities is taken into account by such a model.
The forthcoming analysis will be restricted to the plane stress linear elasticity framework but extends to other situations. Throughout this paper the usual convention of summation of repeated indices is used.

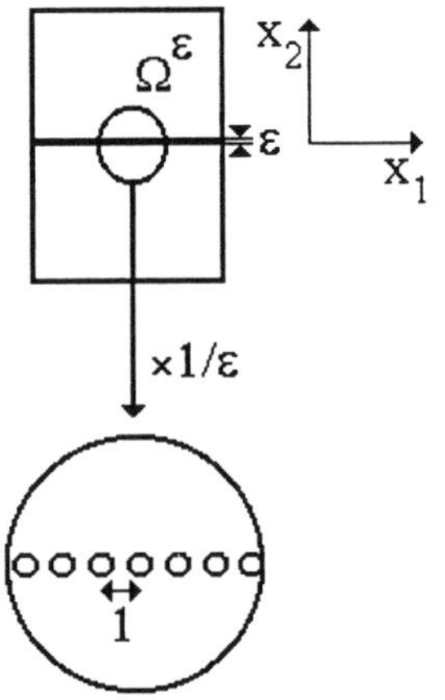

Fig. 1. The domain Ω^ε and the interface microstructure.

2 THE OUTER ASYMPTOTIC EXPANSION

Let us define the domain Ω^ε as the exact structure. It is made of a homogeneous material (with Young modulus E and Poisson ratio ν) separated into two parts by a straight thin strip embedding the interface with its more or less complicated microstructure. The (dimensionless) parameter ε is often chosen as the ratio between the physical thickness of the interface and a characteristic length of the structure. In the present case of microvoids, better than thickness, the appropriate choice for the characteristic dimension of the interface is the distance between the centre of the voids (see fig.1). It is to be noticed that two different materials separated by the interface can be considered as well, it introduces just more complicated notations but no additional difficulties.
The elastic solution $\underline{U}^\varepsilon$ to this perturbed (by the strip) problem is described through two asymptotic expansions. The outer one is defined on the unperturbed domain Ω^0 (roughly speaking it is the limit domain as $\varepsilon \to 0$), as usual when boundary layers are involved, it is assumed to read :

$$\underline{U}^\varepsilon(x_1, x_2) = \underline{U}^0(x_1, x_2) + \varepsilon \underline{U}^1(x_1, x_2) + ... \quad (2.1)$$

The $\underline{U}^j$'s are solutions to problems settled on Ω^0. For this reason the name "outer" is obvious, such a description holds true only away from the location of the perturbation which disappears in Ω^0. Hence, any hypothesis on the precise nature of the interface is useless here. The only assumption contained in this section is that the interface is strong enough to transmit perfectly (at the leading order $\underline{U}^0$) displacements and normal stresses.

The leading term $\underline{U}^0$ in (2.1) is the solution to the unperturbed problem, the interface being considered as perfect. It can be obtained for instance by a finite element approximation without any particularly refined mesh since the problem is settled on the unperturbed structure. The other terms are *a priori* unknown and problems for them can be posed only if additional (*i.e.* matching, see further on) conditions are involved. To this purpose, we define, at any interior point of the interface line, the inner behaviour of the outer terms of (2.1). The ideal line is assumed to be smooth (it is herein the straight line $x_2 = 0$) and thus the elastic solutions are smooth in the vicinity of the interior points of this line. The inner behaviour of these terms is then obtained by a Taylor expansion :

$$\underline{U}^j(x_1, x_2) = \underline{U}^{j\pm}(x_1, 0) + x_2 \frac{\partial \underline{U}^{j\pm}}{\partial x_2}(x_1, 0) + ... \quad (2.2)$$

The index $\pm$ denotes the upper ($x_2 > 0$) and lower ($x_2 < 0$) limits, in particular : $\underline{U}^{0\pm}(x_1, 0) = \underline{U}^0(x_1, 0)$.

3 THE INNER EXPANSION FOR AN INTERFACE WITH MICROVOIDS PERIODICALLY DISTRIBUTED

It is assumed here that the interface is made of circular microvoids (with dimensionless diameter λ) periodically distributed (with period ε) along the line defining the interface at the macro-scale (see section 2). The material parameter :

$$m = 1 - \lambda/\varepsilon \quad (3.1)$$

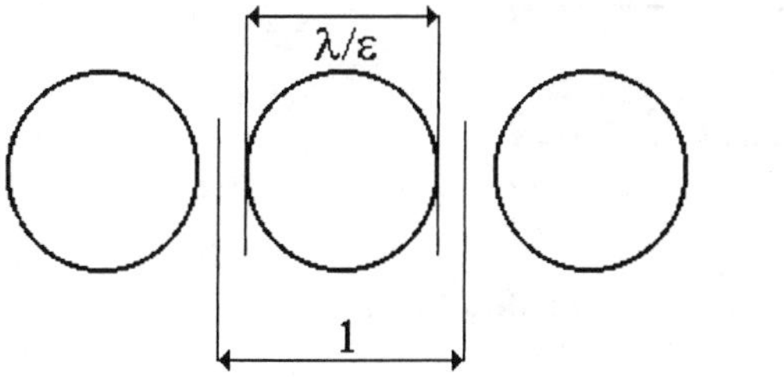

Fig. 2. The solid ligaments part in the interface.

measures the proportion of solid ligaments along the interface. λ/ε is the porosity and can be interpreted as a kind of damage variable.
The inner expansion is obtained by mixing a multiple scalings method and matched asymptotics after the change of variable $y_2 = x_2/\varepsilon$ (Nguetseng & Sanchez-Palencia 1985) :

$$\underline{U}^{\varepsilon}(x_1,x_2) = \underline{V}^0(x_1,y_1,y_2) + \varepsilon\underline{V}^1(x_1,y_1,y_2) + ... \tag{3.2}$$

for $y_1 = x_1/\varepsilon$, contrarily to y_2 this is not *stricto sensu* a change of variable but rather an additional variable. The $\underline{V}^{j'}$s are periodic with respect to this additional variable y_1. They are numerically obtained by a superimposition principle, the $\hat{\underline{V}}^{j'}$s, y_1−periodic functions with an exponential decay at infinity in y_2 (up to a rigid motion), are combined with polynoms in the single variable y_2 which ensure the matching conditions. The calculation of the second (polynomial) parts of the solutions is explicit whereas the periodic parts of the solutions are numerically obtained once for all (independently of the applied loads) by a method close to the one used in homogenization of periodic structures (Sanchez-Palencia 1980). They are computed on an infinite strip (fig.3) or even a semi-infinite one when crack lips are involved (fig.3 and fig.5 below) :

$$\underline{V}^0(x_1,y_1,y_2) = \underline{U}^0(x_1,0) \tag{3.3}$$

$$\underline{V}^1(x_1,y_1,y_2) = y_2\,\frac{\partial \underline{U}^{0\pm}}{\partial x_2}(x_1,0) + \hat{\underline{V}}^1(x_1,y_1,y_2)$$

$$\sim \underline{U}^{1\pm}(x_1,0) + y_2\,\frac{\partial \underline{U}^{0\pm}}{\partial x_2}(x_1,0) \text{ as } \mid y_2 \mid\to\infty \tag{3.4}$$

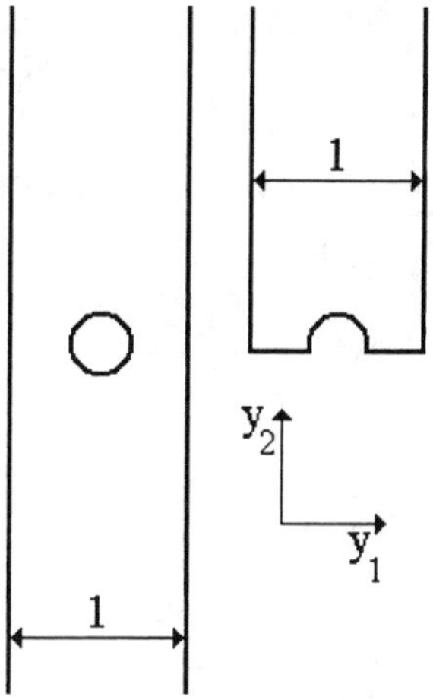

Fig. 3. The inner periodic strip.

$$\underline{V}^2(x_1,y_1,y_2) = y_2\,\frac{\partial \underline{U}^{1\pm}}{\partial x_2}(x_1,0)$$

$$+\frac{y_2^2}{2}\,\frac{\partial^2 \underline{U}^{0\pm}}{\partial x_2^2}(x_1,0) + \hat{\underline{V}}^2(x_1,y_1,y_2)$$

$$\sim \underline{U}^{2\pm}(x_1,0) + y_2\,\frac{\partial \underline{U}^{1\pm}}{\partial x_2}(x_1,0)$$

$$+\frac{y_2^2}{2}\,\frac{\partial^2 \underline{U}^{0\pm}}{\partial x_2^2}(x_1,0) \text{ as } \mid y_2 \mid\to\infty \tag{3.5}$$

4 THE TRANSMISSION CONDITIONS FOR THE SECOND OUTER TERM

The matching rules impose jump conditions to the second outer term $\underline{U}^1$ through the ideal line characterizing the interface in Ω^0. These conditions can be written out formally :

$$[\![\underline{U}^1]\!] \sim \partial(\underline{U}^0)\ ;\ [\![\sigma(\underline{U}^1)]\!].\underline{n} \sim \partial^2(\underline{U}^0) \tag{4.1}$$

It means first that the jump of the displacement field $\underline{U}^1$ is proportional to the first derivatives of the unperturbed displacement field $\underline{U}^0$, and second that the jump of the normal stress associated with $\underline{U}^1$ is proportional to the second derivatives of the unperturbed displacement field (these two conditions are consistent). More precisely, from (3.4) it is clear that $\hat{\underline{V}}^1$ depends linearly on $\partial\underline{U}^0/\partial x_2(x_1,0)$ and in accordance with (4.1) the displacement and stress jump conditions read :

$$[\![\underline{U}^1(x_1,0)]\!] = \lim_{y_2\to+\infty}\hat{\underline{V}}^1 - \lim_{y_2\to-\infty}\hat{\underline{V}}^1 \tag{4.2}$$

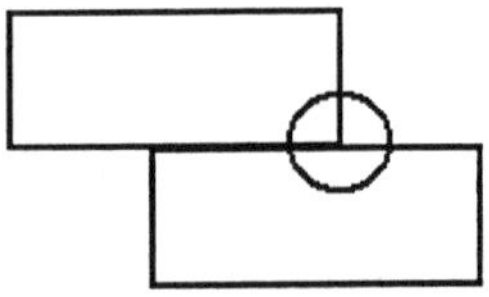

Fig. 4. The single lap assembly.

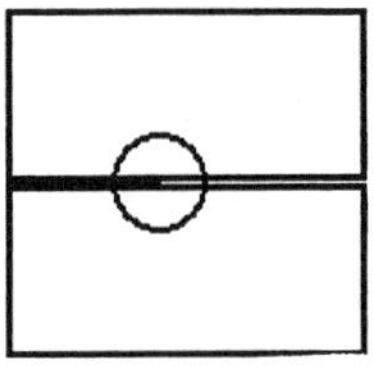

Fig. 5. The interface crack.

$$[\![\sigma_{i2}(\underline{U}^1)(x_1,0)]\!] = \pi\lambda^2 \frac{\partial\sigma_{i1}(\underline{U}^0)}{\partial x_1}(x_1,0) - \frac{\partial}{\partial x_1}\int_Y \sigma_{i1}(\hat{\underline{V}}^1)\, dy \qquad (4.3)$$

The second relation is a consequence of the compatibility condition for the the second term $\hat{\underline{V}}^2$. As a particular feature, the jump of the traction (normal) component of the normal stress is zero (L. 1995, Berrehili 1997) :

$$[\![\sigma(\underline{U}^1)]\!].\underline{n}.\underline{n} = 0 \; i.e \; [\![\sigma_{22}(\underline{U}^1)]\!] = 0 \qquad (4.4)$$

This result underlines the specific role played by the shear component of the normal stress. At the second order, the interface goes on transmitting the tractions but no longer the shear stresses.

5 THE SINGULAR BEHAVIOUR OF THE SECOND OUTER TERM

Obviously relations (4.1) derive from the smoothness assumption (2.2). Thus, for a straight line the above reasoning holds true except at the end points which undergo, in the general case, elastic singularities. At a corner point of a single lap assembly for instance (fig.4) the elastic behaviour is governed by the second term of the following relation expressed in polar co-ordinates r, θ (originating from the corner point) :

$$\underline{U}(x) = \underline{U}^0(0) + kr^\alpha \underline{u}^+(\theta) + ... \qquad (5.1)$$

k is the intensity factor, $0.5 < \alpha < 1$ is the characteristic singular exponent and $\underline{u}^+$ is the associated mode (L. & S. 1987). From (4.1), it comes :

$$[\![\underline{U}^1]\!] \sim r^{\alpha-1} \; ; \; [\![\sigma(\underline{U}^1)]\!].\underline{n} \sim r^{\alpha-2} \qquad (5.2)$$

These are very singular behaviours out of the framework of finite energy solutions which forbid to use the classical energy principles.

The same situation occurs at the tip of a crack (fig.5), in that case the usual crack tip singularity ($\alpha = 0.5$ in (5.1)) governs the behaviour of the unperturbed solution $\underline{U}^0$ and conclusions (5.2) can be drawn identically.
However, using the singularity theory for non homogeneous boundary conditions (L. & S. 1987, L. 1995, 1996, Delisée 1996) provides a solution to the very singular problem by mean of superimposition. In the case of the single lap assembly, there exists a function $\underline{v}(\theta)$ such that its trace on the interface line is the very singular required behaviour, thus the solution splits in two parts :

$$\underline{U}^1(x) = r^{\alpha-1}\underline{v}(\theta) + \hat{\underline{U}}^1(x) \qquad (5.3)$$

The function $\underline{v}(\theta)$ can be numerically calculated, moreover $\hat{\underline{U}}^1$ in (5.3) is solution to a well posed problem. The main reason to this result is that $\alpha - 1$ does not belong to the set of characteristic exponents (L. 1994, 1995). Unfortunately, this last property does not hold in case of a crack tip since $\alpha - 1 = -\alpha = -0.5$ belongs to the set of characteristic exponents and gives rise to a logarithmic term. As a consequence, the outer expansion (2.1) must be revisited in the following form (L. 1996) :

$$\underline{U}^\varepsilon(x) = \underline{U}^0(x) - Ck_I\varepsilon\ln\varepsilon\; \underline{U}^1(x) + k_I\varepsilon\; \underline{U}^2(x) + ... \qquad (5.4)$$

$$\underline{U}^0(x) = \underline{U}^0(0) + k_I\sqrt{r}\; \underline{u}_I^+(\theta) + \hat{\underline{U}}^0(x) \qquad (5.5)$$

$$\underline{U}^1(x) = \frac{1}{\sqrt{r}}\underline{u}_I^-(\theta) + \hat{\underline{U}}^1(x) \qquad (5.6)$$

$$\underline{U}^2(x) = \frac{1}{\sqrt{r}}[C\ln r\; \underline{u}_I^-(\theta) + \underline{v}(\theta)] + \hat{\underline{U}}^2(x) \qquad (5.7)$$

It is assumed for the sake of simplicity that the single mode I $\underline{u}_I^+$ is activated at the crack tip, k_I is the corresponding intensity factor. $\underline{u}_I^-$ is the dual mode to $\underline{u}_I^+$ (L. & S. 1987).
Another slightly different situation occurs at the end of the line at a straight free edge (fig.6). The point is not singular, the exponent to take into account in (5.1) is $\alpha = 1$. Nevertheless, due to

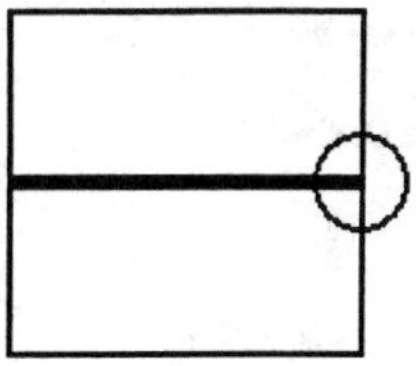

Fig. 6. The straight free edge.

the mismatch between striction in the bulk and in the strip the outer expansion includes a point force term (*) :

$$\underline{U}^{\varepsilon}(x) = \underline{U}^{0}(x) + \varepsilon[C \ln r \ \underline{e}_1 + \underline{v}(\theta) + \hat{\underline{U}}^{1}(x)] + ... \quad (5.8)$$

$\underline{e}_1$ is the outward unit normal to the edge.
It is to be pointed out that the very singular behaviours described above take place in outer expansions and then are valid only at a distance of the singular point, they express mainly a quick growing trend. The role of the micro-structure in (2.1), (5.3)-(5.8) occurs through the constant C and the function $\underline{v}(\theta)$.

6 A NEW INNER EXPANSION IN THE VICINITY OF A SINGULAR POINT

The inner expansion (3.2) was obtained after the change of variable $y_2 = x_2/\varepsilon$, it defines a boundary layer. We are now interested in the vicinity of a point and the full change of variable :

$$y = x/\varepsilon \quad i.e. \ y_1 = x_1/\varepsilon, \ y_2 = x_2/\varepsilon \quad (6.1)$$

Considering the generic singular case (5.1), the change of variable leads to define $\rho = r/\varepsilon$ and the corresponding inner expansion reads :

$$\underline{U}^{\varepsilon}(\varepsilon y) = \underline{U}^{0}(0) + k\varepsilon^{\alpha}[\rho^{\alpha}\underline{u}^{+}(\theta) + \hat{\underline{W}}^{1}(y)] + ... \quad (6.2)$$

This specific form derives from the matching conditions. One has to emphasize on the fact that this is not strictly a classical result since the geometry does not locally depend only on θ (which is the usual assumption for matched asymptotics in the vicinity of singular points (L. & S. 1987, 1992). Each term of the outer expansion has a contribution to the behaviour at infinity of the first significant inner term and there exist difficulties to overcome (L. 1996).

(*) The author is indebted to R. Abdelmoula & J.J. Marigo for discussions on this point.

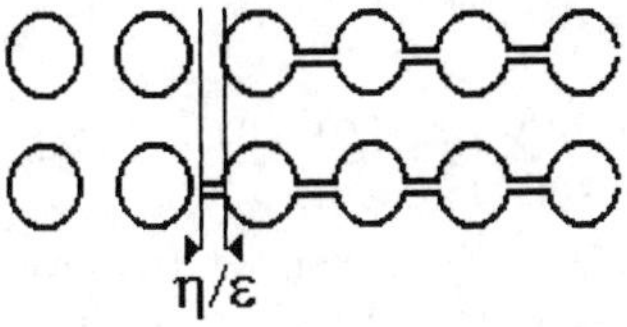

Fig. 7. A small interface crack extension.

In order to go on with matchings, it is necessary to know the behaviour at infinity of $\hat{\underline{W}}^{1}$. As above mentioned, an extension of the classical result leads to the algebraic decay :

$$\hat{\underline{W}}^{1}(y) \sim K\rho^{-\alpha}\underline{u}^{-}(\theta) \quad (6.3)$$

$\underline{u}^{-}$ is the dual mode to $\underline{u}^{+}$ (L. & S. 1987) and K the intensity factor. Eqn. (6.3) derives from the singularity theory for bidimensional elastic problems. Moreover as a consequence such a behaviour gives rise to a higher order term in the outer expansion (the exact rank j of this term depends on α) :

$$f(\varepsilon)\underline{U}^{j}(x) = kK\varepsilon^{2\alpha}[r^{-\alpha}\underline{u}^{-}(\theta) + \hat{\underline{U}}^{j}(x)] \quad (6.4)$$

7 ANALYSIS OF AN INTERFACE CRACK EXTENSION

Let us consider first a pre-existing interface crack and next the same geometry with in addition a small increment of this crack (fig.7). The (dimensionless) extension length η is assumed to be smaller or at most of the same order of magnitude than ε. The corresponding solutions are denoted respectively $\underline{U}^{\varepsilon}$ and $\underline{U}^{\varepsilon\eta}$ and we consider as above that the single mode I is activated.
If the same change of variable (6.1) is used in the two situations, the only change in the inner term (6.2) (with $\alpha = 0.5$) occurs through the function $\hat{\underline{W}}^{1}$ which depends now on η or more precisely on $\zeta = \eta/\varepsilon$:

$$\underline{U}^{\varepsilon\eta}(\varepsilon y) = \underline{U}^{0}(0) + k_I\sqrt{\varepsilon}[\sqrt{\rho}\underline{u}_I^{+}(\theta) + \hat{\underline{W}}^{1\zeta}(y)] + ... \quad (7.1)$$

Following, the most significant change in the outer expansion is due to the coefficient K of (6.3) which depends also on ζ, it is still denoted $K(\zeta)$ and moreover $K(0) = K$:

$$\underline{U}^{\varepsilon\eta}(x) - \underline{U}^{\varepsilon}(x) =$$

$$k_I[K(\zeta) - K(0)]\varepsilon[\frac{1}{\sqrt{r}}\underline{u}_I^{-}(\theta) + \hat{\underline{U}}^{j}(x)] + ... \quad (7.2)$$

The applied loads remaining unchanged, the variation in potential energy (L. 1989, L. & S. 1992) reads :

$$\delta W^{\varepsilon} = W^{\varepsilon} - W^{\varepsilon\eta} =$$

$$\frac{1}{2}\int_{\Gamma}[\sigma(\underline{U}^{\varepsilon\eta})\underline{n}\ \underline{U}^{\varepsilon} - \sigma(\underline{U}^{\varepsilon})\underline{n}\ \underline{U}^{\varepsilon\eta}]ds \qquad (7.3)$$

This integral is independent of the selected contour Γ surrounding the crack extension and starting and finishing on the stress free lips of the crack. From (7.2) and (7.3) and with the help of a biorthogonality property of the singular modes, it comes :

$$\delta W^{\varepsilon} = W^{\varepsilon} - W^{\varepsilon\eta} = k_I^2[K(\zeta) - K(0)]\varepsilon + ... \quad (7.4)$$

However, as already mentioned, the specific geometry implies that it is not a direct consequence of the previous refered results. As a consequence of the contour independence, Γ in (7.3) can be chosen either in the outer or in the inner domain. In the second option it can be as large as needed and then (7.4) is obtained as a limit.

8. THE GRIFFITH CRITERION

The Griffith criterion is based on the definition of a fracture energy per unit surface (length in the present bidimensional case) 2γ. If the released energy is greater or at least equal to the fracture energy then the crack grows. For a fracture length η, this criterion reads :

$$\delta W^{\varepsilon} \geq 2\gamma\eta \qquad (8.1)$$

Of course η in (8.1) is a physiscal crack length, it is the length of fractured material and does not include any void as illustrated on fig.7 (but not on fig.8). Lets k_{Ic} denote the toughness of the material, it is related to 2γ by :

$$2\gamma = k_{Ic}^2 K_F \qquad (8.2)$$

where K_F is the analogous to K in (6.3) in case of a crack propagating in a homogeneous material (L. 1993) (it depends of course on the material properties, it is for instance $(1-\nu^2)/E$ in plane

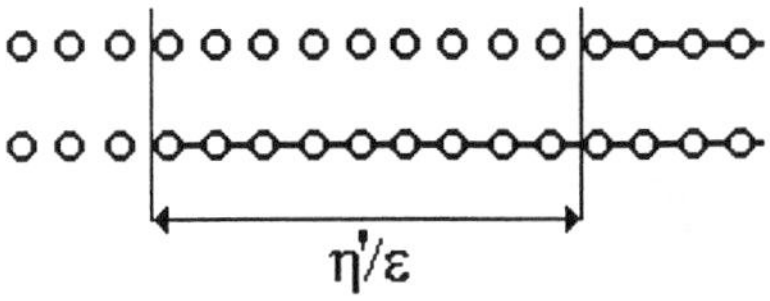

Fig. 8. A long interface crack extension.

isotropic elasticity). Hence, together with (7.4), (8.1) gives at the leading order :

$$k_I^2[K(\zeta) - K(0)] \geq k_{Ic}^2 K_F \zeta \qquad (8.3)$$

This inequality will be discussed below.

9 DISCUSSION - THE EFFECTIVE TOUGHNESS OF THE INTERFACE

There are two degrees of freedom in the above inequality, the intensity of the applied loads (and thus the intensity factor k_I) which trigger the crack growth and the fracture increment length ζ.

Let us first examine the case $\zeta << 1$ (*i.e.* $\eta << \varepsilon$). Under this assumption, η is without any discussion a physical fracture length, and the term $\hat{\underline{W}}^{1\zeta}$ in (7.1) is a perturbation of $\hat{\underline{W}}^{1}$ in (6.2) and can be expanded in powers of ζ (L. & S. 1987). In case of microvoids, the singular exponent to takes into account is $\alpha = 1$ since the small perturbation takes place along a straight edge (the radius of the void is large compared to the perturbation length) and thus :

$$\hat{\underline{W}}^{1\zeta}(y) = \hat{\underline{W}}^{10}(y) + A\zeta^2\hat{\underline{W}}^{11}(y) + ... \qquad (9.1)$$

A is a constant and $\hat{\underline{W}}^{1}$ is denoted $\hat{\underline{W}}^{10}$ for homogeneity reasons. As a consequence :

$$\frac{K(\zeta) - K(0)}{\zeta} = O(\zeta) \to 0 \text{ as } \zeta \to 0 \qquad (9.2)$$

Indeed, from experiments or even from daily custom, such interfaces break under finite applied loads, thus k_I is finite and (8.3) (with (9.2)) implies that ζ cannot be infinitely small. Fracture is made of successive (after each void) spontaneous crack increments. Moreover, this unstable process at the microscopic level has no influence on the crack growth stability at the macroscopic scale.

On the other hand, if the crack increment, denoted now η', becomes large with respect to ε ($\varepsilon << \eta' << 1$), it is no longer a physical fracture length, it embeds voids (fig.8).

The outer and inner expansions must be performed first with respect to the small parameter η' (instead of ε) using the change of variable :

$$y' = \frac{x}{\eta'} \quad ; \quad \rho' = \frac{r}{\eta'} \tag{9.3}$$

Coefficients previously depending on η' (or more precisely on $\zeta' = \eta'/\varepsilon$) are now constant while constant coefficients become functions of ε (or more precisely of $1/\zeta'$ and thus ζ') :

$$\underline{U}^{\eta'}(x) = \underline{U}^0(x) + \eta' k_I \kappa(\zeta') \left(\frac{1}{\sqrt{r}} \underline{u}_I^- + \hat{\underline{U}}^1(x) \right) + \dots \tag{9.4}$$

$$\underline{U}^{\eta'}(\eta' y') = \underline{U}^0(0) + \sqrt{\eta'} k_I \left(\sqrt{\rho'} \underline{u}_I^+ + \tilde{\underline{V}}^{1\varepsilon}(y') \right) + \dots \tag{9.5}$$

In this last expression, $\tilde{\underline{V}}^{1\varepsilon}$ is solution to a problem settled on an inner domain, with a unit crack extension and perturbated by a strip of circular cavities periodically distributed with period $\varepsilon/\eta' = 1/\zeta'$. Thus it can be expanded with respect to this parameter (see (5.4)) :

$$\tilde{\underline{V}}^{1\varepsilon}(y') = \tilde{\underline{V}}^{10}(y') + (\ln\zeta'/\zeta')\, \tilde{\underline{V}}^{11}(y') + \dots \tag{9.6}$$

The stress intensity factors $\kappa(\zeta')$ of $1/\sqrt{\rho'}\ \underline{u}_I^-$ is splitted into :

$$\kappa(\zeta') = K_F + (\ln\zeta'/\zeta') K_{11} + \dots \tag{9.7}$$

$\tilde{\underline{V}}^{10}$ in (9.6) corresponds to $\varepsilon = 0$ (unperturbed but with a unit crack extension) and thus undergoes the intensity factor K_F. K_{11} is associated with the second term of expansion (9.6). Taking into account the two expansions (6.4) and (9.4) leads to :

$$K(\zeta') = \zeta' \kappa(\zeta') \tag{9.8}$$

and then :

$$\lim_{\zeta' - \infty} \frac{K(\zeta') - K(0)}{\zeta'} = K_F \tag{9.9}$$

In this limit case, the Griffith criterion (8.3) becomes :

$$k_I^2 \geq k_{Ic}^2 \frac{\zeta}{\zeta'} \tag{9.10}$$

ζ is the (stretched) physical length of fractured material corresponding to a (stretched) crack increment ζ'. The ratio is the material parameter m defined in (3.1) and then :

$$k_I^2 \geq k_{Ic}^2 m \tag{9.11}$$

This inequality, leading to a simple criterion, was intuitively expected. It means simply that if the microstructure characteristic length is small compared to the length involved in the fracture process, then the microstructure has no specific influence. The effective fracture energy to take into account is the original value 2γ weighted by m, or in other words, the effective toughness is :

$$k_{Ic}^{eff} = k_{Ic}\sqrt{m} \tag{9.12}$$

It is interesting to to note that it arises as a limit case and not as an assumption, it corresponds to a very macroscopic model essentially ignoring the microstructure.

To conclude, let us reexamine the consequences of (9.2). The crack growth is made of successive sudden jumps at each void. We assume that this spontaneous repeated increment length is exactly the width of the ligament separing two voids, *i.e.* the crack jumps from one void to the next one at the microscopic level and thus $\zeta = m$. Eqn. (8.3) can be rewritten :

$$k_I^2 \geq k_{Ic}^2 \frac{K_F}{K(m) - K(0)} m \tag{9.13}$$

which leads to another definition of the effective toughness :

$$k_{Ic}^{eff} = k_{Ic} \sqrt{\frac{K_F}{K(m) - K(0)} m} \tag{9.14}$$

This definition requires the computation of $K(0)$ and $K(m)$ for different values of the diameter λ of the voids. This is performed by a contour integral like (7.3) (L. & S. 1987, Delisée 1996) on the infinite inner domain corresponding to the change of variable (6.1). It must be done with care.

Obviously differences observed in table 1 are not significant. Hence, if the mechanism of jump from one void to the next one is realistic, it seems that the simplified effective toughness (9.12) can be

Table 1. Comparison of the effective toughness for different ligament sizes.

m	k_{Ic}^{eff}/k_{Ic} (9.14)	k_{Ic}^{eff}/k_{Ic} (9.12)
0.7	0.82	0.84
0.5	0.71	0.71
0.3	0.56	0.55
0.1	0.34	0.32

used for a wide range of porosity of the interface. As announced at the beginning of this section, there are two degrees of freedom in (8.3). The discussion was based up to now on ζ, but it can be founded also on k_I. Hypothesis or experiments can be made to provide critical values of the applied loads (and thus of k_I) triggering the fracture. Then (8.3) can be solved with respect to the unknown ζ. Reasonable assumptions (L. 1997) tend to prove, in case of a single void, that $\eta = O(\sqrt{\varepsilon})$ which corroborates the above results.

REFERENCES

Atkinson C. & Chen C.Y. 1996. The influence of layer thickness on the stress intensity factor of a crack lying in an elastic (viscoelastic) layer embedded in a different elastic (viscoelastic) medium (mode III analysis), Int. J. Engng. Sci., 34, n°6: 639-658.

Berrehili Y. 1997. Etude du comportement effectif des matériaux composites à constituants décollés. thèse, Université de Paris-Nord, Villetaneuse.

Delisée I. 1996. Etude de mécanismes de délaminage des composites carbone/epoxy, thèse, Université P. et M. Curie (Paris 6), Paris.

Il'in A.M. 1992. *Matching of Asymptotics Expansions of Solutions of Boundary Value Problems*, Trans. of Math. Monographs, 102, American Mathematical Society.

Leguillon D. 1989. Calcul du taux de restitution de l'énergie au voisinage d'une singularité, *C. R. Acad. Sci. Paris*, 309, série II: 945-950.

Leguillon D. 1993. Asymptotic and numerical analysis of a crack branching in non-isotropic materials, *Eur. J. Mech., A/Solids*, 12, 1: 33-51.

Leguillon D. 1994. Un exemple d'interaction singularité-couche limite pour la modélisation de la fracture dans les composites, *C. R. Acad. Sci. Paris*, 319, série II: 161-166.

Leguillon D. 1995. Concentration de contrainte dans les joints adhésifs, actes du 2ème Colloque national en calcul des structures (Giens 1995), Hermès, Paris: 107-112.

Leguillon D. 1996. Influence de l'épaisseur d'une interface sur les paramètres de rupture, *C. R. Acad. Sci. Paris*, 322, série IIb: 533-539.

Leguillon D. 1997. Asymptotic analysis of a spontaneous crack growth, IUTAM symp. on Non-linear Singularities in Deformation and Flow, Technion, Haifa.

Leguillon D. & Sanchez-Palencia E. 1987. *Computation of singular solutions in elliptic problems and elasticity*, Masson, Paris, John Wiley, New-York.

Leguillon D. & Sanchez-Palencia E. 1992. Fracture in heterogeneous materials - Weak and strong singularities, in *New advances in computational structural mechanics*, P. Ladevèze & O.C. Zienkiewicz eds., Elsevier, Amsterdam: 423-434.

Nguetseng N. & Sanchez-Palencia E. 1985. Stress concentration for defects distributed near a surface, in *Local Effects in the Analysis of Structures*, P. Ladevèze ed., Elsevier, Amsterdam: 55-74.

Ryvkin M., Slepyan L. & Banks-Sills L. 1995. On the scale effect in the thin layer delamination problem, *Int. J. of Fracture*, 71: 247-271.

Ryvkin M. 1996. Mode III crack in a laminated medium, *Int. J. Solids Structures*, 33: 3611-3625.

Sanchez-Palencia E. 1980. *Non- Homogeneous Media and Vibration Theory*, Lect. Notes in Physics, 127, Springer-Verlag, Berlin.

Van Dyke M. 1964. *Asymptotic methods in fluid mechanics*, Academic-Press, New-York.

Wang S.S. & Choi I. 1982. Boundary layer effects in composite laminates. I. Free edge stress singularity. II. Free edge stress solutions and basic characterisitics, *J. Appl. Mech.*, 49, I: 541-548, II: 549-560.

Damage and Failure of Interfaces, Rossmanith (ed.) © 1997 Balkema, Rotterdam, ISBN 90 5410 899 1

Some aspects of stability and bifurcation with moving discontinuities

C. Stolz
Ecole Polytechnique, CNRS URA, Palaiseau, France

R. M. Pradeilles-Duval
DGA-DCE-CREA, Arcueil, France

ABSTRACT: The propagation of moving surface inside a body is analysed within the framework of thermodynamical couplings, when the moving surface is associated with an irreversible change of mechanical properties. The moving surface is a surface of heat sources and of entropy production, intensities of which are related to particular energy release rates defined in terms of Hamiltonian gradients. As examples, we analyse the dynamical evolution of partial damage in a bar, and some problems of bifurcation of evolution in laminates.

1 INTRODUCTION

In the recent past, the propagation of damage has been studied in connection with fracture mechanics. Different approaches based on macroscopic or microscopic description of mechanical degradation of properties have been proposed. In the framework of thermomechanical coupling as in fracture mechanics, the analysis defines two different energy release rates associated with heat production and entropy production. This paper is concerned mostly with a description of damage involved on the evolution of a moving interface along which mechanical transformation occurs.

2 GENERAL FEATURES

At each time, two materials coexist in the structure. The domain Ω is composed of two materials (Ω_1 and Ω_2) with different mechanical characteristics. The interface between them is perfect and denoted by Γ. Material 1 changes into material 2 along Γ by an irreversible process. Hence Γ moves with a normal celerity $c = \phi N$,(N is the normal to Γ outward Ω_2, then ϕ is positive along Γ). When this surface propagates, all mechanical quantities f can have a jump denoted by $[f] = f_1 - f_2$, and any volume average has a rate defined by

$$\frac{d}{dt}\int_\Omega f\, d\omega = \int_\Omega \dot{f}\, d\omega - \int_\Gamma [f]\, c.N\, da$$

We analyse the evolution of Ω under a given loading prescribed on the boundary $\partial\Omega$. Denote by n the unit normal to the boundary $\partial\Omega$ and assume that $\partial\Omega = \partial\Omega_T \cup \partial\Omega_u$, where $\partial\Omega_T$ and $\partial\Omega_u$ are disjoint parts of $\partial\Omega$ on which the stress vector ($\sigma.n = T^d$) and respectively the displacement ($u = u^d$) are prescribed.

At time t, the spatial distribution of the two materials is assumed to be known. The actual state is characterized by the displacement u, from which the strain ε is defined. The elastic behaviour of each phase is given by the free energy density w_i, function of the strain and of the temperature θ. The equations of state are

$$\sigma = \frac{\partial w_i}{\partial \varepsilon}, \quad s = -\frac{\partial w_i}{\partial \theta},$$

where σ is the stress, and s the entropy, the internal energy $(w_i + s\theta)$ is denoted by e_i.

The potential energy of the structure Ω has the following form

$$\mathcal{P}(u, \Gamma, T^d) = \int_{\Omega(\Gamma)} \rho w\, d\omega - \int_{\Omega_T} T^d.u\, da$$

The potential energy represents the global free energy in a thermodynamical description ; the posi-

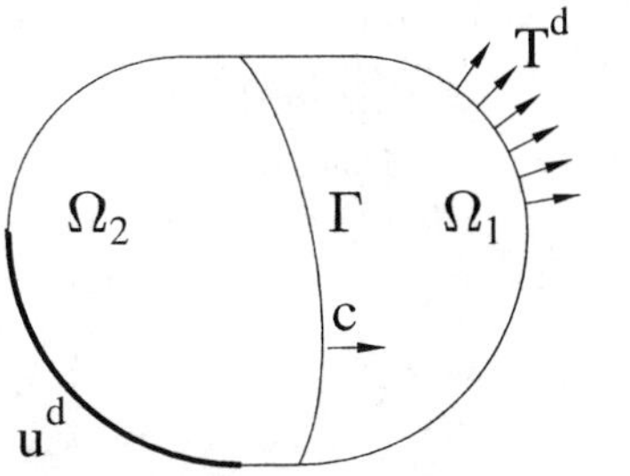

Figure 1: The boundary value problem.

tion of the interface Γ becomes an internal parameter.

The characterization of an equilibrium state is given by

$$\frac{\partial \mathcal{P}}{\partial u} \circ \delta u = 0$$

for all displacement δu kinematically admissible with $\delta u = 0$ over $\partial\Omega_u$. This variational description

$$\int_{\Omega_i} \rho \frac{\partial w_i}{\partial \varepsilon} : \varepsilon(\delta u)\, d\omega - \int_{\partial\Omega_T} T^d.\delta u\, da = 0,$$

implies the set of local equations

- local constitutive relations

$$\sigma = \rho \frac{\partial w_i}{\partial \varepsilon},$$

- equilibrium equations

$$div\sigma = 0,$$

$$[\sigma].N = 0 \text{ over } \Gamma, \ \sigma.n = T^d \text{ over } \partial\Omega_T,$$

- compatibility relations

$$2\varepsilon = \nabla u + \nabla^t u,$$

$$[u] = 0 \text{ over } \Gamma, \ u = u^d \text{ over } \partial\Omega_u.$$

The condition of a perfect interface is taken into account in the continuity of the displacement and of the stress vector along the interface.

Let us consider an evolution of the prescribed loading. The inertia effects are neglected. Some evolutions are induced in the mechanical quantities and a propagation of the interface can occur according to a given evolution law, chosen in a manner that is appropriate to describe the irreversibility of the transformation. The problem is then to characterize the evolution of Γ, that is to determine the normal propagation ϕ, during the time.

Some mechanical quantities are continuous along Γ like the displacement and the stress vector. It is necessary to characterize the evolution of any quantity defined only on Γ. Let us introduce the convected derivative D_ϕ of a function $f(x_\Gamma, t)$ defined along Γ as

$$D_\phi f = \lim_{\Delta t \to 0} \frac{f(x + \phi\nu\Delta t, t + \Delta t) - f(x,t)}{\Delta t}$$

Along Γ, at each time, the displacement and the stress vector are continuous ; then the rates of the displacement v and of the stress vector verify the compatibility equations of Hadamard

$$\begin{aligned} D_\phi[u]_\Gamma &= 0, \\ [D_\phi(\sigma.\nu)]_\Gamma &= 0. \end{aligned}$$

In the plane case, these equation are reduced to

$$\begin{aligned} [\dot\sigma]_\Gamma.\nu + [\frac{d}{ds}(\phi\sigma.t)]_\Gamma &= 0, \\ [v]_\Gamma - \phi[\nabla u]_\Gamma.\nu &= 0. \end{aligned}$$

3 DISSIPATION ANALYSIS

The mass conservation leads to the continuity of the mass flux $m = \rho\phi$, where ρ denotes the mass density. The first law and the second law of thermodynamics give rise to local equations inside the volume and along the moving surface Γ

$$\rho\dot e = \sigma : \dot\varepsilon - divq, \text{over } \Omega,$$

$$m[e] + N.\sigma.[v] - N.[q] = 0, \text{on } \Gamma.$$

Where e is the internal energy density ($e = w + \theta s$), and q is the heat flux associated to the heat conduction. Thanks to Hadamard compatibility equations, the heat power supply is given in terms of a release rate of internal energy

$$N.[q] = G_{th}\phi,$$

$$G_{th} = \rho[e] - N.\sigma.[\nabla u].N.$$

Entropy production The entropy production is given by

$$\int_\Omega (\rho\dot s + \frac{divq}{\theta} - q.\frac{\nabla\theta}{\theta^2}) d\omega + \int_\Gamma (-m[s] + N.[\frac{q}{\theta}]) da \geq 0.$$

The interface is perfect at each time. Under the assumption of separability of the two dissipations, the term inside the volume is reduced to the conduction, and the term along the surface is then

$$D_\Gamma = \frac{\rho[w] - N.\sigma.[\nabla u].N}{\theta}\phi = \frac{G_s}{\theta}\phi$$

where G_s has also the form of a release rate of energy. This quantity has an analogous form to the driving traction force acting on a surface of strain discontinuity proposed by Abeyaratne and Knowles (1990). The criterion which guides the evolution of the interface may be written as function of this quantity.

4 ENERGETICAL INTERPRETATION

The total internal energy of the structure is

$$\begin{aligned} E(u,\theta,T^d) &= \int_\Omega e\, d\omega - \int_{\partial\Omega_T} T^d.u\, da \\ &= \mathcal{P} + \int s\theta\, d\omega. \end{aligned}$$

The kinetic effects are neglected, and the first law of thermodynamics can be written as follows :

$$\frac{dE}{dt} - \frac{\partial E}{\partial T^d}.\dot T^d = -\int_{\partial\Omega} q.n\, da.$$

Taking into account of the momentum conservation, we have

$$\frac{\partial E}{\partial \Gamma} * \dot\Gamma = \int_\Gamma [q].N\, da = -\int_\Gamma G_{th}\phi\, da$$

the second law has the same form as above.

In a thermomechanical coupling, two different energy release rates must be distinguished. One defined in term of variation of the total internal energy gives rise the heat source associated with the moving surface ; the second one gives rise the production of entropy.

Isothermal evolution In that case, we can take the potential energy to describe the whole system. The total dissipation is then given by

$$\frac{d\mathcal{P}}{dt} - \frac{\partial \mathcal{P}}{\partial T^d}.\dot{T}^d = -\int_{\partial\Omega} q.n\, da = -\int_\Gamma G\phi\, da,$$

where $G = \rho[w] - N.\sigma.[\nabla u].N$.

There is only one energy release rate to characterize the propagation, it gives the heat sources and the entropy production due to the propagation of the interface.

These relations can be generalized in the dynamical case, by replacing the internal energy of the system by its Hamiltonian, and can be extended to the case of running cracks,(Stolz 1995, Stolz & Pradeilles-Duval 1996).

5 QUASISTATIC EVOLUTION

In isothermal evolution, complementary relations must be considered to describe irreversibility.

An energy criterion is chosen as a generalized form of the well known theory of Griffith. Then, we assume

$$\begin{aligned} \phi(s) &\geq 0, \text{ if } G(s) = G_c \text{ on } \Gamma, \\ \phi(s) &= 0, \text{ otherwise.} \end{aligned}$$

This is a local energy criterion. At each equilibrium state, the interface Γ, can be decomposed into two subsets where the propagation is either possible or not. Let denote by Γ^+, the subset of Γ where the critical value G_c is reached.

The evolution of the interface is governed by the consistency condition, during the evolution of Γ, at the geometrical point $x_\Gamma(t)$ the criterion is reached

$$G(x_\Gamma(t), t) = G_c$$

then the derivative of G following the moving surface is null

$$D_\phi G = 0.$$

Hence it leads to the consistency condition written for all point inside Γ^+

$$(\phi - \phi^*) D_\phi G \geq 0, \forall \phi^* \geq 0, \text{ over } \Gamma^+.$$

on the complementary part of Γ^+, $\phi = 0$.

By using Hadamard relations the derivative defined above take the final form

$$\begin{aligned} D_\phi G &= [t.\sigma].\nabla v_1.t - N.\dot{\sigma}_2.[\nabla u].N - \phi K \\ K &= t.div_\Gamma \sigma_2.[\nabla u].N - t.\sigma_2.\nabla([\nabla u].N) \\ &\quad + [\sigma : \nabla \varepsilon.N] - N.\sigma.(\nabla\nabla u.N).N] \end{aligned}$$

5.1 Local formulation of the boundary value problem.

During the quasistatic evolution of the system, the loading is increased step by step, so that each state of the body is a state of equilibirum. Then the equilibrium equations are given by

$$\frac{d}{dt}(\frac{\partial \mathcal{P}}{\partial u} \circ \delta u) = 0,$$

This variational description implies the set of local equations

- local constitutive relations

$$\dot{\sigma} = \rho \frac{\partial^2 w_i}{\partial \varepsilon^2} : \dot{\varepsilon},$$

- equilibrium equations

$$div\dot{\sigma} = 0,$$

$$D_\phi([\sigma].N) = 0 \text{ over } \Gamma, \ \dot{\sigma}.n = \dot{T}^d \text{ over } \partial\Omega_T$$

- compatibility relations

$$2\dot{\varepsilon} = \nabla v + \nabla^t v,$$

$$D_\phi([u]) = 0 \text{ over } \Gamma, \ v = v^d \text{ over } \partial\Omega_u.$$

It is a classical problem of heterogeneous elasticity, with non classical boundary conditions defined on Γ.

5.2 Global formulation

In order to perform the solution of the rate boundary value problem from the point of view of existence and uniqueness of the solution which means stability of the equilibrium state and non bifurcation of the evolution, a global formulation is developed now. In fact, the evolution is determined by the functional

$$\begin{aligned} F(v, \phi, \dot{T}^d) &= \int_\Omega \frac{1}{2}\varepsilon(v) : \frac{\partial^2 w}{\partial\varepsilon\partial\varepsilon} : \varepsilon(v)\, d\omega \\ &\quad - \int_{\partial\Omega_T} \dot{T}^d.v\, da \\ &\quad + \int_\Gamma \phi[t.\sigma].\nabla v_1.t + \frac{\phi^2}{2} K\, da. \end{aligned}$$

The solution of the rate boundary-problem defined by the above set of equations is given by :

$$\frac{\partial F}{\partial v} \circ (v - v^*) + \frac{\partial F}{\partial \phi} * (\phi - \phi^*) \geq 0$$

for all v^* kinematically admissible field and $\phi^*(s) \geq 0$ along Γ^+.

The discussion of the stability and bifurcation along an evolution process can be investigate as presented in (Pradeilles-Duval & Stolz 1995).

5.3 Stability and non bifurcation

Let W be the value of F at the solution in the rate of displacement v of the local equations for a given velocity of propagation ϕ

$$div\dot{\sigma} = 0, \ \dot{\sigma} = \frac{\partial^2 w}{\partial \varepsilon \partial \varepsilon} : \varepsilon(v)$$

$$D_\phi(\sigma.N) = 0, \ D_\phi([u]) = 0.$$

Then

$$W(\phi, \dot{T}^d) = F(v(\phi, \dot{T}^d), \phi, \dot{T}^d)$$

The stability of the actual state is determined by the condition of the existence of a solution

$$\delta\phi \frac{\partial^2 W}{\partial \phi \partial \phi} \delta\phi \geq 0, \ \delta\phi \geq 0 \text{ on } \Gamma^+, \delta\phi \neq 0$$

and the uniqueness and non bifurcation is characterized by

$$\delta\phi \frac{\partial^2 W}{\partial \phi \partial \phi} \delta\phi \geq 0, \ \delta\phi \neq 0 \text{ on } \Gamma^+.$$

6 COMPOSITE SPHERES ASSEMBLAGE

In this section, the composite spheres assemblage of Hashin is analyzed, (Christensen & Lo 1979).

The system is composed by the compact assemblage of spheres with external radii in order to fill the whole domain. The microscopic structure is constituted by composite spheres with a core made with material 2 and the shell by material 1, both materials are linear elastic and homogeneous.

As in the general case, materials 1 transforms into material 2 ; the transformation is irreversible and the criterion is a generalized Griffith's criterion based on the energy release rate of the transformation. The concentration of material 2 is denoted by c. Applying the same method than in (Hervé & Zaoui 1991), the assemblage is considered as well-disordered. Using the particular three phase model, the homogeneous equivalent medium denoted by material 0 is unknown. In phase i, the local characteristics are the bulk modulus denoted by k_i and shear modulus by μ_i.In what follows, k_1 is assumed to be larger than k_2.

6.1 Macroscopic behavior with one family of spheres.

There exists only one family of composite spheres in the structure ; then c is the concentration of material 2. Using analytical results obtained in (Hervé & Zaoui 1991), one gets the bulk modulus of the material 0, denoted by k_o

$$k_o = k_1 + c\frac{(k_2 - k_1)}{1 + \frac{3(k_2-k_1)(1-c)}{(4\mu_1+3k_1)}}$$

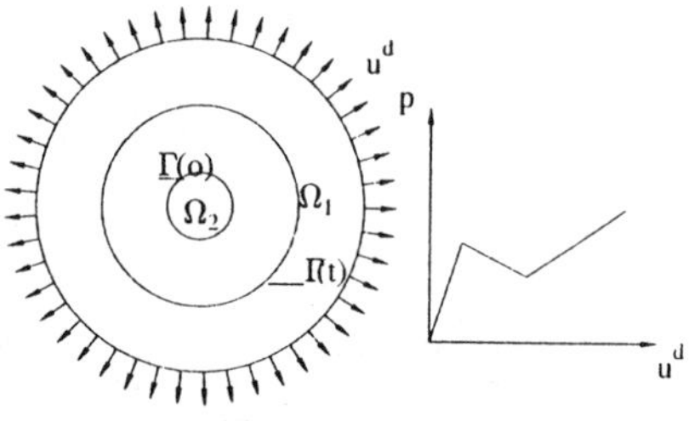

Figure 2: The composite sphere.

On the interface, the energy release rate is

$$G = \frac{\theta_o^2(4\mu_1 + 3k_1)(4\mu_1 + 3k_2)(k_1 - k_2)}{2(4\mu_1 + 3k_2 + 3c(k_1 - k_2))^2}$$

where θ_o represents the uniform strain ($\varepsilon = \theta_o \mathbf{I}$) given at the infinity. When a generalized Griffith criterion is taken into account for the damage transformation, as G reaches the critical value G_c, the ratio c increases such that G remains equal to G_c. The behavior takes the form plotted in figure 2.

6.2 Macroscopic behavior with two families of spheres.

In what follows, we consider the macroscopic evolution of a composite spheres assemblage when two different families coexist in the structure. They are supposed to be perfectly disordered. c_I (respectively c_{II}) denotes the volume fraction of material 2 in the first family (in the second one). If we denoted by G_i the release rate of energy for the family i, we can shown the inequality

$$(G_I - G_{II})(c_I - c_{II})(\mu_1 - \mu_2) > 0.$$

So, the global behavior of the system is the following one : at the beginning the macroscopic behaviour is linear elastic, until the energy criterion is reached and at this moment

- If $\mu_1 > \mu_2$, the difference between the two concentrations $(c_I - c_{II})$ increases until the larger reaches 1.
- If $\mu_1 < \mu_2$, the difference between the two concentrations decreases until there are identical. That means that the two different families become only one.
- If $\mu_1 = \mu_2$, both concentration could increase.

When whole volume is transformed, the behavior is the mechanical behavior of material 2.

Even if the system is composed by only one family, the local response to the loading increment is non unique. In fact, there can exist many kinds of bifurcations : one part of the structure can be damaged (damage localization- no more disorder in the domain). Very well-ordered configurations can appear and specific space distributions of the constituents phases are obtained

One gets order among disorder. If $\mu_1 > \mu_2$ then a new perfectly disordered family can appear along the first one. In that case there is more disorder in the structure.

Here, it is to be underlined that the total dissipation is only due to the change of mechanical characteristic along a moving surface. The macroscopic behavior is dissipative while the components are always in a reversible process. The transformation between the two material corresponds to a volume damage at the macroscopic scale.

7 DYNAMICAL CASE

We must take into account inertia effects. Then, the two thermodynamical principles must be rewritten.

The first law and the second law of thermodynamics give rise to local equations inside the volume and along the moving surface Γ

$$\rho\dot{e} = \sigma : \dot{\varepsilon} - divq \text{ in } \Omega,$$

$$m[e + \frac{v^2}{2}] + N.[\sigma.v] - N.[q] = 0 \text{ over } \Gamma.$$

The displacement is continuous and the momentum is conserved along the moving interface

$$[u] = 0, \ [\sigma].N = m[v].$$

Then the heat power supply is given in terms of a release rate of internal energy

$$N.[q] = G_{th}\phi,$$

$$G_{th} = \rho[e] - N.\bar{\sigma}.[\nabla u].N$$

where $\bar{\sigma} = \frac{1}{2}(\sigma_1 + \sigma_2)$.

Entropy production The entropy production is given again by

$$\int_\Omega (\rho\dot{s} + \frac{divq}{\theta} - q.\frac{\nabla\theta}{\theta^2})\, d\omega + \int_\Gamma (-m[s] + N.[\frac{q}{\theta}])\, da \geq 0.$$

Under the assumption of separability of the two dissipations, the term along the surface is then

$$D_\Gamma = \frac{\rho[w] - N.\bar{\sigma}.[\nabla u].N}{\theta}\phi = \frac{G_s}{\theta}\phi$$

where $G_s = \rho[w] - N.\bar{\sigma}.[\nabla u].N$ has also the form of a release rate of energy.

8 HAMILTONIAN INTERPRETATION

The total Hamiltonian of the structure is

$$H = \int_\Omega \frac{1}{2}\rho v^2 d\omega + \mathcal{P} + \int_\Omega \rho s\theta d\omega$$

The equations of the momentum conservation are then :

$$\frac{\partial H}{\partial p} \bullet \delta p = \int_\Omega v.\delta p \, d\omega,$$

$$\frac{\partial H}{\partial u} \bullet \delta u = -\frac{d}{dt}\int_\Omega p.\delta u \, d\omega,$$

where p is the momentum, these equations lead to the classical equations of movement.

The first law of thermodynamics can be written as follows

$$\frac{dH}{dt} - \frac{\partial H}{\partial T^d}.\dot{T}^d = -\int_{\partial\Omega} q.n \, da$$

and taking into account of the momentum conservation, we have

$$\frac{\partial H}{\partial \Gamma} * \dot{\Gamma} = \int_\Gamma [q].N \, da = -\int_\Gamma G_{th}\phi \, da$$

the second law has the same form as previously.

In a thermomechanical coupling, two different release rates must be distinguished, one defined in term of variation of the Hamiltonian gives rise the heat source associated with the moving surface, the second one is associated with the production of entropy.

Isothermal evolution In that case, we can define another Hamiltonian

$$H = \int_\Omega \frac{1}{2}\rho v^2 d\omega + \mathcal{P}$$

and the total dissipation is then given by

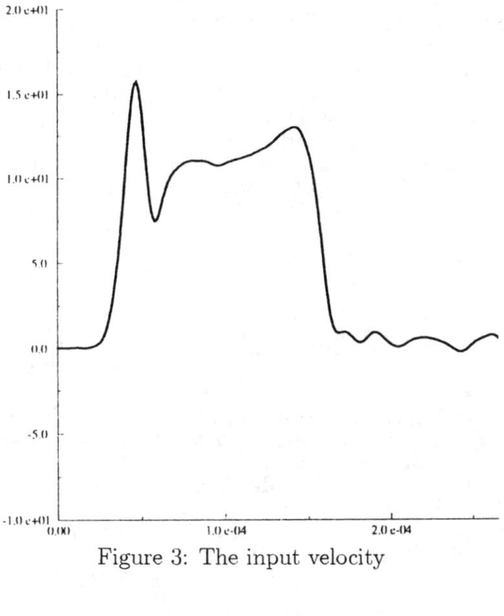

Figure 3: The input velocity

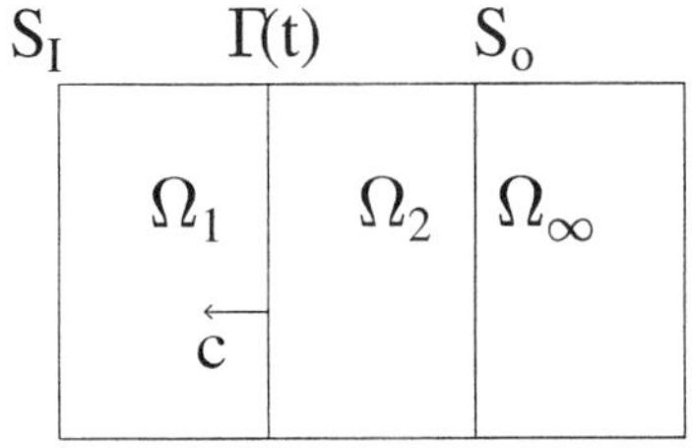

Figure 4: The inpact test

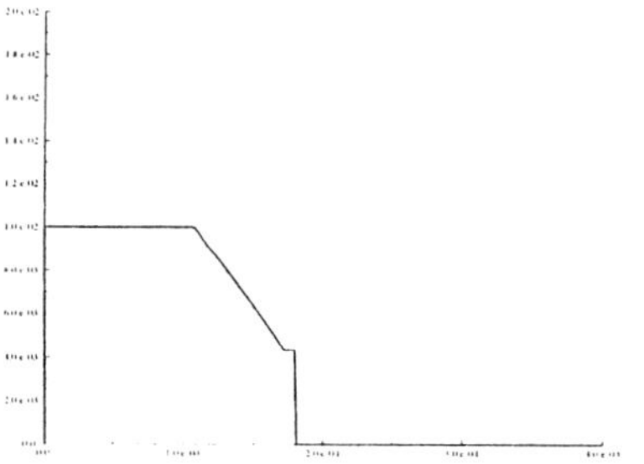

Figure 5: The position of the interface.

$$\frac{dH}{dt} - \frac{\partial H}{\partial T^d}.\dot{T}^d = \int_{\partial\Omega} -q.nda = -\int_{\Gamma} G_{dyn}\phi da$$

where $G_{dyn} = \rho[w] - N.\bar{\sigma}.[\nabla u].N$. In this case, the driving traction acting on the interface of Abeyratne and Knowles is recovered (Abeyaratne & Knowles 1990).

These relations generalize the case of quasistatic evolution, where the potential energy plays the rule of the Hamiltonian. A similar analysis has been given for the case of running cracks,(Stolz 1995, Stolz & Pradeilles Duval 1996).

As an example, a bar composed by two materials is considered, one of which is transformed into the second one when the thermodynamical force G_{dyn} reaches a critical value G_c. The results for such a modelling can be compared with impact test on quasi-brittle material with Hopkinson bars. The results presented on the figures (3,4.5), are obtained for a given velocity history prescribed on the input section S_I. It can be noticed that the interface propagates with a constant celerity in the beginning of the propagation.

9 GENERALIZED MEDIA

The definition of the potential energy, or of the Hamiltonian of the system, is only due to the existence of a free energy. Then, all definition can be generalized for media modelled by beams and plates or shells. The delamination separates the laminate into two laminates. Assume that the laminate is modelled by a beam, the delamination is simulated as an irreversible transformation from a

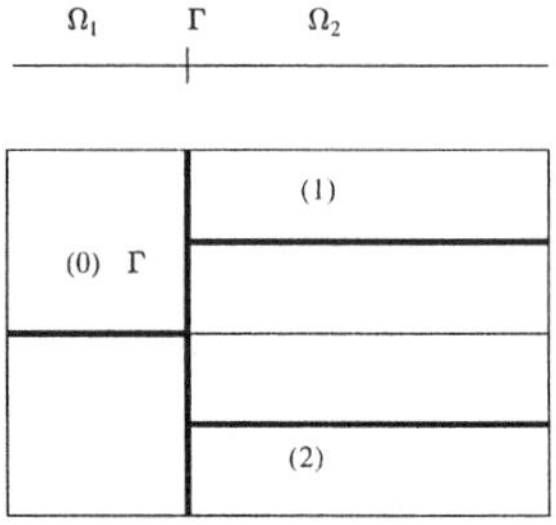

Figure 6: The delamination of laminates as beams assemblage.

sound beam to an assemblage of two beams, as proposed in figure (6).

Along the interface, we have some continuity of the displacement (u, v) and of the rotation ϕ of the section

$$\begin{aligned} u_i &= u_o - h_i\phi_i, \\ v_i &= v_o, \\ \phi_i &= \phi_o. \end{aligned}$$

In this case, the jump of f is defined by $[f] = f_o - (f_1 + f_2)$, The free energy of the beam i is

$$w_i = \frac{E}{2}\{S_i(u')^2 + I_i(\phi')^2\} + \frac{\mu S_i}{8}(v' - \phi')^2$$

where S_i is the section, I_i is the inertia of the section, E is the Young modulus, μ is the shear modulus, and f' is the derivative of f with respect of curvilinear coordinate.

Then the energy release rate is determined by

$$G = -[w] \leq G_c$$

and the functionnal F is

$$\begin{aligned} F(\dot{u}, \dot{v}, \dot{\phi}, c) &= \sum_i \frac{E}{2}[S_i(\dot{u}_i')^2 + I_i(\dot{\phi}_i')^2] \\ &+ \sum_i \frac{\mu S_i}{8}(\dot{v}_i' - \dot{\phi}_i)^2 \\ &+ \frac{c^2}{2}[T\phi'] - P_e(\dot{u}, \dot{v}, \dot{\phi}) \end{aligned}$$

where c is the rate of propagation of the delamination, and P_e the power of the rate of external loading.

The evolution of the system is governed by :

$$\begin{aligned} 0 \leq & \frac{\partial F}{\partial \dot{u}}.(\dot{u} - \dot{u}^*) + \frac{\partial F}{\partial \dot{v}}.(\dot{v} - \dot{v}^*) \\ & + \frac{\partial F}{\partial \dot{\phi}}(\dot{\phi} - \dot{\phi}^*) + \frac{\partial F}{\partial c}(c - c^*). \end{aligned}$$

Hence the problem has the same nature as above. The study of stability and bifurcation can be investigated in the same terms as previously.

10 CONCLUSIONS

This paper shows how local mechanical transformations can influence the global behavior. Even if the determination of a global quantity by average process guarantees the uniqueness of a macroscopic mechanical variable, the localization process seems to be very important because the determination of the macroscopic behavior assumes that the local response is unique when the external loading is changing.

It is to be underlined that, even if at local scale, the components are linear elastic, the macroscopic behavior is no more non dissipative. When initially there exists strains and self equilibrated stresses in the structure, the propagation of the interface will increase their influence on the mechanical macroscopic behavior.

In dynamics, the definition of the Hamiltonian is only defined by a potential energy and a kinetic energy. In the same spirit, the definition of the functional F is governed only by the existence of a potential energy, and by the definition of the thermodynamical forces associated to the propagation. Then all definitions of the thermodynamical forces in terms of energy release rate are valid for all cases, for which a potential energy exists. Some applications have been performed in generalized media, such as beams and plates, to study the delamination of laminates. Some particular cases are then presented.

REFERENCES

Abeyaratne, R., Knowles, J., 1990. On the driving traction acting on a surface of strain discontinuity in a continuum, *J. Mech. Phys. of Solids*, 38, 3, pp.345-360

Pradeilles-Duval, R.M., Stolz, C. 1995. Mechanical transformation and discontinuities along a moving surface, *J. Mech. Phys. of Solids*, 43, 1,pp.91-121

Stolz, C. 1988. Sur les équations de la dynamique des milieux continus anélastiques, *C. R. Acad. Sci. Paris*, t 307, Série II,b, pp.1997-2000

Pradeilles-Duval, R.M., Stolz, C. 1991. Sur les problèmes d'évolution des solides avec changement de phase irréversible, *C.R. Acad. Sci. Paris*, t 313, Série II,b, pp.297-302

Stolz, C., Pradeilles-Duval, R.M. 1996. Approches énergétiques des discontinuités mobiles en dynamique non linéaire, *C.R. Acad. Sci. Paris*, t 322, Série II,b, p.525-532

Stolz, C., 1995. Functional Approach in non linear dynamics, *Arch. Mech*, 47, 3, pp.421-435

Hervé E., Zaoui A. 1991. N-Layered inclusion-based micromechanical modelling. *Int. J. Eng. Sci.*, 31(1):1-10.

Christensen R.M., Lo K.H 1979. Solutions for effective shear properties in three phase sphere and cylinder models. *J. Mech. Phys. Solids*, 27, 3315-330.

Interface degradation and delamination

Damage and Failure of Interfaces, Rossmanith (ed.) © 1997 Balkema, Rotterdam, ISBN 90 5410 899 1

Modelling transverse matrix cracking in laminated fibre-reinforced composite structures

S. Li, S. R. Reid & P. D. Soden
Department of Mechanical Engineering, UMIST, Manchester, UK

ABSTRACT: This paper summarises a recently developed damage model (Li et al., 1996, referred to as LRS hereafter) which describes the effects of transverse matrix cracking in fibre-reinforced laminates of arbitrary layup in the context of continuum damage mechanics. The model is constructed by establishing an appropriate damage representation and a damage growth law. It results in a new laminate theory which describes the deformation of laminates under loads as well as the development of the damage in the form of crack multiplication. This enables practical predictions to be made of the behaviour of fibre-reinforced laminates experiencing transverse matrix cracking.

1 INTRODUCTION

In laminated structures composed of uniaxially fibre-reinforced composite laminae, transverse matrix cracking is one of the most common types of damage arising when such structures are loaded. Many publications have been devoted to micromechanics based investigations of this problem (e.g. Altus & Ishai 1986, Garrett & Bailey 1977, Hashin 1985, Herakovich et al. 1988, Highsmith & Reifsnider 1982, Lewinski & Telega 1996, McCartney 1990 and Swanson 1989) although most of them address only cross-ply laminates. Because of the nature of this type of damage, usually involving a large number of cracks in a representative volume of the material, continuum damage mechanics (CDM) has been employed by several authors (Allen et al., 1987a, 1987b and Talreja, 1985a, 1985b and 1986) to model its effects. A CDM representation of damage expresses mathematically the effects of damage at any given level. Talreja (1985a) used a vectorial damage variable, approached the problem of damage in composites in a systematic way and applied the results directly to the case of damage resulting from matrix cracks. Applications of the theory can be found in his subsequent publications (Talreja, 1985b and 1986). For transverse matrix cracks in uniaxially fibre-reinforced composites, a vectorial damage variable is appropriate and beneficial.

Talreja's (1985a) work appears to be the first attempt of applying the theory of CDM systematically to modelling the effects of damage in fibre-reinforced composite materials. As an early initiative of employing CDM to composite materials Talreja's contribution is remarkable. However, in reviewing it, the following comments can be made:

(1) Talreja treated a whole laminate as a single material with which the damage variable is associated. As a result, all the damage-related material constants are defined for this "material". They depend not only on the properties of the constituent materials in the laminate but also on the layup configuration of the laminate. This makes the implementation of this theoretical approach impractical because of its heavy dependence on experiments.

(2) Another shortcoming of treating a laminate as a material is that it ignores the structural nature of the laminate completely. The position of the cracked lamina in the laminate makes a difference to the behaviour of the laminate. Although such an approach may be acceptable for cases involving only membrane loading and deformation where the layup and the damage distribution have middle-plane symmetry, it is not appropriate when, for instance, bending is present.

(3) In Talreja's (1985a) original work, the damage variable was defined as an unspecified function of crack density (or crack surface area). In his later applications (e.g. Talreja, 1985b) this

function was taken to have a form which results in all the effective material properties being proportional to the crack density. However, from experimental data (Highsmith & Reifsnider, 1982) and from the analyses of cracked laminates (Hashin, 1985), the effective Young's modulus of the cracked laminate in the direction perpendicular to the cracks showed significant nonlinearity with respect to the crack density. The same situation arises with the effective in-plane shear modulus of a cracked laminate. Thus, the applicability of Talreja's damage representation is limited to a small range of crack density in which linearity is a reasonable approximation to a nonlinear curve but this range may not be sufficient for practical problems and, therefore, this aspect too needs to be addressed.

In a recent publication by the authors (LRS) a lamina-based damage representation has been proposed in order to rectify all the above shortcomings.

Having established a lamina-based damage representation, a damage model for transverse matrix cracking can be completed by introducing a damage growth law describing the evolution of damage. The concept of a damage surface has been introduced for establishing damage growth laws in several studies (e.g. Krajcinovic & Fonseka, 1981). This plays a similar role to the yield surface in plasticity. However, as was noted in Krajcinovic & Fonseka (1981), the construction of damage surfaces has suffered from the lack of experimental data and the process has had to be based on a number of arbitrary assumptions. Removing these arbitrary assumptions from the theory is not an easy task and is unlikely to be achieved for general cases in the short term. It therefore seems appropriate at this stage to establish particular damage growth laws for specific problems using the concept of the damage surface.

The growth of damage in the form of crack multiplication in a representative volume is the main feature of damage growth which is considered herein. The damage growth process can be idealized by the problem of generating a third crack between two existing neighbouring cracks. Further cracking like this is controlled by the stress state in the material between the two existing cracks although stress analyses of cracked laminates (Hashin, 1985 and Li et al., 1994) show that the highest stress intensities always appear in local regions around the crack tips at interlaminar interfaces. The high stresses around the crack tips are relevant to the whole damage process, especially to the initiation of other damage modes (Altus & Ishai, 1986) but are unlikely to affect the crack multiplication process and therefore their effects will be ignored here. The region of second highest stress level is in the middle between two cracks and, therefore, a new crack is most likely to appear there. This implies that the increase in crack density is by a doubling process rather than by continuous growth. However, in reality, the process of crack multiplication appears to be defect-sensitive and more often than not cracks are initiated from voids. Due to the random distribution of defects in the material, the cracks will not be regularly spaced even if the nominal global stresses are uniform. The cracks which are generated between existing pairs of cracks in a uniformly loaded lamina will therefore not appear at the same time but will do so separately, so that a continuously varying crack density increase can only be justified in a statistical sense, reflected by an increase in the damage variable. Within the continuum idealization, the crack density, and hence the damage variable, is assumed to be a field variable defined at every point in the material indicating, in a statistical sense, the distribution of cracks in the representative volume surrounding the point.

The damage model has been established along the lines outlined above and it will be presented briefly in the subsequent sections followed by some applications.

2 DAMAGE REPRESENTATION

Given a laminate containing transversely cracked laminae, the major assumption introduced in the damage representation is that the behaviour of all the laminae, cracked and uncracked, can be effectively described by plane-stress states. The implication is that the effects of the out-of-plane stresses including the transverse (through-thickness) direct stress and the two transverse shear stresses can be neglected. These stresses are present in reality as a result of the transverse matrix cracks, especially around the crack tips but are of little significance in the global behaviour of the lamina. This has been justified in LRS in a systematic manner.

In general, constitutive relations can be established with the help of the Helmholtz free energy subject to the constraints imposed by the second law of thermodynamics (Colemen & Noll, 1963). For problems involving small strain and small damage, the Helmholtz free energy can be approximately expressed in the form of a Taylor's expansion truncated at the second order terms of strains and the damage. Use can be made of the integrity basis of invariants of the internal state variables, the strain tensor and the damage vector, to minimise the number of material constants by taking account of all the symmetries the material exhibits (Pipkin & Rivlin, 1959). This procedure has been performed by Talreja (1985a). Still within the same theoretical framework, modifications have been suggested in LRS. After carefully performed manipulations, one obtains the effective material properties of a damaged lamina as

$$\begin{aligned} E_1 &= E^0_1 \\ E_2 &= E^0_2\,(1-\omega) \\ G_6 &= G^0_6\,(1 - k\omega) \\ \nu_{12} &= \nu^0_{12} \\ \nu_{21} &= \nu^0_{21}\,(1-\omega). \end{aligned} \tag{1}$$

where ω is referred to as the *damage parameter*, whose physical meaning is the relative change in the transverse Young's modulus E_2 of the lamina. k is the only damage related material constant. Attempts have been made successfully to determine k theoretically for a given lamina in a laminate in LRS. The above damage representation has been verified in LRS.

The damage representation can now be applied to analyses of laminates with cracked laminae using a conventional laminate theory, e.g. classical laminate theory (CLT), in which the cracked laminae can be replaced by some fictitious homogeneous materials with the effective material properties obtained from the damage representation at defined levels of damage. This will be developed in the next section.

3 CONSTITUTIVE RELATIONS

The stress-strain relation for a lamina, designated by subscript ℓ, in its local coordinate system (aligned with the principal directions of the material) is

$$\sigma_\ell = Q_\ell\, \epsilon_\ell \tag{2}$$

where the stress (tensor) σ_ℓ and strain (tensor) ϵ_ℓ involve only in-plane components since a plane stress state is assumed. In eq.(2),

$$Q_\ell = Q_\ell^0 \quad \text{(if lamina } \ell \text{ is uncracked)} \tag{3}$$

$$Q_\ell = Q_\ell^0 + Q_\ell'\omega_\ell \quad \text{(if lamina } \ell \text{ is cracked)} \tag{4}$$

where Q_ℓ^0 and Q_ℓ' are constant matrices which can be obtained in a straightforward manner from the material properties given in eq.(1).

Denote the stress, strain and the stiffness (tensor) of lamina ℓ in the laminate's global coordinate system by τ_ℓ, γ_ℓ and q_ℓ, respectively. Using the coordinate transformations

$$\tau_\ell = T_\ell\, \sigma_\ell\,, \qquad \gamma_\ell = T_\ell^{-T}\, \epsilon_\ell\,, \qquad q_\ell = T_\ell\, Q_\ell\, T_\ell^T \tag{5}$$

where T_ℓ is the coordinate transformation matrix of lamina ℓ. Relation (2) can be transformed to the global coordinate system as

$$\tau_\ell = q_\ell\, \gamma_\ell\,. \tag{6}$$

In CLT, strains in a lamina are associated with the generalised strains e of the laminate, as a consequence of the Love-Kirchhoff hypothesis, by

$$\gamma_\ell = L_\ell\, e \tag{7}$$

where $e = [\, e_1, e_2, e_3, e_4, e_5, e_6\,]^T$ are the generalised strains as in CLT and L_ℓ is a constant matrix for the ℓ-th lamina if the average stresses over the thickness of the lamina is used to represent the stress state in the lamina as is the case in the subsequent development.

For the laminate, the generalised stress-strain relation is given as

$$s = D\, e \tag{8}$$

where $s = [\, s_1, s_2, s_3, s_4, s_5, s_6\,]^T$ are the stress resultants as in CLT and D is the counterpart of the conventional A, B and D matrix assembly in CLT, which includes the effects of damage at a given level.

A problem involving damage is, in general, a nonlinear one. Conventional solution techniques for nonlinear problem are usually based on an incremental approach and this will be followed in the present study. In order to incorporate this incremental constitutive relation into the laminate analysis described above, it is desirable to formulate an incremental relation between the stress resultants and generalised strains.

The incremental form of the stress, strain and damage relation for lamina ℓ can be obtained from eq.(6) as

$$d\tau_\ell = \frac{\partial \tau_\ell}{\partial \gamma_\ell} d\gamma_\ell + \frac{\partial \tau_\ell}{\partial \omega_\ell} d\omega_\ell = q_\ell d\gamma_\ell + q_\ell' \gamma_\ell d\omega_\ell \tag{9}$$

where $\quad q_\ell' = T_\ell\, Q_\ell'\, T_\ell^T\,.$ (10)

Integrating eq.(9) over the thickness of the laminate in a manner similar to that used in obtaining eq.(8) results in

$$ds = D\,de + \sum_\ell D_\ell' e\, d\omega_\ell \tag{11}$$

where de and ds are the incremental generalised

strains and the stress resultants of the laminate, respectively, and D' can be traced back to Q_ℓ' in eq.(4). Only cracked laminae in the laminate contribute to the second term on the right-hand side of eq.(11) and therefore the summation is only over all the cracked laminae.

Equation (11) relates the incremental stress resultants to the incremental generalised strains but involved also in it are the incremental damage parameters of all the cracked laminae which cannot be determined within the laminate theory. Extra information is required to eliminate these incremental damage parameter from eq.(11). This will be pursued in the next section by considering the damage growth.

4 DAMAGE SURFACE AND DAMAGE GROWTH

Microscopic examinations of cracked laminates show that most cracks span the full thickness of the lamina in which they occur and are arrested by the interfaces between the lamina and the adjacent ones. This suggests that the propagation of cracks through the thickness of the lamina, which is likely to be an unstable process, is not a significant feature in the overall behaviour of a laminate undergoing transverse matrix cracking and, therefore, can be ignored. This leads to the simplification that the behaviour of the material bounded by two cracks in the cracked lamina is determined by the average stresses over the thickness of the lamina, to be referred to hereafter as the stresses in the lamina. Thus, the damage surface for lamina ℓ can be assumed to be defined in terms of the stresses σ_ℓ and the damage ω_ℓ in the lamina. The damage surface introduced in LRS is

$$F(\sigma_\ell) = 1 + h_\ell \omega_\ell^{\eta_\ell} \tag{12}$$

where F can be inherited from a conventional failure criterion while parameters h_ℓ and η_ℓ are two damage growth-related material constants of the ℓ-th lamina which can be explained on the basis of Weibull's size effect theory(Manders et al., 1983 and Wisnom, 1991).

An infinitesimal change of the damage state in lamina ℓ as a result of an infinitesimal change in the stress resultants ds applied to the laminate requires the satisfaction of the following equation so that the internal state of stress, strain and damage remains on the damage surface,

$$\frac{\partial F}{\partial \sigma_\ell}\frac{\partial \sigma_\ell}{\partial s}ds + \frac{\partial F}{\partial \sigma_\ell}\sum_j \frac{\partial \sigma_\ell}{\partial \omega_j}d\omega_j - h_\ell \eta_\ell \omega_\ell^{\eta_\ell - 1} d\omega_\ell = 0 \tag{13}$$

where j covers every cracked lamina reflecting the influence of the damage in lamina j on the internal state in lamina ℓ.

Eliminating ds from eqs.(11) and (13), after a lengthy derivation, the incremental damage parameter of lamina ℓ can be obtained in terms of incremental generalised strain as

$$d\omega_\ell = B_\ell \, de \tag{14}$$

where
$$B_\ell = \frac{\dfrac{\partial F}{\partial \sigma_\ell} Q_\ell T_\ell^T L_\ell}{h_\ell \eta_\ell \omega_\ell^{\eta_\ell - 1} - \dfrac{\partial F}{\partial \sigma_\ell} Q_\ell' T_\ell^T L_\ell e} \, . \tag{15}$$

The incremental stress resultant-generalized strain relation, which has taken account of the growth of damage in all the laminae of the laminate, can then be obtained by substituting eq.(14) into eq.(11) as

$$ds = D_d \, de \tag{16}$$

where
$$D_d = D + \sum_\ell D_\ell' e B_\ell \, . \tag{17}$$

Equation (16) looks similar to the stress resultant-generalised strains relation eq.(8) and indeed it is the incremental form of it. Built into it is the influence of damage growth of laminae experiencing transverse matrix cracking. Matrix D_d gives the tangent stiffness of the laminate at the current deformation and damage state. It can be used in an incremental laminate analysis involving transverse matrix cracking. In such an analysis, when the stress resultant increment ds is applied, the generalised strain increment de can be determined from eq.(16) as

$$de = D_d^{-1} \, ds \, . \tag{18}$$

Substituting this back to eq.(14), $d\omega_\ell$ can be obtained.

It should be pointed out that a negative value produced by eq.(14) for $d\omega_\ell$ indicates unloading in lamina ℓ and eq.(14) should be replaced by

$$d\omega_\ell = 0 \tag{19}$$

because of the irreversible nature of damage. It means that unloading does not follow the same path as that of loading. This behaviour provides the mechanism in the theoretical model which allows energy dissipation as a result of damage development in the material. When unloading is identified in a lamina, the lamina should be excluded from the summation in eq.(17). This operation discounts the

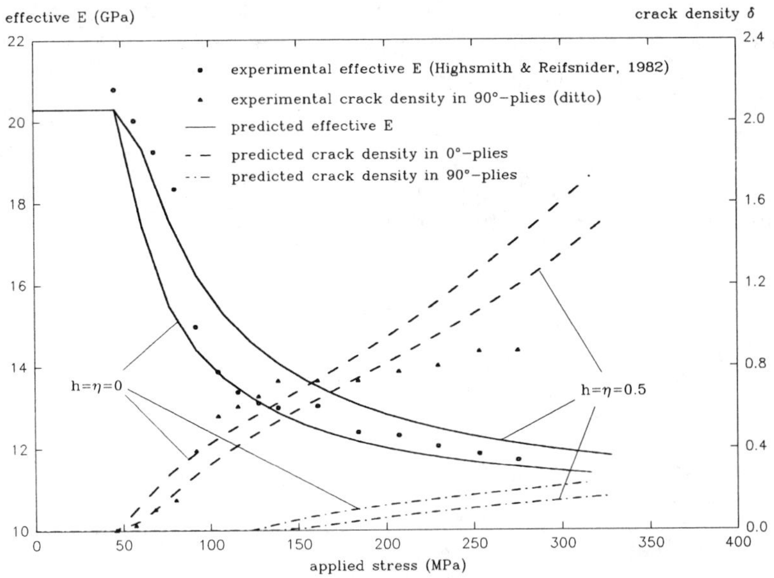

Figure 1 Effective Young's modulus and crack density versus applied stress

contribution of the damage growth in this particular lamina but not the existing damage in it. When the damage in a lamina stops growing, the damage state stays fixed and the effects of fixed damage have been included in the other part of the right-hand side expression, D , of eq.(17).

5 EXAMPLES

The model for the damage in the form of transverse matrix cracking in laminated composites has been presented in the previous sections. Its applicability is not subject to any restriction from the layup of the laminate, such as symmetric or asymmetric, balanced or not, cross-ply or non-cross-ply. In other words, it applies to laminates of uni-directional laminae with arbitrary construction provided that the mode of damage is in the form of transverse matrix cracking. It is also capable of dealing with any combination of loads expressed in terms of stress resultants. Two cases will be studied here, a cross-ply laminate under uniaxial tension and an angle-ply laminate under biaxial tension. When defining the damage surface in the analysis, the maximum stress criterion (Tsai & Hahn, 1980) is employed to define the function F in eq.(12) for the sake of simplicity but there is no restriction in the theory to using any other criterion. The size effects in the context of damage growth has not been involved in any previous treatment of damage of composite materials and, therefore, the parameters h and η are not available. For the purpose of illustration, two sets of values, $h=\eta=0$, corresponding to the case where no size effect is present, and $h=\eta=0.5$ have been chosen on a somewhat arbitrary basis.

5.1 *A cross-ply laminate under uniaxial tension*

A glass/epoxy cross-ply laminate, $[0°/90°_3]_s$, was tested by Highsmith & Reifsnider (1982). It is one of the most frequently cited cases in the literature (Hashin, 1985 and Talreja, 1985b). However, in all the previous analyses, predictions of the behaviour of the laminate were made when the crack density was prescribed and, therefore, the theoretical stress-strain curve has never been reproduced independently of experiments. With the new damage model, this is now possible.

The elastic material constants were given in Hashin (1985) and Highsmith & Reifsnider (1982). In order to construct the damage surface, the strength properties are also required but they have never been provided in the literature. As a crude estimate, an interpolation of the strength properties has been performed assuming them to be proportional to the ratio of the corresponding elastic constants of a similar material (Li et al., 1993). The predicted results for the stiffness of the laminate and the crack density in the cracked lamina versus applied stress curves are shown in Fig.1. The two sets of curves corresponding to $h=\eta=0$ and $h=\eta=0.5$ each give reasonable predictions for the stiffness reduction in different parts

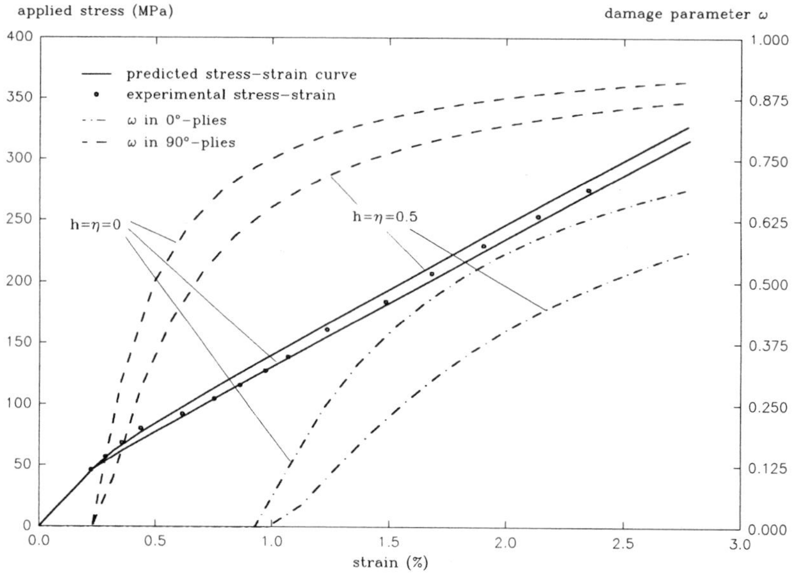

Figure 2 Applied stress and the damage parameter versus strain

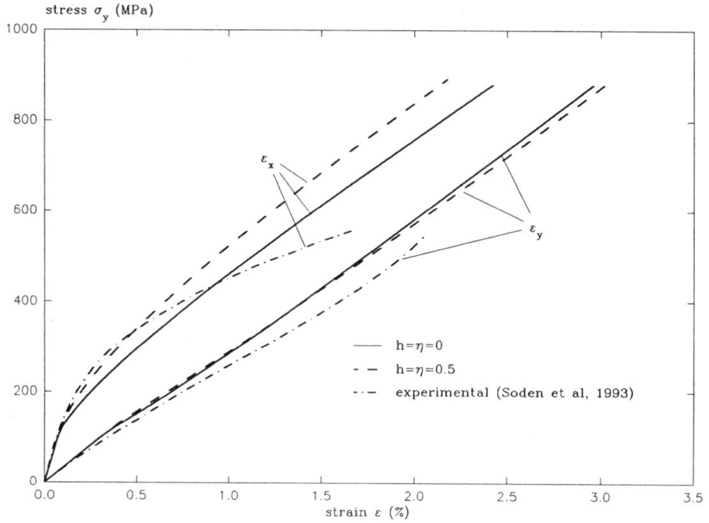

Figure 3 Theoretical and experimental stress-strain curves of a ±55° laminate under biaxial tension at 2:1 ratio

of the curves, the former at higher stress levels and the latter at lower stresses. The predicted crack densities for these two sets of parameters seem to embrace the experiment values reasonably well. In general, with the size effect included, the laminate shows higher stiffness and slower crack density growth than without it. This illustrates the size effects in a straightforward manner. Fig.2 shows the stress-strain curves which are in excellent agreement with the experimental data. Included also in Fig.2 are the predictions of the damage parameters in both the 90° and 0° laminae.

It is interesting to note that the 0° plies on both sides of the laminate start to crack at a stress of about 125MPa. This phenomenon was not reported in

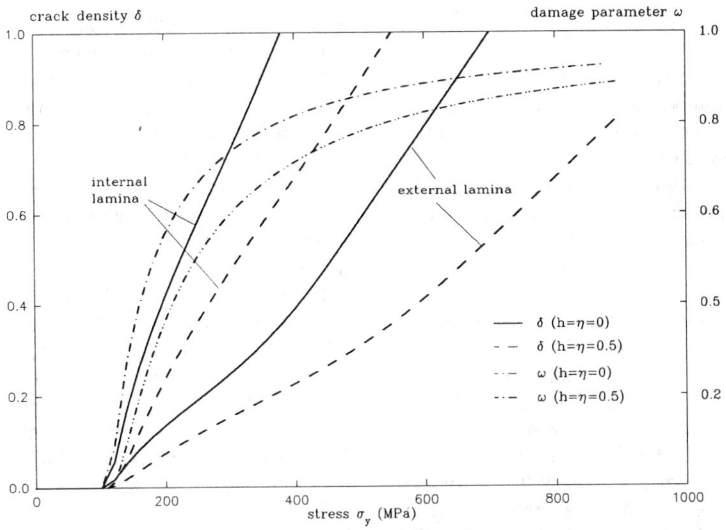

Figure 4 Predicted crack density and damage parameter versus applied load for the ±55° laminate under biaxial tension at 2:1 ratio

Highsmith & Reifsnider (1982) but it is plausible that it occurs as the result of the Poisson's ratio effect. The width of the specimen would affect the result greatly due to the free edge effects. Further experiments are encouraged in order to verify this.

5.2 *An angle-ply laminate under biaxial tensile loads*

A glass/epoxy [+55°/-55°/+55°/-55°] angle-ply laminate is analysed as a second example. The laminate is subjected to biaxial tensile loads applied in a 2:1 ratio. This is the basic loading condition for a cylindrical pressure vessel. Experimental stress-strain data for the internal pressurisation of a closed-ended cylinder are given by Soden et al.. (1993) and the appropriate material properties can be found in Li et al. (1993). Because the laminate layup is antisymmetric, there would be some twisting as a result of in-plane tension if the laminate were free to deform. To simulate a cylindrical pressure vessel, such twisting is suppressed in the analysis.

Stress σ_y (σ_y:σ_x = *2:1*) averaged over the thickness of the laminate is plotted against the strains ϵ_x and ϵ_y in Fig.3, respectively. It can be seen that ϵ_y is not too sensitive to size effects because the behaviour in the *y*-direction of the laminate is fibre-dominated and not sensitive to matrix cracking. ϵ_x behaves differently since in this direction, the matrix in the material plays a significant role. The comparison with the experimental data (Soden, et al., 1993) is encouraging. Without the damage model, a linear prediction would produce straight lines as extensions of the initial linear segments for both ϵ_x and ϵ_y. Neither crack density data nor any other damage measure are available from the experiments. Therefore comprehensive comparisons between experiments and theory are not possible. However the predictions of various aspects of the damage are given in Fig.4. The predicted crack density in each lamina has been non-dimensionalised with respect to the thickness of the lamina. As has been shown in the previous example, the crack density is in general sensitive to size effects. Due to the angle-ply layup of the laminate and the loading, all the laminae behave the same in terms of the stresses and strains referred to the local material axes and the damage parameter. It is interesting to note that the calculated crack density in an external lamina is significantly different from that in an internal lamina. This is due to the different constraints to which an external lamina and an internal lamina are subjected. A crack reduces stresses to a greater extent in an external lamina than in an internal one.

6 CONCLUSIONS

A damage representation for cracked laminates appropriate to the particular mechanism of damage by transverse matrix cracking has been presented. The application of the model has been made and illustrated through the examples analysed. Good agreement with experimental results has been shown.

REFERENCES

Allen,D.H., Harris,C.E. & Groves,S.E. 1987. A theoretical constitutive theory for elastic composites with distributed damage ——— I & II. Int. J. Solids Struct. 23: 1301-1338

Altus,E. & Ishai,O. 1986. Transverse cracking and delamination interaction in the failure process of composite laminates. Comp. Sci. Tech. 26:59-77

Coleman,B.D. & Noll,W. 1963. The thermodynamics of elastic materials with heat conduction and viscosity. Arch. Rational Mech. Anal. 13:167-178

Garrett,K.W. & Bailey,J.E. 1977 Multiple transverse fracture in 90° cross-ply laminates of a glass fibre-reinforced polyester. J. Mater. Sci.12:157-168

Hashin,Z. 1985 Analysis of cracked laminates: a variational approach. Mech. of Mater. 4:121-136

Herakovich,C.T., Aboundi,J., Lee,S.W. & Strauss, E.A. 1988. Damage in composite laminates: effects of transverse cracks. Mech. of Mater. 7:91-107

Highsmith,A.L. & Reifsnider,K.L. 1982. Stiffness-reduction mechanisms in composite laminates. Damage in Composite Materials. ASTM-STP-775, 103-117

Krajcinovic,D. & Fonseka,G.U. 1981. The continuum damage mechanics of brittle materials, Part I and II. J. Appl. Mech. 48:809-824

Lewinski,T. and Telega,J.J. 1996 Stiffness loss of laminates with aligned intralaminar cracks, I & II. Arch. Mech. 48: 245-280

Li,S., Soden,P.D., Reid S.R. & Hinton,M.J. 1993. Indentation of laminated filament-wound composite tubes. Composites. 24:407-421

Li,S., Reid,S.R. & Soden,P.D. 1994. A finite strip analysis of cracked laminates. Mech. of Mater. 18: 289-311

Li,S., Reid,S.R. & Soden,P.D. 1996. A continuum damage model for transverse matrix cracking in laminated firbre-reinforced composites. Submitted for publication

McCartney,L.N. 1990. A three-dimensional stress-transfer model for cross-ply laminates of finite width containing transverse cracks. Proc. IMechE. 4th Int. Conf. FRC'90:19-26

Manders,P.W., Chou,T.-W. Jones,F.R. & Rock,J.W. 1983. Statistical analysis of multiple fracture in 0°/90°/0° glass fibre/epoxy resin laminates. J. Mater. Sci. 18:2876-2889

Pipkin,A.C. and Rivlin,R.S. 1959. The foundation of constitutive equations in continuum physics. Arch. Rational Mech. Anal., 4:129-144

Soden,P.D., Kitching,R., Tse,P.C. and Tsavalas,Y. 1993. Influence of winding angle on the strength and deformation of filament-wound composite tubes subjected to uniaxial and biaxial loads. Composites Sci. Tech., 46:363-378

Swanson,S.R. 1989. On the mechanics of microcracking in fibre composite laminates under combined stress. J. Eng. Mater. Tech. 111:145-149

Talreja,R. 1985a. A continuum mechanics characteriz-ation of damage in composite materials. Proc. R. Soc. Lond. A339:195-216

Talreja,R. 1985b. Transverse cracking and stiffness reduction in composite laminates. J. Composite Mater. 19:355-375

Talreja,R. 1986. Stiffness properties of composite laminates with matrix cracking and interior delamination. Eng. Fract. Mech. 25:751-762

Tsai,S.W. & Hahn,H.T. 1980. Introduction to Composite Materials. Westport CT: Technomic

Wisnom,M.R. 1991. Relationship between strength variability and size effect in unidirectional carbon fibre/epoxy. Composites. 22:47-52

Damage and Failure of Interfaces, Rossmanith (ed.) © 1997 Balkema, Rotterdam, ISBN 90 5410 899 1

Rate-dependent interface models for the analysis of delamination in polymer-matrix composites

A.Corigliano & M.Ricci
Dipartimento di Ingegneria Strutturale, Politecnico di Milano, Italy
R.Frassine
Dipartimento di Chimica Industriale e Ingegneria Chimica, Politecnico di Milano, Italy

ABSTRACT: The present paper focuses on the time-dependent behaviour of delamination properties in PEI (poly-ether-imide) polymer-matrix composites. In order to simulate delamination rate effects, a visco-plastic softening interface law is introduced in which the fracture energy G_c depends on the applied load velocity. A discussion on the identification of interface model parameters and numerical simulations of Double Cantilever Beam tests are presented.

1 INTRODUCTION

Interlaminar crack propagation in polymer-based composite materials has been widely investigated in the past decade, since this type of failure plays a fundamental role in designing damage-resistant structural components. The ability of laminates to withstand crack propagation in an interlaminar plane (i.e. the resin-rich region between adjacent plies), for example, strongly affects the compression strength of a composite panel after impact.

It has been demonstrated that both the toughness of the matrix and the adhesion between matrix and fiber play a role in determining the interlaminar fracture toughness of a laminate, although the resin fracture toughness can only partly be transferred to the composite (Bradley 1989), owing to the constraint set up by the fibres on the development of a large plastic zone.

Furthermore, due to the viscous nature of the polymeric matrices, the interlaminar strain energy release rate of laminates has been found to vary with loading rate and testing temperature by more than one order of magnitude in both thermosetting- and thermoplastic-resin composite materials (see for example Frassine et al. (1993), Friedrich et al. (1989), Aliyu and Daniel (1985)).

The aim of this work is to investigate the rate and temperature dependence of the interlaminar fracture of poly(etherimide)/carbon-fibre unidirectional laminates, with special emphasis on the role played by the polymeric matrix. Fracture toughness vs. crack speed data for the matrix and the composite were experimentally determined (Frassine and Pavan 1995) at different displacement rates for temperatures varying between 23° and 170°C. Tensile tests were also performed on the neat resin.

In order to simulate and interpret experimental data, an approach for the numerical simulation of interlaminar fracture processes which takes rate effects into account is proposed, as recently done in Corigliano et al. (1997). To this purpose the delamination is simulated through the use of a *softening visco-plastic interface law*, coupled with interface finite elements. The interface law, which relates tractions to displacement discontinuities, is such that the energy required to overcome the interface resistance depends on the rate of displacement discontinuity. Hence the variation of interlaminar fracture energy with the loading rate can be reproduced.

The simulation of composite behaviour through the use of time-dependent constitutive laws has been pursued e.g. in Popelar and Kanninen (1980), Nemes and Spéciel (1996), while the use of interface models for the simulation of composite delamination has been recently proposed e.g. in Allix and Ladeveze (1992), Corigliano (1993), Allix et al (1995), Schellekens and de Borst (1993).

The summary of the paper is as follows. In section 2 the experimental evidences which motivate the study are briefly presented; in section 3 the visco-plastic interface model is illustrated, together with a discussion on the identification of material parameters and on the numerical aspects concerning time integration of the interface law and the step-by step procedure for finite element analyses. Examples of 2D finite element simulations concerning DCB delamination tests are presented in section 4 and compared with experimental results.

2 EXPERIMENTAL EVIDENCES

2.1 *Neat resin*

The rate-dependent uniaxial behaviour of the neat PEI (polyetherimide) was examined. Tensile tests at varying displacement rates and temperatures were carried out. Using the time-temperature equivalence postulate valid for most polymers (Ferry 1980) the *master curve* of the maximum stress was obtained that covers several decades of strain rate (fig. 1). A marked increase with the strain rate is observed, which is typical of most polymers.

2.2 *Composite*

Unidirectional laminates were prepared from carbon-fibre /poly(etherimide) prepregs, 0.3 mm thick, with nominal fibre content of 58% by weight (AS4-3K-PEI by Ten Cate AC). The 16-ply laminates were compression moulded at 310°C for 60 min, with a compaction pressure of 10 bar for 30 min. For interlaminar testing, a non-adhering film, 15 μm thick, was inserted at the midplane of the laminate.

Interlaminar fracture tests were conducted following the ESIS protocol (Davies 1992), on Double Cantilever Beam (DCB) specimens 20 mm-wide and 170 mm-long, having an initial crack length of 60 mm. Depending on the moulding conditions the thickness was varying between 3.6 and 4.2 mm.

The crack growth was monitored by observing the crack tip on the specimen's lateral surface. Tests were conducted at constant displacement rates varying from 0.5 to 500 mm/min and temperatures ranging from 23° to 130°C. Notice that these loading rates induce crack propagation velocities much less than the longitudinal wave propagation velocity, i.e. dynamic crack effects can be neglected.

Toughness values obtained at different temperatures were shifted along the logaritmic crack speed axis, using the same time-temperature equivalence postulate already applied for the resin. The *master curve* so obtained, covering several decades of crack speed (Fig. 2) shows a smoothly, slightly increasing trend with crack speed.

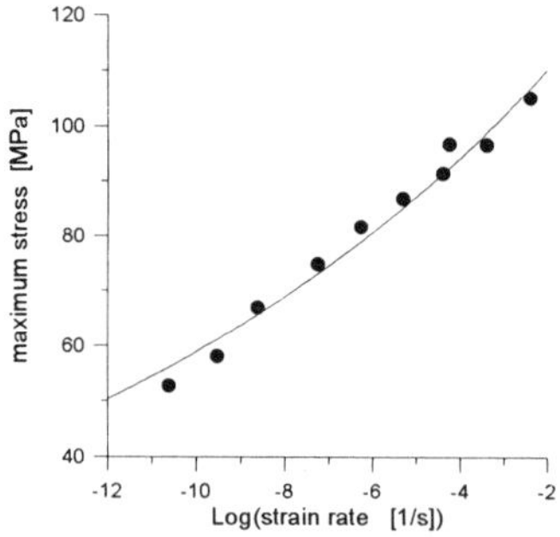

Figure 1 Maximum stress of the resin vs. strain rate *master* curve at T_0=23°C. The line is a power-law least-squares fitting.

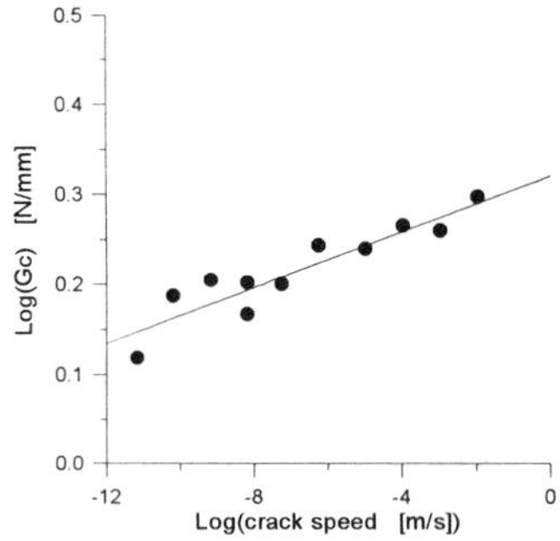

Figure 2 Fracture toughness of the composite vs. crack speed *master* curve at T_0=23°C. The line is a power-law least-squares fitting.

Details are given in Frassine and Pavan (1995).

3 A VISCO-PLASTIC INTERFACE MODEL

3.1 *Model description*

A *visco-plastic* softening interface law is here introduced as a relation between interface tractions **t** and displacement discontinuities $[\mathbf{u}] \equiv (\mathbf{u}_+ - \mathbf{u}_-)$. It is used to simulate decohesion processes, due to delamination in layered composites, which occur along a surface Γ with normal **n**, dividing the body Ω in two parts Ω^+ and Ω^- (see fig. 3).

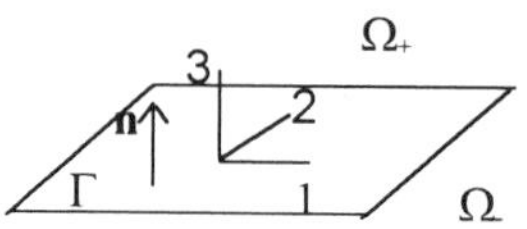

Figure 3. Reference frame for the interface

The relations governing the interface behaviour are as follows:

$$[\mathbf{u}] = [\mathbf{u}]^e + [\mathbf{u}]^{vp} \quad (1)$$

$$\mathbf{t} = \mathbf{K}[\mathbf{u}]^e; \quad \mathbf{K} = diag(K_i) \;\; i = 1,2,3 \quad (2)$$

$$[\dot{\mathbf{u}}]^{vp} = \gamma < f(\mathbf{t},\lambda) >_+^N \frac{\partial f(\mathbf{t},\lambda)}{\partial \mathbf{t}} \quad (3)$$

$$f(\mathbf{t},\lambda) = \sqrt{a_1 t_1^2 + a_2 t_2^2 + a_3 < t_3 >_+^2} - 1 + h\lambda \quad (4)$$

$$\lambda = \int_0^\tau \sqrt{\left([\dot{\mathbf{u}}]^{vp}\right)^T \left([\dot{\mathbf{u}}]^{vp}\right)}\, d\tau' \quad (5)$$

Equation (1) gives the total displacement discontinuity vector $[\mathbf{u}]$ as the sum of an elastic $[\mathbf{u}]^e$ part and a visco-plastic irreversible one $[\mathbf{u}]^{vp}$. In eq. (2) the elastic behaviour is expressed in a decoupled way, through the introduction of a diagonal interface stiffness matrix **K**. The flow rule for the visco-plastic strain rate $[\dot{\mathbf{u}}]^{vp}$ is given in eq. (3), where γ is a *fluidity* parameter. A Perzyna (1966), associate kind of law has been chosen. The symbol $< \bullet >_+$ denotes the positive part of $\bullet$.

The function $f(\mathbf{t},\lambda)$, defining the elastic domain, is given by eq. (4). The positive part of t_3 is introduced in eq. (4) in order to have an *unilateral* effect concerning the interface behaviour in direction 3, normal to the interface (see. fig. 3). The positive variable λ, as defined in eq. (5), is a norm of the cumulated visco-plastic strain rates. λ evolves governing, together with parameter h, the hardening or softening law for the model in point. Here h is assumed to be positive, thus obtaining a softening law in which the current elastic domain, shown in fig. 4, progressively reduces to the negative part of the t_3 axis only.

The response of the interface model in pure mode loading has been evaluated at varying rate of displacement discontinuity and power law parameter N (eq. 3). The following set of parameters has been chosen for the parametric study.

$$K = 100000. \left[N/mm^3\right]; \quad a = 1/60^2 \left[mm^4/N^2\right]$$

$$\gamma = 30. \quad \left[N/mm\,s\right]; \quad h = 47.5 \quad \left[mm^{-1}\right]$$

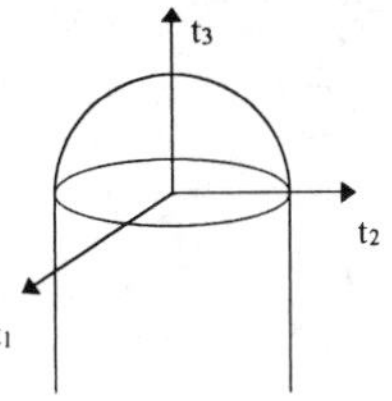

Figure 4. Elastic domain for the interface

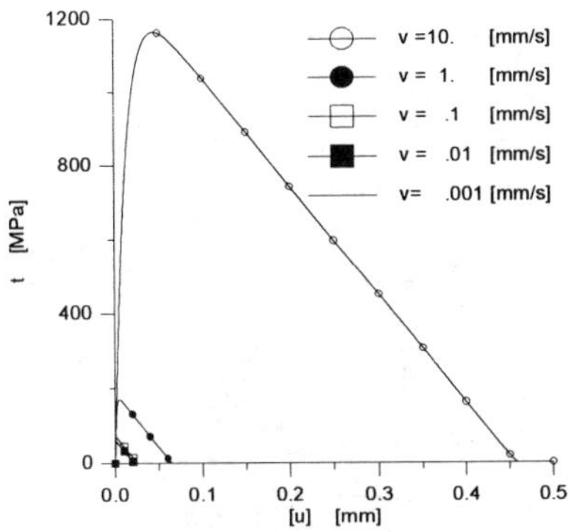

Figure 5a. Traction-displacement discontinuity plots for the interface. $\gamma = 30$ [N/mms]; h= 47.5 [mm^{-1}]; $N = 1$.

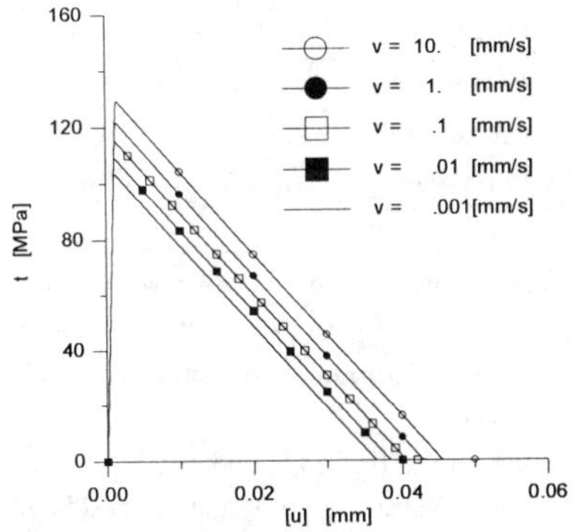

Figure 5b. Traction-displacement discontinuity plots for the interface. $\gamma = 30$ [N/mms]; h= 47.5 [mm^{-1}]; $N = 20$.

In fig. 5 a,b the traction versus total displacement discontinuity plots are shown at varying imposed $v = [\dot{u}]$ in the range $10^{-3} \div 10$ $[mm/s]$ for N=1 and N=20 respectively.

Figures 5 show that the maximum value t_{max} of traction and the area below the traction-displacement discontinuity plot, i.e. the fracture energy G_c, both increase at increasing velocity. Moreover, the influence of the power law parameter N is remarkable. This can be better appreciated in fig. 6a,b

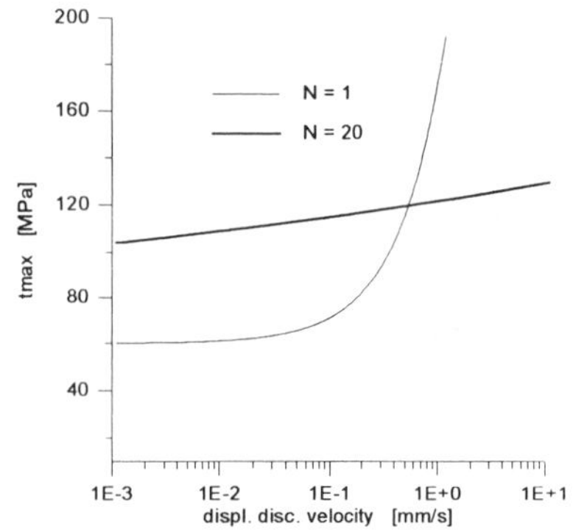

Figure 6a. t_{max} -displacement discontinuity velocity plots for the interface. $\gamma = 30$ [N/mms]; h= 47.5 [mm^{-1}].

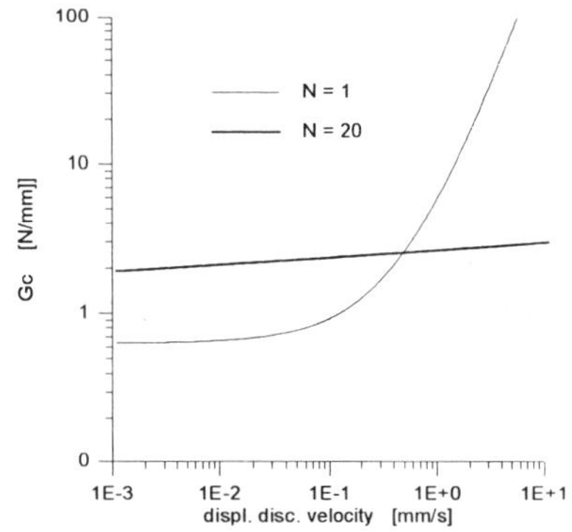

Figure 6b. G_c- displacement discontinuity velocity plots for the interface. $\gamma = 30$ [N/mms]; h= 47.5 [mm^{-1}].

where t_{max} and G_c are plotted versus the displacement discontinuity velocity $v = [\dot{u}]$ for $N = 1$ and $N = 20$ in semi-logarithmic and double logarithmic scale diagrams respectively.

It appears from the above fig. 6a,b that the behaviour of the neat resin (fig. 1) and the composite (fig. 2) can be reproduced only for $N >> 1$.

3.2 *Model parameter identification*

The visco-plastic interface law here discussed depends on parameters K_i, a_i $i = 1,2,3$, γ, N, h.

Assuming that the interface is a resin-reach region, the parameters governing its behaviour can be, at least in the first approximation, determined starting from the properties of the neat resin.

The identification of interface stiffnesses K_i can be carried out as already proposed in recent works (see e.g. Corigliano 1993) starting from the elastic moduli of the resin:

$$K_1 \cong \frac{2G_{13}}{e};\ K_2 \cong \frac{2G_{23}}{e};\ K_3 \cong \frac{E_3}{e} \tag{6}$$

In the above equations: G_{13}, G_{23} are the resin shear moduli for shear in planes 13, 23 respectively; E_3 is the resin Young modulus for direction 3; e is the thickness of the interface.

Parameters a_i are directly linked to the values t_{0i} of tractions corresponding to the onset of yielding in pure mode of the resin:

$$a_i = \frac{1}{t_{0i}^2} \quad i = 1,2,3 \tag{7}$$

In this paper the mode-I DCB delamination specimen is considered; therefore only K_3 and a_3 are required for simulations (cp. Fig. 3). For the material under consideration the following set of parameters has been chosen and introduced in eqs. (6) and (7):

$$E_3 = 3000\ [MPa];\ e = .015\ [mm];\ t_{03} = 60\ [MPa]$$

The last parameters (γ, N and h) govern the evolution of the visco-plastic displacement discontinuity rate through eqs. (3) and (4). They can be identified by fitting the model response to some rate-dependent property of the resin.

Experimental results obtained on the neat resin loaded in tension at different strain rates (see fig. 1), have been interpolated using the following power-law equation:

$$\sigma_{max} = \alpha\dot{\varepsilon}^M;\quad \alpha = 128\ [MPa];\ M = .034 \tag{8}$$

In order to fit the t_{max}-displacement discontinuity velocity plot obtained from the interface model (see fig. 6a) with the experimental curve (8) it is necessary to change the variable $\dot{\varepsilon}$, introducing the interface thickness e as follows:

$$\sigma_{max} = \alpha\dot{\varepsilon}^M = \frac{\alpha}{e^M}[\dot{u}]^M \tag{9}$$

The fitting gives the following interface model parameters

$$\gamma = .18\quad [N/mms];\ h = 33.33\ [mm^{-1}];\ N = 16$$

The experimental and the theoretical curves are shown, in a semi-logarithmic plot, in fig. 7.

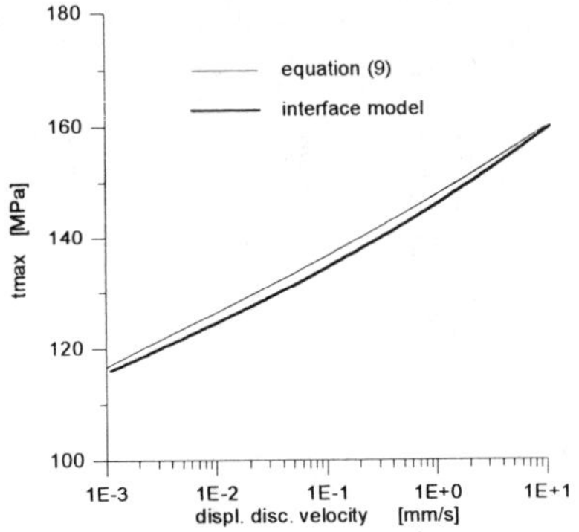

Figure 7. Experimental and modelled t_{max} -displacement discontinuity velocity plots.
$\gamma = .18$ [N/mms]; $h = 33.33$ [mm^{-1}]; $N = 16$.

The consequences of the strong assumption here done (i.e. that the interface behaves like a thin layer of neat resin) are shown in section 4 on the global response of DCB specimens.

3.3 *Time integration of the interface law and finite element formulation*

The visco-plastic interface model presented in section 3.2 is used in step by step finite element analyses. The numerical time integration of the interface law along a time step $\Delta\tau = (\tau_{n+1} - \tau_n)$ is carried out by means of a Runge-Kutta kind algorithm (see also Corigliano et al. 1997). The visco-plastic loading case is governed by equations in box 1:

$$\text{if} \quad (f(\mathbf{t}_n,\lambda_n))^N > 0$$

$$\mathbf{t}_{n+1} = \mathbf{K}\left([\mathbf{u}]_{n+1} - [\mathbf{u}]^{vp}_{n+1}\right)$$

$$[\mathbf{u}]^{vp}_{n+1} = [\mathbf{u}]^{vp}_n + \Delta\tau\left((1-\vartheta)[\dot{\mathbf{u}}]^{vp}_n + \vartheta[\dot{\mathbf{u}}]^{vp}_{n+1}\right),\ \vartheta \in [0,1]$$

$$\lambda_{n+1} = \lambda_n + \Delta\tau\left((1-\vartheta)\dot{\lambda}_n + \vartheta\dot{\lambda}_{n+1}\right)$$

$$[\dot{\mathbf{u}}]^{vp}_{n+1} = [\dot{\mathbf{u}}]^{vp}_n + \left(\left.\frac{\partial[\dot{\mathbf{u}}]^{vp}}{\partial\mathbf{t}}\right|_n\right)^T \Delta\mathbf{t} + \left(\left.\frac{\partial[\dot{\mathbf{u}}]^{vp}}{\partial\lambda}\right|_n\right)^T \Delta\lambda$$

$$\dot{\lambda}_{n+1} = \dot{\lambda}_n + \left(\left.\frac{\partial\dot{\lambda}}{\partial[\dot{\mathbf{u}}]^{vp}}\right|_n\right)^T \Delta[\dot{\mathbf{u}}]^{vp}$$

Box 1. Elastic-visco-plastic case for the time-integrated interface constitutive law.

In box 1 subscripts n and n+1 mean quantities computed at the beginning and at the end of the step respectively; the symbol $\Delta(\bullet)$ means the increment of the quantity $(\bullet)$ in the step, ϑ is an interpolation parameter.
The above equations result in an explicit relation giving the updated interface traction vector in a time step:

$$\mathbf{t}_{n+1} = \mathbf{t}_n + \mathbf{K}^*_n\Delta[\mathbf{u}] - \Delta\mathbf{q}_n \tag{10}$$

with $\mathbf{K}^*_n$ and $\Delta\mathbf{q}_n$ given by:

$$\mathbf{K}^*_n \equiv \left(\mathbf{I} + \Delta\tau\vartheta\mathbf{KAB}\right)^{-1}\mathbf{K} \tag{11a}$$

$$\mathbf{A} \equiv \left(I - \Delta\tau\vartheta\left(\frac{\partial[\dot{\mathbf{u}}]^{vp}}{\partial\lambda}\right)^T_n \frac{\left([\dot{\mathbf{u}}]^{vp}_n\right)^T}{\dot{\lambda}_n}\right)^{-1} ; \ \mathbf{B} \equiv \left(\frac{\partial[\dot{\mathbf{u}}]^{vp}}{\partial\mathbf{t}}\right)^T_n \tag{11b}$$

$$\Delta\mathbf{q}_n \equiv \Delta\tau\mathbf{K}^*_n\left([\dot{\mathbf{u}}]^{vp}_n + \Delta\tau\vartheta\mathbf{A}\left(\frac{\partial[\dot{\mathbf{u}}]^{vp}}{\partial\lambda}\right)^T_n \dot{\lambda}_n\right) \tag{11c}$$

In the numerical simulations here presented, concerning delamination specimens, the interface Γ is thought to belong to an elastic continuum. The discretized form of equilibrium is obtained after time discretization along the time interval and spatial discretization by finite elements. Interface elements are used along Γ. At time instant t_{n+1} equilibrium reads:

$$\mathbf{S}_\Omega\mathbf{U}_{n+1} + \int_\Gamma \mathbf{B}^T_\Gamma\mathbf{t}_{n+1}d\Gamma = \mathbf{P}_{n+1} \tag{12}$$

where $\mathbf{S}_\Omega$ is the elastic stiffness matrix of the continuum part of Ω, $\mathbf{B}_\Gamma$ represents the operator which relates interface displacement discontinuities $[\mathbf{u}]_{n+1}$ to the global displacement vector $\mathbf{U}_{n+1}$ while $\mathbf{P}_{n+1}$ are equivalent nodal loads.

Due to the fact that an explicit time integration rule has been chosen for the interface, the solution in the time step $\Delta\tau = (\tau_{n+1} - \tau_n)$ is straightforwardly obtained after introduction of eq. (10) in eq. (12).

$$(\mathbf{S}_\Omega + \mathbf{S}_{\Gamma n})\Delta\mathbf{U} = \mathbf{P}_{n+1} - \int_\Omega \mathbf{B}^T_\Omega\sigma_n d\Omega - $$
$$-\int_\Gamma \mathbf{B}^T_\Gamma\mathbf{t}_n d\Gamma + \int_\Gamma \mathbf{B}^T_\Gamma\Delta\mathbf{q}_n d\Gamma \tag{13}$$

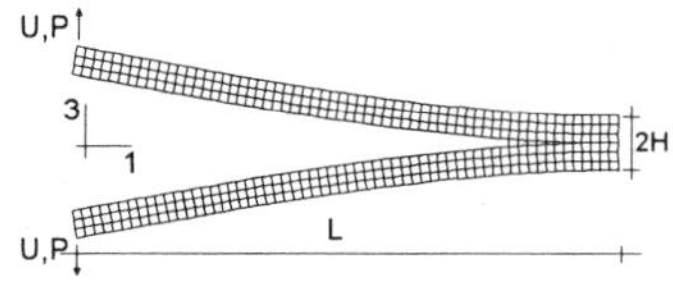

Figure 8. DCB specimen

In the above eq. (13): $\mathbf{S}_{\Gamma n} \equiv \int_\Gamma \mathbf{B}_\Gamma^T \mathbf{K}_n^* \mathbf{B}_\Gamma d\Gamma$ represents the contribution of the interface to the global stiffness matrix; $\mathbf{B}_\Omega$ relates strains to global displacements and σ_n are stresses linearly related to strains in the elastic body Ω.

4 DOUBLE CANTILEVER BEAM SIMULATIONS

The formulation presented in section 3 is here applied to the simulation of DCB tests conducted by Frassine and Pavan (1995), see section 2.2.

The two arms of the specimen are considered to have an elastic, transversely isotropic behaviour, and are discretized with plane-strain four nodes isoparametric elements. The two arms are connected through a line of two-nodes interface elements.

The elastic constants of the two arms are assumed as follows:

$$E_{11} = 84766\,[MPa];\quad E_{33} = E_{11}$$
$$G_{13} = 1000\quad [MPa];\quad \nu_{13} = .035;\quad \nu_{32} = .35$$

Parameters E_{11}, E_{33}, G_{13}, ν_{13}, ν_{32} define the elastic transversely isotropic behaviour of the specimen arms, in the reference frame of fig. 8, 3-2 being the isotropy plane.

4.1 *Simulation based on interface identified from resin properties*

Numerical simulations have been firstly carried out using the interface model parameters identified in section 3.2 in which the interface is assumed to be a thin layer of neat resin.

In fig. 9 the computed load versus load-point displacement plot is compared with the experimental curve for the imposed opening displacement velocity 500 [mm/min].

As it can be observed from the fig. 9, the numerical plot overestimates the fracture energy of the specimen.

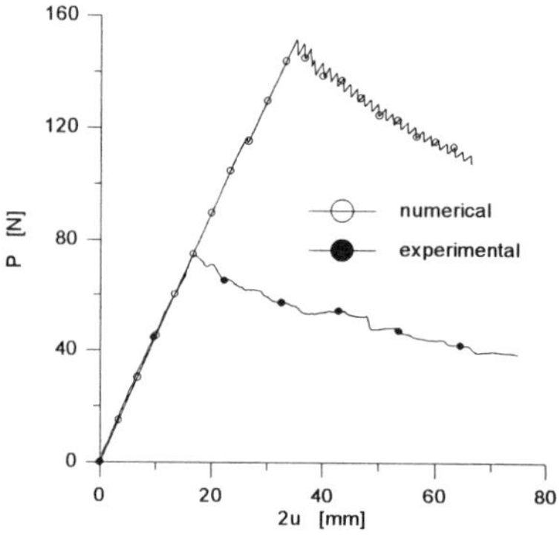

Figure 9. Load-opening displacement responses for a DCB specimen. v=500 [mm/min]; γ=.18 [N/mms]; h = 33.33 [mm^{-1}]; N = 16.

Experimental fracture energy values shown in fig. 2 could also be compared with the calculated G_c-displacement discontinuity velocity plot. To this end, the ratio between the imposed displacement velocity and the displacement discontinuity velocity in points along the interface during the numerical simulation has to be derived. Numerical simulations have shown that this ratio is essentially constant at varying displacement rate.

In fig. 10 the experimental and numerical $G_c - [\dot{u}]$ plots are compared. Figure 10 confirm what already observed from fig. 9. We could therefore conclude that by identifying the interface model starting from neat resin properties, the fracture energy of the composite is overestimated.

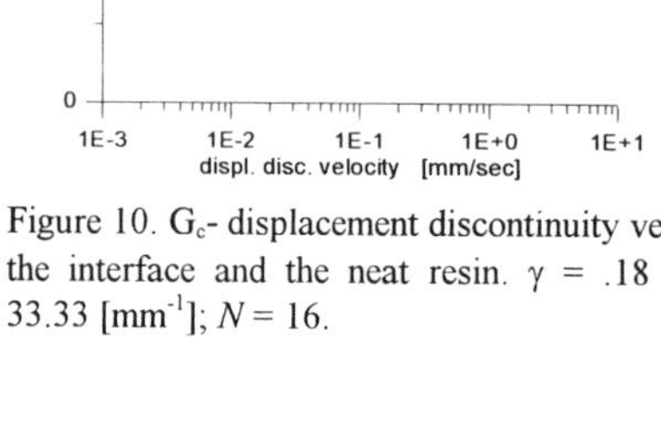

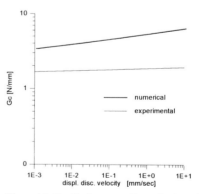

Figure 10. G_c- displacement discontinuity velocity plots for the interface and the neat resin. γ = .18 [N/mms]; h = 33.33 [mm^{-1}]; N = 16.

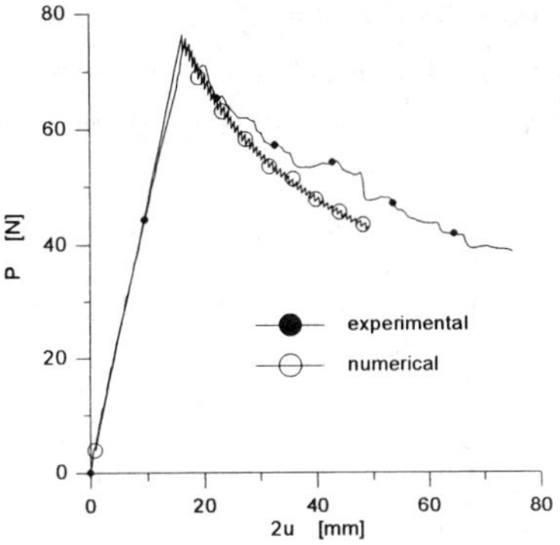

Figure 11. Load-opening displacement responses for a DCB specimen. v=500 [mm/min]; γ= 20 [N/mms]; h = 200 [mm^{-1}]; N = 16.

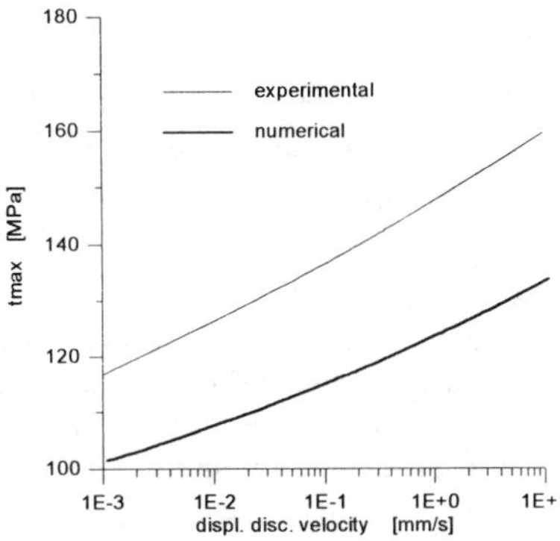

Figure 12a. Experimental and modelled t_{max} - displacement discontinuity velocity plots.
γ = 20. [N/mms]; h = 200 [mm^{-1}]; N = 16.

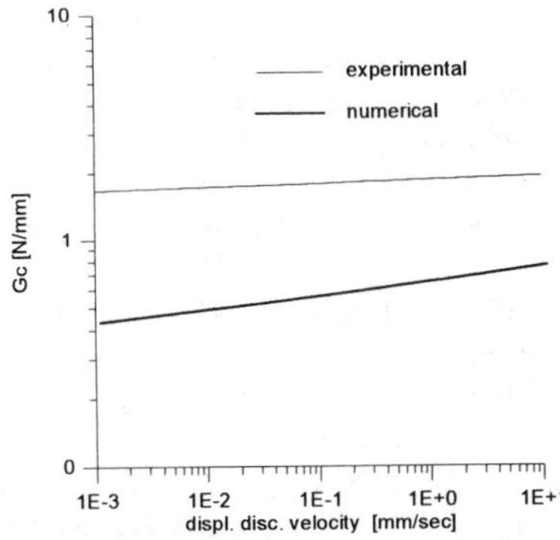

Figure 12b. G_c- displacement discontinuity velocity plots for the interface and the composite.
γ = 20 [N/mms]; h= 200 [mm^{-1}]; N = 16.

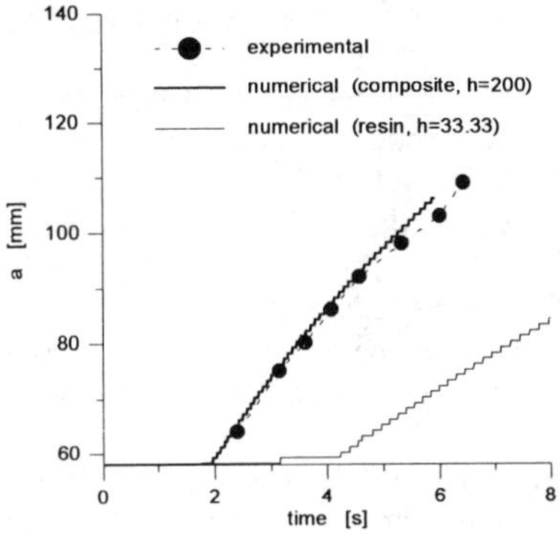

Figure 13. Fracture length-time plots calculated for the DCB test using the two choices of parameters. (v=500 mm/min).

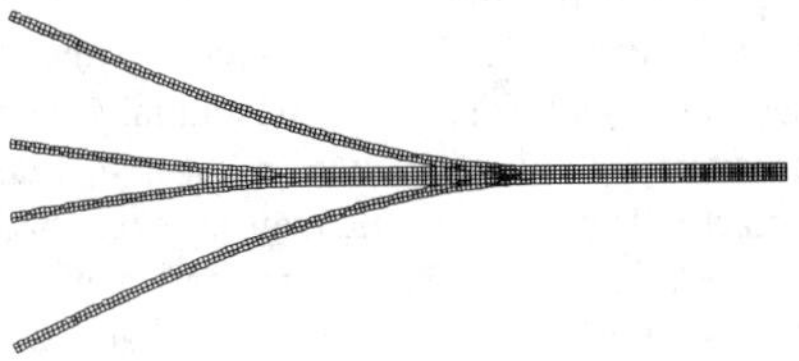

Figure 14. Deformed meshes for a DCB specimen.

4.2 *Simulation based on interface identified from composite global response.*

A second set of interface model parameters has been identified starting from a direct comparison of the numerical and experimental load-versus displacement plots of the DCB test for the opening displacement velocity of 500 mm/min. The value of exponent N has been kept fix, while new values for γ and $\boldsymbol{h}$ have been obtained:

$$\gamma = 20.\ \left[N/mm\,s\right];\quad h = 200.\ \left[mm^{-1}\right]$$

In fig. 11 the result of numerical simulation in terms of load-displacement plot is compared with the experiments.
Experiments and simulations are compared in terms of t_{max} - and G_c- versus velocity plots in fig. 12a,b.
It can be observed that the fracture energy and the t_{max} values are both underestimated in this case.

The numerical and experimental crack length versus time plots are shown in fig. 13 for both choices of interface parameters.

In fig. 14 the deformed mesh of the DCB specimen is shown for two crack length values.

5 CLOSING REMARKS

In the present paper rate-dependent delamination in polymer-matrix composites has been discussed. Numerical simulation of DCB tests have been presented, based on the use of a visco-plastic softening interface law.

The identification procedure of the interface model parameters based on the neat resin properties results in an overestimation of the composite interlaminar fracture toughness. This confirms experimental observations by Bradley (1989) that the resin fracture toughness can only partly be transferred to the composite. It is also interesting to observe that the experimental rate-dependence of the composite fracture toughness is less pronounced than that of the numerical simulation (see fig. 10).

An attempt to identify the model parameters from the overall mechanical behaviour of DCB specimens, on the other hand, gave rise to a substantial lower value of the interface toughness with respect to that of the composite. The latter result may be explained by the presence of some other fracture and damage mechanisms (like fibre bridging, which was actually observed during experiments) implying interaction between adjacent laminae.

Future developments will require finer tuning between experimental and numerical results.

ACKNOWLEDGEMENTS

This research has been carried out in the framework of a research project financed by Murst 40%. The experimental work was carried out by P. Rossi during his Graduation Thesis. Useful discussion with Prof. A. Pavan is gratefully acknowedged.

6 REFERENCES

Aliyu, A.A. & I.M. Daniel, 1985. Effects of strain rate on delamination fracture toughness of graphite/epoxy. In: W.S. Johnson (ed.), *Delamination and debonding of materials*, 336-348 ASTM STP 876, American Society for Testing and Materials, Philadelphia.

Allix, O. and P. Ladeveze 1992. Interlaminar interface modelling for the prediction of delamination. *Int. J. Compos. Struct.*, 22: 235-242.

Allix, O., P. Ladevéze, A. Corigliano 1995. Damage analysis of interlaminar fracture specimens. *Composite Structures*, 31: 61-74.

Bradley, W.L. 1989. Relationship of matrix toughness to interlaminar fracture toughness. in K. Friedrich (ed) *Application of Fracture Mechanics to Composite Materials*: 159-187.

Corigliano A., 1993. Formulation, identification and use of interface models in the numerical analysis of composite delamination. *Int. J. Solids Structures,* 30: 2779-2811.

Corigliano, A, M. Ricci and R. Contro 1997. Rate dependent delamination in polymer-matrix composites. COMPLAS V, 17-20 March 1997, Barcelona: 1168-1175.

Davies, P. 1992. Interlaminar Fracture Testing of Composites. Mode I (DCB) Testing Protocol for ESIS-TC4.

Ferry, J.D. 1980 Viscoelastic Properties of Polymers 3rd Ed., J. Wiley, New York.

Frassine, R., M. Rink and A. Pavan 1993. Viscoelastic effects on intralaminar fracture toughness of epoxy/carbon fibre I. *J. Comp. Mat.* 27: 921-933.

Frassine, R. & A. Pavan 1995. Viscoelastic effects on the interlaminar fracture behaviour of thermoplastic matrix composites: I.Rate and temperature dependence in unidirectional PEI/carbon-fibre laminates. *J.Comp.Sci.Tech.*, 54: 193-200.

Friedrich, K., R. Walter, L.A. Carlsson, A.J. Smiley and J.W. Gillespie Jr, 1989. Mechanisms of rate effects on interlaminar fracture toughness of carbon/epoxy. *J. Mat. Sci.* 24: 3387.

Nemes, J.A. and E. Spéciel 1996. Use of a rate-dependent continuum damage model to describe strain-softening in laminated composites. *Computers and Structures,* 58, 6: 1083-1092.

Perzyna, P. 1966. Fundamentals problems in visco-plasticity. from Recent Advances in Appl. Mech., Academic Press N.Y.

Popelar, C.H. and M.F. Kanninen 1980. A dynamic viscoelastic analysis of crack propagation and crack arrest in a double cantilever beam testing specimen. In: *Crack arrest methodology and applications*, ASTM STP 711, G.T. Hahan, M.F. Kanninen eds., American Society for Testing and Materials, Philadelphia, 5-23.

Schellekens, J.C.J and R. De Borst 1993. A non-linear finite element approach for the analysis of mode-I free edge delamination in composites. *Int. J. Solids and Structures*, 30, 9: 1239-1253

Damage and Failure of Interfaces, Rossmanith (ed.) © *1997 Balkema, Rotterdam, ISBN 90 5410 899 1*

Delamination cracking between plies of different orientation angles in composite laminates

Junqian Zhang
Laboratorium für Technische Mechanik, Universität Paderborn, Germany (On leave from: Department of Engineering Mechanics, Chongqing University, People's Republic of China)

K.P. Herrmann
Laboratorium für Technische Mechanik, Universität Paderborn, Germany

ABSTRACT: Delaminations initiating at the 90°-ply crack tips, and growing along the $\phi/90$ interfaces in $[\cdots/\varphi_p/\phi_m/90_n]_s$ laminates are theoretically investigated. The cracked and delaminated laminates are subdivided by the cross-section separating the laminated and delaminated intervals. A sublaminate-wise first-order shear laminate theory is used to analyze stress and displacement fields in the cracked and delaminated laminates loaded in tension. The strain energy release rate (SERR) for a local delamination normalized by the square of the laminate strain is calculated as a function of delamination length and transverse crack spacing. It is found that the SERR for a delamination is substantially affected by the relative orientation angle of the associated two ply groups.

1 INTRODUCTION

Local delamination, which initiates at matrix ply cracks due to a high interlaminar stress concentration at the crack tips, is one of several important damage mechanisms which contribute to stiffness loss and eventual laminate failure. It is therefore important to be in a position to predict its initiation and growth.

Nairn et al. (1992) used the variational method, which was developed by Hashin (1987) with an application to the stress analysis of cross-ply laminates containing transverse ply cracks, to predict the growth of a delamination induced by matrix cracking.

O'Brien (1985, 1991), using the classical laminate plate theory and by assuming that the laminated portion and the delaminated interval carry loads in series, derived a closed-form equation for the SERR associated with a local delamination. However, the author neglected the effect of matrix cracks which exist before the local delamination occurs. Armanios et al. (1991) developed a sublaminate approach for the analysis of a local delamination including the effect of residual hygrothermal stresses under plane strain conditions by utilizing the first-order shear deformation laminate plate theory. The model was used to predict the onset of local delaminations in T300/934 graphite/epoxy $[\pm 25/90_n]_s$ laminates by utilizing Griffith's energy release rate criterion.

Zhang et al. (1994) derived closed-form expressions for the strain energy release rate and stiffness reduction due to a local delamination where hygrothermal effects were also taken into account. Comprehensive comparisons between this model , Nairn-Hu's variational model and 2-D finite element method (FEM) results have been carried out, suggesting that Zhang-Soutis-Fan's more simple shear-lag model is accurate in comparison with Nairn-Hu's model.

A three-dimensional finite element analysis was performed by Salpekar et al. (1991) to evaluate the energy release rate for local delaminations in glass fiber-reinforced plastic (GFRP) laminates containing 90° matrix cracks of large spacing and loaded in tension. It was observed that the value of the total SERR calculated near the free edge increased with an increasing delamination length and approached O'Brien's closed-form solution for delamination lengths of about four ply thicknesses from the matrix crack. The influence of transverse ply crack spacing on the strain energy release rate associated with a local delamination was investigated numerically by performing a 2-D FEM analysis (Zhang et al., 1994).

The SERR value for a local delamination decreases notably with an increasing matrix cracking.

In this article the dependence of the local delamination on the relative orientation angle of associated two plies is examined in terms of the energy release rate and the laminate lay-ups $[\cdots/\pm\phi_m/90_n]_s$. The cracked and delaminated laminates are sub-divided by the cross-section separating the laminated and delaminated intervals. Analyses of stress and displacement fields are carried out by applying properly the first-order shear deformation laminate theory to each sublaminate. The strain energy release rate for a local delamination is derived as a function of the delamination length and the transverse crack spacing. The effects of the relative ply orientation angle and the transverse crack spacing on the local delamination are examined in terms of the strain energy release rate.

2 STRESS ANALYSIS

Let a symmetric laminate $[\cdots/\varphi_i/\phi_m/90_n]_s$ to be subjected to a tension load, 2N. The matrix cracks are assumed to exist in the 90°-ply group with an uniform crack spacing of 2s; local delaminations initiate and grow from both tips of each matrix crack and span the width of the specimen. Further, for symmetry reasons only one quarter of the repeating interval of the laminate is modeled. Following Armanios's technique (1991) for the subdivision of delaminated and laminated portions, the modeled interval length s is divided into six sublaminates as shown in figure 1. The delamination length at the interfaces of ϕ / 90 is denoted by l. The assumption of the plane strain with respect to the width direction , which to some extent represents the situation of the deformation behavior of the composite laminate interior, is used. The sublaminates are referred to three local co-ordinate systems, respectively, with a common y-axis and their origins at the centers of the left cross-sections of the sublaminates 1, 2 and 3, figure 1.

The displacements in y- and z-directions within each sublaminate are assumed to be of the form

$$v(y,z) = V(y) + z\kappa(y) \tag{1a}$$

$$w = W(y) \tag{1b}$$

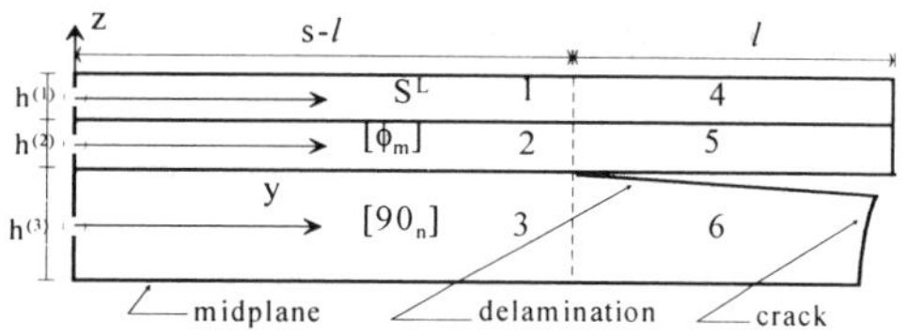

Figure 1. Sub-divisions of a quarter of the repeating interval of a $[\cdots/\varphi_i/\phi_m/90_n]_s$ laminate.

The shear deformation is recognized through the rotation $\kappa(y)$. The equilibrium equations for each sublaminate take the form

$$N_{,y} + T_t - T_b = 0 \tag{2a}$$

$$M_{,y} - Q + \frac{h}{2}(T_t + T_b) = 0 \tag{2b}$$

$$Q_{,y} + P_t - P_b = 0 \tag{2c}$$

where N, Q and M are axial force, shear force and bending moment resultants at a cross-section; P and T denote the interlaminar peel and shear stresses; the subscripts "t" and "b" indicate the top and bottom surfaces; h is the thickness of a sublaminate. Using equations (1a,b), the strain-displacement relations and the in-plane stress-strain relationships of a lamina, the constitutive relationships of a sublaminate in terms of the force and moment resultants read as follows

$$N = A_{22}V_{,y} + B_{22}\kappa_{,y} \tag{3a}$$

$$M = B_{22}V_{,y} + D_{22}\kappa_{,y} \tag{3b}$$

$$Q = A_{44}(\kappa + W_{,y}) \tag{3c}$$

where A_{22}, B_{22}, D_{22} and A_{44} are the extension, coupling, bending and out-of-plane shear stiffnesses, respectively, from the classical laminate theory. Substitution of the equations (3a-c) into the equations (2a-c) gives

$$A_{22}V_{,yy} + B_{22}\kappa_{,yy} + T_t - T_b = 0 \tag{4a}$$

$$(D_{22}-\frac{B_{22}{}^2}{A_{22}})\kappa_{,yy}-A_{44}(\kappa+W_{,y})$$
$$+(\frac{h}{2}-\frac{B_{22}}{A_{22}})T_t+(\frac{h}{2}+\frac{B_{22}}{A_{22}})T_b=0 \tag{4b}$$

$$A_{44}(\kappa_{,y}+W_{,yy})+P_t-P_b=0 \tag{4c}$$

The continuity conditions of displacements at the interlaminar faces between two neighboring sub-laminates read

$$V^{(K)}-V^{(K+1)}=\frac{h^{(K)}}{2}\kappa^{(K)}+\frac{h^{(K+1)}}{2}\kappa^{(K+1)} \quad \text{for } K=1,2,4 \tag{5a}$$

$$W^{(1)}=W^{(2)}=W^{(3)}=W^{(6)}=0 \tag{5b}$$

$$W^{(4)}=W^{(5)} \tag{5c}$$

where superscripts with parentheses indicate sublaminates.

By applying the equations (4a-c) to every sublaminate and by using the equations (5a-c) and the interfacial stress continuity conditions, the governing equations can be obtained in terms of displacement variables, $\kappa^{(K)}$ and $V^{(K)}$ (Zhang et al., 1997). Furthermore, by solving these equations the rotations and the midplane displacements of the sublaminates are arrived at

$$\kappa^{(K)}=\sum_{j=1}^{3}\alpha_j p_j^{(K)}\sinh(\lambda_j y) \qquad \text{for } K=1,2,3 \tag{6}$$

$$V^{(K)}=\sum_{j=1}^{3}\alpha_j\gamma_j^{(K)}\sinh(\lambda_j y)+\alpha_{3+K}y \quad \text{for } K=1,2,3 \tag{7}$$

$$\kappa^{(4)}=\theta_1 e^{\omega y}+\theta_2 e^{-\omega y}+\theta_3 y+\theta_4 \tag{8a}$$

$$\kappa^{(5)}=q(\theta_1 e^{\omega y}+\theta_2 e^{-\omega y})+\theta_3 y+\theta_4 \tag{8b}$$

$$V^{(4)}=k_3(\theta_1 e^{\omega y}+\theta_2 e^{-\omega y})+\theta_6 y+\theta_8 \tag{9a}$$

$$V^{(5)}=-k_1(\theta_1 e^{\omega y}+\theta_2 e^{-\omega y})+\theta_5 y+\theta_7 \tag{9b}$$

where the laminate constants $\gamma_j^{(K)}$, $p_j^{(K)}$, λ_j, ω, q, k_j are functions of the elastic lamina properties and the lay-up parameters; the unknown coefficients α_j, θ_j, which can be determined by the boundary conditions and the continuity conditions between the delaminated and laminated portions (Zhang et al., 1997), are proportional to the applied laminate tensile load, 2N.

3 STRAIN ENERGY RELEASE RATE (SERR)

When the symmetric laminate is subjected to an overall in-plane tensile load, the facts of the periodic configuration of matrix cracks and delaminations require that the rotations of the cross-sections of the constraining sublaminates 4 and 5 are zero at the location y=s, and the midplane in-plane displacements of both layers 4 and 5 should be the same at the cross-section y=s. Consequently, the effective laminate strain is defined by

$$\varepsilon_y=\frac{V^{(4)}(s)}{s}=\frac{V^{(5)}(s)}{s} \tag{10}$$

Moreover, the extension stiffness of a damaged laminate in y-direction is given by

$$2A_{22}^d=\frac{2N}{\varepsilon_y} \tag{11}$$

By using the equations (9a-11) the reduced extension stiffness can be expressed as a function of the delamination length and the transverse crack spacing

$$A_{22}^d=\frac{A_{22}}{1+\chi(\frac{l}{s}+\sum_{j=1}^{3}\gamma_j^{(3)}\bar{\alpha}_j\frac{\tanh(\lambda_j(s-l))}{s})} \tag{12a}$$

$$\Lambda_{22}=\frac{A_{22}-A_{22}^d}{A_{22}^{(3)}}=\frac{(1+\chi)[\frac{l}{s}+\sum_{j=1}^{3}\gamma_j^{(3)}\bar{\alpha}_j\frac{\tanh(\lambda_j(s-l))}{s}]}{1+\chi[\frac{l}{s}+\sum_{j=1}^{3}\gamma_j^{(3)}\bar{\alpha}_j\frac{\tanh(\lambda_j(s-l))}{s}]} \tag{12b}$$

$$\alpha_j=-\bar{\alpha}_j\frac{N}{A_{22}\cosh(\lambda_j(s-l))} \tag{12c}$$

where $\chi = \frac{A_{22}^{(3)}}{A_{22}^{(1)} + A_{22}^{(2)}}$, A_{22} is the initial stiffness of the laminates. Zero or unity value of the parameter Λ_{22} reflects constancy or total loss of the load-carrying capacity of the 90°-plies, respectively.

The strain energy release rate associated with a delamination crack can be found from the compliance equation

$$G^{ld} = \frac{N^2}{2} \frac{\partial C}{\partial l} \tag{13}$$

where C is the extension compliance of the laminate and equal to s / A_{22}^{d}. Further, by combining the equations (11-13), the expressions for the normalized SERR associated with a local delamination is obtained

$$\frac{G^{ld}}{\varepsilon_y^{\ 2}} = \frac{A_{22}^{(3)} s}{2} \frac{\partial \Lambda_{22}}{\partial l} \tag{14}$$

4 RESULTS

The T300/934 graphite/epoxy composite material system is examined here, the basic properties of which are given in Table 1. The results are presented by plotting the normalized strain energy release rate against the normalized delamination length, *l*/t.

Table 1. Material and geometric properties of a lamina

Property	Value
E_{11}	144.8 GPa
$E_{22}=E_{33}$	11.38 GPa
$G_{12}=G_{13}$	6.48 GPa
G_{23}	3.45 GPa
$\nu_{12}=\nu_{13}$	0.3
Single-ply thickness, t	0.132 mm

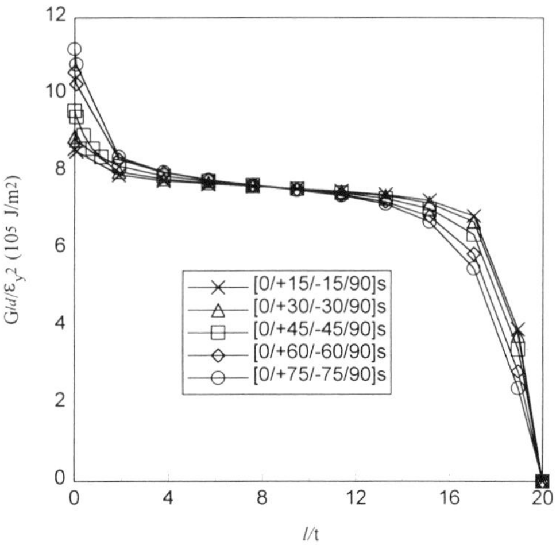

Figure 2. Influence of the relative ply orientation angle on the SERR for a local delamination (s=20t=2.64mm).

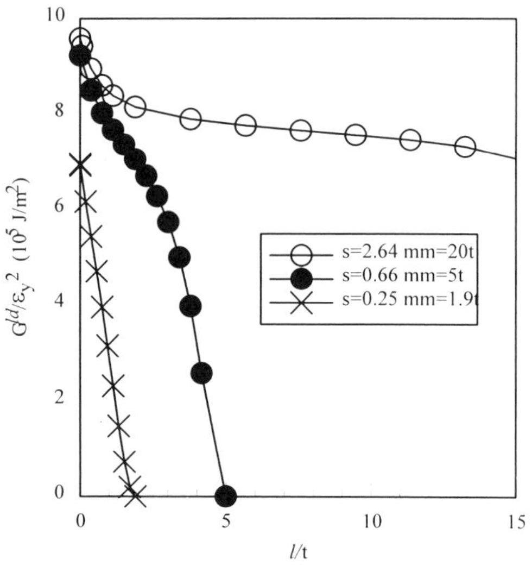

Figure 3. Effect of the 90°-ply crack spacing on the SERR for a -45°/90° delamination in the $[0/\pm45/90]_s$ laminate

Figure 2 presents the predictions of the model for the $[0/\pm\phi/90]_s$ laminates with varying primary constraining ply orientations ranging from 15° to 75°. The SERRs increase considerably with an increasing orientation angle ϕ for a small delamination length (up to about two-ply-thicknesses), but converge to the same steady-state value for large delamination lengths. Furthermore, the SERRs fall quickly to zero when the delamination length approaches the crack spacing.

Figure 3 illustrates the effect of the ply crack spacing on the local delamination. It is important to note that the normalized SERR for the delamination is a function not only of the delamination length but also of the ply crack spacing 2s. The fact that the SERR/delamination length curve for a shorter crack spacing lies below that for a larger crack spacing suggests that in the dense ply cracking configuration there is only a small energy available for an advancing delamination. Moreover, for a long crack spacing, saying 20t, the SERR is expected to reach a steady-state value when the delamination has grown about a length of two single-ply thicknesses 2t. However, for small and intermediate crack spacing, the steady-state is never attained.

5 CONCLUSIONS

1. The strain energy release rate for a local delamination is a function of the delamination length and the ply crack spacing.
2. The delamination driving force, SERR, depend upon the relative orientation angle of two associated plies.
3. The SERR associated with the delamination eventually reaches a steady-state value for large crack spacing, but it never happens for dense crack patterns.

ACKNOWLEDGMENT

This work was financially supported by the Alexander von Humboldt-Foundation of Germany and the National Science Foundation of China.

REFERENCES

1. Armanios, E. A., Sriram, P. and Badir, A. M. (1991) Fracture Analysis of Transverse Crack-Tip and Free-Edge Delamination in Laminated Composites, *Composite Materials: Fatigue and Fracture, ASTM STP* **1110**, 269-286.
2. Hashin, Z. (1987), Analysis of Orthogonally Cracked Laminates under Tension, *J. Appl. Mech.*, **54**, 872-879.
3. Nairn, J. A. and Hu, S. (1992), The Initiation and Growth of Delaminations Induced by Matrix Microcracks in Laminated Composites, *Int. J. Fracture*, **57**, 1-24.
4. O'Brien, T. K. (1985), Analysis of Local Delaminations and Their Influence on Composite Laminate Behaviour, *Delamination and Debonding of Materials, ASTM STP* **876**, 282-297.
5. O'Brien, T. K. (1991), Residual Thermal and Moisture Influences on the Strain Energy Release Rate Analysis of Local Delaminations From Matrix Cracks, *NASA TM-104077, AVSCOM TR-91-B-012*.
6. Salpekar, S. A. and O'Brien, T. K. (1991), Combined Effect of Matrix Cracking and Free Edge on Delamination, *Composite Materials: Fatigue and Fracture, ASTM STP* **1110**, 287-311.
7. Zhang, J., Soutis, C. and Fan, J (1994): Strain Energy Release Rate Associated with Local Delamination in Cracked Composite Laminates, *Composites* **25**, 851-862.
8. Zhang, J., Fan, J. and Herrmann, K. P. (1998) Delaminations induced by constrained transverse cracking in symmetric composite laminates, *to appear in Int. J. Solids & Structures.*

Damage and Failure of Interfaces, Rossmanith (ed.) © 1997 Balkema, Rotterdam, ISBN 90 5410 899 1

On the identification of an interface damage model for the prediction of delamination initiation and growth

O. Allix, P. Ladevèze & D. Lévêque
Laboratoire de Mécanique et Technologie (LMT) / ENS de Cachan, France

L. Perret
Centre National d'Etudes Spatiales (CNES) / Toulouse, France

ABSTRACT: The present study, supported by CNES (French National Center for Space Studies), concerns the identification of an interface damage model devoted to the prediction of delamination. The subject of the identification and validation of the model by means of several tests, for different types of specimens, including holed plates in tension, is discussed. Even though the purpose of this work is quite general, the tests were conducted with M55J/M18 (high-modulus carbon-fibres / epoxy resin) material specimens.

1 INTRODUCTION

Delamination often appears as the result of interactions between different damage mechanisms inside composite laminates, such as fibre-breaking, transverse micro-cracking and the debonding of adjacent layers itself (Highsmith & Reifsnider 1982, Talreja 1985). The analysis of delamination is often split into the study of the onset of delamination and the analysis of the development of an existing delaminated area. Up until now, initiation analysis has involved empirical criteria such as point-stress or average-stress (Kim & Soni 1984). More predicting tools, based on edge effects analysis (Engrand 1982), or singularity computation (Leguillon & Sanchez-Palancia 1987), are used in order to allocate a greater or lesser delamination tendency to different stacking sequences. Most propagation studies on composite laminates involve extensions of Fracture Mechanics (De Charentenay & al. 1984, Davies & al. 1990) usually applied to metallic materials. Recently, some authors tried to extend Fracture Mechanics to the study of the onset of delamination. This was accomplished by the introduction of a minimum length required for the beginning of delamination development (Leguillon 1997).

In order to take the main damage mechanisms into account to be able to predict the behaviour of any structure up to the rupture, an initial step, which has been conducted in other studies, was to define what we call a meso-modelling approach. At the meso-level, the laminate is described as a stacking sequence of non-linear layers and non-linear interlaminar interfaces (see Figure 1). The single-layer model and its identification, including damage and inelasticity, were previously developed (Ladevèze & Le Dantec 1992). In this model, the state of the internal damage variables is prescribed as uniform throughout the thickness of each ply. The interlaminar interface is a two-dimensional entity which ensures traction and displacement transfer from one ply to another. Its mechanical behaviour depends on the angles between the fibers of two adjacent layers. Its primary interest is to allow the modeling of more or less progressive degradation of the interlaminar connection (Allix & Ladevèze 1992).

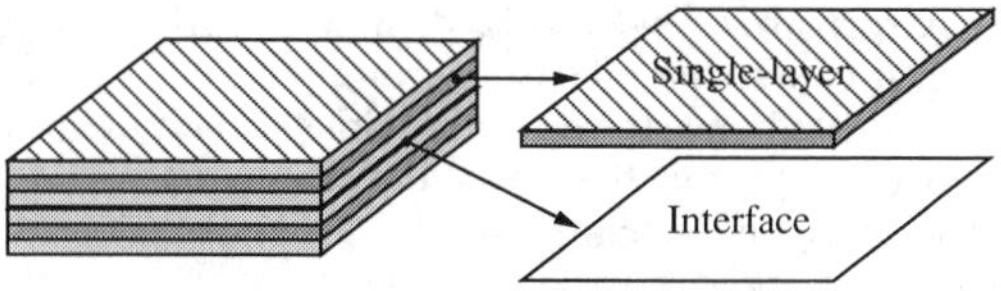

Figure 1. Laminate meso-modelling.

The aim of this work is to define a methodology that allows identifying precisely the few intrinsic characteristics of the interface damage model; characteristics which govern the behaviour of any structure with respect to delamination. This method consists of developing reliable tests with specialized identification software. The most reliable delamination tests are, perhaps, tests on laminated plates with a circular hole. For such specimens, the delamination crack initiation is reproducible and its growth around the hole is often stable within a certain range. Our goal is thus to define precisely a set of holed specimens for which the delamination propagation is relatively easy to use for identification (small number of plies, large delaminated area, etc).

We are just at the beginning of this procedure. Thus, we first identify the interface model by means of standard initiation and propagation delamination tests, for different ±θ interfaces, with θ (θ = 0°, 22.5° and 45°) being the relative direction of the fibers of adjacent plies. The analysis of the standard propagation tests makes use of the links between the critical energy release rates and some of the model's parameters (Allix et al. 1995, Allix & Corigliano 1996). One difficulty encountered is that the damage is not only located at the interface but also concerns the layers. It is thus essential to take into account the part of the energy which is dissipated inside the layers. The interfaces can thus be classified into two categories: the 0°/0° interface, whose behaviour is rather brittle, and the interface at a disorientated angle, whose critical energy release rate is higher. Edge Delamination Tension tests were conducted too for the identification of the other parameters of the interface model. However, such tests are not so easy to analyse since the initiation process in EDT specimens is very unstable.

Afterwards, we verify and improve the identification by means of comparison over the delaminated area for holed specimens submitted to tension. During the test, the damage map is monitored by means of X-Radiography. The difficulty, for identification purposes, is that, due to the complexity of the state of stress, the interpretation of the test requires complex computation. To accomplish this, a software developed at the Laboratory (Allix 1992), which is devoted to the delamination analysis of laminate structures with an initially-circular hole, has been used.

All of the tests presented in this work have been conducted at AÉROSPATIALE, Suresnes (France), with the same M55J/M18 high-modulus carbon-fibre/epoxy resin material. This material is used particularly in space applications because of its good mechanical properties and its dimensional stability. The single-layer model of this material was identified in an earlier work (Allix & Lévêque 1996).

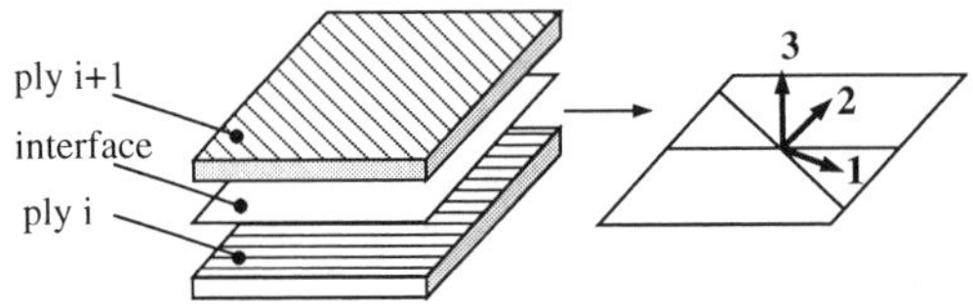

Figure 2. Orthotropic directions of the interface.

2 INTERLAMINAR INTERFACE MODELLING

2.1 *Damageable behaviour modelling*

The effect of the deterioration of the interlaminar connection on its mechanical behaviour is taken into account by means of three internal damage variables. The energy per unit area proposed in (Allix & Ladevèze 1992) is:

$$E_D = \frac{1}{2}\left[\frac{\langle -\sigma_{33}\rangle_+^2}{k_3^0} + \frac{\langle \sigma_{33}\rangle_+^2}{k_3^0(1-d_3)} + \frac{\sigma_{32}^2}{k_2^0(1-d_2)} + \frac{\sigma_{31}^2}{k_1^0(1-d_1)}\right] \quad (1)$$

where k_i^0 is an interlaminar stiffness value and d_i the internal damage indicator associated with its Fracture Mechanics mode, while subscript i corresponds to an orthotropic direction of the interface (Figure 2).

Classically, the damage energy release rates, associated with the dissipated energy ϕ by damage and by unit area, are introduced as:

$$Y_{d3} = \frac{1}{2}\frac{\langle\sigma_{33}\rangle_+^2}{k_3^0(1-d_3)^2}\,; \quad Y_{d1} = \frac{1}{2}\frac{\sigma_{31}^2}{k_1^0(1-d_1)^2}\,;$$

$$Y_{d2} = \frac{1}{2}\frac{\sigma_{32}^2}{k_2^0(1-d_2)^2}$$

$$\text{and} \quad \phi = Y_{d3}\dot{d}_3 + Y_{d1}\dot{d}_1 + Y_{d2}\dot{d}_2$$

with $\phi \geq 0$ to satisfy the Clausius-Duheim inequality.

In what follows, an "isotropic" damage evolution law is described. In this model, as proposed in (Allix & Ladevèze 1996), the damage evolution law is assumed to be governed by means of an equivalent damage energy release rate of the following form:

$$\underline{Y}(t) = \sup\Big|_{\tau\leq t} \left[\left((Y_{d3})^\alpha + (\gamma_1 Y_{d1})^\alpha + (\gamma_2 Y_{d2})^\alpha \right)^{1/\alpha}\Big|_\tau\right] \quad (2)$$

The evolution of the damage indicators is thus assumed to be strongly coupled. γ_1, γ_2 and α are material parameters.

A damage evolution law is then defined by the choice of a material function ω, such that:

$$d_3 = d_1 = d_2 = \omega(\underline{Y}) \text{ if } d < 1;$$

$$d_3 = d_1 = d_2 = 1 \text{ otherwise}$$

One simple case, used for application purposes, is:

$$\omega(\underline{Y}) = \left[\frac{n}{n+1}\,\frac{\langle \underline{Y}-Y_0\rangle_+}{Y_c-Y_0}\right]^n \tag{3}$$

where a critical value Y_c and a threshold value Y_0 are introduced. High values of the n case correspond to a brittle interface.

To summarize, the damage evolution law is defined by means of the six intrinsic material parameters $Y_c, Y_0, \gamma_1, \gamma_2, \alpha$ and n. It will be shown in the next paragraph that Y_c, γ_1, γ_2 and α are all related to the critical energy release rates.

2.2 *Links with Fracture Mechanics*

A simple way of comparing Damage Mechanics with Linear Elastic Fracture Mechanics is to compare the mechanical dissipation yielded by the two approaches. This was performed in (Allix & Ladevèze 1996), and only the results are presented below.

In the case of pure-mode situations, when the critical energy release rate reaches its stabilised value at the propagation denoted by G_c^p, we obtain:

$$G_{cI}^p = Y_c\,;\quad G_{cII}^p = \frac{Y_c}{\gamma_1}\,;\quad G_{cIII}^p = \frac{Y_c}{\gamma_2} \tag{4}$$

For a mixed-mode loading situation, we simply derive a standard LEFM model (Bathias 1995):

$$\left(\frac{G_I}{G_{cI}^p}\right)^\alpha + \left(\frac{G_{II}}{G_{cII}^p}\right)^\alpha + \left(\frac{G_{III}}{G_{cIII}^p}\right)^\alpha = 1 \tag{5}$$

wherein α governs the shape of the failure locus in the mixed mode.

3 FRACTURE MECHANICS TESTS

3.1 *Presentation of the tests*

The tests of crack propagation in interlaminar fracture specimens are usually conducted on beam specimens with an initiated crack at the studied interface. Our specimens are 300 mm long and 20 mm wide. An anti-adhesive film 40 mm long and 25 μm in thickness is inserted at the mid-plane in order to initiate cracking. Each specimen tested is a $[(+\theta/-\theta)_{4s}/(-\theta/+\theta)_{4s}]$ laminate with $\theta = 0°$, 22.5° or 45°, according to the three kinds of $\pm\theta$ interlaminar interfaces. The stacking sequence is equilibrated and symmetric in each arm of the beam in order to suppress any bending/twisting coupling membrane. The mean thickness of a single ply is on the order of

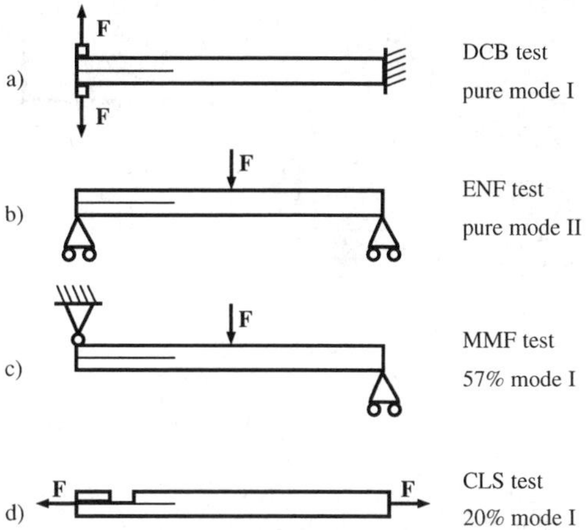

Figure 3. Standard Fracture Mechanics tests.

0.1 mm.

The tests conducted in this work, which were developed at the AÉROSPATIALE facility in Suresnes (France), are the pure-mode I DCB (Double-Cantilever Beam) Test (De Charentenay et al. 1984), the pure-mode II ENF (End-Notched Flexure) test (Davies et al. 1990), and two mixed-mode tests: the MMF (Mixed-Mode Flexure) test (Russell & Street 1985) and the CLS (Cracked-Lap Shear) test (Guédra Degeorges 1993). These various tests are schematically presented in Figure 3. The two mode percentages of the mixed-mode specimens displayed in Figure 3 are calculated based on a unidirectional stacking sequence. In fact, they depend on the geometry of the specimens as well as on the stacking sequence in the multi-directional laminate case (Hwu et al. 1995), yet their variations are sufficiently small to be considered as constants. The tests were conducted on an INSTRON testing machine at ambient temperature (about 25°C) and at an imposed displacement rate. The displacement rate was set at 2 mm min^{-1} in the DCB and CLS tests and at 1 mm min^{-1} in the ENF and MMF tests. In these two latter bending tests, the total useful length is 180 mm (only 100 mm for the ±45° ply-based laminate).

3.2 *Results and non-standard analysis*

The critical energy release rate G_c is classically obtained by deriving the compliance of the specimen, as is usually carried out within the concept of Linear Elastic Fracture Mechanics. This compliance can be analytically computed by use of the Classic Beam Theory, for instance. Here, we derive the critical rates from fitting the experimental compliance. This method is useful since it corrects the measured crack

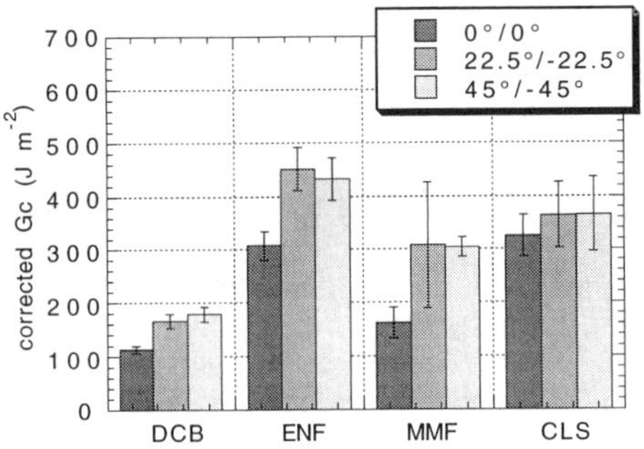

Figure 4. Critical energy release rates at propagation.

length (Williams 1989), which is not easy to accomplish because of the difficulty in locating exactly the end of the crack on each side of the specimen. We applied this method to both the DCB and MMF tests.

Previous studies (Allix et al. 1995, Allix & Corigliano 1996) have highlighted the importance of taking into account the intralaminar damage for an accurate derivation and identification of the local energy release rate. Nevertheless, the portion of the energy dissipated in the layers was not identified. Here, this identification is performed in a simplified manner by making use of: (i) a two-dimensional elastic plate computation, and (ii) a local non-linear re-analysis using for the previously-identified single layer damage and inelastic model. The damage distribution in the laminate thickness is evaluated with the DAMLAM software, developed in the LMT Laboratory, from the data of the generalized loading (at each point of the plate) given by the 2D FE computation. Then, the portion of the energy dissipated inside the layers can be deduced (Allix et al. 1997). We found that this energy is significant only for the ±45° ply-based laminate in the case of the ENF, MMF and CLS tests.

From all the experimental points of three specimens for each test and for each stacking sequence, the mean critical energy release rates are derived and corrected by this method, as presented in Figure 4. From the corrected rates, it seems that the interfaces can be classified into two categories: the 0°/0° interface, whose critical energy release rates are always lower than those of the disorientated-angle interfaces, and this latter kind of ±θ interface, whose critical rates seem to be independent of the angle value. This classification has a physical explanation since the ±θ interfaces are revealed by the wall effect between the fibers of adjacent layers (Billoët et al. 1994), which is not the case for the 0°/0° interface. Moreover, fractographies of the delaminated interfaces show that the interlaminar resin plays a major part in the damage process of ±θ interfaces (Figure 5a), whereas massive full fiber-matrix debonding in the 0°/0° interface case can be observed (Figure 5b), which explains its rather brittle behaviour.

3.3 *Identification of the propagation parameters*

From the corrected critical energy release rates at propagation and from the relationships existing between Fracture Mechanics and Damage Mechanics (4), we deduce the values of the critical energies Y_c and the coupling coefficient γ_1. Without any further information on mode III interlaminar fracture, we can choose $\gamma_2 = \gamma_1$, which is justified at least for a ±45° interface. The identification results are reported in Table 1. For each kind of interface, the parameter α, which governs the shape of the failure locus in the mixed-mode (5), is identified in the normalized plane mode I/mode II (see Figure 6 for instance). It is observed that α is always greater than 1, and we can choose the same parameter α for the two ±θ interfaces ($\theta \neq 0°$).

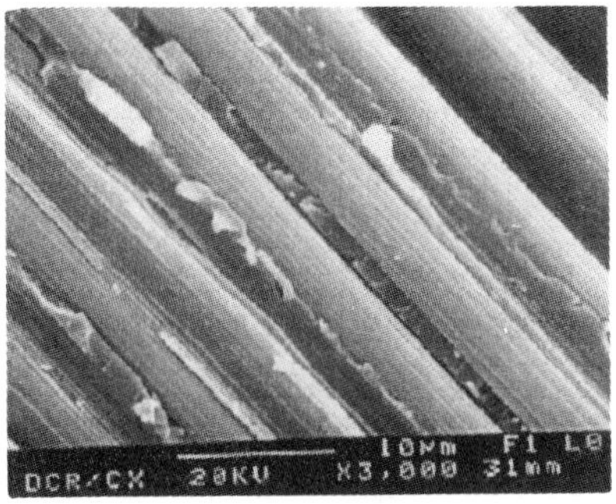

5a. ±45° interface case.

5b. 0°/0° interface case.

Figure 5. Fractography of debonded interlaminar interfaces in mode I DCB test.

Table 1. Interface model parameters.

Interface	Y_c (N mm^{-1})	γ_1	α
0°/0°	0.113 ± 0.007	0.37 ± 0.15	1.59
±22.5°	0.167 ± 0.013	0.36 ± 0.17	1.12
±45°	0.192 ± 0.014	0.44 ± 0.16	1.19

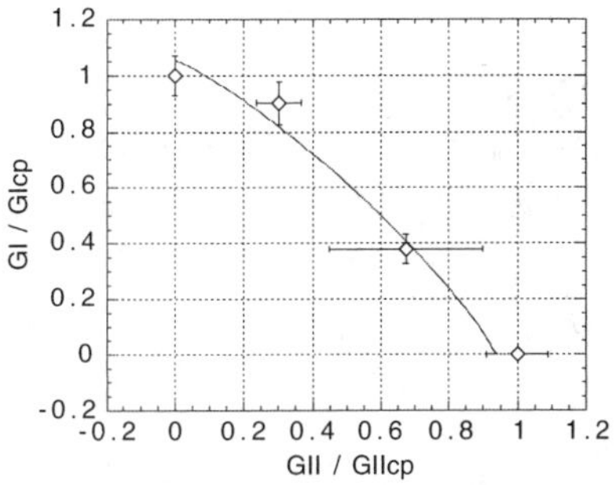

Figure 6. Identification of α for the ±45° interface.

4. EDGE DELAMINATION TENSION TESTS

4.1 *Introduction*

The experimental study of the initiation of delamination often requires EDT (Edge Delamination Tension) specimens (O'Brien 1982). Fracture Mechanics is not well-adapted for the analysis of such a test since the energy release rate vanishes at zero crack length. The meso-modeling concept is then useful when dealing with such a case.

During the initial stage, a review of many EDT tests in the literature has indicated the high frequency of superfluous tests in terms of the type of interface and loading mode (Kim 1989). In selecting the stacking sequences, the level of interlaminar stresses was determined using CLEOPS, a software program dedicated to elastic edge effect computations (Lécuyer & Engrand 1988). For laminates containing some 90° plies, one pragmatic criterion for reducing the transverse cracking is to decrease the relative thickness of these plies (Crossman & Wang 1982). Our laminates tested in tension are presented in Table 3. We can note that the 90°/90° interface is equivalent to a 0°/0° interface.

4.2 *Experiments*

The EDT specimens tested have 16 plies, and measure 30 mm in width and 150 mm in gage length. A bi-directional gage is fastened on each side of the specimen, and acoustic emission is used to detect the beginning of delamination. Optical observations of the edges and X-radiography serve to complete these tests. Each test is conducted at a fixed displacement rate of 0.5 mm min^{-1}.

The longitudinal strain at the beginning of an interlaminar crack and the longitudinal strain at rupture are both read from experimental plots and are displayed in Table 3. The beginning of delamination leads to a brutal rupture of the specimen in the ±22.5° ply-based laminates, whereas the propagation of delamination at the mid-plane of the first laminate is progressive up until rupture.

4.3 *Prediction*

For the computation of delamination initiation, a specialized software EDA has previously been developed in the LMT Laboratory (Daudeville & Ladevèze 1993). It solves the problem posed in a strip perpendicular to the edge, with the damage being located in the interfaces. In order to analyse E.D.T. specimens, the same type of approach was used in (Schellekens & De Borst 1993). Knowing the parameters identified from the last section (see Table 1), we are then faced with having to choose an initial set for the other parameters Y_o, n and the interlaminar stiffness values.

For the energy threshold Y_o, we initially assume that: $Y_o = 0$. Without any further information, we can choose $n = 0.5$, which is the value found for the single-layer. The stiffness values remain to be determined. A $\pm\theta$ interface is assumed to be equivalent to a thin resin layer of two fiber diameters in thickness (Billoët et al. 1994), or about 10 µm. From the properties of a standard epoxy resin, we can set a value of 4.10^5 N mm^{-3} for the normal stiffness and 10^5 N mm^{-3} for the shear stiffnesses. A 0°/0° interface physically resembles a transverse plane of a unidirectional layer. Knowing the rupture stresses measured in previous layer tests, we can identify the stiffness values of this interface from the local instability criterion. For instance, the stiffness value k_3^0 is set at 10^4 N mm^{-3} with $\sigma_{33max} = 41$ MPa.

From the initial set of parameters presented in Table 2 (Set 1), the predicted value of longitudinal strain at the initiation of delamination is displayed in Table 3. For two of the laminates tested, the predicted value is not very close to the experimental results, but in all cases, the delamination locus has been predicted exactly. A second set of parameters from Table 2 (Set 2) yields predictions that are closer to the experimental values (see Table 3) and that could be easily improved. Nevertheless, the discrepancies between computation and experimental results can be explained by the fact that this simulation doesn't take into account the damage occurring inside the layers, which is significant near the edges. Moreover, the

Table 2. Parameter sets (*in italics*: already identified).

Interface	Set n°	Y_c	$\gamma_1\ (=\gamma_2)$	α	k_3^0	$k_1^0\ (=k_2^0)$	n
0°/0°	Set 1	*0.11 N mm⁻¹*	*0.4*	*1.6*	1.10^4 N mm^{-3}	4.10^4 N mm^{-3}	0.5
	Set 2	"	"	"	4.10^5 N mm^{-3}	4.10^5 N mm^{-3}	0.2
± θ	Set 1	*0.18 N mm⁻¹*	*0.4*	*1.2*	4.10^5 N mm^{-3}	1.10^5 N mm^{-3}	0.5
	Set 2	"	"	"	4.10^4 N mm^{-3}	3.10^4 N mm^{-3}	0.5

Table 3. Simulated and experimental longitudinal strain at initiation of EDT specimens.

Laminate	Interface(s)	Mode	$\varepsilon_{rupture}$ (%) (number of specimens)	$\varepsilon_{ini\text{-}experiment}$ (%) (number of specimens)	$\varepsilon_{ini\text{-}prediction}$ (%)	
					Set 1	Set 2
$[0_3/\pm45_2/90]_s$	90°/90°	I	0.54 (3)	0.20 (3)	0.43	0.25
$[\pm22.5]_{4s}$	±22.5° (first)	II	0.64 (5)	0.60 (2)	0.96	0.81
$[0_4/\pm22.5_2]_s$	±22.5°	mixed	0.52 (3)	0.51 (1)	0.50	0.48

beginning of edge delamination is often an unstable phenomenon which is not easy to detect with accuracy.

5 AN INITIAL IDENTIFICATION USING PLATES WITH HOLES

The initial identification with classical tests is not reliable enough. The various problems raised in the preceding sections are essentially of two kinds: the intralaminar damage is not always sufficiently accounted for and the initiation process is unstable and remains not fully understood. According to the authors, more reliable delamination tests would be those conducted on laminated plates with a circular hole. The idea herein is to use the first parameter set identified previously as the initial data for a global identification using plates with holes.

During the test, the damage map is monitored by means of X-ray photography. Also of interest in this test is the valuable information being provided in terms of the shape and size of the delaminated area. The difficulty herein is that, due to the complexity of the state of stresses, interpretation of the test requires complex computations. For such tests, a specialized software DSDM (Delamination Simulation by Damage Mechanics) has previously been developed (Allix 1992) for predicting delamination around initially-circular holes, with the intralaminar damage being taken into account in the analysis.

The laminates tested are the same as those presented in Table 3; a $[0_2/45/0_2/-45/90_2]_s$ was also tested. The specimens are 50 mm in width and 150 mm in gage length, and the hole diameter is 10 mm. The tests were conducted in tension on an INSTRON testing machine at a fixed displacement rate of 0.5 mm min^{-1}. Only the $[0_4/\pm22.5_2]_s$ and the $[0_3/\pm45_2/90]_s$ laminates have exhibited a significant delaminated area around the hole. This finding is surprising because the $[0_2/45/0_2/-45/90_2]_s$ is supposed to be very sensitive to delamination. This was the case for at least the two materials tested in (Trallero 1991) with high strength or intermediate-modulus fibres. This finding serves to demonstrate that the delamination phenomenon is heavily dependent on material properties.

In what follows, we present an initial comparison between the experimental observations in the $[0_3/\pm45_2/90]_s$ laminate loaded in tension and the computation. Figure 7 shows the evolution of the X-revealed damage map near the hole for an increasing applied load. The first damage, appearing at 55% of rupture (Figure 7a), is transverse cracking in 90°-plies near the hole, and matrix cracking in the 0°-plies tangent at the hole and in the fibre direction called "splitting". Delamination only begins at about 80% of rupture (Figure 7b). Just before the rupture (Figure 7c), the delaminated area is always found to be located between the splittings and developed in the 0°-direction with about two hole diameters in length. Micrographies were performed and show that the damage is well-developed in several ways: splittings, transverse cracking not only in the 90°-plies but also in the ±45°-plies, multiple delamination at the 0°/+45°, ±45° and -45°/90° interfaces. From the computation, the splitting can be seen as a shear damage in the 0°-layer (see Figure 8). In fact, when the first 0°-fibres near the hole crack, the local load is transferred by shear in the matrix at the adjacent fibres. The delaminated area computed in the 0°/+45° interface (or a ±22.5° interface oriented at 22.5°) is shown in Figure 9 as an example (the delaminated area corresponds to d = 1). In the same manner, the other

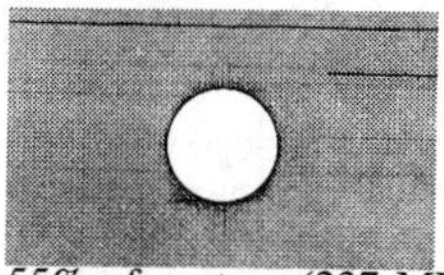

7a. 55% of rupture (237 MPa).

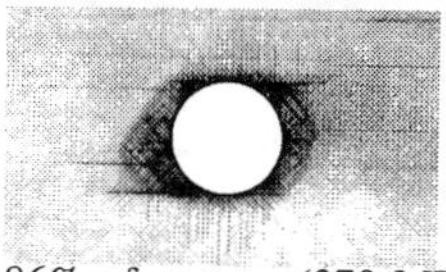

7b. 86% of rupture (370 MPa).

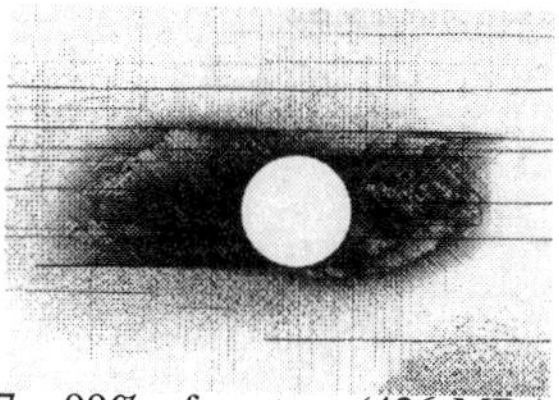

7c. 99% of rupture (426 MPa).

Figure 7. $[0_3/\pm45_2/90]_s$ X-ray damage map.

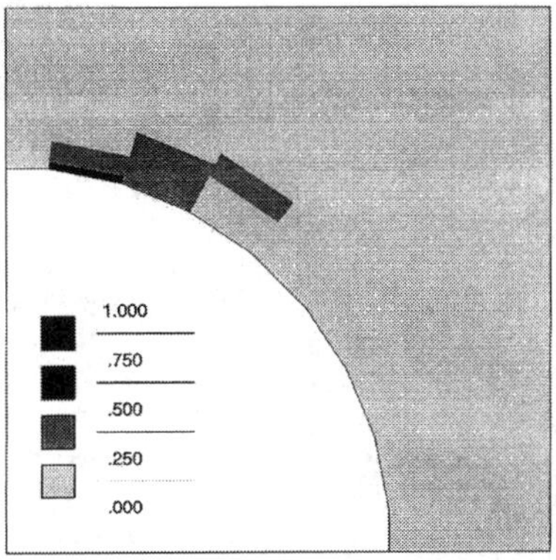

Figure 8. Shear damage indicator computed in the 0°-plies of the $[0_3/\pm45_2/90]_s$ laminate (370 MPa).

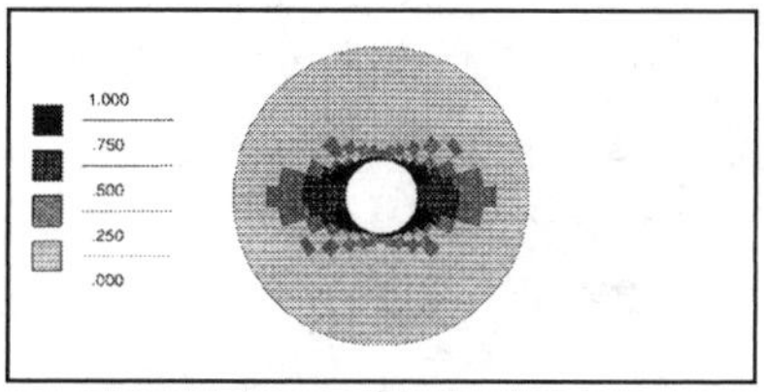

Figure 9. d_3 indicator computed at the 0°/+45° interface in the $[0_3/\pm45_2/90]_s$ laminate (370 MPa).

interfaces – except for the mid-plane – are found to be delaminated. In order to achieve a good comparison, we have not, for the time being, used those parameters identified previously and which have led to the prediction of no delamination. Different explanations for this feature are currently under consideration, and some progress is being hoped for shortly. Nevertheless, this last example shows that it is possible to determine stacking sequences for which, despite the relative brittle behaviour of M55J/M18 material, the onset and the propagation of delamination is sufficiently stable and the size of the delaminated area is wide enough to perform better comparisons between complex computations and experimental test results.

6 CONCLUSIONS

This work represents the first step towards a global identification of an interface damage model dedicated to delamination prediction. An initial set of parameters was determined using standard initiation and propagation tests conducted on M55J/M18 material specimens. We emphasized herein the necessity of considering the dissipative phenomenon due to the damage development inside the layers for both of these tests. Lastly, an initial comparison between computation and holed-specimen tested in tension revealed the value of using such an approach as a standard identification test. The problems discussed in Sections 3 and 4 still have to be resolved and a global identification software along with its accompanying set of reliable tests still has to be developed.

ACKNOWLEDGEMENTS

The authors would like to thank M. Guédra Degeorges for enabling us to conduct these tests at the Centre Commun de Recherche Louis Blériot / AÉROSPATIALE in Suresnes, France.

REFERENCES

Allix, O. 1992. Damage analysis of delamination around a hole. *New Advances in Computational Structural Mechanics*, P. Ladevèze, O. C. Zienkiewicz Eds., Elsevier Science Publishers B. V., p 411-421.

Allix, O. and P. Ladevèze 1992. Interlaminar interface modelling for the prediction of delamination, *Comp. Struct.*, 22:235-242.

Allix, O., P. Ladevèze & A. Corigliano 1995. Damage analysis of interlaminar fracture specimens. *Composite Sructures*, 31:61-74.

Allix, O. & A. Corigliano 1996. Modeling and simulation of crack propagation in mixed-mode interlaminar fracture specimens. *Int. J. of Fracture*, 77:111-140.
Allix, O. & P. Ladevèze 1996. Damage mechanics of interfacial media: basic aspects, identification and application to delamination. In *Damage and Interfacial Debonding in Composites*. Studies in applied Mechanics, 44, Eds. Allen D. and Voyiadjis G., Elsevier, pp. 167-88.
Allix, O. & D. Lévêque 1996. Technical report. *CNES (French National Center for Space Studies)* n°840/CNES/94/1406/00, January 1996.
Allix, O., P. Ladevèze, D. Lévêque & L. Perret 1997. "Identification and validation of an interface model for the delamination prediction", COMPLAS V, *5th Int. Conf. on Comp. Plas., Fund. and App.*, Barcelona, March 1997.
Bathias, C. 1995. Une revue des méthodes de caractérisation du délaminage des matériaux composites. Journée AMAC/CSMA *"Délaminage: bilan et perspectives"*, O. Allix & M.L. Benzeggagh Eds, Cachan, mai 1995.
Billoët, J.-L., T. Ben Zineb and B. Ben Lazreg 1994. Introduction de l'effet de paroi dans l'analyse des contraintes de bords libres pour les plaques stratifiées. *C. R. des JNC 9, AMAC*, Saint-Étienne, J.-P. Favre & A. Vautrin, Editors. pages 833-842, 22-24 Nov. 1994.
Crossman, F.W. and A.S.D. Wang 1982. The Dependence of Transverse Cracking and Delamination on Ply Thickness in Graphite/Epoxy Laminates. *Damage in Composite Materials, ASTM STP 775*, K.L. Reifsnider, Ed., pp 118-139.
Daudeville, L. and P. Ladevèze 1993. A Damage Mechanics Tool for Laminate Delamination. *Comp. Struct.*, 25:547-555.
Davies, P. & al. 1990. Measurement of G_{Ic} and G_{IIc} in Carbon/Epoxy Composites. *Comp. Sci. & Technol.*, 39:193-205.
De Charentenay, F.-X., J.M. Harry, Y.J. Prel and M.L. Benzeggagh 1984. Characterizing the Effect of Delamination Defect by Mode I Delamination Test. *Effect of Defects in Composite Materials, ASTM STP 836*, pp 84-103.
Engrand, D. 1982. Calcul des contraintes de bords libres dans les plaques composites symétriques avec ou sans trou. Comparaison avec l'expérience. *Comptes Rendus des Troisièmes Journées Nationales sur les Composites*, Paris, pp 289-297.
Guédra Degeorges, D. 1993. Principaux essais de délaminage - exploitation. *Une nouvelle approche des composites par la Mécanique de l'Endommagement*, Cachan, pp 229-249.
Highsmith, A.L. and K.L. Reifsnider 1982. Stiffness reduction mechanism in composite material. ASTM-STP 775, *Damage in Composite Materials*, A.S.T.M., 103-117.
Hwu, C., C.J. Kao, and L.E. Chang 1995. Delamination Fracture Criteria for Composite Laminates. *J. Comp. Mater.*, 29:1962-1987.
Kim, R.Y. 1989. Experimental Observations of Free-Edge Delamination. In *Interlaminar Response of Composite Materials*, N.J. Pagano Ed., pp 111-160.
Kim, R.Y. & S.R. Soni 1984. Experimental & analytical studies on the onset of delamination in laminated composites. *J. Comp. Mater.*, 18:70-80.
Ladevèze, P. and E. Le Dantec 1992. Damage modelling of the elementary ply for laminated composites, *Comp. Sci. & Technol.*, 43:257-267.
Lécuyer, F. and D. Engrand 1988. Étude des effets de bords thermomécaniques et hygroscopiques pour les plaques minces multicouches. *JNC 6*, J.-P. Favre and D. Valentin, Editors. pages 777-789. Pluralis.
Leguillon, D. 1997. Le critère de Griffith appliqué à l'amorçage d'une fissure. *Actes du troisième colloque national en calcul des structures*, CSMA/ECN, Presses Académiques de l'Ouest, Giens, mai 1997.
Leguillon, D. and E. Sanchez-Palancia 1987. Effets de bords et singularités dans les matériaux composites. *Annales des Composites, Effets de bords et singularités dans les matériaux composites stratifiés, AMAC*, Chatenay-Malabry, pp 7-20.
O'Brien, T.K. 1982. Characterization of Delamination Onset and Growth in a Composite Laminate. *Damage in Comp. Mat., ASTM STP 775*, Reifsnider, Ed., pp 140-167.
Russell, A.J. and K.N. Street 1985. Moisture and Temperature Effects on the Mixed-Mode Delamination Fracture of Unidirectional Graphite/Epoxy, *Delamination and Debonding of Materials, ASTM STP 876*, W.S. Johnson, Ed., Philadelphia, pp 349-370.
Schellekens, J.C. and R. De Borst 1993. A non-linear finite element approach for the analysis of mode I-free edge delamination in composites. *Int. Journal Solids Structures*, vol. 30-9, 1239-53.
Talreja, R. 1985. Transverse cracking and stiffness reduction in composite laminates. *J. of Comp. Mat.*, 19:355-375.
Trallero, D. 1991. Étude expérimentale et numérique de l'endommagement de structures percees en matériaux stratifiés. *Mémoire CNAM*, Paris, 29 mai 1991.
Williams, J.G. 1989. End corrections for orthotropic DCB specimens. *Comp. Sci. & Technol.* 35:367-376.

Damage and Failure of Interfaces, Rossmanith (ed.) © 1997 Balkema, Rotterdam, ISBN 90 5410 899 1

Examples of delamination predictions by a damage computational approach

Laurent Gornet, Christian Hochard & Pierre Ladevèze
Laboratoire de Mécanique et Technologie (LMT)/ENS de Cachan, France

Lionel Perret
Centre National d'Etudes Spatiales (CNES)/Toulouse, France

ABSTRACT: This study describes, for carbon epoxy laminated composites, some examples of delamination predictions by means of a damage mechanics computational approach. Most of these examples are classical delamination tests for carbon-fiber/epoxy-resin composites. The basic aspects of the damage mechanics approach which remains quite general are also given. Finite element, complete fracture phenomenon predictions of classical delamination tests are developed and compared with experiments.

1. INTRODUCTION

The aim of this paper is to present examples of delamination predictions by means of a damage mechanics computational approach. Three-dimensional non-linear finite element analyses are conducted for standard delamination tests. For this, a previously-defined damage meso-model for composite laminates is used. The interlaminar interfacial deterioration as well as the main inner layer damage mechanisms are taken into account. However, attention is focused herein on modeling the interlaminar connection as an elastic interfacial and damageable medium. A numerical integration scheme for the damage interface constitutive law is proposed. Even though the purpose of this work is quite general, this application concerns the M55J/M18 carbon-fiber/epoxy-resin material which a high modulus carbon-fiber/epoxy-resin material. Examples of comparisons between prediction and experimental results are provided.

All of the experimental results presented in this research work for the adjustment and numerical implementation of the behavior model have been obtained through our joint program with Aerospatiale-Suresnes.

In fact, our aim is to present a finite element delamination analysis of classical delamination tests that include all damage mechanism modeling. Delamination often appears as the result of interactions between different damage mechanisms, such as fiber-breaking, transverse micro-cracking and debonding of the adjacent layers themselves (Talreja 1985, Herakovich et al. 1987, Whithney et al. 1982). Thus, a damage mechanics meso-model, proposed by Ladevèze (1986) and developed in (Ladevèze et al. 1990), which includes both inner layer damage mechanisms and interfacial ones is used herein.

At the meso-level, the laminate is described as a stacking sequence both of inelastic and damageable homogeneous layers throughout the thickness and of damageable interlaminar interfaces. The single-layer model has been identified, the aim is to determine the properties of any structures as regards delamination by knowing only a few characteristics of the interface. The word "interface" denotes here a physical yet two-dimensional medium. At present, applications only concern static loading like in theclassical delamination test.

The single-layer model and its identification, including damage (such as fiber-breaking, transverse cracking and deterioration of the fiber-matrix bond) and inelasticity, were previously developed in (Ladevèze and Le Dantec 1992). The interlaminar interface is a two-dimensional entity which ensures traction and displacement transfer from one ply to another. Its mechanical behavior depends on the angles between the fibers of two adjacent layers (Allix et al. 1995).

Here, we will pay special attention to the basic aspects of the both interlaminar interface model and simulations of standard Fracture Mechanics tests. The analysis of standard propagation tests makes use of the links between the critical energy release rate and some of the interface damage model's parameters. Moreover, the lastest simulations of the Fracture Mechanics tests D.C.B. (Double Cantilever Beam), E.N.F. (End-Notched Flexure), M.M.F. (Mixed-Mode Flexure) and C.L.S., (Cracked Lap

Shear) are presented in connection with Aérospatiale-Suresnes's experimental results (Allix et al. 1996).

These comparisons require a three-dimensional Finite Element tool in order to simulate, up to failure, the behavior of any stacking sequence (Gornet 1996). The use of classical damage modelling for the simulation of failure has led to many theoretical and numerical difficulties which are well-understood and resolved at the present time (Ladevèze 1992). The solution used for laminates, and more generally for composites, is based on the meso-model concept (Ladevèze 1986, 1990). The physical interpretation of this concept is that the state of damage is uniform in each meso-constituent. For example, the damage state is uniform throughout the thickness of each single layer. To be able to perform a complete analysis of the delamination process in all cases, damage models with delay effects are also introduced for the in-plane direction. These damage rate dependence models introduce a characteristic time and then a length scale into the boundary value problem, even though the constitutive description does not contain a material parameter with the dimension of the length. In fact, we obtain a characteristic volume with the use of both damage models with delay effects and the meso-model concept.

2. MESO-MODELING CONCEPT

Let us recall that delamination often appears as an interaction between fiber-breaking, transverse micro-cracking and the debonding of adjacent layers itself. In order to take these mechanisms into account, the first issue is the scale at which they are modeled. For laminates, three different scales can easily be defined: the micro-scale of the individual fiber, the meso-scale associated with the thickness of the elementary ply, and the macro-scale which is the structural one. Due to the low thickness of the elementary ply and the kinematics of the deterioration inside the ply (fiber-oriented), it is possible and worthwhile to derive a material model at the meso-scale. The one proposed in (Ladevèze 1986) is defined by two meso-constituents:

-a single layer, and

-an interface which is a mechanical surface connecting two adjacent layers and dependent on the relative orientation of their fibers.

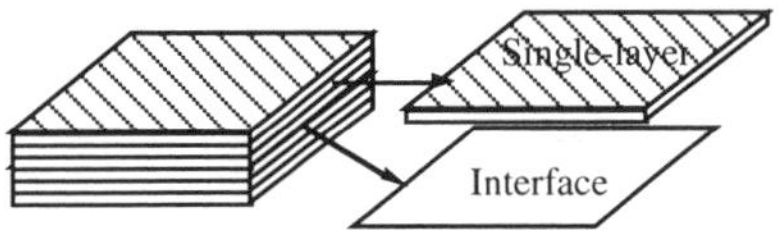

Figure 1. Laminate modeling

A meso-model is then defined by adding another property: a uniform damage state is prescribed throughout the thickness of the elementary ply. This point plays a major role when trying to simulate a crack with a damage model. With this property, Damage Mechanics integrate Fracture Mechanics, i. e. it yields a correct value of the critical energy release rate. Let us recall that in order to be able to perform complete analyses of three-dimensional delamination process cases, damage models with delay effects are introduced for the in-plane direction of both the layer and interface.

One limitation of the proposed meso-model is that the fracture of the material is described by means of only two types of macrocracks, delamination cracks within the interfaces and orthogonal cracks to the laminate with each cracked layer being completely cracked in its thickness.

Let us recall that the single-layer model and its identification, including damage such as fiber-breaking and transverse micro-cracking, as well as inelastic effects were previously developed in (Ladevèze 1986, Ladevèze and Le Dantec 1992).

3. INTERFACE MODELING

3.1. *Elastic modeling of the interface*

The interface model is detailed next. The interlaminar connection is being modeled as a two-dimensional entity that ensures stress and displacement transfers from one ply to another. This model was previously developed in (Allix and Ladevèze, 1994).

In the (N_1, N_2, N_3) axis, the elastic strain energy of the interface may be written as follows:

$$E_d = \frac{1}{2}\int_{\Gamma} [\, k_0[U_3]^2 + k_0^1[U_1]^2 + k_0^2[U_2]^2 \, d\Gamma$$
$$= \frac{1}{2}\int_{\Gamma} \left[\frac{\sigma_{33}{}^2}{k_0} + \frac{\sigma_{13}{}^2}{k_0^1} + \frac{\sigma_{23}{}^2}{k_0^2} \right] d\Gamma$$

where k_0, k_0^1 and k_0^2 are elastic characteristics of the interface. Let us denote the bisectors of the fiber directions by (N_1, N_2). They must necessarily be "orthotropic" directions of the interface, since a $[\theta_1,\theta_2]$ interface is equivalent to a $[\theta_2,\theta_1]$ interface.

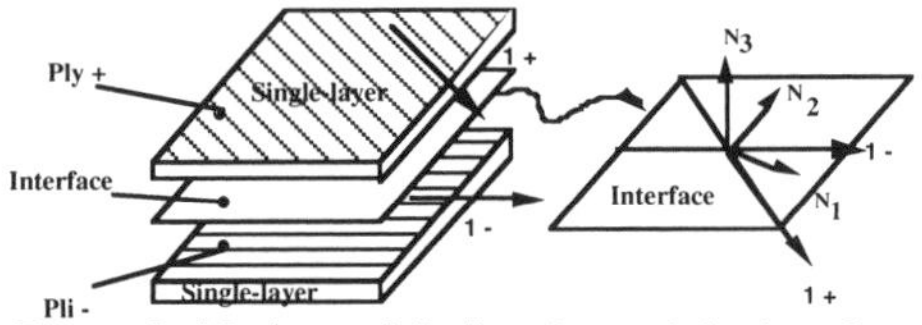

Figure 2. "Orthotropic" directions of the interface

The interpretation of the last relation (1) leads to high stiffness values for the elastic characteristics of the interface. In elasticity, the bond is perfects: it ensures displacement and traction continuities. In the non-linear case, it ensures traction continuity only. The stiffness values of the interface are on the same order of magnitude as an average transverse characteristic modulus of the matrix and of the fiber.

3.2 *Interfacial damage indicators*

The notions and framework that govern the interface damage model are similar to those which are used to derive the layer damage model (Ladevèze 1986, Ladevèze and Le Dantec 1992). The effect of the deterioration of the interlaminar connection on its mechanical behavior is taken into account by means of internal damage variables. The different types of damageable behavior in "tension" and in "compression" are distinguished by splitting the strain energy into "tension-energy " and "compression-energy". More precisely, we use the following expression, proposed in (Ladevèze 1986), for the energy per unit area:

$$E_D=\frac{1}{2}\left[\frac{<-\sigma_{33}>_+^2}{k_3^0}+\frac{<\sigma_{33}>_+^2}{k_3^0(1-d)}+\frac{\sigma_{32}^2}{k_2^0(1-d_2)}+\frac{\sigma_{31}^2}{k_1^0(1-d_1)}\right]$$

Thus three internal damage indicators, associated with the three Fracture Mechanics modes, are introduced (figure 3) .

$$Y_{d3}=\frac{1}{2}\frac{<\sigma_{33}>_+^2}{k^0(1-d_3)^2};$$

$$Y_{d1}=\frac{1}{2}\frac{\sigma_{31}^2}{k_1^0\ (1-d_1)^2};\qquad Y_{d2}=\frac{1}{2k_2^0}\frac{\sigma_{32}^2}{(1-d_2)^2}$$

3.3 *Interfacial damage evolution laws*

In what follows, a classical damage evolution laws is described. It is based on the assumption that the evolution of the different damage indicators is strongly coupled and driven by a unique equivalent damage energy release rate. The following model, proposed in (Ladevèze et al. 1990), considers that the damage evolution is governed by means of an equivalent damage energy release rate of the following form:

$$\underline{Y}\,(t)=\sup_{\tau\leq t}\left[\,(Y_{d3}+(\gamma_1 Y_{d1})+(\gamma_2 Y_{d2}))\,|_\tau\right]$$

This suggests that (i) the evolution in the damage indicators is assumed to be coupled (as for single layers), and (ii) the damage evolution depends (mainly) on the maximal value of the equivalent damage energy release rate. γ_1 and γ_2 are material parameters. In terms of delamination modes, the first term is associated with the first opening mode, and the two others are associated with the second and third modes.

A damage evolution law is then defined by choosing a material function ω, such that:

$d_3 = W(\underline{Y})$ if $d_3 < 1$; $d_3 = 1$ otherwise

$d_1 = \gamma_1\, d_3$ if $d_1 < 1$ and $d_3 < 1$; $d_1 = 1$ otherwise

$d_2 = \gamma_2\, d_3$ if $d_2 < 1$ and $d_3 < 1$; $d_2 = 1$ otherwise

A simple case, used for the purposes of application, is:

$$W(\underline{Y})=\left[\frac{<\underline{Y}-Y_o>_+}{Y_c-Y_o}\right]^m$$

These evolution laws must satisfy the Clausius-Duheim inequality. Classically, the damage energy release rates, associated with the dissipated energy ω by damage and by unit area, are introduced:

$$\omega=Y_d\,\dot{d}_3+Y_{d_1}\dot{d}_1+Y_d\,\dot{d}_2\quad(\omega\geq 0)$$

in which a critical value Y_c and a threshold value Y_o have been introduced. The high values of the "m" case correspond to a brittle interface.

To summarize, the damage evolution law is defined by means of five intrinsic material parameters $Y_c, Y_o, \gamma_1, \gamma_2$ and m. It will be shown next that Y_c, $\gamma_1$and γ_2 are related to the critical energy release rates. The threshold Y_o is introduced here in order to expand the possibility of describing both the creation of a delamination crack and its propagation. As regards the creation of a new delamination crack, the significant parameters are only Y_o and m.

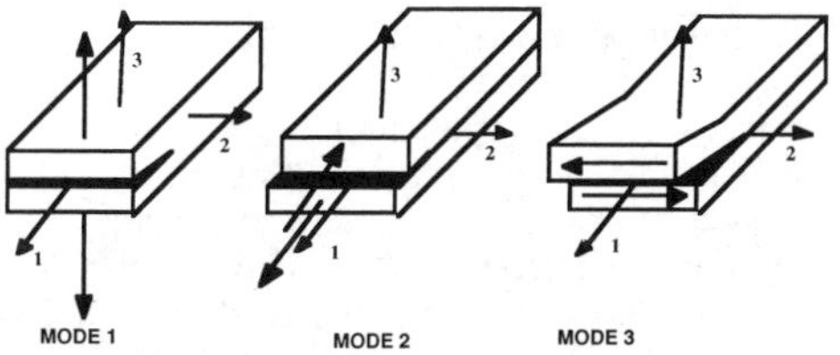

Figure 3. Fracture Mechanics modes

3.4 *Extension with delay effects*

It is well known that classical damage models lead inevitably to strain softening responses. From a fundamental point of view, this effect poses some

severe mathematical difficulties that are now well understood. From a computational perspective, the numerical solution with classical local damage models exhibits a severe mesh dependence in the presence of strain softening, which leads to completely useless results. One way to avoid such difficulties is to use localization-limiter models. A large class of limiter models has been proposed. In order to obtain, in all cases, a consistent model for the description of rupture, a variant of the previous damage model, that introduces delay effects (Ladevèze 1992, 1995) is applied. In quasi-static problems, the use of such damage evolution laws implicitly introduces a length scale into the governing equations of the problem and then avoids the pathological mesh sensitivity for composite structures. Examples are shown in (Gornet 1996). This variant damage model ensures both that the physical variation of the driving force Y does not lead to an instantaneous variation of damage variable d_3 and that the damage rate is bounded.
More precisely, the rate of the damage indicator is defined by:

$$\dot{d}_3 = k < w(Y) - d_3 >_+^n \text{ if } d_3<1; d_3=1 \text{ otherwise}$$

with $w(Y) < 1$; $w(Y) = 1$ otherwise

$$\dot{d}_1 = \gamma_1 \dot{d}_3 \qquad \text{if } d_1<1 \text{ and } d_3<1 \text{ ; } d_1=1$$

$$\dot{d}_2 = \gamma_2 \dot{d}_3 \qquad \text{if } d_2<1 \text{ and } d_3<1 \text{ ; } d_2=1$$

In many practical situations, a model without delay effect is sufficient. This is the case, in particular, for problems where the crack is described by a line. For example, in the case of a D.C.B. (Double-Cantilever Beam) test, results are mesh-independent. But in the general case (three-dimensional delamination shape), the delay effect model must be used to unforced for example the mesh independence.
Remark: In the case of the D.C.B. fracture mechanics test, it is possible to identify the parameter k (inverse of a characteristic time Figure 4) which governs the strain-softening response.

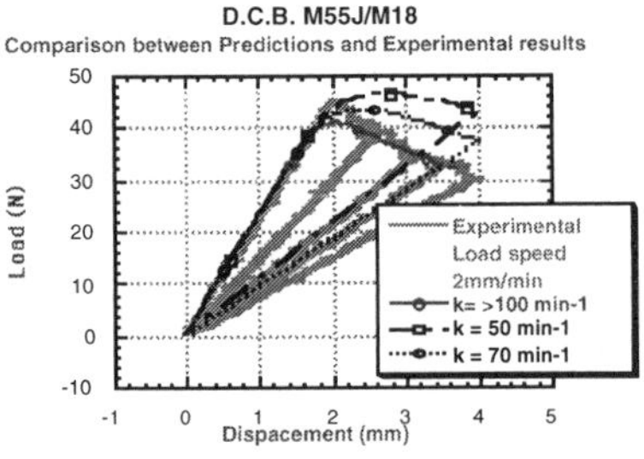

Figure 4. Identification of the damage delay effect's characteristic time parameter (k=100min-1).

3.5 F.E. implementation of the delay effect model

In this work, solid mechanics problems are solved by means of the finite element displacement method. The principe of this method consists of solving, in an incremental way, the discretized equilibrium equations and the material governing equations. A special interface element of zero thickness has been used (Beer 1985). Here, we focus our attention on the implementation used to solve the delay effect damage model in the finite element code. An implicit Euler algorithm is employed, and the non-linear discretized damage equation is solved by the Newton Method. The numerical integration of the constitutive model is summarized in Box 1.

Box 1:
Delay Effect Damage Model Implementation

1. *Compute* $W(Y_{n+1})$
2. *Initialize*

   $i = 0, \quad d_{3_{n+1}}^i = d_{3_n}$
3. *Check the residual:*

   $R_{n+1}^i = d_{3_{n+1}}^i - d_{3_n} - \Delta t\, k_3 < W(Y_{n+1}) - d_{3_{n+1}}^i >^n{}_+$
4. *Check for convergence:*

   If $R_{n+1}^i < Tol$, *Then* $d_{3_{n+1}} = d_{3_{n+1}}^{i+1}$ *Goto* 5.

   Else:

   $$\Delta d_{3_{n+1}} = -R_{n+1}^i \left(\frac{\partial R_{n+1}^i}{\partial d_{3_{n+1}}} \right)^{-1}$$

   $d_{3_{n+1}}^{i+1} = d_{3_{n+1}}^i + \Delta d_{3_{n+1}}$

   $i = i + 1$

   Goto 3.
5. *Compute damages* d_1, d_2:

   $d_{1_{n+1}} = d_{1_n} + \gamma_1 (d_{3_{n+1}} - d_{3_n})$

   $d_{2_{n+1}} = d_{2_n} + \gamma_2 (d_{3_{n+1}} - d_{3_n})$
6. *Compute stress:*

   $\sigma_{33_{n+1}} = k_3^0 (1 - d_{3_{n+1}})[U_3]_{n+1}$

   $\sigma_{23_{n+1}} = k_2^0 (1 - d_{2_{n+1}})[U_2]_{n+1}$

   $\sigma_{13_{n+1}} = k_1^0 (1 - d_{1_{n+1}})[U_1]_{n+1}$
7. *Update and exit.*

4. FRACTURE MECHANICS TESTS

The tests of crack propagation in interlaminar fracture specimens conducted in this work and developed at the Aérospatiale facility in Suresnes (France) are the pure-mode I D.C.B. (Double-Cantilever Beam) test (Whitney et al. 1982, De Charentay et al. 1984, Laksimi et al. 1991), the pure-mode II E.N.F. (End-Notched Flexure) test (Davies et al. 1990) and a mixed-mode tests the M.M.F. (Mixed-Mode Flexure) test (Russell & Street 1985, Carlsson et al. 1986). Standard Fracture Mechanics tests. Our specimens were 300 mm long and 20 mm wide. A 40 mm long anti-adhesive Teflon film was inserted at the mid-plane for the initiation of cracking. These tests were conducted on $[(+\theta/-\theta)_{4s}/(-\theta/+\theta)_{4s}]$ M55J/M18 material laminate with different θ angle values. In this work, we focus our attention solely on the prediction of unidirectional laminates fracture mechanics tests. The mean thickness of a single ply is on the order of 0.1 mm. These tests were conducted with an INSTRON testing machine at ambient temperature. The displacement rate was fixed at 2 mm/min in the D.C.B. test and at 1 mm/min in the E.N.F. and M.M.F. tests. In these latter two bending tests, the total useful length was 180 mm (only 100 mm for the ±45° ply-based laminate).

4.1 *Double-Cantilever Beam Test*

The D.C.B. specimen test is shown in Figure 5. This fracture mechanics test is probably the most widespread in the literature. In this test, the mode I critical energy release rate measured for the unidirectional material is G^p_{cI} = 113 J/m^2. This

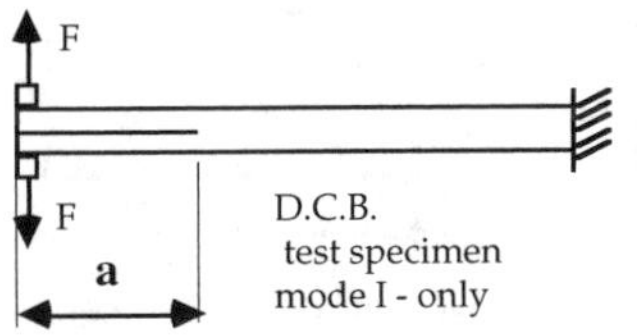

Figure 5. D.C.B. specimen test. The initial crack length is denoted by "a" on the picture.

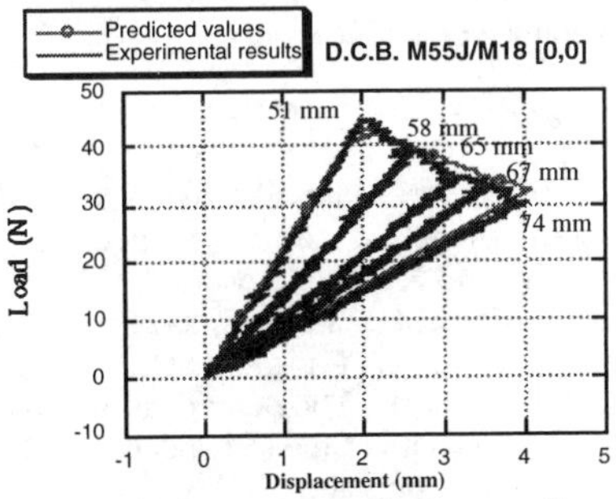

Figure 6. Prediction of a D.C.B. test . Comparison between experimental results and predicted values. The initial crack closure is a= 50 mm. The evolution of the delamination area at the end of the test is 23mm.

value is obtained by deriving the experimental compliance of the specimen. In this mode I test, the links between Linear Elastic Fracture Mechanics and Damage allow identifing the Y_c damageable model parameter. Good correlation is obtained between experimental results and finite element predictions (Figure 6). In particular the lengths of the debonding area are found to be close to one another.

Remarks: Regarding the identification of the damage interface model, the interest of classical Fracture Mechanics tests lies in the possibility of identifing the local critical energy release rate. Such tests are usually analyzed by means of Linear Elastic Fracture Mechanics (LEFM). Nevertheless, in the case of carbon-epoxy laminates, the main assumptions of LEFM are not always satisfied even in the simple case of a D.C.B. specimen. This is true, in particular, in the case of: non-unidirectional stacking sequence and R-curve-like phenomena. In the first situation, inner layer damage mechanisms may be activated, there leading to an apparent energy release rate different from the local interfacial one. In this case, a non-linear damage analysis of the layers and interfaces should be performed.

R-curve-like phenomena appear when the size of the non-linear domain is comparable to that of the specimen. This occurs, for example, in the problem of fiber-bridging. It must be highlighted that the carbon fibers wet oxidative surface treatment levels also influence the interface stiffness, and then the fiber-bridging leads to the R-curve-like phenomena (Albersten et al. 1995). In this case, it also appears to be more worthwhile to use Damage Mechanics rather than LEFM.

4.2 *End-Notched Flexure Test*

The E.N.F. test specimen is shown in Figure 7. This fracture mechanics test is used to obtain the critical energy release rate in mode II (Williams 1990). In this test, the critical energy release rate measured for the unidirectional material is G^p_{cII} = 308 J/m^2. Using both mode I and mode II experimental results, the links between Linear Elastic Fracture Mechanics and Damage allow identifying the γ_1 damageable model parameter.

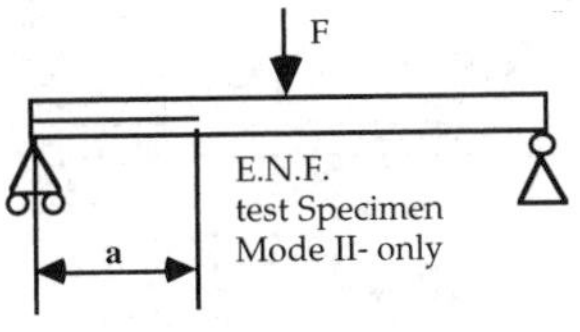

Figure 7. E.N.F. specimen test. The initial crack length is denoted by "a"

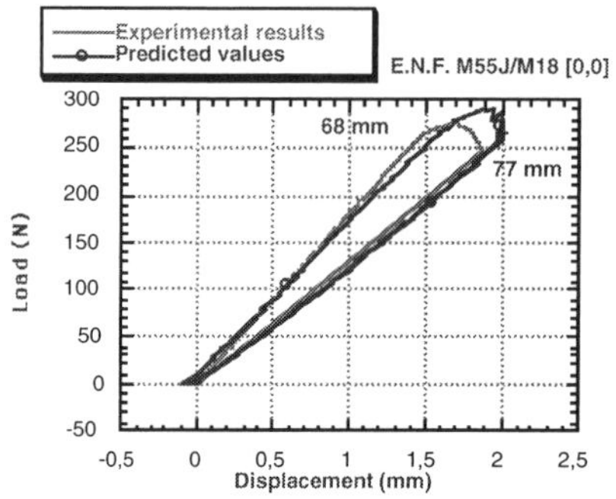

Figure 8. Prediction of an E.N.F. test. Comparison between experimental results and predicted values. The initial crack closure is a=68mm. The evolution of the delaminattion area at the end of the test is 77mm.

The hypothesis ($\gamma_1 = \gamma_2$) is taken without any further experimental information on mode III. The evolution of the damageable area is then refined. Experimental results and finite element predictions are shown in Figure 8. In particular the lengths of the debonding area are found to be close to one another.

4.3 *Mixed-Mode Flexure test*

The M.M.F. test specimen is shown in Figure 9. A mixed-mode critical energy release rate is obtained with this fracture mechanics test. In this mixed-mode test, the mode I is dominant. The critical energy release rate measured for the unidirectional material is G_c^p = 162 J/m^2. This mixed-mode test allows checking the damage model's parameters. The evolution of the damageable area is refined. Experimental results and finite element predicted values exhibit good correlation (Figure 10). In particular, the lengths of the debonding area are found to be close to one another.

4.4 *Cracked Lap Shear*

The C.L.S. (Cracked Lap Shear) test is shown in Figure 11. A mixed-mode critical energy release rate is also obtained with this fracture test (Kyong

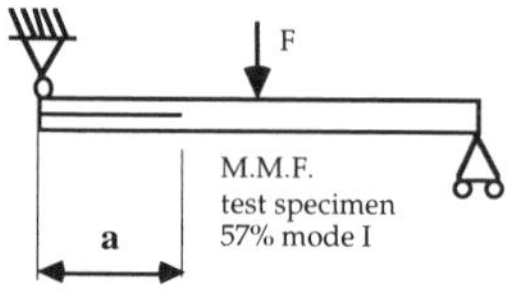

Figure 9. M.M.F. specimen test. The initial crack length is denoted by "a".

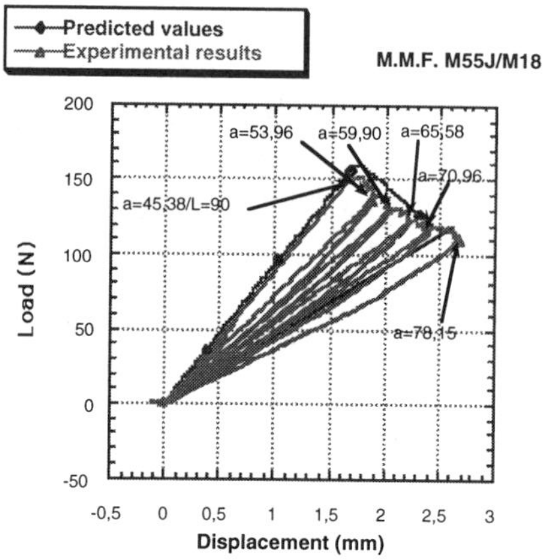

Figure 10. Prediction of an M.M.F. test. Comparison between experimental results and predicted values. The initial crack closure is a=45mm. The evolution of the delaminattion area at the end of the test is 32.77mm.

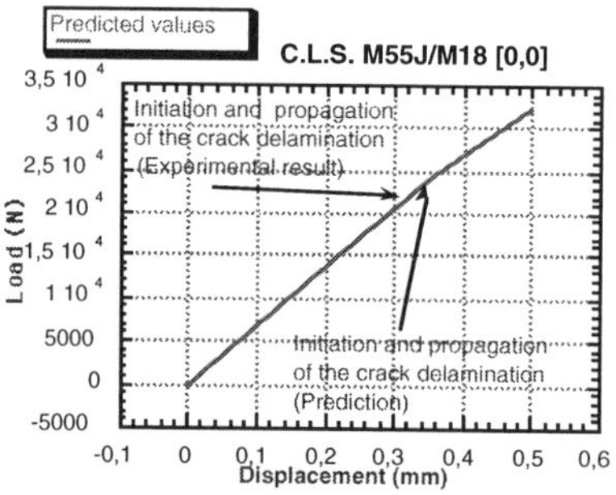

Figure 11. Comparison of the initiation of the crack delamination between experimental results and predicted values.

1994). This mixed-mode test is used to check the interface parameter identification.

4.5 *X-ray results*

After each fracture mechanics test, the experimental delamination shape of the test specimen is highlighted by an X-ray photography. For the unidirectional M55J/M18 material, the X-ray shape is shown in Figure 12 for the D.C.B., E.N.F., M.M.F. and C.L.S. tests. The delamination front is not directly in the width of the test specimens. Near the edge, there is a curvature of the delamination front in all tests. In the case of the C.L.S. test, this shape is not symmetric; this X-ray is not representative of optimum experimental test conditions. The curvature of the delamination front

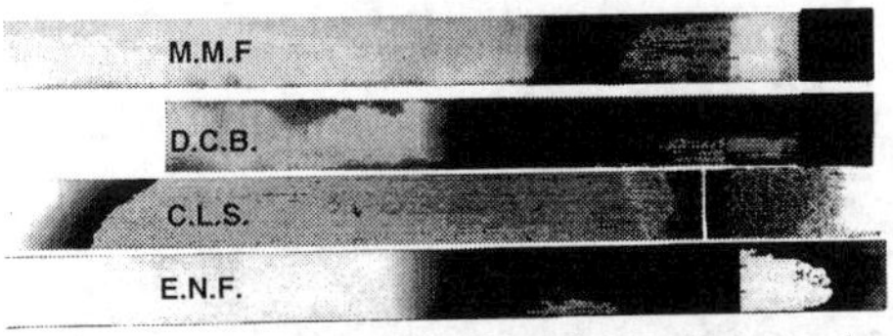

Figure 12. X-ray delamination shape photography.

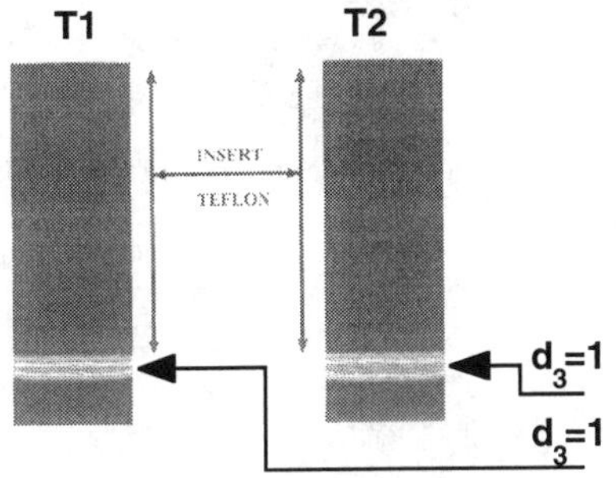

Figure 13. Predicted values of the delamination front of the D.C.B. test.

near the edge is refined by three-dimensional Finite Element predictions generated with the Finite Element code dedicated to the non-linear composite structura analysis which has been developed in this study. The shape of the delamination area is shown for the D.C.B. test in Figure 12. It should be noticed that the curvature of the delamination front is greater for tests conducted with the M55J/M18 material having θ angle values other than 0 degrees.

5. INITIATION TESTS

5.1 *Composite hole tension test*

Let us consider the structural computation example defined in Figure 14. It is a holed plate [+ 22.5° , -22.5°]$_s$ submitted to tension. The loading history is shown in Figure 15. At any point and time, the code is able to yield the "intensity" of the various damage mechanisms up until the ultimate fracture. The main damage mechanism herein is delamination, i.e. the deterioration of the interface [22.5°, -22.5°]. Figure 15 gives the value of the damage variable "d" at times T_1 and T_2. The increase of the delaminated area is very significant. The layer's damage mechanisms are weakly excited (Figure 17).

5.2 *Edge effect analysis*

Numerical simulations of delamination onset near the free edge of carbon-epoxy composite specimens under tension or compression have been previously

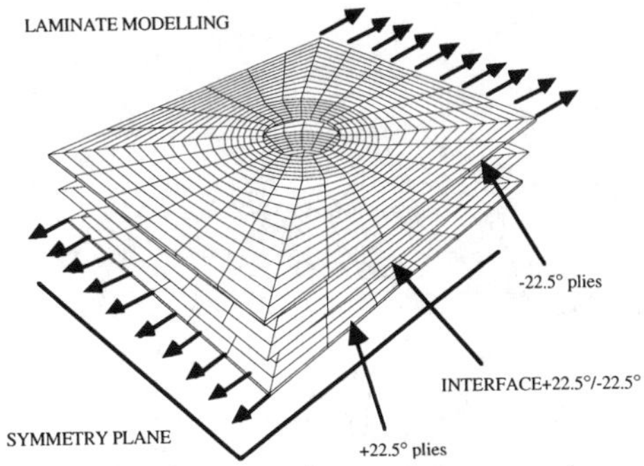

Figure 14. A structural computation example

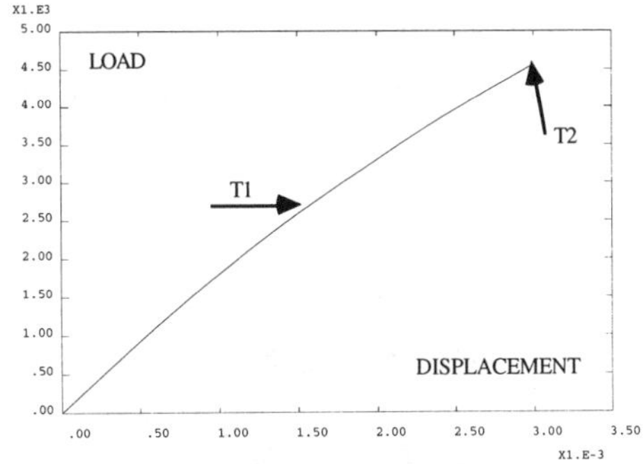

Figure 15. Loading history

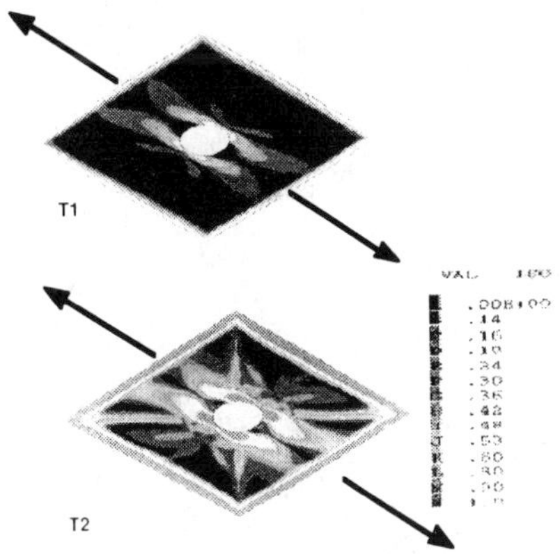

Figure 16. Damage variable d of the interface at times T_1 and T_2.

studied (Daudeville et al. 1995). In this previous study, all of the damage phenomena were modeled as being concentrated on the interface. In this present work, the edge effects are computed by taking into account the damage modeling of both the interface and the layer. As another example of the possibilities of our F.E. code, let us consider the laminate structure defined in Figure 18 where a

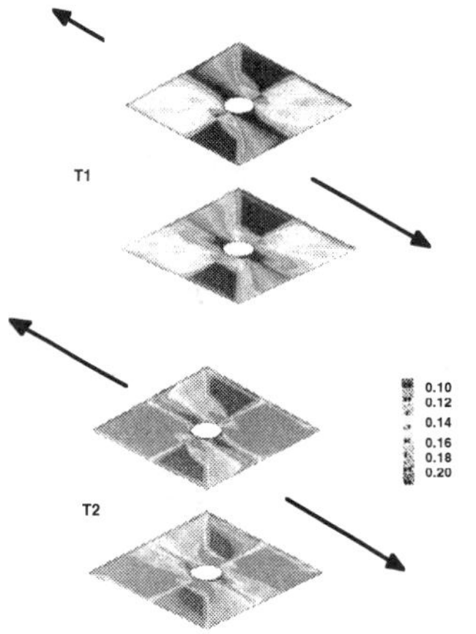

Figure 17. Damage variable d of the layers at times T_1 and T_2

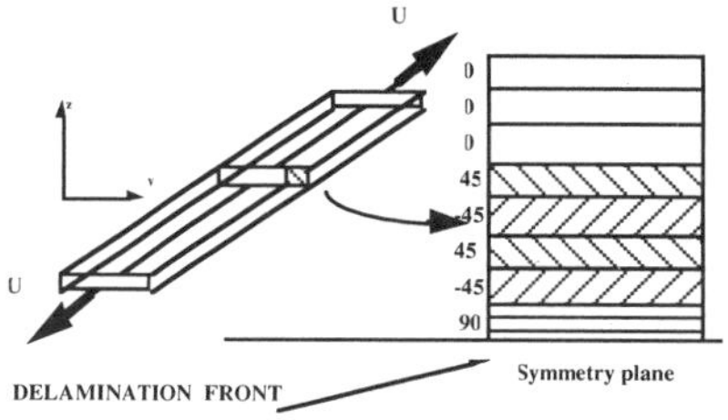

Figure 18. the laminate is a $[0_3, (+-45)_2, 90]$sym M55J/M18 stacking sequence distribution layers.

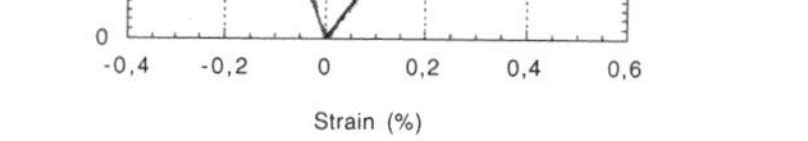

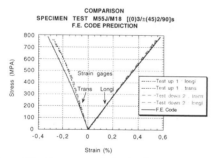

Figure 19. Response of the laminate and the F.E. prediction

tension loading is applied. These initial results have been obtained with the M55J/M18 material's parameters. During the loading history of the structure, a crack in the central interface of the specimen test first appears; at the same time, cracks grow until rupture in the (+-45) layers. The usual mesh sensitivity difficulties are not present. Results are given in Figures 20 et 21.

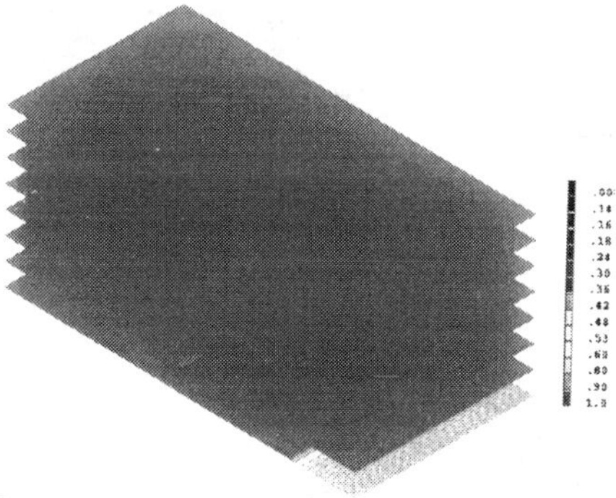

Figure 20. A crack first appears in the central interface.

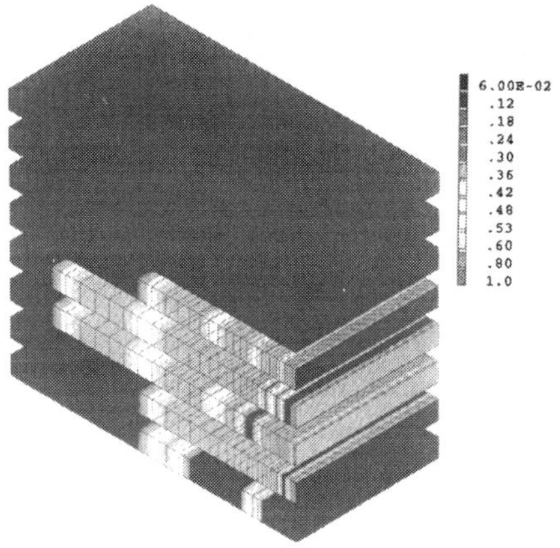

Figure 21. Rupture in the (+- 45) layers of the laminate.

6. CONCLUSION

A meso-damage mechanics modeling of laminates, whose aim is to predict the behavior of any composite structure with respect to delamination through knowing only a few characteristics of the interface, has been detailed. Finite element examples show that this approach is promising in the prediction of delamination under various circumstances. Nevertheless, one of the first challenges will be to build a good interface model which enables predicting, in using a single set of parameters, both the creation and the propagation of a delamination crack over a wide range.

REFERENCES

Albertsen, H., J. Ivens, P. Peters, M. Wevers & I.

Verpoest 1995. Interlaminar fracture toughness of CFRP influenced by fibre surface treatment: Part 1:Experimental results, *Comp. Sc. and Tech.* 54, pp. 133-145.

Allix, O. & P. Ladevèze 1994. A meso-modelling approach for delamination prediction. *In: Fracture and Damage in Quasi-brittle Structures*. Bazant, Z. P., Bittnar, Z., Jirasek, M. Eds, pp. 606-15.

Allix, O., . P. Ladevèze, A. Corigliano 1995. Damage analysis of interlaminar fracture specimen, *J. Composite Structures*, 31(1) , pp. 61-74.

Allix, O., D. Lévêque, L.Perret, 1996, Identification d'un modèle d'interface interlaminaire pour la prévision du délaminage dans les composites stratifiés, *JNC-10, 10th French National Colloquium on Composite Materials*, D. Baptiste and A. Vautrin, eds, Paris, pp. 1041-1052.

Beer, G., 1985. An isoparametric joint/interface element for finite element analysis, *Int.J. for numer. methods eng.*, 21 , pp. 585-600.

Carlsson, L., J. W. Gillospie, R.B.Pipes 1986. On the analysis and design of end-notched flexure (E.N.F.) specimens for mode II testing, *J.Comp.Mat.*, 20 , pp. 594-604.

Daudeville, L., O. Allix, P. Ladevèze 1995. Delamination analysis by damage mechanics: some applications. *Composites Engineering*, 5 (1), pp. 17-24.

Davis P. 1990. Measurement of GIc and GIIc in Carbon/Epoxy Composites, Comp. Sci. & Technol., 39 , pp. 193-205.

De Charentenay, F.-X., J.M. Harry, Y.J. Prel, & M.L Benzeggagh 1984. Characterizing the Effect of Delamination Defect by Mode I Delamination Test. *Effect of Defects in Composite Materials, ASTM STP 836*, pp. 84-103.

Gornet, L. 1996. *Simulation des endommagements et de la rupture dans les composites stratifiés*, Thesis, Université Pierre et Marie Curie Paris 6/LMT/ENS-Cachan, ISBN 2-11-088-9705.

Herakovich, C.T., J., Aboudi, S.W. Lee & E.A., Strauss 1987. Damage in composite laminates: effects of transverse crack, *Mechanics of Materials*, 7, pp. 91-107.

Kyong Y.Rhee 1994. Characterization of delamination behavior of unidirectional graphite/PEEK laminates using cracked lap shear (CLS) specimens. *Composite Structures*, 29, pp. 379-382.

Ladevèze, P. 1986. Sur la Mécanique de l'Endommagement des composites. *In: Comptes rendus des JNC 5*, C. Bathias & D. Menkès eds, Pluralis Publication, Paris, pp. 667-683.

Ladevèze, P. 1992. Towards a Fracture Theory, *Proceedings of the Third International Conference on Computational Plasticity, Part II*, D.R.J. Owen, E. Onate, E. Hinton Ed, Pineridge Press, Cambridge U.K., pp. 1369-1400.

Ladevèze, P., O. Allix & L. Daudeville 1990. Meso-modeling of Damage for Laminate Composites. Application to Delamination. *In: Inelastic Deformation of Composite Materials*,.Springer -Verlag, Dvorak, V.D. Ed, pp. 607-622.

Ladevèze, P. 1992. A damage computational method for composite structures, *J. Computer and Structure*, 44 (1/2), pp. 79-87.

Ladevèze, P. 1995. A damage computational approach for composites: Basic aspects and micromechanical relations. *Computational Mechanics*, 17, pp. 142-150, 1995.

Ladevèze, P. & E. Le Dantec 1992. Damage modeling of the elementary ply for laminated composites, *Composite Science and Technology*, 43-3, pp.257-267.

Laksimi, A., M.L. Benzeggagh, G., Jing, M. Hecini & J.M., Roeland 1991. "Mode I Interlaminar Fracture of Symmetrical Cross-ply Composites", *Comp. Sci. and Tech.*, 41, pp. 147-164.

Russell, A.J. & K.N. Street 1985. Moisture and Temperature Effects on the Mixed-Mode Delamination Fracture of Unidirectional Graphite/Epoxy, *Delamination and Debonding of Materials, ASTM STP 876*, W.S. Johnson, Ed., Philadelphia, pp. 349-370.

Talreja, R. 1985. Transverse cracking and stiffness reduction in composite laminates, *Journal of Composite Materials*, Vol.19, pp. 355-375.

Whitney, J.M., C.E. Browning & W.Hoogsteden 1982. A double cantilever beam test for characterizing mode I delamination of composite materials. *J. Reinforced Plastics and Composites*, 1, pp. 297-313.

Williams, J.G. 1990. Finite displacement correction factors for E.N.F. test", *Composites Science and Technology*, 39 , pp. 279-282.

Damage and Failure of Interfaces, Rossmanith (ed.) © *1997 Balkema, Rotterdam, ISBN 90 5410 899 1*

Interface degradation in ceramics under cyclic loading

K.W.White & L.Olasz
University of Houston, College of Engineering, Tex., USA

J.C.Hay
University of Houston, College of Engineering, Tex., USA (Presently: Oak Ridge National Laboratories, Tenn., USA)

ABSTRACT: The role of the interfaces in the fracture event is studied in monolithic ceramic microstructures. As an alternative approach to R-curves, the toughening of alumina is described via the post-fracture tensile (PFT) test, which has previously been introduced as a method for the direct evaluation of the wake zone traction function. This universal description of the toughening behavior has previously been related to the elements of the microstructure, including thermoelastic constants, grain size and distribution, and grain boundary character. This presentation will address the behavior of the wake process zone microstructure subjected to cyclic loading environments. Where the classical R-curve behavior relates largely to the scale of the grains, the cyclic degradation problem seems to respond to microstructural features of a scale about 1 to 2 orders of magnitude smaller, focusing attention toward the interfacial character.

1 INTRODUCTION

Since the shape of the R-curve influences the stability of crack propagation, the key to improving damage tolerance under typical service conditions is to understand the relationships between microstructure, toughness, R-curve behavior and finally, fatigue behavior. In bulk specimens, an important link between fracture and fatigue presumes that crack growth only occurs when the crack tip stress intensity factor exceeds the inherent material toughness. For crack extension in a wake-toughened ceramic, under either monotonic or cyclic-loading conditions, the applied stress intensity factor, K_a, must exceed the intrinsic toughness, K_o, by K_w, the wake contribution, as given by

$$K_a > K_o + K_w. \qquad (1)$$

Jacobs and Chen (1995) offer a phenomenological account of fatigue crack growth, where, under equilibrium conditions, the reduction in K_w through the degradation of the grain-bridging contribution is offset by the increase in K_w by crack growth, or $dK_w=0$. Mathematically, this treatment indicates that the change in wake shielding, dK_w, is given by

$$dK_w = (\partial K_w/\partial n)\, dn + (\partial K_w/\partial a)\, da = 0 \qquad (2)$$

or

$$(da/dn) = -(\partial K_w/\partial n)\, (\partial K_w/\partial a)^{-1}, \qquad (3)$$

where the crack length is given by 'a' and the number of loading cycles is given by 'n'. The second term on the right-hand side depends on the monotonic crack-growth-resistance curve and the first term describes wake degradation by the dependencies on K_{max} and ΔK. To fully understand the microstructural role, the wake influence under both the monotonic and cyclic conditions must be considered.

1.1 *Bridging Mechanism*

In the decade following 1973 the role of the wake process zone remained unclear. In 1982, however, the renotching experiments of Knehans implicated the wake process zone as the sole source of toughening. Later *in situ* techniques employed methods which allowed for direct observation of the fracture process under SEM (Swanson, 1987; Vekinis, 1990; Rodel, 1990; Lathabai, 1991) and optical (Hay, 1991) microscopy. These works linked the rising R-curve behavior to an interlocking mechanism by which grains effectively bridge the crack, thus, conflicting with previous microcracking theories.

With the reexamination of other monolithic microstructures, the R-curve quickly became the convention for characterizing the toughening properties. By the end of the 1980's, though, careful consideration of the data indicated that the R-curve did not represent a microstructural property; the R-curve depends only upon the crack-opening profile and a stress-displacement relationship describing the local wake loads as a function of crack-face separation. The stress-displacement relationship *is* a unique microstructural characteristic which was quickly adopted for representation of the toughening process.

Studies in reinforced concrete (Wecharatana, 1983) demonstrated that the wake stresses follow a power-law strain-softening behavior described by the form

$$\sigma/\sigma^* = (1-u/u^*)^m \qquad (4)$$

where the exponent, m, characterizes the strain-softening regime, the critical stress, σ^*, is the limiting stress of the mechanical interlocking mechanism, and u^* is a limiting displacement for the active wake. Ceramic studies have shown that the form of equation 4 also applies to ceramics for the strain softening behavior. Of particular interest now, however, the post-fracture tensile (PFT) method developed by Hay and White (1991, 93, 94, 95), offers an experimental means for studying the wake tractions at small crack-face separations. This paper examines the role of subgrain-size features in the fatigue process, supported by previous phenomenological characterizations of the degradation by Paris law parameters.

1.2 *Wake Degradation Phenomenon*

Fatigue in ceramics describes mechanical degradation by various processes. Only recently, though, has cyclic, or mechanical fatigue in ceramics such as alumina (Reece, 1989); Krohn, 1972) or silicon nitride (Jacobs, 1994, 95) received attention. As a part of this new interest the question arises - what role does the microstructure play in the behavior, and specifically, how do the bridging grains contribute to fatigue crack-growth resistance? Ewart and Suresh (1987) suggest that intergranular fracture associated with the polycrystalline samples provided the resistance to rapid crack growth under cyclic fatigue. Fatigue crack growth must undoubtedly be related to the manner in which the wake zone mechanically degrades with each fatigue cycle.

Grain bridging is described by both frictional and elastic constituents, giving rise to observable hysteresis loops in the loading cycles of ceramics with a developed wake zone (Guiu, 1992). When the applied load is lower than that required to overcome the frictional loads, the load-displacement curve exhibits a high stiffness. As the applied load reaches that of the frictional loads, the compliance increases. During unloading the process is reversed. There appears to be substantial evidence to implicate abrasive forces with cyclic fatigue behavior. Microstructural studies by Lathabai et al. (1991) and Jacobs and Chen (1995) showed evidence of wear debris after cyclic loading of grain bridges in the wake region. Phenomenological depictions of the process also appear in the literature to explain the degradation of load-bearing elements leading to crack growth.

Currently, a couple of methods for fatigue characterization appear in the literature. For a decade, research in ceramics has attempted to fit experimental fatigue data to the well known Paris Law for metal fatigue. In a form that includes a K_{max} dependency, this law relates crack growth rates, da/dN, to the range in stress intensity factor, ΔK, used in cyclic loading as: $da/dN = A' (\Delta K)^n (K_{max})^p$. Experimental data from the ceramics literature which use this form exhibit a large dependency on K_{max} ($p \approx 25$) and minimal dependency on ΔK ($n \approx 2$-4). The strong dependency on K_{max} can be rationalized by the sensitivity of bridging stresses to the crack-opening profile. Since the maximum stress intensity factor determines the magnitude of the COD profile, Hu and Mai (1992) suggest that larger COD's reduce the potential for the crack faces fitting back together.

It is clear that the bridging mechanism is displacement controlled, yet the literature has not attempted to directly relate fatigue damage with the change in COD rather than K_{max} or ΔK. Although, Hu and Mai (1992) have offered a characterization of the fatigue phenomenon which indirectly incorporates COD. That work considers a bulk specimen with a long wake zone developed under monotonic loading conditions. When the same specimen is subjected to cyclic loading, where K_{max} is maintained below a critical threshold for crack growth (K_R), wake degradation occurs without the addition of new bridges. The authors characterized wake damage due to cycling by measuring the bulk specimen compliance intermittently. The fatigue process, then, can be characterized by the reduction in wake length and critical COD.

Globally, the cyclic degradation in wake-toughened ceramics will depend on several variables, such as grain size, grain morphology, thermal expansion anisotropy (TEA) stresses, temperature and starting COD. Our goal is to isolate and systematically address each variable to develop a full understanding of their influences on the fatigue characteristics. The post-fracture tensile (PFT) method for experimental measurement of the crack-face loads in the wake zone allows for direct examination of the mechanism under cyclic conditions. The phenomenon we report here controls the cyclic response of this alumina microstructure for one value of starting COD.

2 PROCEDURES

2.1 *Material and Specimens*

The alumina used in this study is a commercial 99.7% alumina, obtained through Johnson Matthey, and is the same as that characterized previously (Hay 1993, 95) The remaining primary constituents are SiO_2, MgO and CaO. The average grain size, found from a polished and etched surface, is approximately

18 μm and a majority of the grains (≈90%) were less than 40 μm.

Short DCB specimen geometry offers two significant advantages over alternatives such as the notched bend bar specimen. First, the DCB geometry is immune to the geometry dependence found in bend bar specimens where the R-curve slopes depend upon the initial crack length. Also, the long cracks obtained from this geometry provide many test specimens for the second half of the experiment (PFT), which characterizes the wake zone.

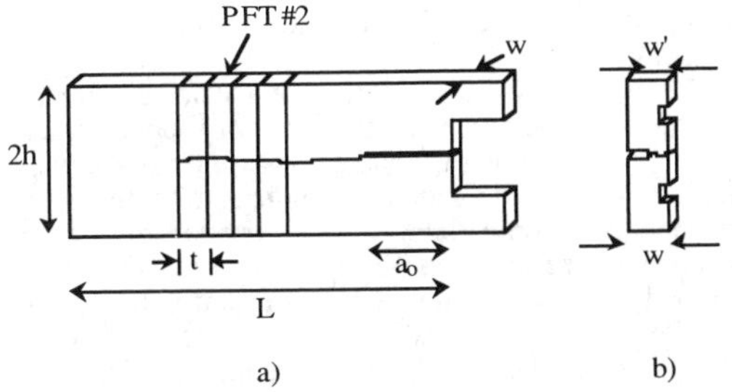

Figure 1. Double cantilever beam (DCB) specimen geometry (a) and the PFT specimen (b) which isolates the wake for direct wake studies.

A half-thickness side groove down the center of the specimen restrains crack growth from the desired fracture plane. The specimens were fractured on an Instron testing machine at a displacement rate of 0.50 μm/min, and crack lengths were observed optically while growing the cracks by approximately 12-15 mm.

2.2 *PFT Experiment*

The second part of the experiment, referred to as the post-fracture tensile (PFT) test, required machining of the tensile specimens from the cracked DCB specimen for characterization of bridging mechanisms. It is important to consider the significance of the PFT technique before proceeding. The PFT technique provides a unique procedure to isolate incremental segments of the crack-wake region of a cracked DCB specimen (Figure 1a). In this figure, the region behind the crack tip is sliced into several strips. Each strip is through-cracked and held together only by bridging ligaments as shown by the schematic in Figure 1b. Two side grooves, in addition to the crack stabilizing groove, facilitate tensile loading on two knife edges.

In the past, the bridging mechanism was studied under quasistatic conditions. Tensile loads were applied under controlled-displacement conditions at a constant displacement rate, and we obtained a stress-displacement relationship by dividing the loads by the cross-sectional area of the intergranular component of the cracked connecting segment. This component has been dealt with in some detail, and appropriately excludes the transgranular fracture surface area from the load-supporting role.

The typical load-displacement curve from a monotonic test contains three distinct behavioral regions, as indicated in Figure 2, for commercial alumina at room temperature. A macroscopic "linear" region characterizes the behavior for loads below approximately 80% of the maximum tensile load (point A). The dashed portion of the curve in Figure 2 was never well characterized by macroscopic displacement measurements, since the actual change in crack face separation, as measured by crosshead motion, is difficult to separate from the nonlinear compliance contributions due to specimen seating at the start of a test. The displacement measuring technique used in this paper faithfully resolves displacements in that range since the displacements are measured directly across the crack face using the interferometer described next.

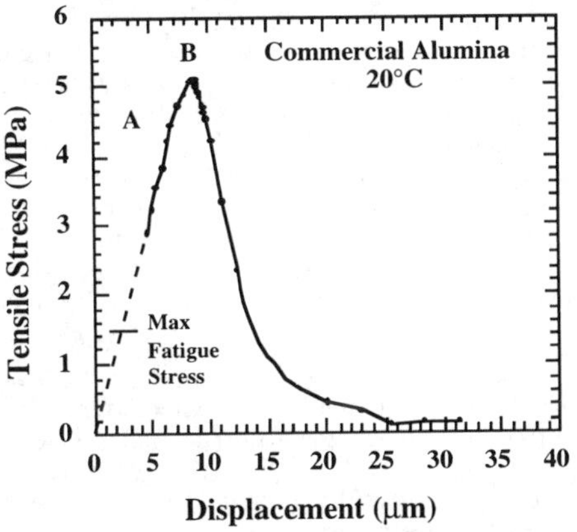

Figure 2. Typical stress-displacement data from the PFT test on commercial alumina at 20°C. Fatigue tests take place in dashed region with peak stresses less than 1.5 MPa.

The linear region is followed by a short nonlinear region, prior to point B, possibly indicating nonrecoverable displacements. A monotonic decrease in load follows the maximum load at point B. Since those regions beyond point B most strongly control R-curve behavior, attention has previously been focused here (Hay 1991, 93, 95). By extrapolating the strain-softening portion of the curve back to the stress axis, we estimated (Hay 1993) that the maximum stress which could be carried by the wake zone in this microstructure was near 50 MPa. This peak stress is roughly 10% of the bulk strength (500 MPa) typically reported for alumina. Although that wake stress may appear low, the stresses actually represent a significant contribution to the load carrying capacity of the specimen. The renotch experiments (Knehans 1982) indicated that the wake zones responsible for R-curve behavior carry up to 1/3 of the total applied load. Following the renotch experiments, the mechanics models discussed in the Introduction provided predictions of the closure stress-displacement curves. Interestingly, those models agreed favorably with the experimental data obtained by the PFT technique. In this paper, the maximum applied stresses are maintained well below point A, as indicated schematically in Figure 2.

Also note that the wake stiffness and peak-load capacity of each specimen depends upon the initial COD conditions, such that PFT specimens from near the crack tip exhibit the greatest peak load and stiffness.

2.3 *Displacement Measurements*

In this study we focus in detail on the bridging mechanism response to small applied displacements through a small number of load-unload cycles. To accomplish this task, the load is maintained well below the macroscopic proportional limit described by point A (Figure 2). Accurate noncontacting measurement of the small changes in displacement were possible using a laser interferometric displacement gage (LIDG). A side view of the PFT specimen is shown in Figure 3 with a representation of the LIDG. Two platinum tabs indented with a Vickers hardness indenter are secured to the PFT specimen above and below the crack plane, using a thermal glue (Crystalbond 509, Aremco Products, Inc., Ossining, NY). When a He-Ne (λ=632 nm) laser beam illuminates the two metal tabs simultaneously, four interference patterns result from the differing path lengths from the indents to the fringe-pattern planes. Only the two fringe patterns in the geometry shown in Figure 3 provide useful displacement information.

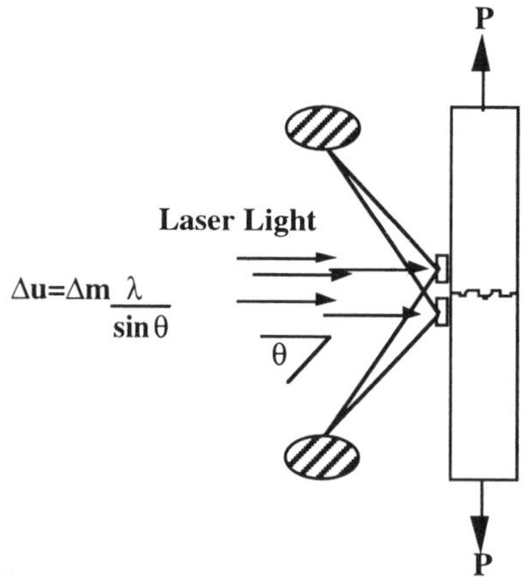

Figure 3. Schematic of the laser interferometric displacement gage (LIDG) for measuring small changes in displacement.

Two remote photodiodes detect changes in the fringe intensities with displacement and generate signals captured by a computer. The wavelength of the laser radiation, λ, the characteristic angle between the incident and reflected beams, θ, and the fringe order, Δm, are used to measure small changes in displacement between the specimen halves. The displacement is determined from

$$\Delta u = \Delta m \, \lambda / \sin\theta. \tag{5}$$

Effects of rigid-body motion are accounted for by using the average fringe order from the two detectors.

We argue that the resolution of the interferometer may actually be much less than the wavelength of the light, provided that the spatial-fringe intensity is well understood. Application of the indents directly to the unattached loading fixtures provides the opportunity to characterize the spatial-fringe intensity. When the unloaded crosshead moves at a constant velocity, the spatial and temporal-fringe shapes coincide, described by the sinusoidal fringe shape (Figure 4) given by

$$y = A\sin(\theta) + B. \tag{6}$$

In the real experiment, the temporal-fringe shape differs from equation (6) due to the applied loads, but the spatial-fringe shape always fits equation (6). Therefore, the measured intensity of an experiment, y(t), and the maximum and minimum fringe intensities (defining A and B) provide a unique phase angle, θ, or partial fringe order, Δm. For the current study, where total displacements take place over a few fringe orders, interpolation of the points between the maximum and minimum fringe points is performed.

An uncertainty for the measurement is difficult to

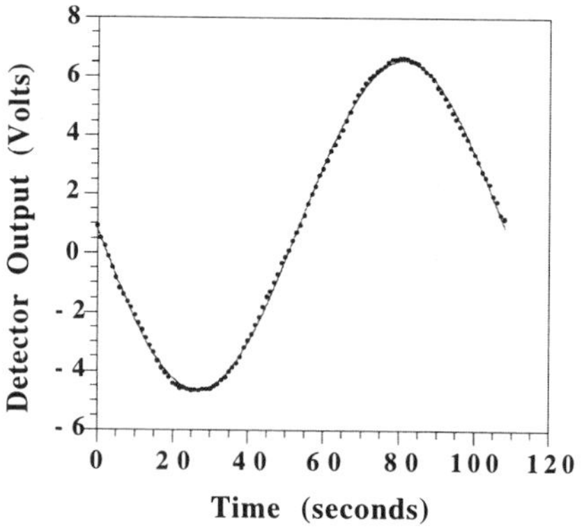

Figure 4. Fringe intensity vs. time while the loading fixtures move apart (no physical contact between the fixtures). The solid line through data points is a sinusoidal curve fit.

define uniquely since it depends on several issues. First, one could make the assumption that the scatter is independent of fringe intensity; that is, the noise is due to the electronics used to measure fringe intensity and not the fringe itself. Second, the sensitivity, given by the slope in Figure 4, implies that the least sensitivity occurs at the positions corresponding to the maximum and minimum fringe intensities. At these points, the noise contributes most significantly to the uncertainty in the displacement measurements. Finally, one must consider the magnitude of noise relative to the fringe peak-to-peak intensities. Since the noise is not a function of the fringe intensity, the greatest uncertainty will be experienced for fringes with small peak-to-peak intensities, while the brightest fringes are expected to provide a lower uncertainty.

2.4 *Cyclic Loading Fixture*

For the ceramic material tested, cyclic loading requires less than 1 μm displacement at the crack interface. Due to the mechanical hysteresis experienced when changing crosshead directions on the universal loading machine, a piezoelectric actuator design was selected to provide the small displacements under controlled-load conditions.

The PFT technique in conjunction with the closed-loop controller and interferometer provides the capability to directly characterize the fatigue behavior of the isolated wake ligaments. For these particular results, PFT specimens have been subjected to small numbers of load-unload cycles to obtain information on the role of subgrain structures in the global bridging mechanism at room temperature.

3 RESULTS AND DISCUSSION

Presently, we focus on characterizing the phenomenon of crack face separation. The role of the starting COD is not clear at this time, however, we anticipate a significant influence over the fatigue behavior, since local stresses and loads vary substantially with contact density. Therefore, our selection of the displacement range followed from consideration of two competing factors, namely, the loss of reliably stable PFT specimens at very small COD's, versus maximizing the test range in a constant contact density regime. The #2 PFT (Figure 1a) falls in the lower end of this range, thereby assuring the capture of the entire event. Future work will consider the effect of starting crack-face separation on the wake degradation by looking at the behavior of several different PFT positions behind the crack tip. However, here we will only communicate the phenomenon related to applied displacements which are small with respect to the grain size, and the associated microstructural response for this particular initial crack-face separation under tensile-tensile loading conditions.

For continuity of the phenomenon, the results of only a single DCB specimen are presented in this paper, however, similar behavior was observed in several specimens having roughly the same starting COD, confirming the validity of the present results. These supporting tests were from DCB samples of equivalent final crack lengths, and therefore, of similar COD profiles, so that the #2 PFT was of the same starting COD. This was confirmed by SEM for each sample.

A load-displacement curve for one load-unload cycle of PFT #2 is presented in Figure 5 where the applied load varied sinusoidally from a minimum of 0.30 N to a maximum of 2.5 N and back to 0.30 N over a period of approximately 400 seconds. The results describe the local processes experienced under tension-tension cyclic loading. Contributions from stored elastic strain energy around the crack tip in a real fracture sample will obviously contribute to compressive, restoring loads in the wake, however, they will depend on the specimen geometry, crack length and load ratio.

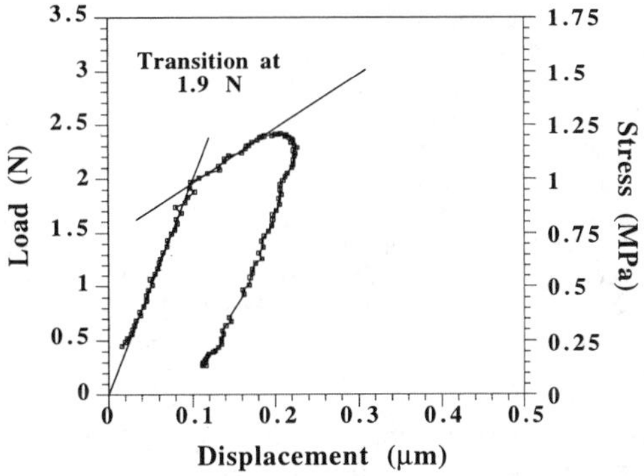

Figure 5. One load-unload cycle for PFT #2 evidencing a change in loading stiffness at 0.1 μm and a nonrecoverable displacement upon unloading.

Note from this figure that the load-displacement relationship follows an approximately linear trace to 1.9 N, or 0.1 μm. This portion of the trace is characterized by a stiffness of roughly 19 N/μm, but at 1.9 N the stiffness abruptly decreases to a stiffness which corresponds closely to the macroscopic stiffness observed in Figure 2. The maximum stresses were intentionally maintained in a range of 1.25 MPa to 1.50 MPa to characterize the transition point. Apparent from recent data, the mechanism responsible for the transition point is displacement controlled and not load controlled.

During unloading, the stiffness is similar to the initial loading stiffness, but there is no evidence of a lower transition point. If the mechanism could be described by classical friction models, we would expect the appearance of a lower transition point and displacements would return to zero (Guiu 1992),

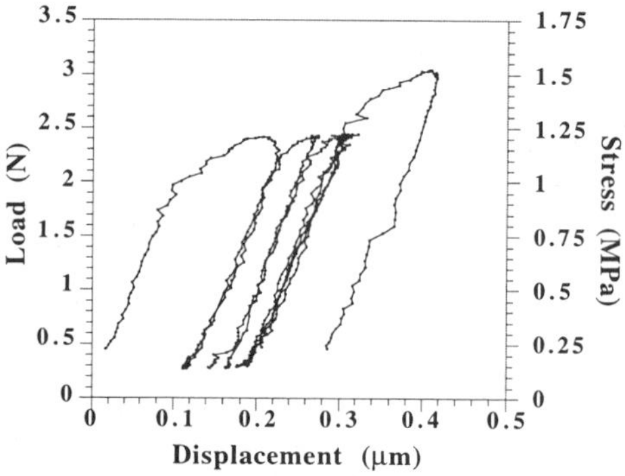

Figure 6. Five load-unload cycles including the data of Figure 5. By the fourth cycle the transition point has increased to 2.5 N. Loading to 3.1 N on the fifth cycle indicates a the transition point is at 2.7 N.

closing the hysteresis loop. Instead, Figure 5 reveals a permanent set at 0.30 N following the unloading portion of the cycle. For this particular load-unload cycle, the specimen experiences a nonrecoverable 0.11 μm residual displacement, where the next loading cycle begins.

Figure 6 depicts the cumulated behavior of the first cycle as well as 4 additional cycles. The load-displacement trace for the second load-unload cycle, which begins with the final point of Figure 5, clearly illustrates differences from the first cycle. One notes that the loading and unloading stiffnesses have not changed significantly from the first cycle, but the transition load has increased from 1.9 N to 2.2 N, providing evidence of a strengthening mechanism, possibly due to some form of a locking mechanism. On this second cycle the maximum load was again limited to 2.5 N, necessitating that the observed increase in the transition load must decrease the contribution to total residual displacement. The increment in residual displacement is approximately 0.04 μm.

Repeated loading and unloading through the third and fourth cycle continues this same trend. The initial and final compliances change very little from cycle to cycle, the transition load increases, and the contributions to the total residual displacement decrease. By the fourth cycle, the transition load has increased to approximately 2.5 N and the specimen exhibits a nearly linear-elastic behavior.

Loading the fifth cycle to a maximum of 3.1 N, caused a reappearance of the transition load, at roughly 2.7 N. The unloading portion of cycle again leads to nonrecoverable displacement, since the specimen displacements exceeded that of the transition point.

To verify the measured displacements from the interferometer, the PFT crack-face separation was observed directly in the SEM after 12 PFT load-unload cycles and compared with the initial COD. At the conclusion of the fracture test, the three-dimensional topography of the fracture surface prohibits the grains from returning to their original positions. Naturally, the residual displacement depends upon the position behind the crack tip, so positions further from the crack tip exhibit larger residual displacements. For this specimen taken from about 1.5 mm from the crack tip the initial crack-face separation was approximately 400 nm.

Figure 6 indicates that 5 load-unload cycles result in approximately 0.28 μm in residual displacement. After 12 cycles, the crack-face separation has increased by approximately 350 nm by monitoring the displacements with the LIDG.

The quasi-static and fatigue processes differ dramatically with respect to their governing mechanisms. Under monotonic loading conditions, the fracture process depends strongly on the combined influences of the COD and the grain size and grain-size distributions. In contrast, the features important to the fatigue mechanism are much smaller than the scale of the microstructure and are independent of the COD as evidenced by our observations that the critical displacement range of 0.1 μm for the present microstructure applies to specimens from different positions behind the crack tip. This is consistent with the literature since fatigue crack growth rates reported for most ceramics (Jacobs 1994, 95; Reece 1989) span many orders of magnitude over a small stress intensity factor range, indicating a characteristically high sensitivity to small changes in crack-opening displacements. These results suggest that the displacement must reach some critical change in displacement (roughly 0.1 μm) to induce damage.

Clearly, wake degradation under cyclic loading emphasizes microstructural features of a much smaller scale than those presently understood to be responsible for the wake behavior under quasistatic conditions. The bridging behavior under quasistatic conditions has been modeled by frictional grain pullout, where critical displacements are on the order of the grain size. When one focuses upon the critical displacements under fatigue conditions, however, the important features are much smaller than the grain size. The critical crack face separation of 0.1 μm for fatigue-induced damage, with support from microstructural evidence described below, implicates secondary mechanisms. Using Equation (6) one discovers that the critical displacements are about 0.5% of the grain size. Although still speculative, we envision a ratcheting mechanism, of sorts, which controls the nonrecoverable displacements. We speculate that grains move irreversibly past features of the 0.1 μm scale, and the unloading process restricts recovery of their initial configuration, and, therefore, the initial crack-face separation.

Figure 7 is offered as fractographic evidence that features of this size exist. The fracture surface generated under monotonic loading (no cyclic history) exhibits several rough patches. The higher magnification of one such area (Figure 7) indicates that the rough patches are crystallographic in nature and exhibit a characteristic size of roughly 0.1 μm.

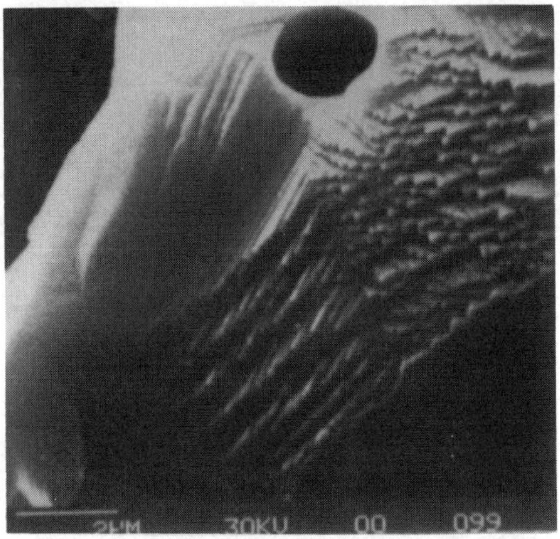

Figure 7. SEM fractograph produced under monotonic loading conditions, illustrating the subgrain-size features which may explain a ratcheting behavior.

Our intention in providing these photographs is not to make the definitive claim that these crystallographic ledges are responsible for the observed phenomenon, but just to suggest that the presence of subgrain-size features may explain the observed behavior.

4 CONCLUSIONS

The post-fracture tensile (PFT) technique was used as a tool for a study of low cycle fatigue by the isolation of the processes and careful consideration of the microstructural role. This study unveiled the existence of a subgrain size mechanism which dominantly controls fatigue in the small displacement regime. The load-unload cycle begins with a high stiffness which decreases substantially when displacements exceed the critical displacement of 0.1 µm. Beyond 0.1 µm, a marked hysteresis loop appears where upon unloading the crack faces do not return to the original crack face separation. During the unloading portion of the cycle, the stiffness differs little from the initial loading stiffness. Repeated cycling results in the transition point moving to a slightly higher load and displacement, possibly implicating a locking mechanism due to loose debris.

SEM images of the crack face separation before and after cycling confirm the LIDG measurements and illustrate the incremental damage which results from small repeated applied displacements. Additionally, fractographic evidence was provided to suggest the existence of the subgrain-size features which may explain the accumulation of damage. In summary, this work contributes significantly to the understanding of why ceramics exhibit a large sensitivity to ΔK in fatigue tests. The features which dominate the R-curve behavior are not the same features which control the fatigue behavior. While the R-curve depends strongly upon grain size features, the degradation of the R-curve through fatigue depends upon the features 0.1 µm in size.

5 REFERENCES

Ewart, L., Suresh, S., "Crack Propagation in Ceramics Under Cyclic Loads," *J. Mat. Sci.*, **22** 1173-1192 (1987).

Guiu, F., Li, M., Reece, M.J., "Role of Crack-Bridging Ligaments in the Cyclic Fatigue Behavior of Alumina," *J. Am. Ceram. Soc.*, **75** [11] 2976-2984 (1992).

Hay, J.C., White, K.W., "Crack Face Bridging Mechanisms in Monolithic $MgAl_2O_4$ Spinel Microstructures," *Acta metall. mater.* [11] 3017-3025 (1991).

Hay, J.C., White, K.W., "Grain Boundary Phases and Wake Zone Characterization in Monolithic Alumina," *J. Am. Ceram. Soc.*, **78** [4] 1025-1032 (1995).

Hay, J.C., White, K.W., "Cyclic Loading and Fracture Process Zone Degradation in Monolithic Ceramics," presented at the 98th annual meeting of the American Ceramic Society, Indianapolis, IN, April 14-17, 1996 (Paper No. SVIII-19-96).

Hay, J.C., White, K.W.," Grain-Bridging Mechanisms in Monolithic Alumina and Spinel," *J. Am. Ceram. Soc.*, **76** [7] 1849-1854 (1993).

Hu, Z-H, Mai, Y-W, "Crack-Bridging Analysis for Alumina Ceramics Under Monotonic and Cyclic Loading," *J. Am. Ceram. Soc.*, **75** [4] 848-853 (1992).

Jacobs, D.S., Chen, I-W, "Cyclic Fatigue in Ceramics: A Balance between Crack Shielding Accumulation and Degradation," *J. Am. Ceram. Soc.*, **78** [3] 513-520 (1995).

Jacobs, D.S., Chen, I-W, "Mechanical and Environmental Factors in the Cyclic and Static Fatigue of Silicon Nitride," *J. Am. Ceram. Soc.*, **77** [5] 1153-1161 (1994).

Knehans, R., Steinbrech, R., "Memory Effect of Crack Resistance During Slow Crack Growth in Notched Al_2O_3 Bend Specimens," *J.Mat.Sci.Let*, **1** [8], 327-29 (1982).

Krohn, D.A., Hasselman, D.P.H., "Static and Cyclic Fatigue Behavior of a Polycrystalline Alumina," *J. Am. Ceram. Soc.*, **55** [4] 208-211 (1972).

Lathabai, S., Rödel, J., Lawn, B.R., "Cyclic Fatigue from Frictional Degradation at Bridging Grains in Alumina," *J. Am. Ceram. Soc.*, **74** [6] 1340-1348 (1991).

Reece, M.J., Guiu, F., Sammur, M., "Cyclic Fatigue Crack Propagation in Alumina Under Direct Tension-Compression Loading," *J. Am. Ceram. Soc.*, **72** [2] 348-352 (1989).

Ritchie, R.O., "Mechanisms of Fatigue Crack Propagation in Metals, Ceramics and Composites:

Role of Crack Tip Shielding," *Mat. Sci. Eng.*, **A103** 15-28 (1988).

Rödel, J., Kelly, J.G., Lawn, B.R., "In Situ Measurement of Bridged Crack Interfaces in the Scanning Electron Microscope," *J. Am. Ceram. Soc.*, 73 [11] 3313-18 (1990).

Swanson, P., Fairbanks, C.J., Lawn, B.R., Mai, Y.W., Hockey, B.J., "Crack-Interface Grain Bridging as a Fracture Resistance Mechanism in Ceramics: I, Experimental Study on Alumina," *J. Am. Ceram. Soc.*, **70** [4] 279-289 (1987).

Vekinis, B., Ashby, M.G., Beaumont, P.W.R., "R-curve Behavior of Al_2O_3 Ceramics," *Acta metall. mater.*, **38** [6] 1151-1162 (1990).

Wecharatana, M., Shah, S.P., "A Model for Predicting Fracture Resistance of Fiber Reinforced Concrete," *Cement and Concrete Research*, **13** 819-829 (1983).

White, K.W., Hay, J.C., "The Effect of Thermoelastic Anisotropy on the R-curve Behavior of Monolithic Alumina," *J. Am. Ceram. Soc.*, **77** [9] 2283-2288 (1994).

Damage and Failure of Interfaces, Rossmanith (ed.) © *1997 Balkema, Rotterdam, ISBN 90 5410 899 1*

Reduction of stiffness characteristics of balanced laminates with intralaminar cracks

T. Lewiński
Faculty of Civil Engineering, Warsaw University of Technology, Poland

J.J.Telega
Institute of Fundamental Technological Research, Polish Academy of Sciences, Poland

ABSTRACT: A new variationally-asymptotic method of modelling effective properties and stress distribution in cracked laminates is applied to assess the loss of the Young, Kirchhoff and Poisson moduli of the $[0^o_m/90^o_n]_s$ laminates with transverse aligned cracks in the 90° layer and to find the stress distribution between interacting cracks and in the vicinity of isolated cracks. The decaying curves describing the loss of the effective Young's modulus in the tensile direction lies slightly over the curves found by Hashin (1985) with the help of similar stress assumptions. The reduction of Kirchhoff's modulus is predicted in the same manner as in Hashin (1985) and Tsai and Daniel (1992). By applying a local stress-type criterion of the onset of nucleation of cracks an interrelation between the crack spacing and the magnitude of applied stress is found and reasonably good fit with experimental data of Highsmith and Reifsnider (1982) is observed.

1. INTRODUCTION

Transverse matrix cracks in composite laminates occur as the first damage mode preceding delamination and fibre breakage. Due to aligned distribution of fibres the intralaminar cracking patterns are proved to be regular as evidenced by Garrett and Bailey (1977) and Highsmith and Reifsnider (1982), although the cracks themselves can be curved, see Groves et al. (1987). The uniform transverse intralaminar cracking results in a decay of the effective stiffnesses of the laminate. Their minimal values, related to the state of saturation of the cracks, can be determined by the method called ply discount by assuming that the cracked laminae do not bear the stresses. A theoretical prediction of a whole curve describing decay of a stiffness up to its minimal value requires an analysis of interaction of the non-cracked and cracked layers.

Under the assumption of a regular and given cracking layout the problem of evaluating the effective moduli of the laminate can be considered within the framework of the theory of homogenization of the stiffnesses of plates, see the three-dimensional approach of Chacha and Sanchez (1992) concerning the unilateral cracks obeying the Signorini's conditions. This method requires solving a nonlinear local problem posed on a three-dimensional periodicity cell. Although formally correct, such analysis can be viewed as unsatisfactory, since it provides no explicit formulae that could be helpful in figuring out the behaviour of the plate with cracks. Therefore, the recent efforts are made towards: i) approximate analysis of the relevant local problem, ii) preceding the homogenization process by forming a two-dimensional (laminate) model. The paper of Gudmundson and Zang (1993) is an example of the former approach.

The methods of type (ii) are based on certain assumptions on the transverse distribution of stresses and/or displacements. Such approximations suppress stress singularities around the crack tips. This is a price to pay for the reduction of the dimension of the problem by one and for having possibility of finding explicit analytical solutions to the problems of tension and shearing of the cross-ply laminates cracked in one direction, cf. Highsmith and Reifsnider's (1982) shear-lag model, Hashin (1985), Nairn (1989), Han and Hahn (1989), Smith and Wood (1990), Tsai and Daniel (1992), McCartney (1992). The averaging methods used in these papers were intuitive, hence there is no guarantee that the effective characteristics introduced there satisfy the known criteria of correctness of the averaging procedure. A correct averaging procedure is provided by the asymptotic homogenization method, see Sanchez-Palencia (1980). In the three-dimensional problem this is the uniquely determined correct method of averaging. In the problems of non-homogeneous plates the asymptotic homogenization can be generated in different ways: i) by the in-plane scaling, ii) by the refined or spatial scaling.

In papers of Lewiński and Telega (1996a-d) new effective models are proposed of three-layer laminates with transverse cracks distributed periodically in the internal layer. The two-dimensional description is obtained by imposing the stress assumptions of Hashin (1985) on the trial stresses in the saddle-point functional of Reissner. The description thus obtained does not coincide with that of Hashin (1985) even in the one-dimensional problem. This model, although two-dimensional, is capable of taking into account the transverse unilateral cracks in the internal layer. According to the method of scaling one finds the effective models:
(h_0, l_0) – for the in-plane scaling, or – (h_0, l) – for the refined scaling. The results of the (h_0, l_0) model produce asymptotes for the results of the model (h_0, l) if the density of cracks $c_d \to \infty$.

The effective model (h_0, l) assumes the form of a thin hyperelastic plate model, its hyperelastic potential being convex and hence the equilibrium problem being well – posed. Application of this modelling for the analysis of $[0^o_m/90^o_n]_s$ composites with aligned cracks in the 90° layer is the subject of papers of Lewiński & Telega (1996c,d). Accuracy of the model has been assessed by comparison with experimental results of Highsmith & Reifsnider (1982), Groves (1986) and Ogin et al. (1985), concerning the decay of the Young modulus. It has been proved that the formula representing the loss of the effective Kirchhoff modulus coincides with that previously found by Hashin (1985) and Tsai & Daniel (1992).

The aim of the present paper is to complete the accuracy analysis of the (h_0, l) model by presenting new fits for the effective Young modulus of the $[0^o_m/90^o_n]_s$ graphite/epoxy laminates and for the effective Poisson ratio of a glass/epoxy laminate with experimental results of Groves (1986) and Smith & Wood (1990). Moreover, the model (h_0, l) makes it possible to formulate a stress criterion for the onset of matrix cracks. A single assumption, that the maximal tensile stress in the 90° layer cannot exceed a certain critical value, has made it possible to find a relation between the crack spacing and the tensile stress. The relation shows good agreement with the experimental results of Highsmith and Reifsnider (1982).

2. EFFECTIVE MODEL OF A CRACKED THREE-LAYER LAMINATE

Consider a three-layer laminate transversely cracked in the central layer of thickness $2c$; the thickness of external layers equals d. Let $x = (x_\alpha)$ parametrize the mid-plane Ω and z is perpendicular to it. The central layer $|z| < c$ is bounded by the faces: $c < z < h = c + d$ and $-h < z < -c$. A part Γ_σ of the boundary of the laminate is subjected to the in-plane loads of density $\overline{N}^\alpha$, the remaining part of the boundary Γ_w being clamped. Distribution of the transverse cracks in the central layer is assumed as Y-periodic, $Y = (0, l_1) \times (0, l_2)$ represents the rectangle of periodicity. The projection of cracks on the mid-plane forms the arcs F lying within the cells Y. The modelling, called further the (h_0, l) approach, is based on

i) relaxing Hashin's (1985) stress assumptions to the form:

$$\sigma^{\alpha\beta}(x,z) = \begin{cases} L^{\alpha\beta}(x)\dfrac{1}{2c}, & |z| < c \\ \left[N^{\alpha\beta}(x) - L^{\alpha\beta}(x)\right]\dfrac{1}{2d}, & \text{otherwise} \end{cases}$$

$$\sigma^{\alpha 3}(x,z) = \begin{cases} -zQ^{\alpha}(x)\dfrac{1}{2c}, & |z| < c \\ \left[z - \text{sign}(z)h\right] Q^{\alpha}(x)\dfrac{1}{2d}, & \text{otherwise} \end{cases} \tag{1}$$

$$\sigma^{33}(x,z) = \begin{cases} (-z^2 + ch)R(x)\dfrac{1}{4c}, & |z| < c \\ (z - \text{sign}(z)h)^2 R(x)\dfrac{1}{4d}, & \text{otherwise} \end{cases}$$

here $\alpha, \beta = 1, 2$.

ii) Displacement assumptions

$$w_\alpha(x,z) = \begin{cases} v_\alpha(x) + \dfrac{3}{2c^2}(c^2 - z^2)u_\alpha(x), & \text{if } |z| < c, \\ v_\alpha(x), & \text{otherwise} \end{cases} \tag{2}$$

$$w_3(x,z) = \begin{cases} \dfrac{z}{cb}w(x), & |z| < c \\ \dfrac{\text{sign}(z)}{b}w(x), & \text{otherwise} \end{cases}$$

with $b = \frac{d}{2} + \frac{c}{3}$; cf. Fig.1.

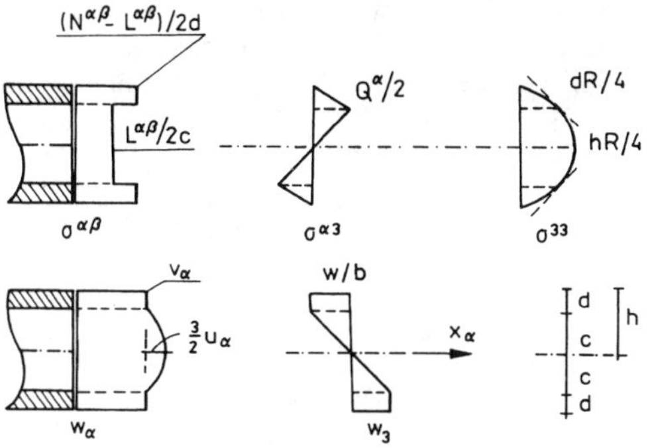

Figure 1. Stress and displacement assumptions

iii) Imposing the assumptions (i)-(ii) onto the trial fields involved in the stress-displacement saddle-point variational principle of Reissner.

iv) Performing the process of smearing-out the cracks in the internal layer by the method of homogenization with the scaling:

$$l_\alpha \to \varepsilon l_\alpha\,, \quad c \to \varepsilon c\,, \quad d \to \varepsilon d\,; \tag{3}$$

$\varepsilon > 0$ is a small parameter.

Within the first order approximation the equlibrium of the laminate with smeared-out cracks is governed by the variational equation

$$\int_\Omega \left(N_h^{11}\tilde{\varepsilon}_{11} + 2N_h^{12}\tilde{\varepsilon}_{12} + N_h^{22}\tilde{\varepsilon}_{22}\right) d\Omega = \\ = \int_{\Gamma_\alpha} \left(\overline{N}^1\tilde{v}_1 + \overline{N}^2\tilde{v}_2\right) ds \tag{4}$$

where $\tilde{\varepsilon}_{\alpha\beta} = \frac{1}{2}\left(\tilde{v}_{\alpha,\beta} + \tilde{v}_{\beta,\alpha}\right)$; $(\cdot)_{,\alpha} = \frac{\partial}{\partial x_\alpha}$ and $\tilde{\boldsymbol{v}} = (\tilde{v}_\alpha)$ represents the kinematically admissible displacements, vanishing on Γ_w. Integration in (4) concerns the uncracked domain.

The overall or effective in-plane forces $N_h^{\alpha\beta}$ depend non-linearly on the macrodeformations $\varepsilon^h_{\alpha\beta} = \frac{1}{2}\left(v^0_{\alpha,\beta} + v^0_{\beta,\alpha}\right)$. There exists a hyperelastic potential W_h of class C^1 and strictly convex such that

$$N_h^{\alpha\beta} = \frac{\partial W_h(\varepsilon^h_{11},\, \varepsilon^h_{12},\, \varepsilon^h_{22})}{\partial \varepsilon^h_{\alpha\beta}}\,. \tag{5}$$

The potential W_h is determined by the properties of the periodicity cell $YF = Y\backslash F$ with the crack F, as follows

$$W_h(x, \boldsymbol{\varepsilon}^h) = \inf\left\{\frac{1}{l_1 l_2}\int_{YF} j(x, \boldsymbol{\varepsilon}^y(\boldsymbol{v}) + \boldsymbol{\varepsilon}^h\,, \\ \boldsymbol{\varepsilon}^y(\boldsymbol{u})\,,\ \boldsymbol{\kappa}^y(\boldsymbol{u}, w)\,,\ w)\, dy_1\, dy_2 | \\ (\boldsymbol{v}, \boldsymbol{u}, w) \in \mathbb{K}_{YF}\right\} \tag{6}$$

with

$$j(x, \boldsymbol{\varepsilon}, \boldsymbol{\gamma}, \boldsymbol{\kappa}, w) = \frac{1}{2}(\boldsymbol{\varepsilon}, \boldsymbol{\gamma}, w)^T \boldsymbol{C}(\boldsymbol{\varepsilon}, \boldsymbol{\gamma}, w) \\ + \frac{1}{2}\boldsymbol{\kappa}^T \boldsymbol{H}\boldsymbol{\kappa}\,, \tag{7}$$

where $\boldsymbol{\varepsilon} = (\varepsilon_{\alpha\beta})$, $\boldsymbol{\gamma} = (\gamma_{\alpha\beta})$, $\boldsymbol{\kappa} = (\kappa_\alpha)$; $\boldsymbol{C}$ and $\boldsymbol{H}$ are tensors of the elastic moduli, their components being fully determined by the elastic properties of the uncracked laminate, see Lewiński & Telega (1996 a,b). Moreover,

$$\varepsilon^y_{\alpha\beta}(\boldsymbol{v}) = \frac{1}{2}\left(\frac{\partial v_\alpha}{\partial y_\beta} + \frac{\partial v_\beta}{\partial y_\alpha}\right)\,. \\ \kappa^y_\alpha(\boldsymbol{v}, w) = v_\alpha - \frac{\partial w}{\partial y_\alpha} \tag{8}$$

for the fields $\boldsymbol{v}$ and w defined on Y. The domain Y is parametrized with Cartesian coordinates y_α. The convex set $\mathbb{K}_{YF}$ reads

$$\mathbb{K}_{YF} = H^1_{per}(Y)^2 \times K_{YF} \times H^1_{per}(Y)\,,$$

$$K_{YF} = \left\{\boldsymbol{v} \in H^1_{per}(YF)^2 | [\![\boldsymbol{v}\cdot\boldsymbol{\nu}]\!] \geq 0 \text{ on } F\right\}$$

where $\boldsymbol{\nu}$ represents a unit vector normal to F.

The condition $\boldsymbol{u} \in K_{YF}$ means that the normal component $u_\nu = \boldsymbol{u}\cdot\boldsymbol{\nu}$ of the displacement field $\boldsymbol{u}$ of the internal layer can suffer a jump on F. If this jump is positive, the crack opens. The jump is zero if the crack is closed. Interestingly enough the analytical formula for the effective potential W_h can be explicitly derived in the important case of aligned cracks.

Apart from the refined scaling (3) one can build the effective modelling on the in-plane scaling: $l_\alpha \to \varepsilon l_\alpha$, $c \to c$, $d \to d$.
The approach based on the scaling (3) is called (h_0, l) and the in-plane scaling leads to the model called (h_0, l_0).

3. CASE OF ALIGNED CRACKS

Consider the laminate with aligned cracks lying parallel to the x_2 axis. The cell Y degenerates to the interval $[0, l]$ and $YF = [0, l/2) \cup (l/2, l]$, since the crack F can be located in the middle of Y. The crack spacing l determines the crack density c_d defined by 1 mm/l or by $2c/l$, see Fig. 2.

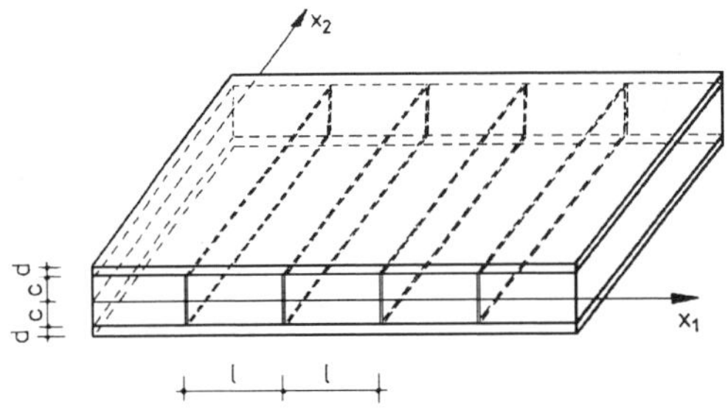

Figure 2. Laminate with aligned cracks

The minimization problem (6) splits into the tension and shearing local problems. The tension problem involves the unknowns: v_1, u_1, w. The shearing problem involves the unknowns: v_2 and u_2.

It turns out that the effective potential W_h depends on $\boldsymbol{\varepsilon}^h$ and on the crack deformation measures

$$\varepsilon_{11}^F = [\![u_1]\!]\frac{1}{l}\,, \qquad \varepsilon_{12}^F = [\![u_2]\!]\frac{1}{2l}\,. \tag{9}$$

The crack is open if

$$E_h = \beta_{11}\varepsilon_{11}^h + \beta_{21}\varepsilon_{22}^h > 0\,;$$

otherwise is closed; $\beta_{\alpha 1}$ are coefficients determined by the characteristics of the uncracked laminate. One can prove that

$$\varepsilon_{11}^F = \begin{cases} 0\,, & \text{for } E_h \leq 0\,, \\ F_{11}(\rho)E_h & \text{for } E_h > 0\,, \end{cases} \tag{10}$$

$$\varepsilon_{12}^F = F_{12}(\rho)\varepsilon_{12}^h\,, \tag{11}$$

where $\rho = \dfrac{l}{2h}$ and $F_{1\alpha}$ are known elementary functions, see Lewiński & Telega (1996c). The effective constitutive relations read

$$\begin{aligned} N_h^{11} &= A\left[\alpha_{11}\varepsilon_{11}^h + \alpha_{12}\varepsilon_{22}^h - \beta_{11}\varepsilon_{11}^F(\varepsilon_{11}^h, \varepsilon_{22}^h)\right]\,, \\ N_h^{22} &= A\left[\alpha_{12}\varepsilon_{11}^h + \alpha_{22}\varepsilon_{22}^h - \beta_{21}\varepsilon_{11}^F(\varepsilon_{11}^h, \varepsilon_{22}^h)\right]\,, \\ N_h^{12} &= B\left[1 - \alpha F_{12}(\rho)\right]\varepsilon_{12}^h\,, \end{aligned} \tag{12}$$

where A, B, α, $\alpha_{\lambda\mu}$ are given moduli.

It turns out that the formula for N_h^{12} coincides with that found by Hashin (1985, Eq. 3.22), Tan & Nuismer (1989) and Tsai & Daniel (1992). For a broad class of glass/epoxy laminates the formulae for the effective stiffnesses determined by (12) yield results very similar to those found by McCartney (1992) and are upper bounds for the results of Hashin (1985), the differences between both models being small, see Lewiński & Telega (1996d).

Let us consider the problem of reduction of the effective Young's modulus E_1^c of the $[0_m^\circ/90_n^\circ]_s$ graphite/epoxy laminate tested in Groves (1986). The experimental results lie between the (h_0, l) predictions and results of Lee et al. (1989). The predictions of Hashin (1985) and of model (h_0, l) practically coincide, cf. Figs. 3-5.

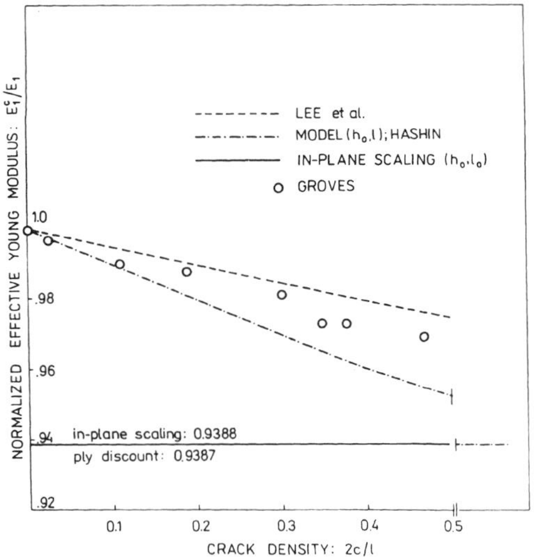

Figure 3. Reduction of the normalized effective Young's modulus (E_1^c/E_1) for the $[0^\circ/90^\circ]_s$ graphite/epoxy laminate. Experimental results of Groves versus theoretical predictions

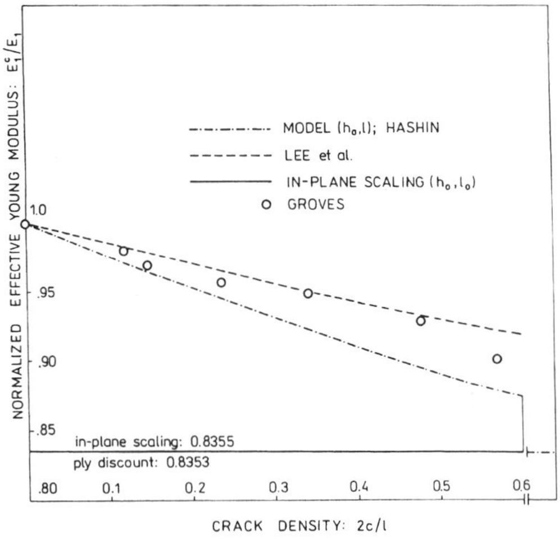

Figure 4. Reduction of the normalized effective Young's modulus (E_1^c/E_1) for the $[0^\circ/90_3^\circ]_s$ graphite/epoxy laminate. Experimental results of Groves versus theoretical predictions

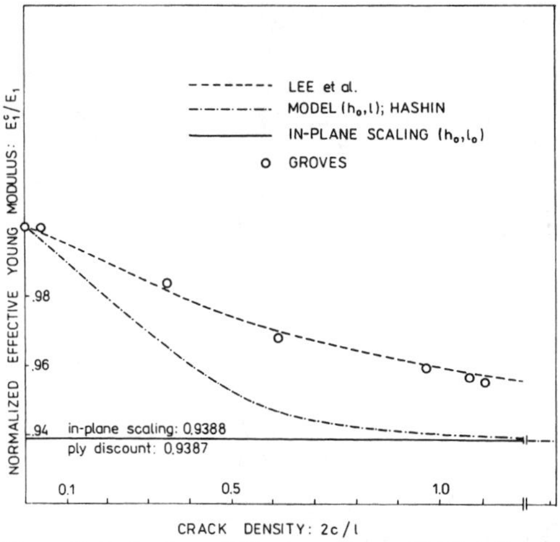

Figure 5. Reduction of the normalized effective Young's modulus (E_1^c/E_1) for the $[0_2^\circ/90_2^\circ]_s$ graphite/epoxy laminate. Experimental results of Groves versus theoretical predictions

Let us refer now to the experimental results of Smith and Wood (1990). The model (h_0, l) provides the lower estimates for decay of Young's modulus E_1^c for the $[0^\circ/90^\circ]_s$ glass/epoxy laminate. In this case the shear-lag predictions lie closer to the experimental data for different samples, cf. signs $(*, \text{o}, \triangle, \times, +)$ in Fig. 6.

The equations (12) determine the formula for decay of the effective Poisson ratio ν_{12}^c, see Figs. 7-8, concerning the $[0_m^\circ/90_n^\circ]_s$ glass/epoxy and carbon/epoxy laminates tested by Smith & Wood (1990).

For the glass/epoxy composites the method (h_0, l) yields acceptable predictions up to the crack density equal ca. 1.7/1 mm. The shear-lag analysis of Smith & Wood (1990) overestimates the experimental results, cf. Fig. 7.

On the other hand the shear-lag model overestimates the experimental data characterizing the $[0^\circ/90^\circ]_s$ carbon/epoxy laminate, see Fig. 8.

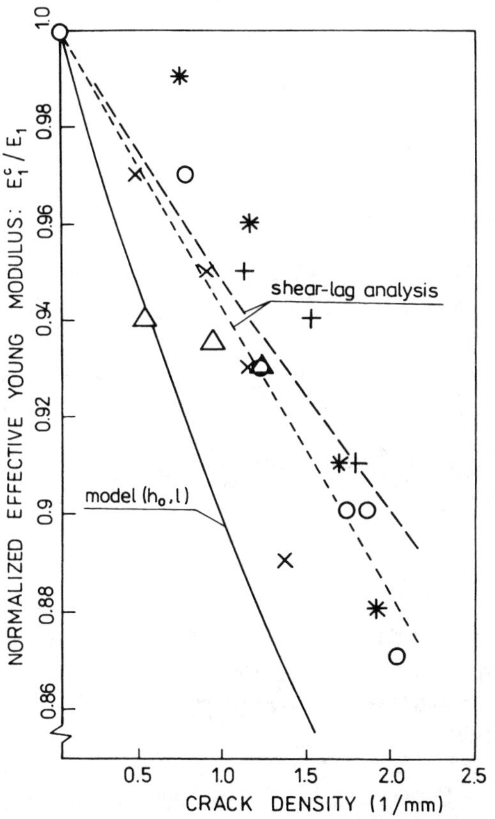

Figure 6. Reduction of the normalized effective Young's modulus (E_1^c/E_1) for the $[0^\circ/90^\circ]_s$ glass/epoxy laminate. Experimental results of Smith & Wood (1990) versus theoretical predictions

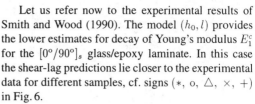

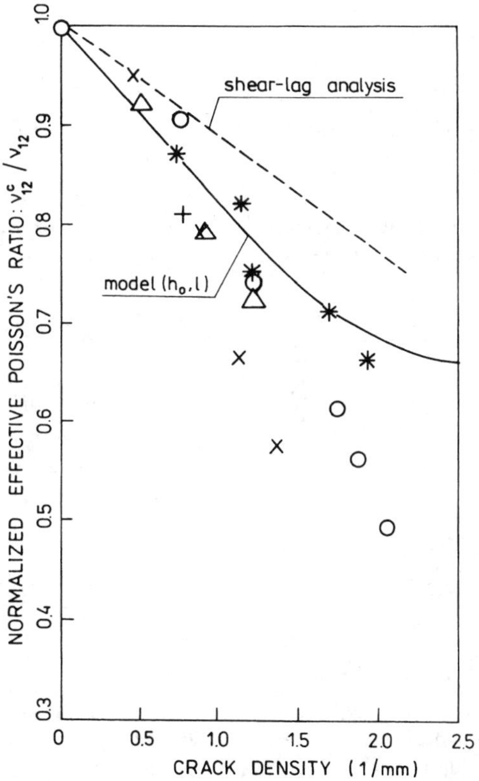

Figure 7. Reduction of the normalized effective Poisson's ratio (ν_{12}^c/ν_{12}) for the glass/epoxy laminate tested in Smith & Wood (1990). Experimental results versus theoretical predictions by the shear-lag model and the (h_0, l) model

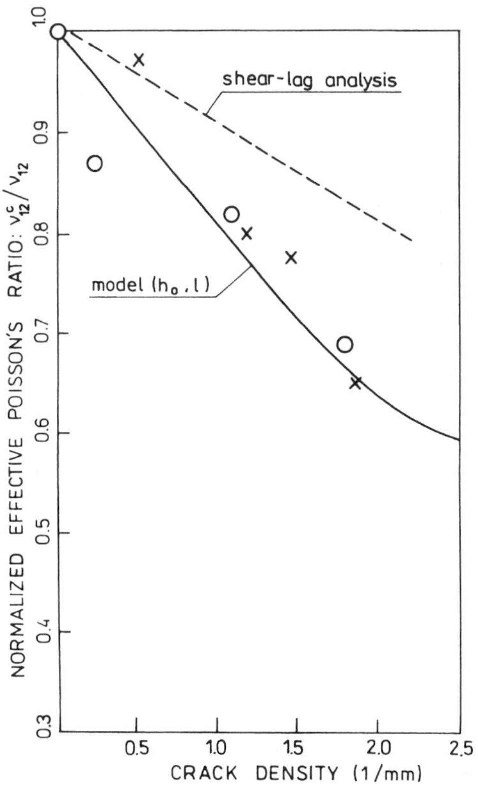

Figure 8. Reduction of the normalized effective Poisson's ratio (ν_{12}^0/ν_{12}) for the carbon/epoxy laminate tested in Smith & Wood (1990). Experimental results versus theoretical predictions by the shear-lag model and the (h_0, l) model

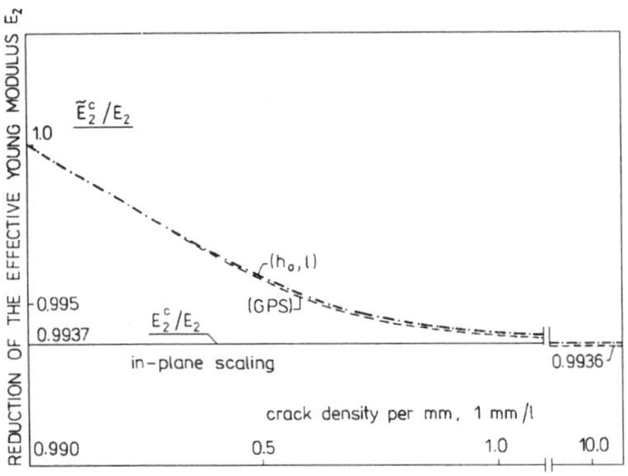

Figure 9. The $[0°/90_3°]_s$ glass/epoxy laminate tested in Highsmith and Reifsnider (1982). Decay of the effective Young modulus E_2^c

The (h_0, l) model is also capable of predicting decay of the Young modulus in the transverse direction. Such decay is usually negligible, cf. Fig. 9 concerning the $[0°/90_s°]_s$ glass/epoxy laminate tested in Highsmith and Reifsnider (1982). The GPS model of McCartney (1992) yields results slightly smaller than those following from the (h_0, l) approach.

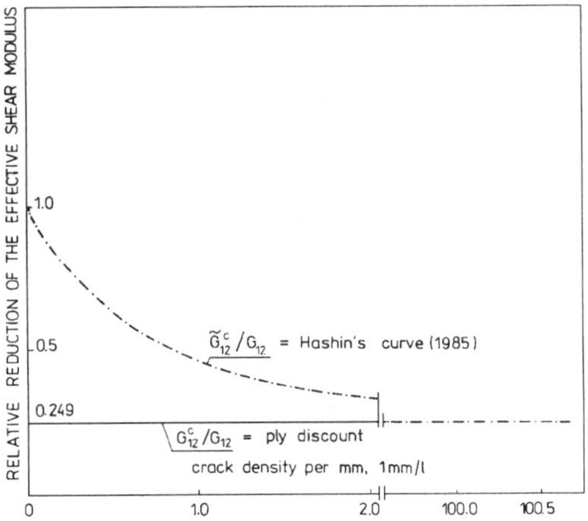

Figure 10. The $[0°/90_3°]_s$ glass/epoxy laminate tested in Highsmith and Reifsnider (1982). Decay of the effective Kirchhoff modulus G_{12}^c

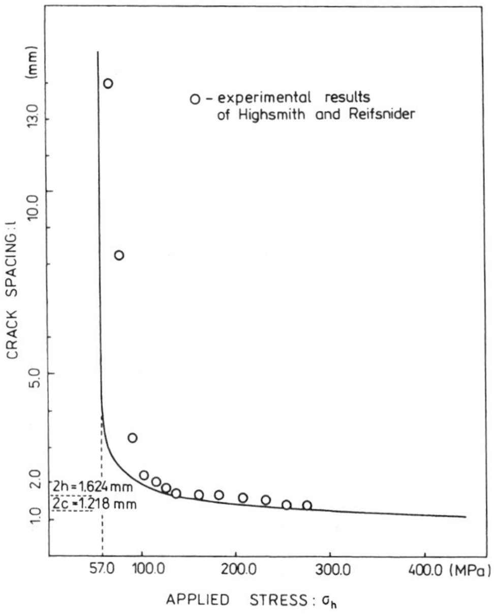

Figure 11. The $[0°/90_3°]_s$ glass/epoxy laminate tested in Highsmith & Reifsnider (1982). Crack spacing l versus the applied stress $\sigma_h = \dfrac{1}{2h} N_h^{11}$ in the case of $N_h^{22} = 0$. Experimental results versus theoretical predictions of the (h_0, l) model

As already mentioned above decay of the Kirchhoff modulus G^c_{12} is evaluated exactly in the same way as derived by Hashin (1985), Tan & Nuismer (1989) and Tsai & Daniel (1992), see Fig. 10 concerning the laminate of Fig. 9.

Let us emphasize that in all cases the results of the conventional (h_0, l_0) model provide the lower asymptotes of the (h_0, l) predictions. This means that the in-plane scaling determines the effective models referring to the case of extremely dense distribution of cracks.

The model (h_0, l) makes it possible to predict distribution of stresses between the interacting cracks and in the vicinity of an isolated crack. These formulae enables one to express a simple fracture criterion: $\max|\sigma^{11}$ (in the central layer)$|$ = critical value.
In this manner one can evaluate the crack spacing l as a function of the applied stress σ_h. Such a relation for the $[0^\circ/90^\circ_3]_s$ glass/epoxy laminate tested in Highsmith and Reifsnider (1982) is shown in Fig. 11.

4. CONCLUSIONS

The model (h_0, l) is capable of predicting the loss of all effective characteristics of the cracked $[0^\circ_m/90^\circ_n]_s$ laminates and of the state of stress between the interacting transverse cracks in the internal layer. The case of cracks in the outer layer has not been solved by this method up till now. The (h_0, l) approach is probably the first model in the literature that joins two features: accuracy of Hashin's stress-based modelling with the rigour of smearing-out the cracks by using the homogenization method.

ACKNOWLEDGEMENT

The work was supported by the Polish State Committee for Scientific Research through the grant No 7 T07 A 01612.

REFERENCES

Chacha, D. & E.Sanchez-Palencia 1992. Overall behaviour of elastic plates with periodically distributed fissures. *Asympt. Analysis* 5: 381-396.

Garrett, K.W. & J.E.Bailey 1977. Multiple transverse fracture in 90° cross-ply laminates of a glass fibre-reinforced polyester. *J.Mater. Sci.* 12: 157-168.

Groves, S.E. 1986. A study of damage mechanics in continuous fibre composite laminates with matrix cracking and interply delaminations. *Dissertation*. Texas A&M University.

Groves, S.E., C.E.Harris, A.L.Highsmith, D.H.Allen, R.G.Norvell 1987. An experimental and analytical treatment of matrix cracking in cross-ply laminates. *Exper. Mech.* 27: 73-79.

Gudmundson, P. & W. Zang 1993. An anlytical model for thermoelastic properties of composite laminates containing transverse matrix cracks. *Int. J. Solids Structures* 30: 3211-3231.

Han, Y.M. & H.T.Hahn 1989. Ply cracking and property degradations of symmetric balanced laminates under general in-plane loading. *Composite Sci. Technol.* 35: 377-397.

Hashin, Z. 1985. Analysis of cracked laminates: a variational approach. *Mech. Mater.* 4: 121-136.

Highsmith, A.L. & K.L.Reifsnider 1982. Stiffness reduction mechanisms in composite materials. In K.L.Reifsnider (ed.), *Damage in Composite Mechanics*: 103-117. ASTM STP 775.

Lee, J.-W., D.H.Allen, C.E.Harris 1989. Internal state variable approach for predicting stiffness reductions in fibrous laminated composites with matrix cracks. *J. Comp. Mater.* 23: 1273-1291.

Lewiński, T. & J.J.Telega 1996a. Stiffness loss in laminates with intralaminar cracks. Part I Two-dimensional modelling. *Arch. Mech.* 48: 143-161.

Lewiński, T. & J.J.Telega 1996b. Stiffness loss in laminates with intralaminar cracks. Part II Periodic distribution of cracks and homogenization. *Arch. Mech.* 48: 163-190.

Lewiński, T. & J.J.Telega 1996c. Stiffness loss of laminates with aligned intralaminar cracks. Part I Macroscopic constitutive relations. *Arch. Mech.* 48: 245-264.

Lewiński, T. & J.J.Telega 1996d. Stiffness loss of laminates with aligned intralaminar cracks. Part II Comparisons. *Arch. Mech.* 48: 265-280.

McCartney, L.N. 1992. Theory of stress transfer in a $0^\circ - 90^\circ - 0^\circ$ cross-ply laminate containing a parallel array of transverse cracks. *J. Mech. Phys. Solids* 40: 27-68.

Nairn, J.A. 1989. The strain energy release rate of composite microcracking: a variational approach. *J. Compos. Mater.* 23: 1106-1129.

Ogin, S.L., P.A.Smith, P.W.R.Beaumont 1985. Matrix cracking and stiffness reduction during the fatigue of a (0°/90°) GFRP laminate. *Compos. Sci. Technol.* 22: 23-31.

Sanchez-Palencia, E. 1980. Non-homogeneous media and vibration theory. Berlin: *Springer.*

Smith, P.A. & J.R.Wood 1990. Poisson's ratio as a damage parameter in the static tensile loading of simple crossply laminates. *Compos. Sci. Technol.* 38: 85-93.

Tan, S.C. & R.J. Nuismer 1989. A theory for progressive matrix cracking in composite laminates. *J. Compos. Mater.* 23: 1029-1047.

Tsai, C.-L. & I.M.Daniel 1992. The behaviour of cracked cross-ply composite laminates under shear loading. *Int. J. Solids Structures.* 29: 3251-3267.

Damage and Failure of Interfaces, Rossmanith (ed.) © 1997 Balkema, Rotterdam, ISBN 90 5410 899 1

Deformation and failure of heat-sealed area in laminated plastic film used for liquid package bags

E. Umezaki & H. Watanabe
Department of Mechanical Engineering, Nippon Institute of Technology, Japan

A. Shimamoto
Department of Mechanical Engineering, Saitama Institute of Technology, Japan

ABSTRACT: This study investigates displacements and failure on a cross section of a heat-sealed area in laminated plastic film used for liquid package bags by a digital image correlation. T-shape specimens of 10mm width which are cut from the bags. The specimens are loaded at a tensile rate of about 2.5mm/min. The images with deformed heat-sealed area are recorded by an 8mm video camera via a microscope, digitized by an image-processing device, and used for obtaining the displacements between the no-load condition and failure of the specimens using a digital image correlation. As a result, a characteristic displacement is found in the heat-sealed area, and the fracture mode is observed to be tension rather than tear.

1 INTRODUCTION

The use of liquid package bags made of plastic film has steadily increased in recent years. However, since a large proportion of these plastics are not recycled, plastic garbage has become a serious environmental problem. Under such circumstances, the reduction in volume and thickness of these wastes to minimize their total quantity is desirable.

Attracting attention in this respect are flexible packages which not only facilitate reduction in volume and thickness but can also be folded into compact sizes for easy recovery. Particularly advantageous among them are liquid packages made of laminated film by which the quantity of plastic used can be reduced by 1/5-1/10 of that used for hard packages. However, one disadvantage is that since a thin laminated film is used, bursting of packages due to temperature changes or shocks during transportation can readily occurs. To overcome this disadvantage, knowledge of the mechanism of bursting is necessary.

Until now, very little has been known about the mechanism and there have been only a few experimental studies on the bursting of packages subjected to impact load and on the relation between the bursting and impact strength of film (Futase et al. 1995a, 1995b). In order to prevent the bursting phenomena, considering that most bursts occur at the heat-sealed area, deformations and stresses on a cross section of this area must be investigated.

From this standpoint, there has been one analytical study of stresses on a cross section of the heat-sealed area using the finite-element method together with linear structure analysis under application of static load (Umezaki et al. 1994a). However, there has been no experimental research on deformations and stresses on the cross section. One of the reasons is that the heat-sealed area is very thin, about 100μm, in thickness. This prevents the use of a traditional extensometer and strain gauge, or a recently developed laser-speckle strain gauge (Yamaguchi et al. 1986, 1988). The only measurement method available is to use a digital image correlation (Peters & Ranson 1982, Chu et al. 1985, James et al. 1990) by which deformations are measured at each pixel on images magnified by a microscope and CCD camera.

In this study, deformations on a cross section of a heat-sealed area in plastic film used for liquid package bags are investigated using a digital image correlation.

2 MEASUREMENT PROCEDURE

2.1 *Specimens*

The laminated film used in this experiment (Figure 1) is commonly used, and consists of nylon, NY, (15μm in thickness), polyethylene, PE, (25μm in thickness) and ethylene vinylacetate copolymer, EVA, (25μm in thickness). The bags made of the laminated film have NY and EVA layers on the outside and inside, respectively, and two sets of PE and EVA layers combined on the heat-sealed sides. The combined layer is called a seal layer.

The package bags of dimensions 80×60×7.8mm as shown in Figure 2 were made of the laminated film and contained about 17ml of water. These bags were heat-sealed on three sides by an automatic packaging machine under the same sealing conditions as those for the usual production of liquid packages. T-shaped specimens of 10mm width as shown in Figure 3 were cut in the longitudinal direction from the bags with a knife as shown in Figure 2.

The tear force acting on the heat-sealed area was applied at the edge of the horizontal part of the specimens as shown in Figure 3. Deformations were measured on the area in the vicinity of point A shown in Figure 3. The measured area was polished using #1500 emery paper to improve the quality of the images.

2.2 *Experimental setup*

Figure 4 shows the equipment used for measuring deformations. The specimens were loaded at a tensile rate of about 2.5mm/min using a wire which is connected to a pulley rotated by a handle. The load was measured with a digital force gauge (dimensions 220×60×35mm) of 196N capacity. The gauge uses a load cell. The value of the load, which is displayed by digits on the upper surface of the case, was recorded with an 8mm CCD camera. An XYZ stage was used to move the loading device so that the heat-sealed area

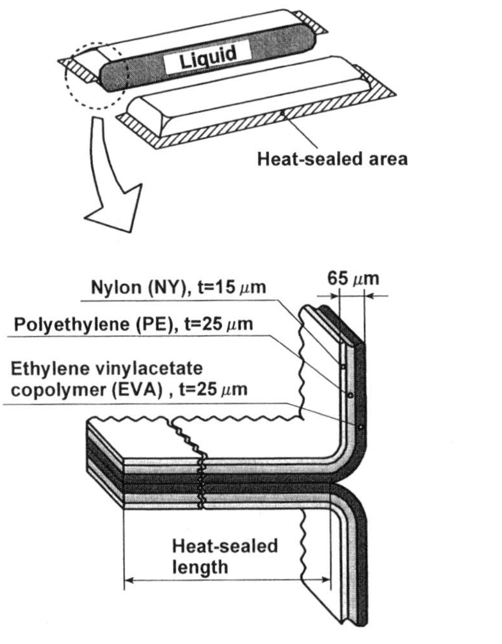

Figure 1. Cross section of laminated plastic film for liquid packages.

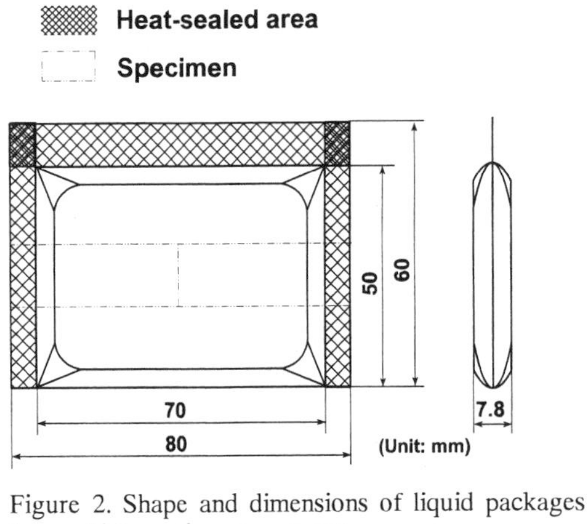

Figure 2. Shape and dimensions of liquid packages from which specimens were cut.

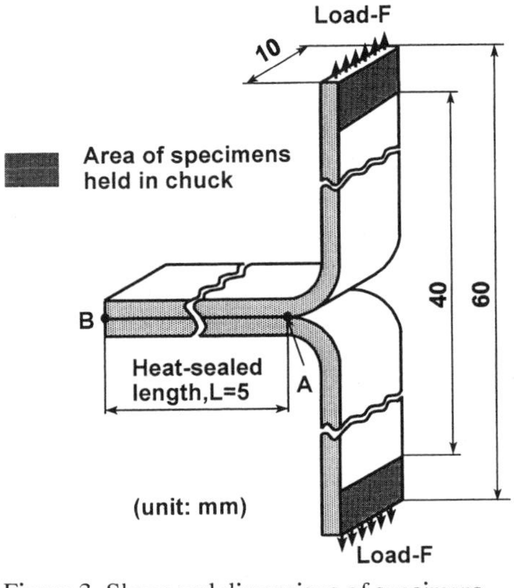

Figure 3. Shape and dimensions of specimens.

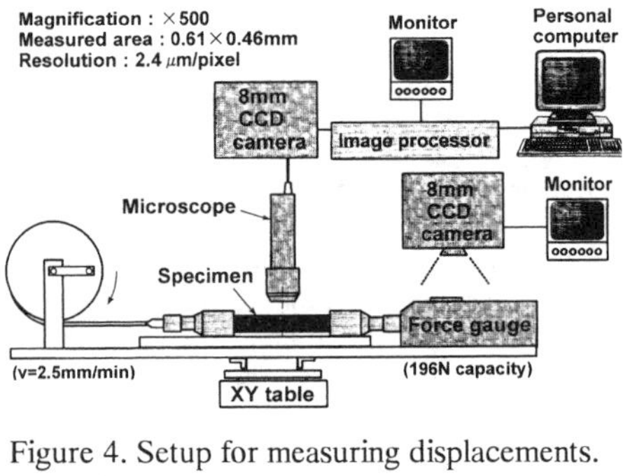

Figure 4. Setup for measuring displacements.

was always located in the photographed zone during the tests. The applied stress, σ_a, is defined by dividing a load, F, by an initial cross-sectional area of 10×0.065mm^2 where the load acts, as shown in Figure 3.

The images with deformed heat-sealed area were stored on 8mm magnetic tape with an 8mm CCD camera connected to an optical microscope of magnification ×500. This system gives an image scaling in the digital system of 2.4μm/pixel, and a photographed area of 0.61×0.46mm. The images recorded on the tape were digitized into an 8-bit (256-level) light-intensity distribution using an image processor. The digital images were stored as a 256×256 pixel array in the image memory of the processor. These were then input into a personal computer for storage in a hard disc unit and for the digital image processing.

2.3 *Digital image correlation*

The digital image correlation utilizes two original digitized images, which are called a reference or undeformed image, R, and a deformed image, D, so as to determine deformations between the images. In this study, the images binarized previously using an adaptive binarization algorithm as R and D were used, in which a threshold value is given to each pixel. The threshold value, $d_{th,o}$, at a noted point, O is written as

$$d_{th,o} = \frac{1}{MN} \sum_{j=1}^{N} \sum_{i=1}^{M} d_{ij} \tag{1}$$

where d_{ij} is one of 256 gray levels at a point (i, j) in an $M \times N$ array of pixels. If a gray level of a noted pixel is greater or less than $d_{th,o}$ obtained from equation (1), then it is replaced by the maximum gray level, white color or the minimum one, black, respectively.

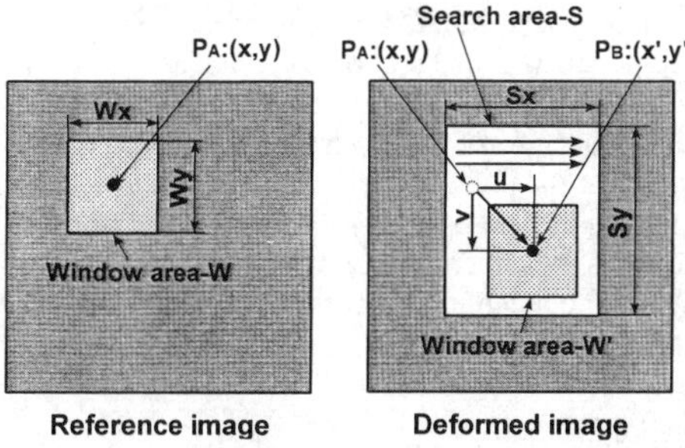

Figure 5. Search of point B on deformed image corresponding to point A on reference image using digital image correlation method.

Consequently, a binarized image obtained.

The algorithm for detecting displacements between two binarized images is as follows. Let W and W' be a window area in R and one of the search area, S, in D, respectively, as shown in Figure 5. Point P_A is located at the center of W, and P_B at the center of W'. Let (x, y) be the coordinates of P_A and (x', y') be the coordinate of P_B, expressed by the equations

$$x' = x + u, \quad y' = y + v \tag{2}$$

where u and v are the components of the pixel offset from R to D.

Firstly, the number of black pixels, W_{Rj} (j=1 to W_y) is counted for each row and W_{Ci} (i=1 to W_x) for each column in W, as shown Figure 6. Next, their numbers are compared to the numbers of black pixels, W'_{Rj} (j=1 to W_y) for each row and W'_{Ci} (i=1 to W_x) for each column in all W' of S by use of the equation (Umezaki et al. 1992, 1993, 1994b)

$$R(u,v) = \sum_{i=1}^{W_x} [W_{C_i} - W'_{C_i}] + \sum_{j=1}^{W_y} [W_{R_j} - W'_{R_j}] \tag{3}$$

Lastly, the values of the displacement components, u and v, at point P_A in R which minimize $R(u,v)$ in equation (3), named an evaluation value, are assumed to be the displacement of the center of W. The point at which the value of $R(u,v)$ is minimum on D is called the corresponding point, P_C. A binarization area of 11×11 pixels and a window area of 23×23 pixels were used.

The above procedure is automatically carried out. However, points at which window areas of 23×23

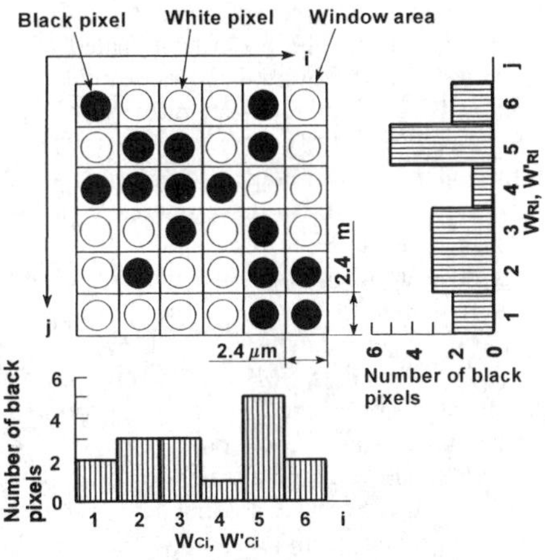

Figure 6. Distribution of black pixels used for investigating correlation between two images.

pixels are put must be selected on a reference image before putting the procedure in practice.

In general, for small deformation, two images obtained at initial and final loads are used as R and D, respectively. However, for large deformations, which is the case treated in this study, there is a greater possibility of being unsuccessful in the identification of the corresponding point between the two images because the surface texture of an evaluation specimen greatly differs from that at initial load. For large deformations, many sets of two images selected in order of the amount of load among those obtained at loads between initial and final loads are used. In this case, a deformed image used prior to the present correlation is used as a reference image for the present correlation. Using this technique, any large deformations can be measured.

3 RESULTS AND DISCUSSION

The images used for measuring displacements were selected from those taken between no-load conditions and failure of the specimens. Twenty-six images selected at intervals of about 1.2s for 32s at σ_a= 11.4MPa to σ_a=16.5MPa were used for measuring the displacements in the heat-sealed area between points A and B as shown in Figure 3. In this case, two images at σ_a=11.4 and 11.8MPa were used as the first image pair, reference and deformed ones, and those at σ_a =11.8 and 12.6MPa as the second pair. The final pair was comprised of two images at σ_a= 16.0 and 16.5MPa. In addition, fifteen images selected at intervals of about 0.7s for 11s at σ_a=8.77MPa to σ_a=11.5MPa were used for the horizontal part near point A of the specimen, as shown in Figure 3. These intervals were determined empirically.

Figure 7 shows examples of images used for measuring the displacements in the heat-sealed area between points A and B. Figure 8 shows examples of images obtained by adaptive binarization of those shown in Figures 7(b) and (d).

The displacements measured at selected points are shown in Figures 9 and 10. In these figures, point A corresponds to that in Figure 3. Figure 9 shows displacement histories of two sets of 7 points selected in the heat-sealed area. In this figure, the starting and final points are shown by symbols "○" and "●" on the images at σ_a=11.4 and 16.5MPa, respectively, and the displacement histories are shown by the broken lines. At the initial stage, each point moved in a direction approximately perpendicular to load. The points near the surface have a tendency to move toward the centerline through points A and B. Thereafter, all the points have a component of motion in the direction of load, and this component at the points near the outside is larger than that at the other positions.

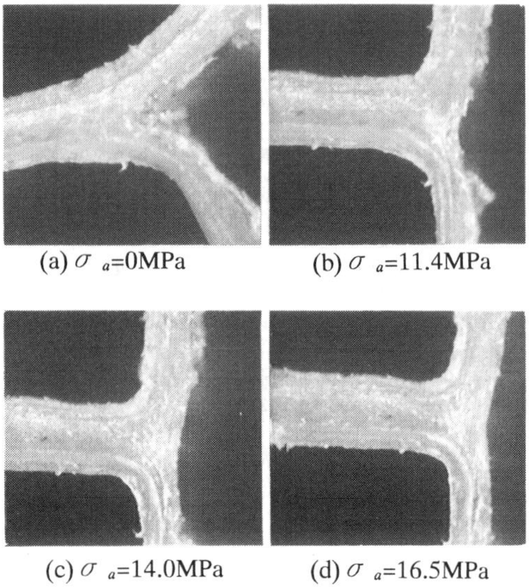

Figure 7. Examples of images taken under loading.

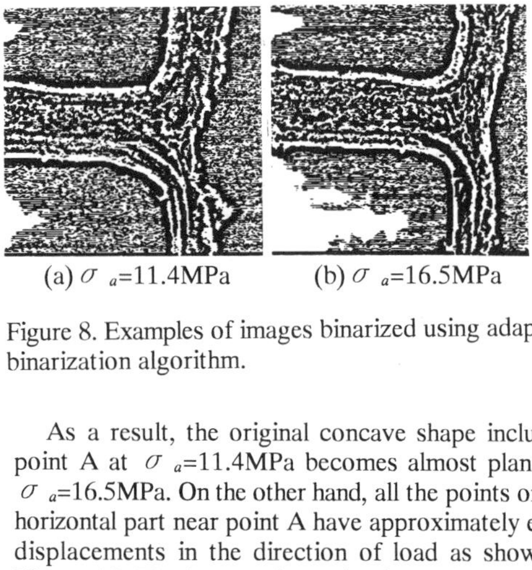

Figure 8. Examples of images binarized using adaptive binarization algorithm.

As a result, the original concave shape including point A at σ_a=11.4MPa becomes almost planar at σ_a=16.5MPa. On the other hand, all the points on the horizontal part near point A have approximately equal displacements in the direction of load as shown in Figure 10. The largest elongation is produced on the seal layer near point A. The fracture initiated at the boundary near the point at which the largest elongation was measured. These results show that the fracture mode of tension rather than tear. Figure 11 shows the diagram of specimen failure at the heat-sealed area.

4 CONCLUSIONS

The deformations on a cross section of a heat-sealed area in laminated plastic film used for liquid package bags were investigated by a digital image correlation.

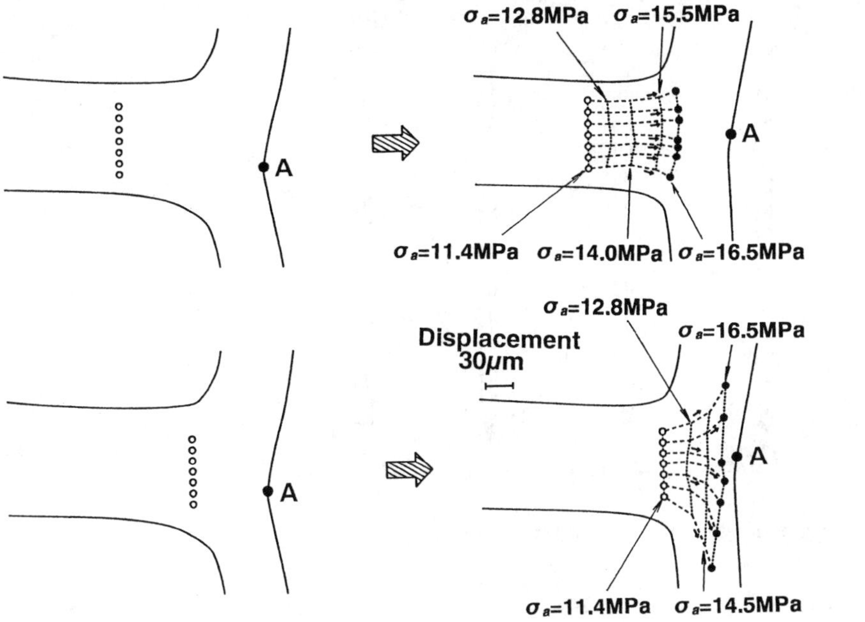

(a) Starting points for measurement of displacements shown by symbols "○" on reference image at σ_a=11.4MPa

(b) Displacement history shown by broken lines and end points shown by symbols "●" on deformed image at σ_a=16.5MPa

Figure 9. Displacements measured at σ_a=11.4MPa to σ_a=16.5MPa.

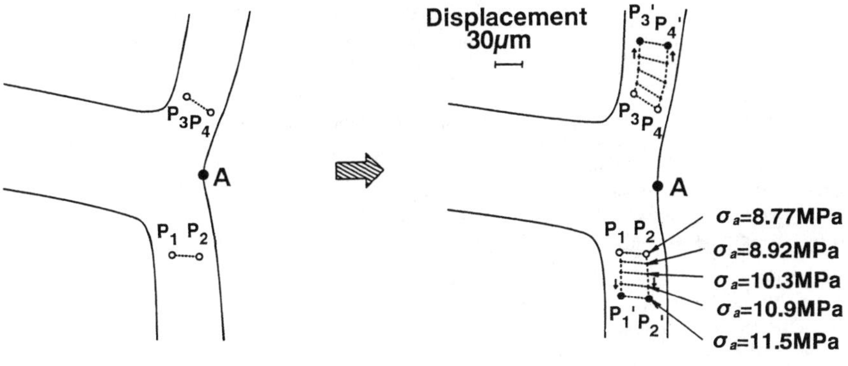

(a) Starting points for measurement of displacements shown by symbols "○" on deformed image at σ_a=8.77MPa

(b) Displacement history shown by broken lines and final points shown by symbols "●" on deformed image at σ_a=11.5MPa

Figure 10. Displacements measured at σ_a=8.77MPa to σ_a=11.5MPa.

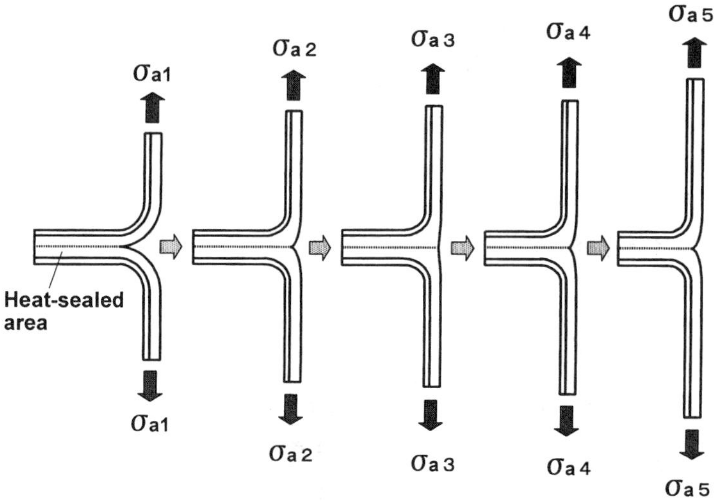

Figure 11. Schematic diagram of specimen failure at heat-sealed area.

The results show a characteristic deformation in the heat-sealed area, and the fracture mode of tension rather than tear.

REFERENCES

Chu, T. C., W. F. Ranson, M. A. Sutton & W. H. Peters 1985. Application of digital-image-correlation techniques to experimental mechanics. *Exp. Mech.* 25(3): 232-244.

Futase, K., A. Shimamoto & S. Takahashi 1995a. Impact strength test on a heat sealed portion of laminated film used for a liquid-filled bag (in Japanese). *J. Japan. Soc. Mech. Eng.* 60(580): 2891-2896.

Futase, K., A. Shimamoto & S. Takahashi 1995b. Changes in internal pressure of composite film packages by their drop impact (in Japanese). *J. Japan Soc. Mech. Eng.* 60(580): 2897-2902.

James, M. R., W. L. Morris & B. N. Cox 1990. A high accuracy automated strain-field mapper. *Exp. Mech.* 30(1): 60-67.

Peters, W. H. & W. F. Ranson 1982. Digital imaging techniques in experimental stress analysis. *Opt. Eng.* 21(3): 427-431.

Umezaki, E., A. Shimamoto & H. Watanabe 1992. Measurement of in-plane displacement in local region by using digital image correlation method. *Proc. 7th Int. Cong. on Exp. Mech.*: 1524-1529, Las Vegas, USA.

Umezaki. E., A. Shimamoto & H. Watanabe 1993. Improvement in speed and accuracy of digital image correlation method. *Proc. Conf. on Advanced Technology in Exp. Mech.*: 173-178, Kanazawa, Japan.

Umezaki, E., M. Shiobara & K. Futase 1994a. Stress analysis of heat-sealed area in plastic film used for liquid package: a fundamental study. In J.F.S. Gomes et al. (eds.), *Recent Advances in Exp. Mech.*: 21-26, Rotterdam: Balkema.

Umezaki, E., A. Shimamoto & H. Watanabe 1994b. Determination of corresponding points with subpixel accuracy in improved digital image correlation, *Proc. 1994 Far East Conf. on Nondestructive Testing*: 191-198, Taipei, Taiwan.

Yamaguchi, I., T. Furukawa, T. Ueda & E. Ogita 1986. Accelerated laser speckle strain gauge. *Opt. Eng.* 25(5): 671-676.

Yamaguchi, I. 1988. Advances in the laser speckle strain gauge *Opt. Eng.* 27(3): 214-218.

Damage and Failure of Interfaces, Rossmanith (ed.) © 1997 Balkema, Rotterdam, ISBN 90 5410 899 1

Delamination buckling and growth at global buckling

K.-F. Nilsson, L.E. Asp & J.E. Alpman
The Aeronautical Research Institute of Sweden, Bromma, Sweden

ABSTRACT: A computational and experimental investigation of buckling and growth of embedded delaminations located at different depths in a composite plate is presented with special emphasis given to the interaction between delamination buckling and global bending of the structure. Delamination growth for all cases occurred at, or slightly below, the global buckling load. The computational model captured qualitatively the observations from the tests with regard to loads for local buckling and crack growth, transverse deflections and shape of crack fronts and stability of growth. A more detailed assessment was difficult to perform due to the scatter in the experimental results.

1. INTRODUCTION

The analysis of delamination buckling and growth requires a combination of kinematically nonlinear structural analysis and fracture mechanics [e.g. Kachanov, 1976, Chai and Babcock 1981, Yin 1985, Whitcomb, 1989]. The complexity of the problem may be significantly reduced if bending and nonlinear kinematics is considered for the delaminated member only. This assumption, which is often referred to as the *thin film assumption*, is well motivated when the delaminated member is thin and located close to the surface. A technically important application is debonding of thin coatings. In more general circumstances, for instance in thin multi-layered structures, delaminations may appear at arbitrary depth (compare Fig. 1).

A finite element based approach used to analyse delamination buckling and growth under rather general circumstances has been presented in a series of papers [e.g. Nilsson and Storåkers, 1992, Nilsson *et al.* 1993, Nilsson 1993, Nilsson and Giannakopoulos 1995]. Delaminations with non-trivial shapes can be analysed and the redistribution of the stresses due to the changing front is accounted for automatically by a moving mesh scheme. In Nilsson *et al.* 1993, results observed from buckling induced delamination growth tests were compared with predictions and an excellent agreement was obtained. The method has also been used to simulate configurational stability of debonded coatings [Nilsson and Giannakopoulos 1995]. Previous investigations have, however, been based on the thin film assumption. The objective of the present study is to extend the methodology to include the effects of global bending. In order to assess the applicability of these extensions, delamination buckling tests of rectangular plates with a single delamination at three different depths have been performed. The paper is organized as follows. A brief description of the structural model is given first. Secondly, a description of the tests and a short presentation of the FE-model is given. Computed results are finally compared with test results with regard to buckling loads, transverse deflections, stability properties and observed crack fronts.

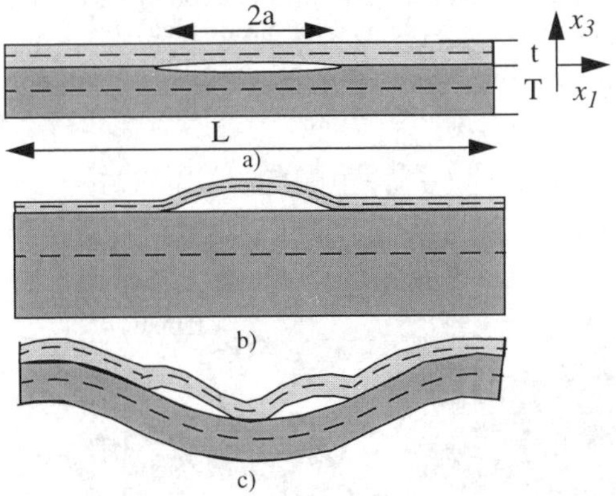

Fig.1 One-dimensional illustration of a structure with a delamination a) undeformed structure with single delamination b) schematic illustration of deformed structure when thin film assumption is adopted, c) deformed structure which bends globally

2. THEORETICAL MODEL

2.1 Kinematical and Constitutive Assumptions

The structure to be analysed consists of a plate of thickness, *t+T*, with a single embedded in-plane delamination of a smooth but otherwise arbitrary front, Γ_D, which encloses an area Ω_D at depth *t*. The thickness of the delaminated member and the total thickness of the structure are assumed small compared to the length of the delamination and entire structure, respectively.

The structure is modelled by two plates with mid-surfaces at $x_3 = t/2$ and $x_3 = -T/2$ respectively as shown in Fig. 1. The displacement of the upper and lower plates are governed by the Reissner-Mindlin assumption, *viz.*

$$\left.\begin{aligned} u_\alpha\left(x_1, \overset{*}{x}_2, \overset{*}{x}_3\right) &= \bar{u}_\alpha(x_1, x_2) + \overset{*}{x}_3\theta_\alpha(x_1, x_2) \\ u_3\left(x_1, \overset{*}{x}_2, \overset{*}{x}_3\right) &= \bar{u}_3(x_1, x_2) \end{aligned}\right\} \quad (1)$$

where 1 and 2 refer to in-plane quantities, 3 is the direction normal to the mid-surface, and $\overset{*}{x}_3$ is the distance from the midsurface. Greek indices run from 1 to 2 and θ_α denotes the rotation of a transverse material fibre. Barred displacements refer to the mid-plane.

Strains are assumed small and rotations moderate. The small strain tensor may then be written as

$$\left.\begin{aligned} e_{\alpha\beta} &= \overset{0}{e}_{\alpha\beta} + x_3{}^* \cdot \kappa_{\alpha\beta} \\ e_{3\alpha} &= (\bar{u}_{3,\alpha} + \theta_\alpha)/2 \end{aligned}\right\} \quad (2)$$

where $\overset{0}{e}_{\alpha\beta} = (\bar{u}_{\alpha,\beta} + \bar{u}_{\beta,\alpha} + \bar{u}_{3,\alpha}\bar{u}_{3,\beta})/2$ is the membrane strain and $\kappa_{\alpha\beta} = (\theta_{\alpha,\beta} + \theta_{\beta,\alpha})/2$ the curvature. The generalized forces conjugate to these quantities can be defined for elastic plates by

$$N_{\alpha\beta} = \frac{\partial W}{\partial \overset{0}{e}_{\alpha\beta}},\ M_{\alpha\beta} = \frac{\partial W}{\partial \kappa_{\alpha\beta}},\ 2Q_\alpha = \frac{\partial W}{\partial e_{3\alpha}} \quad (3)$$

where $W\left(\overset{0}{e}_{\alpha\beta}, \kappa_{\alpha\beta}, e_{3\alpha}\right)$ is the strain energy function.

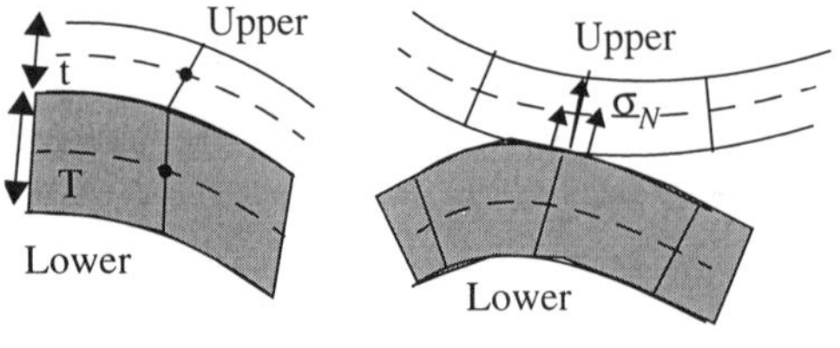

Fig. 2 One-dimensional illustrations of the structural model for a) un-delaminated domain b) delaminated domain with contact

In the *un-delaminated domain* displacement continuity is prescribed along the 'interface' as illustrated in Figure 2a, which results in three constraint equations linking the upper and lower plate

$$\left.\begin{aligned} \bar{u}^U_\alpha - \frac{t}{2}\theta^U_\alpha &= \bar{u}^L_\alpha + \frac{T}{2}\theta^L_\alpha \\ \bar{u}^U_3 &= \bar{u}^L_3 \end{aligned}\right\} \quad (4)$$

where superscripts U and L denote 'Upper' and 'Lower'. In the thin film assumption $\bar{u}_3 = 0$ and $\theta_\alpha = 0$ are imposed except at the delaminated member (upper plate above delamination).

In the *delaminated domain* a contact condition is prescribed where the contact pressure is given by

$$\left.\begin{aligned} \sigma_N &= S \cdot \frac{2d}{T+t} \qquad d < 0 \\ \sigma_N &= 0 \qquad d \geq 0 \\ d &= \left\|\left(\bar{x}^U_i + \bar{u}^U_i\right) - \left(\bar{x}^L_i + \bar{u}^L_i\right)\right\|_2 - \left(-\frac{T+t}{2}\right) \end{aligned}\right\} \quad (5)$$

where σ_N (<0) denotes the resulting contact pressure and *d* the elongation of a line segment linking points in the mid-surface of the upper and lower plate with initially identical in-plane coordinates and *S* is a selected transverse stiffness. Thus, the contact pressure is different from zero when the distance between points with the same coordinates in the upper and lower plate is less than their initial distance. The condition (5) can be implemented into a FE-code, but assumes that the out-of-plane component of the relative displacement is substantially larger than the in-plane component.

The unilateral constraint (5) introduces an additional nonlinearity to the problem and the contact region as well as the contact forces for a given load are computed by an iterative procedure with the two associated contact convergence criteria

$$\int_{\Omega_D} \frac{|d| - d}{|d|} d\Omega \leq \varepsilon_d,\ \int_{\Omega_D} \frac{|\sigma_N| - \sigma_N}{|\sigma_N|} d\Omega \leq \varepsilon_\sigma \quad (6)$$

where ε_d, ε_σ, denote the associated convergence tolerances. The contact algorithm in the thin film case is discussed in some detail in Giannakopoulos, Nilsson and Tsamasphyros (1995). The main difference between the present global bending and thin film contact model is that both plates deflect and the contact area is not a flat surface. Moreover, the thin film contact model did not require that in-plane displacements are small compared to the out-of plane ones.

2.2 Computation of Energy Release rate

Delamination growth of the materials at issue is assumed to be governed by linear elastic fracture mechanics (LEFM) parameters such as the energy release rate or stress intensity factors. The energy release rate at local crack growth, G, can be computed from the discontinuity in field variables across the crack front of a tensor component

$$G = \|P_{nn}\| \tag{7}$$

where double brackets denote the "jump" across the crack front (see Fig. 2a) and

$$P_{nn} = W - N_{n\gamma}\bar{u}_{\gamma,n} - Q_n\bar{u}_{3,n} - M_{n\gamma}\theta_{\gamma,n} \tag{8}$$

is a energy momentum tensor component and n denotes the local normal direction to the crack front. The expression (7) for the energy release rate was first expressed by Storåkers and Andersson (1988) for von Karman's nonlinear plate theory and the extension to the Mindlin theory is given in Nilsson *et al.* (1993).

The nonlinear plate problem may locally be approximated by a linear beam problem by superposing a homogeneous strain field such that the undelaminated region becomes undeformed as illustrated in Fig. 3b. By this procedure the number of unknown load resultants is reduced. Eq. (8) can then be used with the resulting 'beam sectional forces' to compute the energy release rate.

Cracks that propagate along an interface have usually a mixed mode crack tip loading, and the fracture toughness usually shows a strong fracture mode dependence. By solving the linear split beam problem separately, the fundamental fracture modes may be obtained as function of the resulting beam variables. This approach was adopted in Nilsson and Storåkers (1992) for the case of isotropic materials by using closed form results for the split beam problem given by Suo and Hutchinson (1990). In more general cases the split beam problem has to be solved numerically with stiffness properties given by the angle between the crack front normal and principal material axes. In this investigation only the energy release rate will be used for crack propagation.

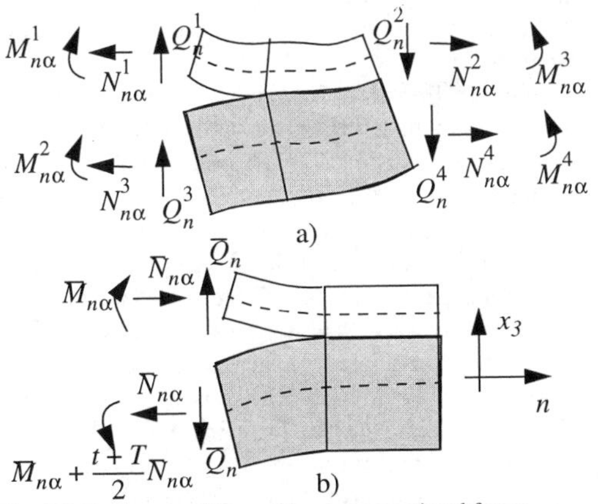

Fig. 3 Split element with resulting beam sectional forces
a) prior to superposition
b) after superposition

3. EXPERIMENTAL PROCEDURE

Cross-ply carbon fibre/epoxy laminates with an implanted artificial delamination were tested. A cross-ply lay-up $[(90°/0°)_{17}/90°]$ was chosen, where the $0°$-direction is parallel to the x_1-direction, (see Fig. 4). The delamination plane is thus bounded by fibres parallel to the major direction of crack growth which prevents out-of-plane kinking of the crack. The laminates were manufactured from HTA/6376C prepreg, by Ciba-Geigy, and cured according to the supplier's recommendations. The artificial delaminations consisted of two, *7.5 μm* thick, stacked circular polyimide films with diameters *55* and *60 mm* respectively. To prevent adhesion, a thin layer of teflon was sprayed between the two polyimide films. The artificial delaminations were placed after prepreg layers three, five and seven, respectively (N=3, 5, 7). The mechanical properties of the HTA/6376C material are $E_{11}=146$ *GPa*, $E_{22}=10.5$ *GPa*, $G_{12}=G_{13}=5.25$ *GPa*, $G_{23}=3.48$ *GPa*, $\nu_{12}=\nu_{13}=0.30$ and $\nu_{23}=0.51$. The nominal ply thickness is *0.130 mm*. The thickness of the laminates were measured to *4.56 mm*, with a standard deviation of *0.08 mm*. The chosen dimensions of composite thickness and width and delamination diameter reflect those typical in aeronautical applications. The fracture toughness of the material is mode dependent with specific mode I and mode II fracture *200 J/m²* and *570 J/m²* respectively [Olsson *et al.* 1996].

Plates dimensions are presented in Fig. 4. A hole with a *1.2 mm* diameter was drilled in the centre of the delamination as to minimize effects of low internal pressure on local buckling. To evaluate uniformity of uniaxial loading all plates were instrumented with 0°/90° strain gauges. Totally, six strain gauges were bonded on each side of the plates.

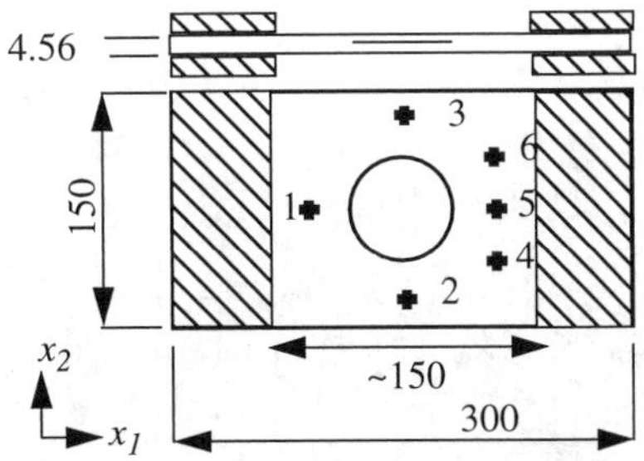

Fig. 4 Schematic top and side view of the delaminated composite plate (lengths in mm). The numbered cross-marks indicate positions of strain gauges around the circular delamination. Hatched areas are clamped.

In the experiments the composite plates were clamped in the load machine with a free length of *145 mm*. As will be further discussed below, the machine clamping was not sufficient to achieve the clamped boundary conditions desired. Therefore, steel plates were bonded on to four composite plates (N=3, 5) with a free length *150 mm*. In all tests the load was applied in the x_1-direction and displacement controlled and at room temperature. A 250 kN servo-hydraulic MTS load frame was used in the experiments. On both sides, out-of-plane displacements were measured in the centre by instrumented dial gauges. A 28 channel data acquisition system, controlled by a IBM PC and using an analogue connection software, Labview 2, recorded the load, in-plane and central out-of-plane displacements and strain gauge reponses. In addition, acoustic emission was employed to detect delamination growth initiation. For this, a Bruel and Kjaer Emission Pulse analyser Type 4429 in conjunction with a Type 8313 AE transducer with a resonance frequency of 200 Hz was used. Determination of the size and shape of the internally growing delamination was accomplished by ultrasonic C-scan.

The composite plates were scheduled to be tested in the following sequence: First local buckling of the delaminated region was to be determined. In the second sequence the plates were to be loaded until global buckling or initial crack growth whichever occurred first. In a final step the plates were to be loaded until failure.

4. NUMERICAL PROCEDURE

The test panels have been analysed with the structural model described above using the commercial FE-code ADINA.

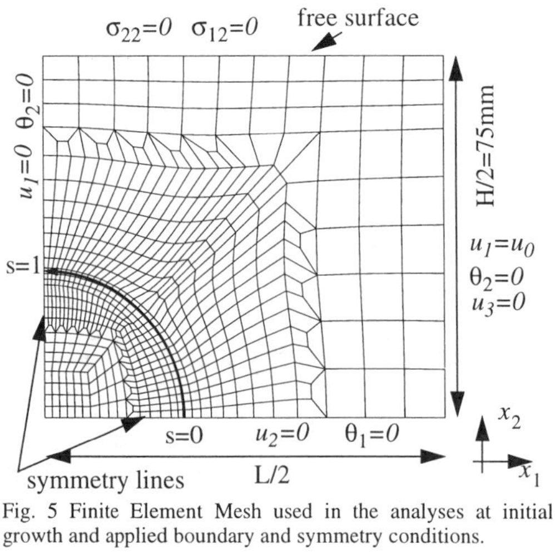

Fig. 5 Finite Element Mesh used in the analyses at initial growth and applied boundary and symmetry conditions.

A plane projection of the adopted FE-mesh with applied boundary and symmetry conditions is shown in Fig. 5. The same mesh has been adopted for the lower and upper part except for a shift of the middle surface. Four-noded mixed interpolation shell elements have been used.

The constitutive parameters were determined by classical laminate theory. The adopted ply thickness is 0.13mm, resulting in a total thickness of 4.55mm. The displacement field of the lower plate element layer in the undelaminated domain is constrained to the upper plates nodewise by the relation (4).

Nodes in the upper and lower part of the delaminated region with the same x_1,x_2-coordinates are joined with a linear elastic spring. Thus the contact condition (5) is modelled nodewise with spring stiffness

$$K_i = A_i E_{33} \tag{9}$$

where K_i replaces S and σ_N is replaced by contact forces R_N. A_i denotes the area associated with the particular node (the sum of all A_i equals the delaminated area). These 'contact' springs have a 'death-birth-option' which makes it possible to assign zero spring stiffness outside the contact zone and to activate and de-activate contact nodes as the contact area changes. The contact zone and contact forces were computed by a predictor correction method where springs with negative reaction forces become inactive and points where interpenetration otherwise occur become active. The contact analysis has converged when the two convergence conditions (6) are fulfilled. In the present investigation $\varepsilon_d = 10^{-3}$ and $\varepsilon_\sigma = 10^{-2}$ were used.

A complete analysis includes the following steps:

- The global buckling load is first determined for the structure by imposing the constraint (4) in the delaminated region and performing the eigenvalue analysis.
- The local buckling load is subsequently determined with due account to contact at buckling following the contact procedure outlined above.
- This is followed by the kinematically nonlinear postbuckling analysis where full Newton method is adopted and where an iterative contact analysis is performed at each load. Once the contact analysis has converged the energy release rate is computed along the front. Load increments are taken automatically such that load increments are small at the local and global buckling load where the tangential stiffness may be very low. Load increments are also adjusted such that the crack growth criterion is attained without significant violation.
- The front may propagate when the crack growth

criterion, ($G = G_c$), has been attained locally nodewise. The front is then advanced by pointwise moving the nodes which have reached the crack growth criterion a small distance in the local normal direction to the front and performing a remeshing. The postbuckling analysis is then restarted at the previous propagation load but with the new updated mesh.

By this approach, the evolution of the delamination propagation is modelled by performing a large number of incremental crack propagations. Note that the number of elements is not increased during propagation. A more detailed description of the moving boundary technique can be found in Nilsson and Giannakopoulos (1995).

5 RESULTS AND DISCUSSION

The experimental results are summarized in Table 1. Superscripts *a* and *b* refer to the two clamping conditions investigated; a) machine clamping of the composite plate with a free length of 145 mm and b) steel plates adhesively bonded to the composite plate with a free length of 150 mm. The specimen numbers gives information of the position of the delamination and consecutive order. Thus, e.g. A3_1 relates to the *first* specimen with a delamination after the *third* layer, N=3. Also, the load-step is given by a third number, e.g. A3_1_1 refers to the first load-step of the A3_1 plate. Aref refers to the tests of two undelaminated plates. Further, P^L_{cr} and P^G_{cr} are the local and global buckling loads at each step, respectively. Finally, P_{growth} is the load at delamination growth and Δa is the subsequent delamination growth in millimetre at each load step. Δa was determined as the average growth in the two delamination growth regions. The cross-head displacement measurements in the load frame have an accuracy of *0.1 mm* only while the imposed end displacements were of order *0.5 mm*. Hence the registered applied displacements were not accurate and are not included in Table 1.

Superscripts *a* and *b* refer to the two clamping conditions investigated; a) machine clamping of the composite plate with a free length of 145 mm and b) steel plates adhesively bonded to the composite plate with a free length of 150 mm. The specimen numbers gives information of the position of the delamination and consecutive order. Thus, e.g. A3_1 relates to the *first* specimen with a delamination after the *third* layer, N=3. Also, the load-step is given by a third number, e.g. A3_1_1 refers to the first load-step of the A3_1 plate. Aref refers to the tests of two undelaminated plates

Table 1: SUMMARY OF TESTS

Spec. no.	P^L_{cr} (kN)	P^G_{cr} (kN)	P_{growth} (kN)	Δa (mm)
Aref_1	_1:none	-110.6	none	none
Aref_2	_1:none	-113.5	none	none
A3_1[a]	_1:-110.5 _2:-12.3 _3:-19.1	-110.5 -110.7 -112.4	-110.5 -109 -110	5 2 fail.
A3_2[a]	_1:-115.0 _2:-10.3 _3:-11.5	-115.0 -110.4 -110.8	-115 -109 -109	7.5 1 fail.
A3_3[a]	_1:pert. _2:-12.6 _3:-12.4	pert. -108.7 -108.9	pert. -108 -108	6 1 fail.
A3_4[b]	_1:-19.4 _2:-17.2 _3:-14.3	-106.9 -106.4 -109.6	none -106.4 -109.6	none 6 fail.
A3_5[b]	_1:pert. _2:-26.2 _3:-26.6	pert. -104.3 -108.7	pert. -104.3 -108.7	4 0.5 3
A5_1[b]	_1:-61.3	-107.8	-107.8	fail.
A5_2[b]	_1:pert. _2:-57.3	-117.0 -100.7	pert. -100.7	20.2 fail.
A7_1[a]	_1:pert. _3:-96.1 _6:-66.8	pert. -110.6 -110.2	pert. -110 -110	none 8 fail.
A7_2[a]	_1:pert. _2:-98.4 _3:-75.4	-115.2 -111.2 -110.1	pert. -110 -109	3.5 3.5 fail.
A7_3[a]	_1:pert. _2:-63.3 _3:-61.3	-106.3 -108.3 -107.3	pert. -108 -107	7.5 2.5 fail.

. In the experiments 'global buckling' load is defined by the asymptotic value of the load for the out-of-plane displacement of the substrate. Local buckling load is defined as the load at which the delamination opened. In practice, the local buckling was difficult to achieve as planned. Local buckling at the first step occurred in four plates only (A3_1, A3_2, A3_4, A5_1), and in two of these (A3_1, A3_2) local buckling was not observed until global buckling. To make the remaining plates buckle locally, an out-of-plane load was applied manually at high compression load (typically close to global buckling load). The perturbed plates are indicated by the abbreviation pert. in Table 1. Delaminations grew at local buckling at the first step in all plates but three (A3_4, A5_1 and A7_1), due to the high compression load. In-plane stiffness of the plates was evaluated by strain gauge measurements and found to agree well to that estimated by the numerical analysis. Furthermore, strain gauge measurements verified symmetry assumptions made in the analysis. Prior to global buckling and delamination growth, the discrepancy between the strain gauge couple 2 and 3, see Fig. 4, was less than five percent. In some cases, at delamination growth and global buckling the gauges indicated the symmetry assumption to be violated as the discrepancy increased to more than 50% in some cases. For strain gauge couples 1 and 5, and 4 and 6 symmetry assumptions could not be evaluated as the gauges were affected by the clamping.

In the C-scan mappings, the actual diameter of the delaminations was close to *55 mm* instead of the expected *60 mm* suggesting some adhesion between the artificial delamination and the composite. Furthermore, the oily teflon spray between the polyimide films may provide a viscous opposition to local buckling.

5.1 Global Buckling Load

The computed global buckling load, P_{cr} (i.e. the reaction load at the clamped edge) as function of the free panel length, L, is shown in Fig. 6.
The computed buckling load at L = *150 mm* is *140 kN* which is higher than the experimentally observed buckling load (~*110 kN*). In order to verify the computational model, the global buckling analysis was also performed by full three-dimensional FE-analysis using the in-house hp FE-code, STRIPE, where each ply was modelled explicitly. The difference in computed buckling load between the plate model and the three-dimensional model was negligible. This lead us to the conjecture that the lower buckling value in the tests was mainly due to non-perfect clamping. This discrepancy was also the reason why the second type of clamping was introduced, but as we se from Table 1, adding the steel plates did not significantly increase the buckling load.

The three-dimensional analysis was also performed by modelling the full length of the panel *(300 mm)* and prescribing $u_3=0$ along upper and lower shaded surfaces in Fig. 4, $u_1=u0$ along the right and $u_1 = 0$ along the left shades surfaces respectively. These boundary conditions were believed to better reflect the actual clamping in the tests. These boundary conditions reduced the buckling load, but only to *135 kN*.

To confirm that the problem was connected with the clamping two Aluminium panels with width and thickness 150.1 and 4.03 mm were tested. The first one was 'clamped' at both end using steel plates and with a free length of 150 mm whereas the second one had one end simply supported. The edge of the simply supported end was worked to a cylindrical shape with a diameter equal to the plate thickness. The computed and tested buckling values were 109 and 83 kN for the plate with two ends clamped. For the simply supported plate the computed value and test value were both 55 kN.

Instead of attempting to model the flexibility of the clamped ends we have adopted an 'apparent length' for the panel in the analyses, such that the computed global buckling load coincides with the one found in the tests. From Fig. 6 we see that L = *170mm* gives a buckling load of *110 kN*. This length has been used in the analyses below.

5.2 Comparison Thin Film vs Global Bending

To assess the applicability of the thin film assumption postbuckling analyses were performed for the circular delamination configuration for all three delamination depths using the global bending model and the thin film model respectively.

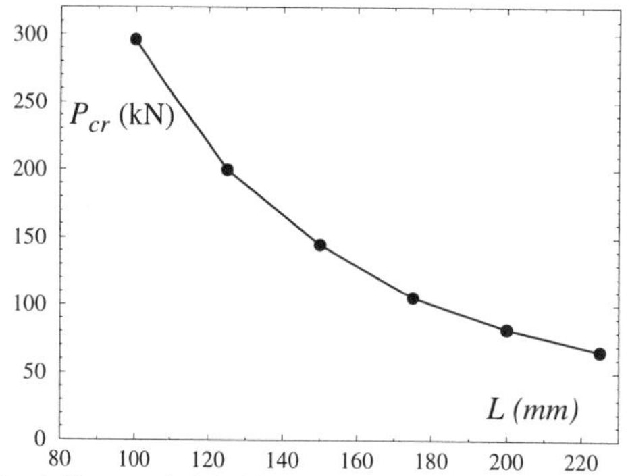

Fig. 6. Computed global buckling load as function of the plate length with material properties and lay-up sequence defined above.

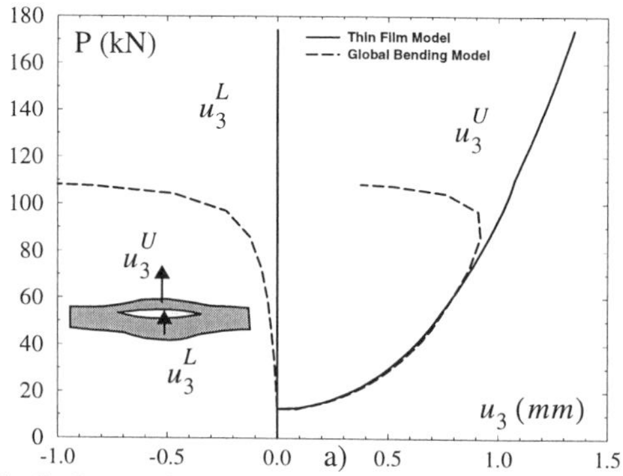

Fig. 7a Computed out-of-plane displacement at centre of a circular delamination as function of applied load for 3 layer delamination as given by the thin film model and the global bending model respectively.

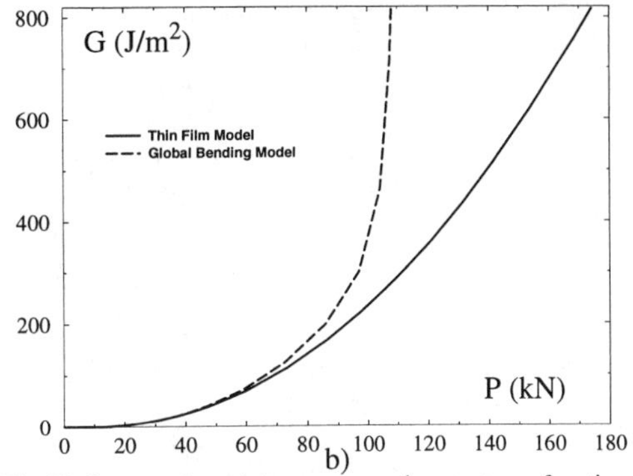

Fig. 7b Computed maximum energy release rate, as function of applied load for 3 layer delamination as given by the thin film model and the global bending model respectively.

Fig. 7a and 7b show the resulting transverse deflection in the centre of the upper and lower part and the maximum energy release rate as function of the load for the 3 layer panel. The difference in displacements and energy release rate between the global and thin film models are very small up to half the global buckling load. As the global buckling load is approached the displacements deviate substantially and the energy release rate increases drastically for the global bending model. In the tests, delamination growth occurred at or slightly below the global buckling load.

5.3 Postbuckling of Initially Circular Delamination

Following the outlined FE-procedure the global and local buckling loads were computed for the three delamination depths followed by a postbuckling analysis where the central deflection along with the energy release rate distribution were recorded as function of the applied load. The computed local buckling loads were *12.5*, *41.3* and *79.5 kN* for the three depths. The global buckling load was *111 kN* for all cases. Fig. 8a-c shows the central deflection in the upper and lower part as function of the applied load for all three depths respectively, along with corresponding deflections from the test panels A3_4_2, A5_1_1 and A7_1_2.

The delaminated layer initially deflects from the thicker lower layer, and then reaches a maximum when the lower layer starts to deflect downward and from then on both layers are deflected downwards with an increasing opening of the delamination.

There is an excellent correspondence between computations and experiments for the 3-layer delamination, whereas for the 5 and 7-layer delamination, the recorded local buckling load is higher than the computed value, however, once the layers have separated from each other, these effects vanish and the correlation between tests and computation is very close. For the 7-layer delamination the correspondence is somewhat worse. Note that in this case in addition to adhesion between layers, there is a noticeable imperfection.

The energy release rate as function of the load attained its maximum value for all depths transverse to the loading direction, (at s = 1 in Fig. 5). The max-

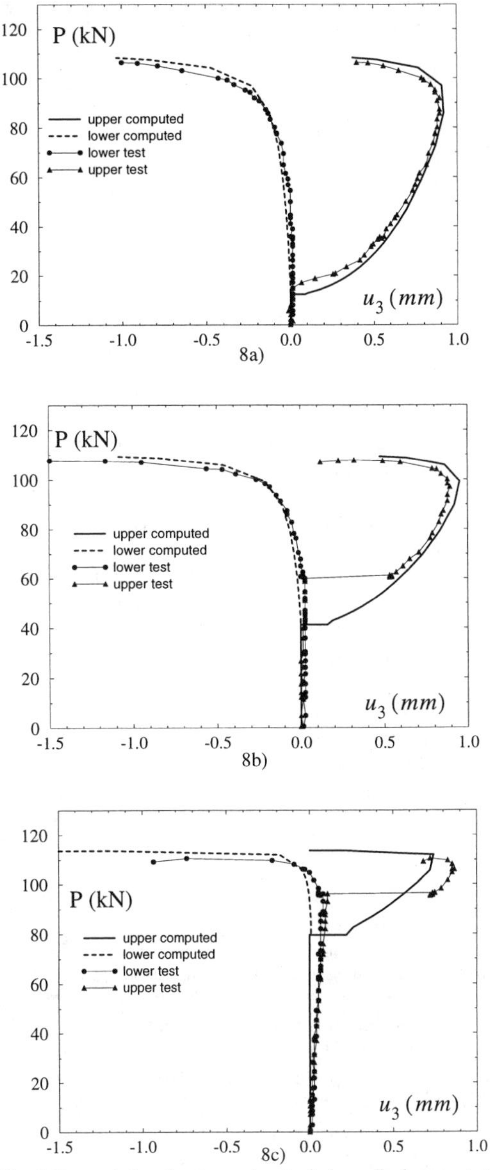

Fig. 8 Computed and measured out-of-plane displacement at centre of circular delamination as function of applied load
a) 3 layer delamination (Test A-3-4-2)
b) 5 layer delamination (Test A-5-1-1)
c) 7 layer delamination (Test A-7-1-2)

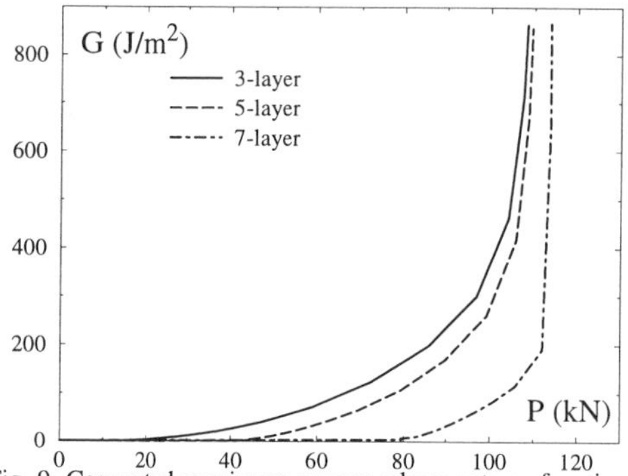

Fig. 9. Computed maximum energy release rate as function of applied load for 3, 5 and 7 layer delamination respectively.

imum energy release rates as function of the applied load for the three depths are shown in Fig. 9. The energy release rate increases drastically at the global buckling load and this feature becomes more pronounced with increasing delamination depth.

The load at which crack growth was initiated for the panels A3_4, A5_1 and A7_1 was *106.8*, *107.8* and *110 kN* respectively. The corresponding energy release rate values read off from Fig. 9 are of the order *500*, *400* and *200* J/m^2. The crack growth resistance for this material has a strong mode dependence as mentioned previously, and a consistent prediction of crack growth would require a separation of the fundamental fracture modes. The shearing mode is usually small at the local buckling, (e.g. Nilsson and Storåkers 1992, Whitcomb 1989) but increases as the load increases. This is a plausible explanation to the apparent increase in critical energy release rate with delamination thickness. If we assume that the mode mixity stays more or less constant as the crack propagates, the values given above may be used in conjunction with a mode independent crack growth criterion to simulate growth. Due to the strong gradient of the energy release rate at the global buckling load, (Fig. 9), growth is expected slightly below the global buckling load regardless of the mode mixity.

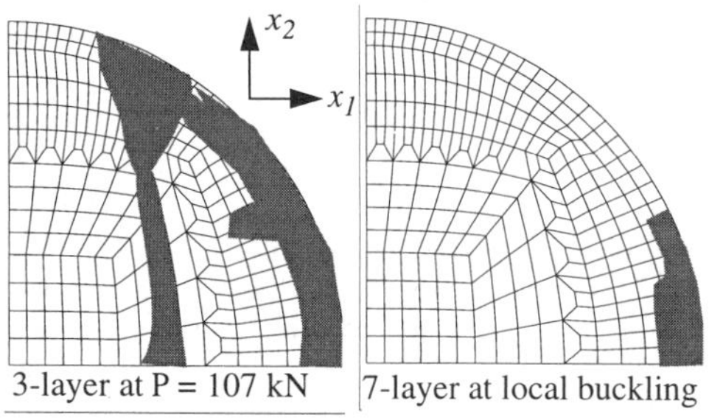

Fig.10 Sketch of computed contact areas

Contact was encountered in the analysis for all three depths already at buckling. The contact area then expanded as the load increased and the contact areas were larger the thinner the delamination. The contact areas at P = 107 kN for the 3-layer delamination and at local buckling for the 7-layer delamination are depicted in Fig. 10.

5.4 Analysis of Delamination Growth

The delamination propagation for the three delamination depths was simulated following the crack growth procedure outlined earlier. A constant critical energy release rate (G_c = 450 J/m^2) was adopted.

Fig. 11 shows the corresponding fronts for the three depths at 7.5 and 15 mm growth in the transverse direction. We see that the growth is more localized the thinner the delaminations. This trend was also observed in the experiments as seen in Fig. 12. The mesh of the delaminated region at 15 mm growth with associated contact zone is shown in Fig. 13 for the 3-layer delamination. The associated distribution of the normalized energy release rate along the front are depicted in Fig. 14.

The applied displacement load for continuous crack growth, (i.e. $G_{max} = G_c$) as function of the growth in

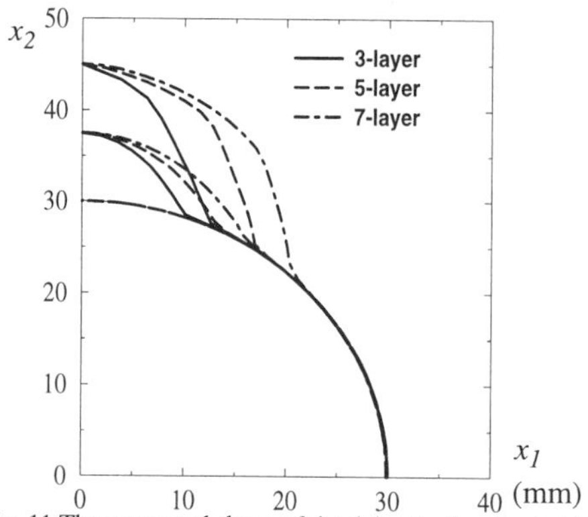

Fig. 11 The computed shape of the delamination front at 7.5 and 15 mm growth for the 3, 5 and 7 layer delaminations respectively. Crack growth criterion: $G = G_c$ and $G_c = 450\ J/m^2$

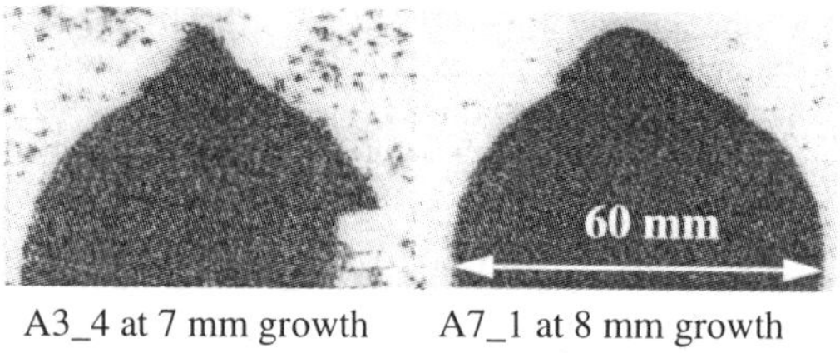

Fig. 12 C-scan of 3 and 7-layer delamination

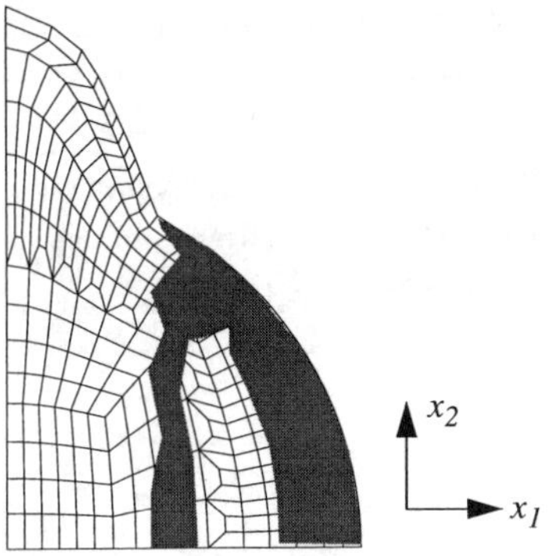

Fig. 13 Mesh at 15 mm growth and associated contact area for the 3-layer delamination.

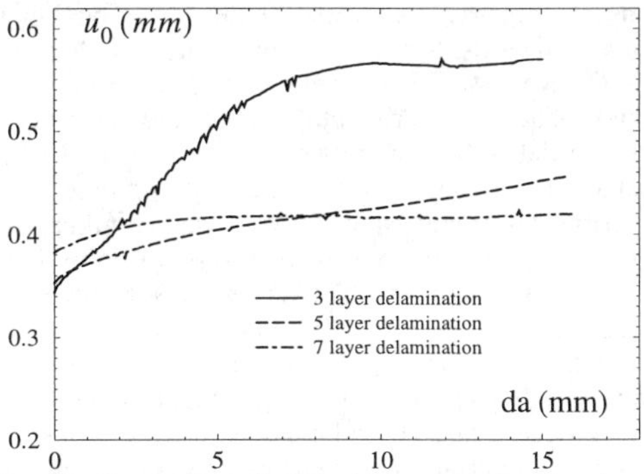

Fig. 15a Computed applied displacement load as function of growth for the 3, 5 and 7 layer delaminations respectively.Crack growth criterions $G = G_c$ and $G_c = 450\ J/m^2$
2

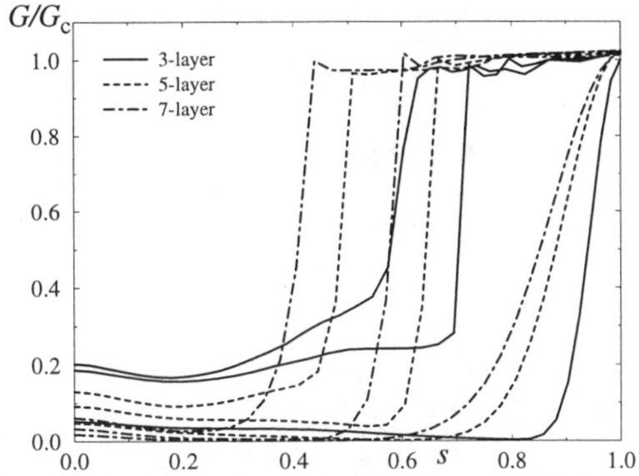

Fig. 14 The energy release rate distribution at initial growth, 25% growth and 50% growth for the 3, 5 and 7 layer delaminations respectively. Crack growth criterions $G = G_c$ with $G_c = 450\ J/m^2$

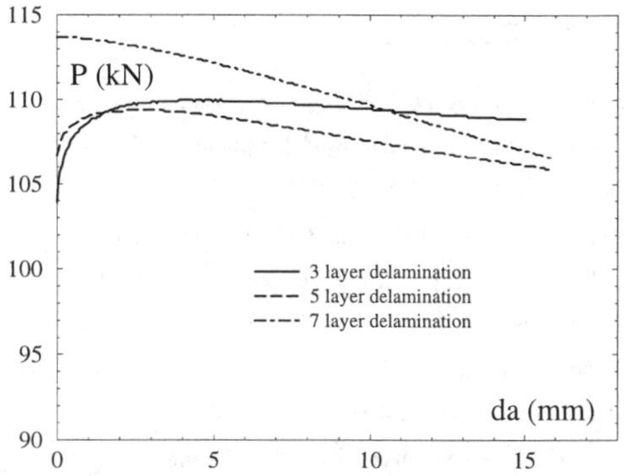

Fig. 15b, Calculated reaction load associated with Fig. 13a.

the transverse direction is presented in Fig. 15a and the associated reaction load is given in Fig. 15b.

Due to uncertainty and scatter in the experimental results, it is perhaps not appropriate to be too effusive in this paper about the capability of the computational method to predict growth stability. Nevertheless, the trends in these figures have some support in the test. Growth for all three depths was sustained at increasing displacement load with roughly the same applied displacement at initiation and where the applied load was increased more for thinner delaminations during growth. The propagation was initially stable as suggested by Fig 15, but became unstable after 5-15 mm growth.

CONCLUDING REMARKS

A numerical and experimental investigation of composite plates with an embedded delamination at three different depths has been presented. The main objective was to study the interaction between delamination buckling and global bending of the plate. Perhaps, the most important result, at least from a practical point of view, is that delamination growth for all cases occurred more or less at the load where the composite plate buckles globally. Consequently structures with delaminations should never be allowed to buckle globally. Another observation is that, at least for the geometries and materials studied here, the so called thin film assumption is not adequate to predict growth, not even for the three layer delamination where the thickness was less than one tenth of the plate. Thus, a delamination buckling analysis should always be accompanied by a global buckling analysis. If delamination growth occurs at loads significantly lower that the global buckling load, say two thirds, then the thin film model may be adopted. At the global buckling load, crack growth parameters increase drastically with the applied load and growth will be predicted fairly accurately irrespective of the sophistication of the fracture model.

The numerical model captured the main observations as regards transverse deflections, buckling loads, critical loads and direction of growth. A more quantitative assessment was, however, difficult to achieve due to experimental scatter and uncertainties. Partial adhesion between delaminated layers lead to large variation in the local buckling load. The experimental global buckling load was significantly lower than the theoretical one due to imperfect clamping. These two problems need to be resolved.

The numerical method handles growth of non-trivial delamination contours, contact at buckling and global bending. In the present investigation the total energy release rate was adopted as the crack growth parameter. Interface crack growth usually shows a strong fracture mode dependence, an efficient mode separation of the fundamental fracture modes is therefore of imminent need.

REFERENCES

Chai, H., Babcock, C.D., 1981, One-dimensional modeling of failure in laminated plates by delamination buckling, *Int.J. Solids Structures*, 17: 1069-1083

Giannakopoulos, A. E., Nilsson, K.-F., and Tsamasphyros, G. 1995 The contact problem at delamination, *J. Appl. Mech.*, 62: 989-996.

Kachanov, L.MM., 1976, Separation failure of composite materials, *Polymer Mech.*, 12: 812-815.

Nilsson, K.-F., and Storåkers, B., 1992, On the interface crack growth in composite plates, *J. Appl. Mech.*, 59, 530-538.

Nilsson, K.-F., Thesken, J.C., Sindelar, P., Giannakopoulos, A. E., and Storåkers, B., 1993,. A theoretical and experimental investigation of buckling induced delamination Growth, *J. Mech. Phys. Solids*, 41: 749-782.

Nilsson, K.-F., and Giannakopoulos, A., 1995, A finite element analysis of configurational stability and finite growth of buckling-driven delamination, *J. Mech. Phys. Solids*, 43: 1983-2021.

Olsson, R., Thesken, J.C., Brandt, F., Jönsson, N., and Nilsson, S., 1996, *Investigations of delamination criticality and the transferability of growth criteria*, FFA-TN 1996-31, The Aeronautical Research Institute of Sweden.

Storåkers, B., and Andersson, B., 1988, Nonlinear plate theory applied to delamination in composites, *J. Mech. Phys. of Solids*, 36: 689-718.

Suo, Z. and Hutchinson, J.W., 1990, Interface crack between two elastic layers. *Int. J. Fracture*, 43: 1-18.

Whitcomb, J.D., 1989, Three-dimensional analysis of a post-buckled embedded delamination, *J. Comp. Mat.*, 23: 862-889.

Yin, W.-L., 1985, Axisymmetric buckling and growth of a circular delamination in a compressed laminate, *Int. J. Solids Structures*, 21: 503-514.

Damage and Failure of Interfaces, Rossmanith (ed.) © 1997 Balkema, Rotterdam, ISBN 90 5410 899 1

A non-osmotic blister growth model in coating systems

T.-J. Chuang
Ceramics Division, National Institute of Standards and Technology, Gaithersburg, Md., USA

T. Nguyen
Building Materials Division, National Institute of Standards and Technology, Gaithersburg, Md., USA

ABSTRACT: A blister growth model is proposed for a coating system consisted of a polymer film applied to a steel substrate exposed to salt solutions. The blister is considered to grow at a constant rate between the coating and the rigid steel substrate. The mechanism of the blister formation is based on corrosion-induced disbondment of the coating at the defect periphery coupled with the stress driven diffusive transport of liquid along the coating/substrate interface at the delamination front. The driving force leading to blister growth is the applied bending moment induced by the in-plane compressive stress of the swelling "buckled" film. By considering the coating as a semi-double cantilever beam loaded by a moment at the periphery, and a distributed load along the beam length due to mass transport, a fifth order ordinary differential equation is derived for the beam "deflection", and the solution is obtained which yields the functional relationship between the blister growth rate and applied bending moment. The predicted blister growth velocity compared favorably with experimental observations on a paint coated steel panel immersed in a 5% salt water solution.

1 INTRODUCTION

The use of organic coatings is the most common and economical means to protect metals against corrosion. Despite great advances in coatings technology in recent years, coatings eventually lose their protective properties wherever corrosion of metals will occur under polymer coatings. One of the most severe degradation modes is the so-called "cathodic blistering" (due to the cathodic half cell reaction of the corrosion process), which occurs when coated panels are exposed to a salted environment. This phenomenon is often observed on metal objects with or without apparent damaged coatings, e.g., a dented car fender or a bridge beam. If the effectiveness of protective coatings is to be increased, it is essential to develop models for quantifying the growth of a cathodic blister. This study presented a model to predict the growth rate of cathodic blisters when coated steel is exposed to salt water.

Cathodic blister initiation is always associated with some form of defects in the coatings, such as artificial scribed mark (Martin et al. 1990), small pores (Funke, 1985), or conductive pathways developed during exposure (Nguyen et al. 1996). Defects allow channels for ions transport from the environment to the metal surface, where corrosion reactions will take place on the metal surface at and around the defects. Corrosion products generated by the cathodic reactions reduce substantially the coating/metal bond strength at these sites (cathodic sites), creating blister nuclei. These sites are formed at or away from the defects depending on the size of the defects. The details of a cathodic blister initiation site and its location are stochastic in nature. It may be an imperfection or weakness in the coating/substrate bonds, a microvoid at the coating/substrate interface, or a defect on the substrate surface. Regardless of the location, an incubation time is always required for the initiation of a cathodic blister on a polymer-coated steel panel. For panels containing no apparent defects or small pores, the time its takes for blisters to occur is much longer for panels contain large defects.

The issue of cathodic blister growth is subjected to debate. There are generally two schools of thought with regards to the driving force. Funke (1985) believed that osmotic pressure is the main mechanism responsible for the growth of a cathodic blister. But in some cases, it is clear (Martin, et al. 1990) that after the blister is formed, a bending moment is set up at the periphery or " tip" of the blister in the film to drive diffusive flux of the cathodic reaction products, leading to the enlargement of the blister. In this case, since the driving force is not from the liquid, rather, it is from the bending moment present inside the coating, it is called non-osmotic disbondment or blister growth.

The present paper adopts the concept of the latter idea wherein it is assumed that a blister is initiated at a cathodic site and reaches a critical nuclei size. It then starts to grow in a steady state at a (unknown, a priori) constant velocity driven by the applied bending moment at the periphery. Under the action of an applied bending moment, the cathodic reaction-containing liquid is driven laterally along the coating/substrate interface to the delamination front from the blister pool, causing the coating to deform. The Fick's law in diffusion links this deflection field with the distributed loads along the beam length. By considering the coating as a semi-double cantilever beam, a differential equation for the unknown deflection can be derived based on the principle of "strength of materials". Details are presented in the following section. The end result is an equation to predict the steady-state blister growth rate as a function of the applied moment, ambient temperature, film thickness, elastic, and diffusive properties. In the discussion section, we test the applicability of the present theory using experimental data given by Martin et al. (1990) in which blister growth rates were measured for an alkyd primer/top coat system applied to a sandblasted steel substrate immersed in a 5% salt solution at room temperature. The agreement between the theory and data shows a promising prospect for the present model.

2 BLISTER INITIATION

Prior to placing into service, a coating/substrate system is usually pre-cured at a higher temperature, e.g., 70°C, thereby introducing a tensile in-plane residual stress in the coating. This stress level can be as high as 15 MPa (Martin et al. 1990). When the coated panel is immersed in a water solution containing salt, water gradually diffuses into the coating causing swelling with a volume expansion. The swelling-induced in-plane stress is so high that can change the residual stress state from 15 MPa in tension into 5 MPa in compression. During exposure, randomly distributed pores are formed in the coatings films, which provide conductive pathways for ions to reach the metal surface. Once ions arrive the metal surface, an electrochemical cell is established where iron is consumed at the anodes and oxygen is reduced to form hydroxyl ions at the cathodes . The generated hydroxyl ions cause a disbondment of the coating from the substrate. When the size of the disbonded area reaches a critical size, buckling of the coating due to the compressive stress occurs, resulting in the formation of the blister.

If w is the buckling distance (i.e., the height of the blister), then a bending moment M_o per unit length of the blister circumference will be introduced at the "tip" of the blister due to the residual compressive stress σ_r:

$$M_o = \sigma_r h w \tag{1}$$

where h is the thickness of the coating, including primer and topcoat. Moreover, if we let σ_f be the cathodic disbondment stress, where σ_f is defined as the tensile strength of the interface below which coating does not separate from the substrate, then at the blister tip where regions of failed and unfailed bonding are separated, the interfacial stress σ_o must always be the disbondment stress σ_f, which is a material constant, namely,

$$\sigma_o = \sigma_{yy}(0) = \sigma_f \tag{2}$$

where subscript o denotes tip location, namely $\sigma_o = \sigma(0)$.

3 BLISTER GROWTH

Once the blister is initiated, it can grow slowly at a steady state velocity V under the action of the applied bending moment M_o located at the periphery of the blister. The present paper aims to predict V as a function of M_o, materials properties, and temperature, so that ultimately the service life of the coating system can be estimated. In order to achieve this goal, a mathematical model is formulated in which Fick's laws and principle of strength of materials are injected to form a well-defined boundary value problem. Those tasks are described in details in the following subsections.

3.1 Liquid transport at interface

Consider mass transport of liquid (electrolyte in this case) along the coating/steel interface at the cathodic delamination front ahead of the blister tip (see Figure 1). When the radius of the blister is large enough, plane-strain conditions in steady state prevail, because the radius of curvature of the periphery in the x-z plane is large in comparison with that of the tip radius in the x-y plane. Accordingly, the problem can be treated as two dimensional on a unit thickness basis in the z-direction. We can then consider the tip as a straight tip

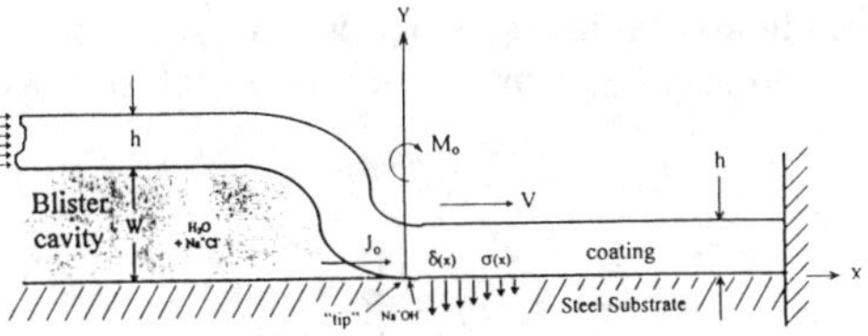

Figure 1. Schematic of the blister growth model

front along the z-axis moving at a velocity V in the x direction with a moving coordinate system with its origin attached to the tip.

The chemical potential at any location in the interface can be expressed as $\mu(x) = -\Omega\,\sigma(x)$ where Ω is the volume of the diffusing species and $\sigma(x) = \sigma_{yy}(x)$ is the normal stresses at location x in the interface. Fick's law then dictates that the diffusive flux be in inverse proportion to the gradient of the chemical potentials. Thus,

$$J(x) = -\frac{D_b\delta_b}{kT\Omega}\nabla\mu(x) = \frac{D_b\delta_b}{kT}\frac{d\sigma}{dx} \qquad \textbf{(3)}$$

where J is the matter flux, $D_b\delta_b$ is the interfacial diffusivity and kT has its usual meaning. At the steady state, the blister volume grows at (w•V) per unit length per unit time. Matter conservation then requires that the matter flux at the tip $J(0)=J_o= wV/\Omega$. Combination with Equation (3) then gives

$$\sigma_o' = \frac{w\ kT\ V}{D_b\delta_b\ \Omega} \qquad \textbf{(4)}$$

where $\sigma' = d\sigma/dx$ is the first derivative of stress, and subscript o denotes tip location at x=0.

3.2 A semi-double-cantilever beam model

Consider a semi-double-cantilever beam located along the positive x-axis with a unit thickness in the z-direction as sketched in Figure 1. The blister is located at $x \le 0$; and the tip at x=0 is moving at a constant speed V in the positive x-direction. The beam is subjected to a bending moment M_o applied at x=0. If we express local deflection of the beam as $\delta(x)$ and local externally applied stress to the beam as $\sigma(x)$, then the principle of the "strength of materials' demands that they must be related by the following equation for the beam:

$$\frac{Eh^3}{12(1-\nu^2)}\frac{d^4\delta}{dx^4} + \sigma(x) = 0 \qquad \textbf{(5)}$$

for $x \ge 0$. Here E is Young's modulus of the coating obtained in the wet (water immersion) state, ν is Poisson's ratio of the coating, and h is the total coating thickness. Meanwhile, at an arbitrary location x, under steady state conditions the local deflection must be related to the flux as $\delta(x) = \Omega J(x)/V$ due to mass conservation requirement. Thus, after combining with equation (3), we have

$$\delta(x) = \frac{D_b\delta_b\Omega}{kT\ V}\frac{d\sigma}{dx} \qquad \textbf{(6)}$$

Upon substituting Eq.(6) into Eq.(5) to eliminate the stress variable, we finally arrive at a fifth order differential equation for the unknown local deflection $\delta(x)$ of the beam:

$$L^5\frac{d^5\delta}{dx^5} + \delta = 0 \qquad \textbf{(7)}$$

where L has a unit in length, and can be expressed by

$$L = \left[\frac{Eh^3 D_b\delta_b\Omega}{12(1-\nu^2)VkT}\right]^{\frac{1}{5}} \qquad \textbf{(8)}$$

The boundary conditions at the blister tip are as follows:

$$-\frac{Eh^3}{12(1-\nu^2)}\delta_o^{(IV)} = \sigma_o \qquad \textbf{(9)}$$

according to Equation (2), and, furthermore,

$$-\frac{Eh^3}{12(1-\nu^2)}\delta_o^{(V)} = \sigma_o' \qquad \textbf{(10)}$$

according to Equation (4). Here the superscripts (IV) and (V) denote fourth and fifth derivatives respectively. Moreover, the shear stress at the tip must vanish. This means that

$$\delta_o^{(III)} = 0 \qquad \textbf{(11)}$$

The differential equation (7), together with the boundary conditions (9-11) forms a well-defined boundary value problem so that a unique solution is guaranteed. This mathematical problem has been solved by Chuang (1975), and the solutions can be expressed as a function of 3 functions superimposed together: exponential; exponential times sines and exponential times cosine functions. Plots of the solutions along the interface indicate that both the stresses and deflections are cyclically varied with decaying amplitudes. Since the bending moment is proportional to the second order derivative of the deflection with respect to x, we finally are able to express M_o at the tip as

$$M_o = L^3\sigma_o' + \frac{1+\sqrt{5}}{2}L^2\sigma_o \approx L^3\sigma_o' + 1.62L^2\sigma_o \qquad (12)$$

Substituting the expressions for L in Eq.(8), σ'_o in Eq. (4) and σ_o in Eq. (2) into Eq. (12), we obtain the relationship between the applied moment at the tip and the steady-state blister growth rate:

$$M_o = A\cdot V^{\frac{2}{5}} + B\cdot V^{-\frac{2}{5}} \qquad (13)$$

where A and B can be expressed as follows in terms of the coating's material properties and temperature:

$$A = \frac{w}{\Omega^{2/5}}\left[\frac{kT}{D_b\delta_b}\right]^{\frac{2}{5}}\left[\frac{Eh^3}{12(1-\nu^2)}\right]^{\frac{3}{5}} \qquad (14)$$

$$B = 1.62\ \sigma_f\left[\frac{Eh^3 D_b\delta_b\Omega}{12(1-\nu^2)kT}\right]^{\frac{2}{5}} \qquad (15)$$

A close examination of Equation (13) indicates that there is a threshold bending moment $(M_o)_{th}$ below which the blister will cease to grow:

$$(M_o)_{th} = 2\sqrt{AB} = 0.73\sqrt{\frac{Ew\sigma_f h^3}{1-\nu^2}} \qquad (16)$$

At this loading level, a minimum blister growth rate is predicted:

$$V_{min} = \left(\frac{B}{A}\right)^{\frac{5}{4}} = 1.83\left(\frac{\sigma_f}{w}\right)^{\frac{5}{4}}\left[\frac{D_b\delta_b\Omega}{kT}\right]\left[\frac{Eh^3}{12(1-\nu^2)}\right]^{-\frac{1}{4}} \qquad (17)$$

below which the blister will not grow. Equation (13) can be written in a non-dimensional form in terms of the normalized moment, m and the blister growth velocity, v:

$$v = \left[m + \sqrt{m^2-1}\right]^{\frac{5}{2}} \qquad (18)$$

for $v \geq 1$ and $m \geq 1$. Here, $v = V/V_{min}$ and $m = M_o/(M_o)_{th}$ are the normalized velocity and moment respectively. Figure 2 plots the predicted blister growth velocity as a function of bending moment applied at the tip. Now, Martin, et al.(1990) has estimated a critical moment for blister initiation which has the following form:

$$M_{ini} = 3\frac{Eh^3 w}{a^2} \qquad (19)$$

where a is the diameter of the blister nuclei at the initiation. It is interesting to compare the bending moment levels of Equations (16) and (19). For instance, if the moment predicted by Equation (16) is less than Equation (19), then the blister will not grow even if it is initiated. On the other hand, if the value calculated by Eq. (16) is larger than by Eq. (19), the blister will grow after its formation.

4 COMPARISON WITH EXPERIMENT

In order to test the validity of the proposed blister growth model, data collected from the experiment performed by Martin, et al. (1990) are used for the prediction. Table 1 lists the relevant material properties

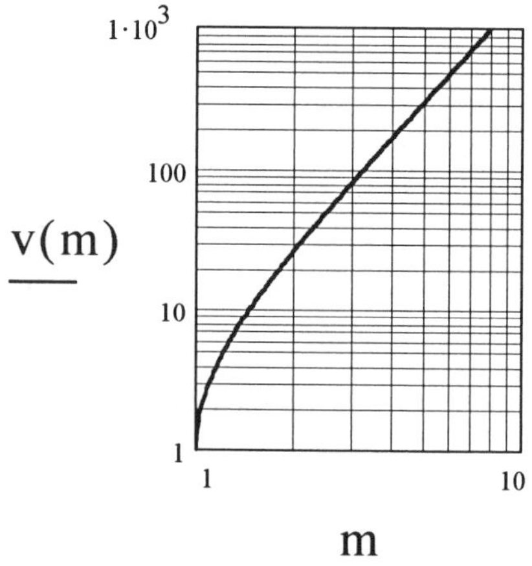

Figure 2. Functional relationship between v and m

Table 1. Values of E, ν, w and σ_r

E(GPa)	ν	h (μm)	w(mm)	σ_r(MPa)
1.0	0.37	145.0	3.0	-5.0

Table 2. Values of $D_b\delta_b$, σ_f, and Ω

$D_b\delta_b(m^3s^{-1})$	σ_f(MPa)	$\Omega(m^3)$	T(K)
$2x10^{-13}$	0.1	10^{-29}	300

Table 3. Applied bending moment and predicted and observed blister growth velocities

M_o(N·m/m)	$V_{predict}$(nm/s)	$V_{measured}$(nm/s)
2.18	1.28	1.86

that are measured, or estimated from data given, by Martin, et al. (1990) and Table 2 provides the other material constants, which were not included in the experiment; they were obtained from open literature. $D_b\delta_b$ was taken from Pommersheim and Nguyen data (Pommersheim & Nguyen, 1995), σ_f was approximated from peel strength results given by Nguyen and Martin (1996) and data on tensile stress and peel strength of polymer/substrate bonds provided by Kinloch (1975), and Ω value was given by King (1970).

Table 3 presents the calculated applied bending moment at the tip (Eq. 12), the predicted blister growth velocity as given by Eq. (18), and the observed velocity given by Martin et al (1990).

It is evident from Table 3 that the velocity predicted by the current theory is in good agreement with the experimental observation. In addition, the calculated threshold applied moment, M_{th} is 0.53 N·m/m, which is about one-quarter of the applied bending moment, and the minimum steady-state blister growth velocity is 0.05 nm/s, which is about thirty times lower than the predicted blister growth rate. Now, the blister nuclei size, a , should be in the same order of magnitude of w, the blister height. If we assume a=5 mm, then the initiation moment required as computed from Eq. (19) is 0.55 N·m/m, which is larger than the threshold level for growth. This means that once the blister is formed, the present theory predicts that the blister will start to grow after initiation.

5 DISCUSSION

Eqs. 13 - 15 indicate that the two materials parameters, σ_f and E, play an important role in the formation and growth of blisters of polymer-coated steel. This section discusses the effects of these two parameters on the growth of a blister formed when a polymer-coated steel panel is exposed to a salt solution. The information should provide a technical base for designing better protective coatings.

5.1 *Significance of σ_f*

For blister formation on a coating system exposed to an aqueous environment, the coating should still adhere to the substrate beyond the periphery of the blister base. On the other hand, in order for a blister to grow, the bonding strength between the coating and the substrate at the tip of the blister must be weaker than the unbroken bonds in the interior. This is supported by the results of Nguyen and Martin (1996), who found that the peel strength of a strong and tough epoxy coating (E=1.7GPa) on a sand-blasted steel substrate decreased from 1.2 kN/m before exposure to a minimum of 0.7 kN/m after 60 days exposure to an alkaline solution at 50 °C. They attributed the loss of adhesion in this region as due to water accumulation at the coating/steel interface. On the other hand, the peel adhesion at the cathodic delamination front was substantially lower, approximately 0.1 kN/m (Alshed et al. 1994). Further, there is no transition between the cathodic-induced delamination front and the water-induced adhesion loss region. Similarly, using a combination of a Kelvin probe apparatus and adhesion test, Stratmann et al. (1989) were able to show that the mechanical adhesion loss at the base of a blister is always behind the cathodic delamination front. Both of these results suggest that the interfacial region beyond the cathodic reaction front is unaffected by the corrosion process, while the interfacial bonding strength at the tip of a cathodic blister is substantially reduced by the corrosion activity.

The above discussion indicates that there is a strong effect of the corrosion process on the disbondment stress at the cathodic blister tip, σ_f. Thus, a knowledge of σ_f above which cathodic blistering does not occur is critical for the development of better protective coatings. Eqs. 13-17 of the model presented here can be used to obtain this value. The relationship in double logarithmic scale between σ_f and V (the blister growth rate) is displayed in Figure 3 for three different types of coatings, ranging from tough (E=1 GPa) to very flexible (E = 0.01 GPa). This figure was obtained using values given in Tables 1 and 2 for the parameters. The results show that, except for the very high values of σ_f, there is a nearly inverse linear relationship between the blistering rate and disbondment stress at the blister tip for a coating applied to a steel substrate subjected to a corrosive environment. For example, for rather strong coatings (E=1 GPa), V decreased from 1.33 nm/s to 0.24 nm/s when σ_f increased from 0.05 MPa to 0.83 MPa. The model also predicts that, for this type of coating, the blister does not grow at all if σ_f is greater than 0.84 MPa. For a flexible coating (E=0.01 GPa), the blister growth rates increased from 245 nm/s to 1378 nm/s when σ_f decreased from 83.0 MPa to a negligible value of 0.05 MPa. Further, for this very flexible coating, a cathodic blister ceases to grow if σ_f is greater than 83 MPa. The results indicate that by measuring the growth rate of cathodic blistering, the critical coating/substrate bonding strength at which a blister does not expand can be estimated.

The facts that cathodic blisters grow rather quickly during exposure to a corrosive environment (Martin et al. 1990) indicated that the bonding strength at the tip of a cathodic blister has been weakened it can no longer resist the applied bending moment stress. Thus, the main question is what are the factors that may affect the bonding strength at the tip of a cathodic blister? To better answer that question, one needs to examine the environment at the tip of a cathodic blister. Evans (1945) classical experiment has shown that when a drop of NaCl solution is placed on a bare steel surface, the center of the drop is the anode while the cathodes are located at the periphery. This happens because of the greater accessability of oxygen at the periphery. The same phenomenon occurs when the same salt solution is placed at the defects through the coating of a polymer-coated steel panel. The anodic process take place at the defect and the cathodic reactions occur on the metal surface underneath the coating, as indicated earlier. At the cathodic sites oxygen is reduced, and, in the presence of cations, such as Na^+ ions, a highly alkaline NaOH solution is formed at the coating/steel (oxide) interface, following the reaction:

$$2H_2O + O_2 + 4Na^+ + 4e^- \rightarrow 4Na^+OH^-$$

In the above reaction, the electrons are supplied by the anodic reactions. It is noted that, in order for the corrosion process to occur there must be a current flow between the anode and the cathode. This means that there must be an electrolyte layer exists at the coating/steel interface within the corrosion cell. In the absence of an external electrical potential applied across the coating, Na^+ ions transport are believed to be along the coating/steel interface. Further, when corrosion occurs, a strong potential gradient exists between the defect and the cathodic sites. This potential gradient accelerates the transport of Na^+ ions

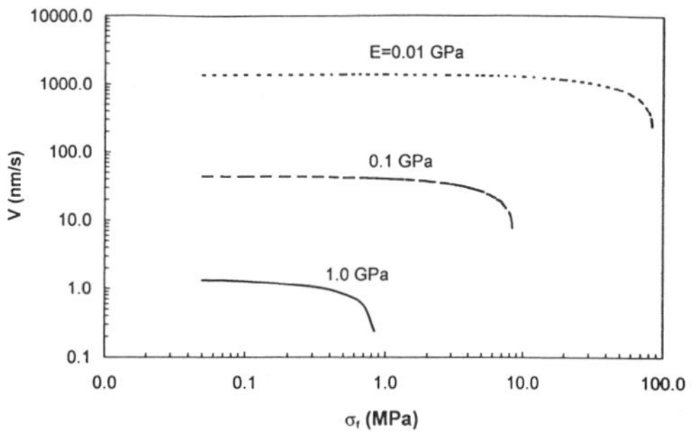

Figure 3. Relationship between σ_f and V for different values of E.

from the defects to the cathodic regions (Starmann, 1994). In addition, Na^+ ions flow also probably produces a potential gradient, similar to that when Cl^- ions permeate to the anodic areas (Sato 1987). Both the corrosion- and Na^+ flow-induced potential gradients should accelerate the lateral transport (electroendosmosis) of water and the corrosion fluid from the blister periphery to the cathodic delamination front. This electroendosmosis transport was demonstrated by Kittleberger and Elm (1945), who showed that 90 % of water uptake into a linseed oil coating was transferred into the coating by the electrical potential gradient. Further, the bending moment stress that causes blistering should also enhance the flow of electrolyte in the blister to the delamination front. All these transport-assisted factors may explain for the high diffusion coefficient value of Na^+ along the coating/steel interface obtained by Pommersheim and Nguyen (1994).

The presence of the alkaline NaOH solution at the cathodic delamination front given in the above reaction suggests that the pH in this region should be very high. Indeed, pH as high as 14 has been measured (Ritter and Kruger, 1983). This high pH at the cathodic sites, which has been proposed as the main cause for the cathodic disbondment of coating/steel systems (Leidheiser,et al. 1983), is probably mostly responsible for the decrease of σ_f at the cathodic blister tip. The degree of adhesion loss depends on a number of variables including surface morphology and treatments, the strength of the molecular interaction at the coating/steel interface, and the pH at the delamination front. The magnitude of the pH generated is, in turn, a function of the rates of diffusion of oxygen, water, and cations to the cathodic sites, the rate of the OH^- ions diffuses away from the cathodic sites, the potential gradient between the cathode and anode, and the volume of the liquid in the regions where OH^- ions are generated.

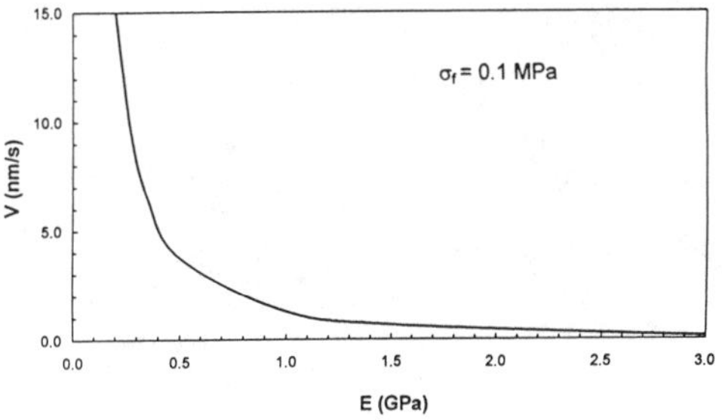

Figure 4. Relationship of V vs. E for $\sigma_f = 0.1$ MPa

Three possible mechanisms have been advocated to explain for the loss of coating adhesion due to the presence of cathodic generated alkaline products: 1) hydrolysis of the coatings (Dickie, 1986), dissolution of the oxide layer (Leidheiser, 1983), and 3) alkaline-induced debonding at the coating/substrate interface (Koeler, 1985). For some coating/steel systems, more than one of these mechanisms is involved, either simultaneously or in stages (Watts, 1989).

5.2 *Significance of E*

The cathodic blistering rate is a function of the coatings Young modulus in the wet state, as given in Eqs. 13-15). The relationship between E and V is presented in Figure 4 for $\sigma_f = 0.1$ MPa. For a coating system having this level of bonding strength at the cathodic blister tip, the blisters grow rapidly from 0.2 nm/s at E = 3 GPa to 15.2 nm/s at E= 0.2 GPa. From this analysis, in order to reduce the cathodic blistering rate, coatings having higher E values should be used. This is consistent with practical observation for the epoxy-coated steel reinforcing bar industry, where stiffer coatings are preferred because they perform better.

6 CONCLUSIONS

Polymer coatings are the most common and economical means to protect metals from corrosion. Despite great advances in coatings technology in the last decade, polymer coatings lose their protective properties and corrosion under coatings do occur. One of the most severe degradation modes of coated steel exposed to salt water is the cathodic blistering. If the effectiveness of protective coatings is to be increased, it is essential to develop models for quantifying the degradation process. This study presented a model to predict the growth rate of blisters formed by the corrosion process. The driving force leading to blister growth is the applied bending moment induced by the in-plane compressive stress of the swelling "buckled" film. The predicted blister growth velocity compared favorably with the experimental observations on a paint coated steel panel immersed in a 5% salt water solution.

REFERENCES

Alsheh, D., Nguyen, T., and Martin, J.W.,1994, Proc. Adhesion Society Meeting, Orlando, Fl, February, 1994, p.209.

Chuang, T.-J., 1975, Models of Creep Crack Growth by Coupled Grain Boundary and Surface diffusion, Ph.D. Thesis, Brown University, Providence, RI.

Dickie, R.A.,1986 in "Polymeric Materials for Corrosion Control," Dickie, R.A. and F.L. Floyd, F.L. (Eds.), *Am. Chem. Soc. Symposium Series 322,* Am. Chem. Soc., Washington, D.C., p.136.

Funke, W.,1985, *Ind. Eng. Chem. Prod. Res. Prod. Dev., 24,* 343.

King, H.W.,1970, in Physical Metallurgy, Ed. R.W Cahn. North Holland, Chapter 2.

Kittelberger, W.W. and Elm, A.C.,1946, *Ind. Eng. Chem. 38,* 695.

Koehler, E.L.,1985, *J. Electrochem. Soc.*, 32, 1005.

Leidheiser, H., Jr., Wang, W. and Igetoft, L., 1983, *Prog. Org. Coat.*, 11, 19.

Martin, J.W., McKnight, M.E., Nguyen, T., and N. Embree, E., 1989, *J. Coatings Technol., 61,* No.772, 39.

Martin, J.W., Embree, E. and Tsao, W., 1990, *J. Coatings Technol., 62,* No.790, 25.

Nguyen, T. and Martin, J. W., 1996, in Durability of Building Materials and Components, Ed. C. S. Jostonrons, EXFN SP on, Vol. P.491.

Pommersheim, J. Nguyen, T., 1995, Cations Diffusion at the Polymer Coating/Metal Interface, *J. Adhesion Sci. and Technol.*, 9, 935.

Ritter, J.J., and Kruger, J., 1981, in "Corrosion Control by Organic Coatings," Leidheiser, H., Jr., (Ed.), Natl. Assoc. Corros. Eng., Houston, TX, p.28

Sato, N.,1987, *Corrosion Sci.*, 27, 421.

Stratmann, M., Feser, R., and Leng, A.,1994, *Electrochimica Acta,* 39, 1207.

Watts, J.F., 1989, *J. Adhesion,* 31, 73.

Damage and Failure of Interfaces, Rossmanith (ed.) © *1997 Balkema, Rotterdam, ISBN 90 5410 899 1*

Crack and debonding at an end of the simple support type constraint for thin plate bending problem

N. Hasebe
Department of Civil Engineering, Nagoya Institute of Technology, Japan

M. Miwa
Technical Research and Development Division, Central Japan Railway Company, Japan

M. Nakashima
Division of Civil Engineering, Obayashi Company, Japan

ABSTRACT: In the case of out of plane loading of a thin plate with a stiffener attached to the boundary , stress concentrations arise at the ends of the stiffener. Therefore, there is a possibility of crack initiation or debonding initiation at the end of the stiffener. If debonding initiates and extends, is there a possibility of crack initiation on the way of debonding? This situation is investigated in this paper. As the stiffener, (as the boundary condition of the thin plate) a simply supported edge is considered.
The analysis is carried out analytically by using rational mapping function and complex variable method. Loading by a uniform bending moment and a concentrated torsional moment is considered. The strain energy release rate criterion is used as the criterion for the direction of crack initiation, crack and debonding initiation.

1 INTRODUCTION

As shown in Fig.1, a thin semi infinite plate with a stiffener is considered. Under some loadings, stress concentrations occur around the ends of the stiffener. Therefore, there is a possibility of crack initiation or debonding initiation at the ends. Debonding is defined as the separation between the plate and the stiffener. It is discussed whether cracking or debonding occurs. Moreover, while the debonding extends, is there a possibility of a crack initiation? These fractures are investigated. The strain energy release rate criterion is used as criterion of crack and debonding initiation. It is also assumed that the strain energy release rate takes a maximum in the direction of crack initiation.
The structural failure as shown in Fig. 1 are often discovered in steel structures, such as bridges and buildings. If the stiffener is allowed to rotate about the x axis, the resistant moment My between the plate and the stiffener may be considered to be zero. The deflection angle along the x axis may also be considered to be zero, since the stiffener is considered to be rigid for the deflection of the thin plate. These conditions for the stiffener correspond to a simply supported edge. The thin plate bending theory is used for the stress analysis. The mixed boundary value problem where the conditions of simply supported edge and free boundary are given is solved. As loading, a uniform bending moment and a concentrated torsional moment are considered (see Fig.1). The case of a rigid stiffener (clamped edge) has been considered by Hasebe et al. (1997).

2 ANALYTICAL METHOD

Classical thin plate bending theory (Savin 1961), complex variable method and a rational mapping function are used for the analysis of the mixed boundary value problem.

Fig.2a shows a half plane with a simply supported edge of length a at a part of the straight boundary on the x axis before crack initiation. Fig.2b shows a half plane with an edge crack of which length and crack angle are b and $\theta\pi(0<\theta<1)$, respectively.
A rational mapping function maps a half plane with an edge crack into the inside of the unit circle on ζ-plane. The rational mapping function was reported in references(Hasebe et al. 1997, Hasebe and Inohara, 1980). The mapping function which maps the semi-infinite region (z-plane) into the inside of a unit circle (ζ -plane) (before a crack initiation as shown in Fig.2a) is given as follows:

$$z = \omega(\zeta) = -a\left(\frac{i}{1-\zeta} + \frac{1-i}{2}\right) \tag{1}$$

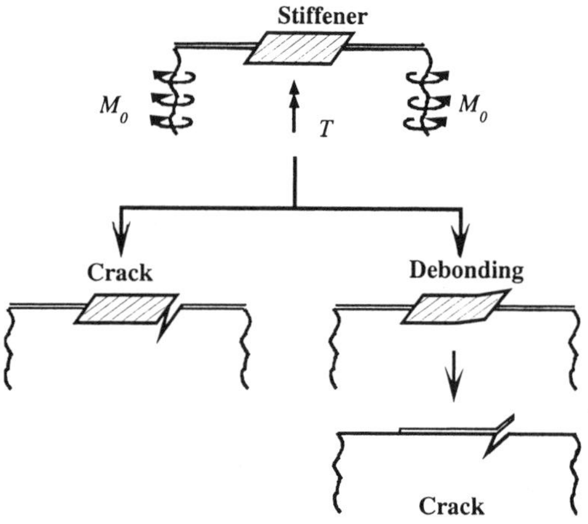

Fig.1 Crack or debonding at the end of stiffener of simple support type

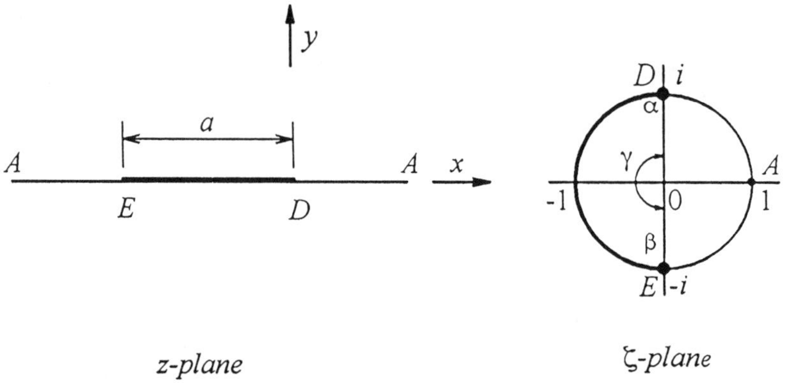

Fig.2a Half-plane before crack initiation

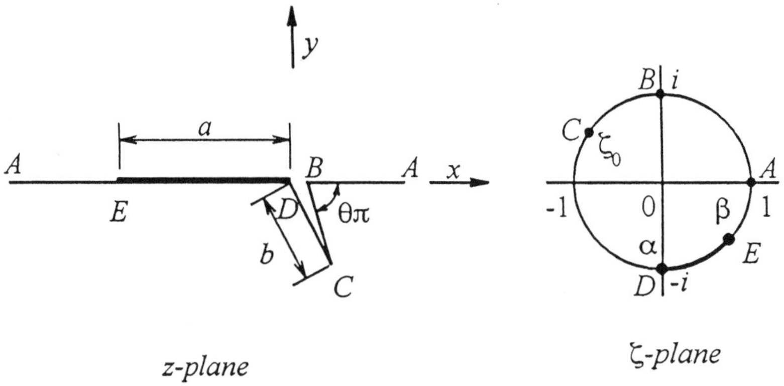

Fig.2b Half-plane after crack initiation

where "a" denotes the length DE for a simply-supported edge. A mapping function which maps the semi-infinite region with a crack length b and at an angle $\theta\pi$ $(0<\theta<1)$ to the x-axis, into the inside of a unit circle as shown in Fig.2b, follows from the use of the Schwarz-Christoffel transformation formula (Hasebe and Inohara 1980):

$$z = \omega(\zeta) = k\frac{(1+i\zeta)^{\theta}(1-i\zeta)^{1-\theta}}{1-\zeta}$$

$$\fallingdotseq \frac{E_0}{1-\zeta} + \sum_{k=1}^{n}\frac{E_k}{\zeta_k-\zeta} + E_c \quad (2)$$

Here k is a constant which defines the length b of the crack, and is given by, $k = b(1-i)i^{-\theta}/\left[2\theta^{\theta}(1-\theta)^{1-\theta}\right]$.

In addition, E_0, E_k, E_c, and ζ_k are complex constants, and n is the total numbers of fractional-expression terms which in this paper is n =24. Also, ζ_0 on the unit circle corresponds to the crack tip C and is given by $\zeta_0 = (1-2\theta+i)/(1-2\theta-i)$. The crack tip C of the rational mapping function (2) is not a strictly sharp corner, but has a very small roundness. Ratio ρ/b of radius ρ of the curvature at the crack tip to the crack length b, is a minimum at θ =0.5, and increases gradually as θ approaches to 0 or 1, but the ratio is very small $\rho/b = 10^{-7} \sim 10^{-12}$, so that the crack tip is quite sharp. This study treats the problem with a condition of $b/a << 1$, which means that the length a of the boundary DE is taken larger enough than the crack length b. Then, the accuracy of the stress intensity factor calculated from this mapping function still increases. While, the lengh a does not explicitly appear in the mapping function (2), it shows up as a parameter in the complex stress functions. The mapping function (1) which maps the uncracked shape corresponds to the special case of $E_k = 0(k=1,2,\cdots,n)$ in (2). The stress function of the thin plate problem using the rational mapping function was reported in Hasebe et al.(1993,1994). Consider a half plane with a simply-supported edge DE of a part of the straight boundary (Figs.2a,b) subjected to a uniform bending moment M_0 acting at infinity of the x-axis and a concentrated torsional moment T acting at infinity of the y-axis. In this case, it may be assumed that T acts on the simply-supported edge. Unlike in the case of a clamped edge (Hasebe et al. 1997), the concentrated torsional moment T is considered as loading instead of a uniform torsional moment at infinity of the x-axis. Since in the case of a uncracked simply-supported edge there is no singular term in the stress function for the uniform torsional moment, no stress concentration is produced at the simply-supported ends. After all, there is no possibility of fracture initiation as a result of the application of a uniform torsional moment. The complex stress function corresponding to a uniform bending moment M_0 for the cracked configuration (Fig.2b), follows from the use of the mapping function (2) as follows (Hasebe et al. 1993):

$$\phi(\zeta) = H(\zeta) + \frac{1}{2\kappa}\sum_{k=1}^{n}\left\{\frac{\bar{A}_k B_k}{\zeta_k-\zeta}\left[1+\frac{\chi(\zeta)}{\chi(\zeta_k)}\right] + \frac{A_k\bar{B}_k\zeta_k'^2}{\zeta_k'-\zeta}\left[1-\frac{\chi(\zeta)}{\chi(\zeta_k')}\right]\right\} + const;$$

$$H(\zeta) = -\frac{M_0}{4D(1+\nu)}\left\{\frac{E_0}{1-\zeta}\frac{\chi(\zeta)}{\chi(1)} - \frac{1}{2\kappa}\sum_{k=1}^{n}\frac{E_k}{\zeta_k-\zeta}\left[1+\frac{\chi(\zeta)}{\chi(\zeta_k)}\right] - \frac{1}{2\kappa}\sum_{k=1}^{n}\frac{\bar{E}_k\zeta_k'^2}{\zeta_k'-\zeta}\left[1-\frac{\chi(\zeta)}{\chi(\zeta_k')}\right]\right\} \quad (3)$$

In the case of a concentrated torsional moment T, $\phi(\zeta)$ is given by (Hasebe et al. 1994):

$$\phi(\zeta) = H(\zeta) + \frac{1}{2\kappa}\sum_{k=1}^{n}\left\{\frac{\bar{A}_k B_k}{\zeta_k-\zeta}\left[1+\frac{\chi(\zeta)}{\chi(\zeta_k)}\right] + \frac{A_k\bar{B}_k\zeta_k'^2}{\zeta_k'-\zeta}\times\left[1-\frac{\chi(\zeta)}{\chi(\zeta_k')}\right]\right\} + const;$$

$$H(\zeta) = \frac{T}{2\pi\kappa D(1-\nu)}\log\frac{-2\chi(1)\chi(\zeta)+(2-\alpha-\beta)\zeta+2\alpha\beta-\alpha-\beta}{1-\zeta} \quad (4)$$

where Poisson's ratio ν, flexural rigidity of the plate D, Plemelj function $\chi(\zeta)$, and the other constants are given as follows:

$$\chi(\zeta) = (\zeta-\alpha)^{0.5}(\zeta-\beta)^{0.5}; \kappa = \frac{3+\nu}{1-\nu}; A_k = \phi'(\zeta_k');$$

$$B_k \equiv \frac{E_k}{\omega'(\zeta_k')}; \zeta_k' \equiv \frac{1}{\zeta_k}. \quad (5)$$

α and β show the coordinates on the unit circle corresponding to simple support D and E, respectively (Fig.2). Complex constant A_k is determined by solving a system of $2n$(=48) linear simultaneous equations with regard to the real and imaginary parts of equations $\phi'(\zeta_k') = A_k(k = 1,2,\cdots,n)$.

Using $E_0 = -ia, E_k = B_k = 0\ (k = 1,2,\cdots,n)$ for (3) and (4), $\phi(\zeta)$ is obtained for the case of acting M_0 and T, respectively,

$$\phi(\zeta) = \frac{iaM_0\chi(\zeta)}{4D(1+\nu)\chi(1)(1-\zeta)} + \text{const.} \quad (6)$$

$$\phi(\zeta) = \frac{T}{2\pi\kappa D(1-\nu)}\times \log\frac{-2\chi(1)\chi(\zeta)+(2-\alpha-\beta)\zeta+2\alpha\beta-\alpha-\beta}{1-\zeta} + \text{const.} \quad (7)$$

The stress components are expressed by the first derivative of $\phi(\zeta)$ of (6) and (7), and they show a singularity and become infinite at $\zeta = \alpha$ and β by

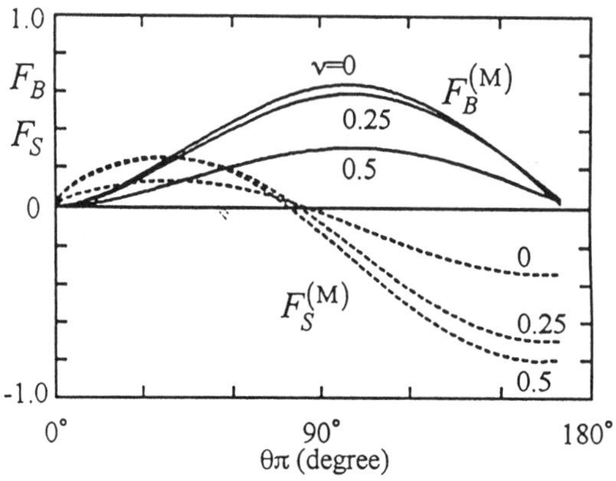

Fig.3a Nondimensional stress intensity factors under uniform bending moment for b/a=0

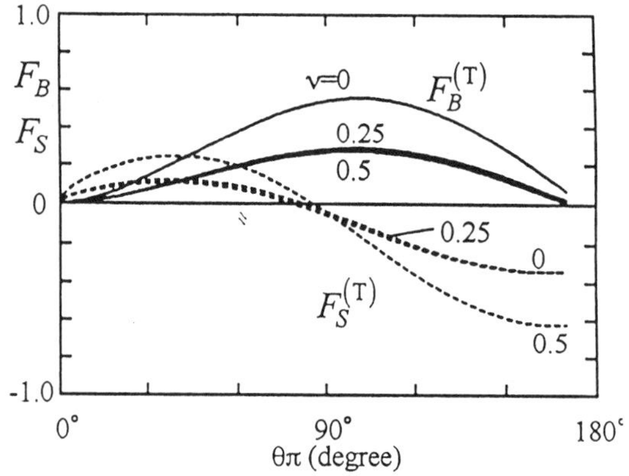

Fig.3b Nondimensional stress intensity factors under concentrated torsional moment for b/a=0

$\chi'(\zeta)$ in $\phi'(\zeta)$, i.e. at the simply-supported ends D and E. Therefore, there is a possibility for crack initiation into the plate from the simply-supported end, or to produce a debonding along the simply-supported edge. By using this stress function, the stress intensity factor and the stress intensity of debonding are calculated and then strain energy release rates are obtained.

STARAIN ENERGY RELEASE RATE

The strain energy release rate for debonding is obtained from the stress intensity of debonding $\tilde{\alpha}_0$ (Hasebe et al. 1997, Hasebe & Salama, 1994) which is given by the following expression

$$\tilde{\alpha}_0 = 2\sqrt{2}D(1+\nu)\exp(-\pi\delta)\frac{|\omega'(\alpha)(\alpha-\beta)|^{\lambda}g(\alpha)}{\omega'(\alpha)(\alpha-\beta)}\times$$

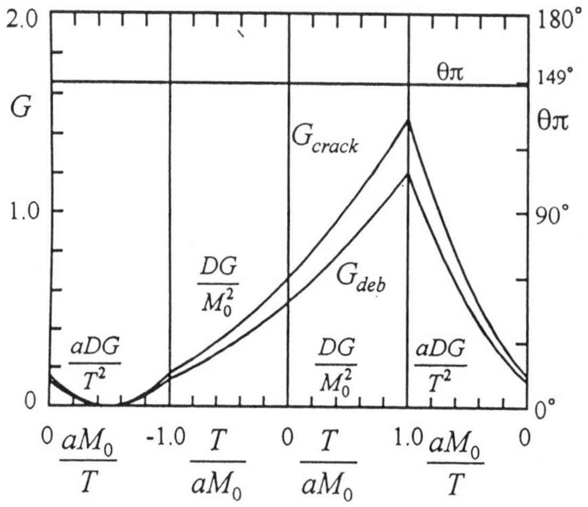

Fig.4 Nondimensional strain energy release rates *Gcrack* and *Gdeb*, and crack angle $\theta\pi$ for ν =0.25

$$\exp\left\{i\lambda\left(\theta_A+\pi+\frac{\gamma}{2}\right)\right\} \tag{8}$$

where $g(\alpha)=\lim_{\zeta\to\alpha}\left[\phi'(\zeta)(\zeta-\alpha)^{\lambda}(\zeta-\beta)^{1-\lambda}\right]$, and θ_A is an angle of the debonded section to the x-axis, and $\theta_A=\pi$ in this case, and γ is $\gamma=\angle\alpha 0\beta$ of the central angle between α and β on a unit circle (see Fig.2a) (Hasebe et al. 1988). The stress intensity of debonding $\tilde{\alpha}_0$ at point D is obtained from (8) with $\lambda=0.5$ (δ=0), and g(α) for M_0 is expressed by the use of (6) in the following form:

$$g(\alpha)=\frac{iaM_0}{8D(1+\nu)\chi(1)}\cdot\frac{\alpha-\beta}{1-\alpha} \tag{9}$$

In the case of T, using (7) follows:

$$g(\alpha)=\frac{-T\chi(1)}{2\pi\kappa D(1-\nu)}\cdot\frac{1}{1-\alpha} \tag{10}$$

For the combined action of M_0 and T, $\tilde{\alpha}_0$ is expressed by the following equation:

$$\tilde{\alpha}_0=\frac{-i\sqrt{a}M_0}{2\sqrt{2}}-\frac{\sqrt{2}i(1+\nu)}{\pi(3+\nu)}\cdot\frac{T}{\sqrt{a}} \tag{11}$$

and the strain energy release rate is given by

$$G_{deb}=\frac{\pi\kappa}{2D(1+\nu)^2}\tilde{\alpha}_0\bar{\tilde{\alpha}}_0=\frac{\pi\kappa a}{16D(1+\nu)^2}\left\{M_0+\frac{4(1+\nu)T}{\pi(3+\nu)a}\right\}^2 \tag{12}$$

Stress intensity factors are calculated for any crack length from the stress function and their definition are shown in reference(Hasebe and Inohara, 1980). Since the strain energy release rate just after crack initiation is needed, stress intensity factors for crack length $b=0$ are obtained by extrapolating the limiting value of b = 0 (Hasebe et al.,1993). The

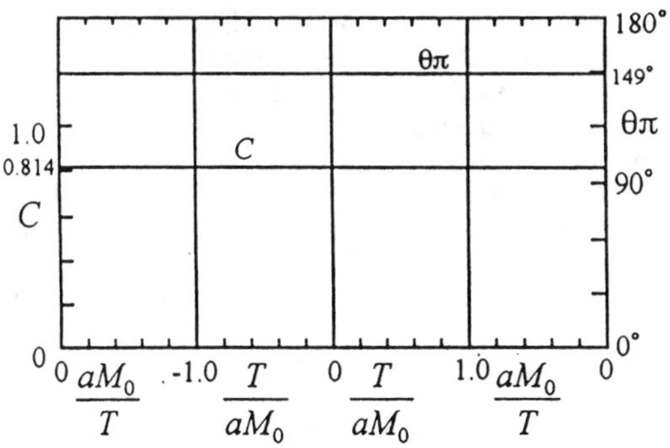

Fig.5 Ratio C and crack angle $\theta\pi$ for ν =0.25

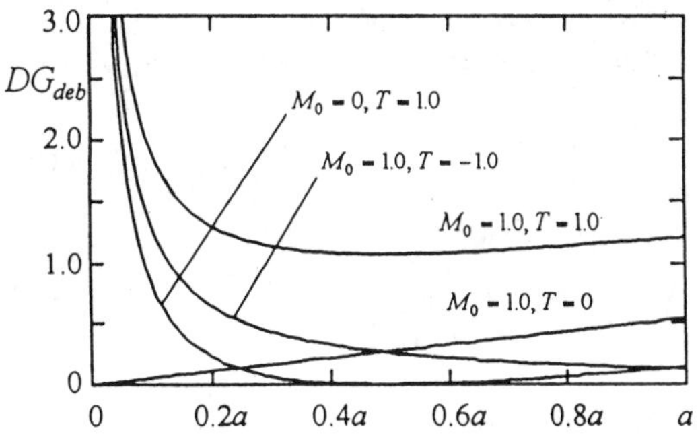

Fig.6 Gdeb during the extension of debonding

limiting value of b = 0 (Hasebe et al.,1993). The following non-dimensional stress intensity factors are shown in Figs.3a and 3b, for a uniform bending moment M_0 and a concentrated torsional moment T,

$$F_B^{(M)} + iF_S^{(M)} = \frac{3+v}{1+v}\cdot\frac{k_B^{(M)} + ik_S^{(M)}}{M_0\sqrt{a}}; \qquad (13a)$$

and

$$F_B^{(T)} + iF_S^{(T)} = \frac{3+v}{1+v}\cdot\frac{\sqrt{a}}{T}\left(k_B^{(T)} + ik_S^{(T)}\right) \qquad (13b)$$

respectively, where the suffix(M) and (T) show cases of loading M_0 and T, respectively. k_B and k_S are the stress intensity factors of the opening and tearing modes, respectively.

The strain energy release rate G_{crack} is obtained from the following equation (see Appendix):

$$G_{crack} = \frac{\pi\kappa}{D(1+\nu)^2}\left|k_B + ik_S\right|^2 = \frac{\pi\kappa a}{D(3+\nu)^2}\times$$

$$\left\{\left(M_0F_B^{(M)} + \frac{T}{a}F_B^{(T)}\right)^2 + \left(M_0F_S^{(M)} + \frac{T}{a}F_S^{(T)}\right)^2\right\} \qquad (14)$$

Values of G_{crack} can be calculated by substituting values F_B and F_S into (14) and changing angle $\theta\pi$ of the crack initiation. Then the angle $\theta\pi$ which yields the maximum value of G_{crack} must be looked for. Figure 4 shows the nondimensional maximum strain energy release rates of crack and debonding which definitions are given in the figure (for example, $aDG/T^2, DG/M_0^2$ etc.) for Poisson's ratio ν=0.25. The horizontal axis shows the ratio of loading such that any magnitude of loading from 0 to $\pm\infty$ can be studied. Also, the initiation angle of the crack is shown in Fig.4 . It is always a constant value 149 ° for any loading. G_{crack} and G_{deb} near $aM_0/T = -0.49$ is zero since singular-terms in the stress functions for M_0 and T cancel each other.

CRITERION OF ENERGY RELEASE RATE

The condition of initiation of debonding and cracking are investigated. Fracture toughness values of debonding and cracking are expressed by $(G_{deb})_{CR}$ and $(G_{crack})_{CR}$, respectively.

The following four cases of failure phenomena produced by the relative magnitude between G_{deb} and $(G_{deb})_{CR}$, and between G_{crack} and $(G_{crack})_{CR}$ are considered:

(i) In the case of $G_{deb} < (G_{deb})_{CR}$ and $G_{crack} < (G_{crack})_{CR}$, neither debond nor crack occurs.

(ii) In the case of $G_{deb} > (G_{deb})_{CR}$ and $G_{crack} < (G_{crack})_{CR}$, a debond initiates, but a crack does not initiate.

(iii) In the case of $G_{deb} < (G_{deb})_{CR}$ and $G_{crack} > (G_{crack})_{CR}$, debond does not initiate, but a crack initiates.

(iv) In the case of $G_{deb} > (G_{deb})_{CR}$ and $G_{crack} > (G_{crack})_{CR}$, there is a possibility of both initiation of debon and cracking. However, in practice, it is observed that either a debond or a crack initiate and the condition must be determined. Therefore, comparing the relative magnitudes by the ratio $C = G_{deb} / G_{crack}$ of strain energy release rates, and by ratio $C_0=(G_{deb})_{CR} / (G_{crack})_{CR}$ of each fracture toughness value, it can be determined which phenomenon will occur. When $C > C_0$, i.e. $G_{deb} /(G_{deb})_{CR} > G_{crack} / (G_{crack})_{CR}$, a debond initiates. When $C < C_0$, i.e. $G_{deb} / (G_{deb})_{CR} < G_{crack} / (G_{crack})_{CR}$, a crack initiates. When $C = C_0$, there is the possibility for simultaneous initiation of a debond as well as a crack. However other conditions may be needed. Case (iv) will be explained in detail. Using (12) and (14), the ratio C is expressed as follows:

$$C = \frac{G_{deb}}{G_{crack}} = \frac{(3+v)^2}{16(1+v)^2}\cdot\frac{\left\{M_0 + \frac{4(1+v)}{\pi(3+v)}\frac{T}{a}\right\}^2}{\left(M_0F_B^{(M)} + \frac{T}{a}F_B^{(T)}\right)^2 + \left(M_0F_S^{(M)} + \frac{T}{a}F_S^{(T)}\right)^2} \qquad (15)$$

If it is found that the stress intensity factors calculated for any value of $\theta\pi$ have the following relationship:

$$\frac{4(1+\nu)}{\pi(3+\nu)} = \frac{F_B^{(T)}}{F_B^{(M)}} = \frac{F_S^{(T)}}{F_S^{(M)}} \tag{16}$$

Equation (15) is expressed as follows:

$$C = \frac{(3+\nu)^2}{16(1+\nu)^2\left(F_B^{(M)^2} + F_S^{(M)^2}\right)} \tag{17}$$

The value of C of (17) is shown in Fig.5 for $\nu = 0.25$. The value C is constant value 0.814 and shown by a straight line which is not dependent on loading.

Next we investigate whether the debonding will arrest during the extension of debonding. Figure 6 shows G_{deb} for $\nu = 0.25$ for some typical load configuration, when the length of a simple support edge changes, (i.e. for various extensions of a debond). In the diagram the horizontal axis is the bond length and D in the vertical axis is the flexural rigidity of the thin plate. It can be distinguished in Fig.6 whether the debonding comes to arrest depends on the loading conditions. For example, when M_0=1.0, and T =0, the value of G_{deb} decreases linearly, thus the debonding stops when $G_{deb} < (G_{deb})_{CR}$ is satisfied. When M_0=1.0 and T=1.0 or M_0=0 and T=1.0, G_{deb} has the minimum value. Therefore when $G_{deb} < (G_{deb})_{CR}$ is satisfied at one point during the extension, debonding stops. When M_0=1.0 and T = − 1.0, G_{deb} increases monotonously. Then the debonding does not stop and reaches to fracture.

The possibility for crack initiation during the extension of the debonding is investigated.

The cases where debonding precedes are cases (ii) and (iv).

In case (ii), the condition of (ii) during the extension is always satisfied, since $G_{crack} = G_{deb} / C$ and C is constant for any loadings. Therefore when the value of G_{crack} increases for some loadings (for example, M_0=1.0 and T=− 1.0), $G_{crack} > (G_{crack})_{CR}$ is satisfied for a certain bond length. Then the situation corresponds to case(iv) as $G_{deb} > (G_{deb})_{CR}$ is also satisfied. However, $C > C_0$ is also satisfied, and a crack does not initiate during the extension of debonding.

CONCLUSION

Loading by a uniform bending moment and a concentrated torsional moment was considered. The uniform torsional moment was not considered as it dose not cause stress concentration around the ends of the stiffener of simple support type. The stress intensity of debonding and the strain energy release rate before crack initiation (see Fig. 2) are expressed by exact expressions (11) and (12), respectively. Also the latter is shown in Fig.4. To consider any magnitude of the loading, the loading ratio was taken and shown on the horigontal axis. The stress intensity factor for crack length $b = 0$ is shown in Fig.3 and the strain energy release rate is shown in Fig.4. A maximum value is attained for the initiation angle of crack 149 ° (case of Poisson's ratio 0.25) which is always constant for any magnitude of uniform bending moment and concentrated torsional moment.

The strain energy release rate criterion was used as the criterion of initiation of cracking and debonding. Comparing the strain energy release rate with the fracture toughness value, it was investigated whether a crack or a debonding will initiate. In order to avoid both debonding and cracking, $(G_{deb})_{CR}$ and $(G_{crack})_{CR}$ of the material and the bond must be chosen larger than G_{deb} and G_{crack} shown in Fig.4 , respectively. When there is a possibility to initiate a debonding and a crack simultaneously, the fracture phenomenon can be investigated by considering the magnitude between the ratio of strain energy release rates and the ratio of fracture toughness values. After a debond has initiated, possible arrest depends on the conditions of the loading, but there is no possibility for a crack to initiate during the extension of the debonding.

As a criterion of initiation of a crack and a debonding, the strain energy release rate criterion was used. The criterion depends on the material and the bond condition. Therefore, in the case where the strain energy release rate criterion can not be used, another criterion must be used. However the investigation as well as this study can be performed.

APPENDIX

Considering work for both sides of a crack surface, which become two times that of the debonding case, the strain energy released ΔU is expressed as follows:

$$\Delta U = \int_{Z_a}^{Z_a-\Delta a} \mathrm{Im}\left[\left(m_y + i\int p_y dx\right)\left(\frac{\partial w}{\partial x} - i\frac{\partial w}{\partial x}\right)\right]dx \tag{A1}$$

Taking the stress intensity factors before and after the extension of the crack as K and K*, respectively, the first derivative of $\Phi(z)$ after the extension can be approximated by the following expressions:

$$\Phi'(z) = \frac{-\overline{K}}{2\sqrt{2}D(1+\nu)}(z - z_a)^{-1/2} \tag{A2}$$

$$\Phi^*(z) = \frac{-\overline{K}^*}{\sqrt{2D}(1+\nu)}(z - z_a + \Delta a)^{1/2} \tag{A3}$$

$$m_y + i\int P_y dx = -\kappa D(1-\nu)\left[\Phi'^+(x) - \Phi'^-(x)\right] \tag{A4}$$

$$\frac{\partial w}{\partial x} - i\frac{\partial w}{\partial y} = \overline{\Phi^{*+}(x)} + \kappa\overline{\Phi^{*-}(x)} \tag{A5}$$

These are substituted into (A1) and integrated. Then $G_{crack} = \lim_{\Delta a \to 0}\left[-\frac{\Delta U}{\Delta a}\right]$ is calculated.

As $K^* \to K$ at $\Delta a \to 0$, the strain energy release rate is expressed by (14).

REFERENCES

Hasebe, N., & S. Inohara, 1980. Stress Analysis of a Semi-Infinite Plate with an Oblique Edge Crack. Ingenieur Archiv, 49: 51-62.

Hasebe, N., S. Tsutui & T. Nakamura, 1988. Debondings at a Semielliptic Rigid Inclusion on the Rim of a Half plane. J. Appl. Mech., 55, 574-579.

Hasebe, N.,M. Miwa, & T. Nakamura, 1993. Second Mixed-Boundary-Value Problem for Thin-Plate Bending. J.Engrg.Mech., ASCE, 119: 211-224.

Hasebe, N., T. Nakamura, & Y. Ito, 1994. Analysis of the Second Mixed Boundary Value Problem for a Thin Plate. J.Appl. Mech., 61: 555-559.

Hasebe, N., & M. Salama, 1994. Thin Plate Bending Problem of Partially Bonded Bimaterial Strips. Arch.Appl.Mech., 64: 423-434.

Hasebe, N., M. Miwa, & M. Nakasima 1997. Branching Problem of a Crack and a Debonding at the End of a Clamped Edge of Thin Plate. Proceedings of Ninth International Conference on Fracture, Sydney : 2219-2226, Pergamon.

Savin, G. N.,1961. Stress Concentration around Hole. Pergamon Press, Oxford England.

Thermal problems

Damage and Failure of Interfaces, Rossmanith (ed.) © *1997 Balkema, Rotterdam, ISBN 90 5410 899 1*

Modelling of temperature dependent failure process in fibre reinforced titanium alloy

H. Assler, M. Gräber & P.W.M. Peters
German Aerospace Research Establishment (DLR), Cologne, Germany

ABSTRACT: Titanium matrix composite (TMC) is a candidate material for application in gas turbines. For this reason it is necessary to know which parameters influence its high temperature strength. In the present investigation modelling of the high temperature strength (from room temperature up to 1000°C) of SCS-6-fibre reinforced IMI834 is performed and compared with experimental results. Modelling is done in two steps. Firstly, three dimensional finite element (FE) calculations are performed to determine the stress distribution around a broken fibre. TMC is a composite with a low fibre/matrix bond strength, so that stress transfer at the fibre/matrix interface mainly takes place as a result of frictional shear stresses. For this reason in the FE calculation two non-linear effects reducing the stress concentration in fibres neighbouring a broken fibre are considered: interfacial sliding and matrix ductility. The stress distribution determined this way is used in a Monte Carlo simulation of the global composite behaviour based on a Weibull strength distribution determined on single fibres. In this simulation influence functions are incorporated to describe the interfering stress fields of fibre breaks. The main reason for a reduction of the tensile strength at increasing test temperature is found to be the strength loss of the matrix. A smaller contribution to the strength reduction results from the reduced stress transfer (leading to a larger ineffective fibre length) at increased test temperature.

1 INTRODUCTION

The today's development of aircraft turbine engines aims at e.g. a reduction of pollutant emission and an increase of efficiency resulting in a reduction of energy consumption. A possibility to reach this aim is to make use of advanced materials. Most important features of materials in use for aircraft turbine components are high temperature stability, oxidation resistance and small specific weight.

Monolithic titanium alloy is an adequate material for the compressor stages of aero-engine gas turbines, although the stiffness and strength improvements realised by fibre reinforcement could significantly extend the range of application. Therefore metal matrix composites, especially SiC-fibre reinforced titanium alloys, are under development to fulfil the above-mentioned requirements.

1.1 *Titanium matrix composites (TMCs)*

Axi-symmetric gas turbine components, such as discs, casings and rings, are most suitable for reinforcement since filament wound embedded continuous fibres can withstand large hoop stresses. If such simple cylindrical arrangements are used to support blading, instead of the bulky disc shapes currently used to resist the centrifugal loading, a significant weight saving could be achieved for such rotating components (Ward-Close & Loader 1995).

In Figure 1 the cross Section of an SiC(SCS-6)-fibre reinforced titanium (Ti-IMI834) alloy with almost ideal hexagonal fibre arrangement is shown.

The SCS-6 fibre, produced by Textron Specialty Materials, has a carbon core and is coated with a thin layer of pure pyrocarbon followed by a SiC/C mixture (in the following designated as carbon layer). The surface coating has two purposes, (a) to essentially heal the irregular surface of the crystalline beta silicon carbide and hence increase its strength and (b) to act as a sacrificial layer to alleviate stress concentration effects resulting from fibre/matrix chemical reaction that occur during high temperature consolidation and application.

A prospective category of titanium alloys for high temperature applications are the near-alpha titanium alloys. IMI834, used in this investigation, is a near-alpha titanium alloy of medium strength and

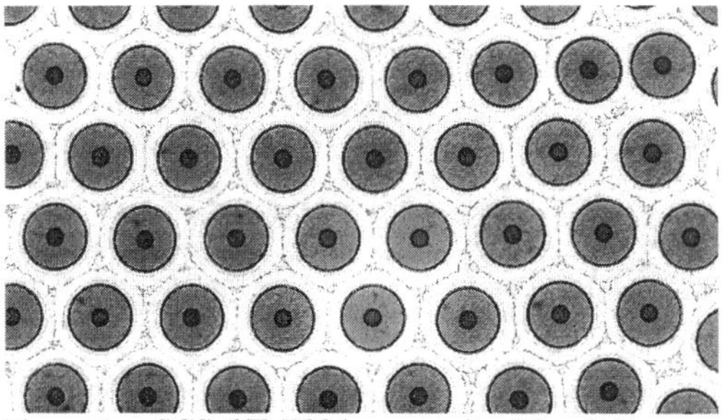

Figure 1. SCS-6/IMI834 composite ($V_f \approx 0.5$)

temperature capability up to about 600°C combined with good fatigue resistance.

1.2 *Processing of TMCs*

The German Aerospace Research Establishment (DLR) developed a special method of processing, which can be divided into two steps. At first SiC-fibres are coated with titanium matrix (e.g. Ti-IMI834, Ti-6Al-4V) by magnetron sputtering, i.e. a single fibre composite is produced. After that bundles of coated fibres are encapsulated in a container of the desired component shape and hot isostatic pressed (HIPed) to obtain the consolidated component. Most important advantages of this type of processing are a variable fibre volume fraction V_f between 0.2 and 0.6, a nearly ideal fibre distribution and a matrix with a small grain size (Leucht & Dudek 1994). A schematic description of the different processing steps with the respective parameters is shown in Figure 2 (Assler & Peters 1997).

1.3 *Aim of this investigation*

A comprehensive investigation of the mechanical properties of unidirectionally SiC-fibre reinforced titanium alloys includes the analysis of the influence of test temperature on the ultimate tensile strength.

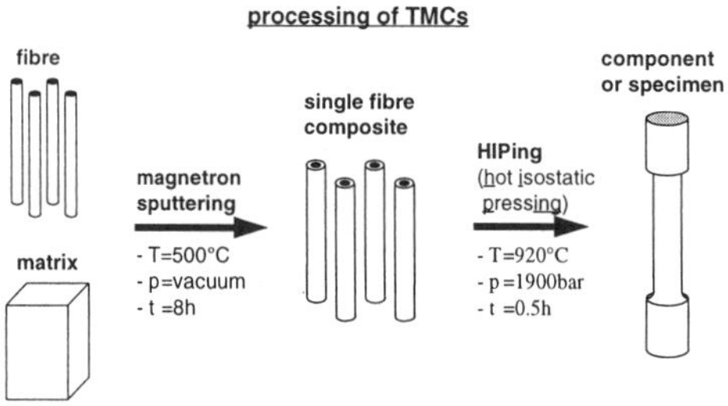

Figure 2. Processing of TMCs

Several methods can be applied to describe this temperature dependence. On one hand simple models (e.g. rule of mixtures) are used to describe the global behaviour of the composite, neglecting local effects like failure of single fibres. More sophisticated models require a more profound knowledge of the parameters, which contribute to the strength. With increasing complexity the results depend on many parameters and the models are suitable to perform studies on the influence of these parameters. An efficient and also simple model is proposed by Curtin (1993).

The model, used in the present work, takes into account:

1. the variability of the fibre strength described by the Weibull distribution.
2. the titanium matrix deforms elastic-plastically (non-linear).
3. the shear stress transfer at the fibre/matrix interface only as a result of frictional interfacial sliding (Coulomb friction). This can be accepted as the bond (shear) strength is low in comparison to frictional shear stresses. Thus in the present investigation the fibre/matrix bond strength is considered to be zero.
4. the thermal residual stresses as a result of cooling down from the production temperature and the different thermal expansion coefficients of fibre and matrix.

In this work only a part of the performed parameter studies are presented, namely that part, which deals with load transfer in the interface.

2 DEVELOPMENT OF THE TENSILE STRENGTH MODEL

A realistic model, which describes the tensile behaviour in fibre direction of unidirectional TMCs, should consider the variability of the fibre strength. This variability gives rise to randomly distributed fibre failures. At increasing load the number of randomly distributed fibre failures increases, whereas close to broken fibres (due to a stress concentration) fibres also can break, leading to a critical damage stage. At this critical damage stage ultimate failure occurs. To simplify the procedure of modelling a subdivision into two models is reasonable.

In the first step (local model: Finite Element Model) the stress distribution in the fibres in the neighbourhood of a broken fibre (local effect) is investigated.

In the second step (global model: Monte Carlo simulation) the global mechanical behaviour of the composite, predicting the increasing number of fibre failures at increasing load, is calculated with the aid of the results of the first step. Stress fields of broken fibres can interfere. A procedure to account for this interference is developed and incorporated in the simulation (Gräber 1997).

2.1 *Local model*

The stress distribution in the neighbourhood of a broken fibre is determined with the aid of the Finite Element Method (FEM). Different models with a fibre volume fraction of V_f=0.3, 0.4 and 0.5 are generated using MSC/PATRAN. Due to symmetry only a sector of 30° out of a composite with hexagonally distributed fibres has to be modelled (see Figure 3).

Close to the broken central fibre the micro-structure is resolved in detail (fibre, matrix, interface), whereas further away from the broken fibre an average material (composite material) with smeared-out properties is considered (Dinter & Peters 1996).

As mentioned the fibre/matrix bond strength is zero, so that shear stress transfer at the fibre/matrix interface can only be realized with the aid of frictional shear stresses. Contact elements (gap elements), obeying Coulomb friction law, are used to model this shear stress transfer (Figure 3: interface 1 and 2). The shear force F_{xz} in these gap-elements measures $F_{xz} \leq \mu \cdot |F_{xx}|$, where F_{xx} is the element radial force at the fibre/matrix interface and μ is the coefficient of friction (μ=0, 0.25, 0.35). Due to the large difference between the thermal expansion coefficients of fibre and matrix ($\alpha_m > \alpha_f$) and the high consolidation temperature (≈930°C) the radial interfacial stresses are high. For a sufficiently large coefficient of friction the element shear force thus can be considerable. Due to the occurrence of interfacial sliding and plastic deformation of the matrix the stress concentration in fibres neighbouring broken fibres is limited. For a better understanding of the mechanical behaviour also the case of a perfect interface bonding (approached by $\mu \rightarrow \infty$) is investigated.

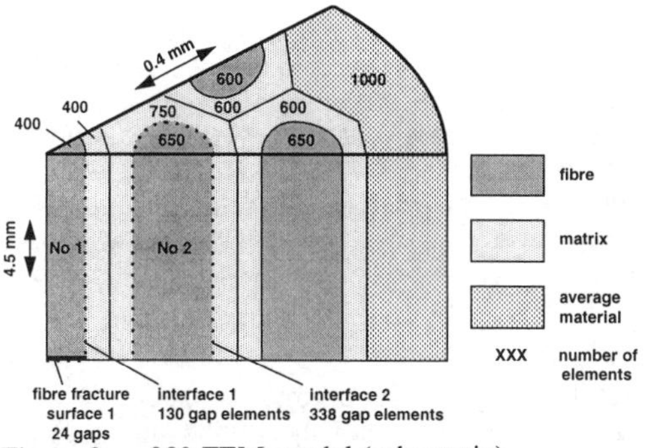

Figure 3. 30°-FEM-model (schematic)

The required stress/strain analyses are carried out making use of MSC/NASTRAN. In eight steps the FE-model is exposed to different levels of loads. In the first load step the thermal residual stresses are introduced by cooling the model from processing temperature (930°C) down to test temperature T of 0, 150, 300, 450 or 600°C. To prevent sliding effects during this step gap-elements (Figure 3: surface 1) are applied to model a closed fibre crack (compressive stress in fibre). During load steps 2 up to 8 a continuous increasing axial displacement (simulation of an axial tensile test), giving rise to an axial strain ε_{ax} of 0.3, 0.5, 0.7, 1.1, 1.3 and 1.5% respectively, is applied. The corresponding calculations are performed for a fractured as well as for an intact central fibre.

In consideration of the test temperature different coefficients of thermal expansion α and the stress/strain relation are chosen for fibre, matrix and average material. The matrix and the average material are assumed to behave elastic-plastically, which is described by a hyperbolic approximation, whereas the fibre behaves linear-elastically. The elastic-plastic behaviour of the unreinforced matrix was determined experimentally at the respective temperatures. The respective poisson ratio ν is assumed to be independent of test temperature T. To estimate the properties of the average material, which should represent a composite, a cylinder model (Hecker et al. 1969) is used instead of the rule of mixtures.

To study the influence of fibre volume fraction V_f several models are produced with the aid of which the influence of other parameters μ: coefficient of friction for sliding in the interface, and test temperature T is analysed.

The following relevant results are determined:

- fibre stresses in case of an intact central fibre $\sigma_f^i = \sigma_f^i(V_f, T, \mu, \varepsilon_{ax})$ and
- fibre stresses in case of a fractured central fibre $\sigma_f^f = \sigma_f^f(V_f, T, \mu, \varepsilon_{ax}, \text{coord}(x, y, z))$.

2.2 *Global model*

The global mechanical behaviour of the composite is predicted with the aid of a Monte Carlo simulation on the basis of the model schematically presented in Figure 4. A hexagonal array of i·j fibres is divided into k segments with each fibre segment surrounded by a matrix segment. It has been shown, that an array of N=10·10·50=5000 fibre/matrix segments, used in the present analysis, is large enough to predict the behaviour of a composite specimen. The length of every segment measures 0.2mm.

In the present analysis the variability of fibre strength is described with the aid of a two parameter Weibull distribution. The Weibull parameters m

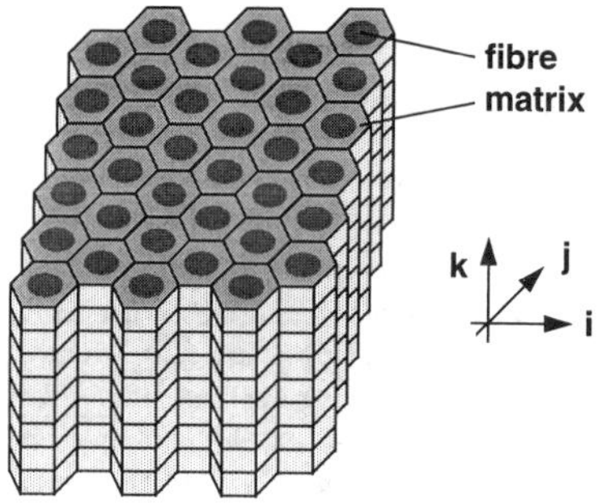

Figure 4. 6·6·8 simulation model

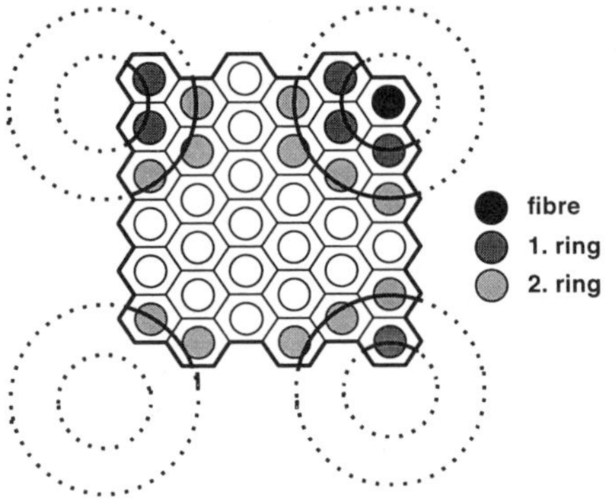

Figure 5. Elimination of free surface

(Weibull modulus) and σ_0 (scale parameter) can be determined evaluating tensile tests of single fibres with a gauge length of l_f. For the numerical generation of N failure stress values $\sigma_{f,n}^{max}$ for the N segments, the Weibull function is inverted and the probability of failure is replaced by N random numbers R_n (n=1..N) between 0.0 and 1.0. This leads to

$$\sigma_{f,n}^{\max} = \sigma_0 \sqrt[m]{\frac{l_f}{l_s} \cdot \ln\left(\frac{1}{1-R_n}\right)} \tag{1}$$

where l_s is the length of a fibre segment and the random number R_n is approached by

$$R_n = \frac{2n-1}{2N} \tag{2}$$

in which n is the segment number in increasing order of strength. The resulting failure stress values $\sigma_{f,n}^{max}$ are then distributed among the fibre segments using (pseudo-) random numbers ($\sigma_{f,n}^{max} \rightarrow \sigma_{\alpha,\beta,\gamma}^{max}$ with α=1 to i, β=1 to j and γ=1 to k).

After assigning the failure stresses to the different fibre segments and choosing the parameters V_f, T, μ, ε_{ax} the simulation of the tensile test starts.
At the beginning of the simulated tensile test all fibre segments are assumed to be intact. Their status changes from „intact" to „fractured" if the equation

$$\sigma_{f,\alpha,\beta,\gamma} \geq \sigma_{f,\alpha,\beta,\gamma}^{\max} \tag{3}$$

becomes true. The following two steps are repeated, until Equation 3 is not fulfiled for any fibre segment.

1. STEP: Making use of the calculated stress distribution near broken fibres, determined with the aid of the local model, stresses are assigned to the fibre segments of the global model. Due to the difference in element length (local model) and segment length (global model) an interpolation of the calculated stresses is necessary. Further an algorithm is developed to suppress a surface effect, which is large for small models. This is realised by assigning fibres of one surface to be neighbours of fibres of the opposite surface as made clear in Figure 5.

Initially the fibre segment failures are probably so wide apart that the stress disturbance fields of the separate segment failures are independent. At increasing number of segment failures the separate stress fields will more and more interfere.
For this reason a function is required that approximates the superposition of the effects of a number of fractured segments on the stresses of those segments that remain intact. This function is defined in terms of two dimensionless stress coefficients that describe the fibre stress changes in a segment due to neighbouring failed segments compared to the axial fibre stress in a composite without failed fibres.

The first coefficient characterizes the influence of a fibre break in segment (δ,η,φ) on the stress in a neighbouring intact fibre segment (α,β,γ). This relation is given by

$$s_{\alpha,\beta,\gamma}^{\delta,\eta,\varphi} = \frac{\sigma_{f,\alpha,\beta,\gamma}^{\delta,\eta,\varphi}}{\sigma_f^i} \tag{4}$$

where σ_f^i is the fibre stress in case of an intact central fibre. The required values are calculated by means of the local model (σ_f^i, $\sigma_f^f \rightarrow \sigma_{\alpha,\beta,\gamma}^{\delta,\eta,\varphi}$). The stress $\sigma_{\alpha,\beta,\gamma}^{\delta,\eta,\varphi}$ in segment (α,β,γ) is caused by a fracture in segment (δ,η,φ). If segment (α,β,γ) is situated in the same fibre like segment (δ,η,φ) ($\alpha=\delta \wedge \beta=\eta$), $s_{\alpha,\beta,\gamma}^{\delta,\eta,\varphi}$ is less than or equal to 1.0 (stress reduction). On the other hand $s_{\alpha,\beta,\gamma}^{\delta,\eta,\varphi}$ is greater than or equal to 1.0 (stress concentration) in case of $\alpha\neq\delta$ or $\beta\neq\eta$.

The second stress coefficient $S_{\alpha,\beta,\gamma}$ is a factor to calculate the unknown stress in segment (α,β,γ) with respect to all significant neighbouring fractured fibre segments with

$$\sigma_{f,\alpha,\beta,\gamma} = S_{\alpha,\beta,\gamma} \cdot \sigma_f^i \,. \tag{5}$$

Three different cases can be distinguished for the value of $S_{\alpha,\beta,\gamma}$:
(a) all fibre segments are intact. This case is the most simple one because only the results for an intact central fibre, which are calculated using the local

model, are required to predict the homogeneous fibre stresses in the simulation model ($S_{\alpha,\beta,\gamma}$=1.0).

(b) few fibre segments break without mutual influence due to a large separation between the broken segments. In this case also the local results for a fractured central fibre are applied. Every single intact fibre segment is influenced by either no or only one segment fracture in its neighbourhood ($S_{\alpha,\beta,\gamma}=s_{\alpha,\beta,\gamma}^{\delta,\eta,\varphi}$).

(c) fibre segments break with mutual stress field influence. With increasing axial strain a distribution of fractured and intact segments in the simulation model is formed. If the stress state of one fibre segment is influenced by at least two different segment breaks a method has to be developed to superpose several overlapping stress distributions. The relation

$$S_{\alpha,\beta,\gamma} = S_{\alpha,\beta,\gamma}^{\alpha=\delta\wedge\beta=\gamma,\varphi} \cdot S_{\alpha,\beta,\gamma}^{\alpha\neq\delta\vee\beta\neq\gamma,\varphi} \quad (6)$$

seems to fulfil the above mentioned requirement. The first factor

$$S_{\alpha,\beta,\gamma}^{\alpha=\delta\wedge\beta=\gamma,\varphi} = \prod_i s_{\alpha,\beta,\gamma}^{\alpha=\delta_i\wedge\beta=\gamma_i,\varphi_i} \quad (7)$$

describes the stress reduction due to i segment breaks in the same fibre as segment (α,β,γ), while the second one

$$S_{\alpha,\beta,\gamma}^{\alpha\neq\delta\vee\beta\neq\gamma,\varphi} = 1 + \sum_j \left(s_{\alpha,\beta,\gamma}^{\alpha\neq\delta_j\vee\beta\neq\gamma_j,\varphi_j} - 1\right) \quad (8)$$

takes the stress increase due to j segment failures in the neighbouring fibres into consideration. The equations mentioned above represent a comprehensive description of the stress state in every fibre segment.

2. STEP: To determine whether or not a segment breaks, the axial fibre stress $\sigma_{f,\alpha,\beta,\gamma}$ of each segment is compared with its corresponding failure stress $\sigma_{f,\alpha,\beta,\gamma}^{max}$. Thus, Equation 3 is evaluated for every fibre segment. If necessary, the status of the fibre segment changes from intact to fractured and repetition of step 1 becomes necessary to correct the stress distribution in the simulation model.

After for each displacement step saturation of fibre failure is reached (Equation 3 becomes false for every fibre segment) the corresponding stresses in the matrix segments are calculated to determine the total stress carried by the composite.

Axial stresses of the matrix segments $\sigma_{m,\alpha,\beta,\gamma}$ are determined, depending on whether the corresponding fibre segment is broken or not. In case a fibre segment is broken, the axial matrix stress is assumed to be equal to the maximum stress of the elastic-plastic matrix material that depends on test temperature. In case a fibre segment is intact, the axial matrix stress is calculated from corresponding axial fibre stress, using a cylinder model. This cylinder model that bases on the same parameters (V_f, T) like the FEM-calculation assumes a linear-elastic fibre and an elastic-plastic matrix under axial load (constant strain state).

The addition of the axial fibre and matrix stresses in a segment plane (γ) according to the rule of mixtures finally yields the composite stress σ_c that corresponds to the applied global composite strain ε_{ax}, of which the lowest value is selected:

$$\sigma_c = \min\left(\sigma_c^{(\gamma)}\right) \quad \text{where}$$

$$\sigma_c^{(\gamma)} = \sum_\delta \sum_\eta \left(V_f \cdot \sigma_{f,\alpha,\beta,\gamma} + \left(1 - V_f\right) \cdot \sigma_{m,\alpha,\beta,\gamma}\right) \quad (9)$$

Equation 9 delivers the composite stress for any axial strain so that the stress strain curve can be determined. The strength follows from the maximum of the stress-strain curve where $d\sigma_c/d\varepsilon_{ax} \leq 0$. In this way the strength can be determined for different values of the parameters V_f, μ, T, m and σ_0.

3 RESULTS

3.1 *Model parameters*

The local FE-model is used to determine the input data for the global model. The number of finite elements of every component and most important dimensions of the FE-model are shown in Figure 3. To improve the accuracy of the calculation the element height at the plane of the fibre fracture (base) are ten times smaller compared to corresponding elements at the opposite plane (top) of the model.

The global model is used to simulate the mechanical behaviour of a composite with a fibre volume fraction of V_f=0.3. Further studies on the influence of fibre volume fractions of V_f=0.4 and 0.5 are not yet completed. As described in Section 2.2 the failure stresses of the fibre segments are assigned to specific values with the aid of random numbers. The strength values are based on the Weibull distribution with m=15 and σ_0=4500MPa, which was determined by tensile testing of fibres (radius R_f=71µm) with a gauge length of l_f=25mm. Ten simulations are performed to calculate the tensile strength and averaged to minimize the influence of randomness on the results.

Table 1 Material properties

	fibre	matrix	T [°C]	σ_m^{max} [N/mm^2]
E	400	115	0	1013
[GPa]			150	850
$\alpha_{0-930°}$	6.5	11.3	300	750
[1E-6/°C]			450	690
ν	0.25	0.30	600	608
			750	480

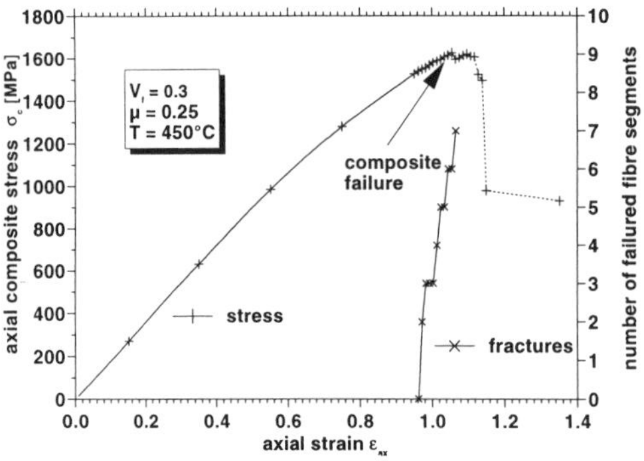

Figure 6. Stress/strain curve

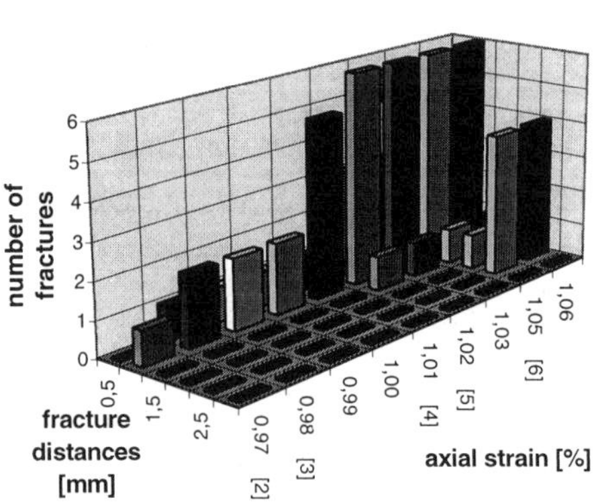

Figure 7. Distribution of fracture distances

Table 1 indicates an extract of the most important material properties, used in this investigation, including the temperature dependent maximum stresses carried by the matrix σ_m^{max}.

3.2 *Parameter studies*

The results of a simulation (V_f=0.3, μ=0.25 and T=450°C) are shown in Figure 6. The axial composite stress (Equation 9) and the number of broken fibre segments, until the maximum stress (=strength, defined by $d\sigma_c/d\varepsilon_{ax} \leq 0$) is reached, are plotted against the axial strain.

For a better insight of the failure process it is useful to investigate the distribution of the distances between the fractured fibre segments (e.g. 4 segment fractures → 3+2+1=6 fracture distances). The distance d_i between broken fibre segments is divided into six different categories with increasing length in steps of 0.5mm. With respect to the fibre volume fraction (V_f=0.3) the minimum distance d_{min} between two fractures can be calculated using the relationship:

$$d_{\min} = R_f \sqrt{\frac{2 \cdot \pi}{\sqrt{3} \cdot V_f}} \approx 0.25\text{mm}\,. \tag{10}$$

The maximum distance between two fractured fibre segments measures due to the elimination of the model free surfaces 5.15mm. For the performed simulation mentioned above (Figure 6) the frequency of occurrence of the different fracture distances are indicated in Figure 7 with respect to the axial strain. The total number of broken fibre segments is also added in brackets. Figure 7 shows that the simulation produces multiple segment failures, which however are not anymore completely randomly distributed. This random distribution of fibre segment failure occurs if the coefficient of friction is set to be zero.

In Figure 8 the calculated average failure stresses of the SCS-6/IMI834 composite for a fibre volume fraction of V_f=0.3 are plotted against the test temperature T.

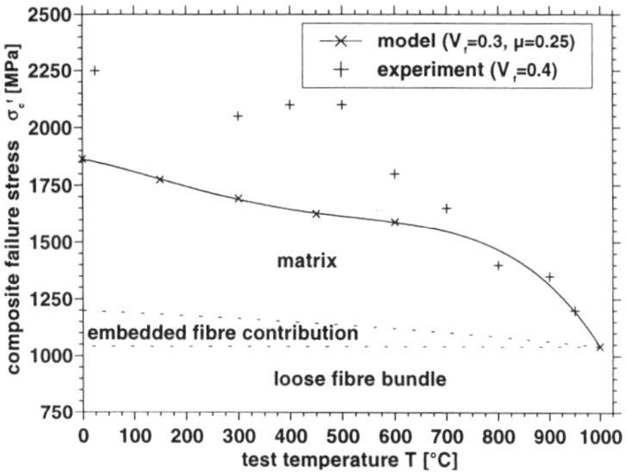

Figure 8. Temperature dependent failure stress

The experimental results, also indicated in Figure 8, were determined on specimens (see Figure 9) with a higher fibre volume fraction ($V_f \approx$0.4, gauge length 30mm, diameter 3.5mm).

At a test temperature of 1000°C the calculation of the failure stress σ_c^f is performed with the aid of the theory of loose fibre bundles (Equation 11) because the tensile strength of the matrix tends towards zero and can be neglected.

$$\sigma_c^f = V_f \cdot \left(\frac{l \cdot m \cdot e}{l_0 \cdot \sigma_0^m} \right)^{-\frac{1}{m}} \tag{11}$$

Simulations are also performed for a (typical compressor operation) temperature of 450°C and different coefficients of friction μ (0, 0.25, 0.35, ∞). The corresponding failure stresses and the number of broken fibres at complete failure n^f are indicated in Table 2. In case there is no stress transfer in the interface (μ=0) the number of broken fibres equals to the value of a loose fibre bundle (n^f=1-exp(-1/m)). In addition to that the ineffective length l_{ineff} (Rosen 1964) in interface 1 (see Figure 3) is also indicated in the same table. This value defines half the length

Table 2. Properties of TMC (V_f=0.3, T=450°C)

μ	σ_c^f [MPa]	n^f[%] (σ_c^f)	l_{ineff} [mm] (ϕ=0.8, ε_{ax}=1.1%)
0.00	1508	6.45	∞
0.25	1626	6-7	3.64
0.35	1650	6-7	3.31
∞	(>1725)	(>15)	1.12

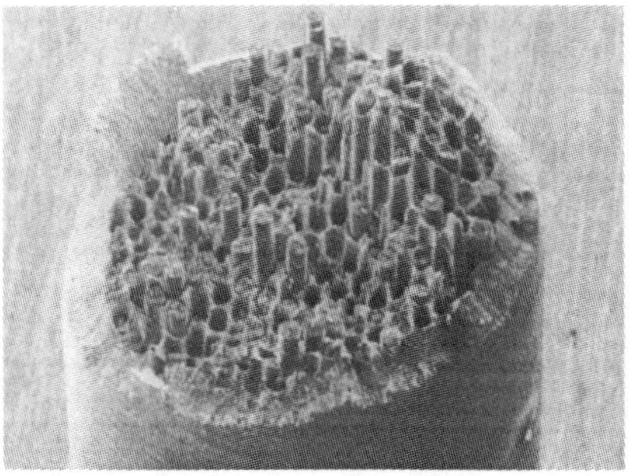

Figure 9. Fracture surface of cylindrical specimen

of a broken fibre, over which the axial fibre stress is less than $\phi \cdot \sigma_f$ (0.8 times the undisturbed fibre stress σ_f).

At increasing values of the coefficient of friction the load carried by the fibres increases. The difference between the two extreme cases (coefficient of friction μ=0 and $\mu=\infty$, approaching the case of perfect bonding) leads to a strength increase of at least 14%. In case of perfect bonding more than 15 fibre segments fail and a significant mutual influence of the stress fields occurs. Due to the fact, that Equation 8 presents only an approximation to determine interfering stress fields, these values are put in brackets for $\mu \rightarrow \infty$.
In the interface of the neighbouring fibres (Figure 3: interface 2) no sliding occurs.

4 DISCUSSION

As described above Figure 6 shows the stress/strain curve as a result of one simulation. At an axial strain of about 0.6% the matrix starts to behave plastically causing a decrease of the modulus. The first two fibre breaks occur at a strain level of $\varepsilon_{ax} \approx 0.97\%$. With increasing number of fibre breaks the modulus continues to decrease. The composite is assumed to fail if the maximum of axial stress is reached ($\sigma_{ax} \approx 1630$MPa, $\varepsilon_{ax} \approx 1.07\%$). At that stress the seventh fibre segment fails and the decrease of composite stress due to fibre failure is larger than the stress increase caused by a higher axial strain.

In real experiments due to dynamic effects during redistribution of stresses the composite fails abruptly. For that reason the second stress maximum ($\sigma_{ax} \approx 1620$MPa, $\varepsilon_{ax} \approx 1.1\%$) is unimportant and does not represent the true mechanical behaviour of the composite. Further the simulation does not consider energy balance so that in contradiction to a real experiment, during which instable fracture occurs, the stress-strain curve can go through a maximum as shown in Figure 6. At a strain level greater than 1.15% most of the fibres are broken and the axial stress of the composite is only born by the matrix.

The frequency of occurrence of different fracture distances are indicated in Figure 7. As expected, the number of broken fibre segments grows with increasing axial strain. With respect to Equation 10 Figure 7 also makes clear that several segment breaks occur, which are not directly neighbouring each other. This fact indicates that a limited number of fibre segments break with mutual stress field influence. For that reason is seems to be allowed to use a simple function that approximates the superposition of the effects of a number of fractured segments on the stresses of those segments that remain intact (see Equation 8). A more thorough evaluation of the superpositioning of stress fields is however necessary.

In Figure 8 the failure stresses of a composite are plotted as a function of the test temperature. It is clear there is a significant influence of temperature on the strength (T=600°C leads to a 15% reduction of composite strength). This is mainly caused by the temperature dependent (a) matrix properties and (b) conditions of load transfer in the fibre/matrix interface. The effect of reduced stress transfer is visible from the difference between loose fibre bundle strength and embedded fibre bundle strength as shown in Figure 8.

The matrix shows a degressive decrease of tensile strength up to 600°C. At higher temperatures (T>600°C) the matrix tensile strength and creep resistance drop progressively.

The strength drop due to the reduced stress transfer in the fibre/matrix interface bases on the residual stresses. With increasing test temperature residual stresses (especially radial stresses at the fibre/matrix interface) caused by the different coefficients of thermal expansion of fibre and matrix decrease. According to Coulomb friction law the shear stresses at the interface also decreases with increasing test temperature. Therefore the ineffective length l_{ineff} increases with increasing test temperature, which leads to a reduction of composite tensile strength.

At the highest test temperature (≈1000°C) the complete elimination of shear stresses at the fibre/matrix interface and of the matrix tensile strength allow to approximate the tensile strength of the composite with the aid of Equation 11 (loose fibre bundle).

Model results based on a fibre volume fraction of $V_f=0.3$ are compared with experimental results ($V_f \approx 0.4$) shown in Figure 8. In spite of the large scatter of the experimental results, which could be caused by the variation of fibre spacing, the model results present the same tendency, especially at temperatures less than 600°C.

Fracture of a fibre introduces a stress concentration in the neighbouring fibres. TMCs possess two possibilities to reduce this stress concentration, i.e. (a) sliding in the interface of the broken fibre and (b) plastic deformation of the matrix. It is shown, that the strength of the composite increases if, by increasing the coefficient of friction, sliding is reduced (Table 2). Simultaneously this leads to an increase of the stress concentration, which however is limited due to the second stress reducing capability, the matrix ductility. Certainly, it can be expected, that the stress reducing capability of the matrix depends on the fibre spacing or matrix volume fraction. Thus the stress born by the fibres can lead to another dependency on the coefficient of friction than given in Table 2 for higher fibre volume fractions with the possibility of a maximum at an intermediate coefficient of friction.

5 CONCLUSIONS

A model is developed, which describes the temperature dependence of the tensile strength of TMCs taking into account: variability of fibre strength, elastic-plastic deformation of titanium alloy, shear stress transfer at the fibre/matrix interface due to frictional sliding and existence of thermally induced residual stresses.

Modelling is performed on the basis of a finite element calculation for the determination of the stress field around a broken fibre. Influence functions are developed to describe the stress field of mutual influencing and multiple fibre breaks. A Monte Carlo simulation is performed to predict the strength of TMCs on the basis of an experimentally determined fibre strength making use of the finite element results for the stress distributions around broken fibres. Parameter studies include a variation of fibre volume fraction and coefficient of friction.

The predicted strength decrease at increasing test temperature shows the same tendency as experimental results for a different fibre volume fraction. The strength decrease mainly is caused by a limited strength reduction of the matrix up to 600°C, whereas above 600°C a severe drop in matrix strength limits the TMC strength.

Further the reduced stress transfer at the fibre/matrix interface at increasing test temperature contributes to the loss of strength. This is caused by the reduction of the residual thermal radial stress at the fibre/matrix interface, which reduces frictional shear stresses and thus increases the ineffective length.

It can also be shown, that similar to a loose fibre bundle, only a few fibres are broken at the strain for ultimate composite failure. For loose fibre bundles this number only depends on the Weibull modulus m ($m>10 \rightarrow$ number of breaks $n_f<10\%$).

Finally the results also indicate, that in reality ($\mu<0.4$) only a few broken fibres with mutual stress field influence exist. Therefore the applied procedure of modelling including the developed procedure to determine the stress concentration around a cluster of several broken fibres is acceptable.

More insight in the failure process of SiC/Ti-composites can be expected after completion of the parameter study on the influence of fibre volume fraction.

6 REFERENCES

Ward-Close, C.M. & C. Loader 1995. Fibre Reinforced Composites. In F.H. Froes & J. Storer (eds), *Recent Advances in Titanium Metal Matrix Composites*: 19-32.

Leucht, R. & H.J. Dudek 1994. Properties of SiC-fibre reinforced titanium alloys processed by fibre coating and hot isostatic pressing. *Mat. Sci. Engn.* A188: 201-210.

Assler, H. & P.W.M. Peters 1997. The Influence of Heat Treatment on the Tensile Strength of Single Fibre Composites (SiC Fibre/Titanium Alloy). *Key Engineering Materials* Vols. 127-131: 1241-1250.

Curtin, W.A. 1993. Ultimate strength of fibre-reinforced ceramics and metals. *Composites* Vol. 24, No. 2: 98-102.

Dinter, J. & P.W.M. Peters 1996. Finite element modelling of the push-out test for SiC fibre-reinforced titanium alloys. *Composites Part A* 27A: 749-753.

Gräber, M. 1997. Entwicklung eines Modells zur Vorhersage der Warmfestigkeit von unidirektional faserverstärkten Titanmatrix-Verbundwerkstoffen *diploma thesis, University of Braunschweig.*

Hecker, S.S., C.H. Hamilton & L.J. Ebert 1969. Elastic-plastic Analysis: A Simplified Approach. *Scripta Metallurgica* Vol. 3: 793-798.

Rosen, B.W. 1964. Tensile Failure of Fibrous Composites. *AIAA Journal* Vol. 2, No. 11: 1983-1991.

Damage and Failure of Interfaces, Rossmanith (ed.) © *1997 Balkema, Rotterdam, ISBN 90 5410 899 1*

Intergranular fracture at low temperatures at polycrystalline materials

G.E.Smith & A.G.Crocker
Department of Physics, University of Surrey, Guildford, UK

P.E.J.Flewitt & R.Moskovic
Magnox Electric, Berkeley Centre, Gloucestershire, UK

ABSTRACT: The proportion of cleavage and intergranular fracture which occurs when high purity α-iron and ferritic steels are tested at temperatures below the ductile-brittle transition has been investigated using both theoretical models and high resolution scanning electron microscopy. Some intergranular fracture is necessary because cleavage cracks in adjacent grains do not meet in a line in their common grain boundary. If it is assumed that cleavage occurs on the three variants of the {100} plane, several different models indicate that the proportion of intergranular fracture should be about 30%. Experiments show that in pure α-iron the proportion may approach 20% but in a C-Mn steel it less than 10%. This discrepancy can be explained if it is assumed that, at least in some grains, cleavage fracture occurs on both {100} and {110} planes. There is some experimental evidence for this suggestion and it is also supported indirectly by preliminary experiments on polycrystalline zinc.

1 INTRODUCTION

At low temperatures materials fracture in a brittle manner. In particular, single crystals fail by cleavage on well-defined crystallographic planes. Similarly, in polycrystalline materials, individual grains usually fail by this mechanism. However, active cleavage planes in adjacent grains do not meet each other in a line in their common grain boundary. Therefore some additional fracture, for example using the alternative brittle failure mode of intergranular fracture, must accompany cleavage. Interpretation of fractographs of materials which have failed at low temperatures has in the past suggested that in practice only a very small amount of intergranular fracture occurs, typically a few per cent. Two-dimensional theoretical models of the fracture process are not inconsistent with this suggestion (Abbott et al. 1994). However, a preliminary study using three-dimensional modelling techniques (Crocker et al. 1996) has indicated that, if intergranular failure is the sole additional mechanism, a much larger proportion is a geometrical necessity. Thus, for example, as much as 30% intergranular fracture is then predicted to occur in polycrystalline alpha-iron and ferritic steels which cleave on the three variants of the {100} family of planes. This result was compared with new experimental measurements on α-iron, tested at temperatures below the ductile-brittle transition, which showed that the proportion of intergranular failure detected at higher magnifications may approach about 20%, (Crocker et al 1996). However, many simplifying assumptions and approximations were made in this earlier work. The present paper describes alternative developments of the theoretical modelling work and additional experimental results on C-Mn steels. It confirms that interfacial failure plays a significant role in low temperature fracture of polycrystalline materials and suggests additional accommodation mechanisms which may be considered in the theoretical models.

2 THE GEOMETRICAL MODEL

In the theoretical model of a polycrystal which has been adopted, each grain is assumed to be a regular tetrakaidecahedron (14-hedron) with six square and eight hexagonal faces. Grains of this shape fit together in a body centred cubic array to fill all space, figure 1. This arrangement is the most efficient way of stacking identical polyhedra together and a polycrystal of this form will have minimum interfacial energy.

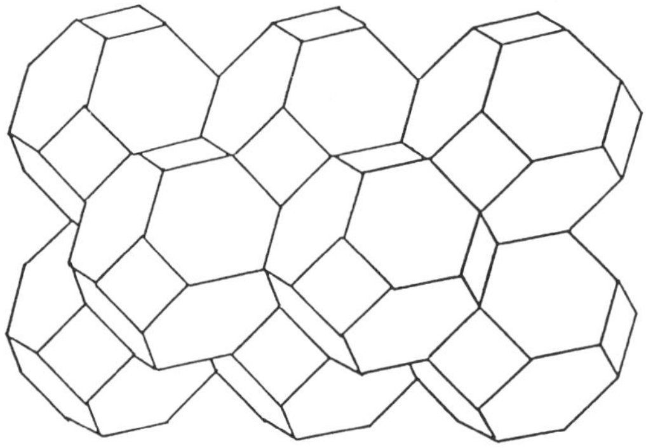

Figure 1. Eight space-filling regular 14-hedra stacked together to form a body cented cubic array.

The model assumes initially that failure occurs entirely by means of brittle fracture, either on cleavage planes or on grain boundaries. No ductile fracture or deviation of cleavage planes from ideal crystallographic orientations is allowed. Fracture may be initiated either at a corner, an edge or a face of a grain or at an interior point. It may then spread outwards either by propagation across grains or along grain boundaries or by means of separate cracks linking together. Two extreme cases may then be identified. If the cleavage fracture energy is very high the failure will be restricted to grain boundaries. It will then be possible to separate the two parts of the polycrystal on either side of the fracture surface. On the other hand, if the cleavage fracture energy is very low, every grain is likely to cleave. However, the traces of these cleavage cracks in the common grain boundary of adjacent grains will not in general be co-incident. Hence if the polycrystal is to be separated into two parts some additional intergranular failure will be necessary. It is the determination of the proportion of intergranular to cleavage failure in this latter case which is being investigated. Clearly the result will depend on many factors including the ratio of intergranular to cleavage surface energies, the spread of intergranular energies resulting in part from impurity segregation, the number of cleavage planes available and their relative energies if they belong to different crystallographic families, and the orientation of grain boundaries and potential cleavage planes relative to the stress axis.

The specific aim of the present application is to obtain a better understanding of low temperature brittle fracture in α-iron and ferritic steels for which cleavage is assumed to occur on the three mutually orthogonal variants of the {100} family of planes. On average these three potential cleavage planes (pcps) meet a grain boundary at 60° to each other. The maximum and minimum angles between the nearest traces of cleavage cracks in two adjacent grains are then 30° and 0°, giving an average of 15°. However the nearest pcp may not be oriented favourably relative to the stress axis so that the second nearest pcp may be adopted, for which the corresponding angle is 45°. If it is assumed that the occurrence of these two cases is inversely proportional to these angles, ie 3:1, the average angle between the traces of the two cracks will be 22.5°.

It is now necessary to consider the different ways in which a cleavage crack can continue to propagate when it meets a grain boundary. Four possibilities arise and these are illustrated schematically in figure 2. In the first mechanism, labelled I, a cleavage crack is nucleated in the second grain at the point at which the first crack meets the grain boundary. To separate the two sides of these fracture surfaces, the area of grain boundary between the traces must then fail. There are clearly two possibilities but normally it will be the smaller area, shown shaded in figure 2(a), which will be selected. In mechanism II a cleavage crack propagates across a grain boundary by first spreading right across its own grain before nucleating a second crack at one of its corners, ie at a point along one of the edges of the common

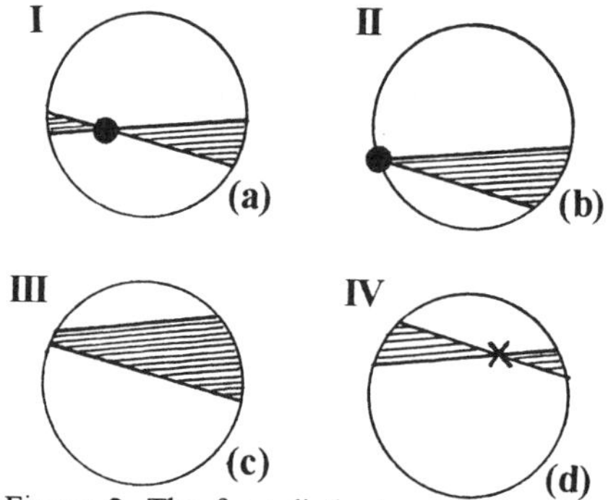

Figure 2. The four distinct mechanisms in which grain boundary failure may link cleavage cracks in adjacent grains. For each mechanism the boundary between adjacent grains is shown as a circle, which represents schematically both the square and the hexagonal faces of a 14-hedron. The part of the boundary which must fail is indicated by shading, the straight edges of which are the traces of the two cleavage planes. In mechanisms I and II it is considered that a crack crosses the boundary whereas in III and IV two cracks meet the boundary from opposite sides.

grain boundary. Again, to separate the grains the included triangular area of grain boundary, shown shaded in figure 2(b), must fail. If cleavage cracks are nucleated independently on either side of a grain boundary two possibilties arise. In mechanism III the traces of the cracks do not intersect in the grain boundary resulting in the need for a quadrilateral shaped area of boundary to fail, as shown in figure 2(c). Alternatively, in mechanism IV, the traces may intersect and the arrangement is shown in figure 2(d).

Using the result that the average angle between the traces of the cleavage planes in the grain boundary is 22.5°, it is now possible to estimate the fraction of the area of a grain boundary which, on average, must fail for each of the mechanisms I to IV. This average will clearly depend on many factors including the shape of the grain boundary and, in

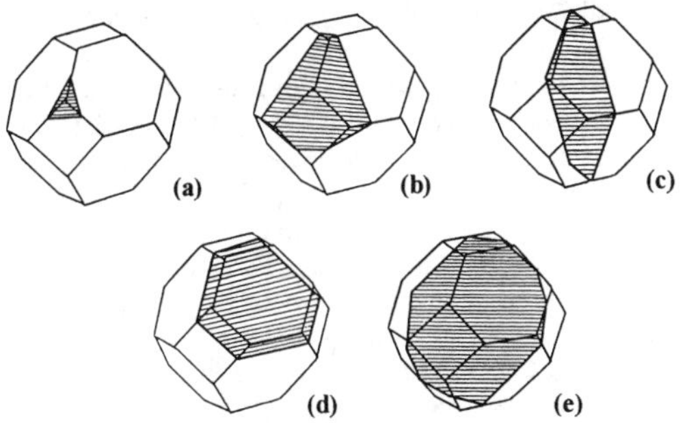

Figure 3. A cleavage plane traversing a 14-hedral grain may intersect any number of grain boundaries from three to ten. Examples are given of cleavage planes, shown shaded, intersecting (a) three, (b-d) six and (e) ten grain boundaries.

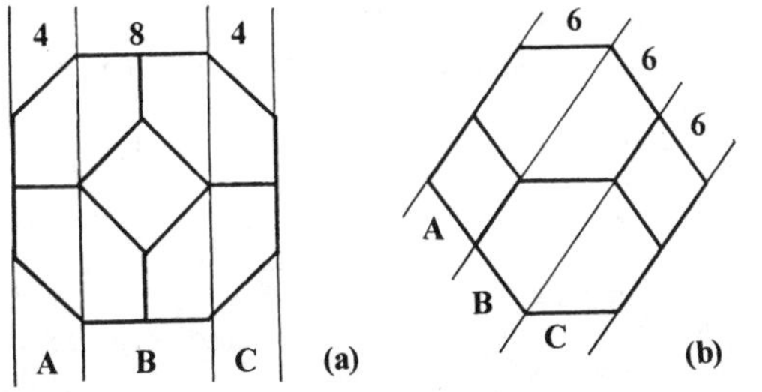

Figure 4. Illustrating the fact that a set of cleavage planes parallel to (a) square and (b) hexagonal faces of a 14-hedron will on average intersect six faces. The extended fine lines represent traces of the intersections. In (a) an equal number of square and octagonal sections are formed whereas in (b) all sections are hexagonal.

the case of mechanisms I and IV, the location of the point of intersection of the two traces. However, using simplifying assumptions, such as a circular grain boundary, it has been possible to deduce approximate average values of 12.5%, 12.5%, 37% and 12.5% for mechanisms I to IV respectively.

The next step is to deduce what proportions of mechamisms I to IV actually occur in practice as brittle fracture spreads across a polycrystalline material. To do this it is first necessary to consider how many grain boundaries a cleavage crack meets, on average, when it spreads across a single grain. The crack then has a polygonal shape and the number of edges will clearly depend on its orientation and position. It has been shown (Crocker & Smith 1996) that for a 14-hedral grain this number may take any value from 3 to 10 and examples of these two extreme cases and three distinct intermediate cases with 6 edges are shown in figure 3. For cracks parallel to planes of symmetry of the 14-hedron, the average is readily shown to be 6, as shown in figure 4 for cracks parallel to the square faces and to the hexagonal faces. For more general cases the average has been computed numerically and the result is always 6 (Leggatt 1997). No general analytical proof of this result has been obtained but it will be assumed that the overall average is indeed 6. Thus the propagation of fracture through a polycrystal may be represented topologically using a diagram consisting of a regular array of hexagonal prisms or, even more conveniently, by a regular array of hexagons.

Consider the case in which the first cleavage crack is initiated at a grain edge and all subsequent grains fail where a propagating crack meets them and not at interior points. Schematically the sequence of events, each spreading crack being indicated by an arc of a circle, is then as shown in figure 5(a). The first crack is initiated in grain 1 at a point on its edge A and immediately causes the two adjacent grains 2 and 3 to crack from the same point. Therefore type II intergranular failure, represented by fine lines, will be necessary in the three boundaries meeting at A. However, because the system has three-fold symmetry, we need only consider the boundary between grains 1 and 2, denoted by 1/2. This boundary meets grain 4 along its edge B and the crack in either grain 1 or grain 2 initiates a further crack in this new grain. It is here assumed that it is the crack in grain 1 so that type II failure occurs in boundary 1/4 and hence type III or IV failure, represented by a bold line, in boundary 2/4. The propagating crack next meets corners C and D and the same situation arises. Thus two of the new boundaries chosen to be

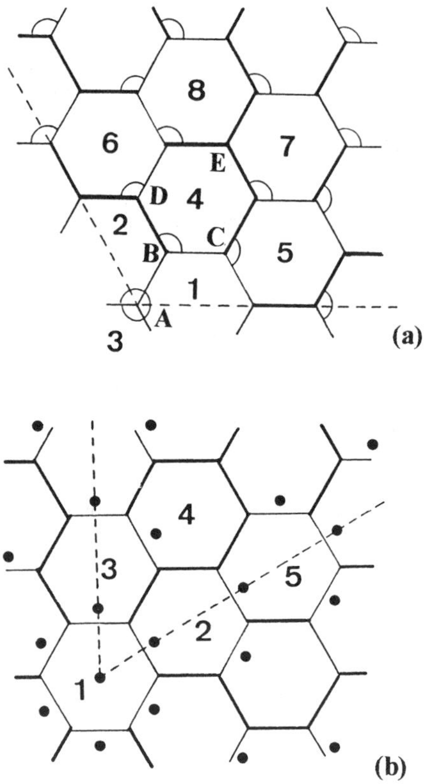

Figure 5. Schematic illustrations of the propagation of a brittle crack across a polycrystal. As each cleavage plane meets on average six grain boundaries, it is possible to represent the polycrystal topologically by an array of hexagonal prisms, cross-sections of which are shown here. Therefore grain boundaries and grain edges are perpendicular to the paper. In (a) the initial crack is assumed to nucleate in grain 1 at a point on grain edge A and to spread into adjacent grains. The arcs of circles indicate the sites, all of which are on grain edges, at which cracks nucleate in each grain. However at each of these edges only one of the incident cracks propagates. Therefore some boundaries fail by mechanism II, represented by fine lines, and some by mechanism III or IV, represented by bold lines. The full pattern of events has 3-fold symmetry about A. In (b) the initial cleavage crack is assumed to nucleate in grain 1 and a dot is used to indicate the nucleation site in each grain. These dots are near the boundaries indicated by fine lines, which are therefore likely to fail by mechanism IV. The remaining boundaries, indicated by bold lines, are remote from nucleation sites and will fail by mechanism III or IV. The full pattern of events has six-fold symmetry about grain 1.

1/5 and 4/6 are of type II and two, 2/6 and 4/5 are of type III or IV. A new situation arises at point E between grains 4, 7 and 8. Cracks have already been nucleated in grains 7 and 8 before the crack front reaches E so that no new crack is formed. Therefore boundary 7/8 is of type III/IV. This situation occurs more frequently as the fracture extends and the conclusion is that about 40% of the grain boundary failures are of type II and 60% of type III or IV. A simple calculation has indicated that on average about two-thirds of this 60% will be of type III and one-third of type IV. Therefore in this case no boundaries fail by mechanism I, 40% by mechanism II, 40% by mechanism III and 20% by mechanism IV. Similar results are obtained if the first crack is assumed to initiate at a grain face, a grain corner or indeed within a grain. Hence assuming that on average 12.5% of the grain boundary area is fractured in mechanisms I, II and IV and 37% in mechanism III it is concluded that overall just over 20% of the area of affected grain boundaries will fail.

To estimate the proportion of intergranular failure as opposed to cleavage failure it is now necessary to deduce the average area of a grain boundary and of a cleavage plane. Assuming the 14-hedral grain to be of unit volume it is readily shown that the average grain boundary area is about 0.4. Also, approximating the 14-hedron by a sphere of unit volume, the average cleavage plane area is about 0.8. Therefore the ratio of these areas is about 0.5. Although on average each cleavage plane will meet six grain boundaries, each of these is shared between two adjacent grains so that the ratio of partially failed grain boundaries to cleavage planes is three. Hence, subject to the constraints and approximations of this geometrical model, it is concluded that the percentage of intergranular failure is approximately 0.5 x 3 x 20%, ie 30%.

Some generalisations of the model will now be explored. First consider the case of all cleavage cracks being initiated within grains and not at grain boundaries. The sequence of events is shown schematically in figure 5(b). The initiation point of the first crack is unimportant, so assume that it is at the centre of grain 1. As this crack spreads, new cracks will be nucleated within the six surrounding grains at weak points, such as carbide particles, which experience large stress concentrations. These points will tend to be near the boundaries with grain 1 and the resulting cracks are likely to meet the initial crack by means of mechanism IV. This is indicated by fine lines in the figure. However, the crack nuclei in grains 2 and 3 are remote from the boundary between them so that either mechanism III or IV will operate, as indicated by a bold line. The crack in grain 4 will be initiated by stress concentrations from either grain 2 or grain 3 and the latter has been selected. Hence boundary 3/4 exhibits mechanism IV and boundaries 4/5 and 4/6 mechanism III or IV. As the fracture surface develops, the proportions of the two types of boundary is found to be one-third of mechanism IV and two thirds of either III or IV. This simple relationship arises because each cell has an initiation site near one edge and, as edges are shared between adjacent cells, there are three edges per cell. Also, as about two-thirds of the III/IV edges have already been shown to be III, the overall result is 44% type III and 56% type IV. Assuming again that 37% and 12.5% of the area of the grain boundary is fractured for these two cases, a similar overall result to that deduced above is obtained. About 23% of the area of affected grain boundaries will fracture which implies about 35% intergranular and 65% cleavage failure.

3 EXPERIMENTAL PROCEDURE & RESULTS

Two ferritic materials were selected: (i) high purity α-iron and (ii) C-Mn steel. The α-iron was in the form of a 3mm diameter bar which contained 230ppm oxygen and 260ppm nitrogen and this was heat treated at a temperature of 1243K for 900s and air cooled to produce a grain size of about 80μm (mean linear intercept). The C-Mn steel plate, composition 0.14C, 1.3Mn, 0.11Si, 0.029S and 0.012P (wt%), was cross rolled and normalised at about 1213K, followed by heat treatment at 873K for 2.16×10^4s, and cooled at $0.003°s^{-1}$ to 523K and finally air cooled. This produced a grain size of about 50μm (mean linear intercept). The α-iron specimens were fractured at a temperature of about 100K and the C-Mn steel specimen was tested using a standard fracture toughness test procedure again at about 100K. The fracture surfaces were examined using a JEOL JSM80A scanning electron microscope operating at an accelerating voltage of 13keV to maximise the observed fractographic detail. The area fractions of cleavage and intergranular fracture were measured from the secondary electron images using standardised imaging conditions at a magnification of 1000X.

Observations of the fracture surfaces of the α-iron specimens have been described in detail previously (Crocker et al 1996). Figure 6 shows that the fracture mode at about 100K is predominantly cleavage with a proportion of inter-granular fracture. This corresponds to about 20% of the fracture surface projected on to the image plane. In the case of the C-Mn steel specimens, again there is a mixture of cleavage and intergranular fracture, figure 7(a), within the predominantly ferritic regions of the microstructure. However, when the fracture surface

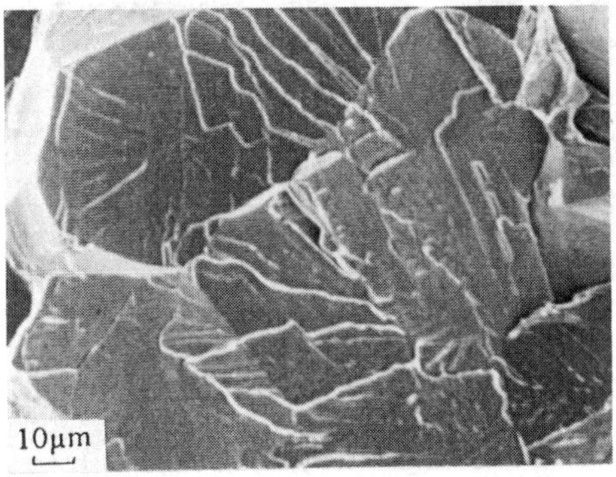

Figure 6. A secondary electron image of an α-iron fracture surface showing cleavage fracture, revealed by river pattern features, with a proportion of intergranular fracture, which totals approxinmately 20%.

is at normal incidence to the electron beam there is a 6.2% intergranular fracture. Despite the fact that the fracture surface is macroscopically planar and normal to the stress axis, on the microscale the topography varies significantly. As a consequence, the same region of the fracture surface, measuring 0.45mm x 0.35mm, was imaged for a range of tilt angles, from 0° to 50°, relative to the incident electron beam. Moreover tilting was undertaken in two orthogonal directions, x and y, and these gave the proportions, of intergranular fracture summarised in table 1.

Table 1. Projected proportion of intergranular fracture in specimens tilted about the x and y axes.

Angle of tilt (±2°)	0°	20°	40°	50°
x-axis (%) (±0.5)	6.2	6.4	6.3	5.0
y-axis (%) (±0.5)	6.2	8.4	8.2	7.6

This shows a variation of the proportion of intergranular fracture about a mean value of about 6.9% with observation angle. Figures 7(a) to (c) are secondary electron images corresponding with part of the sampled fracture surface for 0° and 40° tilt and show the changes in the projected intergranular and cleavage fracture regions. The area in figure 6 represents less than 5% of the total area sampled and has been selected to show a mixture of cleavage and intergranular fracture. The local interactions of the cleavage cracks at the grain boundaries in the α-iron specimens support the theoretical model predictions since they result in a

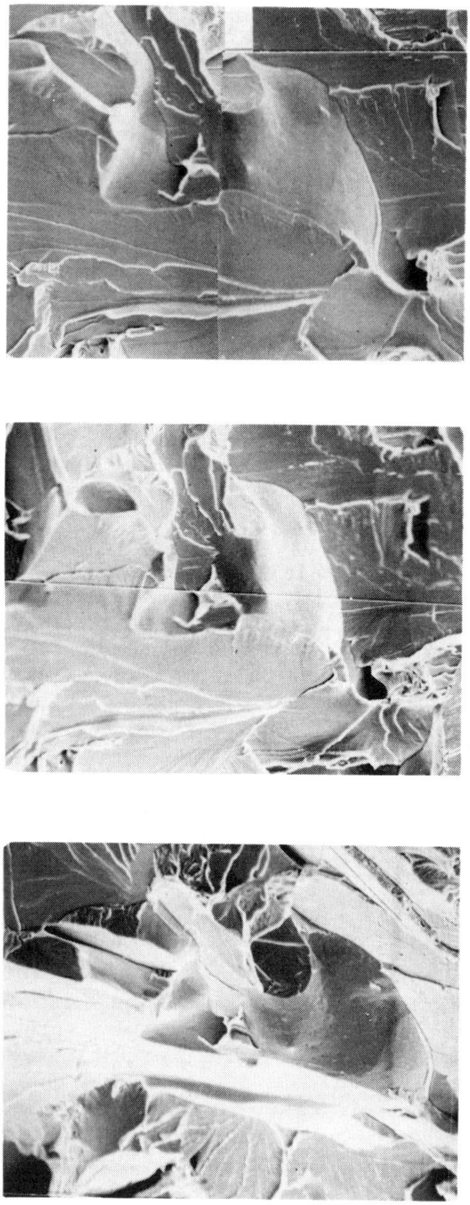

(a)

(b)

(c)

10μm

Figure 7. Secondary electron images of a fracture surface of C-Mn steel showing cleavage and intergranular fracture at (a) normal incidence to the electron beam, (b) 40° tilt in the x direction and (c) 40° tilt in the y direction. Note the lower region of fractograph (c) where tilting the specimen compared with (a) reveals that there is approximately 90° between two cleavage surfaces.

proportion of intergranular fracture. Although similar interactions are observed in the C-Mn steel specimens they are less pronounced.

4 DISCUSSION

This paper has presented new results, obtained from theoretical modelling and fractography, on the amount of intergranular fracture which accompanies cleavage fracture in polycrystalline materials with body centred cubic crystal structures, which fail in a brittle manner at low temperatures. The significance of these results will now be examined. First the theoretical work, subject to the assumptions and approximations adopted, confirms the conclusions of an earlier preliminary investigation (Crocker et al. 1996) that the proportion of intergranular fracture required is at least 30%. Also it is important to recognise that this estimate is concerned entirely with grain boundaries which fail partially in order to link cleavage cracks in adjacent grains. It does not include grain boundaries which fail totally or even from the trace of one cleavage crack to the grain edge.

The experimental work on α-iron (Crocker et al. 1996) has shown that at high magnifications the amount of intergranular failure appears to approach 20%. However, this result was deduced from projected areas and it is anticipated that whereas most cleavage planes will be roughly perpendicular to the stress axis, and hence to the electron beam, many of the partially failed grain boundaries will make a relatively small angle with this direction. Therefore, the true proportion of intergranular fracture is probably larger than 20%. However it appears from the fractographs that some of the grain boundaries have failed completely and are not simply accommodating cleavage cracks in adjacent grains. A direct comparison between the theoretical and experimental results is not therefore valid and in reality the modelling work is probably giving a substantial overestimate.

The new experimental results on C-Mn steel presented in this paper show that for this particular heat treatment the amount of intergranular fracture which occurs is much smaller than that in pure iron. Even allowing for the correction needed for projected areas it is probably less than 10% and again this includes whole grain boundaries which have failed and these are excluded from the model. It is therefore very clear in this case that the theoretical model is giving a large overestimate of the real value and that different mechanisms are needed for α-iron and C-Mn steel. However, these experiments were also intended to investigate how misleading a single projected image of a fracture surface might be. Comparison of the micrographs in figure 7 certainly demonstrates that great care has to be exercised if quantitatively meaningful results are to be obtained.

The idealised theoretical model adopted here provides an estimate of the proportion of intergranular failure arising purely from geometrical constraints. In practice however, as emphasised above, additional areas of grain boundary might fail because they are favourably oriented relative to neighbouring cleavage planes. The extreme situation arises when the grain boundary is perpendicular to the stress axis and all six {100} cleavage planes, three in each grain separated by the boundary, make the same angle of $\cos^{-1}(1/\sqrt{3})$ with this axis. This structure is known as a <111> symmetric twist boundary. In this arrangement, the grain boundary is favoured as long as its energy is no more than three times that of the cleavage planes. Clearly, as this ratio decreases, because for example segregation of impurity elements increases, more and more grain boundaries will fail, either partially or totally, so that the proportion will increase from 30% to 100%, figure 8.

In practice grain boundaries have different structures and impurity concentrations and therefore have a range of energies. This will not effect predictions when the grain boundary energy is greater than three times the cleavage energy as only geometrical constraints are then involved. For lower grain boundary energies, some boundaries will fail with more and some with less difficulty than if they all had identical energies. Hence on average there will again be little effect.

There is no corresponding range of energies for the {100} cleavage planes. However, if additional cleavage planes become active, a dramatic effect on the predictions can arise. Consider first the possibility of the six variants of {110} cleavage, which occurs in molybdenum, replacing the three variants of {100}. Doubling the number of cleavage planes reduces the average angle between neighbouring traces of these planes in the grain boundaries by a factor of two and therefore the predictions of the theory are reduced from about 30% intergranular failure to about 15%. Then, if one variant of either {100} or {110} cleavage occurs in each grain, nine planes are available so that the estimate becomes about 10%. This effect is illustrated schematically in figure 8. Finally, if several cleavage planes can operate together in each grain, fracture can effectively occur on any crystallographic plane. In an extreme case this could be the plane perpendicular to the stress axis in every grain so that no intergranular failure is needed. This is most unlikely if the only available planes are the

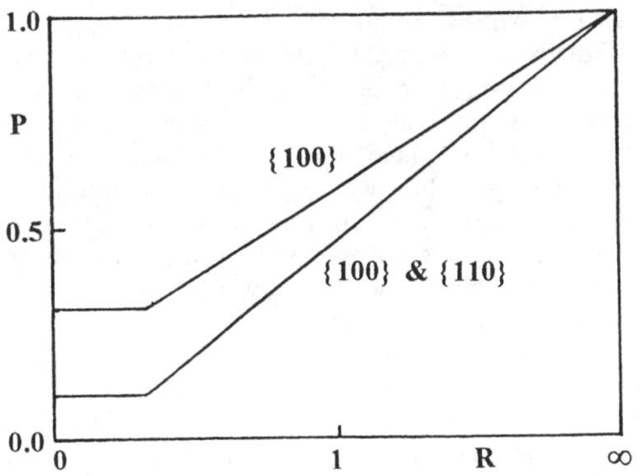

Figure 8. Schematic indication of the variation of the proportion P of intergranular brittle failure which is predicted by the model as the ratio R of the surface energies of cleavage and intergranular fracture increases from zero to infinity. For $R < 0.3$ the plots are horizontal and for $R > 0.3$ they increase gradually to $P = 1$, not necessarily linearly. The upper and lower plots, with P about 0.3 and 0.1 for $R = 0$, are for the case when cleavage is restricted to one variant of {100} and to one variant of either {100}or {110}.

three mutually orthogonal variants of {100} as only one of these can be favourably oriented. However, if {110} is also available, even small amounts of cleavage on these planes could result in a very small proportion of grain boundary failure. A similar conclusion can clearly be reached if some accommodation is achieved by ductile failure.

A preliminary examination of scanning electron micrographs of fracture surfaces in C-Mn steels has indeed revealed the presence of a small amount of ductile fracture. Also, a few grains which have cleaved on two different crystallographic planes have been observed, figure 7(c). It has not however been possible to determine whether these planes are {100}, {110} or both. Further experimental work is therefore required to establish whether the above suggestions explain the present discrepancies between the model and experimental observations. In particular it is intended to etch the fracture surfaces in order to reveal square etch pits in {100} planes and rectangular ones on {110} planes (Honda 1961). In addition, experiments are being carried out, using back-scattered electron diffraction in the scanning elecron microscope combined with stereophotogrammetry. This will allow both the orientation and the surface tilts of the cleavage planes to be measured. Hence a rigorous analysis of the fracture surface leading to a better measure of the true areas of cleavage and intergranular fracture will be provided.

Finally some related preliminary experiments have been performed on polycrystals of zinc (Renshaw 1997). As zinc has only one cleavage plane, the basal (0001) plane, it was anticipated that intergranular failure would play a major role in its brittle fracture. Two types of specimen were used, cast rods and rolled sheet. The rods had large, columnar grains, radiating outwards from the axis. They were fractured using notch bend tests at about 200K. The brittle fracture surfaces revealed about 50% mirror-like cleavage faces and 50% dull intergranular failures. The sheet material was fine grained and assumed to have the basal slip planes within about 30° from the surface in most grains (Chilton et al. 1969). Notched tensile specimens were cut from the sheet and again tested at about 200K. They failed in a brittle manner and the entire fracture surface appeared dull, apparently showing no cleavage failure. It is concluded from these experiments that because zinc only has one cleavage plane a much larger proportion of brittle intergranular failure occurs than in high purity α-iron and C-Mn steel. Also when the single cleavage plane is oriented unfavourably with respect to the stress axis, as in the case of the sheet specimens, only intergranular fracture is observed. This supports the proposals of the present paper that as the number of available cleavage planes increases the amount of intergranular failure decreases.

ACKNOWLEDGEMENTS

This paper is published with the permission of the Director of Technology and Central Engineering, Magnox Electric. The authors are indebted to Spike Leggatt and Andy Renshaw for valuable discussions.

REFERENCES

Abbott, K., R. Moskovic & P. Flewitt 1994. Intergranular fracture in the cleavage fracture temperature range. Mater. Sci. Technol. 10: 813-818.

Chilton, A., J. Stobo, M. Graham & A. Crocker 1969. Grain growth in sheet zinc. Czech. J. Phys. B 19: 103-108.

Crocker, A. & G. Smith 1996. Modelling intergranular and cleavage fracture. In A. Ferro et al. (eds.), *Proc. 7th Int. Conf. Intergranular and Interphase Boundaries, Lisbon, June 1995:* 593-596. Transtec: Zurich.

Crocker, A., G. Smith, P. Flewitt & R. Moskovic 1996. Grain boundary fracture in the cleavage regime of polycrystalline materials. In J. Petit (ed.) *ECF 11: Mechanisms and Mechanics of Damage and Failure*, vol 1: 233-8. EMAS: Warley.

Honda, R. 1961. Occurrence of longitudinal cleavage in stretched silicon iron crystals. Acta Metall. 9: 969-970.

Leggatt, M. 1997. Sections of tetrakaidecahedra. Final year undergraduate project report, Dept. of Physics, University of Surrey.

Renshaw, A. 1997. Brittle fracture of polycrystalline zinc. Final year undergraduate project report, Dept. of Physics, University of Surrey.

Damage and Failure of Interfaces, Rossmanith (ed.) © 1997 Balkema, Rotterdam, ISBN 90 5410 899 1

On some thermoelastic problems of interfacial cracks and rigid inclusions in laminates

A. Kaczyński
Institute of Mathematics, Warsaw University of Technology, Poland

S.J. Matysiak
Institute of Hydrogeology and Engineering Geology, University of Warsaw, Poland

ABSTRACT: This paper attempts a comparative analysis of some plane thermoelastic problems involving an elastic, microperiodic two-layered composite weakened by two types of interface defects: a Griffith crack and a ribbon-like absolutely rigid inclusion. The study is based on the concepts of a new approximate treatment by using the homogenized model with microlocal parameters. Useful solutions with the standard (non-oscillatory) inverse square-root singularity are obtained, so the stress intensity factors are employed as local failure parameters. As the illustration of final results, a simple example is given.

1 INTRODUCTION

Increasing importance of new composite materials consisting of a large number of dissimilar layers in advanced engineering structures requires the study of different aspects of their failure behaviour. Fracture initiated by an interface single flaw is one of the most commonly encountered cases.

In this paper we focus on the problem of determining the temperature, heat flux, stress and displacement plane fields in a bimaterial periodically layered space due to the presence of two types of defects lying on one of the straight interfaces of layers: a Griffith crack and a perfectly rigid ribbon inclusion. The results are obtained in the framework of some homogenized model of the considered microperiodically layered composite (Woźniak 1987, Matysiak & Woźniak 1988) by using a complex representation derived in our earlier papers (see Kaczyński & Matysiak 1988b, 1989). The general problem in question with a thin-walled elastic inclusion of any stiffness was posed and reduced to a complex system of singular integrodifferential equations(Yevtushenko et al. 1995).

As an efficient approach to the analysis of microperiodic layered media (see some achievements in this field given in a survey by Matysiak 1995), the plane linear thermoelasticity with microlocal parameters is applied and briefly reviewed in Section 2. The advantages of the homogenized model used in the paper lie in a relatively simple form of the governing equations, which makes it possible to find the analytical solutions of complex boundary-value problems arising in connection with the presence of interface defects.

An analysis of the results for proper assessment of the strength degradation of laminates due to the defects under consideration is given in Section 3. Characteristics of stress singularity at rigid inclusion or crack tips as well as the effects of the layering are investigated with the context of fracture mechanics. Unlike existing solutions for defects at the interface of materials (Erdogan 1972, Sih & Chen 1981), the solutions obtained display no oscillatory behaviour. An illustrative example is finally presented.

Similar problems dealing with the rectilinear line rigid inclusions in the homogeneous solids were discussed, for example, by Sih 1965, Atkinson 1973, Matysiak & Olesiak 1981, Wang et al. 1985, Tvardovsky 1990. A wide range of several boundary-value problems for isotropic bodies closer to those considered in this paper has been carried out in the monographs by Berezhnitsky, Panasyuk & Staschuk 1983 and Kit & Krivtzun 1983.

2 PROBLEM STATEMENT AND SOLUTION

2.1 *Formulation*

Consider an infinite periodic laminated composite the scheme of its middle cross section is given in Fig.1. A repeated fundamental layer of thickness δ is composed of two homogeneous isotropic materials, denoted by 1 and 2, with thicknesses δ_1 and δ_2 ($\delta_1 + \delta_2 = \delta$), and with different thermomechanical properties characterised by Lamé constants λ_l, μ_l, thermal conductivities k_l and coefficients of volume expansions $3\beta_l/(3\lambda_l + 2\mu_l)$; here and in the sequel,

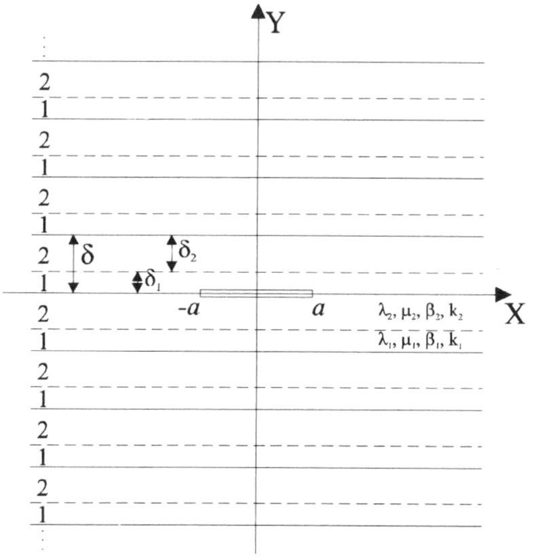

Figure 1. Middle-cross section of periodic composite with an interface defect.

all quantities (e.g. material constants, stresses, etc.) pertaining to the sublayers 1 and 2 will be associated with the index l or (l) taking the values 1 and 2, respectively.

A Cartesian coordinate system (x,y,z) is devised such that the- y axis is normal to the layering and the- x axis coincides with one of the straight interface of materials.

Suppose that this stratified medium is weakened by a Griffith crack (denoted by C) or a line rigid inclusion (denoted by I), occupying the region $|x| \le a$, $-\infty < z < \infty$ in the plane $y = 0$, and is subjected to a certain external loading giving the displacement, stress and temperature fields independent of the coordinate z. In what follows, we consider the plane steady-state thermoelastic problem related to the xy-plane with the interface defect representing a line segment of the length $2a$. Perfect mechanical and thermal bonding between the layers (excluding the interval $< -a,a >$) is assumed. Moreover, the crack surfaces are required to be free of tractions and accordingly the rigid ribbon faces to be free of displacements.

A closed solution of the above posed problem is not possible to obtain because of the complicated geometry and complex boundary conditions. In this study, the specific homogenization procedure called microlocal modelling will be employed in order to seek an approximate solution. We next recall some relevant results from this approach (see detailed treatment described by Matysiak & Woźniak 1988 and Kaczyński & Matysiak 1989).

2.2. *Governing relations*

To determine the following functions of the variables (x, y): the temperature θ, the displacement vector $[U,V]$, the heat flux density vector $\left[q_x^{(l)}, q_y^{(l)}\right]$, the stresses $\sigma_{yy}^{(l)}, \sigma_{xy}^{(l)}, \sigma_{xx}^{(l)}, \sigma_{zz}^{(l)}$, we take into consideration the homogenized model of the treated body characterised by *a priori* given δ- periodic, sectionally linear shape function, defined as

$$h(y) = \begin{cases} y - 0{,}5\delta_1 & \text{for} \quad 0 \le y \le \delta_1 \\ (\delta_1 - \eta y)/(1-\eta) - 0{,}5\delta_1 & \text{for } \delta_1 \le y \le \delta \end{cases} \tag{1}$$

with $\eta = \delta_1 / \delta$. Observe that the values of this function are small but of its derivative, denoted by $h,_l$, are not small taking the value 1 for $l = 1$ and $-\eta/(1-\eta)$ for $l = 2$.

The basis of this approach are some heuristic kinematic postulates leading to the following approximations:

$$\begin{aligned}
&\theta(x,y) = \vartheta(x,y) + h(y)\gamma(x,y) \cong \vartheta(x,y) \\
&U(x,y) = u(x,y) + h(y)p(x,y) \cong u(x,y) \\
&V(x,y) = v(x,y) + h(y)q(x,y) \cong v(x,y) \\
&q_x^{(l)} \cong -k_l\, \vartheta,_x \quad , \quad q_y^{(l)} \cong -k_l\left(\vartheta,_y + h,_l \gamma\right) \\
&\sigma_{yy}^{(l)} \cong \left(\lambda_l + 2\mu_l\right)\left(v,_y + h,_l q\right) + \lambda_l u,_x - \beta_l \vartheta \\
&\sigma_{xx}^{(l)} \cong \left(\lambda_l + 2\mu_l\right)u,_x + \lambda_l\left(v,_y + h,_l q\right) - \beta_l \vartheta \\
&\sigma_{xy}^{(l)} \cong \lambda_l\left(u,_y + v,_x + h,_l p\right) \\
&\sigma_{zz}^{(l)} \cong \lambda_l\left(u,_x + v,_y + h,_l q\right) - \beta_l \vartheta
\end{aligned} \tag{2}$$

Here ϑ and u, v are unknown functions called the macrotemperature and macrodisplacements, respectively, and additional unknown functions γ and p, q stand for the microlocal (thermal and elastic) parameters. Subscipts preceded by a comma indicate partial differentiation with respect to the corresponding coordinates.

Following the microlocal procedure to the macromodelling of the periodic laminated space under consideration, it is arrived at the governing equations and constitutive relations of certain macro-homogeneos medium (the homogenized model), given (after eliminating all microlocal parameters) in terms of macrotemperature and macrodisplacements (in the absence of body forces and heat sources) as follows

$$
\begin{aligned}
&\tilde{k}\,\vartheta_{,xx} + K\,\vartheta_{,yy} = 0\\
&A_2\,u_{,xx} + (B+C)\,v_{,xy} + C\,u_{,yy} - K_2\,\vartheta_{,x} = 0\\
&A_1\,v_{,yy} + (B+C)\,u_{,xy} + C\,v_{,xx} - K_1\,\vartheta_{,y} = 0\\
&q_x^{(l)} = -k_l\,\vartheta_{,x}\quad,\quad q_y^{(l)} = -K\,\vartheta_{,y}\\
&\sigma_{yy}^{(l)} = B\,u_{,x} + A_1\,v_{,y} - K_1\,\vartheta\\
&\sigma_{xy}^{(l)} = C\left(u_{,y} + v_{,x}\right)\\
&\sigma_{xx}^{(l)} = D_l\,v_{,y} + E_l\,u_{,x} - F_l\,\vartheta\\
&\sigma_{zz}^{(l)} = \frac{\lambda_l}{2(\lambda_l+\mu_l)}\left(\sigma_{xx}^{(l)} + \sigma_{yy}^{(l)}\right) - \frac{\mu_l}{\lambda_l+\mu_l}\beta_l\,\vartheta
\end{aligned}
\tag{3}
$$

The positive coefficients appearing in the above relations, describing the material and geometric properties of the composite constituents, are given in the Appendix. It should be emphasized that the condition of perfect bonding between the layers is satisfied (the components $\sigma_{yy}^{(l)}$, $\sigma_{xy}^{(l)}$ and $q_y^{(l)}$ do not depend on l implying the continuity of the stress and heat flux vectors at the interfaces). Let us observe that the components $\sigma_{xx}^{(l)}$, $\sigma_{zz}^{(l)}$, $q_x^{(l)}$ suffer a discontinuity on the interfaces.

2.3 *Solution of the boundary-value problem*

Within the framework of the homogenized model presented above we are faced with the boundary-value problem: find fields $\vartheta(\cdot)$ and $u(\cdot), v(\cdot)$ suitable smooth on $R^2 \setminus (<-a,a> \times \{0\})$ such that $(3)_{1-3}$ hold and satisfy for $|x| < a$ the following global conditions involving a crack C or a rigid inclusion I

$$
\sigma_{yy}^{\pm} = \sigma_{xy}^{\pm} = 0 \ \text{ for } C, \quad u^{\pm} = v^{\pm} \ \text{ for } I \tag{4}
$$

Here and afterwards the quantities assigned with $\pm$ refer to the limiting values as $y \to 0^{\pm}$. Moreover, certain conditions resulting from a given external loading (thermal and mechanical) have to be specified.

A two-stage method for obtaining the solution is used and follows along the same line of reasoning as that in the classical theory of uncoupled thermoelasticity applied to crack problems (Kit & Krivtzun 1983). By means of the superposition principle, the following decomposition of the temperature, displacements and stress tensor can be done

$$
\vartheta = \overset{0}{\vartheta} + \overset{d}{\vartheta},\ (u,v) = \left(\overset{0}{u},\overset{0}{v}\right) + \left(\overset{d}{u},\overset{d}{v}\right),\ \sigma = \overset{0}{\sigma} + \overset{d}{\sigma} \tag{5}
$$

where 0 refers to the problem of the laminated composite without the defect (C or I),loaded by the given external load and d refers to the corresponding perturbed problem in which some fields (tractions, displacements, temperature, heat flux) are imposed on the defect face, equal and opposite to those transmitted across the interface in the absence of the defect, and the stresses and heat fluxes vanish at infinity.

The first problem may be solved by the method of complex potentials devised in our paper Kaczyński & Matysiak 1988b and applied in the case of the action of heat sources by Kaczyński 1993. It is assumed here that its solution on the x-axis is known.

Attention will be drawn next on finding the corrective solution of perturbed problem. To this end, we shall make use of a convenient representation of the governing relations (3) (similar in the form to that of given by Muskhelishvili 1953), derived with the aim of solving interface crack problems in our paper Kaczyński & Matysiak 1989. It is expressed by two potentials Φ and Ω, and thermal potential φ_0 corresponding to thermal loading, taking on the x-axis the following form:

$$
\begin{aligned}
&\overset{d}{\sigma}{}_{yy}^{\pm} - i t_*\,\overset{d}{\sigma}{}_{xy}^{\pm} = \Phi^{\pm} + \Omega^{\mp}\\
&2\mu_*\left[\overset{d}{u}{}_{,x}^{\pm} + i t_*\,\overset{d}{v}{}_{,x}^{\pm}\right] = \kappa_*\,\Phi^{\pm} - \Omega^{\mp} + \\
&\qquad + \beta_*(\varphi_0{}')^{\pm} - \beta'(\overline{\varphi}_0{}')^{\mp}\,;\\
&\overset{d}{\vartheta}(x,y) = 2\,\mathrm{Re}\,\varphi_0{}'(\xi),\quad \xi = x + i\,k_0\,y
\end{aligned}
\tag{6}
$$

In the above, the constants t_*, μ_*, κ_*, k_0 are defined in the Appendix, β_*, β' - in the cited paper.

The primary aim of the further study is to reduce the thermoelastic problem to its mechanical counterpart with given stresses or displacements across the surface of the defect. Hence, a knowledge of the thermal potential $\varphi_0{}'$ describing the temperature perturbed distribution is required. For a thermally insulated problem this potential is determined from the boundary conditions

$$
\begin{aligned}
&\left(\overset{d}{q}_y\right)^{\pm} = -\left(\overset{0}{q}_y\right)^{\pm} \quad \text{for } |x| < a,\\
&\overset{d}{q}_x(\infty) = \overset{d}{q}_y(\infty) = 0
\end{aligned}
\tag{7}
$$

which lead, in view of $(3)_4$ and $(6)_3$, to certain Hilbert problem concerning the function $\varphi_0{}''$ with the line of discontinuity $<-a,a> \times \{0\}$, easily solved by means of Muskhelishvili's method.

We now follow the same technique as developed by Berezhnitsky et. al 1983 for the plane with a single line crack or a rigid inclusion of length $2a$.

Denote the given values of thermal potential φ_0' and thermal stresses and displacements from the solution of the problem without the defect (associated with 0) on the upper and lower faces of the segment $|x| < a, y = 0$ as follows

$$\begin{aligned} &\beta_*(\varphi_0')^{\pm} - \beta'(\varphi_0')^{\mp} \equiv -[S(x) \pm T(x)] \\ &\overset{0}{\sigma}_{yy}{}^{\pm} - i t_* \overset{0}{\sigma}_{xy}{}^{\pm} \equiv -[P(x) \pm Q(x)] \\ &2\mu_*\left[\overset{0}{u}_{,x}{}^{\pm} + i t_* \overset{0}{v}_{,x}{}^{\pm}\right] \equiv -[f'(x) \pm g'(x)] \end{aligned} \tag{8}$$

Taking into account (6) and (4),(5), the basic perturbed problem reduces to finding two single-valued, sectionally holomorphic potentials $\Phi(\cdot)$ and $\Omega(\cdot)$ satisfying the boundary conditions

$$\begin{aligned} &\left.\begin{cases} \Phi^{\pm} \pm \Omega^{\mp} = P(x) \pm Q(x) \\ \kappa_* \Phi^{\pm} - \Omega^{\mp} = f'(x) \pm g'(x) + S(x) \pm T(x) \end{cases}\right. \text{for } |x| < a \\ &\Phi(\infty) = \Omega(\infty) \end{aligned} \tag{9}$$

The generally accepted solution of the above mechanical problem may be written for the crack C and the inclusion I in the common form

$$\begin{aligned} \Phi(\hat{z}) &= \frac{F(\hat{z}) + g_m + g_t}{\sqrt{\hat{z}^2 - a^2}} + G(\hat{z}) \\ \Omega(\hat{z}) &= -\rho_* \Phi(\hat{z}) + 2\rho_* G(\hat{z}) \end{aligned} \tag{10}$$

where the functions (Cauchy integrals) F and G of the generalized complex variable $\hat{z}$ (see, for particulars, Kaczyński & Matysiak 1989) and the constant g_m and g_t are defined by means of known functions $F^*(\rho_*, t)$, $G^*(\rho_*, t)$, $T(t)$ as follows

$$F(\hat{z}) = \frac{1}{4\pi\rho_*} \int_{-a}^{a} \frac{\sqrt{a^2 - t^2}\, F^*(\rho_*, t)}{t - \hat{z}} dt$$

$$G(\hat{z}) = \frac{1}{4\pi\rho_* i} \int_{-a}^{a} \frac{G^*(\rho_*, t)}{t - \hat{z}} dt$$

$$g_m = \frac{\rho + \rho_*}{4\pi i(\rho - \rho_*)} \int_{-a}^{a} G^*(\rho_*, t)\, dt$$

$$g_t = -\frac{\rho_*}{\pi i(\rho - \rho_*)} \int_{-a}^{a} T(t)\, dt \tag{11}$$

provided we set in the problem C and I

$$\begin{aligned} &\left.\begin{aligned} &\rho_* = -1,\ \rho = \kappa_* \\ &F^*(\rho_*, t) = -2P(t),\ G^*(\rho_*, t) = -2Q(t) \end{aligned}\right\} \text{for C} \\ &\left.\begin{aligned} &\rho_* = \kappa_*,\ \rho = -1 \\ &F^*(\rho_*, t) = 2f'(t) + 2S(t), \\ &G^*(\rho_*, t) = 2g'(t) \end{aligned}\right\} \text{for I} \end{aligned} \tag{12}$$

From the general solution obtained above within the framework of the homogenized model one can deduce the results for the local field behaviour in the vicinity of the tips $a^{\pm}$ as well as for the thermal stresses and displacements throughout the periodically layered composite.

3 ANALYSIS

3.1 *Stress intensity factors*

For further discussions concerning the failure analysis a quantity of primary interest is the stress field in the surrounding of the crack (inclusion) tip. It follows from the representation (6) and the solution (10) that the proposed model leads to a typical stress singularity having the classical square-root form, contrary to oscillatory singularity appearing in the interface problems in conventional formulation. The intensification of local thermal stresses may be described therefore in terms of some parameters known from the fracture mechanics as the stress intensity factors (SIFs).

By examining the singular parts of potentials (10) the asymptotic form of the solution in the small vicinity ($0 < r \ll a$) of the tips $a^{\pm}$ on the x-axis is found to be (see a principle established by the conditions (12))

$$\begin{pmatrix} \overset{d}{\sigma}_{yy}(x,0) \\ \overset{d}{\sigma}_{xy}(x,0) \end{pmatrix} = \frac{\rho_* - 1}{2\rho_*} \begin{pmatrix} k_I^{\pm} \\ k_{II}^{\pm} \end{pmatrix} \frac{1}{\sqrt{2r}} + 0(r^0),\ x = \pm a \pm r$$

$$\overset{d}{\sigma}_{xx}(x,0) = -c^{\pm}\frac{3+\rho_*}{2\rho_*}\frac{k_I^{\pm}}{\sqrt{2r}} + 0(r^0), x = \pm a \pm r \quad (13)$$

$$\begin{pmatrix} \overset{d}{u}(x,0) \\ \overset{d}{v}(x,0) \end{pmatrix} = \frac{\rho_* - \kappa_*}{2\mu_*\rho_*}\begin{pmatrix} k_{II}^{\pm} \\ k_I^{\pm} \end{pmatrix}\sqrt{\frac{r}{2}} + 0(r^{3/2}), x = \pm a \mp r$$

where the constants $c^{\pm}$ are given in the Appendix and the SIFs $k_I^{\pm}$, $k_{II}^{\pm}$ (superscipts - and + refer to the left and right-hand crack (inclusion) tips, respectively) are defined by

$$k_I^{\pm} - i\, t_* k_{II}^{\pm} = \frac{1}{2\pi\sqrt{a}}\left[\int_{-a}^{a}\sqrt{\frac{a\pm t}{a\mp t}}F^*(\rho_*,t)\,dt \pm \right.$$

$$\left. \pm i\frac{\rho+\rho_*}{\rho-\rho_*}\int_{-a}^{a}G^*(\rho_*,t)\,dt \mp i\frac{4\rho_*}{\rho-\rho_*}\int_{-a}^{a}T(t)\,dt\right] \quad (14)$$

Now, it is significant to observe that the general character of the above asymptotic relations is similar to that in the homogeneous case. Thus, the $1/\sqrt{r}$ type of stress singularity is preserved and two parameters k_I, k_{II} are the measure of the elevation of stresses due to presence of the interface defect under study. However, the asymptotic angular distribution is different (Kaczyński & Matysiak 1988a). Moreover, the influence of the layering is seen in the dependence of local crack displacements and inclusion stresses on the parameter κ_* pertinent to the composite structure. It should also be noted that the results obtained may be contrasted with the particular results for a homogeneous isotropic body (characterised by the Lamé constants λ, μ, Poisson's ratio ν, thermal conductivity k and thermo-elastic constant β) provided

$$\begin{aligned} &\lambda_1 = \lambda_2 = \lambda,\ \mu_1 = \mu_2 = \mu \\ &k_1 = k_2 = k,\ \beta_1 = \beta_2 = \beta \\ &A_1 = A_2 = \lambda + 2\mu,\ B = \lambda,\ C = \mu \\ &D_l = \lambda,\ E_l = \lambda + 2\mu,\ F_l = \beta \\ &\tilde{k} = K = k,\ K_1 = K_2 = \beta \\ &t_* = 1,\ \mu_* = \mu,\ \kappa_* = \nu \equiv (\lambda + 3\mu)/(\lambda + \mu) \\ &\beta_* = 2\mu\beta/(\lambda + \mu),\ \beta' = 0 \end{aligned} \quad (15)$$

Several problems in this homogeneous case dealing with the interaction between rigid line inclusions and cracks have been extensively treated in a monograph by Berezhnitsky et al. 1983. Due to the same technique it becomes possible to obtain and compare the solutions to corresponding problems considered here.

A simple example dealing with the interface problems in question for a given external loading is given next in order to illustrate the main results. The investigation will be directed primarily to calculation of stress intensity factors.

3.2 *Example: uniform tension and heat flux*

Consider the significant problem of the periodic two-layered space weakened by an insulated interface defect (C or I) and subjected to a one-dimensional uniform heat flow and tension normal to the layering.

Assuming that $q_y(\infty) = -q_\infty$, $\sigma_{yy}(\infty) = p_\infty$, $q_x(\infty) = \sigma_{xy}(\infty) = 0$ and bearing (5),(7),(8) in mind, we can obtain from (12):

$$\begin{aligned} &S(t) = 0,\ T(t) = -\frac{(\beta_* + \beta')q_\infty}{2k_0 K}\sqrt{a^2 - t^2} \\ &G^*(\rho_*,t) = 0 \\ &F^*(\rho_*,t) = \begin{cases} 2p_\infty & \text{for C } (\rho_* = -1) \\ \dfrac{2B\,p_\infty}{A_* - A_-} & \text{for I } (\rho_* = \kappa_*) \end{cases} \end{aligned} \quad (16)$$

and after performing the integration we get from (14)

$$\begin{aligned} &k_I^{\pm} = \begin{cases} p_\infty\sqrt{a} & \text{for C} \\ \dfrac{Bp_\infty\sqrt{a}}{A_* - A_-} & \text{for I} \end{cases} \\ &k_{II}^{\pm} = \pm\frac{(\beta_* + \beta')q_\infty a\sqrt{a}}{2(1+\kappa_*)t_* k_0 K}\cdot\begin{cases} 1 & \text{for C} \\ \kappa_* & \text{for I} \end{cases} \end{aligned} \quad (17)$$

The analysis outlined in the present paper may be successfully used to the investigation of the stresses in laminated medium containing interface cracks or inclusions induced by the action of concentrated forces and/or heat sources. Making use of the results in our papers Kaczyński 1993 and Kaczyński & Matysiak 1995 it is possible to obtain in these cases the stress intensity factors controlling fracture instability.

APPENDIX

1. Denoting by $\eta = \delta_1/\delta$, $b_l = \lambda_l + 2\mu_l$ $(l = 1,2)$,

$b = (1-\eta) b_1 + \eta b_2$, the positive coefficients in (3) are given by the following formulae

$$\tilde{k} = \eta k_1 + (1-\eta) k_2 \quad , K = k_1 k_2 / [(1-\eta)k_1 + \eta k_2],$$

$$A_1 = b_1 b_2 / b \quad , \quad C = \mu_1 \mu_2 / [(1-\eta)\mu_1 + \eta \mu_2],$$

$$B = [(1-\eta)\lambda_2 b_1 + \eta \lambda_1 b_2]/b,$$

$$A_2 = A_1 + 4\eta(1-\eta)(\mu_1 - \mu_2)(\lambda_1 - \lambda_2 + \mu_1 - \mu_2)/b,$$

$$K_1 = [(1-\eta)\beta_2 b_1 + \eta \beta_1 b_2]/b,$$

$$K_2 = \eta \beta_1 \lambda_2 + (1-\eta)\beta_2 \lambda_1 / b +$$
$$+ 2[\eta \mu_2 + (1-\eta)\mu_1][\eta \beta_1 + (1-\eta)\beta_2]/b,$$

$$D_l = \lambda_l A_1 / b_l, \quad E_l = [4\mu_l(\lambda_l + \mu_l) + \lambda_l B]/b_l,$$

$$F_l = (2\beta_l \mu_l + \lambda_l K_1)/b_l$$

2. The constants appearing in the complex representation (6) are given as follows

$$t_* = (A_1/A_2)^{1/4}, \quad \mu_* = A_+ A_- / 2(A_* - A_-),$$

$$\kappa_* = (A_* + A_-)/(A_* - A_-) \quad \text{provided}$$

$$A_* = (A_1 A_2)^{1/4} [(A_+ + 2C) A_- / C]^{1/2} \quad \text{and}$$

$$A_{\pm} = (A_1 A_2)^{1/2} \pm B; \quad k_0 = (\tilde{k}/K)^{1/2}$$

3. The constants $c^{\pm}$ in (13) are defined as

$$c^+ = c^{(1)}, \; c^- = c^{(2)} \quad \text{provided}$$

$$c^{(l)} = 1 + \frac{2\mu_l(2\lambda_l + 2\mu_l - A_+)}{(\lambda_l + 2\mu_l)A_+}$$

REFERENCES

Atkinson, C. 1973. Some ribbon-like inclusion problems. *Int. J. Engng. Sci.* 11:243-266.

Berezhnitsky, L. & Panasyuk, V. & Staschuk, N.G. 1983. *Interaction of rigid inclusions and cracks.* Kiev: Naukova Dumka (in Russian).

Erdogan, F. Fracture problems of composite materials. *Engng. Fracture Mech.* 4:811-840.

Yevtushenko, A. & Kaczyński, A. & Matysiak, S.J. 1995. The stress state of a laminated elastic composite with a thin linear inclusion. *J. Appl. Maths. Mechs.* 59 (4): 671-676.

Kaczyński, A. 1993. Stress intensity factors for an interface crack in a periodic two-layered composite under the action of heat sources. *Int.J.Fract.* 62: 183-202.

Kaczyński, A. & Matysiak, S. J. 1988a. On crack problems in periodic two-layered elastic composites. *Int. J. Fract.* 37: 31-45.

Kaczyński, A. & Matysiak, S. J. 1988b. On the complex potentials of the linear thermoelasticity with microlocal parameters. *Acta Mechanica* 72: 245-259.

Kaczyński, A. & Matysiak, S.J. 1989. Thermal stresses in a laminate composite with a row of interface cracks. *Int. J. Engng. Sci.* 27 (2): 131-147.

Kaczyński, A. & Matysiak, S.J. 1995. Analysis of stress intensity factors in crack problems of periodic two-layered elastic composites. *Acta Mechanica* 110: 95-110.

Kit, G. S. & Krivtzun, M. G. 1983. *Thermoelastic plane problems for bodies with cracks.* Kiev: Naukova Dumka (in Russian).

Matysiak, S. J. 1995. On the microlocal parameter method in modelling of periodically layered thermoelastic composites. *J. Theor. Appl. Mech.* 33 (2): 481-487.

Matysiak, S. J. & Olesiak, Z. 1981. Properties of stresses in composites with ribbon-like inclusions *Mech. Teor. Stos.* 19 (3): 397-408 (in Polish).

Matysiak, S. J. & Woźniak, C. 1988. On the microlocal modelling of thermoelastic composites. *J. of Tech. Phys.* 29 (1): 85-97.

Muskhelishvili, N. I. 1966. *Some basic problems of the mathematical theory of elasticity.* Moskva: Nauka (ed.5, in Russian).

Sih, G. C. 1965. Plane extension of rigidly embedded line inclusions. *Proceedings 9 th Midwestern Mech. Conf.*: 61-79. Wiley.

Sih, G. C. & Chen, E. P. (eds.) 1981. Cracks in composite materials. *Mechanics of fracture* 6. The Netherlands. Martinus Nijhoff.

Tvardovsky, V. V. 1990. Further results on rectilinear line cracks and inclusions in anisotropic medium. *Theor. Appl. Fract. Mech.* 13: 193-207.

Wang, Z.Y. & Zhang, H.T. & Chou, Y.T. 1985. Characteristics of the elastic field of a rigid line inhomogeneity. *Trans. ASME. J. Appl. Mech.* 52: 818-822.

Woźniak, C. 1987. A nonstandard method of modelling of thermoelastic periodic composites. *Int. J. Engng. Sci.* 25: 483-499.

Damage and Failure of Interfaces, Rossmanith (ed.) © *1997 Balkema, Rotterdam, ISBN 90 5410 899 1*

Non-linear analysis of carbon-carbon composite nozzle with interface

S.G. Ivanov & A.A.Tashkinov
Perm State Technical University, Russia

N.V.Tokareva
Perm Military Institute of Ministry of the Home Affairs, Russia

P.G.Udintsev
Urals Research Institute of Composite Materials, Russia

ABSTRACT: Carbon-carbon composite material *3-D CARB*, reinforced at three mutually orthogonal directions is used in nozzle component of a solid fuel rocket engine. The inner surface of the component is subjected to the effect of hot exhaust gas flow, that's why an intensive material loss takes place. The surface layer of *3-D CARB* can be densified with silicon carbide (SiC) to decrease the value of material loss. A densified material has mechanical properties different from an undensified one, and there is an interface between these two materials. A non-linear finite element analysis, used parallel with a linear one, outlined the effectivity of the material densification not only from the point of view of decreasing material loss, but also from the point of view of improving stress state, including the vicinity of interface.

1 INTRODUCTION

Carbon-carbon composite materials (CCCM) have a number of unique properties which allow for instance to use these materials under high temperatures (McAllister & Lachman 1983; Sokolkin et al. 1996). *3-D CARB* is one of such materials, reinforced at three mutually orthogonal directions in ratio 1:1:1. The space between severely anisotropic carbon fibers is filled with a weak carbon matrix. This material is used in a nozzle component of a solid fuel rocket engine. This component is an axisymmetrical body of complex geometry containing three layers: inner layer of *3-D CARB* with an interface, then an intermediate one of heat-protective material (HPM) which preserves the third outer titanium layer from overheating. The inner surface of nozzle component is subjected to the effect of hot exhaust gas flow with extreme temperature up to 3000-3500° C. That's why an intensive material loss takes place. Surface layer of *3-D CARB* is densified with SiC to decrease the value of material loss. An undensified material has density 1650 kg/m^3 and material loss intensity 0.18 mm/s, while the densified with SiC material (noted as *3-D CARB+SiC*) has density of 1750 kg/m^3 and material loss intensity of 0.05 mm/s.

Thus, there is interface between two carbon-carbon materials. Thickness of *3-D CARB+SiC* layer varies in analysis from 0 to 1/3 of the total thickness of both CCCM layers. The mechanical characteristics of the densified material differ from the characteristics of the undensified one. Elastic modulus of the former is about 10% higher, and at the same time tensile strength is about 29% lower, compression strength is about 21% lower than those of undensified *3-D CARB*. That is why it is interesting to compare the stress and strain state of carbon-carbon layers during the nozzle's work both with and without the densified layer. At first we need to obtain an axisymmetric temperature field in structure at various moments of time. Then we are to solve a thermomechanical boundary value problem. Due to anisotropy of orthogonally reinforced composite, this problem cannot be treated as axisymmetrical. Because of the symmetry of material reinforcement in x and y axes the problem should be solved for 1/8 part of the structure. 1/8 part of the axial cross-section of the considered structure is shown in Figure 1.

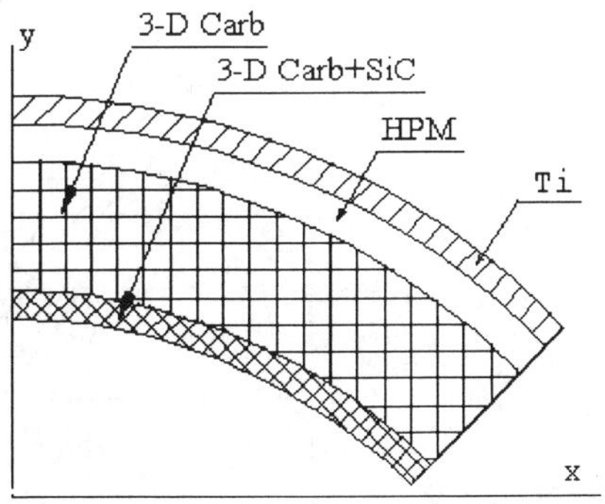

Figure 1. 1/8 part of nozzle cross-section.

2 HEAT CONDUCTION PROBLEM

The temperature field T(r,z,t) in the described compound nozzle component can be evaluated while solving the non-stationary axisymmetrical heat conduction problem. Supposing that the heat flow in materials of the nozzle in axial direction z is much less than the heat flow in radial direction r, we can treat this boundary problem as series of one-dimensional problems in z = const cross-sections. The temperature distribution is found in each cross-section as a solution of equation:

$$c_p \cdot \rho \cdot \frac{\partial T}{\partial t} = \frac{1}{r} \cdot \frac{\partial}{\partial r}\left(\lambda_r \cdot r \cdot \frac{\partial T}{\partial r}\right) \qquad (1)$$

where c_p is specific heat, ρ - density and λ_r - heat conduction coefficient in radial direction. The heat flow at inner (index 1) and outer (index 2) boundary surfaces is defined as :

$$\lambda_r \cdot \frac{\partial T}{\partial r}\bigg|_{r=R_i} = k_i \cdot \left(T\left(R_i, t\right) - T_i\right) \qquad (2)$$

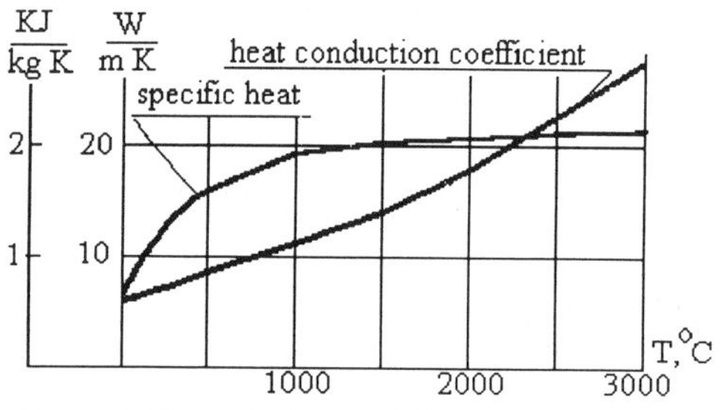

Figure 2. Thermal characteristics of 3-D CARB.

where k_i - heat transfer coefficient and T_i - temperature of gas media near the i - boundary surface (i = 1,2). This coefficient for the critical cross-section of the nozzle was taken as k_1 = 19 KW/m^2/K. The temperature of the inner gas flow T_1 = 3300 ° C, while the outer temperature is supposed to be T_2 = 20° C. Thus the temperature range is very wide and, of course, thermal properties of the materials vary with temperature. Figure 2 indicates the temperature dependences of heat conduction coefficient and specific heat of *3-D CARB*. Note, that the heat conduction problem in such a case becomes non-linear.

While solving the non-linear non-stationary problem (1) , (2) the material loss at the inner surface has been taken into account. Finite element procedure has been applied together with time recurrent scheme. This scheme allows to arrange time step process from initial to required moment of time. Thermal characteristics for other three materials of four-layered structure are set in Table 1.

The *3-D CARB+SiC* layer thickness varied in analysis from 0 to 8 mm, and the total thickness of two CCCM layers was the same for all variants considered, being 25 mm in critical section. The thickness of HPM layer is 7 mm and of titanium layer 5 mm. The designed time of the engine's work is 22.5 s. Examples of calculated temperature distri-

Table 1. Thermal characteristics used in computation.

Property	3-D CARB+ SiC	HPM	Titaniu m
Specific heat, KJ/kg/K	2.0	1.21	0.57
Heat-conduction coefficient, W/m/K	30	0.85	11.3
Density, kg/m^3	1750	1675	4550

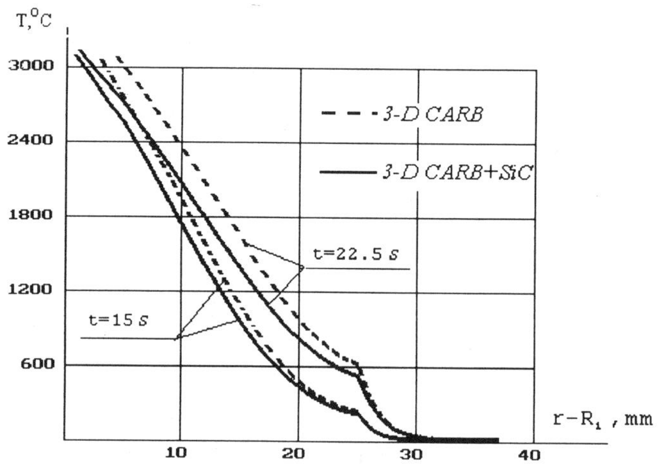

Figure 3. Temperature through thickness distribution.

butions at two moments (15 s and 22.5 s) for two variants - the thickness of the densified with SiC layer being 0 mm and 5mm are shown in Figure 3.

Comparing these temperature distributions, we are to note, that there is more intensive heating of the material by the same time in points with a same coordinate r in case of absence of *3-D CARB+SiC* layer. Thus, the maximum temperature in heat protective material in this case is about 100 degrees higher than in case of the presence of 5 mm densified layer. More intensive material loss leads to more intensive material heating, if we consider not the distance from moving surface, but fixed coordinate value for both cases considered. Note, that during all the designed time of engine's work the titanium shell is at the outer space temperature.

3 LINEAR THERMOELASTIC ANALYSIS

Stress field in structure has been found as a result of solution of the boundary value problem of linear thermoelasticity. Supposing that in examined cross-sections the stress-strain state is close to plane strain state, we have considered the plane boundary value problems for 1/8 part of nozzle cross-sections.

The nozzle component is subjected to internal pressure and temperature field T(r,t) , that has been evaluated in the way it was described in the previous section. In linear analysis we suppose that stresses and strains in each of four materials comply with thermoelastic equations:

$$\begin{Bmatrix}\sigma_x\\ \sigma_y\\ \sigma_z\end{Bmatrix} = D\cdot\begin{bmatrix}1-\nu & \nu & \nu\\ \nu & 1-\nu & \nu\\ \nu & \nu & 1-\nu\end{bmatrix}\cdot\begin{Bmatrix}\varepsilon_x-\alpha\cdot\Delta T\\ \varepsilon_y-\alpha\cdot\Delta T\\ -\alpha\cdot\Delta T\end{Bmatrix}, \qquad (3)$$

$$\tau_{xy} = G\cdot\gamma_{xy}.$$

$$D = E/(1+\nu)/(1-2\nu)\ ; \quad \Delta T = T(r,t) - T_0;$$

T_0 is initial (room) temperature at t = 0. Titanium and HPM are isotropic, and two carbon-carbon materials are anisotropic, having three independent elastic constants. Temperature dependences of Young modulus E of HPM and *3-D CARB* are disposed in Figure 4. A linear approximation between the experimental values has been used. Mechanical characteristics for *3-D CARB+SiC* are known at room temperature only. An elastic modulus E of this material is about 10% higher than that of the initial carbon material. In numerical analysis Young modulus E for densified material was supposed to be 10% higher than for initial material at any temperature.

The Poisson ratio ν of two carbon-carbon materials is considered to be temperature independent and equal to 0.05. The shear modulus G is an independent elastic constant for these orthogonally reinforced materials. It's value for *3-D CARB* at room temperature is 1.96 Gpa, while if material is isotropic with the same values of E and ν, the value of G will be about 7 Gpa. A temperature dependence of shear moduli for two CCCM is supposed to be proportional to the temperature dependence of corresponding longitudinal moduli at compression (Figure 4).

The values of elastic constants and coefficients of linear thermal expansion (CLTE) α for the isotropic materials are given in Table 2. Figure 5 indicates a temperature dependence of CLTE for the both types of CCCM layers.

Finite element method has been used with linear approximation of displacements in triangular elements. It is more descriptive to examine not stresses themselves, but their relations to the corresponding strengths at the same temperature.

The strengths of *3-D CARB* for tension and compression in orthogonal directions are disposed at Figure 6. Some differences in strengths between x,y and z directions are stipulated by technology: at first the fibers of x and y directions are layed , and then the obtained package is sewed by fibers of z direction. Note, that the strengths in temperature range from 1000 to 1500° C are even higher than at room temperature.

A limit shear stress for *3-D CARB* at room temperature is equal to 34.3 MPa. The temperature de-

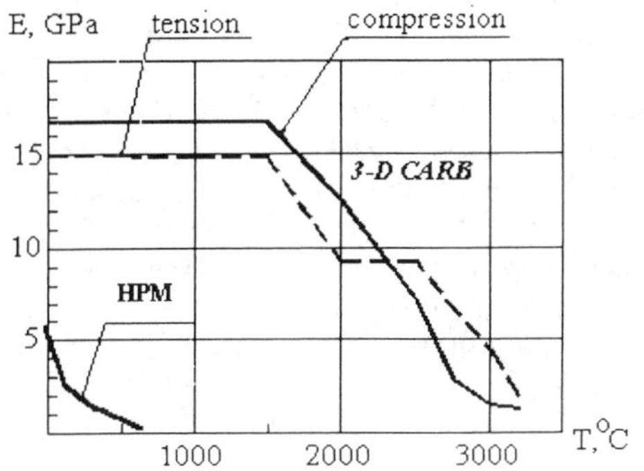

Figure 4. Elastic moduli of 3-D CARB and HPM.

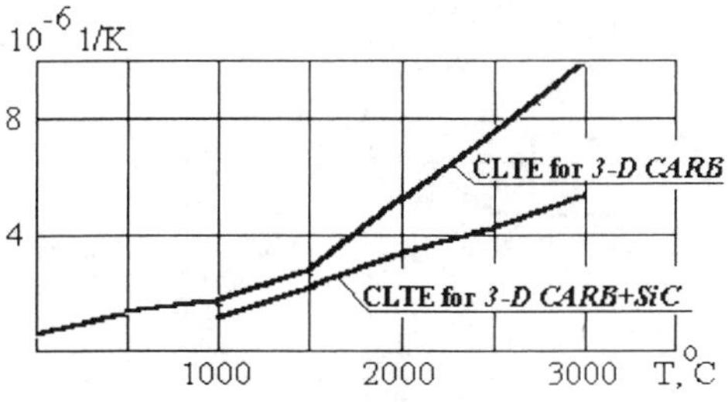

Figure 5. CLTE for carbon-carbon materials.

Table 2. Thermoelastic characteristics of HPM and Titanium.

Property	HPM	Titanium
Young modulus	Figure 4.	110 GPa
Poisson ratio	0.38	0.3
CLTE ,10^{-6} 1/K	11	8.6

pendence of the shear strength is taken as proportional to the temperature dependence of compression strength. The results of FEM analysis show that strength conditions in HPM and titanium layers are satisfied. Though the considered domain has the form of 1/8 of the ring, we still use a cartesian coordinate system, because the main axes of orthotropy for applied materials coincide with this system. Figure 7 illustrates some results of thermoelastic analysis of the nozzle without a densified layer in the most loaded region - the critical section of the nozzle at time of 22.5 s. The isolines of relative stresses (i.e. stresses, related to the corresponding strengths at the same temperature) are depicted only in CCCM layer.

The tensile strengh of *3-D CARB+SiC* material at room temperature is about 29% lower, and compression strength is about 21% lower than those of the undensified *3-D CARB*. That is why it is interesting to know, how these unfavourable changes in mechanical characteristics after the densification with SiC influence the picture of relative stresses. Temperature dependences of the densified material in calculations have been taken proportional to the corresponding dependences of initial material.

Figure 8 illustrates relative stresses in two CCCM layers in the critical section of the nozzle at 22.5 s. The initial thickness of *3-D CARB+SiC* material has been taken 5 mm in this case. Looking at Figures 7 and 8 we see that maximum values of relative compression stresses $\bar{\sigma}_z$ are obtained near the inner surface of the structure. The coordinate dependence for $\bar{\sigma}_z$ is obtained close to axisymmetrical.

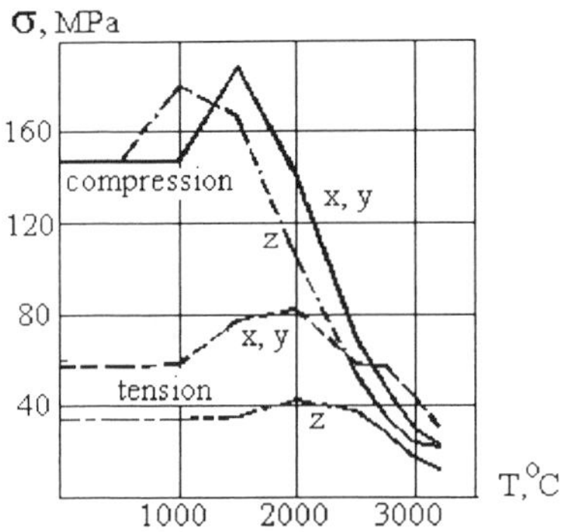

Figure 6. Strength limits for *3-D CARB*

Maximum values of compression stress $\bar{\sigma}_x$ are obtained at x=0, where it coincides with a circumferential one. The shear stress $\bar{\tau}_{xy}$ attains its maximum values in the region adjacent to the radial direction, forming the angle 45° with x and y directions.

Using linear analysis, we have obtained the overestimated values of stresses. Some of them exceed the corresponding strengths in the vicinity of the internal surface during the designed time of the nozzle's work: $\bar{\sigma}_z, \bar{\sigma}_x, \bar{\tau}_{xy}$ in Figure 7, $\bar{\sigma}_z, \bar{\sigma}_x$ in Figure 8.

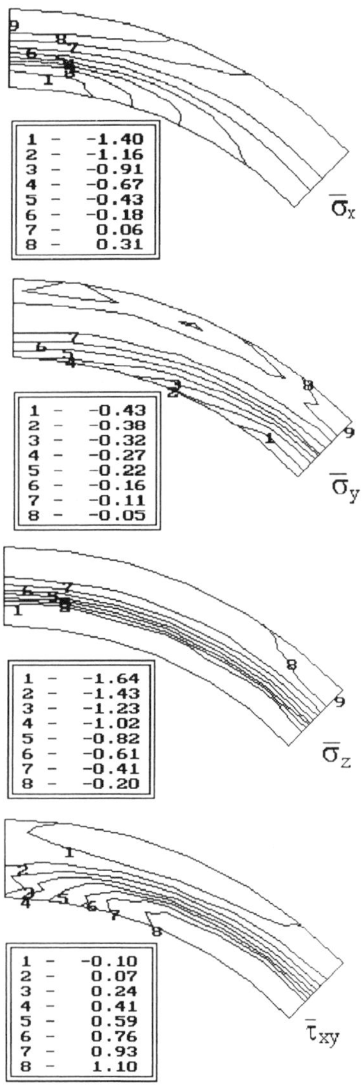

Figure 7. Relative stresses in *3-D CARB* layer in critical cross-section at 22.5 s.

Maximum $\bar{\sigma}_z$ in a case without a densified layer exceeds compression strength more than 1.6 times. Nevertheless, the linear analysis seems to be a good tool to compare different variants of the structure.

The obtained results have outlined the effectivity of CCCM densification with silicon carbide not only from the point of view of decreasing material loss, but also from the point of view of improving stress state, including the vicinity of interface. Notwithstanding the reduction of strength limits of the densified material compared to initial one, the stresses, related to the corresponding strengths in nozzle with a densified layer, are lower, than those in a structure without such a layer. This can be explained mainly by the smaller CLTE of the densified material in comparison to the undensified one (Figure 5).

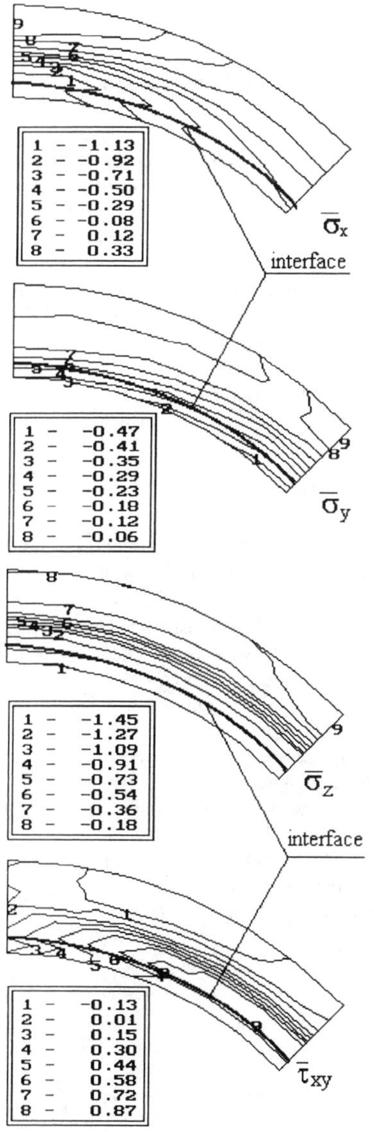

Figure 8. Relative stresses in CCCM layers in critical cross-section at 22.5 s. (initial thickness of the densified layer is 5 mm).

4 NON-LINEAR THERMOMECHANICAL ANALYSIS

The linear stress analysis has shown that namely compressing stresses are critical in these nozzle components. Stress - strain diagrams of uniaxial compression at different temperatures are submitted in Figure 9. We can see that up to the temperature of 2000°C the material can be treated practically as linear. But at higher temperatures the severe non-linearity is inherent to it. Thus we need to solve a non-linear boundary value thermomechanical problem for the structures under consideration. Because of the complex stress state in the points of the structure we are to chose one or another anisotropic plasticity theory for three-dimensional stress state. According to (Pobedrya 1984) an orthotropic material has (generally speaking) six stress invariants which coinside with stress components in the main axes of orthotropy. Let us construct a simple variant of a theory, using the peculiarities of the problem. A secant moduli method has been used for the numerical analysis of the structure. The non-linear boundary value problem is solved as a series of linear problems in which the elastic constants for CCCM materials are defined in each finite element for iteration k as:

$$E_i^{(k)} = \begin{cases} E_i^{tens}, & \varepsilon_i - \alpha \cdot \Delta T > 0; \\ \dfrac{\sigma_i\left(\varepsilon_i^{(k-1)} - \alpha \cdot \Delta T\right)}{\varepsilon_i^{(k-1)} - \alpha \cdot \Delta T}, & \varepsilon_i - \alpha \cdot \Delta T \le 0 \end{cases} \quad (4)$$

where E_i^{tens} - modulus in tension, and i = x, y or z.

While evaluating the elastic moduli of k iteration according to (4) we use the piece-linear approximation of stress - strain diagrams in the necessary interval (Figure 10). Owing to such approximation, we can easily produce a stress - strain dependence for the temperature that occurs in a given finite element. Since the values of tensile stresses in CCCM layers are not large, we deal with a linear stress - strain dependence in tension. A stress - strain diagram in shear at a given temperature is supposed to be proportional to a diagram in compression (Figure 10). For the densified layer stresses are 10% higher at the same strain than those of an undensified layer in any case.

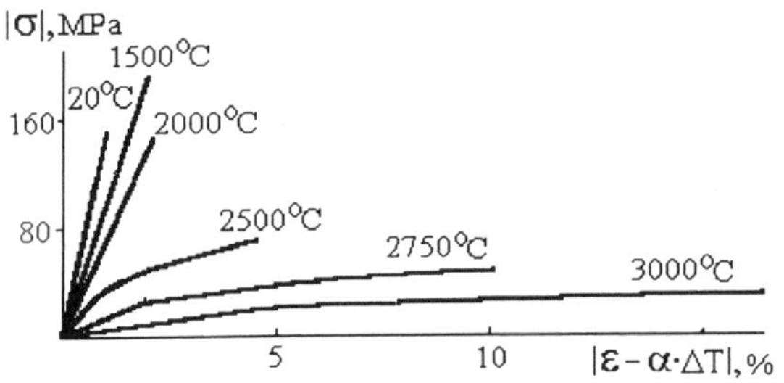

Figure 9. Stress - strain diagrams in compression for *3-D CARB.*

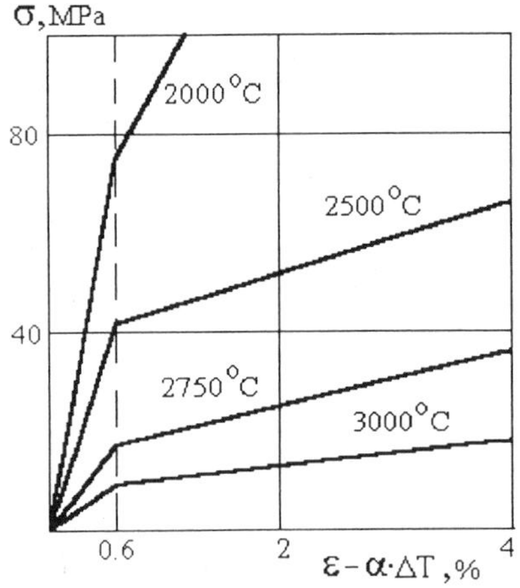

Figure 10. Approximation of compression stress - strain dependences for *3-D CARB.*

For each k iteration we are to solve the linear problem with thermoelastic equations (i = x, y or z):

$$\sigma_i = \sum_{j=1}^{3} D_{ij}^{(k)} \cdot (\varepsilon_j - \alpha \cdot \Delta T);\ \tau_{xy} = G^{(k)} \cdot \gamma_{xy}\ ;$$

$$D_{ii}^{(k)} = \frac{E_i^{(k)} \cdot (1-\nu)}{(1+\nu)\cdot(1-2\nu)};\ D_{ij}^{(k)} = \frac{\nu}{1-\nu}\sqrt{D_{ii}^{(k)} \cdot D_{jj}^{(k)}},$$

$i \neq j$.

5 CONCLUSIONS

Computational procedures were described to evaluate the stress and strain state during the work of the nozzle component containing the three-directional carbon-carbon layer. To decrease the value of material loss the densification of this layer with SiC can be applied. The temperature field has been calculated by FEM solution of the nonlinear non-stationary axisymmetrical heat conduction problem. The linear and nonlinear FEM analyses have been used to solve the thermomechanical problem. Some results concerned a thermoelastic solution are given also in (Tashkinov et al. 1995). Though using the linear analysis we obtain the overestimated values of stresses, it provides a good tool for comparison of different structure variants. A simple variant of anisotropic theory, taking into account the peculiarities of the problem has been constructed to obtain a nonlinear solution. Both the nonlinear and linear analyses of the structure show the effectivity of densifying the CCCM layer not only from the point of view of decreasing material loss but also from the point of view of improving stress state, including the vicinity of interface between densified and undensified carbon-carbon layers.

REFERENCES

McAllister, L.E. & W.L.Lachman 1983. Multidirectional carbon-carbon composites. In A.Kelley & S.T.Mileiko (eds), *Handbook of Composites, vol.4. Fabrication of Composites:* 109-176. Amsterdam:North-Holland.

Pobedrya, B.E. 1984. *Mechanics of composite materials* (in Russian). Moscow: Moscow Univ.Publ.

.Sokolkin, Yu.V, A.M.Votinov, A.A.Tashkinov, A.M. Postnykh & A.A.Chekalkin 1996. *Technology and design of carbon-carbon composites and structures* (in Russian). Moscow : Science, Physmathlit.

Tashkinov, A.A., S.G.Ivanov, P.G.Udintsev, V.Yu.Chunaev & I.N.Rochev 1995. Thermoelastic analysis of the behaviour of carbon-carbon structure components with densified surface layer (in Russian). *Transactions of PermSTU. Mechanics*.2: 66-74.

Damage and Failure of Interfaces, Rossmanith (ed.) © *1997 Balkema, Rotterdam, ISBN 90 5410 899 1*

Prediction of damage and failure in thermal barrier coatings

Kwai S. Chan, Yi-Der Lee & Gerald R. Leverant
Southwest Research Institute, San Antonio, Tex., USA

Thomas J. Fitzgerald & John G. Goedjen
Westinghouse Electric Corporation, Orlando, Fla., USA

ABSTRACT: This paper summarizes the development of a life prediction model for treating failure of thermal barrier coatings (TBC) in gas turbine applications. The model has been formulated based on consideration of the formation of a wing crack at the tip of a shear crack and its propagation near the interface of the TBC and the thermally grown oxide. The effects of thermomechanical fatigue, oxidation, bond coat creep, sintering, curvature, and interface roughness on the propagation of the wing cracks have been treated mechanistically to obtain the form of the TBC life equation. Material constants in the life equation are, however, evaluated from experimental data. Application of the TBC life model to predicting failure of air-plasma-sprayed TBCs indicates that the coating life can be reasonably predicted by the model.

1 INTRODUCTION

Many hot-section components in gas turbine engines are coated with ceramic thermal barrier coatings (TBCs) that are intended to increase the operating temperature and efficiency of the engines. In the coating process, the Ni-base superalloy substrate is coated first with a bond coat on top of which a ceramic TBC is then applied. A thin oxide layer is usually formed between the bond coat and the TBC during the coating process. As a result, a TBC system consists of a four-layered structure that includes the TBC top coat, a thermally grown oxide (TGO), an oxidation resistant bond coat (BC), and a Ni-base superalloy substrate.

Several damage and failure mechanisms are possible in the TBC system when subjected to temperature and strain cycling; they include thermomechanical fatigue (Miller 1984, Miller 1987, DeMasi et al. 1989, Meier et al. 1991, Meier et al. 1991, Cruse, et al. 1988), bond coat oxidation (Miller 1984, Miller 1987, DeMasi et al. 1989, Meier et al. 1991, Meier et al. 1991, Cruse, et al. 1988), fracture in the TBC (DeMasi et al. 1989, Strangman et al. 1987, DeMasi-Marcin et al. 1989), oxide (Strangman et al. 1987), or at the oxide/TBC interface (Brindley 1996). Creep and stress relaxation in the bond coat (Brindley 1996, Brindley 1995), sintering in TBC (Brindley 1996), roughness of the TBC/TGO interface (Evans et al. 1983, Chang et al. 1987), as well as interdiffusion of Al from the bond coat to the substrate are also important (Brindley 1996). Most of these damage mechanisms are time- and temperature-dependent processes whose synergism (e.g., thermomechanical fatigue and oxidation) is detrimental to the integrity of the TBC coating (Miller 1984, Miller 1987, DeMasi et al. 1989, Meier et al. 1991, Meier et al. 1991, Cruse, et al. 1988). An accurate prediction of time- and temperature-dependent damage accumulation and failure in TBC coating is, therefore, important for the safe and efficient utilization of gas turbine engines. A number of life prediction models have been developed to treat failure of TBC resulting from thermomechanical fatigue and bond coat oxidation (Miller 1984, Miller 1987, DeMasi et al. 1989, Meier et al. 1991, Meier et al. 1991, Cruse, et al. 1988, Strangman et al. 1987). However, time-dependent damage processes such as bond coat creep, sintering of microcracks in TBC, and changes in the roughness of the TBC/TGO interface during high-temperature exposure are not addressed in the current models.

The development of a life prediction methodology for treating several time-dependent damage mechanisms in TBC coating is summarized in this paper. The methodology is developed based on micromechanical treatments of relevant damage mechanisms in the TBC to obtain the general form of the damage evolution and life expressions.

Material constants in these equations, however, are obtained from experimental data so that accurate and reliable life predictions of engineering components are feasible. The synergistic effect of thermomechanical fatigue and oxidation under time-dependent degradation processes such as TBC sintering, bond coat creep, and roughness of the bond coat/oxide interface is also considered in the derivation of the coating life equation. Application of the proposed model to predicting the useful life of air-plasma-sprayed TBC is presented and compared against experimental data.

2 SUMMARY OF MODEL DEVELOPMENT

The development of a life prediction model for treating the spallation of TBC coatings is summarized here. The model is developed for a TBC system that consists of a Ni-based superalloy substrate coated with a bond coat and a TBC top coat. A thin alumina oxide layer is formed between the bond and the TBC top during the coating process, resulting in a four-layered structure as shown schematically in Figure 1. The TBC system is subjected to an externally applied cyclic stress, temperature, and strain. The surface of the TBC system is at temperature T_{TBC} and subjected to a heat flux, Q_{flux}, that causes a temperature gradient across the layered structure. The interface between the oxide layer and the TBC is generally nonplanar. Both the oxide thickness and the interface roughness increase with time at elevated temperatures.

Motivated by experimental observations (DeMasi et al. 1989, DeMasi-Marcin et al. 1989), the TBC is considered to contain microcracks that could slide by shear when subjected to compressive loading. Sliding of the shear cracks leads to the development of wing cracks at the tips of the shear cracks. Spallation of TBCs is treated in terms of fatigue crack growth of these wing cracks when the TBC system is subjected to cyclic thermomechanical loads. The wing cracks are considered to either lie near the TBC/oxide interface or reside within the oxide layer, depending on the type of TBC. For air-plasma-sprayed TBC considered in this paper, damage accumulation and failure occur at or near the oxide/TBC interface (Brindley 1996, DeMasi et al. 1989). The driving force for fatigue crack growth of the wing cracks are the stress ranges associated with the stress field of the shear crack and the radial stress range originating from curvature, waviness of the TBC/TGO interface, and bond coat creep, as illustrated in Figure 1. The growth of the wing crack is described in terms of a power-law (Paris

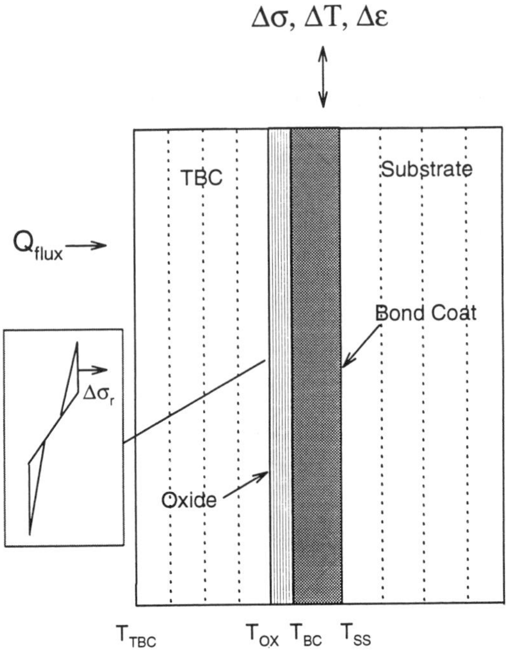

Figure 1. A schematic of a four-layered TBC system consisting of a TBC top coat, a thermally grown oxide (TGO), a bond coat (BC), and a Ni-base superalloy substrate. Fatigue crack propagation of shear-induced wing cracks leads to the spallation of TBC near the TBC/oxide interface.

equation). Integrating the fatigue crack growth equation from an initial crack length the critical crack length at fracture gives the low cycle fatigue (LCF) life equation. After sintering of microcracks in the TBC and the radial stresses due to curvature, interface roughness, and bond coat creep are incorporated, the final form of LCF life relation for TBC spallation is given by

$$\Delta \epsilon_{mech} N_f^b = c \Lambda_t^m (1 - \Lambda_o) \quad (1)$$

where $\Delta \epsilon_{mech}$ is the mechanical strain range; N_f is the cycles-to-failure (spallation) of the TBC; $(1-\Lambda_o)$ is the oxidation damage term; Λ_t is a time-dependent degradation term that includes microcrack sintering in the TBC, curvature, interface roughness, and bond coat creep; and b, c, and m are material constants that are evaluated from experimental data. Both Λ_o and Λ_t evolve with time and thermomechanical loading histories; their development is described by appropriate evolution equations for individual degradation mechanisms.

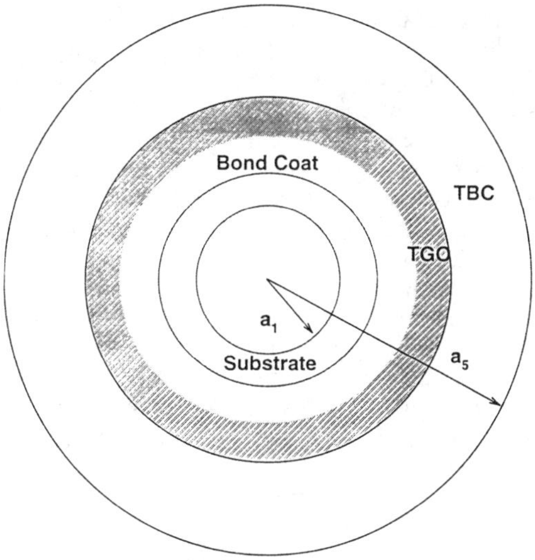

Figure 2. A schematic of a four-layered TBC cylinder used to model the effects of curvature, interface roughness, and bond coat creep on thermally-induced radial stress at the oxide/TBC interface and on the spallation of the TBC.

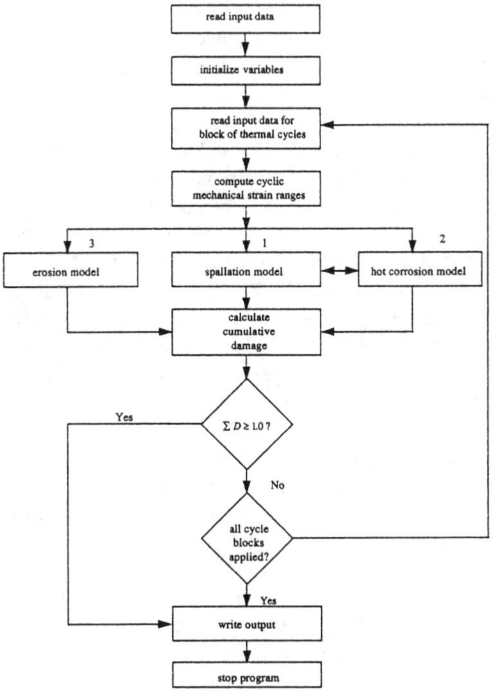

Figure 3. An overview of the TBC Life Prediction Model (TBCLPM).

The effects of curvature and the growth stress associated with oxide formation have been modeled by considering a four-layered cylinder, as shown in Figure 2. These effects of curvature are analyzed by prescribing thermal strain in the TBC and a transformation strain in the oxide layer. A thermal strain is prescribed in the oxide layer. The stress distributions in each of the four layers are then analyzed as elastic media and obtained by solving the radial force and displacement equations, while imposing compatibility at each of the layer interfaces. The result shows that the radial stress due to a geometric curvature is small but not negligible. On the other hand, the growth stress is large but is almost stationary, and its range during thermal cycling, being two orders of magnitude smaller than the curvature term, can be ignored. The result obtained for the curvature effect is also applied to treat interface roughness by replacing the geometric radius of curvature by the local radius of curvature of the wavy TBC/TGO interface. The resultant radial stress on the wing crack is the sum of the curvature and the interface roughness terms.

The four-layered cylinder model is also used for treating stress relaxation due to bond coat creep at elevated temperatures. In this formulation, the bond coat is assumed to exhibit primary creep according to a simple power-law as described by

$$\epsilon_c = A(\sigma/E)^n t^s \tag{2}$$

where ϵ_c is the creep strain; σ is the stress; E is Young's modulus; t is time of creep; n is the creep exponent; s is a time constant which is < 1 for primary creep, but is unity for steady state creep; and A is a constant. Using this material law, stresses in the bond coat calculated based on the elastic analysis of the four-layered cylinder are allowed to relax as a function of time under constant strain conditions. As the stresses in the bond coat are relaxed, stresses in the TBC, oxide, and substrate are lowered as well. As a result, bond coat creep exerts a significant effect on the magnitude of the radial stress at the TBC/TGO interface and, therefore, on the spallation and the life of the TBC.

The life prediction model has been programmed into a computer code dubbed TBC Life Prediction Model (TBCLPM), whose overall structure is summarized in Figure 3. The computer model calculates the mechanical strain range based on input of material parameters and thermal and loading histories. The calculated mechanical strain

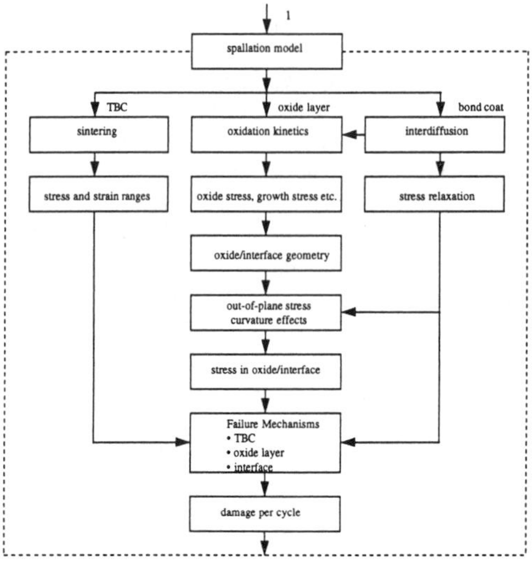

Figure 4. A summary of degradation mechanisms treated in the spallation model in the TBCLPM code.

range is then used in one of three degradation models to calculate the value of damage for the current thermomechanical cycle. The current model treats failure by spallation only; damage by hot corrosion and erosion are not considered. The damage per cycle, ΔD, is calculated as the inverse of the cycles-to-failure, N_f, for the corresponding mechanical strain range through Eq. (1). A linear damage rule given by

$$D = \sum \Delta D = \sum \frac{1}{N_f} \geq 1 \qquad (3)$$

is used as the criterion for the onset of spallation and for computing the TBC life. The various degradation mechanisms treated in this model are summarized in the flow chart shown in Figure 4.

3 MODEL APPLICATION TO AIR-PLASMA-SPRAYED TBC

As part of NASA's Hot Section Technology Program, Pratt & Whitney Aircraft conducted a number of burner rig tests on air-plasma-sprayed TBC to generate coating life data to aid TBC life prediction model development (DeMasi et al. 1989, Meier et al. 1991). In addition to baseline data, additional verification experiments involving more complicated thermal histories that emphasized strain cycling or oxidation damage were also performed. The TBC life baseline data were used for evaluating some of the material constants in the life prediction model developed in this program. The remainder of the model constants were evaluated from experimental data generated by Westinghouse. Furthermore, the TBC life data for the strain-emphasized and oxidation-emphasized thermal cycles from the HOST program were used for independent evaluation of the present model. Because of limited space, only important results from the strain-emphasized and oxidation-emphasized thermal cycles are highlighted here. Details of the model development, determination of material constants, and model calculations are presented elsewhere.

A typical strain-emphasized thermal cycle is shown in Figure 5(a), while an oxidation-emphasized thermal cycle is shown in Figure 5(b). The difference between the two types of thermal cycle is a temperature hold at the peak temperature in the oxidation-emphasized thermal cycle. The minimum temperature (T_{min}) and the maximum temperature (T_{max}) at the interface of the TBC/TGO for the verification tests in the HOST Program (DeMasi et al., 1989) are shown in Table 1, with the corresponding times for heat-up, hold, and cold down. Using the thermal histories as input, the TBCLPM calculated the mechanical strain ranges in various elements of the layered TBC system during a thermal cycle. Figure 6 presents a typical result of the distribution of the calculated mechanical strain range in the TBC system after one cycle. In most instances, the largest mechanical strain range occurred in the TBC element located adjacent to the TBC/oxide interface. Once the mechanical strain range was calculated, the cycles-to-failure were obtained through Eq. (1) and the cumulative damage was computed via Eq. (3). Figure 7 shows the damage evolution curve calculated for a strain-emphasized thermal cycle compared to that for an oxidation-emphasized cycle. The TBC life was calculated as the cycle at which D reached unity.

Six different thermal history cycles were calculated and the results are compared against experimental data in Table 1. Since none of the experimental data were used to obtain the material constants in the model, Table 1 presents an independent comparison between model and experiment. Table 1 shows the types of thermal cycle, which include strain-emphasized and oxidation-emphasized cycles; some of the strain-emphasized cycles contain two different levels of strain ranges in series. The minimum and maximum

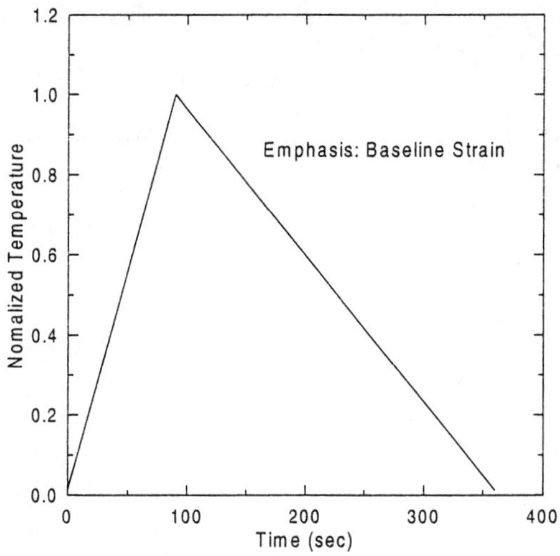

(a)

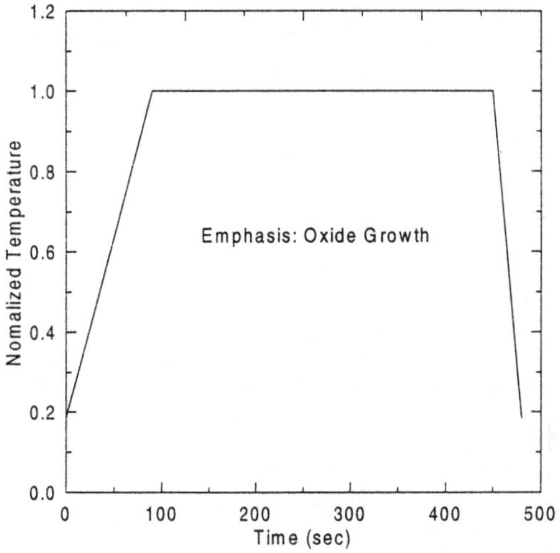

(b)

Figure 5. Temperature histories of thermal loading cycles used in verification experiments: (a) strain-emphasized thermal cycle, and (b) oxidation-emphasized thermal cycle.

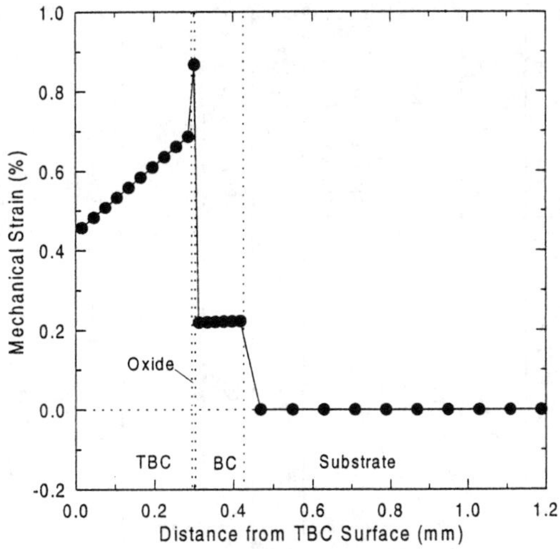

Figure 6. Distribution of mechanical strains in the TBC at the maximum temperature.

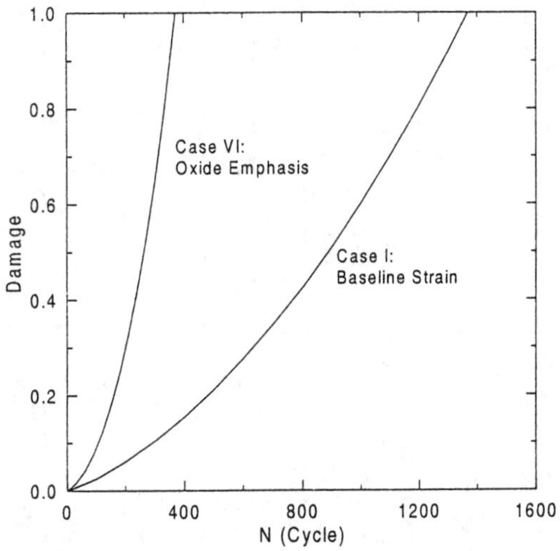

Figure 7. Calculated damage evolution curves for the strain-emphasized and the oxidation emphasized cycles.

temperatures at the TBC/TGO interface are indicated, together with the times for heat-up, hold at the peak temperature, and cool down. The calculated mechanical strain range, the oxide thickness at the onset of TBC failure, the calculated and the actual TBC lives (DeMasi et al. 1989) are all presented in Table 1. Model calculations are presented for cases with and without the time-dependent damage term, Λ_t. In the former case, failure of TBC was due to a combination of thermomechanical fatigue, oxidation, and time-dependent degradation mechanisms such as sintering, interface roughness, and bond coat creep. In the latter case, failure was due to thermomechanical fatigue and oxidation only. As indicated in Table 1, the inclusion of time-dependent damage in the calculation led to a lower calculated TBC life and a better agreement with the experimental data compared to the calculation without the time-dependent damage term. A comparison of

Table 1. Comparison of calculated and observed TBC lifes for six verification tests. Experimental data are from DeMasi et al. [4]

		Interface Temperature							TBC Life		
									Model Prediction (Cycles)		
Case	Emphasis	T_{min}, °C	T_{max}, °C	Heat-up Time (min)	Hold Time (min)	Cool-down Time (min)	Mechanical Strain Range (%)	Oxide Thickness (μm)	Without Λ_t	With Λ_t	Experimental Results (Cycles)
I	Baseline Strain	38	1149	1.5	0	4.5	0.72	2.31	5510	1369	1513
II	Oxide	481	1138	1.5	6	0.5	0.63	12.9	9719	1212	431
III	Mixed (strain)	28	1077	1.5	0	4.5	0.63	1.732	1310	1310	1310
	(strain)	28	1154	1.5	0	4.5	0.77	1.95	980	157	665
IV	Mixed (strain)	29	1137	1.5	0	4.5	0.74	1.749	602	602	602
	(strain)	28	1102	1.5	0	4.5	0.68	3.05	21425	2003	267
V	Baseline Strain	38	1149	1.5	0	4.5	0.76	1.85	1289	503	570
VI	Oxide	217	1136	1.5	6	0.5	0.72	5.70	2110	371	1000

model predictions with both the time-dependent damage and oxidation terms against the verification data is shown in Figure 8 with the baseline data from which the model constants were obtained. Figure 8 shows that the baseline data exhibit a relatively large scatter; part of the scatter in the baseline data appears to arise from the various time-dependent degradation processes in the TBC. The pair of open and closed diamond symbols joined by a dotted line represents the model prediction and verification data for the same test condition. Currently, the model prediction is within a factor of four of the experimental result in the worst case, which is within the scatter of the baseline data. Figure 9 shows a comparison of the measured and predicted TBC lives for the baseline data. It is noted that some of the model constants were obtained using the baseline data. As a result, the comparison indicates how well the model reproduces the experimental data only. Figure 9 also shows a comparison of the current model against results of TBCLIF model reported by Cruse et al. (1988) and DeMasi et al. (1989). The comparison indicates a slight improvement by the current model. The standard error of the predicted lives for the baseline data is ± 893 cycles for the current model (TBCLPM) compared to ± 5659 cycles for the TBCLIF model. Comparison of the observed and predicted TBC lives for the verification experiments is shown in Figure 10. In this case, both the current model (TBCLPM) and the TBCLIF model are within a factor of three of the measured TBC lives. The standard error of the predicted lives is ± 847 cycles for the current model and is ± 991 cycles for the TBCLIF model (Cruse et al., 1988; DeMasi et al., 1989). Effort is ongoing to improve further the accuracy of the current (TBCLPM) model.

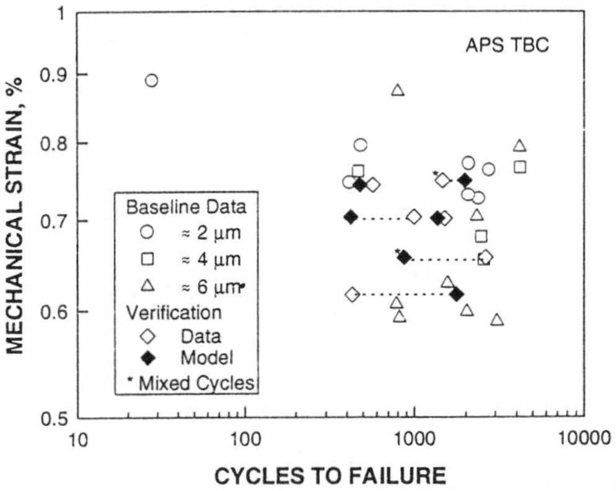

Figure 8. A comparison of the model prediction with both the oxidation and time-dependent damage terms against the verification and the baseline data from which the model constants were obtained.

4 SUMMARY

The development of a TBC life prediction model for land based gas turbine applications is summarized in

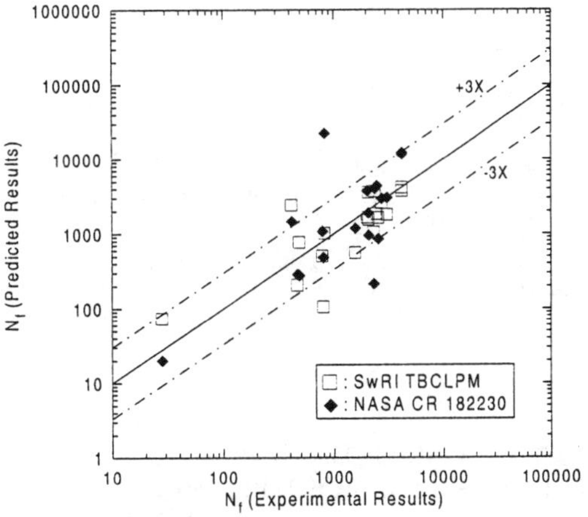

Figure 9. A comparison of measured and predicted TBC lives of the baseline data for the current model (TBCLPM) and for the (TBCLIF) model of Cruse et al. (1988) and DeMasi et al. (1989).

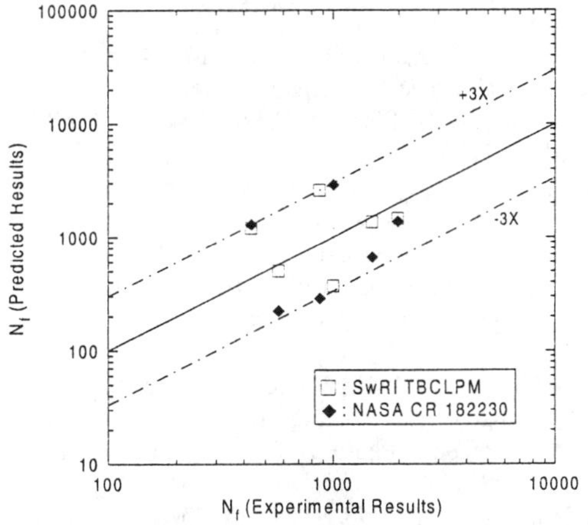

Figure 10. A comparison of measured and predicted TBC lives of the verification data for the current model (TBCLPM) and for the (TBCLIF) model of Cruse et al. (1988) and DeMasi et al. (1989).

this paper. Spallation of the TBC is treated as a thermomechanical fatigue process under the adverse influence of oxidation, bond coat creep, and sintering. The degradation mechanisms are treated in the strain-life relation that governs the spallation of TBC, which also incorporates the effects of curvature and interface roughness. A linear damage rule is used to describe the accumulation of fatigue damage for arbitrary thermomechanical strain cycles. Application of the TBC life model to air-plasma-sprayed TBC has been demonstrated with favorable results.

ACKNOWLEDGEMENT

This work was supported by the ATS Program, Department of Energy through Contract No. DE-AC05-95OR22242. The clerical assistance of Ms. P. A. Soriano in the preparation of this paper is appreciated.

REFERENCES

Brindley, W. J. 1995. *Proceedings of TBC Workshop*. NASA-CP 3312: 189-202.

Chang, G. C., W. Phucharoen & R. A. Miller 1987. Thermal expansion mismatch and plasticity in thermal barrier coating. NASA CP-2493 (10): 357-368.

Cruse, T. A., S. E. Stewart & M. Ortiz 1988. Thermal barrier coating life prediction model development. *J. Engineering for Gas Turbines and Power* 110: 610-616.

DeMasi, J. T., K. D. Sheffler & M. Ortiz 1989. Thermal barrier coating life prediction model development—Phase I Final Report. NASA CR-182230.

DeMasi-Marcin, J. T., K. D. Sheffler & S. Bose 1989. ASME Paper 89-FT-132.

Evans, A. G., G. B. Crumley & R. E. Demaray 1983. On the mechanical behavior of brittle coatings and layers. *Oxidation of Metals* 20(5-6): 193-216.

Meier, S. M., D. M. Nissley & K. D. Sheffler 1991. Thermal barrier coating life prediction model development—Phase II Final Report. NASA CR-189111.

Meier, S. M., D. M. Nissley, K. D. Sheffler & T. A. Cruse 1991. Thermal barrier coating life prediction model development. ASME Paper 91-GT-40.

Miller, R. A. 1984. Oxidation based model for thermal barrier coating life. *J. Amer. Cer. Soc.* 67(8): 517.

Miller, R. A. 1987. Progress towards life modelling of thermal barrier coatings for aircraft gas turbine engines. ASME Paper 87-ICE-18.

National Council 1996. Coatings for high-temperature structural materials trends and opportunities. National Academy Press: 26-33, 72-77.

Strangman, T. E., A. Liu & J. Neumann 1987. Thermal barrier coating life-prediction model development final report. NASA CR-179648.

Failure of fibrous composites

Damage and Failure of Interfaces, Rossmanith (ed.) © *1997 Balkema, Rotterdam, ISBN 90 5410 899 1*

Multiple fiber breaks in a fiber composite with a creeping matrix

Irene J. Beyerlein & S. Leigh Phoenix
Department of Theoretical and Applied Mechanics, Cornell University, Ithaca, N.Y., USA

ABSTRACT: This paper investigates the influence of pre-existing damage on the local creep response of a planar unidirectional fiber composite with elastic reinforcement and a Newtonian viscous matrix or interface under steady axial tension. We perform micromechanical calculations for the time-dependent stress and strain redistribution in the fiber and matrix around various combinations of transversely aligned and staggered fiber breaks. Calculations for large numbers of breaks are possible through the use of our efficient computational mechanics technique, called viscous break interaction (VBI), which is built on shear-lag theory. The results elucidate key relationships and differences between the time variation in peak stress concentrations and break opening displacements for cracks versus spatially staggered breaks, and quantify how the interactive effects of matrix creep and microstructural length scales influence the time-scales of stress redistribution. Favorable comparisons to continuum fracture mechanics are made.

1 INTRODUCTION

Long term tests on fiber composites indicate that the onset of stress-rupture or tertiary creep is due to accelerated nucleation and coalescence of fiber breaks, void or cavity formation in the matrix, and linkage of interfacial cracks. The process determining lifetime appears to involve numerous fiber breaks particularly at the beginning and end. Moreover, it is widely recognized that when and where creep damage progresses is governed by the time-varying stress redistribution around the current state of damage. In order to model creep damage evolution as would naturally occur in a large composite, a *physically realistic* and *fast* micromechanical stress analysis is needed. In this work, we employ our recently developed and efficient computational mechanics technique, called viscous break interaction (VBI), to examine the time-evolution of the fiber tensile stresses and displacements in and around both transversely staggered breaks and aligned breaks.

The key difference between VBI and other numerical schemes, such as finite element and spring network models, is that the computation time is tied to the volume of damage (e.g. number of fiber breaks) rather than the composite volume itself. To date a few 'damage-dependent' computational mechanics techniques have been developed for large scale simulations of failure in elastic fiber composites (Beyerlein et al. 1996, Beyerlein & Phoenix 1997a, Ibnabdeljalil & Curtin 1996). In particular, Beyerlein et al. (1996) showed that the 2-D elastic Hedgepeth shear-lag model produced stress fields consistent with linear elasticity and linear elastic fracture mechanics for long rows of aligned breaks modeling a transverse crack. In the VBI stress analysis of this paper, the matrix is assumed to creep in shear as a linearly viscous (Newtonian) material and the fibers remain the primary load bearing component with time independent, elastic properties. Such a model applies at elevated temperatures to polymer matrices or to advanced ceramics composites with a solid glassy matrix.

Several previous models involving viscoelastic matrices have been developed but for much smaller numbers of fibers or fiber breaks than treated here (Lifshitz & Rotem 1970, Gusev & Ovchinskii 1984, Lagoudas et al. 1989). These works show that an important consequence of the viscoelasticity is broadening of the effective load transfer or overload lengths on those fibers neighboring breaks, approximately at a rate proportional to the square root of the matrix creep compliance. We show similar results for much larger aligned break arrays or cracks than treated before, but also find significantly different features for non-aligned, widely spaced, fiber breaks.

2 MODEL FORMULATION

2.1 *Composite lamina and governing equations*

Figure 1 shows the central region within the model planar unidirectional fiber composite, which is infinitely long and infinitely wide, and contains a

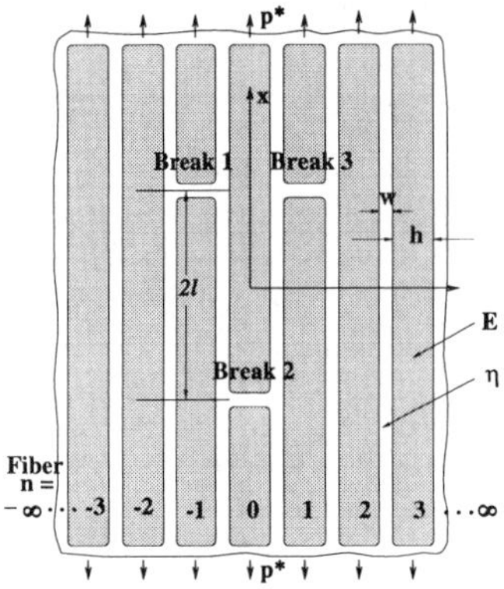

Figure 1. Region in infinitely large fiber composite lamina.

few staggered fiber breaks. The center fiber is $n = 0$, the fibers to the right are $n = 1, 2, \ldots, \infty$ and those to the left, $n = -1, -2, \ldots, -\infty$. The matrix bay to the *right* of fiber n is matrix bay n. Also indicated on Figure 1 is E, the Young's modulus of the fibers; η, the matrix viscosity; w, the matrix width; h, the fiber diameter. We also let h be the lamina thickness, and A, be the fiber cross-sectional area. The applied far field tensile load is denoted as p^* per fiber.

For problems studied here, the matrix creep compliance $J(\mathcal{T})$, ignoring any initial elastic response, is a linear function of time:

$$J(\mathcal{T}) = J_v(\mathcal{T}/\mathcal{T}_c) \tag{1}$$

where $\mathcal{T}$ is time and $\mathcal{T}_c$ is the characteristic time constant for matrix relaxation. Note that when $\mathcal{T} = \mathcal{T}_c$, the creep strain is the constant J_v, which has dimensions [stress^{-1}], reminiscent of the elastic compliance. According to (1), the matrix is a linear Newtonian viscous material with constant viscosity $\eta = \mathcal{T}_c/J_v$ or in other words, the matrix strain $\gamma_n(x,\mathcal{T})$ is a linear function of matrix shear stress $\tau_n(x,\mathcal{T})$:

$$\tau_n(x,\mathcal{T}) = \eta \partial\gamma_n(x,\mathcal{T})/\partial\mathcal{T}. \tag{2}$$

where $\gamma_n(x,\mathcal{T})$ is approximated as the differential displacement between the two flanking fibers n and $n+1$ over w, i.e. $\gamma_n(x,\mathcal{T}) = (u_{n+1}(x,\mathcal{T}) - u_n(x,\mathcal{T}))/w$. For the above constitutive behavior (2), we obtain the following dimensionless shear-lag equation for $P_n(\xi,t)$, which encompasses equilibrium, constitutive behavior, and compatibility:

$$\frac{\partial^2 P_n(\xi,t)}{\partial \xi^2} + \frac{\partial}{\partial t}\left(P_{n+1}(\xi,t) - 2P_n(\xi,t) + P_{n-1}(\xi,t)\right) = 0, \tag{3}$$

where we have applied the following normalizations for the axial coordinate x and time $\mathcal{T}$ (Beyerlein et al. 1997):

$$\xi = \frac{x}{\sqrt{\dfrac{wEAJ_v}{h}}}, \qquad t = \frac{\mathcal{T}}{\mathcal{T}_c} = \frac{\mathcal{T}}{\eta J_v}, \tag{4}$$

and for the fiber displacements $u_n(\xi,t)$ and loads $p_n(\xi,t)$:

$$U_n(\xi,t) = \frac{u_n(x,\mathcal{T})}{p^*\sqrt{\dfrac{wJ_v}{AhE}}} \tag{5}$$

and

$$\frac{\partial U_n(\xi,t)}{\partial \xi} = P_n(\xi,t) = \frac{p_n(x,\mathcal{T})}{p^*}. \tag{6}$$

By (6) the normalized far field load P is unity. Dimensionless versions of the matrix shear stresses, strains, and strain rates, $T_n(\xi,t)$, $\Gamma_n(\xi,t)$, and $\partial\Gamma_n(\xi,t)/\partial t$, can also be determined (Beyerlein et al. 1997).

2.2 *VBI technique*

The first task is to solve for the fiber and matrix displacements and loads due to an isolated fiber break under the boundary conditions of $P = 0$ far field, a unit compressive load -1 on the fiber break ends, and an initial condition of zero displacement at the break. To distinguish this unit solution from the general solution, we use the symbols V_n and L_n instead of U_n and P_n for the fiber displacements and loads. It can be shown (Beyerlein et al. 1997) that the general solution to (3) subject to these boundary conditions involves first coupling dependence on time and length using the similarity variable $z = \xi/\sqrt{t}$, applying this similarity transformation to (3), and then applying a discrete Fourier transform on n. $L_n(\xi,t)$ and $V_n(\xi,t)$ become:

$$L_n(\xi,t) = -\frac{1}{2}\int_0^{\pi} C_\theta \cos(n\theta)\,\mathrm{erfc}(C_\theta|z|)\,d\theta, \tag{7}$$

and

$$V_n(\xi,t) = \mathrm{sgn}(\xi)\frac{1}{2}\sqrt{t}\int_0^{\pi} \cos(n\theta)\left\{\frac{1}{\sqrt{\pi}}\exp(-C_\theta^2 z^2) - C_\theta|z|\,\mathrm{erfc}(C_\theta|z|)\right\}d\theta, \tag{8}$$

where $C_\theta = \sin(\theta/2)$ and $\mathrm{sgn}(\xi) = 1$ for $\xi \geq 0$ and

$\mathrm{sgn}(\xi) = -1$ for $\xi < 0$.

The isolated break solutions (7) and (8) are then used to calculate the solution for multiple fiber breaks, $r \geq 1$, in the composite under the same boundary and initial conditions. The first key concept is that the tensile load at any fiber point (n,ξ) at t is a weighted sum of all loads transmitted by the individual r breaks since application of the load at time zero up to t. The following expressions result (Beyerlein et al. 1997) for the time-dependent fiber loads or stress concentrations $F_n(\xi,t)$ at every point (n,ξ) and t:

$$F_n(\xi,t) = \sum_{j=1}^{r} \Big[\Big(\int_{0+}^{t} L_{n-n_j}(\xi-\xi_j,t-\tau)\frac{\partial \mathcal{K}_j(\tau)}{\partial \tau}d\tau + L_{n-n_j}(\xi-\xi_j,t)\mathcal{K}_j(0^+)\Big)\Big] \quad (9)$$

A formula for the displacements $W_n(\xi,t)$ is obtained by using $V_{n-n_j}(\xi-\xi_j,t)$ in place of $L_{n-n_j}(\xi-\xi_j,t)$ in (9). The unknown weighting functions, the $\mathcal{K}_j(t)$'s, are obtained from the knowledge that the load on each end of the r fiber breaks is the prescribed compressive load -1, which leads to the following system of r equations:

$$-1 = \sum_{j=1}^{r} \Big[\Big(\int_{0+}^{t} \Lambda_{ij}(t-\tau)\frac{\partial \mathcal{K}_j(\tau)}{\partial \tau}d\tau + \Lambda_{ij}(t)\mathcal{K}_j(0^+)\Big] \quad (10)$$

where $i = 1, \dots, r$, and $\Lambda_{ij}(t) = L_{n_i-n_j}(\xi_i-\xi_j,t)$ from (7). Lastly, a uniform tensile load field $P = 1$ is superimposed to obtain the desired solution; that is, $P_n(\xi,t) = L_n(\xi,t) + 1$ and $U_n(\xi,t) = V_n(\xi,t) + \xi$. Solving for the $\mathcal{K}_j(t)$'s exactly requires integration of $[\underline{\Lambda}(s)s]^{-1}$ by use of the Laplace inversion Bromwich integral formula, which is generally intractable. In one approach (Method I), we solve (9) and (10) numerically, wherein error enters only through discretization of the convolution (10) and can be minimized by considering small time steps (Beyerlein et al. 1997).

Another method we use, designated as Method II, is to simply solve for $\{\boldsymbol{K}(t)\}$ based on the current values of $[\Lambda(t)]$, and so the evolving history of $\{\boldsymbol{K}(t)\}$ is ignored. Though Method II is more crude, we later show it yields a good approximation of Method I. So given t and r breaks, the following system of r equations is solved for the r weighting functions $\mathcal{K}_j(t)$:

$$\{\boldsymbol{K}(t)\} = [\Lambda(t)]^{-1}\{\boldsymbol{P}\} \quad (11)$$

where $\{\boldsymbol{P}\}$ is the r dimensional vector whose entries are -1, and $[\Lambda(t)]$ is the $r\times r$ matrix of the respective influence functions between the fiber breaks evaluated at time t. At any fixed time t, we simply sum over the current proportions of the loads transmitted from each fiber break j multiplied by its current $\mathcal{K}_j(t)$ to obtain:

$$F_n(\xi,t) = \sum_{j=1}^{r} \mathcal{K}_j(t)\, L_{n-n_j}(\xi-\xi_j,t) \quad (12)$$

and $W_n(\xi,t)$, but with $V_{n-n_j}(\xi-\xi_j,t)$ in place of $L_{n-n_j}(\xi-\xi_j,t)$ in (12).

3 RESULTS

3.1 *Isolated break.*

Interesting comparisons can be made between results from VBI and results for a linear elastic matrix found in previous work (Hedgepeth 1961, Hikami and Chou 1990, Beyerlein et al. 1996). Figure 2 shows the overload profiles on the neighboring intact fibers $s = 1, \dots, 4$ from the break versus $\xi/\sqrt{t}$. Also shown for comparison is the time-independent elastic shear-lag prediction (Beyerlein et al. 1996), which is plotted vs ξ. Notably $P_n(\xi,t)$ achieves close correspondence with the elastic case when $z = \xi$ or $t = 1$ (corresponding to $J(\mathcal{T}) = J_v$ in (1)). In fact, the peak stresses along $z = \xi = 0$ are the same (1.333 for $s = 1$, 1.061 for $s = 2$, 1.026 for $s = 3$, and 1.014 for $s = 4$). At the break site, $n = 0$ and $\xi = 0$, an isolated fiber break separates exactly as $(\pi t)^{1/2}/2$, when the matrix is viscous, while for a linear elastic matrix, the break displacement is $\pi/4$ under the same load (Hedgepeth 1961). Thus at $t = \pi/4$, we obtain exact correspondence in break displacements.

Of particular interest in failure progression is the load transfer length, $l_c(\mathcal{T})$ or the fiber axial distance on $s = 1$, such that $p(l_c(\mathcal{T}))/p^* \approx 1.00$ (defined before the 'dip' and on one side of the break). Using normalization (4) and noting from Figure 2 that at $z_c \approx 1.0$, $P(\xi,t) \approx 1.0$, we find that $l_c(\mathcal{T})$ increases in $\sqrt{\mathcal{T}}$ and inversely in $\sqrt{\eta/w}$:

$$l_c(\mathcal{T}) = \sqrt{\mathcal{T}}\sqrt{\frac{EA}{(\eta/w)h}}. \quad (13)$$

Comparisons with the elastic case in Figure 2 suggest that beyond time $t = 1$ (or $\mathcal{T} = \mathcal{T}_c$), $l_c(\mathcal{T})$

begins to grow longer than that predicted for a linear elastic matrix.

3.2 *Large transverse cracks.*

Many of the time-dependent features of a single break also appear for a row of r aligned and contiguous breaks, forming a 'crack'. (Figure 3 shows one side of a central crack containing a total of $r = 2N + 1$ fiber breaks.) Figure 4 shows the load profiles on the first intact fiber ahead of a row of $r = 2N + 1$ fiber breaks or $P_{N+1}(\xi,t)$ versus $\xi/t^{1/2}$, as predicted by the VBI technique. Similarly the overload region grows proportional to $z_c t^{1/2}$, wherein z_c increases with r. Also in Figure 4, the peak stress at the crack plane $P_{N+1}(0,t)$ is time-independent and equal to that predicted by the elastic, time-independent shear-lag model. The peak stress enhancement factor, denoted as K_r, is well approximated by (Beyerlein et al. 1996):

$$K_r \approx \frac{\sqrt{\pi}}{2}\sqrt{r+1}. \tag{14}$$

An asymptotic expression for the crack opening displacement $U_n(0,t)$ (Figure 3), is an equation for an ellipse (Beyerlein & Phoenix 1997b):

$$\left(\frac{\sqrt{\pi}\,U_n(0,t)}{2\,(N+1)}\right)^2 + \left(\frac{n}{N+1}\right)^2 = 1. \tag{15}$$

Figure 5 compares the time-scaled opening displacements $U_n(0,t)/\sqrt{rt}$ in a quarter plane of the crack, i.e. $0 \le n/N \le (N+1)/N$ as predicted by VBI and the approximation (15) for $N = 5$, 10, and 25. The dots representing the numerical results correspond to the actual fiber break locations and fall nicely onto the analytical approximation curve, forming an ellipse. The time-scale of $U_n(0,t)$ in Figure 5 suggests that at all points across the crack face, the crack opens proportional to $\sqrt{t}$, and therefore maintains its elliptical shape in the creep process. In summary, when the matrix is linearly viscous, the overloads and opening displacements produced by an r-sized crack grow proportional to $t^{1/2}$, while the peak stresses occur along $\xi = 0$ ahead of the crack and remain constant.

3.3 *Process zone fiber breaks.*

Thus far, we have examined the stress redistribution around multiple fiber breaks which are aligned. In this section, using the VBI technique, we compute and show how the fiber tensile stresses in and around transversely staggered fiber breaks redistribute in time. Take for example, the three staggered fiber breaks in Figure 1, in which break 2 and break 1 or 3 are separated longitudinally by a dimensionless distance $2L$, where l is rendered dimensionless to L through (4).

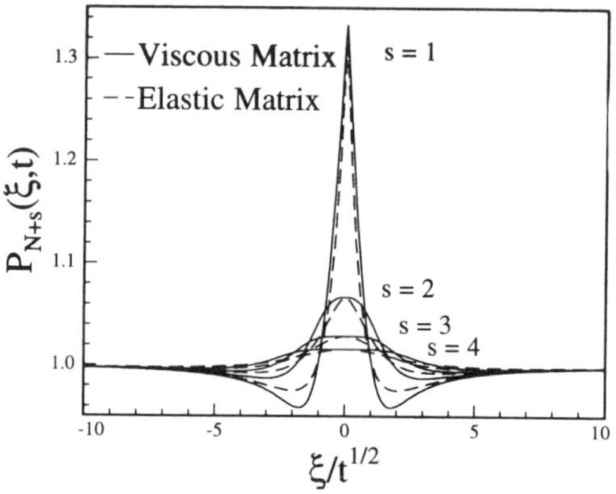

Figure 2. Stress concentration profiles in fibers adjacent to a single break.

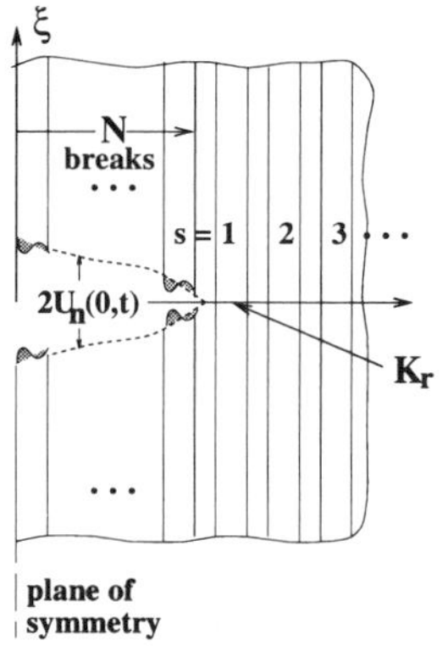

Figure 3. One side of central crack containing $r = 2N+1$ fiber breaks.

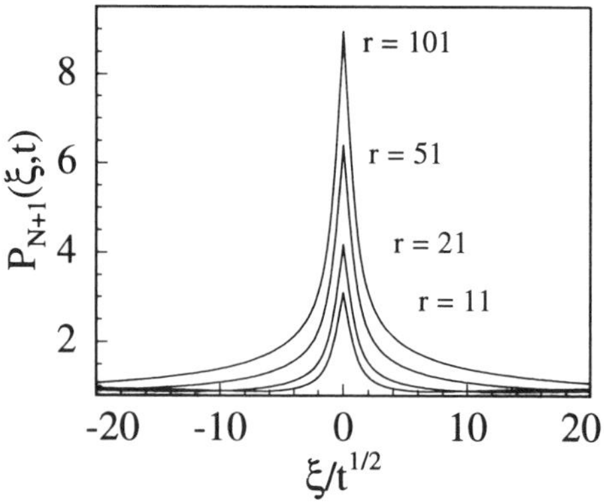

Figure 4. Stress concentration profiles in 1st intact fiber next to r-sized crack.

Figures 6 and 7 show the tensile stress concentration profiles, $P_n(\xi,t)$, on the center broken fiber $n = 0$ and on the adjacent intact fiber $n = 2$,

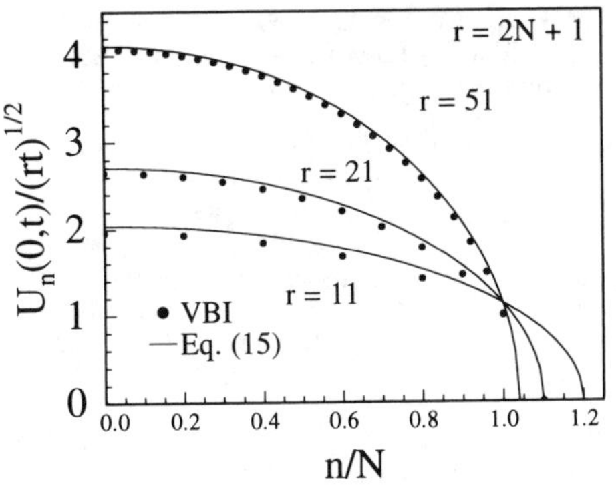

Figure 5. Crack opening displacements.

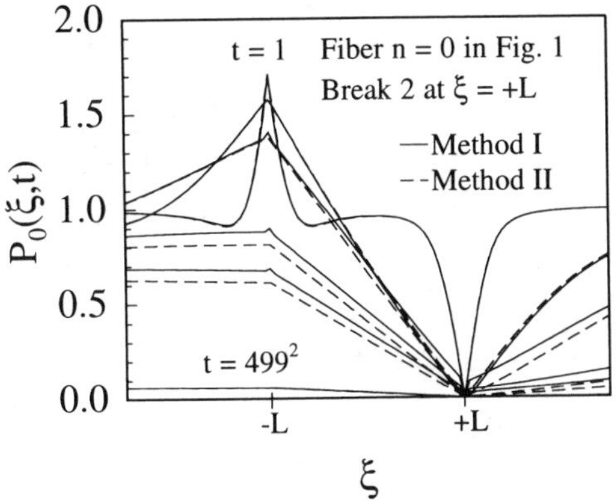

Figure 6. Time evolution of stresses in broken fiber.

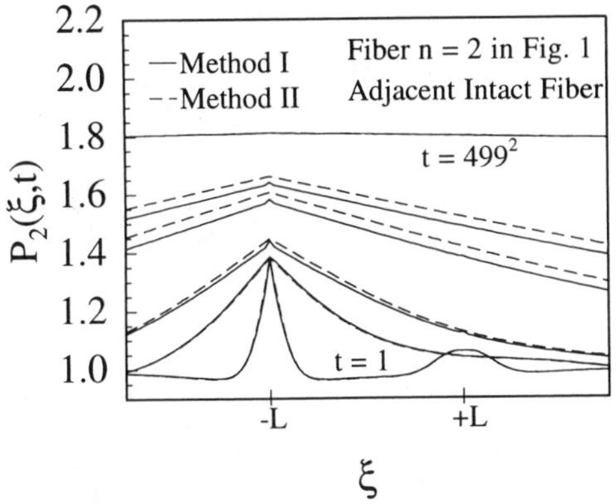

Figure 7. Time evolution of stresses in intact fiber.

respectively, at various times using Methods I and II. To minimize error in Method I, time increments leading to small changes in the stress state were used in numerically integrating (9) and (10) for $F_n(\xi,t)$. As shown, the quick and conservative approach Method II yields nearly the same stress

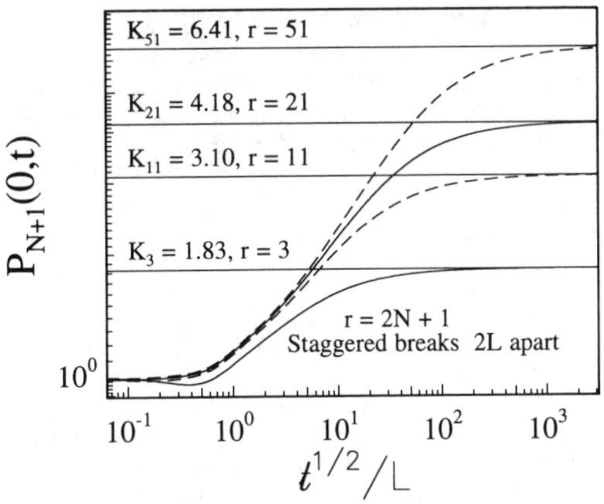

Figure 8. Time evolution of stress concentration in 1st intact fiber next to r-sized process zone.

predictions as Method I at short times, e.g. $t \leq 81$, and asymptotically long times, e.g. $t = 499^2$.

In Figures 6 and 7 we see that at $t = 1$, the stress fields of these breaks are non-interacting. Later in time, the peak stresses in fiber $n = 0$ which contains a break, decay as well as the rate at which the tensile stresses build up from this break. At the same time, the amount of tensile stress transferred from these 3 broken fibers to the adjacent unbroken fibers $n = 2, 3, 4, \ldots$ gradually increases with time. For instance for fiber $n = 2$, the maximum tensile stress increases from 1.33 at $t = 1$ to 1.8 at $t = 499^2$, and will eventually approach a maximum of 1.83, the elastic prediction for three aligned breaks. However, unlike an elastic matrix composite, the fiber stress distributions elsewhere are more dispersed.

To further demonstrate this effect, we extend this transversely staggered break pattern from $r = 3$ breaks to $r = 11$, 21, and 51 breaks and monitor the stress concentration $P_{N+1}(0,t)$ at $\xi = 0$ on the first intact fiber. Figure 8 is log-log plot of this stress evolution versus $t^{1/2}/L$. Significantly, for this and similar symmetrical, staggered configurations, we find that $P_n(\xi,t)$ scales approximately with $t^{1/2}/L$. This scaling implies large separations between breaks lead to a slower rise in the stress concentrations with time. Initially the stress fields of the breaks are non-interacting at $\xi = 0$, and so the stress concentration is unity. Later in time, $P_n(0,t)$ increases up to a magnitude as if an r-sized crack had formed. During the growth period the fibers go through a phase where they slide with constant velocity. This limit K_r, indicated on Figure 8 is calculated exactly and is well-predicted by (14). Recall that when $t = 1$ or $\mathcal{T} = \mathcal{T}_c$, the elastic Hedgepeth shear lag result closely matches the linear viscous matrix result. Therefore, if $L = 10$ or in real units, $l = 10(wEAJ_v/h)^{1/2}$ (widely spaced breaks), then the stress predicted by the elastic solution would correspond to $P_n(0,t)$ at $t^{1/2}/L$ = 1/10, which is unity for all r. However, if $L = 10^{-2}$ (closely spaced breaks), then the elastic prediction corresponds to the stress concentrations where the $P_n(0,t)$ curves intersect $t^{1/2}/L = 10^2$.

4 CONCLUSIONS

In this work, we use our recently developed micromechanical technique, VBI, to compute the stresses and displacements in a composite lamina, wherein the matrix creeps in shear as a viscous medium and that the fibers, the primary load bearing component, have time-independent, elastic properties. Using VBI, investigations focus attention on the effects of large rows of aligned or transversely staggered fiber fractures, how the resulting fiber and matrix stresses redistribute in time due solely to linearly viscous matrix creep, and how this evolution can lead to new fiber (delayed) fiber breaks. The results demonstrate that it is possible that the local fiber tensile stresses could rise at rates depending on microstructural length scales, such as crack length and break spacing. The motivation for VBI is for use in Monte Carlo simulations of creep damage accumulation through sequential fiber breaks, so that one may perform micromechanical creep analyses and estimate the stress rupture lifetime of composites.

REFERENCES

Beyerlein, I. J. & S. L. Phoenix 1997a. Statistics of fracture for an elastic notched composite lamina containing Weibull fibers-Part I and II. *Engng. Fract. Mech.*, In press.

Beyerlein, I. J. & S. L. Phoenix 1997b. In preparation.

Beyerlein, I. J., S. L. Phoenix & R. Raj 1997. Time evolution of stress distributions around arbitrary arrays of fiber breaks in a composite with a viscoelastic matrix. *Int. J. Solids Structures*, submitted.

Beyerlein, I. J., S. L. Phoenix & A. M. Sastry 1996. Comparison of shear-lag theory and continuum fracture mechanics for modeling fiber and matrix stresses in an elastic cracked composite lamina. *Int. J. Solids Structures* 33:2543-2574.

Gusev, Y. S. & A. S. Ovchinskii 1984. Computer modeling of failure processes of fiber-reinforced composite materials under the effect of a constantly acting tensile load. *Mech. Comp. Mater.* March-April, No. 2:196-203.

Hedgepeth, J. M. 1961. Stress concentrations in filamentary structures. *NASA TN D-882* .

Hikami, F. & T. W. Chou 1990. Explicit crack problem solutions of unidirectional composites: elastic stress concentrations. *AIAA J.* 28:499-505.

Ibnabdeljalil, M. & W. A. Curtin 1996. Size effects on the strength of fiber-reinforced composites. To appear in the *Int. J. Solids Structures* .

Lagoudas, D. C., C. Y. Hui & S. L. Phoenix 1989. Time evolution of overstress profiles near broken fibers in a composite with a viscoelastic matrix. *Int. J. Solids Structures*, 25: 45-66.

Lifshitz, J. M. & A. Rotem 1970. Time-dependent longitudinal strength of unidirectional fibrous composites. *Fibre Sci. Technol.* 3:1-20.

Damage and Failure of Interfaces, Rossmanith (ed.) © *1997 Balkema, Rotterdam, ISBN 90 5410 899 1*

Fracture and interface properties of a composite with long aligned fibers

J. Botsis
Department of Mechanical Engineering, Swiss Federal Institute of Technology, Lausanne, Switzerland

ABSTRACT: Results on fracture, fatigue crack growth and debonding on compact tension specimens of a specially made composite material are reported. The specimens consisted of an epoxy matrix and layers of equally spaced long aligned glass fibers. The experimental data showed that for a range of fiber spacing λ, the composite's strength σ_A, scaled with the fiber spacing in the form of $\sigma_A\sqrt{\lambda}$ = constant. In all fatigue experiments, crack arrest was observed after a few fiber layers. Analysis of the experimental results was aimed at quantifying the effects of the fibers ahead of the crack tip and the contributions of the bridging fibers to the total stress intensity factor. The debonding in the bridging zone was evaluated using a one dimensional analysis. The model was calibrated with the debonding on the first row of fibers, in a typical specimen, and consequently used to describe debonding on specimens with different fiber spacing. In spite of the assumptions adopted, the model described debonding well.

1. INTRODUCTION

Understanding the strength characteristics and fracture behavior of composites with long aligned fibers has been the subject of intense investigation in the past several years. This is demonstrated by a large number of analytical investigations that have been proposed to describe matrix fracture, crack bridging and interface properties in various composite systems (for a review see Kerans et al., 1989). These analyses have played an important role in our efforts to characterize the response of composite materials.

Mechanical tests used to determine the behavior of composites are usually performed on commercial or model composites and include, among others, the fiber pullout and pushout tests (Kerans and Parthasarathy 1991; Herrera-Franco and Drzal 1992). It has been difficult, however, to reveal basic mechanisms that contribute to strength and fracture through testing of commercial composite materials and to relate interface behavior from pullout or pushout tests to the overall in situ behavior of these materials. Thus, it has been a formidable task to assess the contributions of crack fiber - interaction, tractions in the bridging fibers, interface debonding and sliding, energy dissipation in the matrix material, etc., to the overall response of composite materials. From a physical standpoint, some of these processes are related to micromechanics and others to macromechanics thus, involving a wide range of scales. The interactions of these mechanisms are the main obstacle in understanding and modeling fracture in a large class of modern composite materials.

To better understand and characterize the physical processes taking place during fracture in composites, experimental and numerical studies on specially made composites were undertaken (Botsis and Beldica 1994; Botsis et al. 1994; Zhao and Botsis 1996; Beldica and Botsis 1996). This approach offers certain advantages. By controlling the fiber architecture, the properties of the constituent materials and the interface, different types of behavior can be simulated experimentally. Thus, some of the pertinent parameters (fiber size, spacing, interface, etc.,) affecting strength, crack growth and debond evolution can be identified and thoroughly characterized. Results from such studies can be an important supplement to efforts aimed at modeling the response of composites particularly when a continuum approximation of the reinforcement is not appropriate. Such cases may arise when studying: (a) crack initiation, (b) driving forces of relatively short cracks, (c) the area around the tip of a long crack, (d) criteria for fracture, etc.

In this paper, we report results of research related to strength, fracture evolution and interface properties as a function of fiber spacing. The fibers were stronger than the matrix and the strain-to-failure of the matrix was less than that of the fibers. The combination of specimen geometry and matrix used resulted in a brittle fracture. Moreover, the system was sufficiently simple to allow for an in situ observation of crack growth, debonding, any dissipative mechanisms in the

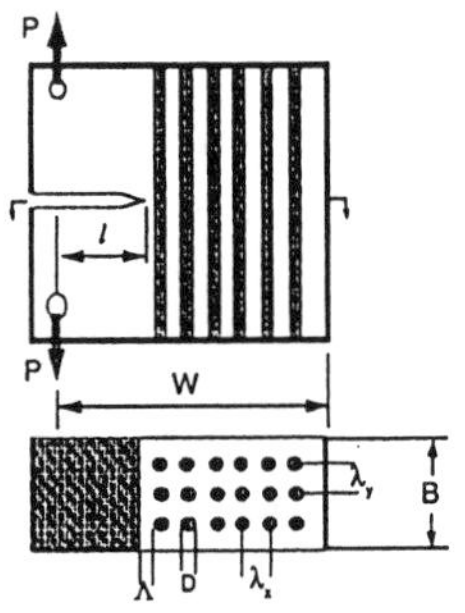

Figure 1. Schematic of the compact tension specimen for ramp and fatigue testing

matrix material and crack front changes due to the reinforcement. Following the introduction, the experimental methods are outlined in Sec. 2. The results and discussion on strength, fatigue crack growth, stress intensity factors and debonding are presented in Sec. 3.

2. EXPERIMENTAL METHODS

Compact tension (CT) specimens were used in the experimental studies. The composite specimens consisted of an epoxy matrix and glass fibers with 0.4 mm in diameter. The Young's moduli of the constituent materials were, E_m = 3.5 GPa and E_f = 72.50 GPa. The specimens were machined from composites blocks prepared in a specially designed mold. A typical schematic of a specimen used in the present studies is shown in Figure 1. Details of specimen preparation and experimental methods can be found elsewhere (Botsis and Beldica 1994; Zhao and Botsis 1994).

Strength characteristics of the composite material were determined using CT specimens that were pulled to fracture. The fiber spacing in these specimens was $\lambda_x \approx \lambda_y \approx \lambda \approx 1$, 1.5, 2 and 2.5 mm (Figure 1). The thickness was determined from $B = 2\lambda_x + 5$. The rest of the dimensions were evaluated according to the ASTM standards for toughness measurements, i.e., $W = 4B$ and $l = 0.5W$. The specimens were tested under tension with a cross-head speed of 1 mm/min.

An important parameter in the fracture response of the composite specimens is the distance Λ of the crack tip to the first layer of fibers (Figure 1). In the present studies this distance was kept the same and equal to about two fiber diameters.

The fatigue experiments were load controlled with a sinusoidal waveform function of frequency ν = 1 Hz. The maximum loads of the fatigue cycle were P_{max} = 265, 310, and 355 N, while the minimum load was kept the same, P_{min} = 10 N.

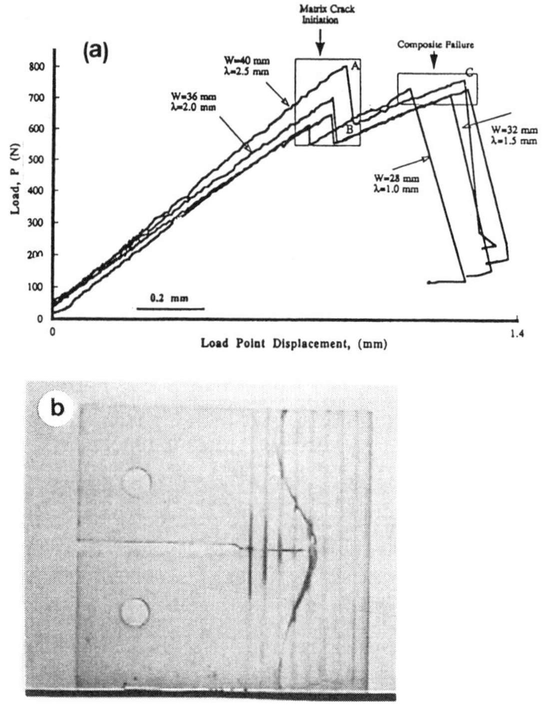

Figure 2. (a) Load-displacement curves for composite specimens with different fiber spacing, (b) typical failure mode of composite specimens

All fatigue experiments were performed on a dual servohydraulic Instron Mechanical Testing System at room temperature and laboratory environment.

3. RESULTS AND DISCUSSION

3.1 *Strength*

Load displacement curves of the CT specimens are shown in Figure 2 for four fiber spacing.

It is interesting to note here that after the initial linear part, a sudden drop in the loading occurred followed by an increase and a final decrease of the load leading to specimen fracture. To gain further understanding on the fracture behavior of the CT specimens, the crack tip was monitored with a video recording unit which was attached to an optical microscope. These observations revealed that during the first loading phase (O-A, Figure 2) the crack grew by about 0.2 mm and stopped well before the peak load P_A. Afterwards the crack jumped for about 3 fiber spacing (point B). Final failure occurred at loading P_C (Figure 2b). For all fiber spacing examined in these studies, the shapes of load displacement curves were similar. Moreover, the load P_A depended on the fiber spacing while P_C was independent of the fiber spacing.

To quantify the maximum stress level and its dependence on fiber spacing, a nominal strength σ_A, was taken in the form

$$\sigma_A = \sigma_A^{tension} + \sigma_A^{bending}$$
$$\frac{P_A}{B(W-l)} + \frac{6P_A[l + 0.5(W-l)]}{B(W-l)^2} \qquad (3.1)$$

where W, B and l are defined in Figure 1. Note that the stress σ_A is easily obtained from the beam theory when it is assumed that the neutral axis is at the middle of the ligament. Moreover, this stress level can be used to calculate a stress intensity factor (Tada et al. 1973)

$$K_I = \sigma_A \sqrt{W-l}F \qquad (3.2)$$

where F is a dimensionless correction factor due to specimen geometry. Strength values σ_A, obtained from the CT specimens with different fiber spacing are shown in Figure 3a as a function of fiber volume fraction V_f. Note that the data are well described by a linear relationship. The intercept corresponds to the strength of the plain matrix specimen which was about 27 MPa. Although the strength data are well correlated with V_f, the volume fraction does not always represent the local morphology (especially around the crack tip) because it is a volume average parameter. Therefore, the effects of fiber spacing on strength can not be easily examined. In our previous work (Botsis et al., 1994), strength data of specimens with one layer of long aligned fibers were correlated with the fiber spacing. Using dimensional analysis arguments, a relationship of the following form was obtained

$$\sigma_c \sqrt{\lambda} = \kappa \qquad (3.3)$$

Here σ_c is the strength of the specimen and κ is a constant which is proportional to the matrix fracture toughness. Using the same arguments (Zhao and Botsis 1996), the strength data obtained from the CT specimens were expressed in terms of the fiber spacing.

Experimental measurements of strength plotted against $1/\sqrt{\lambda}$ for the range of fiber spacing tested in this work are shown in Figure 3b (note that $\lambda \approx \lambda_x \approx \lambda_y$). Interestingly the data are well described with a relationship similar to (3.3).

3.2 *Crack propagation*

Crack propagation in the CT specimens with $\lambda_x = \lambda_y = 2.5$ mm for three different maximum loads plotted against the crack length l, obtained from surface measurements, is shown in Figure 4. In all cases, the time to crack initiation largely depended upon the applied load. After initiation, a significant decrease in the crack speed was observed. The rate of crack speed decrease was a function of the applied load. While a monotonic decrease was seen in the experiment with the highest load, a phase of almost constant speed was observed in the experiment with the lowest load. In all three experiments, crack arrest occurred at about the same crack length (Figure 4).

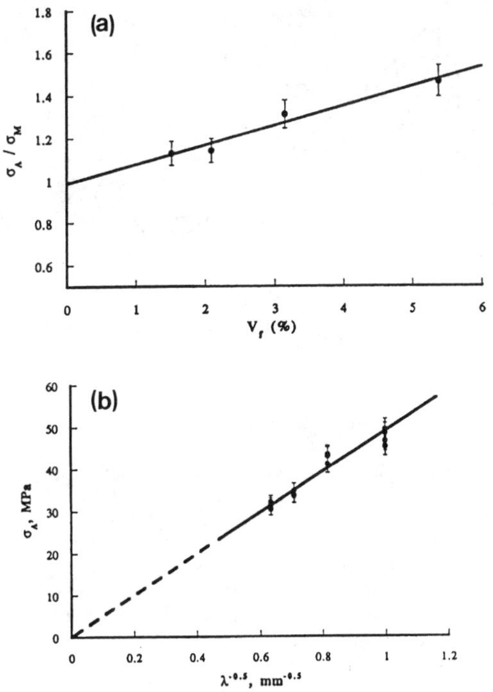

Figure 3. (a) Normalized strength vs. fiber volume fraction, (b) strength as a function of inverse square root of fiber spacing

In all fatigue experiments, the crack front was not straight. Instead, a curved crack front was seen with the curvature being much larger when the crack front was near to a fiber. Within the resolution of the observations, fiber debonding was the dominant mechanism of energy dissipation. At a particular fiber, debonding started when the center of the crack front ran into a fiber. Although fiber friction and deformation may have also contributed to energy dissipation, they were not recorded in the present studies. The decrease in crack speed can be explained in terms of the effects of the reinforcement on the stress field at the crack tip and the bridging effects due to the fibers in the wake of the crack.

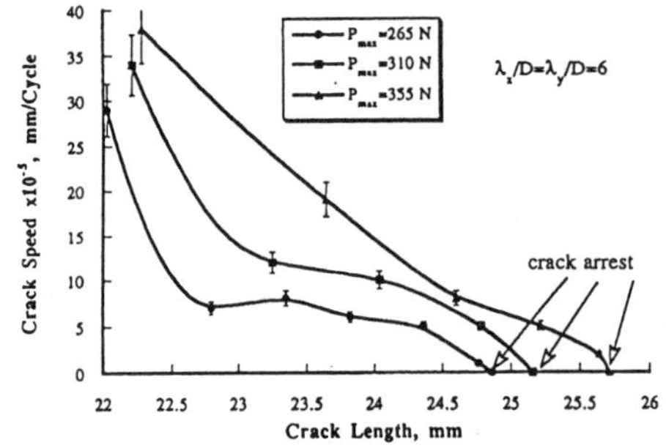

Figure 4. Crack speed vs. crack length for three different loads

3.3 *Stress Intensity factors*

Analysis of the experimental results on strength, debonding and crack growth was aimed at understanding and quantifying the effects of the fibers ahead of the crack tip on the stress and strain fields and the contributions of the bridging fibers on crack growth, debonding and overall stability of the fracture process. Towards this end, two complementary directions were undertaken. In the first one a numerical analysis, using boundary element method, of the problem was pursued in order to understand and quantify the level of crack-fiber interaction in terms of the elastic properties and the pertinent geometrical parameters (Beldica and Botsis 1996). Typical results of a two dimensional analysis on the effects of a stiffer reinforcement ahead of the crack tip on the stress intensity factor K_r, are shown in Figure 5 (K_0 is the stress intensity factor of the unreinforced specimen).

In the second one a semi-experimental method was used to obtain the tractions on the fibers using experimentally measured crack opening displacements (COD) and appropriate influence functions (Zhao and Botsis 1996).

To calculate the driving force for fracture of a specimen reinforced with fibers, the tractions on the bridging fibers should be available. With these forces known we can calculate a total stress intensity factor and also describe the debonding along the fibers in the bridging zone using simplified analysis similar to fiber pull-out tests as a first approximation.

One method to obtain the bridging tractions in a linear elastic system is to use experimentally measured COD profiles and tools of fracture mechanics. To use the method we need accurate measurements of COD profiles at different crack lengths. At this point in time, however, complete COD measurements are not available. Thus, to apply the method for tractions in a CT specimen, we used a linear profile for the points located in the middle of the fiber rows,

$$\Delta = \Delta_{max}(1 - x / l) \qquad (3.4)$$

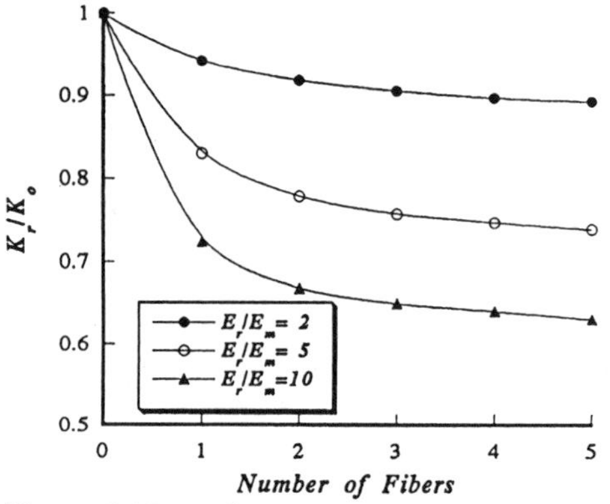

Figure 5. Normalized stress intensity factor vs. number of fibers ahead of the crack tip (distance of the crack tip to the first layer of fibers is two fiber diameters)

where Δ_{max} is the COD at zero crack length. Although the profile given by Eq. 3.4 is an assumption, it is realistic when we stay away from the vicinity of the crack tip. Using experimental values of Δ_{max} for each crack length and the appropriate pin load, forces in three consecutive rows of fibers were calculated and are shown in Figure 6. In the simulations the crack lengths were considered at the middle of two consecutive fiber rows.

With the tractions on the bridging fibers known a total stress intensity factor K_l, can be easily calculated. Assuming that for a crack bridged by fibers the principle of superposition applies and that the level of residual stresses, due to specimen preparation was negligible, K_l is expressed as

$$K_l = K(P_0; l) - \sum_{i=1}^{n} K_i^f(P_i, c_i; l) \qquad (3.5)$$

Here $K(P_0; l)$ is the stress intensity factor due to

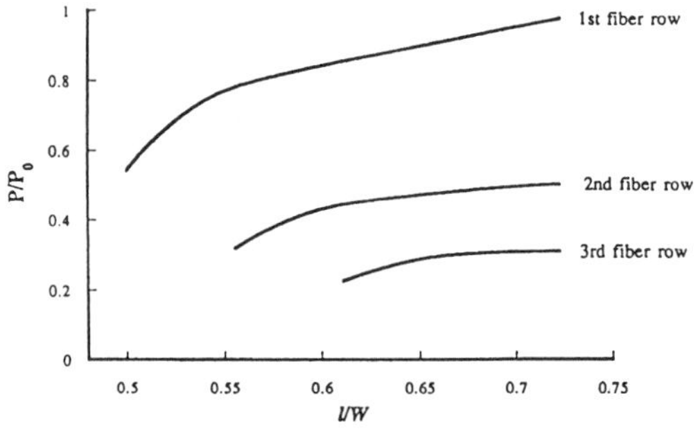

Figure 6. Forces in the bridging zone as a function of normalized crack length

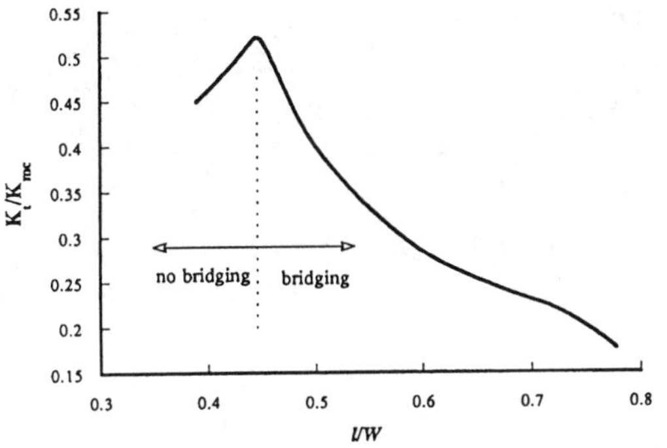

Figure 7. Total stress intensity factor plotted against the crack length

the applied pin load P_0, on a specimen with an unbridged crack of length l and $K_i^f(P_i, c_i; l)$ is the contribution to K_t of the i-th fiber, in the bridging zone, expressed as the effect of a closure force P_i acting at a distance c_i from the loading point.

Using the force distribution shown in Figure 6, a total stress intensity factor was calculated using Eq. 3.5. The plot in Figure 7 shows that the values of K_t decrease with crack length, consistent with the crack growth data in Figure 4.

3.4 *Debonding analysis*

The experimental data for the analysis were taken from CT specimens with different fiber spacing tested under ramp loading conditions (Figure 2a) and are shown in Table I (Zhao and Borsis 1996). The specimen geometries and the experimental measurements are shown in Table I (note that for the specimens with W = 32 and 36 mm, two sets of data are shown that correspond to two different tests). Referring to Figure 2a, one sees that after the initial linear part (O - A), a sudden drop in the loading occurred (A - B) followed by an increase (C) and a final decrease of the load leading to specimen fracture. The bridging zone was established after the first load drop. Note also that due to symmetry, debonding in each fiber of a certain fiber layer (Figure 1) across the specimen thickness was the same.

To describe debonding of the fibers in the bridging zone we used a one dimensional analysis model for fiber pull out (Kerans and Parthasarathy 1991). According to this model, the debonding along a fiber is

$$l_d = \frac{r}{2\mu k}\ln\left(\frac{P^* - P_d}{P^* - P_a}\right) \qquad (3.6)$$

Table I. Data from compact tension specimen used in debonding analysis

Specimen	W28	$W32_1$	$W32_2$	$W36_1$	$W36_2$	W40
W, mm	28	32	32	36	36	40
λ, mm	1.0	1.5	1.5	2.0	2.0	2.5
P_A, N	615	654	616	705	660	758
δ_A, mm	0.771	0.839	0.773	0.842	0.773	0.798
P_B, N	550	562	531	560	530	560
δ_B, mm	0.774	0.846	0.787	0.856	0.786	0.834
P_C, N	732	712	732	730	750	870
δ_C, mm	1.076	1.198	1.204	1.250	1.340	1.734
V_f(%)	5.38	3.14	3.14	2.09	2.09	1.51

where r is the fiber radius; k is a coefficient that depends on the elastic constants of the matrix and fiber; μ is the friction coefficient; P_d is a force for debond growth; P^* is a critical force equal to $P^* = -\sigma_0 \pi r^2 / k$ with σ_0 being the normal stress to the fiber due to shrinkage of the matrix and mismatch between matrix and fiber. Note that are three-parameters, μ, P_d and P^* (or σ_0) in this debonding model.

Implicit to using this model is the assumption that debonding in the bridging zone is similar to that in a fiber pull-out test. The only conceptual difference here is that σ_0 is a stress that also depends on fiber-fiber interaction and is expected to vary with fiber spacing.

To apply the model described above, the forces in the fibers should be evaluated and the three parameters should be known. To obtain the bridging tractions, we used the semi-experimental method (Zhao and Botsis 1996). The pin load P_{BC}, was taken as the average of P_B and P_C and the corresponding load point crack opening δ_{BC} as the average of δ_B and δ_C (Figure 2a). With regard to the COD profile, a linear variation of COD at fiber spacing midpoints was assumed. The Young's modulus of each specimen was taken as the system modulus obtained from the specimen compliance method. With all this information the bridging forces, on each fiber layer in the bridging zone, can be obtained and used in the debonding analysis.

In our specimens, three to four layers of fibers were in the bridging zone and three fibers within each layer. Accordingly, the force in each fiber P_a was taken equal to one third of the forces in the corresponding layer. To calibrate the model, ranges of $\mu[0.3-0.7]$ and $P_d[5-20N]$ were taken and the value of P^* was varied to fit

Table II. Debonding measurements, model description and fitting parameters

Debonding (mm)	W28	W32-1	W32-2	W36-1	W36-2	W-40
1st fiber, d_1	1.75-1.75	3.1--3.10	2.5--2.50	4.1--4.10	4.4--4.40	7.5--7.50
2nd fiber, d_2	1.0--0.92	2.0--1.75	1.4--1.44	2.0--2.36	2.5--2.36	4.5--3.79
3rd fiber, d_3	0.1--0.23	0.5--0.55	0.25-0.32	0.5--0.44	1.0--0.97	3.0--2.55
4th fiber, d_4,	-----	----	----	----	----	1.5--1.39
P_d, N	10	10	10	10	10	10
μ	0.6	0.6	0.6	0.6	0.6	0.6
P^*,N	754	605	730	544	518	466
σ_0 , MPa	52	42	50	38	36	32

the debonding on the first layer of fibers. Using the data of the CT specimen with fiber spacing λ =2.5 mm, the values of the model parameters for which the experimental debonding on the first fiber coincided with that of Eq. 3.6 were μ=0.6, P_d=10N and σ_0=32 MPa. These parameters were next used in Eq. 3.6 to describe debonding on the rest of the fibers in the bridging zone of the same specimen. To obtain an idea on the effects of the forces in the bridging zone on debonding we applied a force distribution changed by about 10%. The corresponding changes in debonding, according to the model used herein, were less than 12%.

Since all specimens were of the same materials and prepared under identical conditions, we assumed that the interface was the same in all specimens. Thus, in specimens with different fiber spacing, μ and P_d should be the same and σ_0 be different due to the effects of fiber spacing. Accordingly, μ and P_d were kept the same in each fiber spacing and the value of σ_0 was evaluated by identifying the debonding length on the first fiber, in each specimen, with the value given by Eq. 3.6. Subsequently, the debonding on the rest of the fibers in each bridging zone was described using Eq. 3.6.

According to the procedures mentioned above, debonding in specimens with different fiber spacing were evaluated and the results are shown in Table II.

The load P^* indicates to a certain degree the interactions between the fibers. From the definition, $P^* = -\sigma_0 \pi r^2 / k$, the value of the normal stress σ_0, is equal to $-P^* k / \pi r^2$ (the minus sign means compression). For the best fit, the values of σ_0 are shown in Table II. One sees that the smaller the fiber spacing, the higher the normal stress. According to experimental data (Daniel 1995), for a glass-epoxy system, the initial normal stress due to the shrinkage and mismatch is in the order of 10 MPa. The higher σ_0 in our data is attributed to the fiber-fiber interactions and to processing conditions. With regard to the coefficient of friction, its value should be measured independently. However, the value used to fit the data (μ = 0.6), is reasonable and compares with reported values on various systems (Warner 1995).

4. CONCLUSIONS

a. For a range of fiber spacing λ, strength σ_A, was correlated with the fiber spacing in the form $\sigma_A\sqrt{\lambda} = \kappa$ where κ is a constant. This results are similar to those reported by Botsis, et al. (1994) on a monolayer specimen and other combinations of fiber-matrix systems (Gaffney and Botsis 1995).

c. Crack arrest was observed in all fatigue crack propagation experiments performed in the present studies provided that no fiber fracture occurred in the bridging zone. Within the resolution of the observations no fiber failure was observed in the bridging zone.

d. The reduction in the stress intensity factor due to the layers of fibers ahead of the crack tip depended on the elastic properties of the fibers. The analysis also indicated that the first two layers of fibers produced the most important effect. The contribution of the subsequent fibers was much smaller.

f. A semi-experimental method was used to evaluate the tractions in the bridging zone. Simulations indicated that the total stress intensity factor decreased with the crack length, consistent with the experimental results on crack propagation. A simple fiber pullout analysis seems to describe debonding well in the bridging zone.

REFERENCES

Beldica, C. & J. Botsis 1996. Experimental and numerical studies in model composites, Part II: numerical results. *International Journal of Fracture*, 82: 175.

Botsis, J. & C. Beldica 1994. Strength characteristics and fatigue crack growth in a composite with long aligned fibers. *International Journal of Fracture* 69: 27.

Botsis, J., C. Beldica & D. Zhao 1994. On strength scaling of composites with long aligned fibers. *International Journal of Fracture* 63: 375.

Daniel, I. (1995) private communication.

Gaffney, K & J. Botsis 1995. To be published

Herrera-Franco, P. J. & L. T. Drzal 1992. Comparison of methods for the measurement of fibre/matrix adhesion in composites. *Composites* 23:2.

Kerans, R. J., R. S. Hay, N. J. Pagano & T. Parthasarathy 1989. The role of fiber-matrix interface in ceramic composites. *Ceramics Bulletin* 68: 429.

Kerans, R. J. & T. Parthasarathy 1991. Theoretical analysis of the fiber pullout and pushout tests. *Journal of the American Ceramic Society* 74: 1585.

Tada, H., P. C.Paris & G. P. Irwin 1973. *The Stress Analysis of Cracks Handbook*,: Del Research Corporation, Hellertown, PA.

Warner, S. B. 1995. *Fiber Science*: Prentice Hall.

Zhao, D. & J. Botsis 1996. Experimental and numerical studies in model composites. Part II: experimental results. *International Journal of Fracture* 82: 153.

Damage and Failure of Interfaces, Rossmanith (ed.) © *1997 Balkema, Rotterdam, ISBN 90 5410 899 1*

Stress analysis in a three phase material with fibre, coating and matrix

W.Wu & I.Verpoest
Department of Metallurgy and Materials Engineering, Katholieke Universiteit Leuven, Belgium

Janis Varna
Department of Materials and Production Engineering, Lulea University of Technology, Sweden

ABSTRACT: A new model for the stress state analysis in the single fibre specimen with 3 phases, fibre, coating, and matrix, is developed. The stress distributions in radial and axial directions are obtained using the principle of minimum complementary energy. Assumptions made to simplify the stress analysis are: a) axial stress distribution is uniform across the fibre cross-section; b) axial stress distribution is uniform across the coating thickness; c) the non-uniform axial stress distribution in the matrix is given by a decreasing function containing a free parameter which is obtained by minimization procedure. The developed model is used in parametric analysis to determine the effect of the coating properties and geometry on the rate of the stress transfer and shear stress concentration at the fibre end.

1. INTRODUCTION

In order to study the effect of the fibre coating on the behaviour of a composite, an understanding is needed of the influence of varying the coating stiffness and thickness on the stress transfer at the fibre ends, of the shear stress redistribution on the coating properties etc. Since all these phenomena are very clearly displayed in the single fibre fragmentation test, the following analysis will focus only on this test. The analysis aims at the determination of the elastic stress state at a perfectly bonded interface between the fibre and matrix, but it also serves as the base for a more advanced analysis including both debonded and bonded interfaces.

An advantage of an analytical model is the simplicity of parametric analysis, in particular involving a change of the geometry (coating thickness). We therefore focus on the derivation of a 2-D axisymmetric analytical model.

In the literature for 2-D models, a relatively early 2-D micromechanical model of 2-phases (Whitney & Drzal 1987) has been modified (Dong et al. 1996) to evaluate the microfracture behaviour near a fibre end in short fibre-reinforced thermoplastic composite, including an interphase. However, such a model will have the same serious defects as in Whitney's model. The major defect is that the selection of the stress function was guided by the questionable 1-D shear-lag solution.

Among all existing two phase models (2-D) which can be extended to a 3-phase composite, the model developed by the current authors (Wu, Verpoest & Varna 1997) seems to be the best: it includes a non-uniform stress distribution in radial and axial directions, thus does not need any uncertain parameter, R as in most of the existing analytical models (Nairn 1992). It exactly satisfies all equilibrium equations. The best approximate stress state is extracted by applying the principle of minimum complementary energy. In the present paper we develop a rigorous 3-phase model based on the same approach. The assumptions for fibre and matrix are: a) axial stress distribution in the fibre is uniform across the thickness of the fibre; b) axial stress distribution is uniform across the coating thickness since the coating layer is very thin relative to the fibre radius; c) the non-uniformity in the radial direction of the axial stress distribution in the matrix is described by a decreasing function containing a free parameter which is obtained as a result of the application of the minimum complementary energy principle. The validity of the developed model was checked by FEM calculation and then used for parametric analysis in order to understand the effect of the coating geometry and stiffness on the axial stress built-up and on the maximum shear stress concentrations.

2. ANALYSIS OF THE STRESS STATE

We consider a single fibre fragment with an isotropic coating, embedded in an infinite matrix.

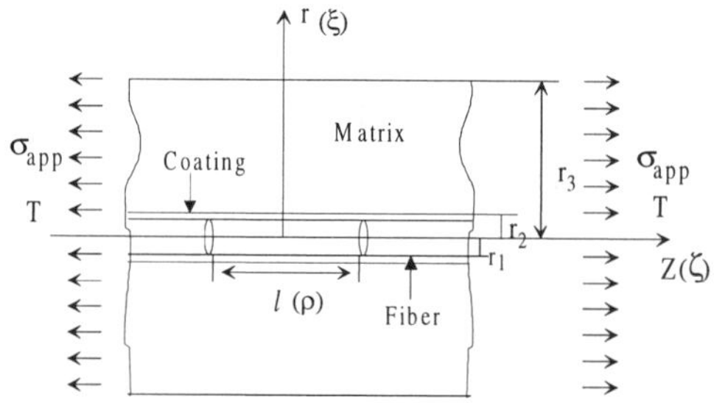

Fig. 1: A single fragment of length l, composed of fibre, coating and matrix, is subjected to mechanical (σ_{app}) and thermal (T) loading. Nondimensional coordinates are $\xi=r/r_1$, $\zeta=z/r_1$ and the fragment length is $\rho=l/2r_1$.

One fibre fragment between two breaks is shown in Fig. 1. σ_{app} is the applied axial stress in the matrix and T is the temperature difference between the room temperature and the stress-free temperature. r_1, r_2 and r_3 are the radius of fibre, coating and matrix, respectively. The notations for material properties: E_A and E_T are the axial and transverse modulus of the fibre, ν_A is the axial and υ_T is the transverse Poisson's ratio of the fibre, α_A and α_T are the axial and transverse thermal expansion coefficients of the fibre. E_c is the modulus, ν_c is the Poisson's ratio and α_c is the thermal expansion coefficient of the coating. E_m is the modulus, ν_m is the Poisson's ratio and α_m is the thermal expansion coefficient of the matrix.

The interfacial stress continuities and the stress boundary conditions along the radial and axial directions are

$$\tau_{rz,f}(r_1)=\tau_{rz,c}(r_1) \qquad \tau_{rz,c}(r_2)=\tau_{rz,m}(r_2) \tag{1}$$

$$\sigma_{rr,f}(r_1)=\sigma_{rr,c}(r_1) \qquad \sigma_{rr,c}(r_2)=\sigma_{rr,m}(r_2) \tag{2}$$

$$\tau_{rz,m}(r \Rightarrow \infty)=0 \qquad \sigma_{rr,m}(r \Rightarrow \infty)=0 \tag{3}$$

$$\sigma_{zz,f}(z=\pm l/2)=\tau_{rz,f}(z=\pm l/2)=0 \tag{4}$$

where $\tau_{rz,j}$, $\sigma_{rr,j}$ and $\sigma_{zz,j}$ denote the shear, radial and axial stresses. Subscripts, j = f, c and m denote the fibre, coating and the matrix, respectively.

The stress components can be obtained from stress functions, $\Psi(\zeta,\xi)$ and $\Omega(\zeta,\xi)$

$$\sigma_{rr}=\frac{1}{\xi}\frac{\partial\Omega}{\partial\xi}+\frac{\partial^2\Psi}{\partial\zeta^2} \qquad \sigma_{\theta\theta}=\frac{\partial^2\Omega}{\partial\xi^2}+\frac{\partial^2\Psi}{\partial\zeta^2} \tag{5}$$

$$\sigma_{zz}=\frac{\partial^2\Psi}{\partial\xi^2}+\frac{1}{\xi}\frac{\partial\Psi}{\partial\xi} \qquad \tau_{rz}=-\frac{\partial^2\Psi}{\partial\xi\partial\zeta} \tag{6}$$

where $\xi=\frac{r}{r_1}$, and $\zeta=\frac{z}{r_1}$ are nondimensional coordinates.

In the following notations, $\psi_{i,j}$ are arbitrary functions of ζ only and ϕ is arbitrary function of ξ only. In $\psi_{i,j}$, i is the function number while subscripts, j = f, c and m denote the fibre, coating and the matrix, respectively. ψ_f, ψ_c and σ_{app} are the far field axial stresses in the fibre, coating and the matrix respectively. Far field stresses are the stresses with the same external loading described in Fig. 1 and without fibre cracks. ψ_f, ψ_c and σ_{app} are independent on both z and r. The explicit expressions for all far field stress components are not available in the literature and have been derived by using stress functions (5)-(6) with the boundary/interface conditions (1)-(4) plus interfacial continuities of both axial and radial displacements. These expressions are presented in the Appendix.

We first assume that in the transversely isotropic fibre and the isotropic coating, the axial perturbation stresses, due to cracks, are uniform over the radial coordinate ξ, but are dependent on the axial coordintae ζ via the function $\psi_{0,f}(\zeta)$ and $\psi_{0,c}(\zeta)$, respectively

$$\sigma_{zz,f}=\psi_f-\psi_f\psi_{0,f}(\zeta) \tag{7}$$

$$\sigma_{zz,c}=\psi_c-\psi_f\psi_{0,c}(\zeta) \tag{8}$$

whereas in the isotropic matrix, the perturbation axial stress is not only dependent on the axial coordintae ζ but also decays with the increase of the radial coordinate. We use a general expression as

$$\sigma_{zz,m}=\sigma_{app}-\psi_f\psi_{0,m}(\zeta)\left[\phi''(\xi)+\frac{\phi'(\xi)}{\xi}\right] \tag{9}$$

In Eqs. (7)-(9), each axial stress is the superposition of the far field and perturbation stresses. Using a similar procedure as our previous work of a two phase model (Wu, Verpoest & Varna 1997), we substitute Eqs. (7)-(9) into the first one of Eq. (6) to find the stress functions Ψ for the three phase. Functions Ω in Eq. (5) is used only for producing the far field radial and hoop stresses, based on our previous conclusion (Wu, Verpoest & Varna 1997) for a two phase model. Then, we substitute the derived Ψ into Eqs. (5)-(6) to obtain the other stresses. In the fibre cylinder:

$$\tau_{rz,f} = \frac{\psi_f \xi \psi'_{0,f}}{2} - \frac{\psi'_{1,f}}{\xi} \quad (10)$$

$$\sigma_{rr,f} = \sigma_{fr} - \frac{\psi_f \xi^2}{4} \psi''_{0,f} + \psi''_{1,f} \ln\xi + \psi_{2,f} \quad (11)$$

$$\sigma_{\theta\theta,f} = \sigma_{f\theta} - \frac{\psi_f \xi^2}{4} \psi''_{0,f} + \psi''_{1,f} \ln\xi + \psi_{2,f} \quad (12)$$

where σ_{fr} and $\sigma_{f\theta}$ are the far field radial and hoop stresses (linked to Ω only) in the fibre. In the coating cylinder:

$$\tau_{rz,c} = \frac{\psi_f \xi \psi'_{0,c}}{2} - \frac{\psi'_{1,c}}{\xi} \quad (13)$$

$$\sigma_{rr,c} = \sigma_{cr} - \frac{\psi_f \xi^2}{4} \psi''_{0,c} + \psi''_{1,c} \ln\xi + \psi_{2,c} \quad (14)$$

$$\sigma_{\theta\theta,c} = \sigma_{c\theta} - \frac{\psi_f \xi^2}{4} \psi''_{0,c} + \psi''_{1,c} \ln\xi + \psi_{2,c} \quad (15)$$

where σ_{cr} and $\sigma_{c\theta}$ are the far field radial and hoop stresses (linked to Ω only) in the coating. In the matrix cylinder:

$$\tau_{rz,m} = \psi_f \psi'_{0,m} \phi' - \frac{\psi'_{1,m}}{\xi} \quad (16)$$

$$\sigma_{rr,m} = \sigma_{mr} - \psi_f \psi''_{0,m} \phi + \psi''_{1,m} \ln\xi + \psi_{2,m} \quad (17)$$

$$\sigma_{\theta\theta,m} = \sigma_{m\theta} - \psi_f \psi''_{0,m} \phi + \psi''_{1,m} \ln\xi + \psi_{2,m} \quad (18)$$

where σ_{cr} and $\sigma_{c\theta}$ are the far field radial and hoop stresses (linked to Ω only) in the matrix. All the first terms in Eqs. (10)-(18), except for the shear stresses, are the far field stresses and the rest are the perturbation stresses. The far field shear stress is zero in the three cylinders.

The requirement for finite stresses at ξ=0 forces $\psi'_{1,f} = \psi''_{1,f} = 0$ in Eqs. (10)-(12). $\psi''_{1,m}$ in Eqs. (17)-(18) must vanish because $\sigma_{rr,m}$ and $\sigma_{\theta\theta,m}$ must be finite at $\xi \to \infty$. We can further prove that $\psi'_{1,m}$ must be zero (Wu 1997). Using Eqs. (1), (10), (13), we can express $\psi'_{1,c}$ in terms of $\psi'_{0,f}$ and $\psi'_{0,c}$ and using Eqs. (2), (11), (14), (17), we can express both $\psi_{2,f}$ and $\psi_{2,c}$ in terms of $\psi''_{0,f}$, $\psi''_{0,c}$ and $\psi''_{0,m}$ ($\psi_{2,m}$ vanishes as explained after Eq. (21)). Thus, we have expressed all ζ-dependent functions by $\psi_{0,f}$, $\psi_{0,c}$ and $\psi_{0,m}$ and their derivatives. The force balance in the three phases requires

$$2\pi\int_0^{r_1} \sigma_{zz,f}\,rdr + 2\pi\int_{r_1}^{r_2} \sigma_{zz,c}\,rdr + 2\pi\int_{r_2}^{\infty} \sigma_{zz,m}\,rdr$$
$$= 2\pi\int_0^{r_1} \psi_f\,rdr + 2\pi\int_{r_1}^{r_2} \psi_c\,rdr + 2\pi\int_{r_2}^{\infty} \sigma_{app}\,rdr \quad (19)$$

Using Eqs. (7)-(9), we achieve

$$\psi_{0,f}\int_0^1 \xi d\xi + \psi_{0,c}\int_1^{\xi_2} \xi d\xi + \psi_{0,m}\int_{\xi_2}^{\infty} (\xi\phi')' d\xi = 0 \quad (20)$$

Eq. (20) is the expression for the total axial perturbation force and the second integral must be finite. It forces $\xi\phi' \underset{\xi\to\infty}{\Rightarrow} A^*$, where A* is an arbitrary constant. It can be proved (Wu 1997) that $A^*=0$:

$$\underset{\xi\to\infty}{\xi\phi'=0} \quad (21)$$

Note that the second one of Eq. (1) can be satisfied automatically by differentiating Eq. (20) with regard to the axial coordinate. Eq. (3) is also satisfied if $\psi_{2,m}$=0 because Eq.(21) implies $\phi(\xi \Rightarrow \infty) = 0$.

To decrease the mathematical complicity, we search, in addition to Eq. (21), one more equation to express $\psi_{0,c}$ and $\psi_{0,m}$ in terms of only $\psi_{0,f}$. Considering that physically both the coating and matrix should have a similar changing rate of axial perturbation stress along the z direction, we assume

$$\psi_{0,m} = A_1 \psi_{0,c} \quad (22)$$

where A_1 is an arbitrary constant (determined below). Using Eqs. (20) and (22), we achieve

$$\psi_{0,c} = A_2 \psi_{0,f} \quad \text{and} \quad \psi_{0,m} = A_3 \psi_{0,f} \quad (23)$$

where

$$A_2 = \frac{1}{2A_1\xi_2\phi'(\xi_2) + 1 - \xi_2^{\,2}} \quad \text{and } A_3 = A_1 A_2 \quad (24)$$

Using Eq. (23) and denoting $\psi_{0,f}$ by the shorter notation ψ, we obtain the final stress expressions (only containing ψ and its derivatives) in the fibre:

$$\sigma_{zz,f} = \psi_f - \psi_f \psi \quad (25)$$

$$\tau_{rz,f} = \frac{\psi_f \xi \psi'}{2} \quad (26)$$

$$\sigma_{rr,f} = \sigma_{fr} - \left[\frac{\xi^2}{4} - A_5\right] \psi_f \psi'' \quad (27)$$

$$\sigma_{\theta\theta,f} = \sigma_{fr} - \left[\frac{\xi^2}{4} - A_5\right]\psi_f \psi'' \tag{28}$$

In the coating:

$$\sigma_{zz,c} = \psi_c - A_2 \psi_f \psi \tag{29}$$

$$\tau_{rz,c} = \left[\frac{A_2 \xi}{2} - \frac{A_4}{\xi}\right]\psi_f \psi' \tag{30}$$

$$\sigma_{rr,c} = \sigma_{cr} - \left[\frac{A_2 \xi^2}{4} - A_4 \ln\xi - A_6\right]\psi_f \psi'' \tag{31}$$

$$\sigma_{\theta\theta,c} = \sigma_{c\theta} - \left[\frac{A_2 \xi^2}{4} - A_4 \ln\xi - A_6\right]\psi_f \psi'' \tag{32}$$

In the matrix:

$$\sigma_{zz,m} = \sigma_{app} - A_3 \psi_f \psi (\phi'' + \frac{\phi'}{\xi}) \tag{33}$$

$$\tau_{rz,m} = A_3 \psi_f \psi' \phi' \tag{34}$$

$$\sigma_{rr,m} = \sigma_{mr} - A_3 \psi_f \psi'' \phi \tag{35}$$

$$\sigma_{\theta\theta,m} = \sigma_{m\theta} - A_3 \psi_f \psi'' \phi \tag{36}$$

where

$$A_4 = \frac{A_2 - 1}{2} \tag{37}$$

$$A_5 = \frac{A_2\left(\xi_2^{\,2} - 2\ln\xi_2 - 1\right) + 2\ln\xi_2 + 1 - 4A_3\phi(\xi_2)}{4} \tag{38}$$

$$A_6 = \frac{A_2\left(\xi_2^{\,2} - 2\ln\xi_2\right) + 2\ln\xi_2 - 4A_3\phi(\xi_2)}{4} \tag{39}$$

The boundary conditions at the fibre crack surface, Fig.1, are given by Eq. (4). From Eqs. (25)-(26) we have

$$\psi(\pm\rho) = 1 \qquad \text{and} \qquad \psi'(\pm\rho) = 0 \tag{40}$$

where $\rho = l/2r_1$ is the fragment aspect ratio.

A_1 in Eq. (22) can be determined by some compatibility condition in an average sense. We use condition that the average axial displacements along the axial direction reach the same at the coating/matrix interface for both the coating and matrix. It leads to

$$\begin{aligned} &\int_{-\frac{1}{2}}^{\frac{1}{2}} \varepsilon_{zz,c}(r = r_2)dz = \int_{-\frac{1}{2}}^{\frac{1}{2}} \varepsilon_{zz,m}(r = r_2)dz \\ &\Rightarrow \int_{-\rho}^{\rho} \varepsilon_{zz,c}(\xi = \xi_2)d\zeta = \int_{-\rho}^{\rho} \varepsilon_{zz,m}(\xi = \xi_2)d\zeta \end{aligned} \tag{41}$$

where ε_{zz} is the axial strain. Using Hook's law, Eqs. (29)-(36) in Eq. (41) and making use of $\int_{-\rho}^{\rho}\psi'' d\zeta = 0$ (Eq. (40), we achieve

$$A_1 = \frac{E_m}{E_c}\frac{1}{\phi''(\xi_2) + \dfrac{\phi'(\xi_2)}{\xi_2}} \tag{42}$$

To obtain the best solution for $\psi(\zeta)$, we use the principle of minimization of complementary energy. The general expression of complementary energy Γ is

$$\begin{aligned} \Gamma = &\sum_{i=f,c,m} \int_{V_i} \frac{1}{2}[\sigma][K][\sigma]dV \\ &+ \sum_{i=f,c,m} \int_{V_i} [\sigma][\alpha]TdV - \int_S [P][U]dS \end{aligned} \tag{43}$$

where [K] is the compliance matrix, $[\alpha]$ is the vector of thermal expansion coefficients and $[\sigma]$ is the stress vector in Voigt notation. V_i (i = f, c, m) denotes the volume and S is the body surface subjected to a known displacement of [U]. [P] is the traction vector. The last term in (43) is zero (Wu 1997) for the current problem. On substituting the perturbation terms of Eqs. (25)-(36) into Eq. (43), and performing the ξ-integration only, we achieve the following expressions for the part of Γ related to the perturbation stresses:

$$\Gamma = \pi r_1^3 \psi_\infty^2 \int_{-\rho}^{\rho} \begin{pmatrix} C_{00}\psi^2 + C_{02}\psi\psi'' \\ + C_{22}\psi''^2 + C_{11}\psi'^2 \end{pmatrix} d\zeta \tag{44}$$

where the constants C_{ij} are defined as $C_{ij} = C_{ij}^{\,f} + C_{ij}^{\,c} + C_{ij}^{\,m}$

In the fibre

$$C_{00}^f = \frac{1}{2E_A} \qquad C_{02}^f = \left(-\frac{\upsilon_A}{E_A}\right)\left[\frac{1}{4} - 2A_5\right]$$

$$C_{11}^f = \frac{1}{16G_A}$$

$$C_{22}^{f} = \left(\frac{1-\upsilon_T}{E_T}\right)\left[\frac{1}{48} - \frac{A_5}{4} + A_5^{\,2}\right] \tag{45}$$

In the coating

$$C_{00}^{c} = \frac{A_2^{\,2}\left(\xi_2^{\,2}-1\right)}{2E_c}$$

$$C_{02}^{c} = \frac{A_2 \upsilon_c}{E_c}\left[\begin{array}{l}\frac{A_2\left(1-\xi_2^{\,4}\right)}{4} + A_4\left(\begin{array}{l}1-\xi_2^{\,2}+\\ 2\xi_2^{\,2}\ln\xi_2\end{array}\right)\\ +2A_6(\xi_2^{\,2}-1)\end{array}\right]$$

$$C_{11}^{c} = \frac{1}{G_c}\left[\begin{array}{l}\frac{A_2^{\,2}\left(\xi_2^{\,4}-1\right)}{16} - \frac{A_2A_4\left(\xi_2^{\,2}-1\right)}{2}\\ +A_4^{\,2}\ln\xi_2\end{array}\right]$$

$$C_{22}^{c} = \frac{1-\upsilon_c}{E_c}\left\{\begin{array}{l} A_2\left[\begin{array}{l}\frac{A_2\left(\xi_2^{\,6}-1\right)}{48} - \\ A_4\left(\frac{4\xi_2^{\,4}\ln\xi_2 + 1 - \xi_2^{\,4}}{16}\right)\\ -A_6\left(\frac{\xi_2^{\,4}-1}{4}\right)\end{array}\right]\\ +A_6^{\,2}\left(\xi_2^{\,2}-1\right)\\ +A_4\left[\begin{array}{l}\left(\begin{array}{l}2\xi_2^{\,2}\ln\xi_2\\ +1-\xi_2^{\,2}\end{array}\right)\\ \left(A_6 - \frac{A_4}{2}\right)\\ +A_4\left(\xi_2\ln\xi_2\right)^2\end{array}\right]\end{array}\right\} \tag{46}$$

In the matrix

$$C_{00}^{m} = \frac{A_3^{\,2}}{E_m}\left[\int_{\xi_2}^{\infty}\left(\xi\phi''^2 + \frac{\phi'^2}{\xi}\right)d\xi - \phi'(\xi_2)^2\right]$$

$$C_{02}^{m} = \frac{4\upsilon_m A_3^{\,2}}{E_m\phi'(1)^2}\left(\int_{\xi_2}^{\infty}\phi'^2\xi d\xi + \phi(\xi_2)\phi'(\xi_2)\xi_2\right)$$

$$C_{11}^{m} = \frac{A_3^{\,2}}{G_m}\left(\int_{\xi_2}^{\infty}\phi'^2\xi d\xi\right)$$

$$C_{22}^{m} = \frac{2(1-\upsilon_m)A_3^{\,2}}{E_m}\int_{\xi_2}^{\infty}\phi^2\xi d\xi \tag{47}$$

The standard procedure for the minimization of Γ produces the homogeneous equation

$$\psi'''' + p\psi'' + q\psi = 0 \tag{48}$$

where $p = \frac{C_{02} - C_{11}}{C_{22}}$ and $q = \frac{C_{00}}{C_{22}}$. The solution of Eq. (48) that satisfies the boundary condition (40) is available in (Wu, Verpoest & Varna 1997) The solution for ψ is expressed in term of exponential functions of ζ. It was shown (Hashin 1985) that by making use of Eq. (48), the minimum value of the perturbation complementary energy of Eq. (44), corresponding to ψ of Eq. (48), can be expressed

$$\Gamma = -2\pi r_1^3\psi_\infty^2 C_{22}\psi'''(\rho) \tag{49}$$

Note that as yet all expressions, such as C_{22}, $\psi'''(\rho)$ etc., depend still on an arbitrary function $\phi(\xi)$. Thus this function might be used for further minimization of the complementary energy Γ in (49). We assume a simple exponential decrease of perturbation stresses in radial direction by using

$$\phi(\xi)=e^{-\lambda\xi} \tag{50}$$

where λ is an arbitrary real parameter that is larger than zero. From Eq. (49), we have $\Gamma_{min} = \Gamma_{min}(\lambda)$ and the problem is reduced to the minimum estimation as a function of one variable, λ. The physical support for the exponential form of $\phi(\xi)$ is that starting at a short distance from the fibre break, all the perturbation stress components in Eqs. (33)-(36) must decay with the same rate along radial direction. As the different-order derivatives of ϕ are required in Eqs. (33)-(36), Eq. (50) is a rational selection. The study of the size of the perturbation zone along the radial direction has been made for the two phase case (Wu, Verpoest & Varna 1997) and the numerical results compared with FEM also supported the selection of Eq. (50).

3. RESULTS AND DISCUSSION

In this section we redefine the nondimensional coordinate ζ such that the origin is at the left crack rather than midway as in Fig. 1. We plot the results based on this coordinate system. The properties of the fibre, the coating and the matrix are listed in Table 1. The model fragment length is 30x the diameter of the fibre. Fig. 2 compares the predictions from the present model and an axisymmetric finite

Table 1: mechanical properties for a carbon fibre, a coating and an epoxy.

Property	Carbon fibre	Coating	Matrix
E_A, E_c or E_m (MPa)	238000	500	3000
E_T (MPa)	16000		
G_A, G_c or G_m (MPa)	35000	186.57	1119.4
ν_A, ν_c or ν_m	0.2	0.34	0.34
ν_T	0.25		
α_A, α_c or αm (ppm/°C)	-0.36		40.0
α_T (ppm/°C)	18.0		
diameter (mm)	0.007		

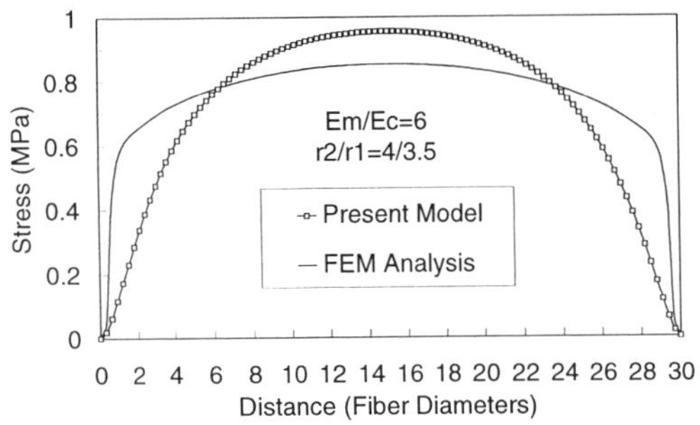

Fig. 2. The axial tensile stress in the fibre, σ_{zz}, as a function of axial distance (in units of fibre diameter). The curve is the stresses calculated by a ratio, 6, of matrix stiffness (E_m) to coating stiffness (E_c). The stresses are normalised by far field axial stress.

element analysis. We use an extremely high ratio, 6, of matrix stiffness (E_m) to coating stiffness (E_c) in order to test the model's capability. This ratio is also high enough to model a practical two phase composite system where a cracked fibre, surrounded by the matrix and average composite, is used instead of a cracked fibre, surrounded by coating and matrix as in Fig. 1. Finite element calculations have been carried out by using the Ansys software, version 5.1. The element chosen is a "2-D, 8-Node, structure solid" element, which is used as an axisymmetric element. Due to the assumption of constant axial fibre stress in the analytical model, the plotted finite element stress is the average axial stress over the fibre cross-section. Although the analytical model leads to a higher_peak stress and a lower initial increasing rate of the axial stress than FEM, there is general agreement between both analyses if one checks the average (throughout the fragment length) axial fibre stresses. The analytical model predicts an average stress of 117.61 MPa and FEM gives 123.45 Mpa. Some difference between two analyses can be improved by a more complicated (but still closed-form analytical) mathematical treatment in the

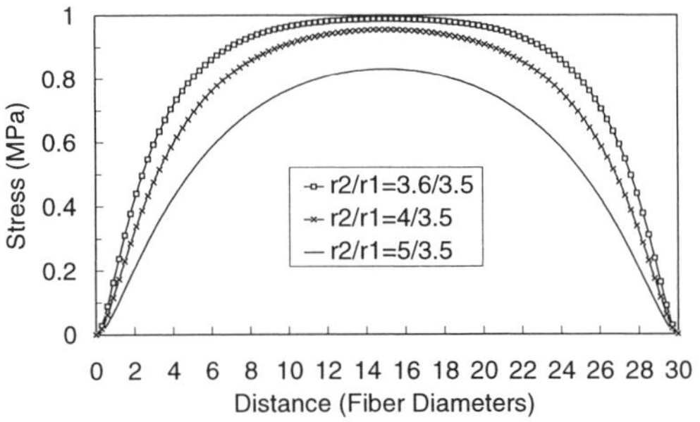

Fig. 3. The axial stress (σ_{zz}) distribution in the fibre, as a function of axial distance from the fibre break. The curves are the stresses calculated by the different ratios of the coating radius r_2 to the fibre radius r_1 at the same stiffness ratio (E_m/E_c=6). The stresses are normalised by far field axial stress

matrix. It has been shown for two phase case in two papers (Wu, Verpoest & Varna 1997) and (Wu, Verpoest & Varna submitted). In general, we consider that the present model is useful for readily practical use because it is more accurate and rigorous than the three phase analytical models which are known to us. Moreover, the accuracy increases with the decrease of the ratio of E_m to E_c.

The effect of the coating thickness on the stress transfer efficiency is plotted in Fig. 3. The curves are the stresses calculated for different ratios of the coating radius r_2 to the fibre radius r_1 at the same stiffness ratio (E_m/E_c=6). Fig. 3 indicates that the built-up speed of the axial fibre stress as well as the value of the curve peak remarkably decrease with the increase of the coating thickness, if the stiffness of the coating is much lower than that of the matrix. Thus, the coating thickness has an important influence on the stress transfer efficiency.

The stress concentration in the matrix due to the fibre breaks are studied in Fig. 4 and 5 where the shear stresses at the fibre/coating (ξ=1) and coating/matrix ($\xi=r_2/r_1$=4/3.5) interfaces are plotted along the axial direction. The curves are the stresses calculated by the same ratio of matrix stiffness (E_m) to coating stiffness (E_c) at the same ratio of r_2/r1. In this case, we assume the same applied stress, σ_{app}=2 Mpa in the matrix as we intend to figure out the change of the shear stress concentration within the matrix due to the addition of coatings.

We use the peak values in both Figures as the index of stress concentration due to the fibre breaks. As expected, the stress concentration at both interfaces increases with the increasing coating stiffness. However, if one compares the peak stress of E_m/E_c=1 in Fig. 4 (being the representative for a two phase sample) with that of E_m/ E_c=0.5 in Fig. 5, it is found that both values are very close. It concludes

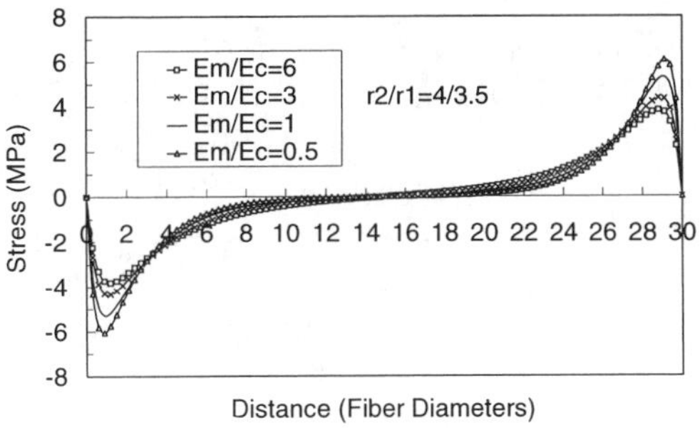

Fig. 4. The shear stress (τ_{rz}) at the fibre/coating interface (ξ = 1), as a function of axial from the fibre break. The curves are the stresses calculated by the different ratios of matrix stiffness (E_m) to coating stiffness (E_c). The stresses are normalised by the far field axial stress (σ_{app}=2 MPa) in the matrix.

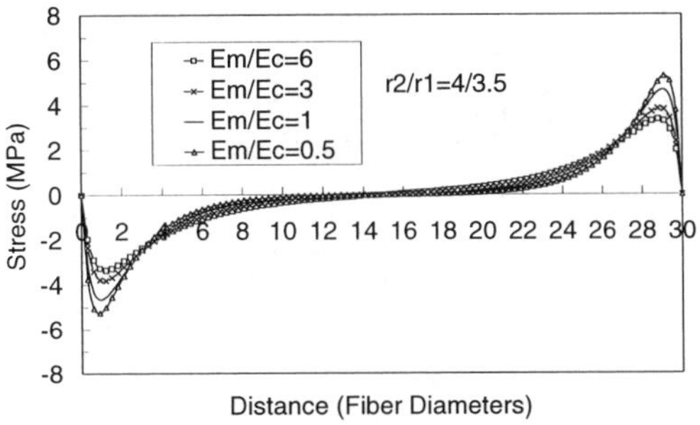

Fig. 5. The shear stress (τ_{rz}) at the coating/matrix interface (ξ = r_2/r_1=4/3.5), as a function of axial distance. The curves are the stresses calculated by the different ratios of matrix stiffness (E_m) to coating stiffness (E_c). The stresses are normalised by the far field axial stress (σ_{app}=2 MPa) in the matrix.

that the maximum shear stress within the matrix in the presence of a high stiffness coating might be equal or lower than maximum shear stress within the matrix in the absence of the coating, if a optimal combination of coating stiffness and thickness is realised. Hence it is possible to use a coating to achieve both decreasing stress concentration in matrix and increasing stress transfer efficiency. Consequently the composite system will achieve a better fracture toughness if the coating itself is tougher than the matrix, while keeping up a sufficient stress transfer. Figs. 4-5 show that the present model is capable to provide a quantitative guidance on the optimal design of a coating interphase for a certain fibre/matrix system with a perfectly bonded interface.

Some discussions: (1) the calculated λ values in the analytical model reach the same value for all fragments of longer than 30d. It means that the rate of the stress decrease in z-direction is the same for all fragments in this aspect ratio region (ρ > 30). (2) another important application of the model is to model a practical two phase composite system where a cracked fibre, surrounded by the matrix and average composite. (3) The total energy change caused by fibre breaks might be calculated using the present model with a high accuracy as the strain energy can be obtained from the rather accurate complementary energy of the present approach. (4) A new approach to derive micromechanical models for a debonded/bonded mixture interface have been presented for a two phase case (Wu, Verpoest & Varna submitted) where higher level mathematical variational analysis was also performed. The present work forwards to build a micromechanical model with a partially debonded interface for a three phase case.

4. CONCLUSION

The objective of the present research was to develop an enough accurate and still readily-used analytical model for the prediction of the stress state in a perfectly bonded interface system with a cracked fibre, an interphase and a matrix of an infinite radius. The stress state could be determined without using an uncertain parameter like that in the shear lag analysis and other closed-form models in literature. The stress-state estimation was performed by using the principle of minimum complementary energy. The analysis provided a more reliable prediction of the stress state.

ACKNOWLEDGEMENTS

This work was funded as a Brite EuRam project under the title 'Plasma polymerised coatings and interphases for improved performance carbon fibre composites', Contract No BRE2-0453. In addition to the named authors this collaboration also involves groups within Department of Engineering Materials, University of Sheffield, UK and Institut für Kunststoffverarbeitung, RWTH Aachen, Germany.
The text presents research results of the Belgian Programme on Interuniversity Poles of attraction initiated by the Belgian State, Prime Minister's Office, Science Policy Programming. The scientific responsibility is assumed by its authors.

REFERENCES

1. Cox, H.L., The elasticity and strength of paper and other fibrous materials, *Br. J. Appl. Phys.*, 3 (1952) 72-9.
2. Dong, Y.S., Nobuo, T., Tadashi, S. & Kazuo, N.,

Approximate analysis of the stress state near the fibre ends of short fibre-reinforced composites and the consequent microfracture mechanisms, *Composites* 27A (1996) 357-64.
3. Hashin, Z., Analysis of cracked laminates: A variational approach, *Mech. Mater.* 4 (1985) 121-37.
4. Nairn, J.A., A variational mechanics analysis of the stresses around breaks in embedded fibers, *Mech. Mater.* 13 (1992) 131-57.
5. Rosen, B. W., Mechanics of composite strengthening, *In Fiber composite materials*, ASM Publication, Metals Park, Ohio, 1965, 37-86.
6. Sabat, P. J., *Evaluation of the fiber-matrix interfacial shear strength in fiber reinforced plastics. Virginia Tech. center for Adhesion Science* Report VPI/CAS/ESM-86-1, Virginia Polytechnic Institute and State University, 1986.
7. Wu, W., Verpoest, I. & Varna, J., An improved analysis of the stresses in a single fibre fragmentation test. Part I: 2-Phase Model, *Compos. Sci. Technol.*, in press.
8. Wu, W., *Ph.D. Thesis*, Katholieke Universiteit Leuven, Leuven, 1997.
9. Wu, W., Verpoest, I. & Varna, J., A novel axisymmetric variational analysis of the stress transfer into fibre through a partially debonded interface, submitted to *Compos. Sci. Technol.*
10. Whitney, J.M. & Drzal, L.T., Axisymmetric stress distribution around an isolated fiber fragment, *Toughened Compos.*, ASTM STP 937, ed. N. J. Johnston, Philadelphia, 1987, 179-96.

APPENDIX: FAR FIELD STRESS EXPRESSIONS

$$\psi_f = \frac{\left[\begin{array}{l}\dfrac{AE_c}{2\nu_c E_m} + \dfrac{BE_A}{2\nu_A E_m} - \dfrac{\nu_m}{E_m} - \\ \dfrac{H}{2E_m}\left(\dfrac{2}{\nu_c} - \dfrac{E_A(1+\nu_c)}{\nu_A E_c} - \dfrac{E_A(1-\nu_T)}{\nu_A E_T}\right)\end{array}\right]\sigma_{app} + \left[\begin{array}{l}(\alpha_m - \alpha_c)\left(1 + \dfrac{AE_c}{2\nu_c}\right) - H\left(\alpha_T - \alpha_c - \dfrac{\alpha_c}{\nu_c} + \dfrac{\alpha_m}{\nu_c}\right) \\ -\dfrac{E_A(\alpha_A - \alpha_m)}{2\nu_A}\left(B + \left(\dfrac{1+\nu_c}{E_c} + \dfrac{1-\nu_T}{E_T}\right)H\right)\end{array}\right]T}{H\left(-\dfrac{\nu_A}{E_A} + \dfrac{1+\nu_c}{2\nu_A E_c} + \dfrac{1-\nu_T}{2\nu_A E_T}\right) + \dfrac{B}{2\nu_A}}$$

where

$$A = \frac{1-\nu_c}{E_c} + \frac{(1+\nu_c)}{E_c}\left(\frac{r_1}{r_2}\right)^2 - \frac{(1+\nu_m)}{E_m}\left[\left(\frac{r_1}{r_2}\right)^2 - 1\right]$$

$$B = \left(-\frac{1+\nu_c}{E_c} + \frac{1+\nu_m}{E_m}\right)\left(\frac{r_1}{r_2}\right)^2 \qquad H = \left(\frac{A}{2\nu_c} - \frac{\nu_c}{E_c}\right)\frac{E_c\nu_c}{1-\nu_c^2}$$

$$\sigma_{fr} = \sigma_{f\theta} = \frac{1}{2}\left[\frac{\psi_f}{\nu_A} - \frac{E_A\sigma_{app}}{\nu_A E_m} + \frac{E_A(\alpha_A - \alpha_m)T}{\nu_A}\right]$$

$$\psi_c = \frac{\left[\begin{array}{l}-\dfrac{\nu_A}{E_A}\psi_f + \dfrac{1}{E_m\nu_c}\sigma_{app} + \left(\dfrac{1+\nu_c}{E_c} + \dfrac{1-\nu_T}{E_T}\right)\sigma_{fr} \\ +\left(\alpha_T - \alpha_c - \dfrac{\alpha_c}{\nu_c} + \dfrac{\alpha_m}{\nu_c}\right)T\end{array}\right]E_c\nu_c}{1-\nu_c^2}$$

$$\sigma_{cr} = \left[1 - \left(\frac{r_1}{r}\right)^2\right]\frac{1}{A}\left[\begin{array}{l}\dfrac{\nu_c}{E_c}\psi_c - \dfrac{\nu_m}{E_m}\sigma_{app} - B\sigma_{fr} \\ +(\alpha_m - \alpha_c)T\end{array}\right] + \left(\frac{r_1}{r}\right)^2\sigma_{fr}$$

$$\sigma_{c\theta} = \left[1 + \left(\frac{r_1}{r}\right)^2\right]\frac{1}{A}\left[\begin{array}{l}\dfrac{\nu_c}{E_c}\psi_c - \dfrac{\nu_m}{E_m}\sigma_{app} - B\sigma_{fr} \\ +(\alpha_m - \alpha_c)T\end{array}\right] - \left(\frac{r_1}{r}\right)^2\sigma_{fr}$$

$$\sigma_{mr} = -\sigma_{m\theta} = \left[\left(\frac{r_2}{r}\right)^2 - \left(\frac{r_1}{r}\right)^2\right]\frac{1}{A}$$
$$\left[\frac{\nu_c}{E_c}\psi_c - \frac{\nu_m}{E_m}\sigma_{app} - B\sigma_{fr} + (\alpha_m - \alpha_c)T\right] + \left(\frac{r_1}{r}\right)^2\sigma_{fr}$$

Damage and Failure of Interfaces, Rossmanith (ed.) © 1997 *Balkema, Rotterdam, ISBN 90 5410 899 1*

Finite element modelling of damage development during longitudinal tensile loading of coated fibre composites

E. Jacobs & I. Verpoest
Department of Metallurgy and Materials Engineering, Katholieke Universiteit Leuven, Belgium

ABSTRACT: This paper presents the results of finite element calculations for determining the stress state near a broken coated fibre in unidirectional carbon fibre - epoxy matrix composites. The influence of coating thickness, coating stiffness and assumed crack pattern on the stresses in the broken fibre and in adjacent, unbroken fibres is evaluated and their possible impact on composite longitudinal tensile strength is discussed. For this analysis, a two-dimensional axisymmetric model was developed and all the elastic constants of the coated fibre composite were separately calculated and used as input data in the model.

1. INTRODUCTION

Resistance to damage initiation in composites can be substantially improved by aiming at a composite with a very high fibre-matrix adhesion. This leads e.g. to a strong increase in the transverse strength of unidirectional composites. Composite final failure however is often determined by the accumulation of damage inside the material. Too strong fibre-matrix bonds lead to a decrease in longitudinal tensile strength and interlaminar fracture toughness. In the former case, this is due to increased stress concentrations in adjacent, non-broken fibres whereas in the latter a decrease in energy absorption due to less extensive fibre bridging is the reason (Ivens et al. 1995).

It is believed that an optimum combination of interlaminar fracture toughness / longitudinal tensile strength and transverse tensile strength can be achieved by introducing a tough interlayer, provided that this third phase adheres very well to both fibre and matrix and has a very high energy absorption capacity. To try to fulfil the latter requirement, various options are possible : a plastically deforming coating, a coating with well-chosen stiffness properties in order to achieve a beneficial stress redistribution inside the composite or a change in the failure pattern.

In the present analysis, our attention has been predominantly focused on the influence of an existing fibre coating on the longitudinal tensile strength of unidirectional carbon fibre - epoxy matrix composites. Their possible effects on composite transverse strength and interlaminar fracture toughness are sometimes referred to.

2. LONGITUDINAL TENSILE STRENGTH

Detailed longitudinal tensile strength calculations of unidirectional composites require the use of statistical models. All these models rely on information concerning the stress state near one or more adjacent broken fibres in the material.

2.1 Strength model parameters

The three important input parameters, derived from the beforementioned stress state calculation, are the *ineffective length* δ, determining the distance from the broken fibre end to the location at which 90% of the nominal value of the axial fibre stress has been reached, the *positive affected length* PAL, indicating the distance in an adjacent fibre over which the stress exceeds the nominal stress, and the *stress concentration factor* K, that quantifies this stress increase (Fig.1).

In order to be able to focus more on the influence of an existing interlayer, while not being to much dependent on computational power, a simplified two-dimensional finite element analysis has been developed (Jacobs & Verpoest 1996). The model geometry is presented in Fig. 2 for the case of a fibre volume fraction of 60%. As a part of the model represents a composite area with equivalent three phase composite properties, these data had to be calculated for different transversely isotropic fibre-coating matrix systems.

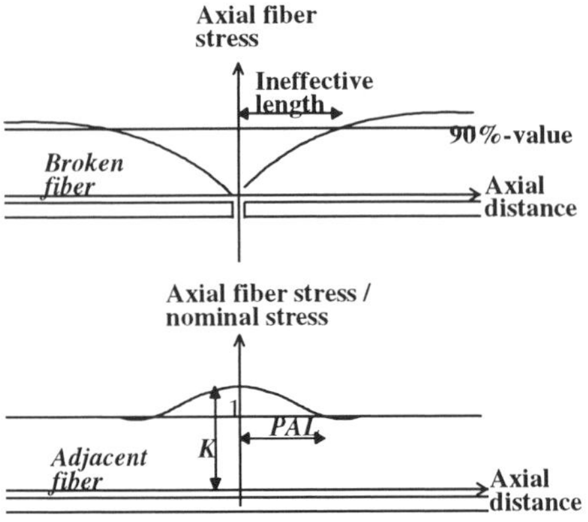

Fig.1 : Model parameters

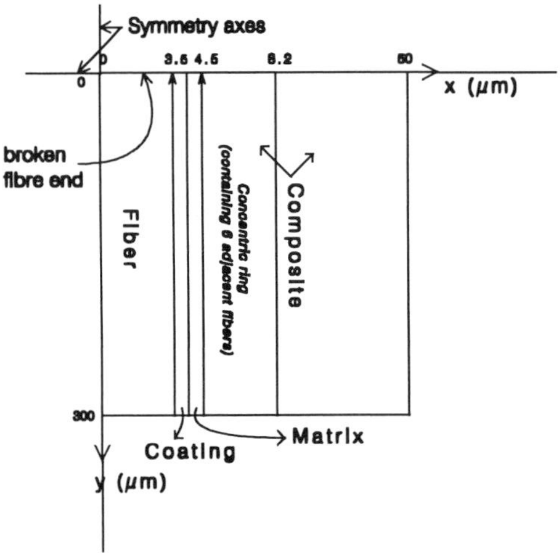

Fig.2: Finite element model geometry.

2.2 Elastic properties

The elastic behaviour of transversely isotropic material systems is completely characterised by 5 independent constants. Different sets of 5 independent elastic constants can be determined. The following 5 are often chosen: the longitudinal Young's modulus E_l, the longitudinal Poisson ratio ν_{lt}, the longitudinal shear modulus G_l, the transverse plain strain bulk modulus k_t and the transverse shear modulus G_t.

In the case of a 2-phase system, the improved lower bound expressions of these 5 constants, as calculated by the Bounding Energy method, are widely recommended for use (Chou 1993). These expressions are in a closed form and the solutions for E_l and ν_{lt} are equal to the rule of mixture result with a correction term. However, as the closed-form expressions of E_l, ν_{lt}, G_l and k_t fully coincide with the results of two other analytical models, i.e. the Doubly (Triply) Embedded model (Hermans 1967) and the Concentric Cylinder Assemblage model (Hashin & Rosen 1964), the latter two models can be equally used for extension into 3-phase expressions of these 4 elastic constants.

3-phase results of E_l and ν_{lt} were calculated, using the simple rule of mixture equation. It was not found worthwhile to adapt the almost negligible two phase correction term for this equation, to that for a three phase system. A closed form solution for the three phase longitudinal shear modulus of the composite $G_{l,c}$ is quite easy to derive; the calculation principle, based on the Doubly (Triply) Embedded model (Hermans 1967), has already been reported by several authors (Aronhime 1986, Wilczynski 1990).

$$G_{l,c} = G_m * \left(\frac{(I - II)}{(I + II)} \right) \tag{1}$$

$$\text{where: } I = (2*\alpha+1)^2 * (G_i + G_f)*(G_m + G_i) + (G_i - G_f)*(G_m - G_i)$$

$$II = ((2*\alpha+1)^2 * (G_i + G_f)*(G_m - G_i) + (G_i - G_f)*(G_m + G_i))*V_f*(2*\alpha+1)^2$$

$$\alpha = \left(\frac{1}{2}\right)*\left(\sqrt{\frac{(1 - V_m)}{V_f}} - 1\right)$$

(l=longitudinal, c=composite, f=fibre, m=matrix, i=interlayer, V_x=volume fraction of phase x)

Finding an expression for the transverse plain strain bulk modulus $k_{t,c}$ is somewhat more complex. Based on the Concentric Cylinder Assemblage model (Hashin & Rosen 1964), we derived a three phase, closed form solution for this constant. The final form is given in Equation 2.

$$k_{t,c} = \left(\frac{a1*b1 + (1+C)*V_m*k_{t,m}*G_{t,m}}{a1*V_m + (1+C)*b2} \right) \tag{2}$$

$$\text{where: } a1 = k_{t,i}*C - G_{t,i}$$

$$b1 = k_{t,m} + G_{t,m}*(1 - V_m)$$

$$b2 = G_{t,m} + k_{t,m}*(1 - V_m)$$

$$C = \left(\frac{1 - V_m}{V_f}\right)*\left(\frac{G_{t,i} + k_{t,f}}{k_{t,i} - k_{t,f}}\right)$$

(t=transverse, c=composite, f=fiber, m=matrix, i=interlayer, $G_{t,x}$=transverse shear modulus of phase x, V_x=volume fraction of phase x)

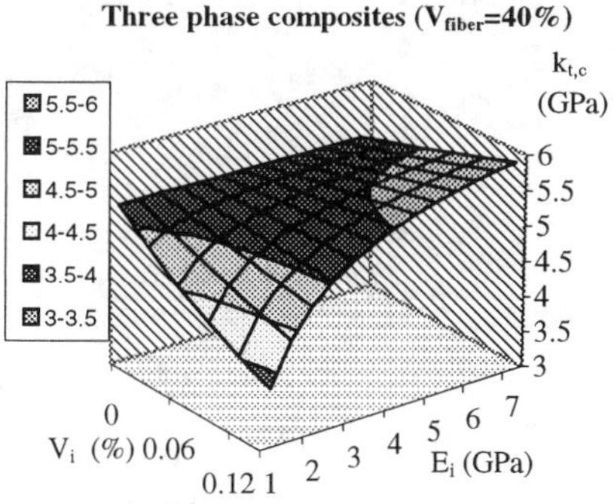

Fig.3: Composite k_t-values vs coating properties

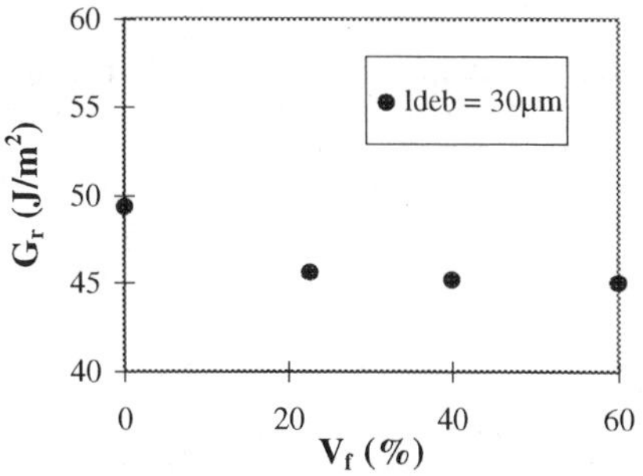

Fig.4: Energy release rate at debond extension vs fiber volume fraction in the case of a 30μm long fibre-matrix debond and an applied strain of 1.5%.

In the case of a unidirectional carbon fibre - epoxy matrix composite with component properties $E_{l,f}$=230GPa, $E_{t,f}$=13.8GPa, $\nu_{lt,f}$=0.2, $\nu_{tt,f}$=0.37, G_l=18GPa, E_m=3.0GPa, ν_m=0.35, ν_i=0.35, the variation of the composite transverse plain strain bulk modulus with varying coating stiffness (E_i=1-7GPa) and coating volume fraction (V_i=0-0.15%) is schematically shown in Fig.3.

Finally, for the transverse shear modulus G_t, we extended the Christensen solution (Christensen & Lo 1979) to a three phase system instead of deriving a closed form three phase lower bound expression.

2.3 Finite element calculations

2.3.1 Results without interfacial debond

The authors have previously shown that, in the simplified case of a single fibre crack without debonding or plastic deformation of coating or matrix, the existence of an interlayer markedly influences the *ineffective length* and the *positive affected length* data; on the contrary, it does almost not change the *stress concentration factor* in an adjacent, non-broken fibre (Jacobs & Verpoest 1996).

In this case of very strong fibre-coating bonds and a coating with very high fracture toughness, the longitudinal tensile strength will be low because of the stress concentrations in adjacent non-broken fibres, causing rapid composite failure. The corresponding finite element results consequently point out that tailoring the interlayer stiffness and thickness will probably not have a big impact on the composite longitudinal tensile strength; only energy absorbing mechanisms such as plastic deformation or changes in failure pattern could lead to improvements.

2.3.2 Parametric study of energy release rate values at interfacial debond extensions for a 2 phase system

A next step involves the introduction of damage in the model geometry. As interfacial debonds very often occur after fibre cracking in carbon fibre-epoxy systems, this damage form is further looked at. In order to get some insight into the influence of several parameters - such as the fibre volume fraction, the fibre stiffness to matrix stiffness ratio, the applied strain and the debond length - on the interfacial crack propagation, a parametric study of these variables has been performed for a two phase system. The starting model parameters include a fibre volume fraction of 40%, a fibre Young's modulus of 230GPa, a matrix Young's modulus of 3.0GPa, an applied strain of 1.5% and a debond length of 30μm. In further simulations, one or two of the previous parameters have been varied and their influence on the potential energy release rate for a unit extension of the debond G_r calculated. The remaining elastic properties of fibre and matrix are the same as previously mentioned for the calculation of composite transverse plain strain bulk moduli.

The influence of the fibre volume fraction appears to be negligible between V_f=20% and V_f=60%, i.e. the radial distance between the broken fibre segment and its adjacent fibres doesn't severely influence the energy release rate at the interface (Fig.4).

In contradiction to the stress transfer built up, which is controlled by the ratio E_f/E_m, the G_r-values are mainly determined by the *fibre* Young's modulus (Fig.5); in other words, the potential amount of strain energy, stored in the fibre prior to debond extension, determines the expected energy release rate. This conclusion is also reinforced by the results, obtained at different levels of applied strain: $G_r \sim \varepsilon_{br}^2$ (Fig.6).

Finally, energy release rate values at fibre/matrix debond extension as a function of the assumed debond length are given in Fig. 7 at ε =1.8%. The biggest variations in energy release rate occur in the case of very small debonds. The potential amount of energy release decreases with higher debond lengths.

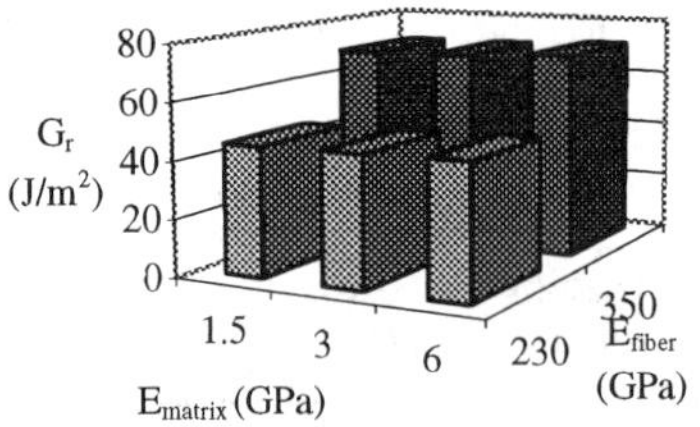

Fig.5: Energy release rate at debond extension vs fibre/matrix stiffness in the case of a 30μm long fibre/matrix debond, a fibre volume fraction of 40% and an applied strain of 1.5%.

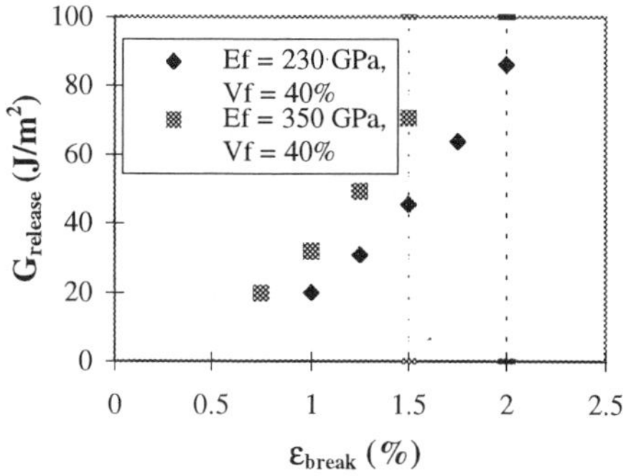

Fig.6: Energy release rate at debond extension vs applied strain in the case of a 30μm long fibre/matrix debond and an epoxy matrix with E_m=3.0GPa.

Consequently, if a fibre/matrix debond exists, debond growth at higher applied strains - if it occurs - is a stable process.

2.3.3 Prediction of fibre/matrix debond length at fibre fracture for a 2-phase system

For a given interfacial toughness, the actual debond length at fibre fracture and the debond growth at increasing loading strains can be calculated with the use of energy release rate data at different interfacial debond lengths. For comparison purposes, actual debond lengths are first calculated in the case of a two phase material. The component elastic properties are again the same as previously mentioned for the calculation of the transverse plain strain bulk moduli. For high strength fibres with a Young's modulus of 230 GPa, fibre breaking strain is about 2%. Taking this value into account, the applied strain at which a single fibre fractures has been arbitrarily chosen at 1.8%.

In this analysis, an energy balance between the released and the required amount of energy at fibre fracture and subsequent damage development (=debonding) has been used to calculate the initial

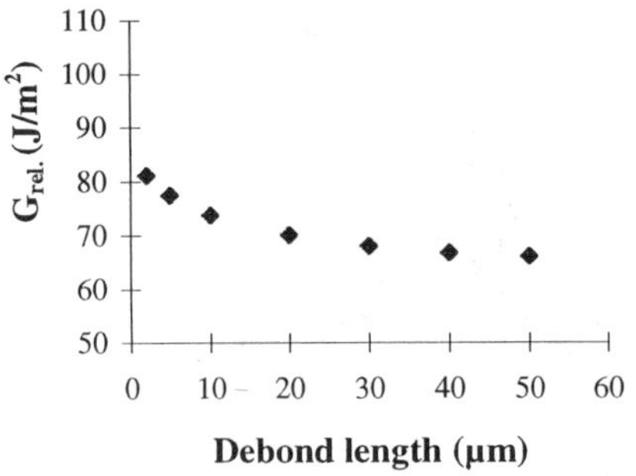

Fig.7: Energy release rate at fibre/matrix debond extension vs debond length at the moment of extension for ε=1.8%..

interfacial debond length. It is assumed that the excess of energy, released after fibre fracture, will be totally consumed by the development of an interfacial debond.

$$E_{required} = E_{released}$$

where : $E_{required} = \gamma_f * A_f + G_{ft} * (2\pi r_f) * l_d$

$$E_{released} = \Delta U_{fc} + \sum_i G_i * (2\pi r_f) * \Delta l_i$$

γ_f = surface energy of the carbon fibre

A_f = fibre surface area

G_{ft} = interfacial fracture toughness

r_f = fibre radius

l_d = debond length

ΔU_{fc} = energy release at fibre fracture

G_i = energy release rate at debond extension

Δl_i = incremental debond extension

By using the results for energy release rate values at fibre/matrix debond extension, as given in Fig.7, taking γ_f = 4.2 J/m² (Dresselhaus 1988) and ΔU_{fc} = $2.68*10^{-8}$ J (finite element result), initial debond lengths for different values of the interfacial fracture toughness can be determined (Table 1).

2.3.4 Extension to 3-phase systems and calculation of strength input parameters δ, PAL and K.

Similar calculations have been performed for coated fibre composite systems, i.e. with a coating stiffness of 1.5GPa and 6.0GPa (coating thickness=0.5μm). The elastic constants of the beforementioned unidirectional composites are shown in Table2.
Fig. 8 presents the energy release rate values at fibre/coating debond extension as a function of the

Table1: Debond lengths vs fracture toughness values.

G_{ft} (J/m^2)	100	150	200
$l_{in.\ deb.}$ (μm)	43.1	17	9.8

Table2: Independent elastic constants of unidirectional coated fibre composites.

$E_{coating}$ (GPa)	1.5	3.0	6.0
$G_{l,comp.}$ (GPa)	1.91	2.33	2.63
$G_{t,comp.}$ (GPa)	1.58	1.82	2.02
$E_{l,comp.}$ (GPa)	93.62	93.8	94.17
$E_{t,comp.}$ (GPa)	4.65	5.39	5.94
$\nu_{lt,comp.}$	0.29	0.29	0.29

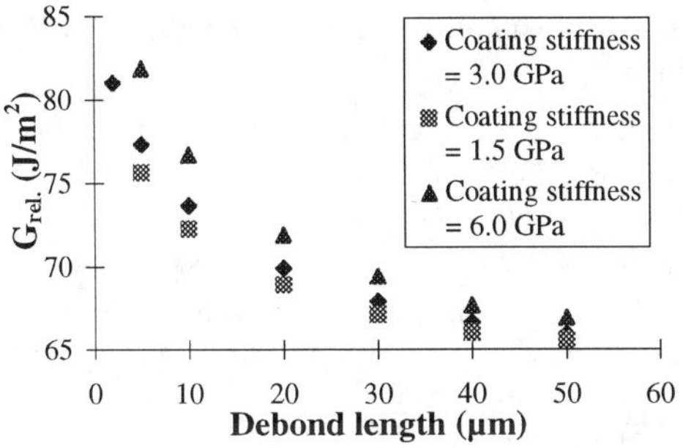

Fig. 8: Energy release rate at fibre/coating debond extension vs debond length for different coating stiffness values.

assumed interfacial debond length for the three systems. The change in energy release rate within the evaluated domain (1.5-6.0 GPa) is rather small.

Data of initial debond lengths for an interfacial fracture toughness value of 150 J/m^2, calculated as explained in chapter *2.3.3*, are shown in Table3. It should be noticed that, in the case of an interlayer with E_i=3.0 GPa, the composite is simply a 2-phase system. The results show that, if the same type of interfacial failure is observed in the different unidirectional composites (=debond at the outer fibre radius), the initial debond length varies only slightly.

In the case of a very strong fibre/coating bond, the released energy at fibre fracture causes a crack extension into the coating layer and can further lead to a debond, growing at the coating-matrix interface. For a coating with a thickness of 0.5μm and a stiffness of 6.0 GPa, the energy release rate at coating/matrix debond extension is calculated as a function of different debond lengths and compared

Table3: Initial debond lengths vs coating stiffness.

$E_{coating}$ (GPa)	1.5	3.0	6.0
$l_{in.deb.}$ (μm)	18.1	17.0	15.6

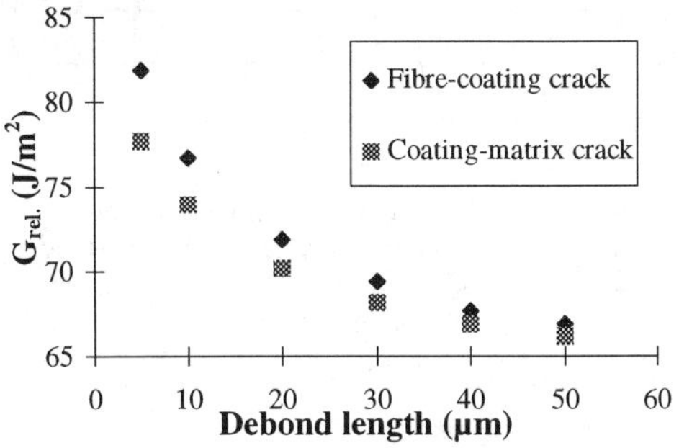

Fig.9: Energy release rate at debond extension vs debond length for different cracking patterns.

with the previous results for a fibre/coating debond (Fig. 9).

The variation in G_r-values is rather small. However, if the fibre crack extends into the interlayer, the excess of energy after fibre fracture and radial interlayer cracking is much larger, leading to a longer initial value for the interfacial debond. Assuming a coating/matrix interfacial fracture toughness value of 150 J/m^2 and a coating toughness of 200 J/m^2, the initial debond length becomes 20.9μm (>15.6μm).

As the energy release rate value continuously decreases as a function of the debond length for all the above composite systems, a possible debond extension in these systems, due to a further increase of the applied load, should be stable. The required G_{ft} =150 J/m^2 will however not be reached before composite failure (Fig.6: G_r=90 J/m^2 at ε=2% for a two phase system) and we can hence assume that the debond lengths will remain unchanged until composite failure.

Using therefore the calculated values of the initial debond length - for a fibre/coating debond with E_i=1.5, 3.0 and 6.0 GPa and a coating/matrix debond with E_i=6.0 GPa - ineffective length δ, positive affected length PAL and maximum stress concentration factor K are determined. The case of a coating with E_i=3.0GPa represents the 2-phase system. Results are summarised in Table4. If a debond at the outer fibre radius develops at fibre fracture, the existence of an interlayer does not change profoundly the stress concentration factor. Stiffer coatings however lead to a reduction in ineffective length and positive affected length. Only in the case of a coating/matrix crack with E_i=6.0

Table4 : Results for strength model parameters.

$E_{coating}$ (GPa)	Crack pattern	δ (μm)	PAL (μm)	K (-)
1.5	fibre-coating	98.8	61.4	1.061
3.0	fibre-coating	85.2	54.3	1.063
6.0	fibre-coating	76.3	48.9	1.063
6.0	coating-matrix	84.7	57.4	1.054

GPa, the maximum stress concentration factor is substantially lower. As, in this case, the ineffective length is almost the same as that for the 2-phase system, further improvements in longitudinal tensile strength could be obtained.

3. CONCLUSIONS

Finite element calculations of the stress state near a broken coated fibre show that the introduction of a low or high modulus coating does not lead to a substantial change in interfacial debond length and maximum stress concentration factor, assuming a debond at the outer fibre radius and assuming the same level of interfacial adhesion. A stiffer coating reduces the ineffective length and positive affected length due to better stress transfer into the carbon fibre and hence offers a possibility to improve the longitudinal tensile strength.

In the case of strong fibre/coating bonds, the debond can occur at the weaker coating/matrix interface. For a coated fibre composite with a coating thickness of 0.5μm and a coating stiffness of 6GPa, the maximum stress concentration factor appears to be substantially lower than that of a two phase system, whereas both ineffective length values are approximately equal. Consequently, this type of fibre coating and debond pattern provide a second way of improving the longitudinal tensile strength as compared to the uncoated system.

This work is funded by the Brite-Euram project "Plasma polymerised coatings and interphases for improved performance carbon fibre composites", contract No BRE2-0453. Research continued by the first author in the framework of his IWT-grant. The text presents research results of the Belgian Programme on Interuniversity Poles of attraction initiated by the Belgian State, Prime Minister's Office, Science Policy Programming. The scientific responsibility is assumed by its authors.

4. REFERENCES

1. Aronhime M.T., Marom G., Elastic properties of fiber-reinforced composites modified with an interlayer, Report from the Casali Institute of Applied Chemistry, Hebrew University of Jerusalem,1986.
2. Chou, T.-W., Materials Science and Technology, Vol. 13: Structure and properties of composites, VCH, 1993, p. 406-425.
3. Christensen, R.M., Lo, K.H., Solutions for effective shear properties in three phase sphere and cylinder models, *J. Mech. Phys. Solids*, 27, 1979, p. 315-330.
4. Dresselhaus, M.S., et al., Graphite fibers and filaments, Springer-Verlag Berlin Heidelberg, Germany, 1988, p.120-130.
5. Hashin, Z.V.I., Rosen, B.W., The elastic moduli of fiber-reinforced materials, *J. Appl. Mech.*, June 1964, p. 223-232.
6. Hermans, J.J., The elastic properties of fiber reinforced materials when the fibers are aligned, *Proc. K. Ned. Acad. Wet. B*, 70,1,1967.
7. Ivens, J., Albertsen, H., Wevers, M., Verpoest, I., Peters, P., Interlaminar fracture toughness of CFRP influenced by fibre surface treatment, part 1: experimental results, *Comp. Sci. Technol.*, 54, 1995, p. 133-145.
8. Jacobs, E., Verpoest, I., Finite element modeling of the stresses near a broken coated fibre in a unidirectional composite, In '*Proc. of ECCM-7*', Woodhead Publishing Ltd, England, Vol. 1, 1996, p. 21-26.
9. Wilczynski, A.P., A basic theory of reinforcement for unidirectional fibrous composites, *Comp. Sci. Technol.*, 38, 1990, p. 327-337.

Dynamics and impact

Damage and Failure of Interfaces, Rossmanith (ed.) © *1997 Balkema, Rotterdam, ISBN 90 5410 899 1*

Dynamic stresses in elastic and elastic-plastic solids with cracks

A. Rivinius & J. Ballmann
Lehr- und Forschungsgebiet für Mechanik, RWTH Aachen, Germany

ABSTRACT: The paper deals with a numerical scheme for the propagation of stress waves in solids with cracks. Both elastic and elastic-plastic materials with cracks and/or interfaces are considered by solving the nonlinear, hyperbolic system of partial differential equations for elastic-plastic waves in solids. The integration is carried out by a two step Godunov-type method. In the first step the fluxes at the cell interfaces are worked out by a finite volume scheme, followed by an updating procedure in the second step to calculate the stress and velocity components in the cell centers. Our main concern are the interaction processes of waves with crack tips and their significance for crack initiation and crack arrest. Especially Rayleigh and Stoneley waves, travelling along crack surfaces and material interfaces, respectively, are supposed to play a very important role in dynamic fracture mechanics. Numerical results will be presented for a stationary interface crack in an elastic bi-material and a comparison with an experiment in an inelastic, homogeneous body, including the stages of crack initiation and arrest, will be discussed.

1 INTRODUCTION

Strong dynamic loading or unloading is typically characterized by stress waves such as longitudinal and shear waves, producing Rayleigh waves along the crack surfaces and eventually Stoneley waves along the interface. Besides other models, the dynamic stress intensity factors can be assumed as quantitative criteria for start or arrest of crack growth. Studying those problems numerically, it is very important that all phenomena are reproduced correctly by the numerical scheme.

Different explicit characteristic-based one- and two-step methods have been developed and improved (Lin and Ballmann, 1995) to treat the propagation of wave fronts with correct speeds and amplitudes. Key words are the accurate modelling of the elastic-plastic stress path, the consideration of the contact – no-contact problem along the crack surfaces (Zhang and Ballmann, 1996) and the improvement of the numerical resolution by restriction of the computational domains to the region of physical relevance by using artificial, non-reflecting boundary conditions (Zhang and Ballmann, 1997).

Our method of choice for the bi-material problems is an explicit bicharacteristic scheme for the non-singular part of the solution domain and a modified Zwas-two-step scheme for subdomains with singular points (Niethammer and Ballmann, 1995). By this way almost all non-physical perturbations, e.g. numerical oscillations at wave fronts, numerical cutting traces from singular points, spurious reflections from artificial boundaries, could be avoided. In problems where elastic-plastic yielding may occur we prefer a Godunov-type two-step method in the complete solution domain. In the vicinity of singular points an additional term derived from the bicharacteristic scheme (Lin and Ballmann, 1993) is applied to minimize numerical oscillations. These methods have successfully been applied to elastic bi-material and elastic-plastic homogeneous bodies to reproduce the important part especially of the Rayleigh waves travelling along crack faces and of the Stoneley waves arising along interfaces under certain conditions. They are now being extended to crack problems in elastic-plastic bi-materials and to the crack growth and arrest phenomena in homogeneous elastic-plastic media.

In this paper we will confine ourselves to the latter method. We will present numerical results in elastic as well as in elastic-plastic materials. We will discuss a stationary interface crack with particular emphasis on the generated Rayleigh and Stoneley waves. In the second example we will compare our numerical results for a wave impact problem in an elastic-plastic medium with the associated experiment (Prakash et al., 1992).

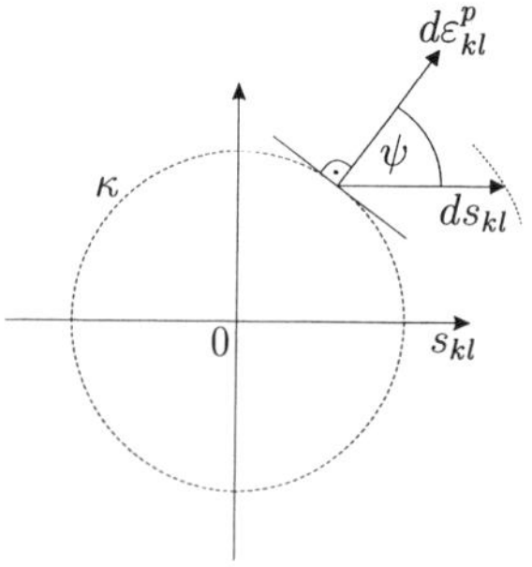

Figure 1: Drucker's hypothesis.

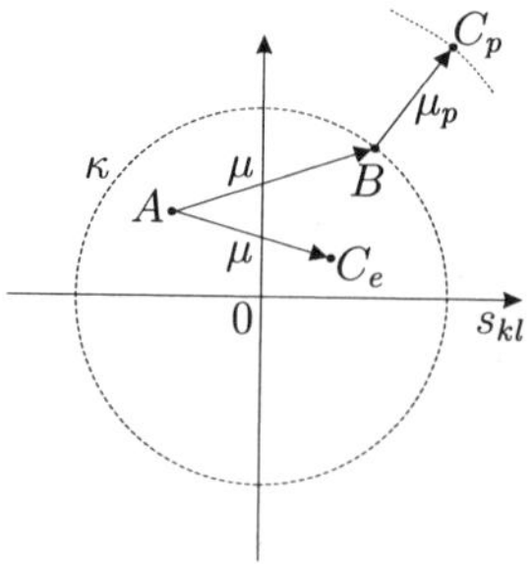

Figure 2: Proposed elastic-plastic stress path.

2 CONSTITUTIVE RELATIONS

The elastic-plastic constitutive relation we use in our numerical code (Lin and Ballmann, 1993) is based on Druckers hypothesis (see figure 1). The increments of a plastic strain $d\varepsilon^p_{kl}$ are assumed to be normal to the instantaneous yield surface κ and the angle ψ between the plastic strain increment $d\varepsilon^p_{kl}$ and the increment of the deviatoric stresses ds_{kl} to be acute but not known, i.e. $|\psi| \leq \frac{\pi}{2}$. Adopting the von Mises yield condition

$$\frac{1}{2} s_{\lambda\nu} s_{\lambda\nu} - \kappa^2 = 0, \tag{1}$$

the following relation can be obtained:

$$d\varepsilon^p_{kl} = \frac{h}{2\mu}\frac{s_{kl}}{\kappa} d\kappa, \tag{2}$$

where

$$h = \begin{cases} \dfrac{\mu}{\mu_p(\kappa)} - 1, & if\ d\kappa > 0 \\ 0, & if\ d\kappa = 0 \end{cases} \tag{3}$$

is the so-called plastic factor, μ and μ_p are the material shear modulus at the elastic and plastic stage, respectively. For work-hardening materials μ_p is constant, for power-law-hardening ones, obeying $\frac{\sigma}{\sigma_0} = \left(\frac{\varepsilon}{\varepsilon_0}\right)^\alpha$ and $\sigma_0 = \kappa_0$ with ε_0 denoting the elastic limits of stress and strain, respectively, and α denoting the hardening exponent, we have

$$\mu_p = \frac{\alpha\kappa_0}{\varepsilon_0}\left(\frac{\kappa}{\kappa_0}\right)^{\frac{\alpha-1}{\alpha}}. \tag{4}$$

Lin and Ballman proposed the angle ψ to be zero for a numerical stress path, i.e. the deviatoric stress increment to be normal to the yield-surface, too. Together with the assumption of pure elastic volume change of the material, i.e. $de^p_{kl} = d\varepsilon^p_{kl}$, where de^p_{kl} are the plastic part of the deviatoric strain increments de_{kl}, the elastic-plastic contitutive relation follows as

$$de_{kl} = de^e_{kl} + de^p_{kl} = \frac{1+h}{\mu} ds_{kl}. \tag{5}$$

The stress path (see figure 2) is determined by the starting point A inside the yield-surface and the end-point C_p outside of it. The line from the original yield-surface point B to the point C_p is a part of a radial ray through the origin of the deviatoric stress plane:

$$s^A_{kl} \xrightarrow{\mu} s^B_{kl} \xrightarrow{\mu_p} s^{C_p}_{kl}. \tag{6}$$

Numerical experiments for one-dimensional problems have shown a very good agreement of this proposed stress path with the exact solution (given by Ting and Nan, 1969).

3 GOVERNING EQUATIONS

With the assumptions of small strains and isotropy the governing equations for an elastic-plastic solid under plane strain conditions without body forces can be written in this matrix form:

$$\mathbf{A}\frac{\partial \mathbf{w}}{\partial t} = \frac{\partial \mathbf{f}}{\partial x_1} + \frac{\partial \mathbf{g}}{\partial x_2}, \tag{7}$$

$$\mathbf{A} = \text{diag}\left(\rho, \rho, \frac{1}{3K}, \frac{1+h}{2\mu}, \frac{1+h}{2\mu}, \frac{1+h}{\mu}\right), \tag{8}$$

$$\mathbf{w} = \begin{pmatrix} \dot{u}_1 \\ \dot{u}_2 \\ \sigma_{11}+\sigma_{22}+\sigma_{33} \\ \sigma_{11}-\sigma_{33} \\ \sigma_{22}-\sigma_{33} \\ \sigma_{12} \end{pmatrix}, \tag{9}$$

$$\mathbf{f} = \begin{pmatrix} \sigma_{11} \\ \sigma_{12} \\ \dot{u}_1 \\ \dot{u}_1 \\ 0 \\ \dot{u}_2 \end{pmatrix}, \quad \mathbf{g} = \begin{pmatrix} \sigma_{12} \\ \sigma_{22} \\ \dot{u}_2 \\ 0 \\ \dot{u}_2 \\ \dot{u}_1 \end{pmatrix}, \tag{10}$$

where x_1 and x_2 are the Cartesian coordinates, t the time; ρ denotes the density of the solid, K is the bulk modulus. Since the last-mentioned two coefficients are constant, the first three equations are always linear, wheras the last three are nonlinear due to h being a function of the instantaneous yield stress κ.

This system reveals four characteristic wave speeds, those of the elastic longitudinal and shear waves c_1 and c_2 and those of the fast and slow plastic waves c_f and c_s, respectively:

$$\begin{aligned} c_1 &= \sqrt{\frac{3K+4\mu}{3\rho}}, & c_2 &= \sqrt{\frac{\mu}{\rho}}, \\ c_f(\kappa) &= \sqrt{\frac{3K+4\mu_p}{3\rho}}, & c_s(\kappa) &= \sqrt{\frac{\mu_p}{\rho}} \end{aligned} \tag{11}$$

with $c_1 > c_f > c_2 > c_s$.

4 NUMERICAL SCHEME

4.1 *Inner point*

Our numerical method is a two-step Godunov-type method in finite volume discretisation, derived from the HEMP code (Wilkins, 1964). Our computational domain is covered by a regular mesh of quadrilateral cells of the area $A_{(i)}$, cell values are denoted by $\mathbf{w}^n_{(i)}$ at time level t^n, see figure 3. The fluxes $\mathbf{w}^{n+\frac{1}{2}}_{(\bar{j})}$ at the grid points $(\bar{j})$, are worked out in the first step at time level $t^{n+\frac{1}{2}} = t^n + \frac{\Delta t}{2}$ from the values in the four adjoining

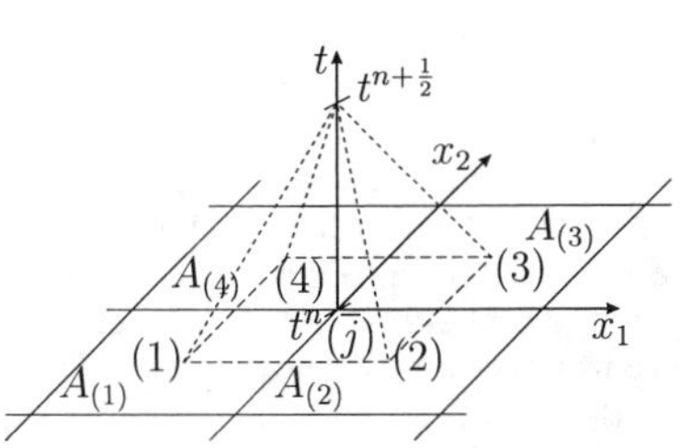

Figure 3: Flux calculation at an interior grid point.

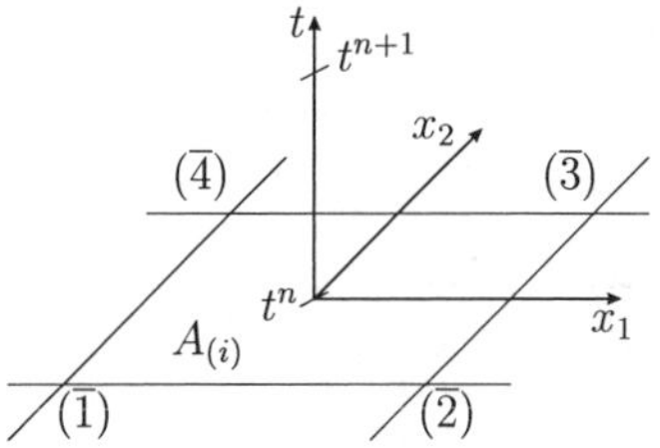

Figure 4: Updating of the cell values.

cells:

$$\sum_{i=1}^{4} \left(\frac{A_{(i)}}{4} \int_{\mathbf{w}^n_{(i)}}^{\mathbf{w}^{n+\frac{1}{2}}_{(\bar{j})}} \mathbf{A}\, d\mathbf{w} \right) = \frac{\Delta t}{4} \cdot [\\ (x_{2(4)} - x_{2(2)})(\mathbf{f}^n_{(3)} - \mathbf{f}^n_{(1)}) - \\ (x_{2(3)} - x_{2(1)})(\mathbf{f}^n_{(4)} - \mathbf{f}^n_{(2)}) - \\ (x_{1(4)} - x_{1(2)})(\mathbf{g}^n_{(3)} - \mathbf{g}^n_{(1)}) + \\ (x_{1(3)} - x_{1(1)})(\mathbf{g}^n_{(4)} - \mathbf{g}^n_{(2)})]. \tag{12}$$

Due to the nonlinear character of the system described in equations (7) to (10) an iteration like Newton's iterative method has to be applied in the case of plastic yielding (Lin and Ballmann, 1993). In the second step the fluxes at cell vertices are used to update the values in the shared cell at time level t^{n+1} (see figure 4):

$$\int_{\mathbf{w}^n_{(i)}}^{\mathbf{w}^{n+1}_{(i)}} \mathbf{A}\, d\mathbf{w} = \frac{\Delta t}{2A_{(i)}} \cdot [\\ (x_{2(\bar{4})} - x_{2(\bar{2})})(\mathbf{f}^{n+\frac{1}{2}}_{(\bar{3})} - \mathbf{f}^{n+\frac{1}{2}}_{(\bar{1})}) - \\ (x_{2(\bar{3})} - x_{2(\bar{1})})(\mathbf{f}^{n+\frac{1}{2}}_{(\bar{4})} - \mathbf{f}^{n+\frac{1}{2}}_{(\bar{2})}) - \\ (x_{1(\bar{4})} - x_{1(\bar{2})})(\mathbf{g}^{n+\frac{1}{2}}_{(\bar{3})} - \mathbf{g}^{n+\frac{1}{2}}_{(\bar{1})}) + \\ (x_{1(\bar{3})} - x_{1(\bar{1})})(\mathbf{g}^{n+\frac{1}{2}}_{(\bar{4})} - \mathbf{g}^{n+\frac{1}{2}}_{(\bar{2})})]. \tag{13}$$

Again the yield condition has to be taken into account for the nonlinear part of the governing equations.

On a mesh of quadratic cells the described method will provide a CFL number of 1 in every cell for signals transported with the elastic longitudinal wave c_1.

4.2 *Crack tips*

Due to the mixed boundary value problem

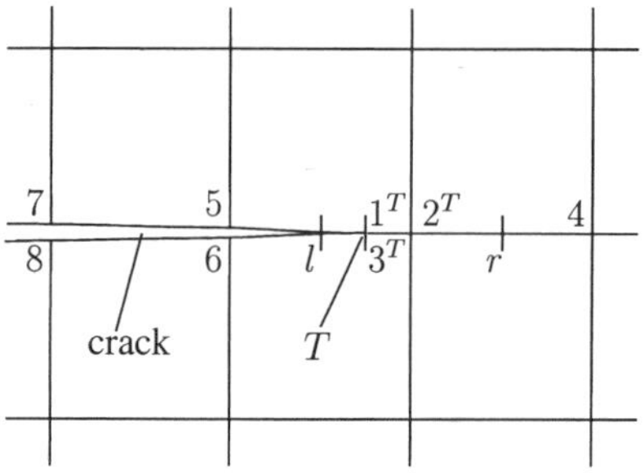

Figure 5: Sketch of the mesh around a crack tip.

arising at crack tips the stresses are singular. In the case of crack propagation the particle velocities are singular, too. These singularities depend just on the mathematical model, since no real material can support such a situation. To overcome this problem we consider the crack tip as the border of crack surfaces to inner points, see figure 5. The fluxes at the grid point nearest to the crack tip T, 1^T, 2^T and 3^T as long as T moves between l and r, are composed by the values obtained under two or three different boundary conditions, depending on whether the crack line is a symmetry line or not, respectively. In case of symmetry that grid point is treated as well as a crack surface point 1^T with dynamic boundary conditions and as an inner point 2^T obeying the symmetry conditions. In other cases values from both crack faces, 1^T and 3^T, and from an inner point 2^T are obtained. While the crack tip is stationary it lies on a grid point of the computational domain, i.e. the crack tip T is identical to the points with a superscript T. Then the average values of 1^T and 2^T in case of symmetry or else 1^T, 3^T and doubled weighted 2^T compose the fluxes for the updating procedure. If T propagates between l and r, additional weighting functions with the distance of T from the nearest grid point as a parameter are used. So the domain influenced directly by the crack tip is shifted with the propagating crack tip causing small wave radiations and numerical oscillations in every time step, which could be reduced by applying a gradual nodal release method (Atluri and Nishioka, 1985). To clarify this consider the instant when the crack tip just passes point r, so that all numbered points in the figure 5 have to be shifted by one cell to the right. At the new points 1^T and 3^T we do not prescribe the real boundary conditions at once, instead the values at the former inner point 4 are used as artificial conditions and reduced linearly to the correct ones until points 7 and 8 are shifted to that position. This procedure counteracts the effects of grid splitting.

In our present researches the stress intensity factor (SIF) is used as fracture criterion. Due to its restriction to small scale yielding conditions the SIF is a quantitatively good criterion just in elastic problems, whereas in elastic-plastic materials other criteria are being introduced in our further research. The calculation of the SIF is carried out for stationary as well as for running cracks by a least-squares method over the spatial derivatives of the stress components (given by Freund, 1990).

4.3 *Interfaces*

In the case of non-homogeneous media such as bi-materials the different material properties have to be taken into account. Obviously the above described method is applicable without any change except at interface points which need a special treatment.

Assume an interface in x_1-direction between two materials glued or welded together. Then the jump or continuity conditions at points of the interface are as follows:

$$\begin{aligned} [\![\sigma_{22}]\!] &= [\![\sigma_{12}]\!] = 0, \\ [\![\dot{u}_1]\!] &= [\![\dot{u}_2]\!] = 0. \end{aligned} \tag{14}$$

The other unknown components, i.e. the normal stresses σ_{11} and σ_{33}, may jump across the interface. In our numerical code the values at these grid points are determined by considering both sides of the interface as homogeneous boundaries and calculating the values iteratively until the jump conditions are fulfilled.

5 NUMERICAL RESULTS

5.1 *Interface crack*

In order to emphasize the significance of the high energetic Rayleigh and Stoneley waves results for a numerical experiment in a bi-material with interface crack are presented. The mentioned waves are surface waves propagating along crack surfaces and interface boundaries, respectively. They are generated when a shear wave hits these types of boundaries. Table 1 shows the chosen material parameters.

The experimental setup is sketched in figure 6. The specimen is assumed to be infinite with a centered crack. At the lower far boundary of the aluminium part the medium is impacted by the Heavyside load σ^L, which causes a longitudinal wave c_1 to travel towards the crack and interface line. The calculation was carried out on a mesh of 900x1000 cells including non-reflecting boundary layers of 100 cells at the upper, right and lower boundaries. At time level $t = 0$ the interaction process of this wave front with crack and interface starts. Longitudinal waves $c_1^{A,S}$ and shear waves $c_2^{A,S}$ are radiated from the crack tip

Table 1: Bi-Material properties.

	Aluminium	Steel
mass density $\rho\ [\frac{kg}{m^3}]$	2710	7800
Poisson's ration ν [1]	0.333	0.319
longitudinal wave speed $c_1\ [\frac{m}{s}]$	6614	6132
shear wave speed $c_2\ [\frac{m}{s}]$	3332	3203
Rayleigh wave speed $c_R\ [\frac{m}{s}]$	3105	2977

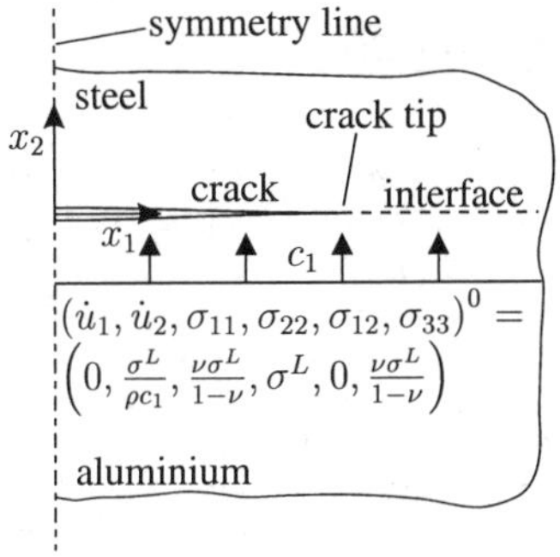

Figure 6: Sketch of the experimental setup.

into both materials. Figure 7 shows the generated wave patterns in a shaded representation of the von Mises equivalent stress at the time when the longitudinal wave c_1^A in aluminium has just reached the symmetry line. At both crack surfaces the shear waves $c_2^{A,S}$ change into Rayleigh waves $c_R^{A,S}$. At the interface boundary a Stoneley wave c_{St} can be observed. Additionally, several von Schmidt waves $c_{vS}^{A,S\,(1,2,3)}$, marked by dashed lines, travel into both materials. The latter wave type is a head wave front enveloping waves of one type which are generated by a disturbance running faster than the produced waves themselves. The von Schmidt waves denoted by a superscript (1) are formed by shear waves generated at the crack surface or the interface line by the faster longitudinal waves in the same material. They propagate with the speed of shear wave. The superscript (2) refers to head waves of shear wave speed, too, caused by the faster longitudinal waves in the respective other material across the interface boundary. $c_{vS}^{S\,(3)}$ is made up of longitudinal waves in the steel part produced by the faster aluminium longitudinal wave across the interface. The angle of these von Schmidt waves with the x_1-direction is the arcsin of the ratio speed of

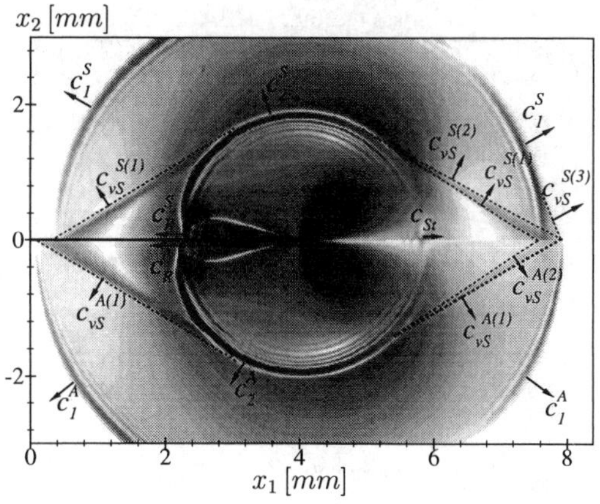

Figure 7: Wave pattern in a bi-material with interface crack.

its producing disturbance over speed of the generated wave type.

5.2 *Crack initiation and arrest*

In the following we are going to compare our numerical results for an impact loading setup (see figure 8) with the experimentally measured data (Prakash et al., 1992). The experiment involves the plane strain loading of a plane crack by a square tensile pulse with the duration of approximately 1 μs. The specimen was a metal disk 62.5 mm in diameter and 8 mm thick. A planar crack on its half midplane was prepared. The loading was produced by catapulting a second disk against the cracked one at very high speed and thus generating a compressive wave in both disks. The stress magnitude of the wave depends on the impact velocity and the material properties, the impact duration on the thickness of the impacting disk, since the compressive wave reflects at the free surface as a tensile wave which then unloads the impact when arriving at the contact line. In the cracked disk the compressive wave travels over the crack line almost without being affected, reflects at the free surface as a tensile pulse and then loads the crack when arriving again at the crack line. The interaction of the pulse with the crack tip leads to crack initiation and sudden crack arrest. The waves radiated from the crack tip, as indicated in figure 8, finally load the free surfaces where the normal velocity on four monitoring points is measured by means of a laser interferometer system. As long as no disturbances from the outer boundaries of the specimen arrive at the middle portion the conditions of plane strain are fulfilled.

Table 2 shows the material parameters and the experimental data describing the impact load and measurement points.

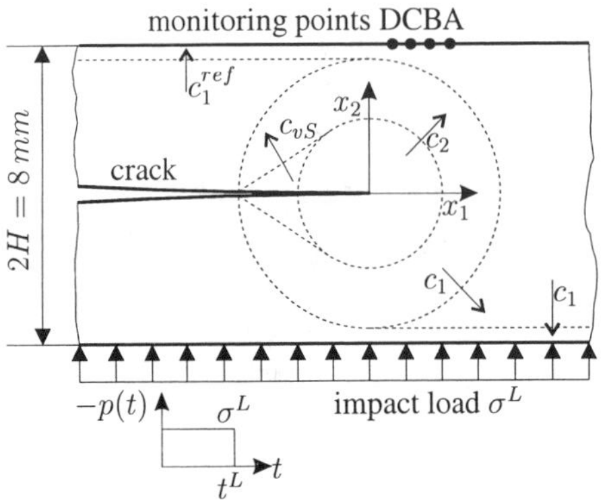

Figure 8: Experimental setup.

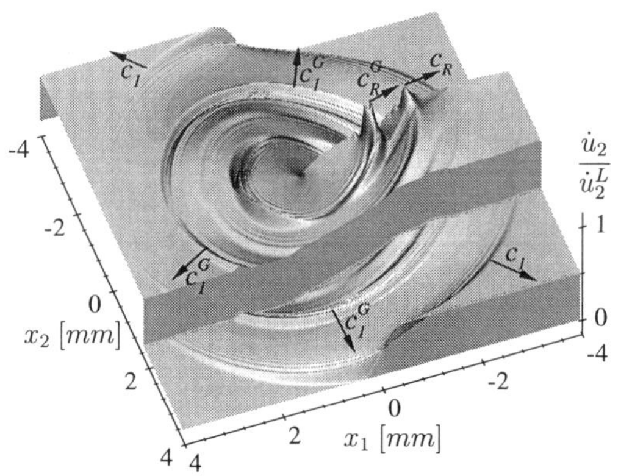

Figure 9: Velocity field in x_2-direction when the load pulse arrives at the free surface.

Table 2: Material properties and experimental data.

mass density ρ	=	$7600\,\frac{kg}{m^3}$
Poissin's ratio ν	=	0.3
longitudinal wave speed c_1	=	$5893\,\frac{m}{s}$
shear wave speed c_2	=	$3124\,\frac{m}{s}$
Rayleigh wave speed c_R	=	$2903\,\frac{m}{s}$
impact load σ^L	=	$1941.0\,MPa$
impact velocity $\dot{u}_2^L$	=	$85.4\,\frac{m}{s}$
impact duration t^L	=	$0.969\,\mu s$
crack initiation time t^*	=	$190.5\,\mu s$
crack propagation Δx_1	=	$0.109\,mm$
x_1^A	=	$0.53H$
x_1^B	=	$0.41H$
x_1^C	=	$0.29H$
x_1^D	=	$0.17H$
$x_2^A = x_2^B = x_2^C = x_2^D$	=	H

The most interesting observation in the experiment was the appearance of a sharp spike with a duration of less then $80\,ns$, which seems to be related to the onset of crack growth. Prakash et al. modelled this spike in an elastic-viscoplastic material by the sudden formation of a small traction free hole at the crack tip. In our linear-elastic investigations this sharp peak could successfully be modelled by assuming the crack tip to propagate with the Rayleigh wave speed (Lin, 1996). But the calculated velocities at the monitoring points show just a qualitatively good agreement with the measured data. The normal velocity field for this model is presented in figure 9 at the time, when the first longitudinal wave reflected from the crack surface (c_1^{ref} in figure 8) arrives at the free boundary. Superscript G denotes waves generated by crack growth. From the crack tip a circular logitudinal wave c_1 is radiated. At the free crack surface the shear wave causes a Rayleigh wave. The shear wave is superimposed with a second longitudinal wave c_1^G generated by the crack initiation and arrest. Moreover a second Rayleigh wave c_R^G due to crack growth propagates along the crack surface. Our calculation resolves this second Rayleigh wave as a two pointed peak, the first one caused by the energy release at crack initiation, the second one by the energy accumulation at crack arrest.

Finally we present results for elastic-plastic models, too, see figures 10 and 11. The represented quantity is the normal velocity at the four monitoring points. Figure 10 refers to the case when the crack stays stationary, in figure 11 the crack undergoes the stages of initiation, running very shortly with the Rayleigh wave speed and arrest. Our numerical results were obtained for three elastic-plastic hardening laws and for an elastic material. They are compared with a numerical, viscoplastic simulation of Prakash et al. for the stationary crack and with their experiment for the non-stationary crack. In all elastic-plastic models we assumed an elastic limit $\kappa_0 = 700\,MPa$. The plastic shear modulus μ_p was set to $\frac{1}{16}\mu$ for the work-hardening, the hardening exponent α to 3 for the power-law hardening constitutive relation.

The calculations were carried out on a mesh of 1000x800 cells including non-reflecting boundary layers of 100 cells at both artificial x_1-borders of the computational domain. Due to the formation of a plastic zone around the crack tip reflections occur at the elastic-plastic interfaces which lead to higher velocities at the monitoring points. So, a much better

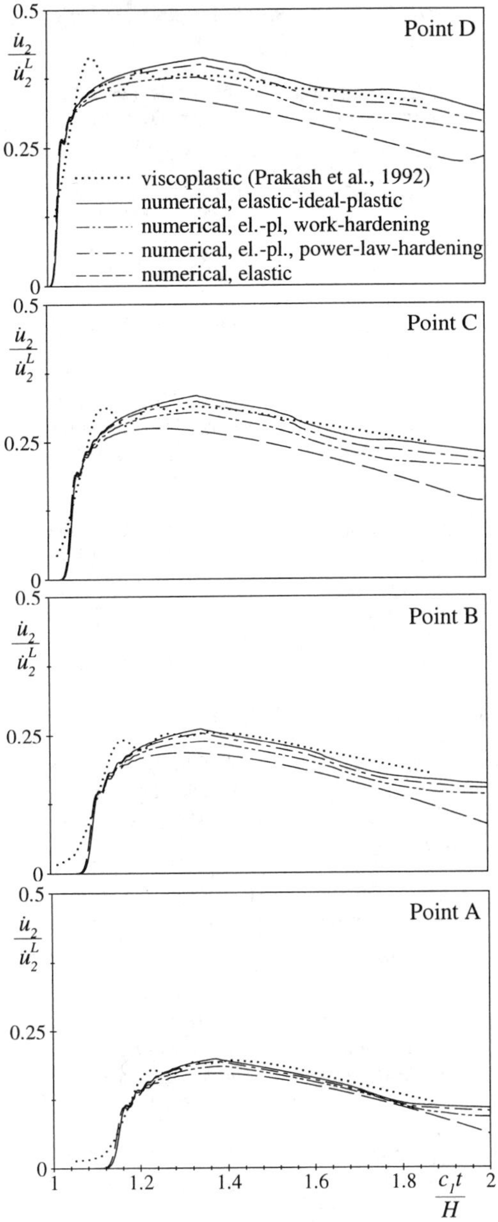

Figure 10: Normal velocities at the monitoring points in comparison with the numerical, viscoplastic simulation of Prakash et al., 1992, for the crack staying stationary.

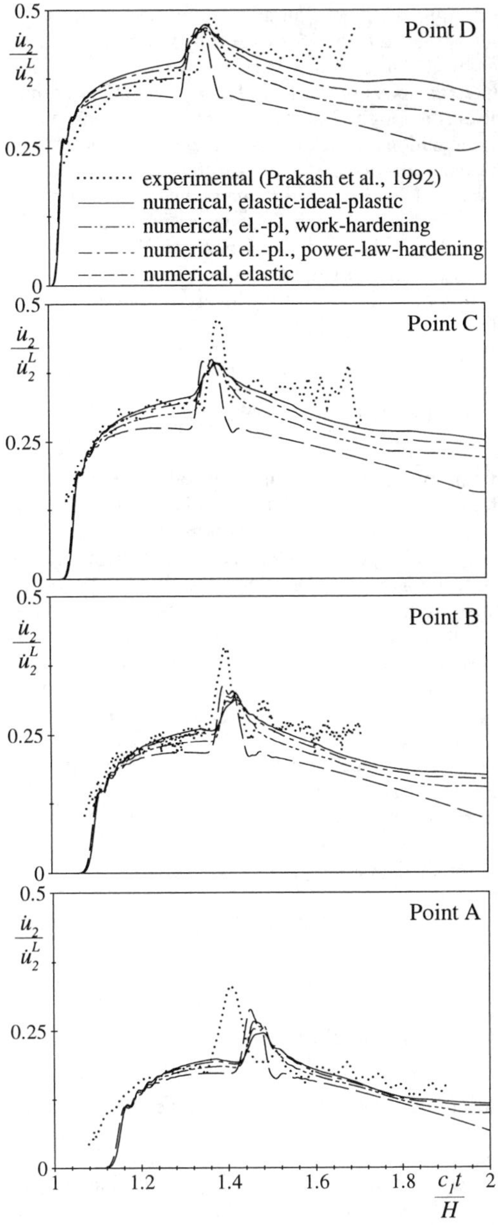

Figure 11: Normal velocities at the monitoring points in comparison with the experiment of Prakash et al., 1992, for the crack undergoing initiation and arrest.

and up to the crack initiation time even almost perfect agreement with the experimental data could be achieved. On the other hand the spike is now wider, since the speed of the fast plastic wave is lower than the elastic longitudinal wave speed. Other reasons for the deviations from the experimental results may be the used elastic-plastic constitutive relation and the numerical load pulse not matching the experiment. Possibly the experimental, compressive load pulse did interact with the crack line when travelling across.

But nevertheless these results attest our method to be a good tool for further investigations in elastic-plastic fracture mechanics.

6 SUMMARY

In this paper we have investigated crack problems in elastic bi-materials and elastic-plastic homogeneous materials by a finite volume method. By using non-reflecting boundary conditions the computational domain could be reduced to the domain of interest around the crack tip improving the numerical resolution in order to resolve the significant role of stress waves interacting with crack tips in fracture mechanics. Results for a numerical experiment in a bi-material medium with an interface crack were discussed to show the significance of surface waves at free and interface boundaries, i.e. the Rayleigh and Stoneley waves, respectively, on the time history of the stress intensity factor used as the fracture criterion. Furthermore we demonstrated the efficiency of this method in a comparison of numerical results in a cracked, elastic-plastic, homogeneous medium with the associated experiment including the stages of initiation and arrest.

ACKNOWLEDGEMENT

The authors would like to express their thanks to the Deutsche Forschungsgemeinschaft, Germany, for the financial support of the reasearch project "The significance of elastic-plastic waves for the non-stationary crack propagation in solids" under grant no. Ba 661/17-1.

REFERENCES

S.N. Atluri, T. Nishioka. Numerical Studies in Dynamic Fracture Mechanics. *International Journal of Fracture*, 27:245–261, 1985.

L.B. Freund. *Dynamic Fracture Mechanics.* Cambridge University Press, 1990.

X. Lin, J. Ballmann. Numerical Method for Elastic-Plastic Waves in Cracked Solids, Part 1: Anti-Plane Shear Problem, Part 2: Plane Strain Problem. *Archive of Applied Mechanics*, 63:261–282, 63:283–295, 1993.

X. Lin, J. Ballmann. Elastic-Plastic Waves in Cracked Solids under Plane Stress. *Applied Mechanics Review*, 46(12), 1993. Proceeding of IUTAM Symposium – Nonlinear Waves in Solids, 1993.

X. Lin, J. Ballmann. Improved Bicharacteristic Scheme for Two-dimensional Elastodynamic Equations. *Quaterly of Applied Mathematics*, 53:383–398, 1995.

X. Lin. *Numerical Computation of Stress Waves in Solids.* Akademie Verlag, Berlin, 1996.

R. Niethammer, J. Ballmann. Numerical Simulation of Shock Waves in Linear-Elastic Plane Plates with Curved Boundaries and Material Interfaces. *International Journal of Impact Engineering*, 16:711–725, 1995.

V. Prakash, L.B. Freund, R.J. Clifton. Stress Wave Radiation From a Crack Tip During Dynamic Inititiation. *Transactions of the ASME*, 59:356–365, 1992.

T.C.T. Ting, N. Nan. Plane Waves Due to Combined Compressive and Shear Stresses in a Half Space. *Journal of Applied Mechanics*, 36:189–197, 1969.

M.L. Wilkins. Calculation of Elastic-Pastic Flow. In *Methods in Computational Physics, Advances in Research and Applications*, 1964.

Y.G. Zhang, J. Ballmann. An Explicit Finite Difference Procedure for Contact-Impact Analysis of Crack Edges. *Archive of Applied Mechanics*, 66:493–502, 1996.

Y.G. Zhang, J. Ballmann. Two Techniques for the Absorption of Elastic Waves Using an Artificial Transition Layer. *Wave Motion*, 25:15–33, 1997.

Damage and Failure of Interfaces, Rossmanith (ed.) © 1997 Balkema, Rotterdam, ISBN 90 5410 899 1

Two optical techniques applied to the investigation of intersonic interface fracture

R.P.Singh, A.J.Rosakis & O.Samudrala
Graduate Aeronautical Laboratories, California Institute of Technology, Pasadena, Calif., USA

A.Shukla
Dynamic Photomechanics Laboratory, Department of Mechanical Engineering, University of Rhode Island, Kingston, R.I., USA

ABSTRACT: This paper describes experimental observations of various phenomena characteristic of dynamic intersonic decohesion of bimaterial interfaces. Two separate but complementary optical methods are used in conjunction with high speed photography to explore the nature of the large scale contact and shock wave formation at the vicinity of running cracks in two different bimaterial systems. Theoretical predictions of crack tip speed regimes where large scale contact is implied are confirmed. Also, the theoretically predicted shock wave emanating from the intersonically propagating crack tip is observed. Direct visual evidence is also obtained for *another* traveling shock wave emanating from the end of the intersonically moving contact zone.

1 INTRODUCTION

In homogeneous materials, the observation of crack growth speeds greater than the shear wave speed, $v>c_s$, is limited to cases when the loading is applied directly to the propagating crack tip. For remotely loaded cracks, energy considerations make it impossible for the crack tip speed to exceed the Rayleigh wave speed of the material (Broberg, 1960 and Freund, 1990). Thus, the only experimental observations of intersonic or supersonic crack tip speeds, $v>c_s$ or $v>c_l$, in a laboratory setting have been on crack growth along weak crystal planes in single crystals of Potassium Chloride, where the crack faces were loaded by laser induced expanding plasma (Winkler *et. al*, 1970 and Curran *et. al*, 1970). Indirect observation of intersonic shear rupture ($c_s<v<c_l$) has also been reported for crustal earthquakes (Archuleta, 1982). These observations have motivated extensive theoretical work in the area of high speed shear fracture in homogeneous materials (Burridge, 1973, Burridge *et. al*, 1979, Freund, 1979, Broberg, 1985 and 1989, Bykovtsev and Kramarovskii, 1989, and Aleksandrov and Smetanin, 1990).

In bimaterial systems, however, intersonic crack propagation along bimaterial interfaces is possible even under remote loading conditions (Liu *et. al*, 1993, Lambros and Rosakis, 1995a, 1995b, 1995c and Singh and Shukla, 1996). Under these conditions only a *finite amount* of energy has to be supplied to the crack tip to maintain extension as the propagational speed approaches the lower of the two Rayleigh wave speeds (Yang *et. al*, 1991).

Despite these initial attempts, the phenomenon of intersonic crack propagation is still more or less unexplored. The experimental evidence is quite limited and still there does not exist a completely physically realistic theoretical model for the intersonically propagating interfacial crack. In view of these limitations, the current study presents valuable experimental observations on interface failure in the intersonic regime and interprets these observations based on currently available theory.

2 EXPERIMENTAL TECHNIQUES

The two techniques of coherent gradient sensing (CGS) and photoelasticity were employed independently to study intersonic crack propagation along a bimaterial interface subjected to impact loading. Both these optical techniques provide real-time, full-field information and are ideally suited to investigate dynamic fracture events when used in conjunction with high speed photography. However, the two techniques have their own advantages and limitations and provide complementary information.

The two techniques of coherent gradient sensing (CGS) and photoelasticity have been employed extensively to investigate a variety of solid mechanics problems including dynamic fracture. The details of analyzing the CGS optical method can be found in several previous articles including Tippur *et al.* (1991) and Rosakis (1993) and will not be repeated here for the sake of brevity. Similarly Dally and Riley (1991) provide detailed information regarding the technique of photoelasticity.

For the case of plane stress and a transmission configuration CGS fringe patterns are interpreted in terms of two dimensional stress field approximations. For points outside the near-tip three dimensional region the fringes can be related to the in-plane gradients of $\hat{\sigma}_{11}+\hat{\sigma}_{22}$ as, (Rosakis, 1993),

$$c_\sigma h \frac{\partial\left(\hat{\sigma}_{11}+\hat{\sigma}_{22}\right)}{\partial x_\alpha}=\frac{n^{(\alpha)} p}{\Delta}, \quad n^{(\alpha)}=0, \pm 1, \pm 2, \ldots \qquad (1)$$

where, c_σ is the stress optical coefficient of the transparent material, h is the specimen thickness and $\hat{\sigma}_{11}$ and $\hat{\sigma}_{22}$ are thickness averages of the in-plane stress components in the plate, $n^{(\alpha)}$ represents the integer identifying fringes observed for shearing along the x_α-direction, p is the grating pitch, and Δ is the grating separation.

The generation of isochromatic fringe patterns is governed by the stress optic law. For the case of monochromatic light, the condition for the formation of fringes is expressed as,

$$\hat{\sigma}_1-\hat{\sigma}_2=\frac{N f_\sigma}{h}, \qquad (2)$$

where, $\hat{\sigma}_1-\hat{\sigma}_2$, is the principal stress difference of the thickness averaged stress tensor, $\hat{\sigma}$, f_σ is the material fringe value, h is the specimen thickness and N is the isochromatic fringe order.

The bimaterial specimens were subjected to one-point bend type impact loading, as shown in figure 1, using a cylindrical steel projectile launched by a gas-gun. This impact resulted in dynamic crack growth along the bimaterial interface. The dynamic failure event was imaged using a high speed camera. For the CGS experiments a rotating mirror type high speed camera (Cordin Co., model 330A) was employed at an interframe time of 1.2μs. Meanwhile, a Cranz-Schardin spark-gap camera was used for the photoelastic experiments at an interframe time of 4μs.

The bimaterial specimens used in these experiments consisted of a transparent polymer bonded directly to a metal. The material

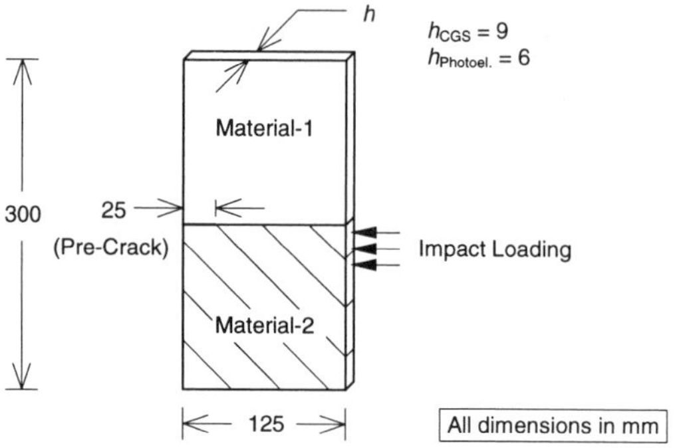

Figure 1. Bimaterial specimen subjected to impact.

combinations were chosen so that there would be a large mismatch in the mechanical properties across the interface and this would intensify the dynamic effects. The particular choice of the transparent polymer was also dictated by the particular needs of the experimental technique employed, i.e. CGS or photoelasticity. For the case of CGS, the transparent half was Poly-methylmethacrylate (PMMA) while the metal half was AISI 4340 steel. For the photoelastic experiments, the transparent half was Homalite-100, a polyester resin that exhibits stress induced birefringence, while the metal half was 6061 aluminum. Throughout this paper, the transparent polymer side of the specimen will be referred to as material-1 and the metal side as material-2.

As demonstrated by the mechanical properties of the bimaterial constituents (Singh *et. al*, 1997), either of the PMMA/steel or the Homalite-100/aluminum combinations result in a significant mismatch of mechanical properties and, most importantly for dynamics, wave speeds across the bimaterial interface. Moreover, both the material combinations behave very similar to an elastic/rigid bimaterial system. Thus, one can expect intensified interfacial effects during the dynamic fracture event.

Bonding for the PMMA/steel interface is described by Tippur and Rosakis (1991) while that for the Homalite-100/aluminum interface by Singh and Shukla (1996). The issue of bond strength and toughness has also be addressed by the same authors.

3 RESULTS AND DISCUSSION

A typical sequence of CGS interferograms from a one point bend experiment on a PMMA/steel specimen is shown in figure 2. Note that fringes are

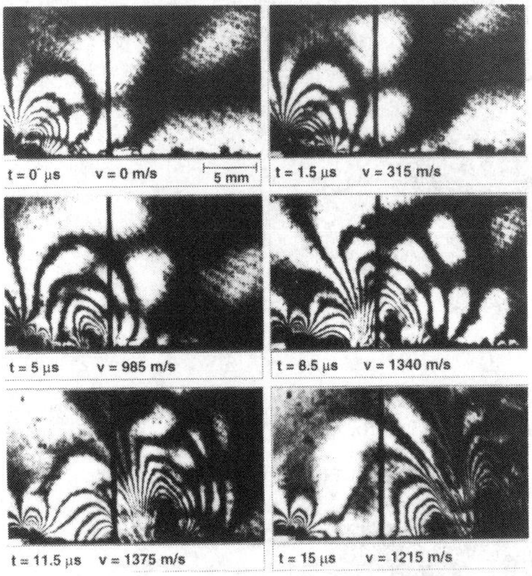

Figure 2. CGS fringes obtained for dynamic crack growth along a PMMA/steel bimaterial interface.

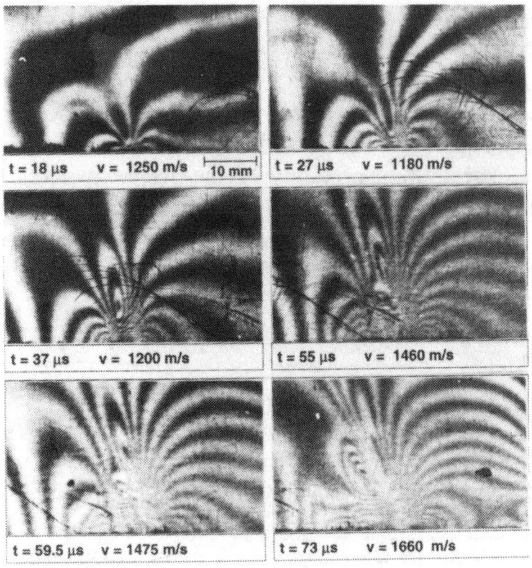

Figure 4. Isochromatic fringes obtained for crack growth along a Homalite-100/aluminum interface.

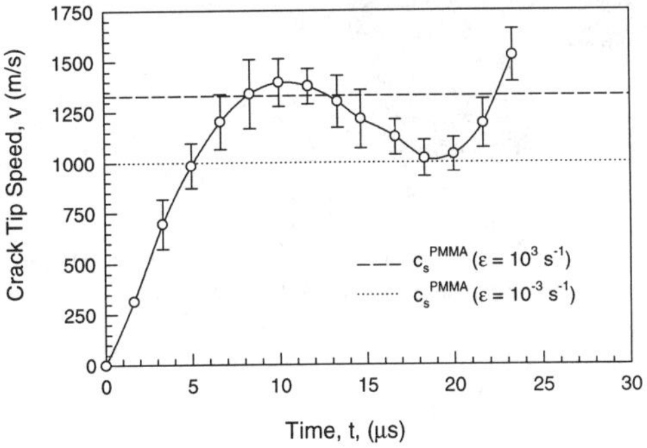

Figure 3. Crack tip speed history for dynamic crack growth along a PMMA/steel bimaterial interface.

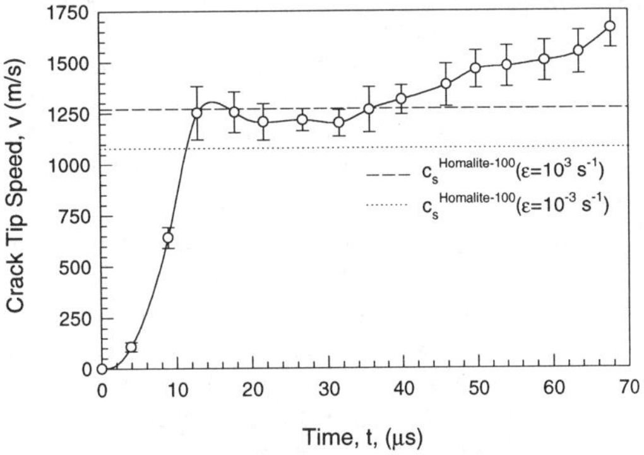

Figure 5. Crack tip speed history for dynamic crack growth along a Homalite-100/aluminum interface.

observed only in the transparent (PMMA) side. The instantaneous location of the crack tip is known from each frame and this was used to determine the crack tip speed. The crack tip speed history obtained from one such experiment is plotted in figure 3. After initiation the crack tip accelerated very rapidly to beyond the upper extreme of the shear wave speed of PMMA, c_s^{PMMA}. Then the crack tip speed oscillated between the upper and lower extremes of c_s^{PMMA} for about 15 μs before accelerateing further. Accelerations were of the order of 10^8 m/s^2, which establishes the highly unstable nature of this event.

Tests conducted with Homalite-100/aluminum specimens, using of photoelasticity, yield similar observations. Figure 4 shows a typical set of isochromatic fringe patterns obtained for dynamic crack propagation along a Homalite-100/aluminum bimaterial interface. The history of the crack tip speed is plotted in figure 5. As observed for the PMMA/steel case, the crack tip accelerated rapidly after initiation to beyond the upper extreme of the shear wave speed of material-1 (Homalite-100). Thereafter, the crack tip speed stayed at this value for about 20 μs, after which it accelerated further.

The terminal crack tip speeds that have been observed for these experiments were about 140% of the upper extreme of c_s^{PMMA} for the PMMA/steel interface and about 130% of the upper extreme of $c_s^{Homalite-100}$, for the Homalite-100/aluminum interface. Nevertheless, the dilatational wave speed of material-1 (PMMA or Homalite-100) was not exceeded in either case. Crack growth in this speed regime is termed as being *intersonic*.

Crack propagation in the intersonic regime has a direct effect on the nature of the fringe patterns observed. At first the fringes are smooth and continuous, while the crack tip is still subsonic, as shown in the first few frames in figures 2 and 4. Moreover, the forward and rear fringe loops focus at a *single point* along the interface, i.e. the crack tip. In later frames, however, the fringes become squeezed and elongated normal to the interface. Finally, in the intersonic crack growth regime, the fringes in the center of the two lobed fringe pattern intercept the bond line over a finite area between the two main lobes, which is evident in the last frame in figure 2 and the last three frames in figure 4. This effect is seen clearly in figure 6. This fringe pattern is caused by large scale contact of the crack faces along *d*. This large scale contact of crack faces was first observed by Lambros and Rosakis (1995c), theoretically confirmed by Liu *et. al* (1995), and also observed by Singh and Shukla (1996). From the numerical simulation point of view, Xu and Needleman (1996) have confirmed the existence of a contact zone area when the crack tip speed exceeds the lower of the two Rayleigh wave speeds.

Another direct consequence of intersonic crack propagation is the formation of a mach wave (or line-of-discontinuity) in the stress field surrounding the crack tip. The propagating crack tip acts as a source of shear and dilatational stress waves which radiate out into the material and establish the stress field surrounding the crack tip. If the crack tip propagates faster than the shear wave speed then the spreading out of the shear waves is limited and a mach wave forms. The existence of such mach waves was predicted by Liu *et. al* (1995) and experimentally confirmed by Singh and Shukla (1996). The mach wave is observed in the form of discontinuous isochromatic fringe contours as shown in figure 7.

It was noted earlier that both the PMMA/steel and Homalite-100/aluminum bimaterial systems can be modeled as an elastic/rigid approximation. Now, consider a crack propagating intersonically along an elastic/rigid interface, as shown in figure 8. Liu *et. al* (1995) have shown that the stress field around the crack tip can be expressed as,

$$\sigma_{ij} = \frac{\mu A_0}{1+\alpha_l^2\hat{\alpha}_s^2}\left\{\frac{\Sigma^a_{ij}}{r_l^q} + \frac{\Sigma^b_{ij}}{(\eta_1+\hat{\alpha}_s\eta_2)^q}H(\eta_1+\hat{\alpha}_s\eta_2) + \frac{\Sigma^c_{ij}}{(-\eta_1-\hat{\alpha}_s\eta_2)^q}H(-\eta_1-\hat{\alpha}_s\eta_2)\right\}, \quad (3)$$

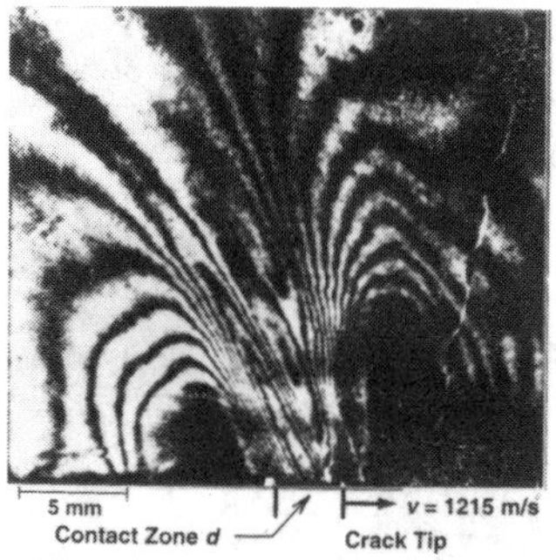

Figure 6. Enlarged view of CGS fringes showing the area of crack face contact.

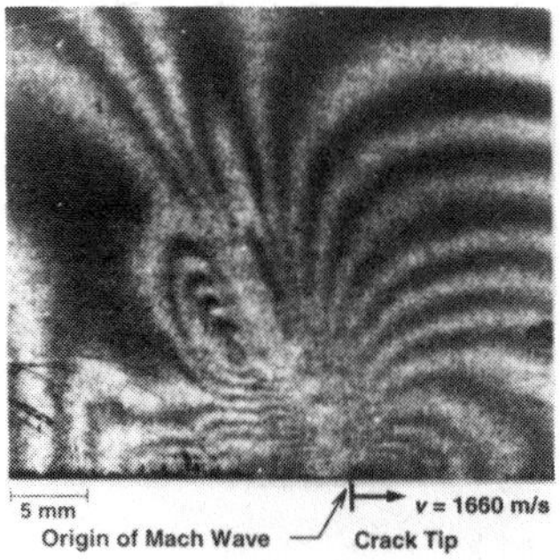

Figure 7. Discontinuities of isochromatic fringe contours representing the formation of a mach wave.

$$\alpha_l^2 = 1-\frac{v^2}{c_l^2}, \qquad \hat{\alpha}_s^2 = \frac{v^2}{c_s^2}-1. \quad (4)$$

where, v is the crack tip speed; μ, c_l and c_s are the shear modulus, dilatational wave speed and shear wave speed, respectively, of material-1 and $H(\bullet)$ is the Heaviside unit step function. Also, the functions Σ^a_{ij}, Σ^b_{ij} and Σ^c_{ij} are functions of θ_l, the crack tip speed, v, and the wave speeds of material-1, c_l and c_s. The scaled polar coordinates are defined as,

$$r_l = \sqrt{\eta_1^2+\alpha_l^2\eta_2^2}, \theta_l = \arctan(\alpha_l\eta_2/\eta_1) \quad (5)$$

The strength of the crack tip singularity is given as,

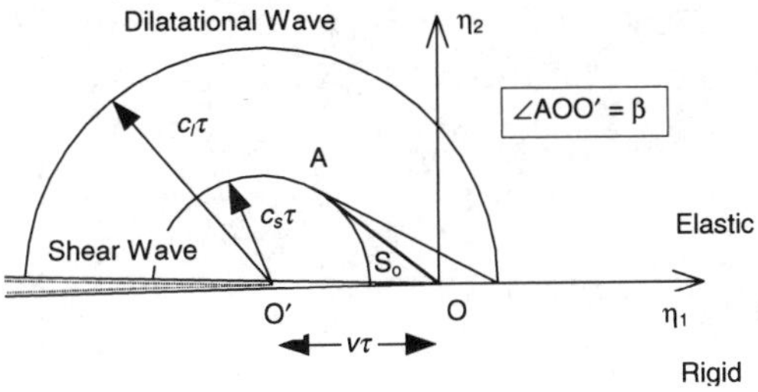

Figure 8. A crack propagating intersonically along an elastic/rigid interface.

$$q(v)=\frac{1}{\pi}\arctan\left\{\frac{\alpha_l\hat{\alpha}_s\left[4-\left(1-\hat{\alpha}_s^2\right)^2\right]}{4\alpha_l^2\hat{\alpha}_s^2+\left(1-\hat{\alpha}_s^2\right)^2}\right\} \tag{6}$$

The singularity exponent, $q(v)$, has a value of $q(v) = 0$ at $v = c_s$ and increases monotonically with crack tip speed till it reaches a maximum value at $v=\sqrt{2}c_s$. With further increase in the crack tip speed the exponent decreases monotonically back to the value $q = 0$. Note that $q(v)$ remains less than 0.5 for the entire speed range considered. This limit on the maximum value of $q(v)$ implies that energy flux into the moving crack tip is always zero irrespective of the crack tip speed in the intersonic crack growth regime.

This analysis is very useful in explaining several key features of intersonic crack propagation along bimaterial interfaces. Across the head wave front S_o, see figure 8, the components of stress and particle velocity are discontinuous. Therefore, for the case of intersonic crack growth, an entire singular line of infinite jumps in stress and particle velocity appears in the body. The singularity across this line is the same as that at the crack tip. The line originates from the propagating crack tip and radiates out into the elastic solid. This is the line-of-discontinuity that appears in the isochromatic fringe patterns obtained in the photoelastic experiments. The angular orientation of this line, with respect to the interface, can be expressed in terms of the crack tip speed, v, and the shear wave speed of the elastic material (or material-1), c_s, as,

$$\tan\beta=\left(\frac{v^2}{c_s^2}-1\right)^{-1/2}. \tag{7}$$

The orientations of the line-of-discontinuity determined from the experimental isochromatic

Table 1. Experimentally measured and theoretically predicted orientations of the line-of-discontinuity.

Frame No.	$v/c_s^{\text{Homalite-100}}$	β_{Theory}	$\beta_{Experiment}$
13	1.16	59.5°	63°
14	1.19	57.5°	55°
15	1.21	55.7°	53°
16	1.30	50.3°	48°

fringe patterns were compared with the angles predicted by the above equation and are listed in Table 1. The correspondence between the experimentally observed and theoretically predicted angles is excellent and substantiates the fact that the experimentally observed line-of-discontinuity is indeed the theoretically predicted mach wave. Further evidence of the line-of-discontinuity is presented in the numerical simulations of Xu and Needleman (1996). Liu *et. al* (1995) also showed that there is no energy dissipation when the singular line S_o moves through the elastic material.

Consider the normal tractions along the interface at an arbitrary distance a ahead of the moving crack tip, $\sigma_{22}(a, 0^+, t)$, and the crack opening displacement at the distance a behind the moving crack tip, $u_2(a, 0^+, t)$. Then it can be shown (Liu *et. al*, 1995) that if the crack tip speed is in the range $c_s < v < \sqrt{2}c_s$, $\sigma_{22}(a, 0^+, t)$ and $u_2(a, 0^+, t)$ have opposite signs. This implies that when the normal traction ahead of the crack tip is positive, crack face penetration into the rigid substrate is predicted. Now, positive normal tractions ahead of the crack tip are required to facilitate interface rupture and crack face penetration is physically impossible. Hence, in the crack tip speed range $c_s < v < \sqrt{2}c_s$ the crack faces would come into contact behind the propagating crack tip. This accounts for the large scale contact of crack faces observed experimentally in the CGS interferograms when the crack tip speed was indeed in the range $c_s^{PMMA} < v < \sqrt{2}c_s^{PMMA}$.

When crack face contact does indeed occur the asymptotic solution is no longer valid and the problem must be revisited under different crack face boundary conditions. Nevertheless, despite this limitation the solution does provide considerable conceptual insight into the intersonic crack growth phenomenon.

Same qualitative observations as above have been made for a more realistic elastic/elastic analysis for intersonic bimaterial crack growth (Huang *et. al*, 1996). Here again traveling shock waves have been

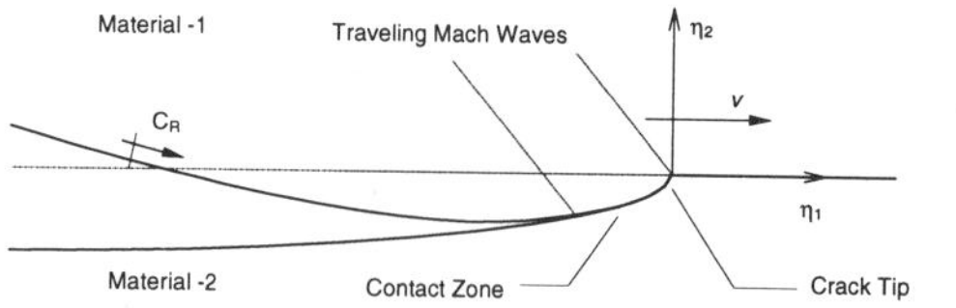

Figure 9. Schematic of crack face openings for a intersonically propagating interface crack.

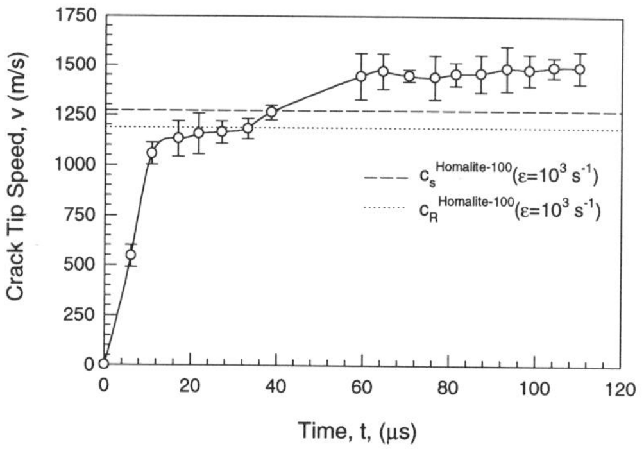

Figure 10. Crack tip speed history for dynamic crack growth along a Homalite-100/aluminum interface.

predicted. Also, for certain crack tip speed range large scale contact was implied by the solution. However, the details of the speed range where contact is predicted is slightly modified depending on the relative magnitudes of wave speed mismatch of the two elastic solids.

Given the experimental observations presented in the previous section and the asymptotic analyses of Liu *et. al* (1995) and Huang *et. al* (1996) it is believed that when the crack tip is propagating in the regime $c_s < v < \sqrt{2}c_s$, with respect to material-1, the crack faces behind the moving crack tip would be in contact. This contact zone would propagate along with the moving crack tip as illustrated in figure 9.

To further investigate the phenomenon of intersonic crack growth along a bimaterial interface another series of experiments was conducted using a wider specimen. This specimen was identical to the geometry shown in figure 1 except that it was twice as wide, which provides a longer path of propagation for the interface crack. This increased path of crack propagation allowed for a fuller development of the near-tip stress field in the intersonic regime and also reduced the interaction of the propagating crack with the specimen boundaries. Thus, the intersonic regime could be explored in greater detail.

The specimen was impact loaded in the one point bend configuration as before and the failure process was observed using photoelasticity in conjunction with high speed photography. In these experiments, only the technique of photoelasticity was employed since we were interested in observation of the line-of-discontinuity, which would not show up in CGS interferograms. The history of the crack tip speed corresponding to this experiment is plotted in figure 10. As for the previous experiments, the crack rapidly accelerated to the shear wave speed (upper extreme) of Homalite-100, $c_s^{Homalite\text{-}100}$. Thereafter, the speed accelerated further until it reached about 120% of $c_s^{Homalite\text{-}100}$. Subsequently, the crack tip speed stabilized at this value and further propagation occurred in a "steady-state" manner without any further crack tip acceleration or deceleration.

As the crack tip was propagating intersonically, the formation of a mach wave was observed in the form of a line-of-discontinuity emanating from the propagating crack tip, as shown in figure 11. Also, since the crack tip was propagating in the $c_s < v < \sqrt{2}c_s$ speed regime large scale contact of the crack faces was observed behind the moving crack tip. This contact zone is characterized by fringes that run parallel to the interface, as shown in figure 11. A *secondary* mach wave was also observed in addition to the previously discussed mach wave (the *primary* mach wave that originated from the propagating crack tip). This originated from the trailing edge of the contact zone and represents the singularity that occurs when the crack faces separate at the end of the contact zone. The trailing

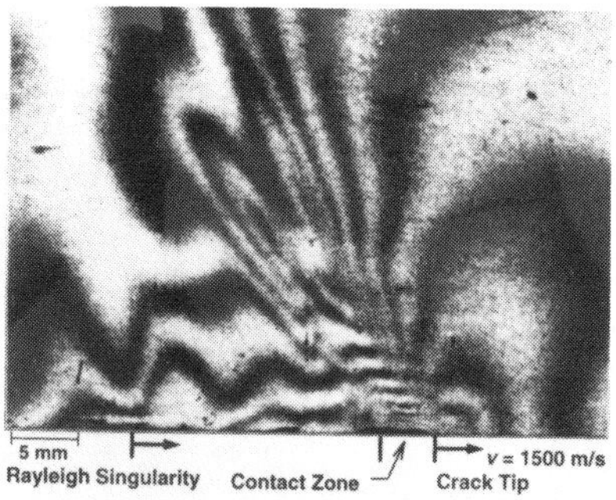

Figure 11. Details of the isochromatic fringe patterns around the intersonically propagating crack tip showing the *primary* and *secondary* mach waves as lines-of-discontinuity, the dynamically moving contact zone, and the Rayleigh disturbance.

edge of the contact zone propagated with the same speed as the moving crack tip. Thus, the size of the contact zone remained relatively constant at around 2-3 mm during the "steady-state" intersonic propagation phase of the interfacial crack tip. This was further confirmed by the equal angles of inclination of the primary and secondary lines-of-discontinuity. Hence, this test confirms the mechanics of interfacial crack propagation in the $c_s < v < \sqrt{2}c_s$ crack tip speed regime, as illustrated in figure 9.

The history of the crack tip speed shown in figure 10 clearly demonstrates that "stable" crack propagation is possible in the $c_s < v < \sqrt{2}c_s$ regime. This is in contrast with the analysis of Liu *et. al* (1995) which postulates that this crack tip speed regime is inherently unstable and the crack would accelerate beyond $\sqrt{2}c_s$. However, the asymptotic analysis does not take into account the dissipation that would be introduced by the dynamically moving contact zone. It is believed that incorporating the energy dissipated by the propagating contact zone will shed greater light on the regimes of "stable" intersonic crack propagation along bimaterial interfaces. Indeed since the crack tip energy release rate is identically zero in this regime, frictional contact is the only mechanism of energy dissipation in the entire system.

4 CLOSURE

This paper describes the first experimental observations of various phenomena characteristic of dynamic intersonic decohesion of bimaterial interfaces. The optical techniques of coherent gradient sensing (CGS) interferometry and photoelasticity, were employed in conjunction with high speed photography, in separate yet complementary experiments, to explore intersonic interfacial crack propagation in two different bimaterial systems, namely, PMMA/steel and Homalite-100/aluminum.

Using the two techniques the nature of large scale contact and shock wave formation at the vicinity of running cracks in the two bimaterial systems is explored. It is confirmed that large scale contact does indeed occur when the crack tip speed is in the $c_s < v < \sqrt{2}c_s$ regime, as implied theoretically. Also, direct visual evidence is obtained for shock waves emanating from the intersonically moving crack tip *and* the end of the intersonically moving contact zone.

5 ACKNOWLEDGMENT

The authors would like to acknowledge the support of the National Science Foundation through a joint grant no. CMS-9424113 to the University of Rhode Island and the California Institute of Technology (Dr. O. Dillon, Scientific Officer). A. J. Rosakis and J. Lambros would also like to acknowledge the support of the Office of Naval Research under grant no. N00014-95-1-0453 (Dr. Y. Rajapakse, Scientific Officer).

6 REFERENCES

Aleksandrov, V. M. and Smetanin, B. I. 1990 Supersonic cleavage of an elastic strip. *PMM U.S.S.R.* **54**(5) 677-682.

Archuleta, R. J. 1982 Analysis of near-source static and dynamic measurements from the 1979 Imperial Valley earthquake. *Bull. Seismological Soc. Am.* **72**(6) 1927-1956.

Broberg, K. B. 1960 The propagation of a Griffith crack. *Ark. Fys.* **18** 159.

Broberg, K. B. 1985 Irregularities at earth-quake slip. *J. Tech. Phys.* **26**(3-4) 275-284.

Broberg, K. B. 1989 The near-tip field at high crack velocities, *Int. J. Fract.* **39**(1-3) 1-13.

Burridge, R. 1973 Admissible speeds for plane-strain shear cracks with friction by lacking cohesion. *Geophys. J. R. Soc. Lond.* **35** 439-455

Burridge, R., Conn, G. and Freund, L. B. 1979 The stability of a rapid mode II shear crack with finite cohesive traction. *J. Geophys. Res.* **85**(B5) 2210-2222.

Bykovtsev, A. S. and Kramarovskii, D. B. 1989 Non-stationary supersonic motion of a complex discontinuity. *PMM U.S.S.R.* **53**(6) 779-786.

Curran, D. R., Shockey, D. A. and Winkler, S. 1970 Crack propagation at supersonic velocities, II. Theoretical model. *Int. J. Fract.* **6**(3) 271-278.

Dally, J. W. and Riley, W. F. 1991 *Experimental stress analysis*. McGraw Hill Inc.

Freund, L. B. 1979 The mechanics of dynamic shear crack propagation. *J. Geophys. Res.* **84**(B5) 2199-2209.

Freund, L. B. 1990 *Dynamic Fracture Mechanics.* Cambridge University Press, Cambridge.
Huang, Y., Liu, C. and Rosakis, A. J. 1996 Transonic crack growth along a bimaterial interface: An investigation of the asymptotic structure of near-tip fields. *Int. J. Solids Struct.* **33**(18) 2625-2645.
Lambros, J. and Rosakis, A. J. 1995a Dynamic decohesion of bimaterials: Experimental observations and failure criteria. *Int. J. Solids Struct.* **32**(17/18) 2677-2702.
Lambros, J. and Rosakis, A. J. 1995b On the development of a dynamic decohesion criterion for bimaterials. *Proc. R. Soc. Lond.*.
Lambros, J. and Rosakis, A. J. 1995c Shear dominated transonic interfacial crack growth in a bimaterial-I. Experimental observations. *J. Mech. Phys. Solids*, **43**(2) 169-188.
Liu, C., Lambros, J. and Rosakis, A. J. 1993 Highly transient elastodynamic crack growth in a bimaterial interface: Higher order asymptotic analysis and experiments *J. Mech. Phys. Solids* **41**(12) 1887-1954.
Liu, C., Huang, Y. and Rosakis, A. J. 1995 Shear dominated transonic interfacial crack growth in a bimaterial-II. Asymptotic fields and favorable velocity regimes. *J. Mech. Phys. Solids* **43**(2) 189-206.
Rice, J. R. 1988 Elastic fracture mechanics concepts for interfacial cracks. *J. Appl. Mech.* **55** 98-103.
Rosakis, A. J. 1993 Two optical techniques sensitive to gradients of optical path difference: The method of caustics and the coherent gradient sensor (CGS). *Experimental Techniques in Fracture*, 327-425.
Singh, R. P. and Shukla, A. 1996 Subsonic and intersonic crack growth along a bimaterial interface. *J. Appl. Mech.* Vol. 63, pp. 919-924, 1996.
Tippur, H. V., Krishnaswamy, S. and Rosakis, A. J. 1991 A coherent gradient sensor for crack tip measurements: Analysis and experimental results. *Int. J. Fract.* **48**, 193-204.
Tippur, H. V. and Rosakis, A. J. 1991 Quasi-static and dynamic crack growth along bimaterial interfaces: A note on crack-tip field measurements using coherent gradient sensing. *Expt. Mech.* **31**(3) 243-251.
Winkler, S., Shockey, D. A. and Curran, D. R. 1970 Crack Propagation at Supersonic Velocities, I. *Int. J. Fract.* **6**(2) 151-158.
Xu, X.-P. and Needleman, A. 1996 Numerical simulations of dynamic crack growth along an interface. *Int. J. Fract.* **74** 289-324.
Yang, W., Suo, Z. and Shih, C. F. 1991 Mechanics of dynamic debonding. *Proc. R. Soc. Lond.* **A433**, 679-697.

Damage and Failure of Interfaces, Rossmanith (ed.) © 1997 Balkema, Rotterdam, ISBN 90 5410 899 1

Rayleigh pulse interaction with partially contacting dissimilar interfaces

K. Uenishi, H. P. Rossmanith, R. E. Knasmillner & C. Böswarth
Institute of Mechanics, Vienna University of Technology, Austria

ABSTRACT: An experimental and numerical investigation of the interaction of a Rayleigh pulse with a partially contacting interface is presented. The interface is subjected to static normal and shear pre-stresses. Utilizing dynamic photoelasticity in conjunction with high speed cinematography, the evolution of time-dependent isochromatic fringe patterns associated with Rayleigh pulse-interface interaction is experimentally recorded. For the numerical studies, the finite difference wave propagation simulator SWIFD is used for a quantitative analysis of the problem under different combinations of contacting materials. The effect of acoustic impedance ratio of the two contacting materials on the wave patterns is discussed. As a case example, the mechanism of damage concentration caused by the 1995 Hanshin (Kobe), Japan, earthquake is considered based on the results obtained by the model study.

KEYWORDS: Acoustic impedance ratio, Contact mechanics, Dynamic faulting, Dynamic photoelasticity, Earthquake rupture mechanism, Finite difference method, Interface instability, Interface slip, Mohr-Coulomb condition, Rayleigh wave interaction.

1 INTRODUCTION

Rayleigh and interface waves are considered to play a crucial role in damage and failure of interfaces between two dissimilar or even similar materials, and it is known that these waves can induce separation (delamination) of the interface surfaces (Lambros & Rosakis 1995a, b, Liu et al. 1995).

The most familiar interface waves involving separation of interface surface are Schallamach waves which occur when two media of large differences in rigidity slide past one another (Schallamach 1971). Comninou and Dundurs (1977, 1978a,b) showed mathematically that a dynamic wave involving separation can stably propagate along an interface when two elastic media are compressed and simultaneously sheared. However, the validity of their mathematical solution was questioned by Freund (1978) from an energy point of view, and the solution has mostly been ignored (Anooshehpoor & Brune 1994).

From the results obtained by frictional tests in the laboratory, Brune et al. (Brune et al. 1993, Anooshehpoor & Brune 1994, Brune 1996) propose that dynamic rupture on a geological fault can be triggered by interface waves with separational sections propagating along the fault. It is suggested that the excitation of Rayleigh waves on a rupture surface can lead to pulses of separation. However, this mechanism has not been confirmed in a conclusive manner (Turcotte 1997).

In this contribution, the basic mechanisms of dynamic interface instability caused by a Rayleigh (R-) pulse will be investigated (Uenishi et al. 1997, Rossmanith & Uenishi 1997). First, the fundamental characteristics of a R-pulse will be summarized in terms of stress field and particle motion. Second, the results of laboratory model experiments utilizing dynamic photoelasticity in conjunction with high speed cinematography will be presented. It will be shown that a R-pulse can trigger instability of a partially contacting interface between similar materials and that a static shear pre-stress has an influence on the initiation of interface instability. Third, the problem will be numerically simulated using the finite difference simulator SWIFD (Rossmanith & Uenishi 1995, 1996) and the dynamic interaction process will be quantitatively assessed. The effect of acoustic impedance ratio of the two contacting materials on the dynamic interaction process will be discussed. Finally, the results obtained by the model investigation will be applied to explain the damage concentration caused by the 1995 Hanshin (Kobe), Japan, earthquake.

2 CHARACTERISTICS OF A RAYLEIGH PULSE

In an elastic material a R-pulse propagates in a non-dispersive fashion, carrying the energy concentrated in a shallow layer adjacent to the surface with low geometrical damping (Rayleigh 1885, Lamb 1904). The stress and displacement field associated with a plane R-pulse of arbitrary shape can be analytically represented in terms of one complex potential (Cardenas-Garcia 1983). Knowing the stress field from the complex potential, one can draw isochromatic fringe patterns (contours of maximum in-plane shear stress) pertaining to a R-pulse. A typical theoretically predicted isochromatic fringe pattern is shown in Figure 1(1) where the R-pulse, produced by a concentrated line load as a result of a detonating line charge, propagates from left to right along the free surface of the half-space. As indicated by the high fringe density in Figure 1(1), the disturbance is largely confined to a thin layer adjacent to the free surface. The particle movement associated with the same R-pulse is shown in Figure 1(2) where one recognizes the push and pull normal particle movement on the free surface which becomes important in contact problems.

(1)

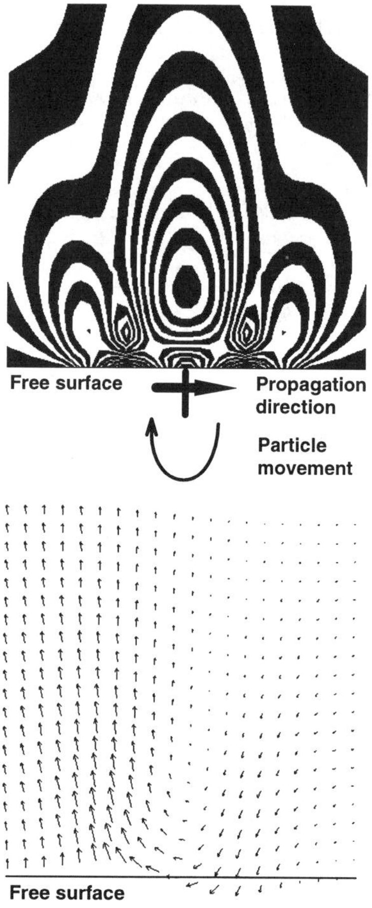

(2)

Figure 1. A Rayleigh pulse propagating from left to right along a free surface: (1) isochromatic fringe patterns (contours of maximum shear stress); and (2) vector representation of particle movements.

3 EXPERIMENTAL INVESTIGATIONS

3.1 *Experimental Set-up*

Dynamic photoelasticity in conjunction with high speed cinematography is used to analyze the interaction between a R-pulse and a statically pre-loaded, non-welded partially contacting interface.

The experimental model is schematically shown in Figure 2. The model consists of two plates of Araldite B which are in contact. The dimensions of the plates are selected so as to prevent reflected waves from impinging upon the region of contact and altering the results. The upper surface of the plate 2 (lower plate) is given a very blunt double wedge cut such that only the central section of the surface would initially be in contact with the upper plate (plate 1). Three contact configurations are investigated:

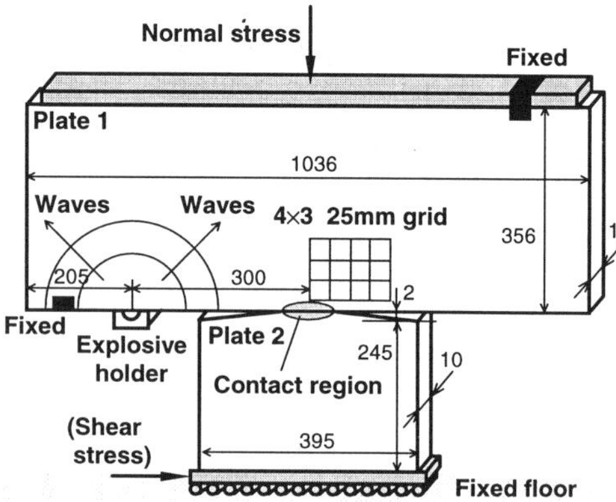

Figure 2. The experimental model set-up for Rayleigh pulse interaction investigation [all lengths in mm].

(1) one long contact of 55mm length;
(2) two short contacts of 12.5mm length each; and
(3) three short contacts of 12.5mm length with two gaps (symmetrically arranged) each.

The three geometrically different contact configurations are schematically shown in Figure 3. The contact region is not glued (non-welded) and during the interaction process the contact could diminish and be enforced depending on the relative position of the R-pulse with respect to the contact region. The two plates are statically pre-loaded in compression and in shear. The static shear stress is either zero (Exp. #1), positive (Exp. #2), or reverse (Exps #3 to #5) (see Fig.3).

A typical stress wave pattern was generated in plate 1 by the detonation of a small amount of explosive (240mg of PbN_6). The pulse length of the incident R-pulse is 75mm. A Cranz-Schardin type multiple spark gap camera is used to record the isochromatic fringe patterns which are produced by circularly polarized monochromatic light.

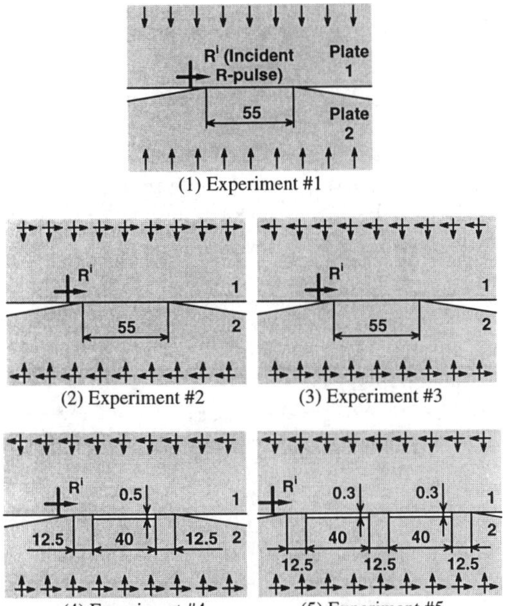

Figure 3. Five experiments of Rayleigh pulse interaction with a partially contacting interface [all lengths in mm]:
(1) Compression, no shear; one long contact;
(2) Compression with positive shear; one long contact;
(3) Compression with reverse shear; one long contact;
(4) Compression with reverse shear; two short contacts; and
(5) Compression with reverse shear; three short contacts.

3.2 *Results*

In the following, the results obtained from each experiment will be presented and discussed individually.

3.2.1 *Experiment #1: compression, no shear; one long contact*

In this experiment the interaction of a R-pulse with a non-welded interface, loaded under static compression only, is investigated. Figure 4 shows a sequence of three isochromatic fringe patterns of the dynamic wave interaction process. The R-pulse propagates from left to right and interacts with the contact region. The time scale *t* indicates the time elapsed from the instance of maximum stress amplification at the lhs edge of the contact region.

Figure 4(1) pertains to the event where the incident R-pulse impinges upon the lhs edge of the contact region. It is clearly seen in Figure 4(1), that the strength of stress singularity expressed by the fringe order about the lhs edge is larger than that about the rhs edge where the dynamic effect is negligible in this phase. This stress amplification about the lhs edge is due to the particle motion on the free surface [Fig.1(2)] where the leading part of the incident R-pulse induces a back- and downward movement of the particles [Fig.4(1)]. The surface particles which are already in contact move together towards the lower plate 2, and thus increase the stresses about the lhs edge.

Whereas the first part of the retrograde motions in the incident R-pulse is attributed to the stress amplification about the lhs edge, the trailing part of the R-pulse induces back- and upwardly oriented movement [Fig.4(2)], which may lead to a stress reduction with possible cancellation if the surfaces separate during a later stage of the interaction process, because the particles are now receding and open the interface. In fact, in Figure 4(2), the structure of the wave interaction patterns changes considerably. The fringe patterns about the lhs diffraction edge show very small stress singularity.

Theoretically, the R-pulse does not exist in the contact region and there must be other kinds of generalized interface pulses which carry the energy across and along the contact region. As the incident (generalized) R-pulse approaches the rhs edge of the contact region, as seen in Figure 4(3), partial wave energy transmission occurs across the interface into the lower plate 2. Due to the separational movements of surface particles, in the trailing part of the R-pulse there are no corresponding fringes in the lower plate 2.

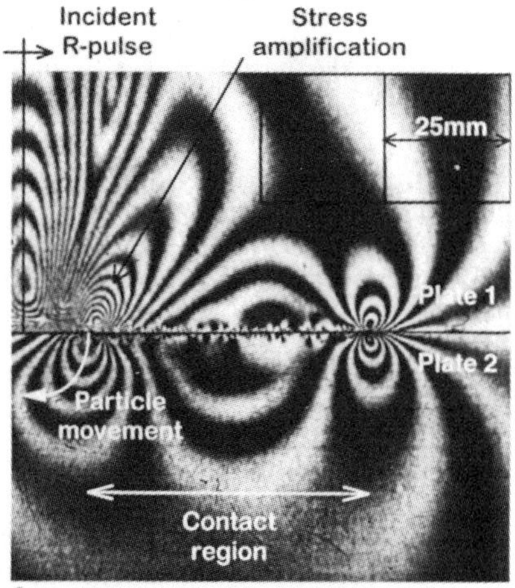

(1) $t = 0\mu s$

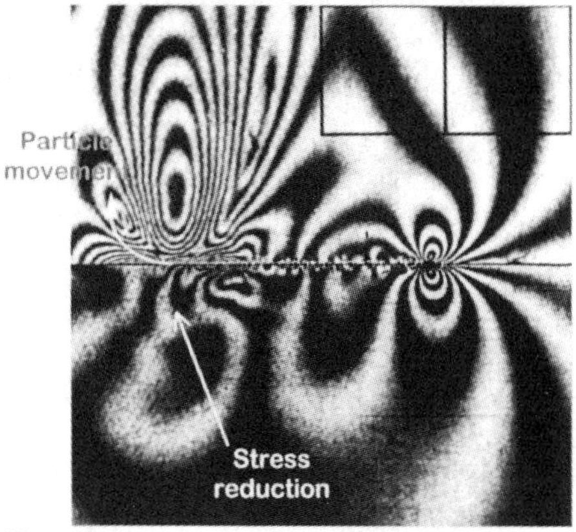

(2) $t = 13\mu s$

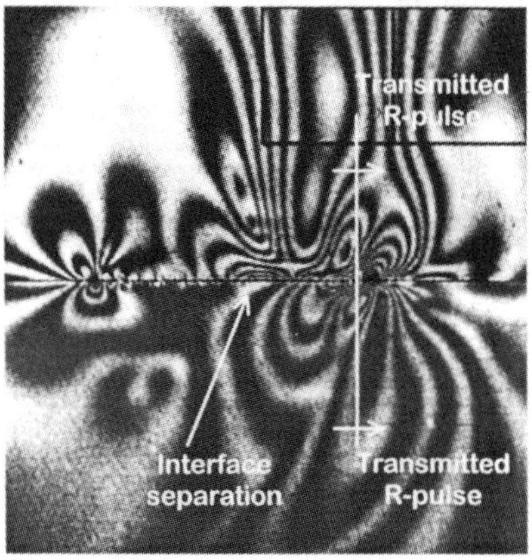

(3) $t = 53\mu s$

Figure 4. Sequence of experimentally obtained snapshots of isochromatic fringe patterns.
Experiment #1: compression, no shear; one long contact.

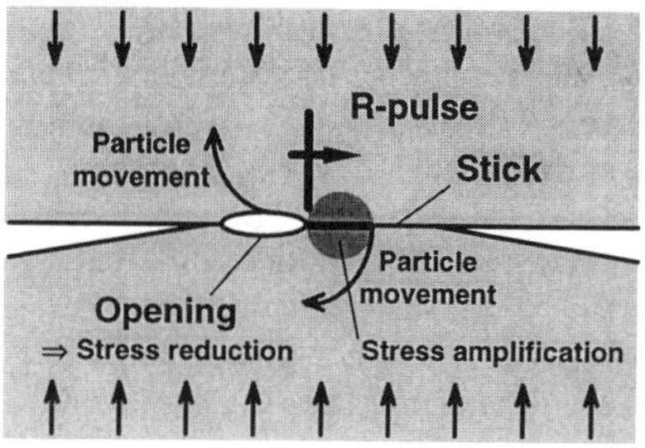

Figure 5. Interface opening caused by a Rayleigh (R-) pulse.

In summary, the retrograde motion, especially the down- and upward movement of the particles in the incident R-pulse, plays an important role in the interaction process as shown in Figure 5.

3.2.2 *Experiment #2: compression with positive shear; one long contact*

In the second series of experiments a positive shear stress is superimposed onto the static compressive load. The level of additional shear stress is set such that the contact region is at the limiting state, i.e. one still has stick contact conditions where slip will be induced at the slightest increase of the shear stress (increase in the same positive direction). In Figure 6, two experimentally obtained dynamic isochromatic fringe patterns are shown where the basic wave interaction process is the same as in Experiment #1 with some differences in the details. The following differences can be identified:

- The static shear stress induces an inclination of the fringes associated with the static contact stress singularity at the edges of the contact region [Fig.6(1)];
- The impinging R-pulse causes a relatively weak stress amplification about the lhs diffraction edge [compare Fig.6(1) with Fig.4(1)]. This weakened amplification with less energy transmitted across the interface [Fig.6(2)] indicates that interface slip has been initiated at an earlier stage during the wave interaction process. As the shear traction level is already at the critical limit any additional shear exerted due to particle movement within the R-pulse will trigger local slip within the contact area. When the incident R-pulse approaches the contact region and imparts a particle movement into the opposite direction of travel of the R-pulse the positive shear stress along the contact region is increased and slip will occur.

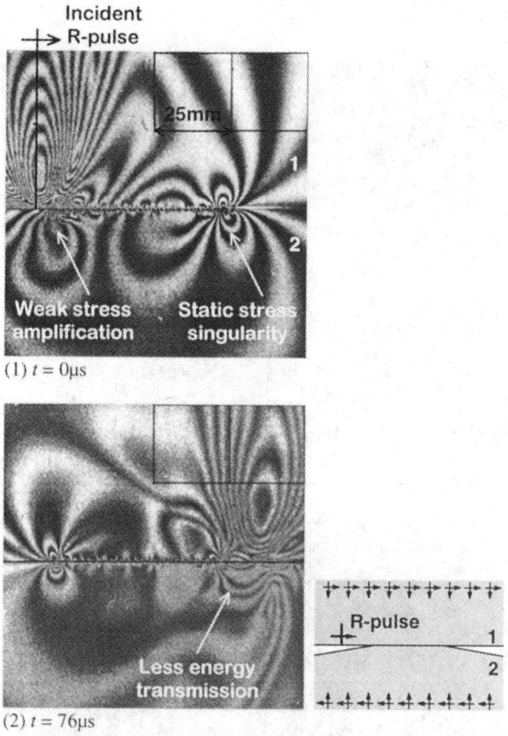

Figure 6. Dynamic isochromatic fringe patterns.
Experiment #2: compression with positive shear; one long contact.

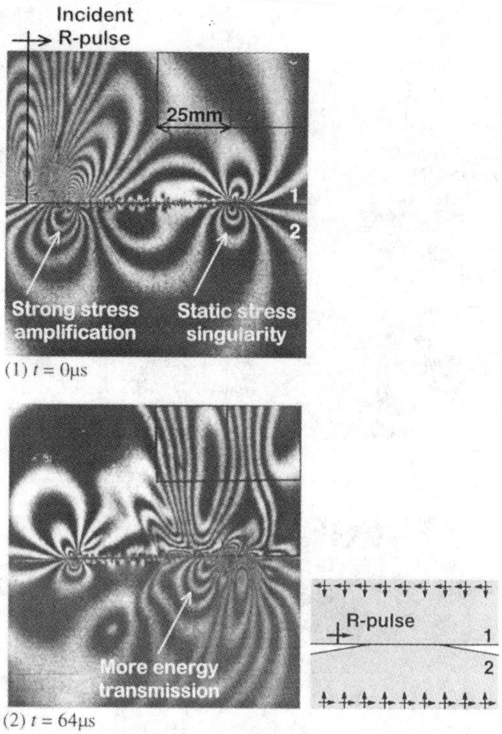

Figure 7. Snapshots of isochromatic fringe patterns.
Experiment #3: compression with reverse shear; one long contact.

3.2.3 *Experiment #3: compression with reverse shear; one long contact*

In the third series of experiments the static compressive load is retained and the direction of the shear traction is reversed. As before, the static pre-load combination between compressive stress and shear traction has been chosen such that the contact conditions in the contact region induce a limiting state (stick) and any increase of the shear traction (in the same reverse direction) will cause slip in the contact region.

Figure 7 shows two experimentally obtained snapshots of isochromatic fringe patterns. The general structure of the wave interaction patterns is the same as in the previous two experiments. However, the following differences can be noted:

- The reverse shear pre-stress causes an inclination of the contact edge stress singularity patterns as shown in Figure 7(1). The inclination is in the opposite direction as compared to the one associated with static positive shear stress [Fig.6(1)];
- Larger stress amplification occurs upon incidence of the R-pulse about the lhs diffraction edge [compare Fig.7(1) with Fig.6(1)]. This large stress intensification can be explained by considering the direction of static shear stress and the retrograde movement of particles in the R-pulse: as they act in opposite directions immediate local slip is suppressed and, therefore, more energy can be transmitted across the contacting zone. However, when the incident R-pulse is diffracted at the entrant edge and a complicated state of stress is generated in the volume adjacent to the contact region, as shown in Figure 7(2), no energy from the trailing part of the incident R-pulse is transmitted across the contact into the lower plate. This suggests that slip can only be initiated if the incident R-pulse is very strong, in fact strong enough to over-compensate the reverse shear pre-stress. Interface slip may thus be inhibited and delayed, and more energy is transferred into the lower plate across the "welded" interface [Fig.7(2)].

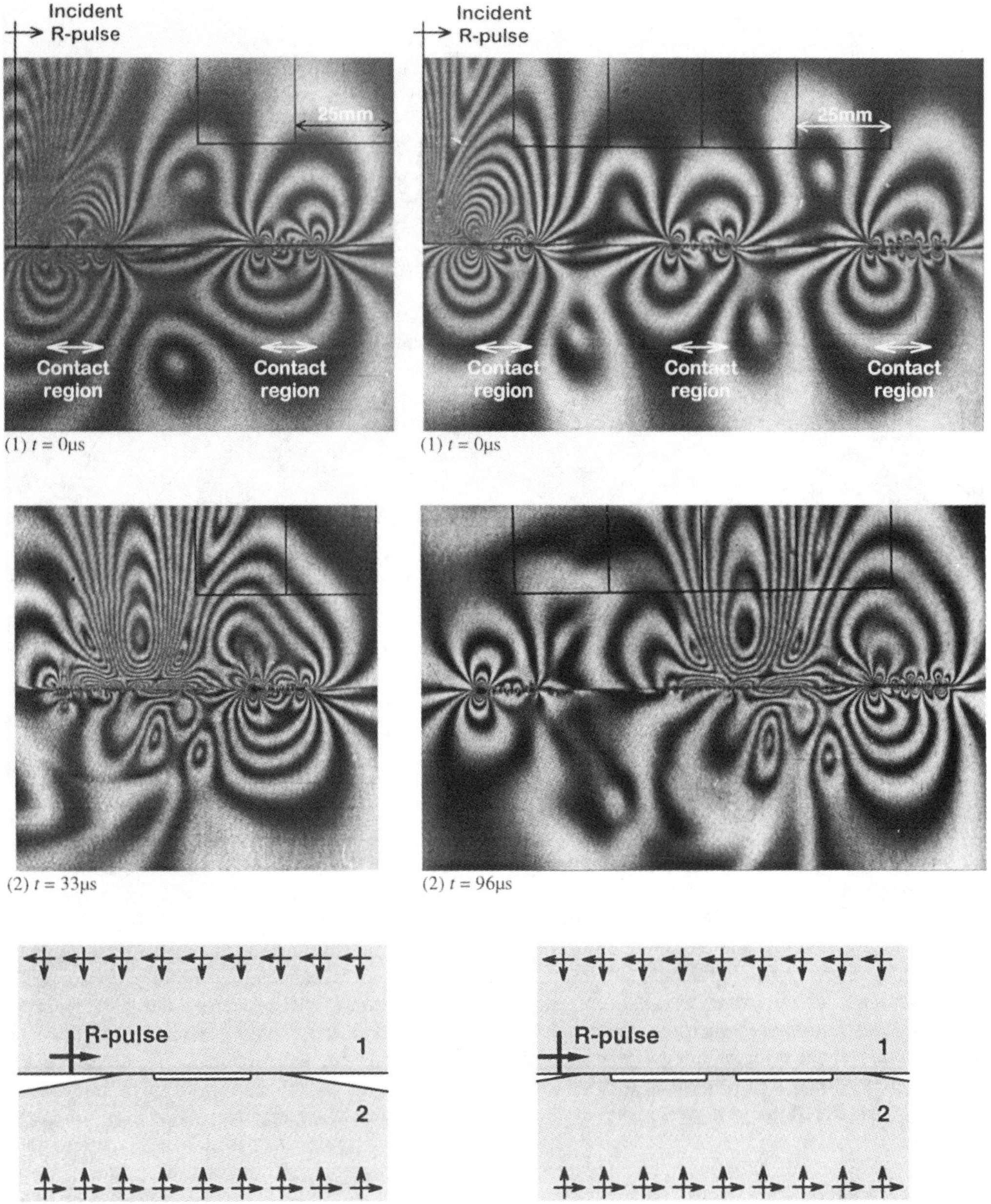

Figure 8. Snapshots of isochromatic fringe patterns obtained by experiments.
Left column: Experiment #4: compression with reverse shear; two short contacts; and
Right column: Experiment #5: compression with reverse shear; three short contacts.

3.2.4 *Experiments #4 and #5: multiple contact*

In order to investigate the influence of number of contacts on R-pulse scattering, a double [Exp. #4 in Fig.3(4)] and triple [Exp. #5 in Fig.3(5)] contact systems are investigated. Smaller contact lengths are chosen.

In Figure 8, isochromatic fringe patterns pertaining to two particular phases of the interaction process allow a comparison between similar dynamic situations occurring in the double (lhs column) and triple (rhs column) contact experiment. The phases shown pertain to the stages where (1) the maximum stress amplification at the lhs edge of the first contact can be observed; and (2) the transmitted R-pulse approaches the last contact section.

The overall characteristic features of the dynamic interaction patterns, i.e. stress amplification and reduction due to the R-pulse, basically appear to be similar in both experiments. However, the amplitude of the transmitted R-pulse in the upper plate shown in Experiment #5 (2) (i.e. after passing through two contacts) is smaller than that in Experiment #4 (2) (i.e. after the interaction with one contact). This indicates that each individual interaction reduces the energy of the incident/transmitted R-pulse by generating reflected R-pulses and bulk waves in the upper and lower plates.

4 NUMERICAL SIMULATIONS

The dynamic R-pulse interaction problems are numerically investigated using the finite difference wave propagation simulator SWIFD (Rossmanith & Uenishi 1995, 1996). The contact problems are considered under plane stress conditions, and the R-pulse is assumed to interact with a contact region which is characterized by a Mohr-Coulomb friction criterion with the coefficient of friction equal to 0.3. The interface is pre-stressed and the tensile strength of the interface is set at a very low level.

4.1 *Pulse energy partition*

It is informative to evaluate the relative amount of wave energy transmitted across and reflected at the contact area during the interaction process. However, there are no expressions of energy parameters for the surface waves, and even in the simplest cases, they are given in a rather complicated form obtained by direct integration over a certain section of the wave structure (Biryukov *et al.*, 1995). Therefore, a method to calculate the R-pulse energy from isochromatic fringe pattern or displacement data has been developed and used to evaluate the pulse energy partition. The results are schematically shown in Figure 9 for the statically compressed, however not sheared, one long contact between two acoustically identical materials. It is indicated that more than one half of the energy initially contained in the incident R-pulse has been radiated in the form of bulk waves into the far-field and only 37% of the total energy is transmitted in the form of a new R-pulse R^t_1 along the free surface. Note, that the energy carried by the reflected R-pulses is negligibly small compared with that contained in the transmitted R-pulses.

Table 1 shows the relative energy contained in the upper-transmitted R-pulse, R^t_1, after the interaction with one long contact. The results are shown for three different static shear pre-loading conditions. It is indicated that positive (reverse) static shear pre-loading does increase (decrease) the energy transmitted along the contact region in plate 1, respectively. These results are consistent with the experiments (Figs 6 and 7) where less (more) energy radiation in the form of bulk waves has been observed under positive (reverse) static shear stress, respectively.

4.2 *Influence of the acoustic impedance ratio*

During the course of numerical investigations, the influence of acoustic impedance ratio $(\rho c_P)_2 / (\rho c_P)_1$ is also studied. Here, ρ is the mass density, c_P is the longitudinal (P-) wave speed, and the subscripts 1 and 2 correspond to the upper and lower material, respectively. The acoustic impedance ratio of the two contacting materials plays a crucial role in wave transmission and reflection at the interface (see e.g. Rinehart 1975). The investigation is performed for one long contact under static compression only.

Figure 10 shows the numerically generated snapshots of isochromatic fringe patterns taken at the same timing, 60μs after the maximum stress amplification at the lhs edge of the contact region. Figure 10(1) pertains to the case where the incident R-pulse speed lies between the shear (S-) and P-wave speeds of the lower material, $(c_S)_2<(c_R)_1<(c_P)_2$, showing a shear-type (S-) Mach wave.

If the lower material is sufficiently soft, $(c_S)_2<(c_P)_2<(c_R)_1$ [Fig.10(2)], both longitudinal- (P-) and shear-type (S-) Mach waves are generated and the disturbance in the lower material is largely confined to the region behind the P-Mach wave front [the triangle in Fig.10(2)].

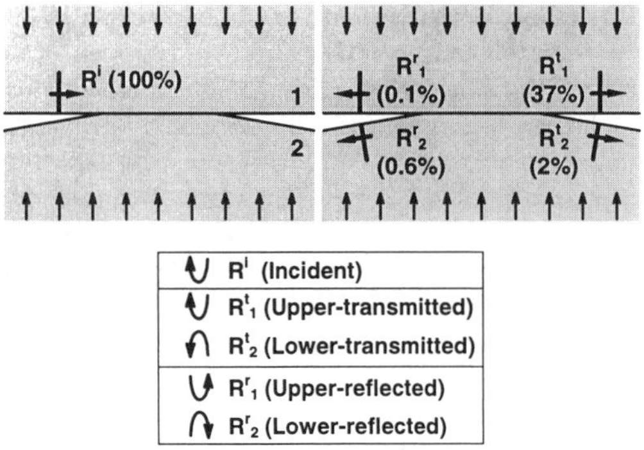

Figure 9. Partition of the energy contained in the incident R-pulse for the case where the two contacting materials are mechanically identical.

Table 1 Relative energy in the upper-transmitted R-pulse after the interaction with one long (55mm) contact.

Static shear	Reverse shear	No shear	Positive shear
Relative energy contained in R^t_1	29%	37%	41%

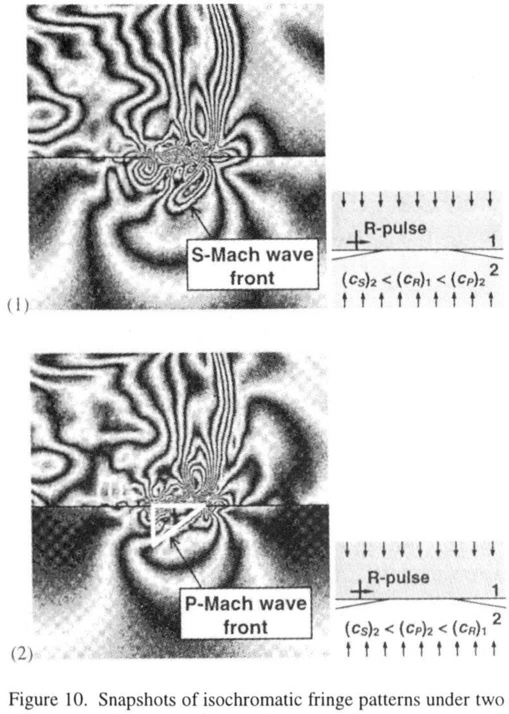

Figure 10. Snapshots of isochromatic fringe patterns under two different combinations of contacting materials.
(1) $(c_S)_2 < (c_R)_1 < (c_P)_2$; and (2) $(c_S)_2 < (c_P)_2 < (c_R)_1$.

5 APPLICATIONS OF THE MODEL

5.1 *Seismological interpretation of the model*

From a seismological point of view, the model and the results obtained in the previous chapters can be interpreted as follows:

In reality, the crustal rock adjacent to geological faults is fragmented by secondary faults over a wide range of scales. In modeling earthquakes, one consequence of this complexity is the introduction of the concept of asperities. The regions of very high slip, or asperities, are considered to be important in earthquake hazard analysis because the failure of the asperities radiates most of the high-frequency seismic energy into the far-field. The geometrical explanation for asperities reflects the fact that faults are not perfectly planar and on all scales rough, containing jogs or steps (Lay & Wallace 1995).

This study has shown that a relatively large amount of energy is radiated in the form of bulk waves from a contact region into the far-field. Interface slip is found only inside the initially contacting regions. These observations suggest that – when scaled up – the contact regions can be regarded as asperities on a geological fault. Therefore, the energy partition patterns obtained from the laboratory and numerical models are of practical importance in evaluating the influence of asperities.

5.2 *The 1995 Hanshin (Kobe), Japan, earthquake*

On 17 January 1995, at 5:46 a.m. local time, an earthquake of moment magnitude 6.9 struck the region of Kobe and Osaka (Hanshin region) in west-central Japan. This region is Japan's one of the most populated and industrialized area, with a population of some 10 million. Seismic inversion (Wald 1995, 1996) suggests that the rupture started at a shallow depth on a fault system running from Awaji Island through the city of Kobe and propagated bilaterally: along the Suma/ Suwayama faults toward the city of Kobe [Fig.11(1)] and along the Nojima fault. Strong ground shaking motions lasted for about 20 seconds and caused damage over a 100km radius from the epicenter, but Kobe and its neighboring region were most severely affected.

One of the puzzling phenomena observed in Kobe is the strip of the most severely damaged zone [the Japan Meteorological Agency Intensity 7 zone marked in dark gray in Fig.11(1)]. This strip, approximately 20km long, is located close, but not parallel, to the Suma/Suwayama faults. Damage due to liquefaction was hardly observed inside this area, and it is suggested that the damage was caused directly by the seismic waves.

As the Suma/Suwayama faults dip steeply, nearly 90°, and a near-source, SH directivity pulse from strike-slip faulting was recorded in the city of Kobe (Nakamura 1995, Toki et al. 1995), a two-dimensional model, which includes a plane perpendicular to the fault plane, is considered appropriate for a first order analysis of the rupture mechanisms of the Hanshin earthquake.

A region of relatively large slip (asperity) was found beneath the lhs edge [in Fig.11(1)] of the strip. A concentrated shear disturbance arrived and the resulting large (particle) velocity response was recorded in central Kobe (Wald 1995, 1996). This suggests that the rupture-induced shear wave was of a Mach wave type.

As discussed above, an asperity on a fault corresponds to a contact region in the model. In Figure 10(1), a S-Mach wave is observed in the numerical simulation where an interface is located between dissimilar materials [$(c_S)_2 < (c_R)_1 < (c_P)_2$]. In the simulation material 1 corresponds to the acoustically harder region in the foothills of the Rokko Mountains, where soils are very shallow or rock outcroppings are found. There the damage tended to be relatively minor. Material 2 fits the acoustically softer region which includes the "damage strip" where primarily soft alluvial soils prevail.

During the course of dynamic interaction, each particle in the materials will experience a history of velocity and acceleration. The maximum values of these quantities are of practical importance and, hence, the peak particle velocity (PPV) and the peak particle acceleration (PPA) have been selected as the design parameters in many applications such as blasting in mines, engineering seismology.

Figure 11 shows the distributions of PPV [Fig.11(2)] and PPA [Fig.11(3)] obtained by the numerical simulation [$(c_S)_2 < (c_R)_1 < (c_P)_2$]. It is interesting to note that the region of high PPV (or PPA) is located in a narrow band, similar to the shape of the "damage strip" in Kobe [Fig.11(1)]. The angle between the interface (fault) and the region of high PPV (PPA) is controlled by the S-Mach wave generated during the interaction process [Fig.10(1)]. For the two-dimensional numerical simulation, this angle is approximately 45°. For the Hanshin earthquake, the measured angle between the fault and the "damage strip" is 18°. The difference in these angles is possibly due to the local geological as

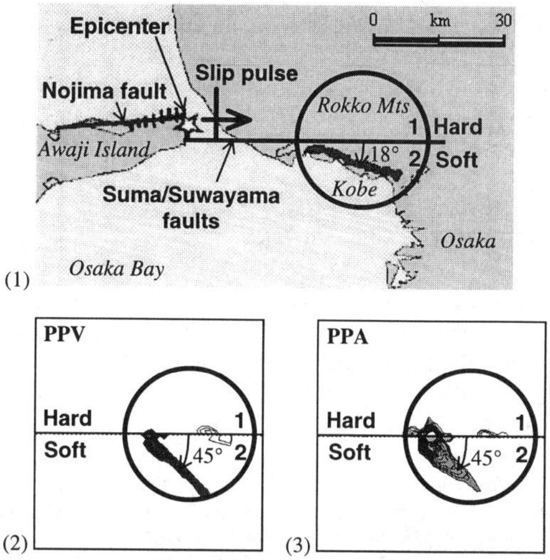

Figure 11. Comparison of the Hanshin (Kobe) earthquake and the numerical simulation [for the case $(c_S)_2 < (c_R)_1 < (c_P)_2$]. (1) the Japan Meteorological Agency Intensity 7 (the most severely damaged) zone (dark gray region) associated with the Hanshin earthquake; and contours of (2) high PPV (peak particle velocity) and (3) high PPA (peak particle acceleration) obtained by the numerical simulation.

well as the three-dimensional effect on the real fault rupture process. However, the result indicates that a simple two-dimensional model may be able to provide the information about the earthquake rupture and the ensuing dynamic wave phenomena, although for more sophisticated study three-dimensional analysis is required.

6 CONCLUSIONS

The purpose of this study was to obtain an improved understanding of the interaction between a Rayleigh (R-) pulse and a partially contacting interface between similar and dissimilar materials. The experimental and numerical investigations have given a clear insight into the basic mechanisms of the R-pulse-induced instability of a statically pre-stressed interface. The particle motion in a R-pulse can be attributed to the observed phenomena. It is hoped that the knowledge gained from this work will assist in obtaining a better understanding of the earthquake rupture mechanisms. It should also be noted that due to a vast variety of application areas of Rayleigh pulses, the basic phenomena observed in this study can be of practical importance in various other fields of engineering.

ACKNOWLEDGMENTS

This work was financially sponsored by the Austrian National Science Foundation (FWF) through Research Project No P10326-GEO.

REFERENCES

Anooshehpoor, A. & Brune, J.N. 1994. Frictional heat generation and seismic radiation in a foam rubber model of earthquakes. *Pure Appl. Geophy.* 143: 735-747.

Biryukov, S.V., Gulyaev, Yu.V., Krylov, V.V. & Plessky, V.P. 1995. *Surface acoustic waves in inhomogeneous media.* Berlin: Springer-Verlag.

Brune, J.N. 1996. Particle motions in a physical model of shallow angle thrust faulting. *Proc. Indian Acad. Sci. (Earth Planet. Sci.)* 105: L197-L206.

Brune, J.N., Brown, S. & Johnson, P.A. 1993. Rupture mechanism and interface separation in foam rubber models of earthquakes: a possible solution to the heat flow paradox and the paradox of large overthrusts. *Tectonophysics* 218: 59-67.

Cardenas-Garcia, J.F. 1983. *On Rayleigh waves and Rayleigh wave extension of surface micro-cracks.* Ph.D. Thesis: Mechanical Engineering Department, University of Maryland.

Comninou, M. & Dundurs, J. 1977. Elastic interface waves involving separation. *J. Appl. Mech.* 44: 222-226.

Comninou, M. & Dundurs, J. 1978a. Can two solids slide without slipping? *Int. J. Solids Structures* 14: 251-260.

Comninou, M. & Dundurs, J. 1978b. Elastic interface waves and sliding between two solids. *J. Appl. Mech.* 45: 325-330.

Freund, L.B. 1978. Discussion: elastic interface waves involving separation. *J. Appl. Mech.* 45: 226-228.

Lamb, H. 1904. On the propagation of tremors over the surface of an elastic solid. *Phil. Trans. Roy. Soc. London* A203: 1-42.

Lambros, J. & Rosakis, A.J. 1995a. Shear dominated transonic interfacial crack growth in a bimaterial – I. Experimental observations. *J. Mech. Phys. Solids* 43: 169-188.

Lambros, J. & Rosakis, A.J. 1995b. Development of a dynamic decohesion criterion for subsonic fracture of the interface between two dissimilar materials. *Proc. R. Soc. Lond.* A451: 711-736.

Lay, T. & Wallace, T.C. 1995. *Modern global seismology.* San Diego: Academic Press.

Liu, C., Huang, Y. & Rosakis, A.J. 1995. Shear dominated transonic interfacial crack growth in a bimaterial – II. Asymptotic fields and favorable velocity regimes. *J. Mech. Phys. Solids* 43: 189-206.

Nakamura, Y. 1995. Waveform and its analysis of the 1995 Hyogo-Ken-Nanbu earthquake. *JR Earthquake Information* 23c, Railway Technical Research Institute.

Rayleigh, J.W.S. 1885. On waves propagated along the plane surface of an elastic solid. *The Proceedings of the London Mathematical Society* 17: 4-11.

Rinehart, J.S. 1975. *Stress transients in solids.* Santa Fe: HyperDynamics.

Rossmanith, H.P. & Uenishi, K. 1995. *SWIFD user's manual version 1995.* Vienna: Fracture and Photomechanics Laboratory, Institute of Mechanics, Vienna University of Technology.

Rossmanith, H.P. & Uenishi, K. 1996. PC software assisted teaching and learning of dynamic fracture and wave propagation phenomena. In H.P. Rossmanith (ed.), *Teaching and Education in Fracture and Fatigue*: 253-262. London: E & FN Spon.

Rossmanith, H.P. & Uenishi, K. 1997. Fault dynamics – dynamic triggering of fault slip. To appear in: *Proceedings of International Symposium on Rock Stress* (RS Kumamoto '97, Kumamoto, Japan, 8-10 October 1997).

Schallamach, A. 1971. How Does Rubber Slide? *Wear* 17: 301-312.

Toki, K., Irikura, K. & Kagawa, T. 1995. Strong motion data recorded in the source area of the Hyogo-Ken-Nanbu earthquake, January 17, 1995, Japan. *J. Nat. Disas. Sci.* 16: 23-30.

Turcotte, D.L. 1997. Earthquakes, fracture, complexity. In: J.R. Willis (ed.), *Proceedings of the IUTAM Symposium on Nonlinear Analysis of Fracture*: 163-175, Dortrecht: Kluwer Academic Publishers.

Uenishi, K., Rossmanith, H.P. & Knasmillner, R.E. 1997. Interaction of a Rayleigh pulse with non-uniformly contacting interfaces. To appear in: *The Proceedings of the International Conference on Materials and Mechanics '97* (20-22 July 1997, Tokyo).

Wald, D.J. 1995. A preliminary dislocation model for the 1995 Kobe (Hyogo-Ken Nanbu), Japan, earthquake determined from strong motion and teleseismic waveforms. *Seism. Res. Let.* 66: 22-28.

Wald, D.J. 1996. Slip history of the 1995 Kobe, Japan, earthquake determined from strong motion, teleseismic, and geodetic data. *J. Phys. Earth* 44: 489-503.

Damage and Failure of Interfaces, Rossmanith (ed.) © 1997 Balkema, Rotterdam, ISBN 90 5410 899 1

Reflection and refraction of plane stress waves at dissipative interfaces

A. Daehnke
CSIR Division of Mining Technology, Johannesburg, South Africa

H. P. Rossmanith
Vienna University of Technology, Austria

ABSTRACT: When stress waves interact with joints and fractures in a rock mass, the incident energy is reflected from and refracted (transmitted) across the interface. Knowledge of the amplitudes of reflected and transmitted waves is of importance to understand stress wave propagation through discontinuous rock, and to assess the stability and ultimately the safety of structures in rock loaded transiently.

In this paper the dynamic behaviour of joints containing fluids or thin layers of comparatively soft material are analysed. A generalised Kelvin model is used to simulate the interactions of incident *P*- and *SV*-waves. It is found that the amplitudes of the reflected and refracted waves are dependent on the joint stiffness and viscosity, as well as the frequency of the incident wave. The energy dissipation of the viscous interface is analysed and parameters influencing the maximum dissipation are quantified and discussed.

1 INTRODUCTION

The theory of rock dynamics has been considerably advanced this century by seismology and earthquake engineering, leading to further understanding of wave propagation effects along free surfaces and along the interfaces between different material layers in the crust of the earth. Within the mining context, research in this field has been spurred in the last three decades particularly by the technical difficulties of deep level mining operations and the associated occurrence of seismicity and rockbursts. Although the research effort has accelerated over the past decade, these dynamic events continue to account for the single largest cause contributing towards the toll of injuries and fatalities suffered by the workforce during mining operations.

An important general aspect of rock dynamics is the interaction of stress waves with geological or mining induced discontinuities and fractures. Knowledge of this field is necessary to assess the propensity of wave trapping and energy channeling in geological strata surrounding mining excavations. Furthermore, it gives valuable insight towards understanding the performance of blasting, percussion drilling and the development of non-explosive rock breaking equipment such as impacting and ripping tools.

It is of interest to consider the following specific aspects of stress wave interactions with rock mass discontinuities, which have to be assessed by the engineer engaged in mining design and operations (see Figure 1):

- *Mining Induced Seismicity:* Mining through highly stressed rock in deep level mines is often accompanied by considerable seismic activity. High geological and mining induced stresses can lead to sudden movements on existing discontinuities (e.g. faults) or on geological inhomogeneities (e.g. intrusive dyke structures). The resulting seismic waves interact with mining excavations leading to loss of stability with falls of ground or with dynamic formation of new fractures and the explosive failure of brittle rock.

- *Blasting:* In hard rock mining, the most common technique of excavating rock is in the form of blasting. Blast induced stress waves can cause strong dynamic loading, which can have an unpredictable effect on the subsequent stability of the excavation. The blast waves interact with planes of weakness and may trigger further movements on faults, where these triggered movements may lead to rockfalls and damage to the excavations.

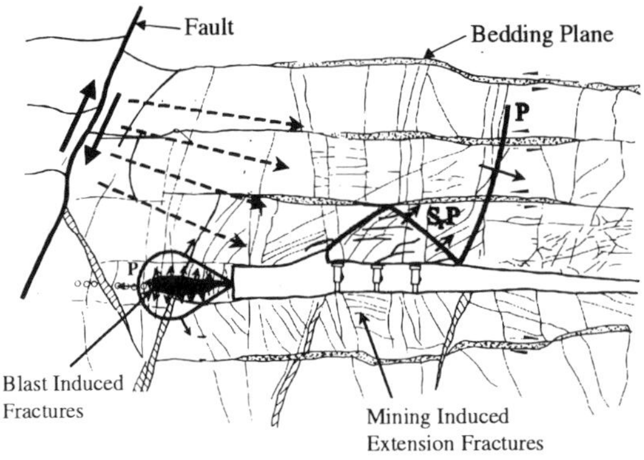

Figure 1: Stress wave interactions with rock mass discontinuities.

It is evident that there is a need to quantify and gain insights into stress wave interactions with rock mass discontinuities. Ongoing research at the Fracture and Photomechanics Laboratory of the Vienna University of Technology has centered on various aspects of rock dynamics, such as stress wave initiated fault rupture, wave interactions with mining excavations and stress wave propagation though fractured and jointed rock. In this paper some new results concerning the wave interactions with dissipative rock mass interfaces are presented.

2 PREVIOUS WORK

The basis of the present work was established at the turn of the century by Knott (1899) and Zoeppritz (1919), who were amongst the first to study, by analytical means, the reflection and refraction of planes waves. Since then various authors (Mindlin 1960, Koefoed 1962, Tooley et al. 1965, Kendall and Tabor 1971, Schoenberg 1980, Pyrak-Nolte et al. 1990a,b, Myer et al. 1995, Daehnke and Rossmanith 1997) investigated increasingly complex interface models simulating the dynamic behaviour of rock mass discontinuities.

Of note is the work by Schoenberg (1980), who studied wave propagation across an interface with slip linearly related to the stress traction, which is continuous across the interface, i.e. an interface incorporating an elastic spring (hereafter referred to as the interface spring model). The physical justification of this model is that fractures and joints in rock appear as a planar collection of void spaces and asperities in contact. The presence of these void spaces collectively defines a thin zone with effective normal and shear stiffness, where the stiffness can range from close to zero for open fractures (thereby reflecting most of the incident energy), to high stiffness values for smooth, closed fractures (thereby transmitting the incident waves). This model was subsequently verified by Pyrak-Nolte et al. (1990a,b) who investigated longitudinal and shear wave transmission across natural fractures, and successfully correlated laboratory findings with the analytical results of the spring model. The reflection and transmission amplitudes of stress waves interacting with a spring model were related to actual rock joint surface properties by Daehnke and Rossmanith (1997).

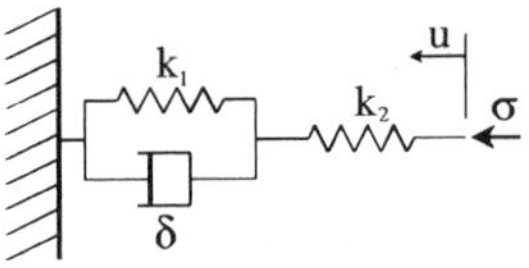

Figure 2: The generalised Kelvin model.

3 THE GENERALISED KELVIN MODEL

The objectives of the work reported here are to extend the interface stiffness model to a generalised Kelvin model. The components of the Kelvin model are shown in Figure 2, and the relation between stress (σ), displacement (u), stiffness (k_1, k_2) and viscosity (δ) is given by Equation (1).

This model is suitable to simulate the dynamic behaviour of fluid filled rock joints, as well as discontinuities containing comparatively soft material.

$$\sigma = \delta\left(\dot{u} - \frac{\dot{\sigma}}{k_2}\right) + k_1\left(u - \frac{\sigma}{k_2}\right) \tag{1}$$

The width of rock joints and fractures encountered in mining operations is typically of the order of a few mm. This width is negligible compared to the wavelengths of seismic and blast induced stress waves. Hence, for the analysis conducted here, the width of the interface is not taken into account. A further assumption is that the mass of the interface filler material is negligible.

4 REFLECTION AND REFRACTION OF WAVES AT A KELVIN MODEL INTERFACE

One of the major aspects of wave interaction with an

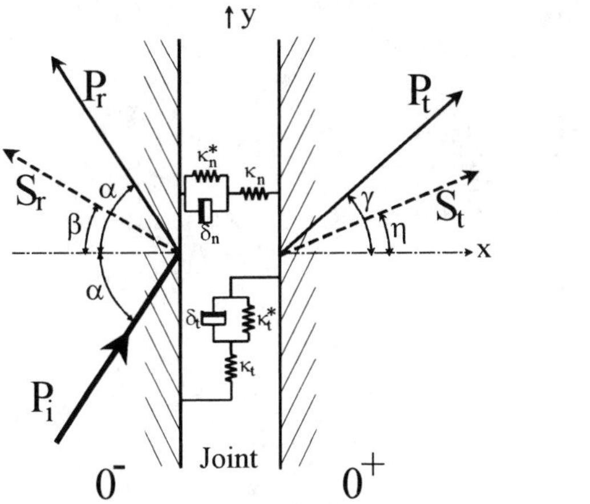

Figure 3: Wave normals (rays) resulting from a P-wave interacting with a Kelvin model interface.

interface is that, for a single incident wave, additional waves are created due to mode conversion, giving rise to a reflected P- (P_r) and a SV-(S_r) wave, and a transmitted P- (P_t) and a SV-(S_t) wave. Figure 3 shows the wave field resulting from an incident P-wave interacting at an angle α with an interface.

As shown in Figure 3, the interface is modelled applying a Kelvin model both in a normal and tangential direction. Currently no data exist regarding the dynamic material properties of actual joint surfaces, which could be used to determine the parameters κ and δ. Here, the stiffness of the two springs acting in a normal (κ_n) and tangential (κ_t) direction are taken as equal, i.e. $\kappa_n = \kappa_t$, as well as $\kappa_n = \kappa_n^*$ and $\kappa_t = \kappa_t^*$. Similarly, the viscosity acting in a normal and tangential direction is set equal, i.e. $\delta_n = \delta_t$.

The purpose of the present analysis is to conduct a parametric study in order to establish trends and the effect of varying viscosity and stiffness for various angles of wave incidence. In doing so the reflection and transmission coefficients associated with this rock joint model are established.

To solve the problem of wave interaction with a Kelvin model interface, the following boundary conditions must be satisfied:

(i) continuity of normal stress,

(ii) continuity of shear stress,

(iii) the normal stress, and

(iv) the shear stress are related to the relative joint surface displacement and velocity (Equation 1, with $\sigma = \sigma_x$ or $\sigma = \tau_{xy}$).

The mathematical form of the interface conditions is given by Equations (2):

(i) $\sigma_x^- = \sigma_x^+$

(ii) $\tau_{xy}^- = \tau_{xy}^+$

(iii) $$\sigma_x^{-/+} = \delta_n\left((\dot{u}_x^- - \dot{u}_x^+) - \frac{1}{k_n}(\dot{\sigma}_x^- - \dot{\sigma}_x^+)\right) + k_n\left((u_x^- - u_x^+) - \frac{1}{k_n}(\sigma_x^- - \sigma_x^+)\right)$$

(iv) $$\tau_{xy}^{-/+} = \delta_t\left((\dot{u}_y^- - \dot{u}_y^+) - \frac{1}{k_t}(\dot{\tau}_{xy}^- - \dot{\tau}_{xy}^+)\right) + k_t\left((u_y^- - u_y^+) - \frac{1}{k_t}(\tau_{xy}^- - \tau_{xy}^+)\right) \quad (2)$$

where

$$\sigma_x^- = \sigma_x^{Pi} + \sigma_x^{Si} + \sigma_x^{Pr} + \sigma_x^{Sr}, \qquad \sigma_x^+ = \sigma_x^{Pt} + \sigma_x^{St},$$
$$\tau_{xy}^- = \tau_{xy}^{Pi} + \tau_{xy}^{Si} + \tau_{xy}^{Pr} + \tau_{xy}^{Sr}, \qquad \tau_{xy}^+ = \tau_{xy}^{Pt} + \tau_{xy}^{St},$$
$$u_x^- = u_x^{Pi} + u_x^{Si} + u_x^{Pr} + u_x^{Sr}, \qquad u_x^+ = u_x^{Pt} + u_x^{St},$$
$$u_y^- = u_y^{Pi} + u_y^{Si} + u_y^{Pr} + u_y^{Sr}, \qquad u_y^+ = u_y^{Pt} + u_y^{St},$$

and the values for $\dot{\sigma}_x^-$, $\dot{\sigma}_x^+$, $\dot{\tau}_{xy}^-$, $\dot{\tau}_{xy}^+$, $\dot{u}_x^-$, $\dot{u}_x^+$, $\dot{u}_y^-$ and $\dot{u}_y^+$ follow analogously. Here σ_x^X, τ_{xy}^X and $u_{x,y}^X$ (X = P_i, S_i, P_r, S_r, P_t and S_t) are the normal stress, shear stress and displacement components, respectively, of the incident (P_i, S_i), reflected (P_r, S_r) and refracted (P_t, S_t) waves.

The reflection and transmission coefficients, defined as the ratio of the reflected and transmitted wave amplitude versus the incident wave amplitude, can be calculated for each wave type by satisfying the interface conditions and simultaneously solving Equations (2).

4.1 *Incident P-wave*

Figure 4 gives the reflection and transmission coefficients for an incident P-wave interacting with a Kelvin type interface. The curves are given for normalised values of stiffness ($k_{n,t}$) and damping ($d_{n,t}$), which are defined as (Pyrak-Nolte et al. 1990, Gu et al. 1996):

$$k_{n,t} = \frac{\kappa_{n,t}}{\omega\sqrt{E\rho}} \quad \text{and} \quad d_{n,t} = \frac{\delta_{n,t}}{\sqrt{E\rho}}, \quad (3)$$

where $\kappa_{n,t}$ and $\delta_{n,t}$ are the effective stiffness and damping values acting in a normal and tangential direction, E is Young's modulus, ρ is the material density and $\omega = 2\pi f$, where f is the frequency of the incident periodic wave. The results are given for

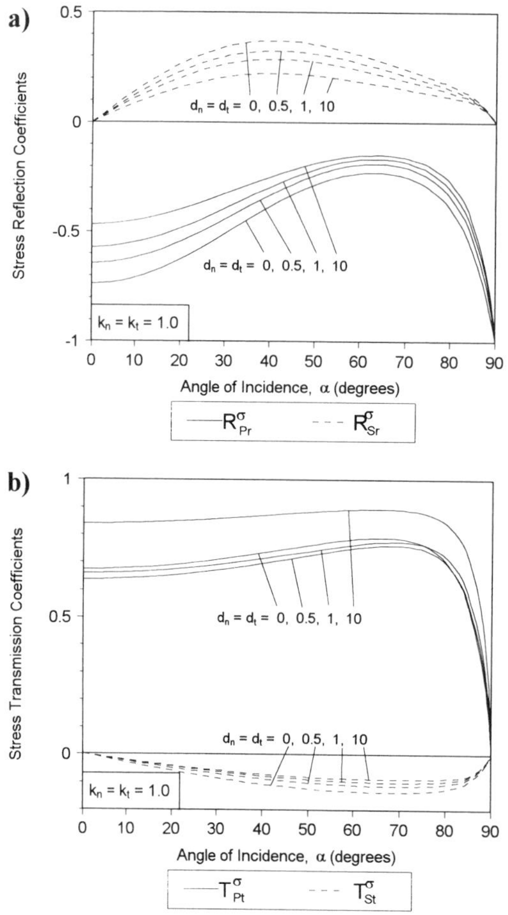

Figure 4: Reflection a) and transmission b) stress coefficients for an incident *P*-wave.

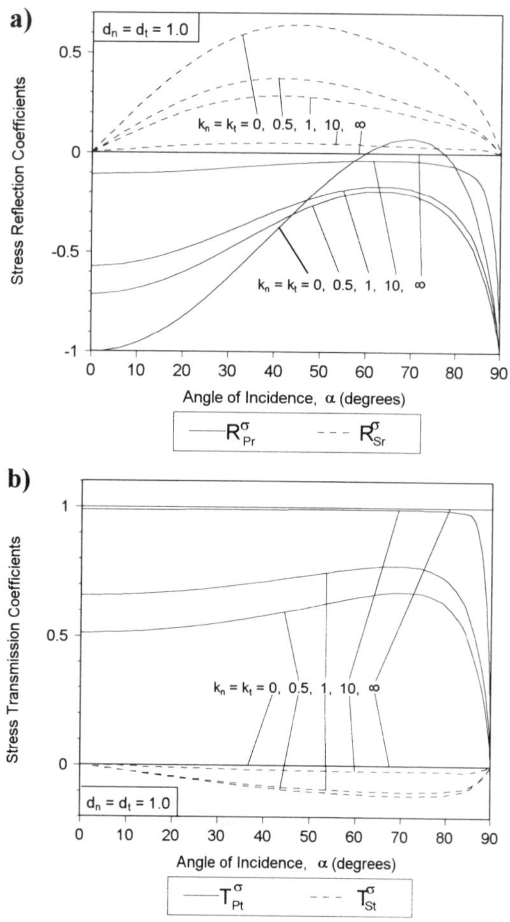

Figure 5: Reflection a) and transmission b) stress coefficients for various values of k_n and k_t.

Poisson's ratio $\nu = 0.25$, which is a reasonable approximation for most rock types. The rock on either side of the interface is assumed to have identical acoustic impedances, i.e. $\rho^- c_P^- = \rho^+ c_P^+$.

As is evident from Figure 4 a), for a constant value of $k_n = k_t = 1.0$, the stress amplitude of reflected wave decreases as the viscosity increases. Conversely, the amplitude of transmitted waves increases with increasing viscosity.

Figures 5 a) and b) give the reflection and transmission coefficients, respectively, for the case where the viscosity is kept constant ($d_n = d_t = 1.0$), and k_n, k_t are varied from zero to infinity. From the curves and the definition of k_n and k_t (see Equation 3) it is seen that at decreasing wave frequency and/or increasing joint stiffness ($k_{n,t} \rightarrow \infty$) the amplitudes of the reflected *P*- and *SV*-wave decrease, whereas the transmitted amplitudes increase. In the limit ($k_{n,t} = \infty$), all the incident energy is transmitted. If $k_{n,t} = 0$ the interface acts as a free boundary, and all the incident energy is reflected.

These results are expected, as a high frequency wave is more likely to be reflected at a viscous interface, whereas low frequency waves are more easily transmitted.

4.2 *Incident SV-wave*

The reflection and transmission coefficients for an incident *SV*-wave interacting with a viscous interface are given in Figures 6 a) and b). The curves

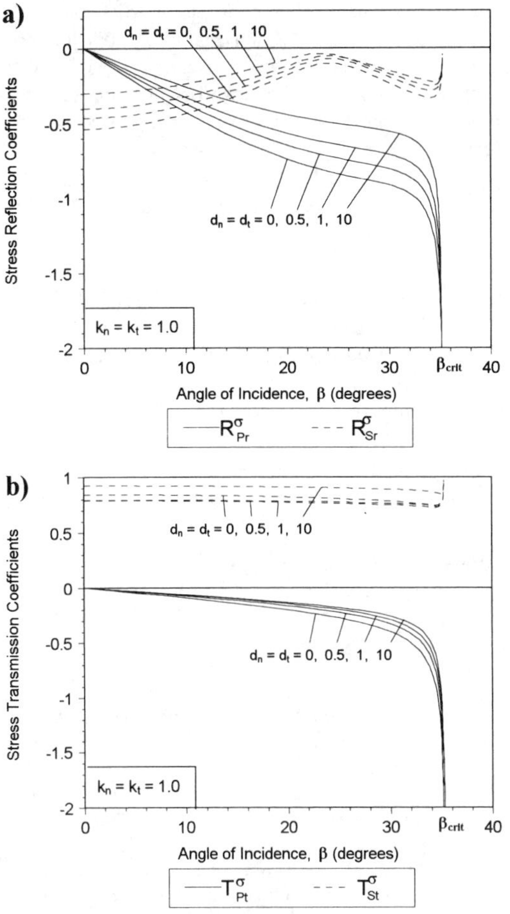

Figure 6: Reflection a) and transmission b) stress coefficients for an incident *SV*-wave.

are given for various values of normalised damping (defined in Equation 3), whilst the condition is set that $k_n = k_t = 1.0$. The coefficients are given for angles of incidence ranging from $\beta = 0^{\circ}$ to $\beta = \beta_{crit}$, where β_{crit} is the critical angle of incidence. For $\beta > \beta_{crit}$ the incident *SV*-wave is totally reflected and the reflection and transmission coefficients become complex valued. The case of total reflection is not dealt with in this paper.

As observed in the case of an incident *P*-wave, with increasing viscosity less energy is reflected and more energy is transmitted.

5 ENERGY DISSIPATION

An important aspect of viscous interfaces is that energy is dissipated during wave interaction. This energy is absorbed by the viscous material and can lead to further interface disintegration and rock failure.

Figure 7 shows the percentage of energy dissipated during wave passage across an interface, where $k_n = k_t = 1.0$ and the normalised viscosity is varied from $d_n = d_t = 0 \rightarrow \infty$. At zero and infinite viscosity, the dashpot is in effect inactive, and no energy is dissipated. At intermediate values energy is dissipated, with a maximum of 24 % and 20 % for a normally incident *P*- and *SV*-wave, respectively, where $d_n = d_t = 1.0$.

Table 1 lists the peak energy dissipation values for incident *P*- and *SV*-waves at various angles of incidence. Also shown are the normalised stiffness and damping values at which maximum dissipation occurs.

A three dimensional representation of the energy dissipation characteristics of viscous interfaces is given in Figures 8 and 9 for normally incident *P*-

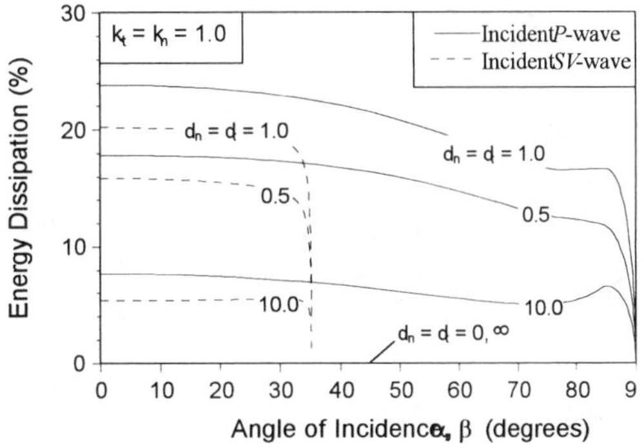

Figure 7: Dissipation of incident *P*- and *SV*-waves.

Table 1: Dissipation maxima for *P*- and *SV*-waves.

Incident *P*-wave:

α	Max. Dissipation	$k_{n,t}$	$d_{n,t}$
0°	25 %	0.775	1.095
22.5°	24.7 %	0.725	1.025
45°	23.7 %	0.582	0.822
67.5°	21.3 %	0.378	0.535
90°	0	–	–

Incident *SV*-wave:

β	Max. Dissipation	$k_{n,t}$	$d_{n,t}$
0°	25 %	0.447	0.632
30°	22.4 %	0.447	0.632

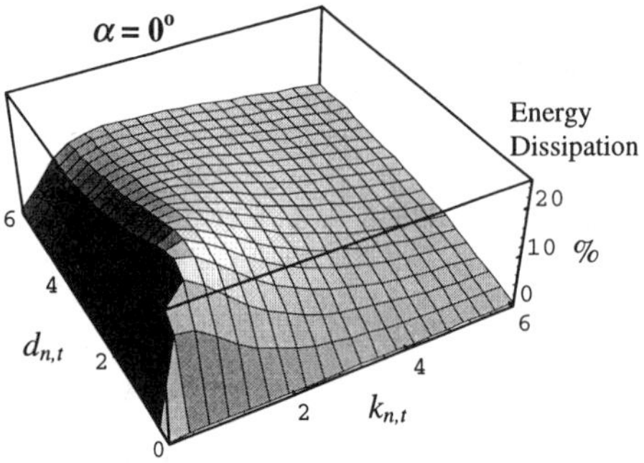

Figure 8: Energy dissipation of an incident *P*-wave.

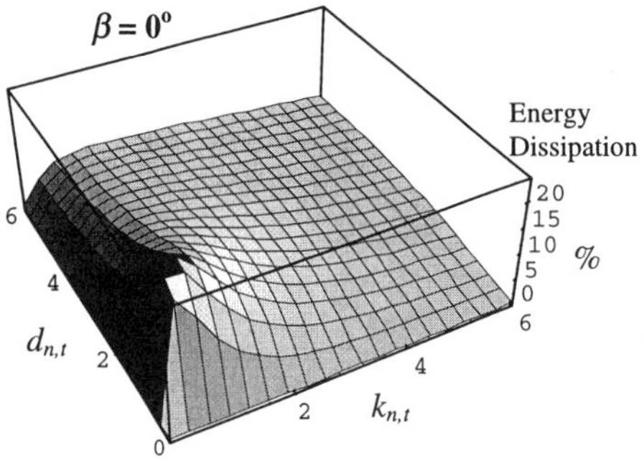

Figure 9: Energy dissipation of an incident *SV*-wave.

($\alpha = 0^o$) and *SV*-waves ($\beta = 0^o$), respectively. A prominent peak exists at which the maximum dissipation occurs.

6 CONCLUSIONS

The generalised Kelvin model is suitable to investigate the dynamic behaviour of interfaces modelling fluid filled joints, as well as discontinuities containing comparatively soft material.

The reflection and transmission characteristics of this model are dependent on the joint stiffness, viscosity and the frequency of the incident wave. With increasing stiffness and viscosity, the amplitude of the reflected waves decreases, while more energy is transmitted. Increasing wave frequency results in more energy being reflected, and a reduced portion is transmitted.

For the Kelvin model analysed here, a maximum of 25 % of the energy can be dissipated during wave passage across the interface. It is shown that the energy dissipation is relatively insensitive to the angle of wave incidence.

Values of wave frequency, joint stiffness and damping are given at which maximum dissipation occurs. Knowledge of this parameter window (at which maximum energy is absorbed) is important to assess the interface integrity during passage of seismic and blast induced stress waves.

ACKNOWLEDGEMENTS

The first author would like to acknowledge the South African Department of Mineral and Energy Affairs for permission to publish. The second author thanks the Austrian National Science Foundation for financial support of this project, No. P10326 GEO.

REFERENCES

Gu, B., Suarez-Rivera, R., Nihei, K.T. & Myer, L.R. 1996. Incidence of plane waves upon a fracture. *J. of Geophys. Research* 101(B11): 25337-25346.

Daehnke, A. & Rossmanith, H.P. 1997. Reflection and transmission of plane stress waves at interfaces modelling various rock joints. *Int. J. for Blasting and Fragmentation* 1(2), (in press).

Kendall, K. & Tabor, D. 1971. An ultrasonic study of area of contact between stationary and sliding surfaces. *Proc. R. Soc. London* 323: 321-340.

Knott, C.G. 1899. Reflexion and refraction of elastic waves with seismological applications. *Philos. Mag.* 48: 64-97.

Koefoed, O. 1962. Reflection and transmission coefficients for plane longitudinal incident waves. *Geophys. Prosp.* 10: 304-351.

Mindlin, R.D. 1960. *Waves and vibrations in isotropic elastic planes.* In *Structural Mechanics,* J.F. Goodier & W.F. Hoff (eds.), Pergamon, New York.

Myer, L.R., Hopkins, D., Peterson, J.E. & Cook, N.G.W. 1995. Seismic wave propagation across multiple fractures. In *Fractured and Jointed Rock Masses*, L.R. Myer, N.G.W. Cook, R.E. Goodman & P. Tsang (eds.): 105-109. Rotterdam: Balkema.

Pyrak-Nolte, L.J., Myer, L.R. & Cook, N.G.W. 1990a. Transmission of seismic waves across single natural fractures. *J. Geophys. Research* 95(B6): 8617-8638.

Pyrak-Nolte, L.J., Myer, L.R. & Cook, N.G.W. 1990b. Anisotropy in seismic velocities and amplitudes from multiple parallel fractures. *J. Geophys. Research* 95(B7): 11345-11358.

Schoenberg, M. 1980. Elastic wave behaviour across linear slip interfaces. *J. A. Soc. Am.* 68(5): 1516-1521.

Tooley, R.D., Spencer, T.W. & Sagoci, H.F. 1965. Reflection and transmission of plane compressional waves. *Geophysics* 4: 552-570.

Zoeppritz, K. 1919. *Nach. d. Königl. Gesell. d. Wissen. z. Göttingen, Math.-Phys.*: 66-84. Berlin.

Damage and Failure of Interfaces, Rossmanith (ed.) © *1997 Balkema, Rotterdam, ISBN 90 5410 899 1*

Flexural vibrations of sandwich beams with deteriorating interfaces

C. Adam & F. Ziegler
Department of Civil Engineering, Vienna University of Technology, Austria

ABSTRACT: Analysis of inelastic slip developments in the physical interfaces of sandwich beams under severe dynamic loading is presented in the context of a multiple field analysis. Materials are considered in the regime of rate-dependent plasticity and are subjected to accumulated damage. Inelastic defects of the material are considered in the linear elastic background beam by a second strain field (eigenstrains). Layerwise continuous and linear in-plane displacement fields are implemented, and as such, model both the global and the local inelastic response of sandwich beams. An effective cross-sectional rotation is defined and the complex problem of a layerwise laminate theory reduces to the simpler case of a homogenized shear-deformable beam, with effective stiffness and boundary conditions. Superposition applies in the linear elastic background in an incremental formulation. Linear methods, as those based on Green's functions, are used to account for the given as well as for the resultants of the imposed strain fields.

1 INTRODUCTION

In various engineering applications structural components are designed of layers with a proper choice of the material properties. These components can be beams, plates or shells. In many cases the mechanical assumption of rigid contact between the layers is reasonable, e.g. if the layers are connected continuously by means of strong adhesives. For the analysis, basically two classes of theories can be distinguished: the equivalent-single-layer theories and the layerwise laminate theories (Reddy 1993). The latter category is derived by admitting a separate displacement field within the individual layers of the composite, see e.g. Swift & Heller (1974), Heuer (1992). Alternatively, the extension of homogeneous beam and plate theories is based on one displacement expansion throughout the thickness of the laminate that results in equivalent-single-layer theories, e.g. Whitney & Pagano (1970), Reddy (1984). Such theories cannot accurately model laminates made of dissimilar material layers.

In some frequently used structures, such as in layered wood systems connected with nails or layers that are glued with weak adhesives, rigid bonds between the layers cannot be achieved. Due to relative deformation of the connection an interlayer slip occurs, that significantly can affect both strength and deformation of the layered structure. Assuming that the Bernoulli-Euler hypothesis holds for each layer separately, and a linear constitutive equation between the horizontal slip and the interlaminar shear force applies, a sixth-order initial-boundary value problem results. Linear analysis of layered beams with partial or flexible connection is well established by Goodman & Popov (1968), Girhammar & Pan (1993) and Adam et al. (1997). An extension to steel and concrete composite beams with a nonlinear shear force-slip relationship is undertaken by Mistakidis et al. (1994). Alternatively, if the interlayer joints are modeled as additional layers, a proper layerwise laminate theory can be used to analyze such composites.

In the present paper an inelastic layerwise laminate theory, developed by Adam & Ziegler (1997a, 1997b, 1997c), is adopted to discuss the flexural response of symmetrically designed sandwich beam structures where inelastic slip may develop in the physical interfaces. According to the layerwise laminate theories, the governing equations are derived by the application of the Timoshenko theory of shear-deformable beams to each individual layer. The continuity of the transverse shear stress across the interfaces is considered by definition of the interlaminar shear stress, and hence, by means of the generalized Hooke's law. An effective cross-sectional rotation is introduced subsequently, which reduces the complex problem to the simpler case of an equivalent homogeneous beam with effective stiffness and with a corresponding set of boundary conditions.

The material of the physical interfaces behaves elastic-viscoplastic in the applied range of dynamic loading. Thereby, fracture as a consequence of plastic flow is included. This kind of damage may be smeared, which is done in continuum damage mechanics (Kachanov 1986). Within a multiple field approach the additional inelastic fields of strain can

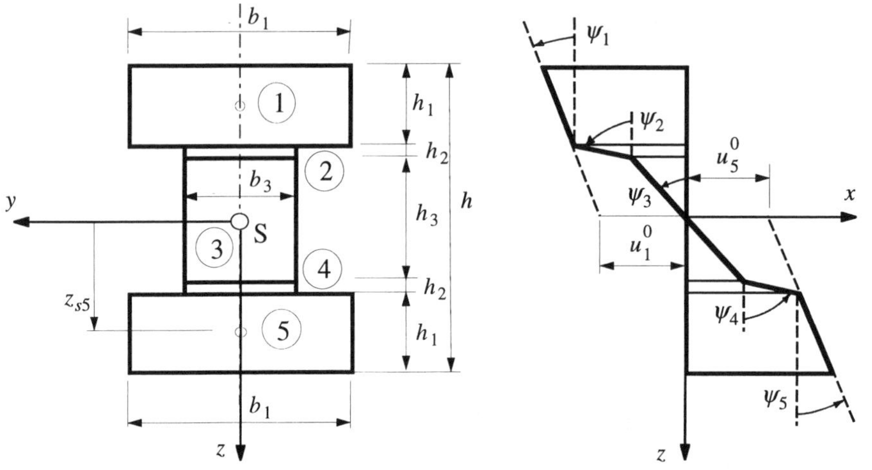

Figure 1. Cross-section and horizontal displacements.

be interpreted conveniently as eigenstrains acting in the background beam (Irschik & Ziegler 1995).

The nonlinear problem turns incrementally in two linear problems, where the first one is simply the response of the associated linear elastic background structure to the given external loads. The second part accounts incrementally for the effects of the physical nonlinearities in the structure. Since the response in both cases is linear within a given time step, solution methods of the linear theory of flexural vibrations are applied.

2 GOVERNING EQUATIONS

Symmetrically designed three layer composite beams of span l in principal bending about the y-axis are considered. Each layer has a constant rectangular cross-sectional area A_i and modulus of elasticity E_i. The adhesive physical interfaces are modeled as additional layers. The investigation is limited to materials, which exhibit the same behavior in tension and compression. The governing equations are derived by applying the assumptions of Timoshenko's theory of shear deformable beams to each individual layer. Consequently, the displacement field in the i-th layer is assumed to be of the form (Yu 1995),

$$u_i = u_i^{(0)} + z\,\psi_i\ ,\quad w_i = w\ ,\qquad i = 1, \ldots, 5\ , \tag{1}$$

where u_i represents the horizontal displacement at vertical distance z from the central axis, $u_i^{(0)}$ is the portion of u_i at $z = 0$, ψ_i denotes the cross-sectional rotation of the i-th layer and w is the lateral deflection common to all layer axes, see Figure (1). The components $u_i^{(0)}$ $(i = 1, 2, 4, 5)$ can be expressed in terms of the cross-sectional rotations in order to satisfy the interface displacement continuity relations and the symmetry requirements (Heuer 1992),

$$u_1^{(0)} = -u_5^{(0)} = u_2^{(0)} + (h_2 + \frac{1}{2}h_3)(\psi_1 - \psi_2)\ ,$$
$$u_2^{(0)} = -u_4^{(0)} = \frac{1}{2}h_3(\psi_2 - \psi_3)\ ,\quad u_3^{(0)} = 0\ , \tag{2.1}$$

$$\psi_1 = \psi_5\ ,\quad \psi_2 = \psi_4\ . \tag{2.2}$$

In Eqs. (1) and (2) the numbers 1 and 5 refer to the upper and lower face, respectively, 3 corresponds to quantities of the core, and 2 and 4 refer to the physical interfaces. h_2 and h_3 denote the thickness of the physical interfaces and of the core, respectively.

Within a geometrically linearized theory, total strains are given by the gradients

$$\varepsilon = u_{i,x}\ ,\quad \gamma = u_{i,z} + w_{,x}\ , \tag{3}$$

and they are directly derived from equations (1) and (2). The comma in the subscripts indicates partial differentiation.

The normal stress component σ_{zz} is neglected. The remaining stress components $\sigma_j \equiv \sigma_{xx\,j}$, $\tau_j \equiv \tau_{xz\,j} \equiv \tau_{zx\,j}$, of the interface layers are related to the strain and shear angle by the generalized Hooke's law, in general the rate form applies,

$$\dot{\sigma}_j = E_j(\dot{\varepsilon}_j - \dot{\bar{\varepsilon}}_j)\ ,\quad \dot{\tau}_j = G_j(\dot{\gamma}_j - \dot{\bar{\gamma}}_j)\ ,\quad j = 2, 4\ , \tag{4}$$

where G_j is the shear modulus and the imposed fields $\bar{\varepsilon}_j$ and $\bar{\gamma}_j$ denote the inelastic strain and inelastic shear angle, respectively, which are assumed to be constant across the small thickness of the adhesive layers.

The equations of motion are derived by considering the free-body diagram of an infinitesimal beam element, loaded by a given transverse force per unit length $q(x,t)$ and an external moment per unit length $m(x,t)$. Conservation of momentum in the z- and x-direction renders, after some algebra, the following coupled set of partial differential equations (Adam & Ziegler 1997a),

$$\sum_{i=1}^{5} \left\{ E_i \left[A_i (u_{i,xx}^{(0)} - \bar{\varepsilon}_{i,x} \delta_{ij}) + B_i \psi_{i,xx} \right] - \rho_i (A_i \ddot{u}_i^{(0)} + B_i \ddot{\psi}_i) \right\} = 0 , \quad (5.1)$$

$$\sum_{i=1}^{5} \left[S_i (\psi_{i,x} + w_{,xx} - \bar{\gamma}_{i,x} \delta_{ij}) - \mu_i \ddot{w} \right] + q = 0 , \quad (5.2)$$

$$\sum_{i=1}^{5} \left\{ E_i \left[B_i u_{i,xx}^{(0)} + C_i (\psi_{i,xx} - \frac{2}{h_2 + h_3} \bar{\varepsilon}_{i,x} \delta_{ij}) \right] - S_i (\psi_i + w_{,x} - \bar{\gamma}_i \delta_{ij}) - \rho_i (B_i \ddot{u}_i^{(0)} + C_i \ddot{\psi}_i) \right\} + m = 0 , \qquad j = 2, 4, \quad (5.3)$$

with

$$A_i = b_i h_i , \quad B_i = b_i \int_{a_i}^{a_{i+1}} z \, dz , \quad (6.1)$$

$$C_i = b_i \int_{a_i}^{a_{i+1}} z^2 \, dz , \quad S_i = \kappa^2 G_i A_i , \quad (6.2)$$

and

$$a_1 = - a_6 = - h / 2 , \quad a_2 = - a_5 = - h_3 / 2 - h_2 , \quad (7.1)$$

$$a_3 = - a_4 = - h_3 / 2 \ . \quad (7.2)$$

In Eqs. (5) δ_{ij} is the Kronecker delta. The factor κ^2 denotes a shear coefficient. The proper choice of its value is discussed by Yu (1995).

An additional set of equations is obtained when the transverse shear stress is specified to be continuous across the interface. Two types of approximations are acknowledged in the literature. If the "correct" shear stress, that is its value expressed by means of the law of conservation of momentum, is used, the in-plane equilibrium of the faces is automatically satisfied. The resulting beam theory is of the sixth order (Di Taranto 1965, Mead 1982). Alternatively, a simplified boundary value can be derived by prescribing the shear stress continuity according to the generalized Hooke's law (Heuer 1992, Adam & Ziegler 1997a),

$$G_j (\psi_j + w_{,x} - \bar{\gamma}_j) = G_k (\psi_k + w_{,x}) , \qquad j = 2, 4 ; \ k = 1, 3, 5 \ . \quad (8)$$

In analogy to the Timoshenko's theory for homogeneous beams also Eq. (8) exhibits the simplified assumption that the shear stress distribution is uniform throughout the layer.

Furthermore, both, the longitudinal as well as the rotatory inertia are neglected,

$$\ddot{u}_i^{(0)} = 0, \quad \ddot{\psi}_i = 0 \ , \quad (9)$$

thus, limiting the analysis to the lower frequency band of structural dynamics.

Eliminating the cross-sectional rotations ψ_i and the horizontal displacements $u_i^{(0)}$ from Eqs. (5) together with Eqs. (2) and (8) renders a fourth-order differential equation for the deflection w (Adam & Ziegler 1997a, 1997b),

$$B_e w_{,xxxx} - \mu \frac{B_e}{S_e} \ddot{w}_{,xx} + \mu \ddot{w} = q - \frac{B_e}{S_e} q_{,xx} + m_{,x} - B_e \bar{\kappa}_{e,xx} + B_e \bar{\gamma}_{e,xxx} \ . \quad (10)$$

Eq. (10) can be interpreted as the equation of a homogeneous shear-deformable background beam with effective flexural stiffness B_e, effective shear stiffness S_e and mass per unit length μ forced by the given loads q and m and by the two fields: the effective inelastic curvature $\bar{\kappa}_e$ and the effective inelastic shear angle $\bar{\gamma}_e$. The distribution and time evolution of the latter are to be determined by considering the constitutive relations. The effective properties and the effective resultant eigenstrains of the beam of Figure 1 are given by

$$\mu = \sum_{i=1}^{5} \rho_i A_i , \quad B_e = \sum_{i=1}^{5} E_i C_i , \quad S_e = \frac{B_e}{\alpha} \delta , \quad (11.1)$$

$$\bar{\kappa}_e = \frac{2 E_2 C_2 \bar{\varepsilon}_2}{B_e (h_2 + h_3)} , \quad \bar{\gamma}_e = \frac{\beta}{B_e} \bar{\gamma}_2 , \quad (11.2)$$

with the abbreviations

$$\delta = \kappa^2 G_3 \sum_{i=1}^{5} A_i , \quad D_i = A_i z_{si} , \quad (12.1)$$

$$\beta = E_2 (D_2 h_3 + 2 C_2) - 2 E_1 D_1 h_2 , \quad (12.2)$$

$$\alpha = G_3 \sum_{i=1}^{5} \frac{E_i}{G_i} C_i + E_1 D_1 [h_3 (\frac{G_3}{G_2} - 1)$$

$$+ (2 h_2 + h_3) (\frac{G_3}{G_1} - \frac{G_3}{G_2})] \qquad (12.3)$$

$$+ E_2 D_2 h_3 (\frac{G_3}{G_2} - 1) .$$

In (12) z_{si} denotes the vertical coordinate of the centeroid of the i-th layer.

In order to obtain a complete analogy to the homogeneous shear-deformable beam, an effective cross-sectional rotation ψ_e is defined by Heuer (1992), which is connected to the layer deformation by means of the total shear force and the bending moment,

$$Q = S_e (\psi_e + w_{,x} - \bar{\gamma}_e) , \quad M = B_e (\psi_{e,x} - \bar{\kappa}_e) , \qquad (13)$$

relations that lead to the following expression for ψ_e (Heuer 1992, Adam & Ziegler 1997b),

$$\psi_e = \frac{1}{B_e} [\alpha \psi_3 + (\alpha - B_e) w_{,x}] + \bar{\gamma}_e . \qquad (14)$$

By means of Eq. (13) and (14) the equation of motion (10) can be separated to form a set of two second order equations,

$$S_e (\psi_e + w_{,x}) - B_e \psi_{e,xx} = m - B_e \bar{\kappa}_{e,x} + S_e \bar{\gamma}_e , \qquad (15.1)$$

$$\mu \ddot{w} - S_e (\psi_{e,x} + w_{,xx}) = q - S_e \bar{\gamma}_{e,x} . \qquad (15.2)$$

Equations (15) describe the higher order problem of a composite beam with inelastic interface slip in full analogy to the lower order engineering theory of a homogeneous inelastic shear beam.

Classical homogeneous boundary conditions are specified in analogy to the homogeneous shear-deformable beam (Adam & Ziegler 1997a).

- Simply supported end:

$$w = 0 , \quad M \equiv EJ_e (\psi_{e,x} - \bar{\kappa}_e) = 0 . \qquad (16)$$

- Free end:

$$M \equiv EJ_e (\psi_{e,x} - \bar{\kappa}_e) = 0 ,$$

$$Q \equiv S_e (\psi_e + w_{,x} - \bar{\gamma}_e) = 0 . \qquad (17)$$

- Partly clamped end:

$$w = 0 , \quad \psi_3 = 0 . \qquad (18)$$

Note, that the clamped-end boundary condition for which $\psi_1 = \psi_2 = \psi_3 = 0$ cannot be formulated in a consistent manner within this effective beam theory.

The coupled set of differential equations (15) is solved considering the actual boundary conditions for the deflection w and the effective cross-sectional rotation ψ_e in an incremental procedure. Subsequently, the cross-sectional rotations of the core and of the faces are to be determined. Decomposition of equation (14) yields the cross-sectional rotation of the core,

$$\psi_3 = \frac{1}{\alpha} [B_e (\psi_e - \bar{\gamma}_e) - (\alpha - B_e) w_{,x}] . \qquad (19)$$

The cross-sectional rotations of the faces and of the interlayer joints are calculated from Eq. (8),

$$\psi_1 = \psi_5 = \frac{G_3}{G_1} (\psi_3 + w_{,x}) - w_{,x} , \qquad (20.1)$$

$$\psi_2 = \psi_4 = \frac{G_3}{G_2} (\psi_3 + w_{,x}) - w_{,x} + \bar{\gamma}_2 . \qquad (20.2)$$

3 CONSTITUTIVE RELATIONS

From the constitutive equations the necessary amount of equivalent eigenstrains $\Delta\varepsilon$, $\Delta\gamma$ has to determined. In the following, elastic-viscoplastic materials are considered that are based on yield surface $F \geq 0$, see Fotiu (1990),

$$F = \frac{\tilde{\eta}_e^2}{k^2} - 1 , \qquad (21)$$

with the following quantities for the state of plane stress $\sigma \neq 0$, $\tau \neq 0$,

$$\tilde{\eta}_e^2 = \tilde{\sigma}^2 + 3 \tilde{\tau}^2 . \qquad (22)$$

In (21) k is the radius of the static yield surface $F = 0$. The rate dependent plastic strain and plastic shear angle are derived from F by a viscoplastic law that is similar to that of Perzyna (1963),

$$\dot{\varepsilon}^p \equiv \dot{\varepsilon}^p_{xx} = \Gamma k \langle \Phi(F) \rangle \frac{\partial F}{\partial \sigma} , \quad \dot{\gamma}^p = 2 \Gamma k \langle \Phi(F) \rangle \frac{\partial F}{\partial \tau} ,$$

$$\Phi(F) = F^m , \qquad (23)$$

where m and Γ are viscosity parameters. The brackets $\langle . \rangle$ have the following meaning: $\langle \Phi(F) \rangle = 0$ for $F \leq 0$ and $\langle \Phi(F) \rangle = \Phi(F)$ for $F > 0$.

The flow rule Eq. (23) determines the plastic strain rate, the plastic dilatation rate and the plastic shear angle,

$$\dot{\varepsilon}^p = 2\,\lambda\,\tilde{\sigma}\,,\ \dot{\gamma}^p = 12\,\lambda\,\tilde{\tau},\ \lambda = \frac{\Gamma}{k}\frac{\langle \Phi(F)\rangle}{(1-D)}\,. \tag{24}$$

Ductile damage is introduced via the concept of effective stress, see also Lemaitre (1992), by substituting the effective stress, e.g. $\tilde{\sigma} = \sigma/(1-D)$. The damage parameter D, $0 \le D \le 1$, of value 0 denotes non-fractionated material, an amount of 1 denotes material failure. Ductile damage is approximated by the following evolution equation,

$$\dot{D} = D\,(1-0.5\,D)\left[\alpha_1\,\tilde{\sigma}^2 + \alpha_2\,\tilde{\tau}^2\right]^n \dot{\varepsilon}^p_e\,, \tag{25}$$

where α_1, α_2 and n are material constants, ε^p_e denotes the equivalent plastic strain,

$$\varepsilon^p_e = \sqrt{\left[(\varepsilon^p)^2 + \frac{1}{3}\,(\gamma^p)^2\right]}\,. \tag{26}$$

For the incremental formulation the values of Eqs. (24) and (25) at t_{a+1} are expanded into Taylor series at t_a, retaining only terms up to the first order. The equivalent increment of the eigenstrain $\Delta\bar{\varepsilon}$ is obtained from the equivalence

$$\Delta\left[(1-D)\,(\varepsilon - \varepsilon^p)\right] = \Delta\varepsilon - \Delta\bar{\varepsilon}\,. \tag{27}$$

Subsequently, Eq. (27) is solved for the eigenstrain increment,

$$\Delta\bar{\varepsilon} = (1 - D_a - \Delta D)\,\Delta\varepsilon^p + (D_a + \Delta D)\,\Delta\varepsilon + (\varepsilon_a - \varepsilon^p_a)\,\Delta D. \tag{28}$$

In a similar procedure the increment $\Delta\bar{\gamma}$ is determined,

$$\Delta\bar{\gamma} = (1 - D_a - \Delta D)\,\Delta\gamma^p + (D_a + \Delta D)\,\Delta\gamma + (\gamma_a - \gamma^p_a)\,\Delta D. \tag{29}$$

Eqs. (28), (29) exhibit the coupling of the elastic and the plastic strain fields due to ductile damage as pointed out by Dafalias (1977). Note that Eqs. (28), (29) are still activated during unloading.

4 NUMERICAL SOLUTION

In the present analysis the solution of the coupled set of equations of motion (15) is derived by superposition of two linear elastic contributions,

$$w = w^e + w^*\,,\ \ \psi_e = \psi^e_e + \psi^*_e\,, \tag{30}$$

where w^e, ψ^e_e are the deflection and effective cross-sectional rotation, respectively, of the elastic homogenized background beam due to the given loads q and m, while w^*, ψ^*_e are the deflection and effective cross-sectional rotation, respectively, produced in the homogeneous background by the imposed effective curvature $\bar{\kappa}_e$ and by the imposed effective shear angle $\bar{\gamma}_e$. The response w^e, ψ^e_e thus is evaluated in advance by the well-known procedure of linear elastodynamics. In simple cases that linear response of the linear elastic background beam is given in analytic form for the whole observation time. Since the distribution of imposed fields of eigenstrains in the background beam is not known in advance and depends on the current state of overall stress and strain, w^*, ψ^*_e have to determined incrementally by stepping the time and updating the strength of the effective eigenstrains iteratively in each time step.

In general, the solution of the equations of motion (15) exists in a modal form. This representation of the solution, and especially of its derivatives, is not desirable because the quasistatic response may contain singularities and discontinuities due to sudden changes in the load history, which are poorly modeled through a finite modal series approximation. Hence, a separate treatment of the quasistatic and of the complementary dynamic response gives higher numerical accuracy for the same number of summations. Further, the analysis is based on an integral equation formulation. The complementary dynamic increments of the deformation due to effective eigenstrains are obtained from the following integral equation (Adam & Ziegler 1997a),

$$\Delta w^*_D(x) = \int_l \int_0^{\Delta t} \left[\tilde{M}^q_D(\xi,x,\Delta t-\bar{\tau})\,\Delta\ddot{\bar{\kappa}}_e(\xi,\bar{\tau}) + \tilde{Q}^q_D(\xi,x,\Delta t-\bar{\tau})\,\Delta\ddot{\bar{\gamma}}_e(\xi,\bar{\tau})\right] d\tau\, d\xi + \Delta w^{0*}_D(x)\,, \tag{31.1}$$

$$\Delta\psi^*_{e\,D}(x) = \int_l \int_0^{\Delta t} \left[\tilde{M}^m_D(\xi,x,\Delta t-\bar{\tau})\,\Delta\ddot{\bar{\kappa}}_e(\xi,\bar{\tau}) + \tilde{Q}^m_D(\xi,x,\Delta t-\bar{\tau})\,\Delta\ddot{\bar{\gamma}}_e(\xi,\bar{\tau})\right] d\tau\, d\xi + \Delta\psi^{0*}_{e\,D}(x)\,. \tag{31.2}$$

In (31) the complementary dynamic Green's functions $\tilde{M}^q_D(\xi,x,\Delta t)$ and $\tilde{Q}^q_D(\xi,x,\Delta t)$ are the bending moment and the shear force in the cross-section at ξ and at time Δt that are produced by an instantaneous transverse unit force in x applied at time $\Delta t = 0$, respectively. A superscript m denotes to Green's functions due to external unit moments. The complementary dynamic Green's functions of undamped vibrations of such a homogeneous shear beam with effective properties are given by infinite modal series, see e.g. Adam & Ziegler (1997a). The functions Δw^{0*}_D, $\Delta\psi^{0*}_{e\,D}$ represent the portion of the deformation referring to the initial conditions to be prescribed at the beginning of the time interval $\Delta t = 0$, at time instant t_a (Adam 1997),

$$\Delta w_D^{0*}(x) = \mu \int_l [\, \dot{\tilde{w}}^q(\xi,x,\Delta t)\, \overset{*}{w}_D(\xi,t_a) + \tilde{w}^q(\xi,x,\Delta t)\, \dot{\overset{*}{w}}_D(\xi,t_a)\,]\, d\xi\,, \tag{32.1}$$

$$\Delta \psi_{eD}^{0*}(x) = \mu \int_l [\, \dot{\tilde{w}}^m(\xi,x,\Delta t)\, \overset{*}{w}_D(\xi,t_a) + \tilde{w}^m(\xi,x,\Delta t)\, \dot{\overset{*}{w}}_D(\xi,t_a)\,]\, d\xi\,. \tag{32.2}$$

Integral equations (31) are solved assuming a linear variation of the imposed effective eigenstrains $\Delta \kappa_e$, $\Delta \gamma_e$ within the time increment $\Delta t = t_{a+1} - t_a$. Such a separation of functional time and space dependence is possible for sufficiently small increments of the eigenstrain resultants: thus a spline function $g(\bar{t})$, commonly a polynomial, is selected that must be at least of the first order

$$\Delta \bar{\kappa}_e(\xi,\bar{t}) = \Delta \bar{\kappa}_e(\xi)\, g(\bar{t})\ ,\quad \Delta \bar{\gamma}_e(\xi,\bar{t}) = \Delta \bar{\gamma}_e(\xi)\, g(\bar{t})\ , \qquad \bar{t} = t - t_a\ , \tag{33}$$

$$g(\bar{t}) = \begin{cases} 1\,, & \bar{t} \geq \Delta t \\ \bar{t}\,/\,\Delta t\,, & 0 \leq \bar{t} \leq \Delta t \\ 0, & \bar{t} \leq 0 \end{cases}\ . \tag{34}$$

With the linear time shape function $g(\bar{t})$ the time integration in Eqs. (31) can be carried out explicitly within Δt using the properties of the Dirac delta function.

In a first step, i.e. in the first time interval, the linear elastic increments $\Delta \boldsymbol{w}^e$, $\Delta \boldsymbol{\psi}_e^e$ due to the given loads are determined. The second step consists of the iterative computation of the increments of the effective eigenstrains $\Delta \bar{\boldsymbol{\kappa}}_e$, $\Delta \bar{\boldsymbol{\gamma}}_e$. The length of the physical interfaces equal to the length l is divided in elements and the discrete values of the strain increments $\Delta \boldsymbol{\varepsilon}_2$, $\Delta \boldsymbol{\gamma}_2$ are determined. The constitutive equations (28), (29) render a first estimate of the eigenstrain increments $\Delta \bar{\boldsymbol{\varepsilon}}_2^{(1)}$, $\Delta \bar{\boldsymbol{\gamma}}_2^{(1)}$. Further, effective quantities of the eigenstrain distribution $\Delta \bar{\boldsymbol{\kappa}}_e$, $\Delta \bar{\boldsymbol{\gamma}}_e$ are determined elementwise, Eq. (11.2). Substituting these amounts into the corresponding solution equations, static part and Eq. (18), yields a first estimate $\Delta \boldsymbol{w}^{*(1)}$, $\Delta \boldsymbol{\psi}_e^{*(1)}$, and consequently the total deformation $\Delta \boldsymbol{w}^{(1)} = \Delta \boldsymbol{w}^e + \Delta \boldsymbol{w}^{*(1)}$, $\Delta \boldsymbol{\psi}_e^{(1)} = \Delta \boldsymbol{\psi}_e^e + \Delta \boldsymbol{\psi}_e^{*(1)}$. These results are the initial values for the second iterative step. The procedure is repeated and continued in time. The algorithm was found stable in many applications.

5 ILLUSTRATIVE EXAMPLE

The proposed multiple field procedure is applied to a simply supported beam with rectangular cross-section composed of three layers and two adhesive interlayers. The eigenfrequencies ω_n and the normalized mode shapes $\Phi^{(n)}$, $\Psi^{(n)}$ of such a simply supported shear deformable beam with effective stiffness can be found e.g. in Magrab (1979),

$$\omega_n = \alpha_n^2 \sqrt{\frac{B_e}{\mu\, \beta_n}}\ ,\quad n = 1, \ldots, \infty\ , \tag{35.1}$$

$$\alpha_n = \frac{n\,\pi}{l}\ ,\quad \beta_n = 1 + \alpha_n^2 \frac{B_e}{S_e}\ , \tag{35.2}$$

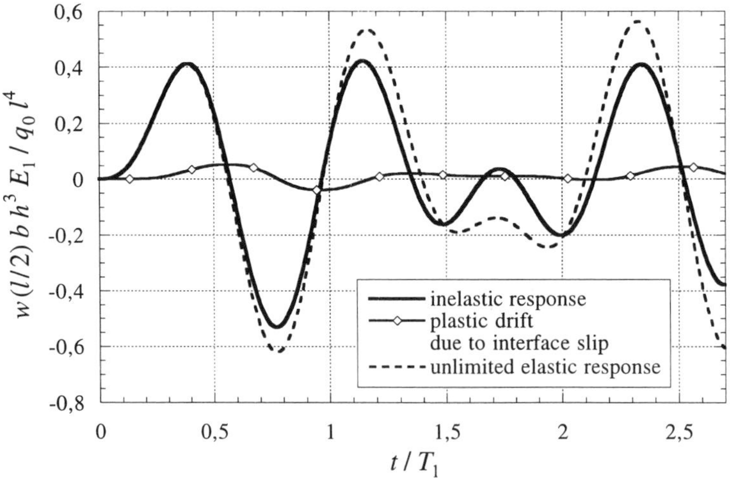

Figure 2. Lateral deflection and plastic drift at mid-span

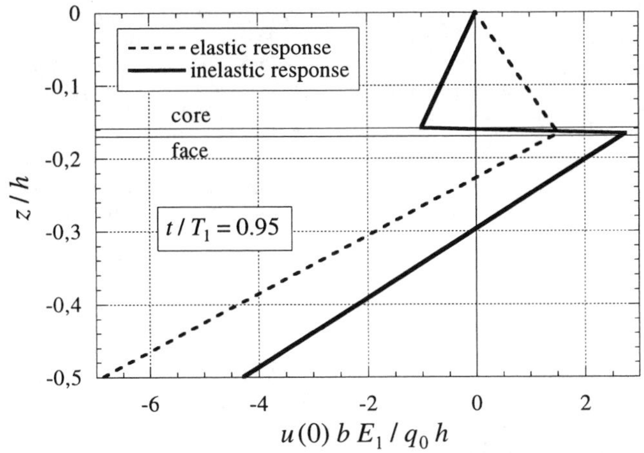

Figure 3. Horizontal displacement at $x = 0$ for specified time instant.

$$\Phi^{(n)}(x) = \frac{1}{\sqrt{l/2}} \sin \alpha_n x \ , \tag{36.1}$$

$$\Psi^{(n)}(x) = -\frac{\alpha_n}{\beta_n \sqrt{l/2}} \cos \alpha_n x \ . \tag{36.2}$$

The formation of interface slip is modeled by considering the ductility of the adhesive layers in the form a material of the ideally elastic - perfectly viscoplastic type, Eq. (24), while, for simplicity's sake, the core and the faces are assumed to remain unlimited elastic. The numerical results are compared with those predicted when the adhesive interlayers behave elastic, likewise to the core material, in the applied range of loading. In all subsequent calculations the mechanical properties of the composite are characterized by the following parameters: longitudinal wave speed $\sqrt{E_j/\rho_j} =$ 5040 m/s, Poisson's ratio $\nu_j = 0.3$, ductility parameter $E_k/k_k = 1000$, viscosity parameters $\Gamma_k T_1 = 4.96 \cdot 10^{-4}$ and $m_k = 0.4$, and the ratios $E_j/E_3 = 23.0$, $\nu_j/\nu_3 = 0.75$, $\rho_j/\rho_3 = 1.94$, $E_3 = E_k$, $\nu_3 = \nu_k$, $\rho_3 = \rho_k$ $(j = 1, 5;\ k = 2, 4)$, and with $T_1 = 2\pi/\omega_1$ denoting the linear fundamental period. Prescribing Young's modulus of the faces by $E_1 = 68.9 \cdot 10^9\ N/m^2$ renders the foregoing parameters such that they refer to a laminate beam, whose faces consist of aluminum and the core of a thermoplastic material. The material of the interface layers has the same elastic stiffness as the core, however, with a much lower yield stress. The dimension of the beam is determined by the ratios $h_1 = h_5 = h/3$, $h_3 = 19h/60$, $h_2 = h_4 = h/120$, $b_i = b$ $(i = 1, .., 5)$, $h/b = 1.5$, and $l/h = 4$. The shear coefficient is chosen to be $\kappa^2 = 0.83$ so that the linear fundamental frequency of the equivalent shear deformable beam and that derived by the plane stress theory takes on the same value. At time instant $t = 0$ the composite is subjected to a distributed load according to a sine-half wave over the span, which varies harmonically in time,

$$q(x,t) = q_0 \sin \frac{\pi x}{l} \sin \nu t \ , \ t \geq 0 \ . \tag{37}$$

The ratio of excitation frequency versus linear fundamental eigenfrequency is chosen to be $\nu/\omega_1 = 1.58$, the amplitude of the applied force is determined by the non-dimensional ratio $\underline{q}_0 = q_0 l^2/(E_1 b h^2) = 52.2 \cdot 10^{-4}$.

Figure 2 shows the normalized time evolution of the plastic drift that is due to the interface slip and of the inelastic mid-span deflection. For comparisons sake the unlimited elastic response is also included. Phase shift and damping by energy dissipation due to the inelastic slip can be observed. Also the distribution of the horizontal displacement at the boundary $x = 0$ (Figure 3) illustrates the jump-effect of the inelastic slip.

CONCLUSIONS

In the present paper, the dynamic response of sandwich beams, where inelastic slip develops in the physical interfaces, is incrementally determined by means of a multiple-field analysis. Materials are considered in the regime of rate-dependent plasticity and are subjected to accumulated damage. Inelastic defects of the material are considered in the linear elastic background beam by a second field of strains (eigenstrains). Layerwise continuous and linear in-plane displacement fields are implemented, and as such, model both the global and the local inelastic response of sandwich beams. The adhesive physical interfaces are modeled as additional layers. An effective cross-section is defined and the complex problem of a layerwise laminate theory reduces to the simpler case of a homogenized shear deformable beam, with effective stiffness and boundary conditions.

REFERENCES

Adam, C. 1997. Dynamics of linear elastic Timoshenko beams with eigenstrains. *Asian Journal of Structural Engineering*: in press.

Adam, C., Heuer, R. & A. Jeschko 1997. Flexural vibrations of elastic composite beams with interlayer slip. *Acta Mechanica*: in press.

Adam, C. & F. Ziegler 1997a. Forced flexural vibrations of elastic-plastic composite beams with thick layers. *Composites Part B* 28B: 201 - 213.

Adam, C. & F. Ziegler 1997b. Composite beam dynamics under conditions of inelastic interface slip In H.A. Mang & F. Rammerstorfer (eds), Proc. IUTAM Symp. on *Discretization Methods in Structural Mechancs II,* Vienna, in press.

Adam, C. & F. Ziegler 1997c. Dynamic response of elastic-viscoplastic sandwich beams with asymmetrically arranged thick layers. In Y.A. Bahei-El-Din & G.J. Dvorak (eds), Proc. IUTAM Symp. on *Transformation Problems in Composite and Active Materials*, Cairo, in press.

Dafalias, Y.F. 1977. Elastic-plastic strain coupling within a thermodynamic strain space formulation in plasticity. *J. Non-Linear Mech.* 12: 327 - 337.

Di Taranto, R.A. 1965. Theory of vibratory bending for elastic and viscoelastic layered finite-length beams. *J. Appl. Mech.* 32: 881 - 886.

Fotiu, P.A. 1990. Die dynamische Wirkung dissipativer Prozesse in Materialien mit Mikrodefekten mit Anwendungen auf Schwingungsprobleme inelastischer Plattentragwerke. Doctoral thesis (in german), Technical University Vienna.

Fotiu, P.A., Irschik, H. & F. Ziegler 1991. Micromechanical foundations of dynamic plasticity with applications to damaging structures. In O. Brüller , V. Mannl and J. Najar (eds), *Advances in Continuum Mechanics*: 338-349. Berlin: Springer.

Girhammar, U.A. & D. Pan 1993. Dynamic analysis of composite members with interlayer slip. *Int. J. Solids Structures* 30: 797 - 823.

Goodman, J.R. & E.P. Popov 1968. Layered beam systems with interlayer slip. *J. Struct. Div., ASCE* 94: 2535 - 2547.

Heuer, R. 1992. Static and dynamic analysis of transversely isotropic, moderately thick sandwich beams by analogy. *Acta Mechanica* 91: 1 - 9.

Irschik, H. & F. Ziegler 1995. Dynamic processes in structural thermo-viscoplasticity. *Appl. Mech. Rev. AMR* 48: 301 - 316.

Kachanov, L.M. 1986. *Introduction to continuum damage mechanics*. Dordrecht: Martinus Nijhoff.

Lemaitre, J. 1992: *A Course on Damage Mechanics*. Berlin: Springer.

Magrab, E.B. 1979. *Vibrations of elastic structural members*. Germantown, Maryland: Sijthoff & Noordhoff.

Mead, D.J. 1982. A comparison of some equations for the flexural vibration of damped sandwich beams. *J. Sound Vibration* 83: 363 - 377.

Mistakidis, E.S., Thomopoulos, K., Avdelas, A. & P.D. Panagiotopoulos 1994. On the non-monotone slip effect in the shear connectors of composite beams. *Int. J. Eng. Analysis and Design* 1: 395 - 409.

Perzyna, P. 1963. The constitutive equations for rate sensitive plastic materials. *Quart. Appl. Math.* 20: 321 - 332.

Reddy, J.N. 1984. A simple higher-order theory for laminated composite plates. *J. Appl. Mech.* 51: 745 - 752.

Reddy, J.N. 1993. An evaluation of equivalent-single-layer and layerwise theories of composite laminates. *Composite Structures* 25: 21 - 35.

Swift, G.W. & R.A. Heller 1974. Layered beam analysis. *J. Engng. Mech. Div. ASCE* 100: 267 - 282.

Whitney, J.M. & N.J. Pagano 1970. Shear deformation in heterogeneous anisotropic plates. *J. Appl. Mech.* 37: 1031 - 1036.

Yu, Y.-Y. 1995. *Vibrations of elastic plates*. New York: Springer.

Ziegler, F. 1995. *Mechanics of solids and fluids*. 2nd ed. New York: Springer.

Damage and Failure of Interfaces, Rossmanith (ed.) © 1997 Balkema, Rotterdam, ISBN 90 5410 899 1

Impact response of a composite plate with a delamination

S.K. Datta
Department of Mechanical Engineering, University of Colorado, Boulder, Colo., USA

J. Zhu & A.H. Shah
Department of Civil and Geological Engineering, University of Manitoba, Winnipeg, Man., Canada

ABSTRACT: Effect of a near-surface delamination on the dynamic response of a laminated plate has been analyzed in this paper. The numerical technique used to model the dynamic response is combines the boundary integral representation and a multi-domain approach. The response is studied in the frequency domain and then in the time domain by using a fast Fourier transform (FFT). The time-harmonic Green's function used in the boundary integral equation (BIE) is evaluated by combining a stiffness method and the modal summation technique. Results presented show that the delamination causes significant changes in the dynamic response and can be used for ultrasonic nondestructive evaluation of delamination defects.

1 INTRODUCTION

Wave propagation and scattering in laminated composite plates are of interest for ultrasonic nondestructive evaluation of defects, material characterization, and studying dynamic response. In recent years considerable progress has been in modeling propagation of ultrasonic guided waves in composite plates. A comprehensive review of current work has been published recently (Chimenti 1997). Scattering of guided waves by cracks in homogeneous and composite plates has been studied before (Al-Nassar et al. 1992, Karunasena et al. 1991.) A hybrid technique combining near-field finite element discretization and far-field modal expansion was used. The problem of scattering by a crack in a plate due to a line source of excitation was investigated using a combination of finite element representation of a finite region containing the crack and a boundary integral solution of the external field (Paffenholz et al. 1990, Liu et al. 1991.) Scattering by a near-surface delamination in a laminated plate has also been analyzed by the same method (Datta et al. 1992, Ju & Datta 1992.) In this paper, scattering by a near-surface delamination in a uniaxial laminated graphite-epoxy plate has been studied using a boundary element method (BEM.)

The primary difficulty associated with the finite element method for crack or delamination problems is the stress field singularity at its tip(s). Accurate approximation of the singularity requires refined discretization, which is computationally time consuming. Working solely with function values at the domain boundary a boundary integral formulation allows circumventing the singularity. However, since the boundary integrals involve singular kernels, the integrations have to be performed with care and the overall accuracy of the boundary element method is largely dependent on the precision with which the various integrals are evaluated. Much research has been done and several techniques have been suggested to accurately evaluate the singular integrals. In the usual application of the BEM, the Green's functions are full space fundamental solutions, and most of the techniques to treat the singular integrals are based on the explicit, analytical expressions of the displacements and stresses as $\sim \ln r$ and $\sim r^{-1}$ for two-dimensional problems, and $\sim r^{-1}$ and $\sim r^{-2}$ for three-dimensional problems, where r is the distance between the source and receiver points.

When BEM is applied to the wave-scattering problem in composite laminates, the application of infinite space Green's function is extremely difficult.

Although Green's functions for of layered media can be expressed as wave-number integrals, the numerical evaluation of these integrals is a difficult task, even for isotropic media (Xu and Mal 1987.) Also, when these Green's functions are used in BEM the time-consuming computations have to be repeated from the beginning for any change in source or receiver points. To overcome this difficulty, the Green's functions in the present paper are computed by using a stiffness method (Dong and Huang 1985.) A multidomain technique developed recently (Zhu et al. 1996) has been used to Cauchy Principal Value (CPV) singular and weakly singular equations arising in the boundary integral equations. It is shown that the BEM used here to obtain the scattered wave fields gives excellent results that agree well with those given by the hybrid method (Zhu et al. 1995.)

2 FORMULATION OF THE PROBLEM

Consider a linearly elastic body of volume V bounded by a regular surface S. The displacement at any point in the frequency domain can be expressed as a boundary integral equation,

$$c_{ij}(\xi)u_j(\xi) = \int_S [G_{ij}(\mathbf{x},\xi)t_j(\mathbf{x}) - H_{ij}(\mathbf{x},\xi)u_j(\mathbf{x})]dS_x + \int_V G_{ij}(\mathbf{x},\xi)f_j(\mathbf{x})dV_x \quad (1)$$

where, u_i and t_i are the displacement and traction vectors, respectively, ξ and $\mathbf{x}$ are respectively the field and source points, and $\mathbf{f}$ is the body force. The Green's displacement tensor $\mathbf{G}$ represents the displacement at point $\mathbf{x}$ due to a unit harmonic point force at ξ. It is assumed that the time dependence is of the form $e^{-j\omega t}$ $(j=\sqrt{-1})$, ω being the circular frequency. $\mathbf{H}$ is the traction at $\mathbf{x}$ due to the same force. The tensor c_{ij} is the well known discontinuity term.

2.1 *Governing equations*

A stiffness method (Dong and Huang 1985) is used to express the Green's tensor $\mathbf{G}$. In this technique, each lamina is subdivided into several sublayers and the displacement at a point within each sublayer is expressed as a quadratic function of a depth variable defined within the sublayer and the generalized coordinates in this representation are the displacements at the top, middle, and the bottom of the sublayer.

It is assumed that each sublayer material is orthotropic with the symmetry axes x, parallel to the sublayer, and z, perpendicular. The stress and strain components for the i-th sublayer are then related by the relations, assuming plane strain conditions,

$$\begin{bmatrix}\sigma_{xx}\\ \sigma_{zz}\\ \sigma_{xz}\end{bmatrix} = \begin{bmatrix}D_{11} & D_{13} & 0\\ D_{13} & D_{33} & 0\\ 0 & 0 & D_{55}\end{bmatrix}\begin{bmatrix}\varepsilon_{xx}\\ \varepsilon_{zz}\\ \varepsilon_{xz}\end{bmatrix} \quad (2)$$

where σ_{ij} and ε_{ij} are the stress and strain components, respectively, and D_{ij} are the elastic stiffness constants. By applying the principle of viral work to each sublayer, a set of approximate differential equations of motion can be established. The equations for the entire plate are then obtained by assembling all the sublayer equations (Karunasena et al. 1991.) In matrix form, the equation is

$$\mathbf{F} = (-\mathbf{K}_1\partial^2/\partial x^2 + \mathbf{K}_2\,\partial/\partial x + \mathbf{K}_3 - \omega^2\mathbf{M})\,\mathbf{Q} \quad (3)$$

where the vectors $\mathbf{Q}$ and $\mathbf{F}$ represent the nodal displacements and the tractions applied at the interfaces of the plate, respectively. The sizes of $\mathbf{Q}$ and $\mathbf{F}$ are $M \times 1$ and the sizes of $\mathbf{K}_i$ (I=1,2,3) and $\mathbf{M}$ are $M \times M$, where $M = 2 \times (2N + 1)$, N being the total number of sublayers in the plate. The matrices $\mathbf{M}$, $\mathbf{K}_1$, and $\mathbf{K}_3$ are real and symmetric, whereas $\mathbf{K}_3$ is real and antisymmetric.

Applying the Fourier transform with respect to x as,

$$\tilde{f} = \int_{-\infty}^{\infty} f(x)e^{-jkx}dx \quad (4)$$

to (3), the governing equation in the transformed domain is found to be

$$\tilde{\mathbf{F}} = (k^2\mathbf{K}_1 + jk\mathbf{K}_2 + \mathbf{K}_3{*})\tilde{\mathbf{Q}} \quad (5)$$

where k is the wavenumber in the x direction, $\tilde{\mathbf{F}}$ and $\tilde{\mathbf{Q}}$ are the Fourier transformed $\mathbf{F}$ and $\mathbf{Q}$, and $\mathbf{K}_3{*} = \mathbf{K}_3 - \omega^2\mathbf{M}$.

2.2 *Eigenvalue problem*

By setting $\mathbf{F} = \mathbf{0}$, the equation for the eigenvalues and eigenvectors is obtained from Eq.(5) and can be arranged in the form,

$$\mathbf{E}_1\Phi_m^R = k_m \Phi_m^R,\ \Phi_m^L\mathbf{E}_2 = k_m\Phi_m^L \tag{6}$$

where

$$\mathbf{E}_1 = \begin{bmatrix} \mathbf{0} & \mathbf{I} \\ \mathbf{K}_1^{-1}\mathbf{K}_3^* & j\mathbf{K}_1^{-1}\mathbf{K}_2 \end{bmatrix}$$
$$\mathbf{E}_2 = \begin{bmatrix} \mathbf{0} & \mathbf{K}_1^{-1} \\ \mathbf{K}_3^* & j\mathbf{K}_2\mathbf{K}_1^{-1} \end{bmatrix} \tag{7}$$

Φ_m^R and Φ_m^L are right and left eigenvectors, respectively, and can be written in the partitioned forms,

$$\Phi_m^L = < \tilde{\mathbf{Q}}_m^L \quad k_m \tilde{\mathbf{Q}}_m^L >$$
$$\Phi_m^R = \begin{bmatrix} \tilde{\mathbf{Q}}_m^R \\ k_m\tilde{\mathbf{Q}}_m^R \end{bmatrix} \tag{8}$$

Here $\mathbf{Q}_m^L$ and $\mathbf{Q}_m^R$ have sizes 1×M and M×1, respectively.

2.3 *Green's function in frequency domain*

Following the modal summation technique (Liu and Achenbach 1995) and making use of the orthogonality of the left and right eigenvectors, we obtain the displacement in the transformed domain as,

$$\tilde{\mathbf{Q}} = \sum_{1}^{2M} k_m\tilde{\mathbf{Q}}_m^L\tilde{\mathbf{F}}\tilde{\mathbf{Q}}_m^R / \{(k_m - k)B_m\} \tag{9}$$

where

$$B_m = \tilde{\mathbf{Q}}_m^L\tilde{\mathbf{Q}}_m^R + k_m^2\,\tilde{\mathbf{Q}}_m^L\mathbf{K}_1\tilde{\mathbf{Q}}_m^R \tag{10}$$

The displacement Green's functions for the plate in the spatial domain due to a load at (x_0, z_0) can be obtained by applying the inverse Fourier transformation to equation (9) as,

$$\mathbf{G}(x\,;x_0, z_0) = 1/2\pi \int_{-\infty}^{\infty} \sum k_m\tilde{\mathbf{Q}}_m^L\mathbf{F}_0\,\tilde{\mathbf{Q}}_m^R e^{jk(x-x)} / \{(k_m - k)B_m\}\,dx \tag{11}$$

where $\mathbf{F}_0$ is a constant vector representing the amplitude of the external force. It is noted that in equation (11), $\mathbf{Q}_m^L$, $\mathbf{F}_0$, $\mathbf{Q}_m^R$, and B_m do not contain k explicitly. Also, of the 2M eigenvalues half (M) correspond to the waves traveling or decaying from the source point towards infinity, white the other half correspond to the waves traveling towards the source or increasing in amplitude. Applying Cauchy's residue theorem and choosing the M modes that travel or decay away from the source, we obtain

$$\mathbf{G}(x\,;x_0, z_0) = \sum k_m\tilde{\mathbf{Q}}_m^L\mathbf{F}_0\,\tilde{\mathbf{Q}}_m^R e^{jk(x-x_0)}/B_m \tag{12}$$

Once the displacements are known from (12), the tractions can be obtained by using the constitutive equations.

3 NUMERICAL IMPLEMENTATION

For numerical evaluation of equation (1), the boundary is represented by a series of elements connected to boundary nodes. With the spatial discretization, writing Eq. (1) for each of the nodes, and allowing field point ξ to coincide sequentially with all the nodal points of the boundary, the global system of equations ia obtained to be,

$$\mathbf{Hu} = \mathbf{Gt} + \mathbf{b} \tag{13}$$

where, matrix $\mathbf{H}$ contains the c_{ij} tensor and the traction kernel integrals, and $\mathbf{u}$ and $\mathbf{t}$ are the displacements and tractions at the nodes, resp. $\mathbf{b}$ is due to the last integral in equation (1) containing the body force. As mentioned before, most of the existing methods to treat the CPV of the singular traction kernel integrals and the weakly singular displacement kernel integrals are based on the properties of the full space Green's functions. These cannot be used here. A new multidomain technique was developed (Zhu et al. 1996) to circumvent the difficulties. Details of the technique will be omitted here and only the numerical results are presented below.

3.1 *Numerical results*

The method discussed above is used to investigate the elastodynamic displacements in a uniaxial graphite-epoxy plate of thickness 2H (5.08 mm) with a delamination, as shown in Figure 1. A vertical load is applied on the top surface of the plate at a distance 3.5H to the left of the origin.

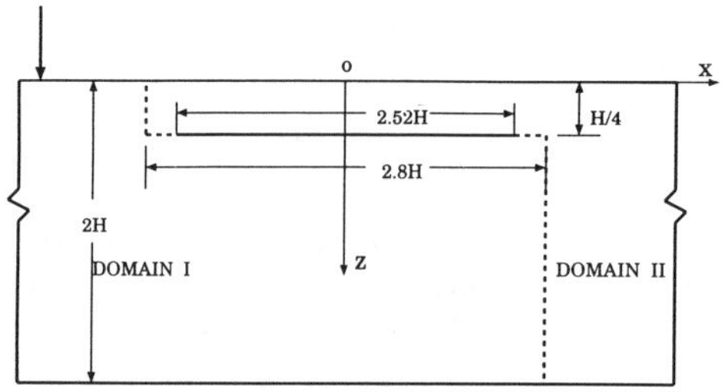

Figure 1.Geometry of the composite plate with delamination.

The relevant material constants characterizing the uni-axial graphite-epoxy laminated plate are (in GPa):
$D_{11} = 160.7$, $D_{33} = 13.92$, $D_{13} = 6.44$, $D_{55} = 7.07$.
The density is $\rho = 1.8$ g/cm^3. Thus, the longitudinal (c_p) and the shear (c_s) wave speeds in the x-direction are respectively 9.45 mm/μs and 1.98 mm/μs. The normalized cut-off frequencies, Ω ($=\omega H/c_s$), for the symmetric S_1 and S_2 modes and the antisymmetric A_1 and A_2 modes are found to be 2.20, 3.14, and 1.57, 4.4, respectively.

In the first step, the plate is divided into sixteen sublayers to compute the eigenvalues and eigenvectors. The multidomain technique (Blanford et al. 1981) is used in implementing the BEM to avoid the difficulties associated with the calculation of the singular and weakly singular integrals. As shown in Figure 1, the plate is divided into two domains by the crack and the fictitious boundaries (dashed lines.) The boundary element mesh is composed of 124 quadratic elements and 204 nodes. Eight crack-tip singular elements are used.

For validation of the method, the problem is also solved by a hybrid method (Karunasena et al. 1991), in which the plate is divided into two domains: the interior domain that contains the delamination and that is bounded by two vertical boundaries at x = -1.4H and x = 1.4H, and the exterior domain composed of the semi-infinite regions x > 1.4H and x < -1.4H. The interior domain is modeled by finite elements and the field in the exterior domain is represented as modal sums. The finite element mesh is composed of 392 quadratic elements and 1280 nodes. The vertical displacements at different depths at two sections of the plate are shown in Figure 2. Here Ω is taken to be 1. It is noted that due to the fairly large finite element discretized region, the accuracy of the hybrid method degenerates with higher frequency. The results in the far field are better, as expected.

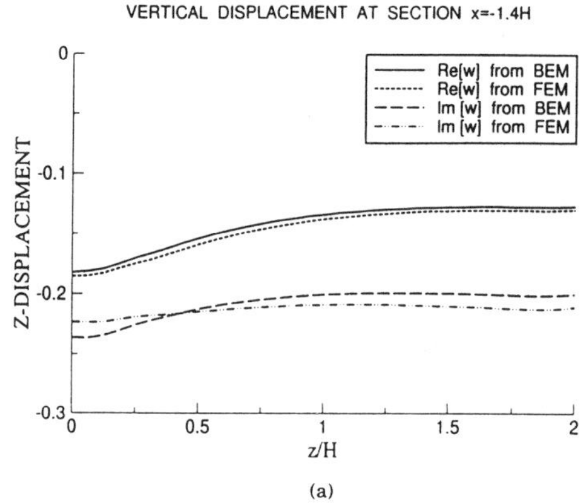

(a)

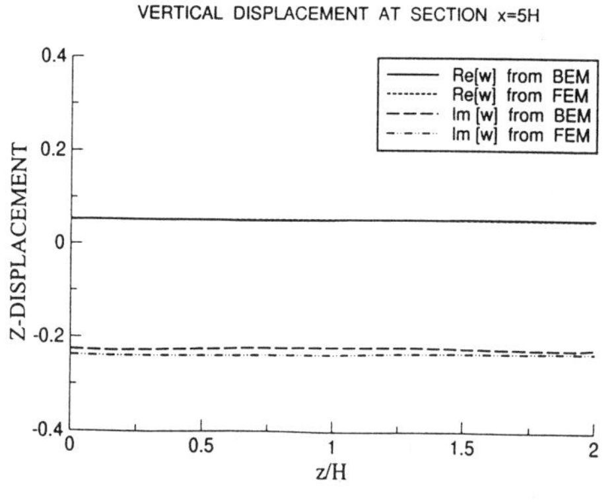

(b)

Figure 2. Vertical displacements through the thickness at two sections of the plate for $\Omega = 1$. (a) x=-1.4H, (b) x=5H.

Figures 3(a) and 3(b) show the frequency response of the top surface of the plate between x=-1.5H and x=1.5H. So this length is right above the delamination. Comparison of the two shows interesting differences in the response between the two cases. The response at the surface above the delamination shows sinusoidal oscillations at high frequencies that are absent when there is no delamination. In the latter case, it is seen that the response shows two pronounced peaks, one at $\Omega = 2.2$, the cut-off frequency of the S_1 mode, and at $\Omega = 4.4$, the cut-off frequency of the A_2 mode. In the presence of delamination, a peak appears at the cut-off frequency of the A_1 mode ($\Omega = 1.57$). This is shown in Figure 4.

The two peaks shown in Figure 4 at x = -0.63H (~ L/4) and x = 0.63H can be used to estimate the delamination length L. Figure 5 shows the frequency

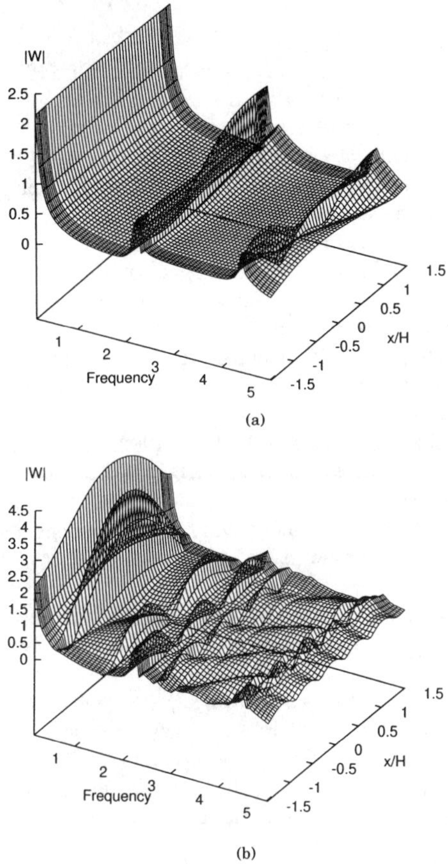

Figure 3. Frequency and spatial dependence of the vertical displacement at points on the top surface above the delamination. (a) without delamination and (b) with delamination.

spectrum of the vertical displacement at x = 0. It is seen that the response with delamination has several sharp peaks, one of which is at a low frequency, Ω = 0.55, which corresponds to the fundamental mode of bending vibration of a simply supported plate of length 2.52H and thickness H/4. This feature was discussed earlier (Keer et al. 1984, Cawley and Theodorakopoulos 1989, and Datta et al. 1992.) This can be used to predict the crack size or depth. The other peaks can be identified with the resonance frequencies of the plate.

Transient response of the top surface was also considered. In this case the load was taken to have time dependence,

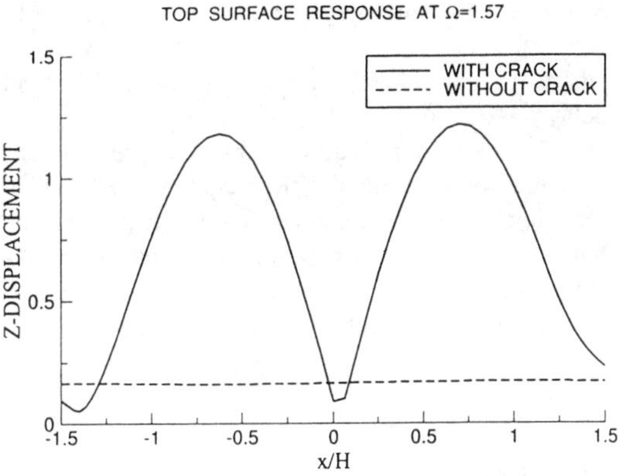

Figure 4. Response of the top surface at $\Omega = 1.57$.

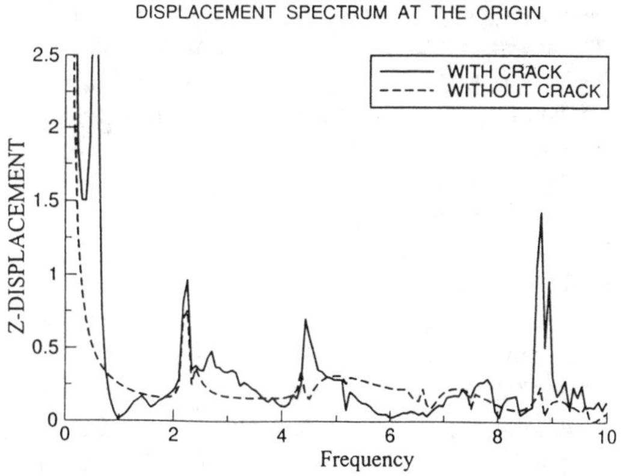

Figure 5. Displacement spectrum at the origin.

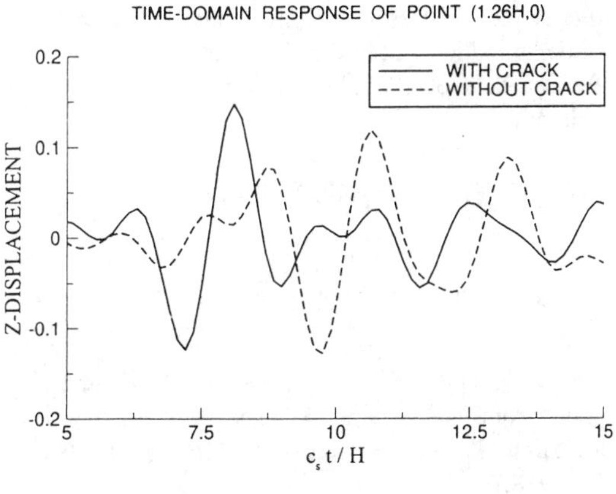

Figure 6. Transient responseof point (1.26H, 0).

$$f(\tau) = 2e^{-(\tau-\tau\ )\ /2} \sin(\Omega_c\tau)/\sqrt{2\pi} \tag{14}$$

where the normalized time is $\tau = c_s t/H$. The time delay $\tau = 3.0$ and $\Omega_c = 3.14$.

It is seen from Figure 6 that the predominant response is due to the Rayleigh wave that is traveling at a speed of 1.8 mm/μs. In the presence of delamination the response is much more complicated because of the interaction with the tips of delamination and multiple reflections between the top surfaces of the plate and the delamination.

4 CONCLUSIONS

Impact response of a laminated composite plate with a near surface delamination has been studied in this paper. A boundary element method coupled with modal representation of guided waves in the plate has been used to obtain the frequency and time domain responses. The Green's functions needed for the boundary element solution are obtained by modal summation. A multi-domain technique developed earlier by the present investigators has been employed to evaluate the Cauchy Principal Value integrals and the weakly singular integrals arising in the boundary integral representation. The frequency dependence of the vertical displacement at the top surface due to a line load shows resonance peaks that are related to the cut-off frequencies of the plate as well as to the plate region between the top surfaces of the plate and the delamination. These can be used to find the depth and the length of delamination. The transient response of the plate at the top of the delamination shows complicated interaction between waves diffracted by the two tips of delamination and multiple reflections between the free surface of the plate and the top surface of delamination. This study demonstrates the significant effects a weak interface has on the dynamic response of a composite plate.

5 ACKNOWLEDGMENT

The work reported here was supported in part by a grant from the Natural Science and Engineering Research Council of Canada (NSERC grant # 0GP-0007988.)

REFERENCES

Al-Nassar, Y.N., S.K. Datta & A.H. Shah 1991. Scattering of Lamb waves by a normal rectangular strip weldment. *Ultrasonics*. 29:125-132.

Blanford, G.E., A.R. Ingraffea & J.A. Liggett 1981. Two-dimensional stress intensity factor computations using the boundary element method. *Int. J. Numer. Methods Eng.* 17:387-404.

Cawley, P. & C. Theodorakopoulos 1989. The membrane resonance method for nondestructive testing. *J. Sound Vib.* 130:299-311.

Chimenti, D.E. 1997. Guided waves in plates and their use in material characterization. *Appld. Mech. Rev.* 50:247-284.

Datta, S.K., T.H. Ju & A.H. Shah 1992. Scattering of an impact wave by a crack in a composite plate. *J. Appl. Mech.* 59:596-603.

Dong, S.B. & K.H. Huang 1985. Edge vibrations in laminated composite plates. *J. Appl. Mech.* 52:433-438.

Ju, T.H. & S.K. Datta 1992. Pulse propagation in a laminated composite plate and nondestructive evaluation. *Comp. Engrg.* 2:55-66.

Karunasena, W.M., A.H. Shah & S.K. Datta 1991. Plane-strain-wave scattering by cracks in laminated composite plates. *J. Engrg. Mech.* 117:1738-1754.

Keer, L.M., W. Lin & J.D. Achenbach 1984. Resonance effects for a crack near a free surface. *J. Appl. Mech.* 51:65-70.

Liu, G.R. & J.D. Achenbach 1995. Strip element method to analyze wave scattering by cracks in anisotropic plates. *J. Appl. Mech.* 62:607-613.

Liu, S.W., S.K. Datta & T.H. Ju 1991. Transient scattering of Rayleigh-Lamb waves by a surface-breaking crack: comparison of numerical simulation and experiment. *J. Nondestr. Eval.* 10:111-126.

Paffenholz, J., J.W. Fox, X. Gu, G.S. Jewett, S.K. Datta & H. Spetzler 1990. Experimental and theoretical study of Rayleigh-Lamb waves in a plate containing a surface-breaking crack. *Res. Nondestr. Eval.* 1:197-217.

Xu, P.C. & A.K. Mal 1987. Calculation of the Green's functions for a layered viscoelastic solid. *Bull. Seism. Soc. Am.* 77:1823-1837.

Zhu, J., S.K. Datta & A.H. Shah 1995. Modal representation of transient dynamics of laminated plates. *Comp. Engrg.* 5:1477:1487.

Zhu, J., A.H. Shah & S.K. Datta 1996. The evaluation of Cauchy principal value integrals and weakly singular integrals in BEM and their applications. *Int. J. Num. Methods Eng.* 39:1017-1028.

Damage and Failure of Interfaces, Rossmanith (ed.) © *1997 Balkema, Rotterdam, ISBN 90 5410 899 1*

Impact damage and post impact residual strength of pultruded composite beams

T.J.Chotard & M.L.Benzeggagh
Université de Technologie de Compiègne LG2mS., UPRES, CNRS, France

ABSTRACT: This paper reports detailed results about impact damage and residual failure behavior of some glassfibre/polyester pultrusion structures. The impact aspects studied include the damage analysis and the impact behavior of these shapes. The influence of test parameters such as impact velocity and impactor size was emphasized as well as the type of damage observed. Four-point flexural tests where performed on box and 'U'-beam sections in order to determine their post-impact residual strength. The acoustic emission was used to follow the evolution of damage during flexural tests. Several types of failure mode were observed on damaged and undamaged specimen. A strong correlation between the profile geometry and the post impact failure mode of these beams was found.

1 INTRODUCTION

Over the past ten years, analysis of fiber reinforced composite materials working processes shows pultrusion gaining an ever-increasing share of the market (Sumerak et al, 1985). Pultrusion is one of the fastest and most cost-effective process by which composite can be manufactured. Recently, the uses of pultruded composites have proliferated to include a number of new structural applications. These composites structures have desirable properties in corrosive and chemical environment. For all these reasons, pultruded composite structures are being more and more competitive with steel and concrete for construction. With the increasing possibilities of pultruded composites into construction industry, the need of full characterization of these materials become essential (Sims et al, 1987, Bank, 1987, Bank et al 1987, Bank, 1989, Barbero et al, 1993, Davalos et al, 1996). As these sections are frequently loaded in flexure, there is a strong dependence of performance on section shapes as well as on material composition (Chotard et al, 1996b). Today, knowledge about pultruded composites increasingly takes their dynamic properties into account and especially the impact behavior of these structures (Svenson et al, 1992,1993, Chotard et al 1996a).

In this study, several types of Glass / polyesters pultruded beams were tested under low velocity impact solicitation with different parametric configurations. both destructive and non-destructive damage analysis were carried out to evaluate the influence of impact velocity and impactor diameter on damage shape. Afterward, 4-point flexural tests were performed on these structures in order to determine their residual flexural strength.

2 EXPERIMENTAL

2.1 Materials

Four pultruded structures; called Pul1B, Pul1G, Pul3G and Pul4G, were examined in this work. Two geometrical types have been investigated, UPN geometry for Pul1 and Pul3 and boxbeam for Pul4. (see fig 1a & b). Several types of matrix (resins+filler) were also investigated (B and G). Two different types of Pultruded stacking sequence were considered. These sequences were constituted with a Continuous strand Mat (M) and Unidirectional Roving fibers (R). The thickness of these sequences were 5 mm for the M/R/M sequence and 8 mm for the M/R/M/R/M sequence. The localization of the different reinforcements in the cross section is shown in Figure 1a..

The matrix set is constituted with a mixture of two different kinds of polyester unsaturated isophtalic resins named here for confidential reasons resin I and resin II. In addition to this matrix mix, several type of additives (fillers, pigment and other retardant or absorbers) are joined to the matrix composition in order to prevent environmental degradation (UV's aggression etc.). The separation of the fillers from the matrix was performed to evaluate the type and the proportions of filler in the matrix. The two different identified filler sets called here α and β are quit similar in composition (majority of calcium carbonates) but different in quantities in the global matrix set. A summary of these data is presented in Table 1.

Table 1 : Materials configurations

Profile	Sequence / Vf	Profile geometry	Filler group
Pul1B (white matrix)	M/R/M/R/M 0.468±3%	UPN 90x35x8x8 mm	α
Pul1G (grey matrix)	M/R/M/R/M 0.458±2%	UPN 90x35x8x8 mm	β
Pul3G (grey matrix)	M/R/M 0.438±2%	UPN 50x55x5x5 mm	β
Pul4G (grey matrix)	M/R/M 0.462±3%	Boxbeam 50x50x5 mm	β

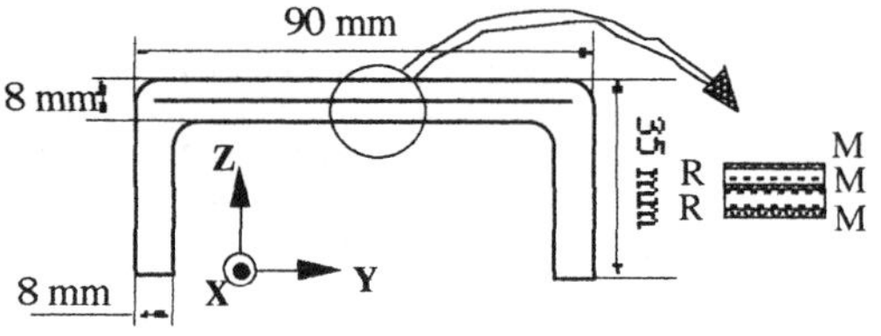

Figure 1a: Pul1 cross section and pultrusion sequence

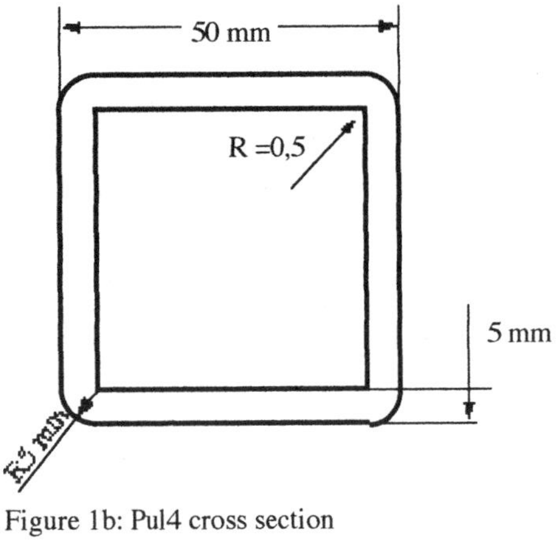

Figure 1b: Pul4 cross section

The difference of filler quantities can have a great influence on the material failure modes under static solicitation such as tension and bending (Chotard & al, 1996b).

2.2 Impact testing

Low velocity impact tests were conducted using a conventional drop weight fixture (see fig.2). Here, several impactor weights with different diameter hemi-spherical noses were allowed to fall freely from a height varying from 0 to 2m. The global fixture was modified to allow clamped device configuration and to

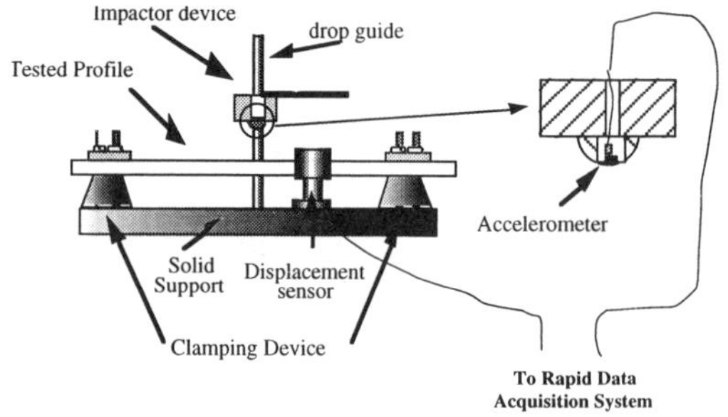

Figure 2 : Details of Drop-weight fixture

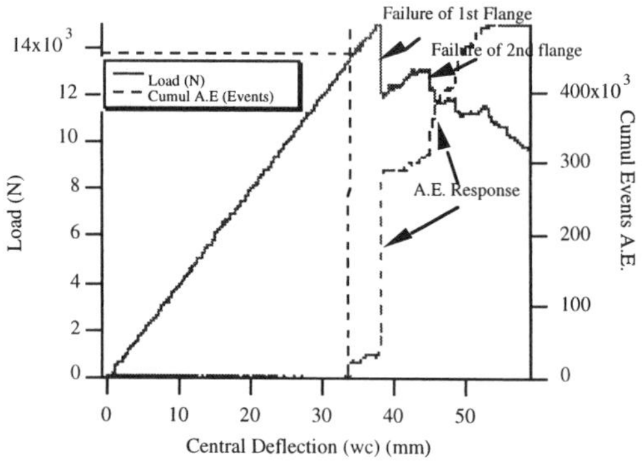

Figure 3: A.E response of Pul1G specimen during 4-point bending test.

perform long span impact tests. The acceleration while impact duration is measured by an Endevco 2225 accelerometer affixed to the impactor nose which can operate in the range of ±20000 g. The displacement of the mobile weight is measured by a magnetic displacement sensor clamped on the solid support. Both the specimen length (1200mm) and the profile clamped configuration (lateral impact, 900 mm span between clamping device) were chosen in order to be representative of the industrial use of these structures (safety structures, load bearing structures in civil engineering).

Iso-velocity and iso-energy tests were carried out on clamped specimen. The span between clamping device was 900 mm. Three levels of energy (50J, 80J and 110 J) and three levels of impact velocity (4m/s, 5m/s and 6m/s) were considered for impact tests. In addition, three different diameters of impactor (Big Impactor (BI) = 50 mm, Medium Impactor (MI) = 25 mm and Small Impactor (SI) = 12.5 mm) were used in order to evaluate the influence of the diameter nose on damage shapes.

2.3 Four-point bending tests

Flexural tests were performed on both undamaged and impacted specimen on an Instron 1186 universal testing machine equiped with four point bending fixture at crosshead speed of 2mm/min. Bending tests

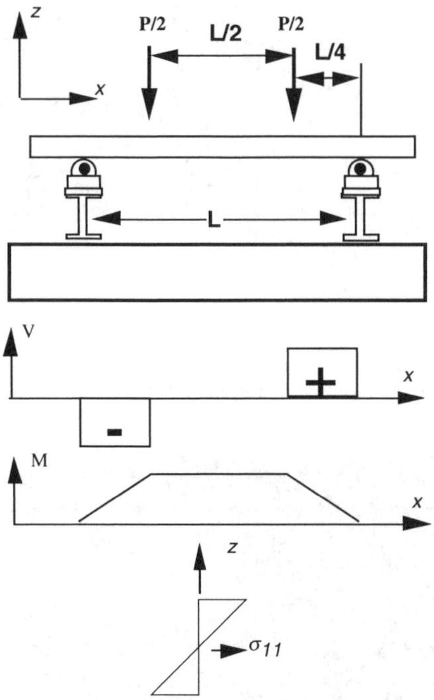

Figure 4 : Load configuration for 4 points bending test with Shear force and Bending moment diagrams and longitudinal stress distribution through-the-thickness.

were also undertaken on a series of 3 specimen of each profile. The acoustic emission technique was used to follow the evolution and the type of damage encountered during the test (e.g. fig. 3).

The total span (L) was chosen to be the same than impact configuration (900 mm). The flexural test configuration (four-point bending test) was chosen in order to have both a pure bending moment in the central part of the tested profile and a pure shear solicitation (negative and positive) at the extremities. This load configuration is presented in figure 4.

3 RESULTS AND DISCUSSION

3.1 Impact damage

Damage identification was carried out with both destructive and several non destructive evaluation (NDE) methods such as US C-Scan and X-ray tomography. These techniques allow the impact damage shape through-the-thickness of the specimen to be identified and the delamination area to be determined (De belleval et al, 1996). The ultrasonic cartography has been correlated with destructive testing. In order to better determine the shape and the extend of internal damage, a post-mortem analysis was carried out. Several transverse cross sections were cut in the specimen at different distances from the impact point. The fissuration was then visualized with dye. A very good correlation was found between real and C-Scan representation of internal damage extension (Potel et al, 1996). The next figure shows the typical damage shape for Pul1B specimen impacted at 110J / 6m/s with Big Impactor (BI).

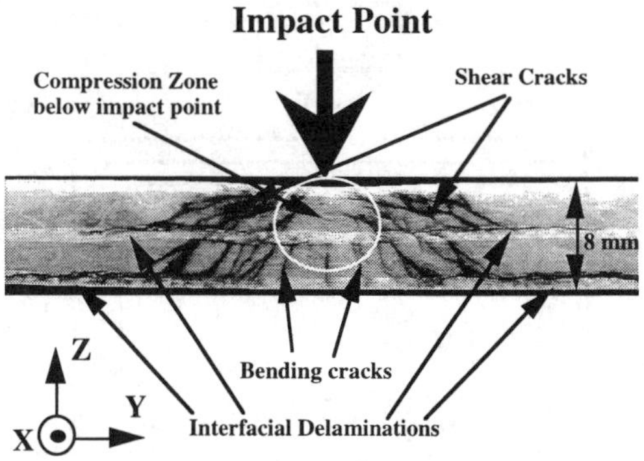

Figure 5 : Damage shape on Pul1B specimen impacted at 110 J with a big impactor (Diameter = 50 mm) (taken below impact point)

As we can see in this figure, the damage shape is different from side to side of the intermediate mat layer. In the upper part, the cracking pattern (α crack near from 45°) shows that a majority of shear solicitation due to high shear stress field is taking part in this zone near the edge of the impactor. Moreover, we can observe a small undamaged and compacted area localized right below the impact point. This area is strongly influenced by the compression stress field generated by the impactor burst. In the lower part, The type of effective solicitation is different. The cracking direction is more vertical. This orientation is specific to a mode I solicitation due to the transverse bending of the profile. Furthermore, large delamination areas are observed. These areas take place at the interface between Mat and Roving layers at intermediate location and in the lower part of the specimen. These locations seem to be the weak points

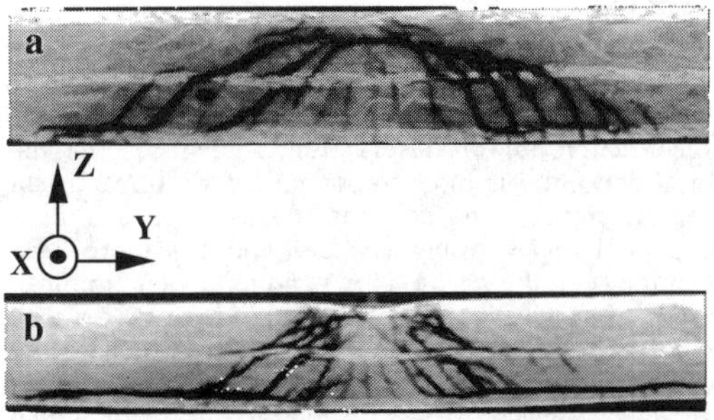

Figure 6: Views of transverse sections, taken below impact point, of 80 J / 5m/s impacted Pul1B specimen (a: with Big Impactor, b: with Small Impactor).

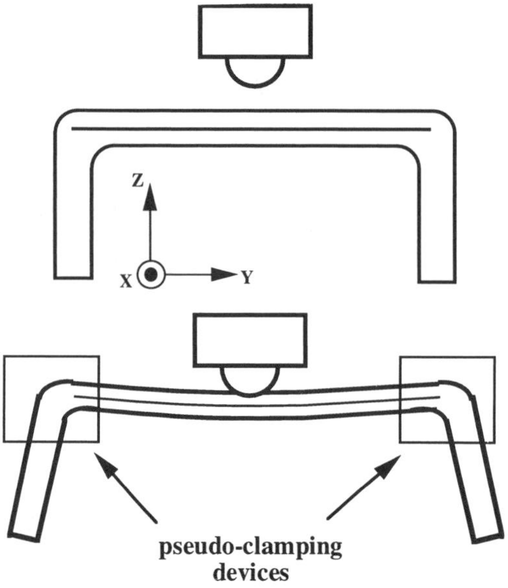

Figure 7 : Schematic representation of the clamping role played by the corner of the profile.

of the material. Although that Mat layers stop the moving forward of the intralaminar cracks initiated in the Roving layers, the multiplication of these interfaces in the material could affect the global integrity of the structure after impact.

Figure 6 shows the influence of the impactor size on the damage shape of 80 J impacted Pul1B specimen. These pictures have been taken from transverse cross sections cut below the impact point. As we can see, the damage pattern is very different for SI and BI impact. In both pictures, the delamination areas are present at the same place as explained before, but it seems that a larger delamination area occurs when Pul1B profile is impacted by a small impactor (SI).These observations have been already reported (Chotard & al, 1996 a) and correlated with energy absorption quantification during this kind of impact tests. Others information can be extracted from these pictures. The compression zone is still present for 80J SI impact but not in the 80J BI impact. This area has been replaced by a web of shear cracks in the upper zone. Furthermore, the delamination zone is localized nearer the lower Mat / Roving interface for the 80J / SI than for the 80J / BI specimen. From this observation, we can say that the local deflection is more important for SI impact than for BI impact. In fact, the shape is loaded in a structural way with big impactor and in a more local fashion with the small. This bending solicitation lead to a high tensile one. With the tensile strength mismatch between Roving layer (transversally considered) and Mat layer, a high shear stress field appears between these two different layers which leads to delamination near the rear face of the specimen. In addition, the geometry of the specimen

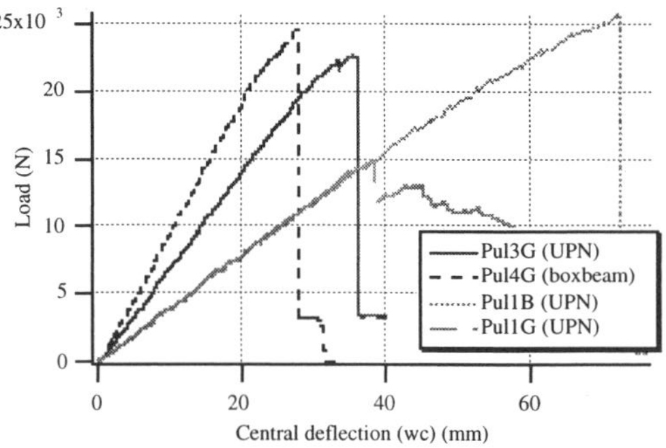

Figure 8 : Load-displacement curves of undamaged profiles under flexural solicitation.

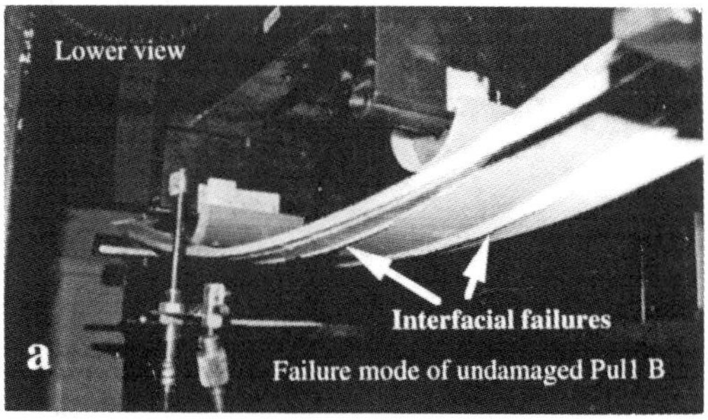

Figure 9 a: Failure modes of undamaged Pul1B specimen

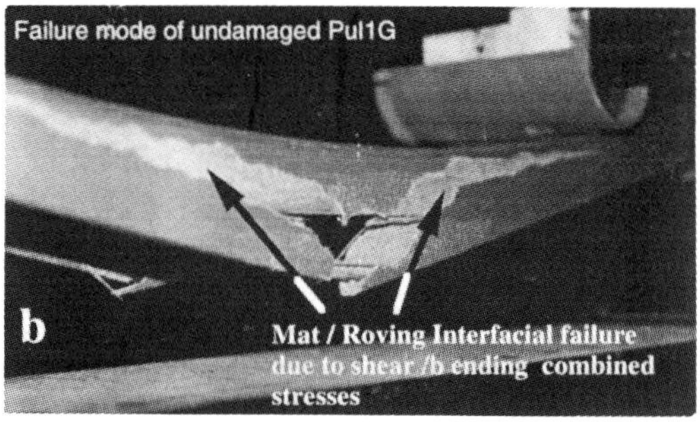

Figure 9 b: Failure modes of undamaged Pul1G specimen.

have a great influence on this shear field. Indeed, the corners of the profile play the role of a pseudo clamping system as showed in Figure 7

We will see later that this difference in damage shapes could have important effects on both residual flexural performance and behavior of these structures. The SI generated a more localized damage instead of the BI one which develops a more extended delamination zone near impact point. This zone will be more sensitive to a buckling mode generated by a compressive solicitation

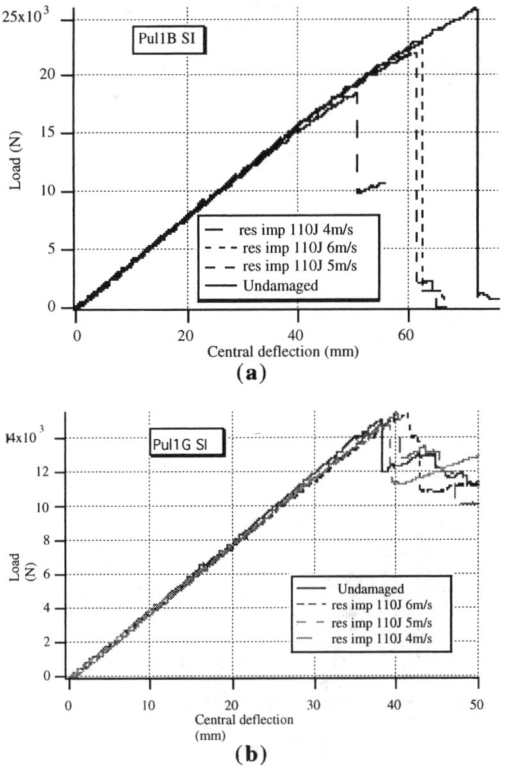

Figure 10 : Residual Load-deflection curves for Pul1 B (a) and Pul1G (b) (Impact at 110 J with SI at varying impact velocities)

3.2 Residual Flexural tests

Firstly, flexural tests were performed on undamaged specimens in order to determine their initial flexural behavior. The next figure presents the load-displacement curves of the four tested shapes.

In this figure, we can easily see that the geometry of the considered profiles plays an important role in the flexural behavior of these structures. Moreover, the open / close geometry aspect, for example close geometry for Pul4G and open geometry for Pul3G, has to be taken into account to explain the failure modes of these shapes. Here, we can observe a major difference in the value of the maximum load between Pul1G and Pul1B specimen. This difference is confirmed by the two types of failure mode encountered in these two profiles both for impact damaged and undamaged specimens (see fig. 9 a & b)

Here, although the tested profiles got strictly the same geometry, the observed failure modes are totally different for Pul1B and Pul1G specimens. The same type of failure modes were observed for both these profiles in residual flexural tests.

The residual behavior of the Pul1B profile seems to be more influenced by impact load than the Pul1G one. The reduction of the residual strength reached a magnitude of 30 % for an impact event at 4 m/s with a small impactor. This reduction is less important for higher velocities (near 10% at 5 and 6 m/s). This observation has also been done for other impactor sizes. This characteristic has not been noted for the Pul1G profile. Indeed, all the post-impact residual tests did not report any reduction of the flexural strength for this shape. This could be explain by the catastrophic shear failure of the flanges which dominate the global behavior of this profile. It seems that, as already noted, the weak point of this structure is the bad Mat/Roving interface strength under combined bending / shear solicitation (e.g. fig. 9 b). This could be explained by the difference in the matrix composition between the B matrix and the G matrix. A recent study (Chotard et al, 1996b) has emphasized the influence of resin additives on the failure behavior of pultruded composite materials.

These curves show the influence of the impact velocity for a given energy level and a given impactor size (here, small impactor, SI). The flexural failure

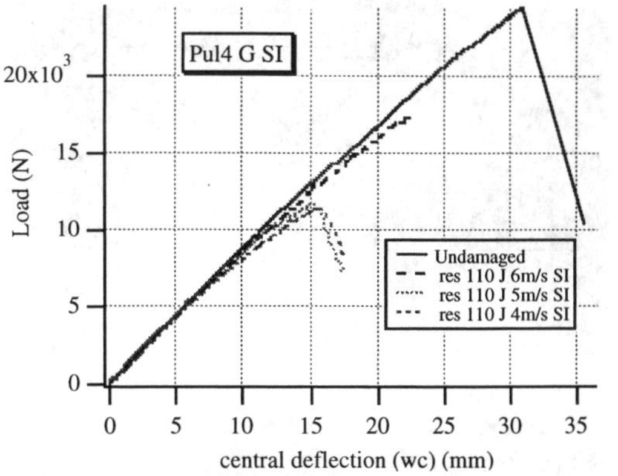

Figure 11 : Residual Load-deflection curves for Pul4 G (Impact at 110 J with SI at varying impact velocities)

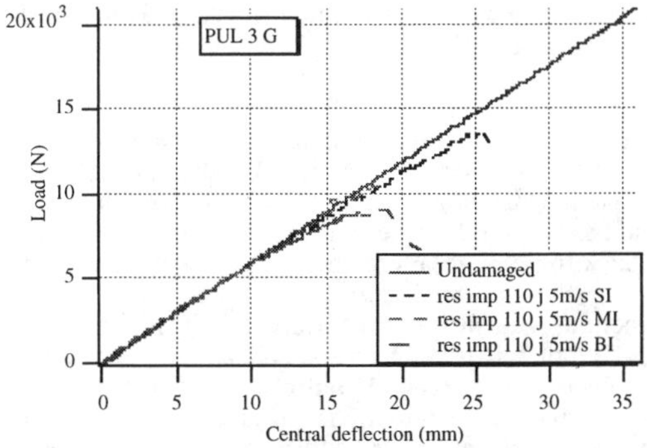

Figure 12 : Residual Load-deflection curves for Pul3 B and G (Impact at 110 J at 5m/s with varying impactor diameters(BI, MI and SI))

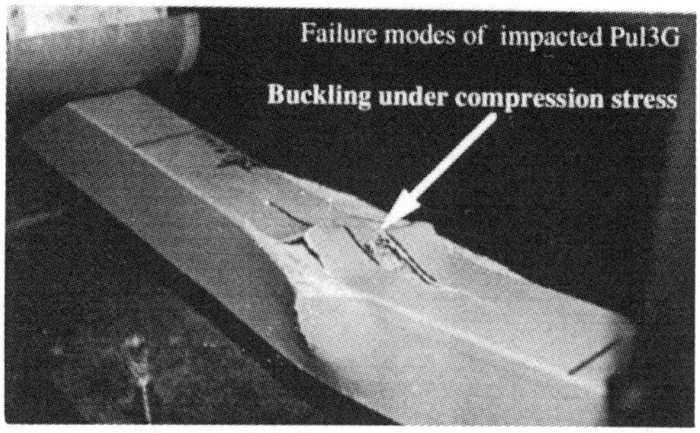

(a)

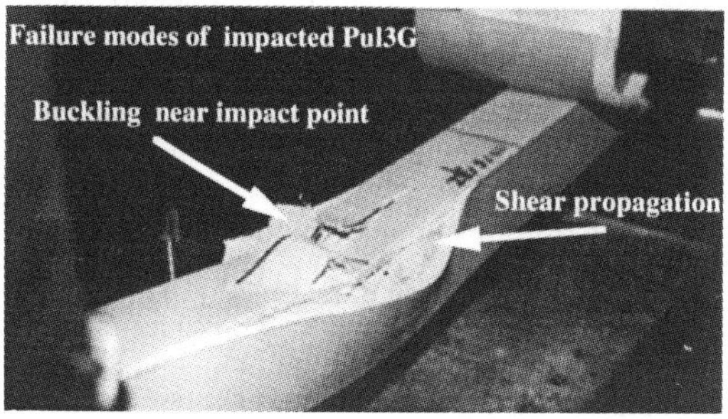

(b)

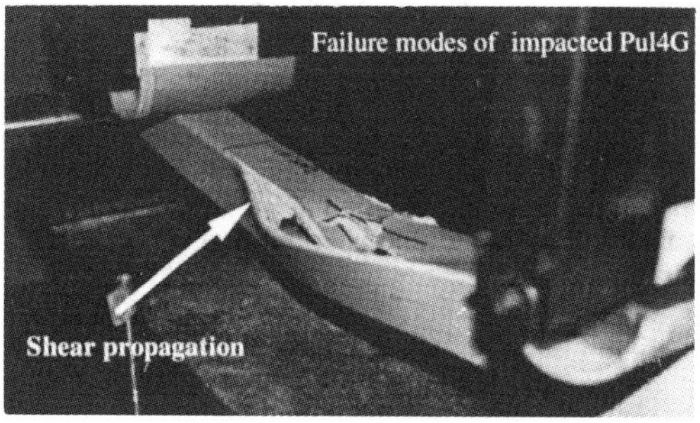

(c)

Figure 13 : Typical failure mode of 110 J impacted Pul3G (a & b with SI) specimens and 110 J impacted Pul4G (c with BI).

mode of the undamaged and impact damaged Pul4G specimen are very different. As the undamaged sample fails under a compression mode just below the load bearing system, the impacted specimen develops a buckling failure mode near the impact location during flexural test. This observation leads to a important conclusion : contrary to Pul1G profile, Pul4G structure under bending solicitation is strongly influenced by the type of impact load. The reduction of the flexural strength is around 50% for SI impact.Another parametric effect can be emphasized: the impactor diameter

The same observation than Pul4G can be done for Pul3G concerning the influence of the impactor

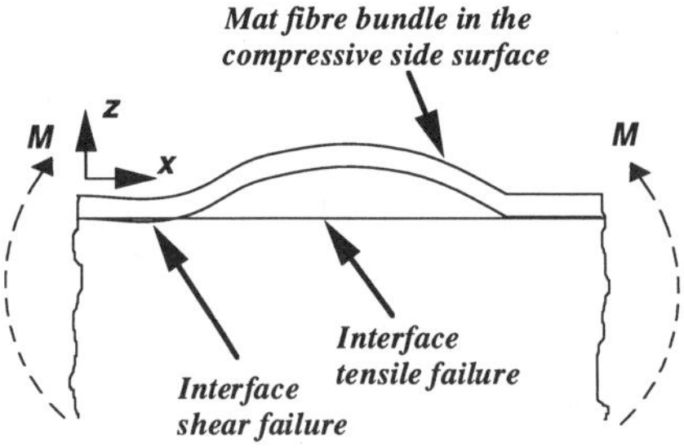

Figure 14 : Schematic representation of post impact failure modes near the impact point for both Pul3G and Pul4G profiles

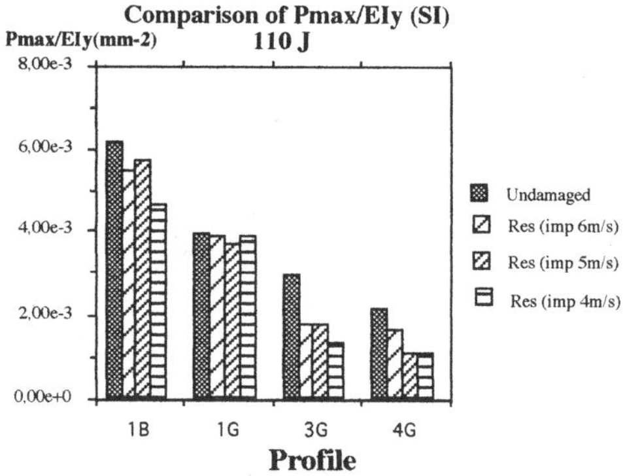

Figure 15: Comparison of Pmax/EIy parameter for undamaged and 110 J (SI) impact damaged profiles

diameter on the residual strength of the structure. It appears that, as previously proposed, the Big Impactor causes worse damage to the structure than the small one. As explained before, Although the total delamination area is less important for a BI than a SI (Chotard et al 1996a), this delamination area is localized at middle thickness or in the upper side of the profile section. Due to the vicinity of the compression stress field during a flexural test, this delamination zone may be more sensitive to a compression solicitation which can leads to a local instability such as buckling or shear failure mode (e.g. fig 13 & fig. 14)

In order to compare the residual strength of the different shapes, the maximum load obtained with bending tests have been normalized by dividing the maximum flexural load by both the longitudinal modulus and the second moment of inertia of the structure. The next figure shows this comparison.

This graph was obtained for a given energy (110 J) and a impactor size (SI). Many observations can be done regarding this diagram. Firstly, impact velocity have a great influence on residual performance of Pul1 B, Pul3G and Pul4G shapes.

Secondly, the failure modes observed under bending solicitation for undamaged specimens are the same as the impacted ones for Pul1B and Pul1G profiles. They are different for Pul3G and Pul4G shapes. The post impact failure modes under flexural loading for these structures are closely related to the impact damage previously introduced.

Finally, regarding these results, it seems that the worst configuration for residual evaluation is the lowest impact velocity with the biggest impactor diameter (50 mm). It is important to note that no reduction of the flexural strength was observed for both 50 and 80 J iso-energy tests. This observation tends to demonstrate that an minimum energy threshold exists and when this level is overshot, the global integrity of the structure is severely reduced, near 50 % of reduction of the normalized flexural strength for Pul3G and Pul4G (e.g. fig. 15)

4 CONCLUSION

In impact tests, the influence of the impactor size has been emphasized. The Big Impactor (BI) produces less delamination than the small one. However, the location of this delamination area seems to have an higher influence on residual properties of these structures than the damage extension.

The Small Impactor (SI) generate a local loading leading to a material solicitation. At the same time, the big one have a more structural effect. This global solicitation allows a bigger part of the impacted energy to be accommodate by the structure.

Concerning the post-impact behavior, it has been found that the impact velocity is also an important parameter which could have notable influence both on residual performance and on residual failure modes of these shapes under bending loading.

The lowest velocity (4m/s) associate to the highest impactor size (BI) seems to be the worst impact configuration for residual behavior of these beams.

Different failure modes were observed during residual strength tests. For Pul1B and Pul1G, no changes in the bending failure modes were noted between impact damaged and undamaged tested specimens. For both Pul3G and Pul4G profiles, flexural residual tests report an important change in the failure modes (evolution from an axial compressive mode for undamaged specimen to a combined shear/buckling failure for the impacted one) and in the performance (50% of reduction for the normalized residual strength).
It is important to note that although the impact velocities are relatively close, a notable dynamic effect, due to impact velocity, is revealed by the residual testing.

ACKNOWLEDGMENT

The authors want to thank the DCP Company for their material support.

5 REFERENCES

Bank .L.C. 1987. Shear Coefficients for thin-Walled Composite Beams, *Comp Struct..* 8, 47-61.

Bank.L.C & Bednarczyk.P.J. 1987. Deflection of Thin walled Fiber-reinforced Composite Beams *Proc of Snd Tech Conf, Am Soc for Comp*, 553-562. Nework, Delaware, Technomic Publishing company,

Bank.L.C. 1989. Flexural and shear moduli of full section fiber reinforced plastic (FRP) pultruded beam. *J of Test and Eval*, 17, N°1, 40-45.

Barbero.E.J., Lopez-Anido.R & Davalos.J.F. 1993. On the mechanics of thin-walled laminated composite beams, *J of Comp Mater*, 27: 806-829.

Chotard.T.J. & Benzeggagh.M.L. 1996 a. The influence of velocity and impact energy on industrial glass/polyester pultruded profile. *Proc of JNC 10 (in french)*, 1: 307-317 .Paris.

Chotard.T.J. & Benzeggagh.M.L. 1996 b. Influence of resin additives on failure behavior of pultruded glass/polyester composites shapes. *J of Mater Sciences Lett.* (In Press).

Davalos.J.F, Salim.H.A., Qiao.P, Lopez-Anido.R.& Barbero.E.J. 1987. Analysis and design of pultruded FRP shapes under bending, *Composites Part B*, 27B: 295-305.

De Belleval J.F, Potel.C., Chotard.T.J., & Benzeggagh.M.L. 1996. Ultrasounds and their model for composite materials characterization. *Proc of 3th Int.Conf of Comp.Eng (ICCE 3)*..211-212.

Potel.C., Chotard.T.J., De Belleval J.F & Benzeggagh.M.L. 1996. Characterization of composite materials by ultrasonic methods: modelisation and application to impact damage. *Composites Part B*. Submitted.

Sims.G.D, Johnson.A.F & Hill.R.D. 1987. Mechanical and structural properties of a GRP Pultruded section, *Comp Struct*, 8: 173-187.

Sumerak .J.E & Martin.J. 1985. Pultruded products-new capability on the horizon. *Proc of Advan Comp Conf.*: 133-138.

Svenson,A.L, Hargrave,M.W & Bank,L.C. 1992. Impact performance of glass-fibre Composite materials for Roadside safety structures. *Proc of Advanced Composite materials in bridges and structures*:.559-568.

Svenson,A.L, Hargrave,M.W & Bank,L.C. 1993. Impact behavior of Pultruded composites. *Proc of 48th An Conf, Composites Institute, The Society of Plastics Industry*, February 8-11, Session 21D:1-6.

Damage and Failure of Interfaces, Rossmanith (ed.) © *1997 Balkema, Rotterdam, ISBN 90 5410 899 1*

Damage behaviors in CFRP laminates due to repeated impacts of low energy

Hyung-Seop Shin
Department of Mechanical Engineering, Andong National University, Korea

Ichiro Maekawa
Department of Mechanical Systems Engineering, Kanagawa Institute of Technology, Atsugi, Japan

ABSTRACT: A simple experimental method for repeated impacts of low-energy was devised to provide some insights into the impact damage behavior of CFRP laminates. The macroscopic failure mode and the internal damage as a consequence of repeated impacts were investigated, at four different levels of incident impact energy, with the laminated specimens prepared from different stacking sequences. The results indicated that there existed a critical incident impact which led to the delamination and the critical size of delamination which produced the degradation of residual strength. The delamination behavior in the laminated specimens depended largely upon the level of incident impact energy. The stiffer the specimen, the larger the delamination damage induced in the interior of specimen.

1 INTRODUCTION

Since composite materials possess superior properties such as high specific strength, stiffness and excellent fatigue resistance, their applications to structural components of aircraft have been increasing. However, laminated composite materials are very susceptible to damage induced by the impingement of rain-drops, hailstones, debris, hard particles and tools dropped during maintenance. The impact of particles will produce some damage and eventually result in the degradation of the designed strength in composite structures. Until now, many attempts have been carried out to investigate damage caused by a low velocity impact of drop-weight(Abrate 1991). These attempts represented that the damage induced by an impact could be classified as matrix cracking, delamination, fiber breakage and debonding at the interfaces of fiber and matrix(Grag 1988). Generally, in laminated composite materials, it is known that the delamination is a major damage mode and closely relating to the strength degradation of specimen. Therefore, considering a long-term usage of composite materials in structural components, the investigation of the damage behavior caused by the repeated impacts of particle with low energy becomes important(Gweon & Bascom 1992, Prayogo et al 1996).

The objective of this study is to investigate the damage behavior induced in CFRP laminates under repetition of a steel ball impact with a low energy and to provide some insights into the impact damage behavior of CFRP laminates. A simple experimental apparatus is devised for the repeated impact testing of low-energy. The macroscopic failure mode and the internal damage as a consequence of repeated impacts are examined with an optical microscope and a C-scan method, at some different levels of incident impact energy in the laminated specimens with different stacking sequences. After impact tests, the residual strength is evaluated by compression and flexural tests, respectively.

2 EXPERIMENTAL PROCEDURES

2.1 *Specimen*

The CFRP laminates, where 12 layers of prepregs were laminated and hot-pressed to the size of 350x350x1.5mm, were supplied as a specimen. In prepregs, carbon fibers (TORAY T300) were reinforced to the thermo-setting epoxy resin(TORAY #2500), and the volume fraction of epoxy resin was 36%. Two kinds of stacking sequence were introduced in this study. They were called specimens C and D, respectively, having the following stacking sequences,

Specimen C : $[0^{\circ}_{2}/(+45^{\circ}/-45^{\circ})_{2}]_{s}$

Specimen D : $[(+45^{\circ}/-45^{\circ})_{2}/0^{\circ}_{2}]_{s}$

The size of specimen was small as compared with the usually used one in the drop weight impact test, and the dimension was 70x15x1.5mm. In this case, the fiber direction of 0° layer was selected to the longitudinal axis of specimen. The mechanical properties of each specimen are shown in Table 1. The specimen C, in which 0° layers were located at its outside, showed a larger value in both bending stiffness and compressive strength than the specimen D where 0° layers were located at the inside region. A steel ball of 4.8mm in diameter (0.45g in mass) was used as an impactor.

2.2 *Repeated impact experiment*

A schematic illustration of the setup for repeating impacts used in this experiment is shown in Fig. 1. The apparatus is similar to an air gun, having a shortened length of barrel. In the set-up, the specimen was located above the barrel and both ends of the specimen were fixed firmly to the supporting block, with a span of 50mm. The operating principle is very simple: the steel ball is fired up and impacts the center of the specimen, then the ball rebounds into the barrel after contact The rebounded ball is fired after an interval of 2 seconds. Thus, repeated impacts of the same kinetic energy

Table 1. Mechanical properties of laminated specimens.

Specimen	Young's modulus (GPa)	Bending rigidity ($\times 10^5$N/m)	Compressive strength (MPa)
C	53*	3.0	271
D	53*	0.79	112

* represents the value calculated by the rule of mixtures.

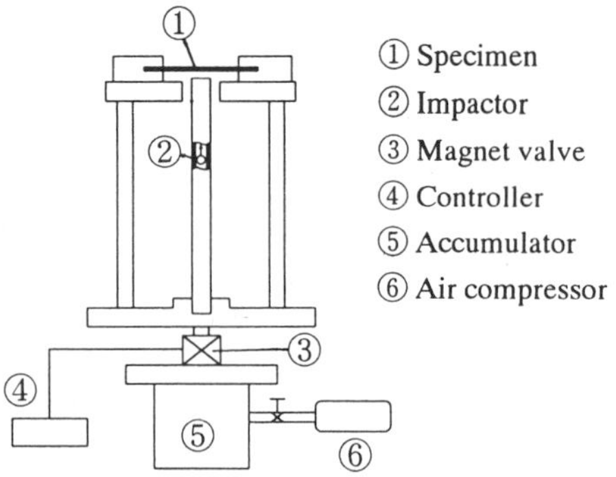

Fig. 1 Experimental set-up of a repeated impact apparatus.

are obtained having nearly the same contacting area(Shin 1991).

The impact velocity of an impactor can be obtained by measuring the flight time of impactor between two diodes located in front of barrel, at a single impact test. In real repeated impact tests, to provide the desired impact velocity, a specific value of pressure was selected from the relationship between the pressure of accumulator and the velocity of impactor which was determined previously through a single impact test. The kinetic energy of the impactor in a single impact test is defined as an incident impact energy E_{in}. In this study, the value of E_{in} was limited to the ranges 0.08-0.8J(which corresponds to the impact velocity of 18-60m/s).

2.3 *Evaluation methods of damage*

After the single and repeated impact tests, the damage at the surface of the specimen was examined using an optical microscope, and the interior damage was investigated using ultrasonic image scanning equipment (C-scan; Hitachi AT5000, frequency 10kHz), respectively. In order to investigate the damage morphology developed along the thickness direction of the specimen, the specimen was sectioned longitudinally through the center of the impacted part, and the section was polished using alumina powders of grain size 1μm.

After the impact test, the residual strength was evaluated to investigate the influence of damage induced by repeated impacts on the degradation of strength and stiffness in each specimen. However the testing method was different according to the stacking sequence of specimens. In specimen C, a compression test was carried out, and a test jig (the gage length of specimen receiving compressive load was 30mm) was used to prevent the occurrence of buckling during compression,. But in the case of specimen D, a four point bending test was used because it was impossible to evaluate the residual strength exactly through the compression test. During the four point bending test, a test fixture with an inner span of 30mm and an outer span of 60mm was used. The impacted part was located at the tensile side of specimen due to bending. The residual strength was determined from the maximum value of load on the load-displacement curve obtained in each test.

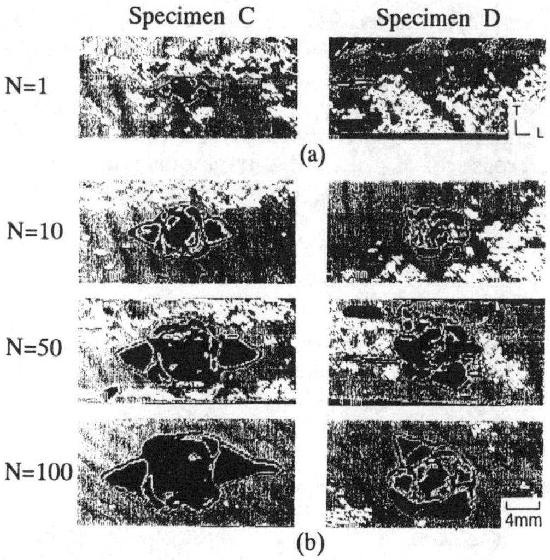

Fig. 2 C-scan images of internal damage after (a) a single impact and (b) repeated impacts (at E_{in}=0.34J).

3 EXPERIMENTAL RESULTS AND DISCUSSION

3.1 *Characteristics of damage induced due to a single impact*

In order to characterize the damage and fracture behaviors in CFRP laminates by repeated impacts, first the damage caused by a single impact of steel ball was examined. A crater appeared at the surface of the impact side and a deformation (bulging) on the opposite side of the specimen when the incident impact energy, E_{in} reached at 0.34J for specimen C, and 0.5J for specimen D, respectively. Moreover, with increasing of E_{in}, there also occurred various types of damage such as delamination, matrix cracking and fiber breaking.

The damage that occurred in the interior of the specimen was examined by using ultrasonic image scanning equipment. Fig. 2(a) shows a representative example of planar distribution of delamination occurred at the interface of layers by a single impact in the case of E_{in}=0.34J. In this figure, different delamination patterns according to stacking sequences can be seen. In the case of specimen C, the delamination occurred mainly along the 0° layer, but for specimen D it had a nearly circular shape representing that the delamination spreaded along $\pm 45^o$ direction. It could be thought that these differences in delamination patterns according to stacking sequence were resulted from the fact that the delamination at the interfaces between layers was propagated along the fiber direction in just the upper layer(Ishikawa et al 1992).

As increasing E_{in}, the damaged region became larger. The projected area of delamination in each specimen was measured from the ultrasonic image. It had been known that there existed a linear relation between the total delaminated area among layers of laminates and the projected delamination area in thin laminated specimens (Abrate 1991). Figure 3 shows the relationship between the projected area of delamination and the incident impact energy. When E_{in} exceeded 0.22J, the delamination was initiated at the interiors of

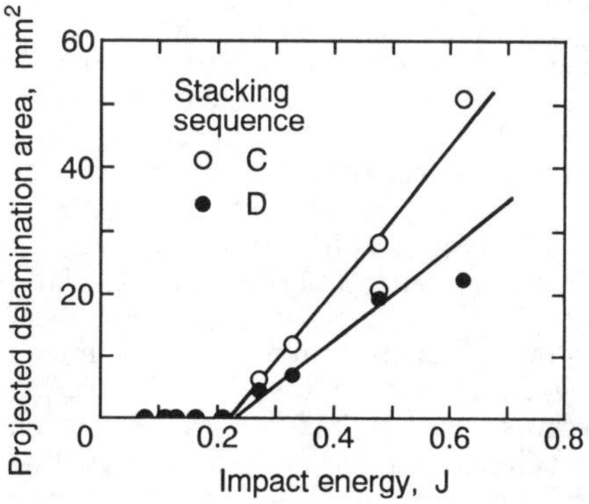

Fig. 3 Relationship between projected delamination area and incident impact energy for a single impact.

both specimens. And, the energy at that time was defined as a critical impact energy to the initiation of delamination, E_c. At the impact exceeding E_c, the projected delamination area increased linearly with increase of E_{in}, and the behavior was similar to the other results of laminated composites (Hong et al. 1989). Also, there existed the effect of stacking sequence, and the specimen C represented a larger projected delamination area than specimen D at the same value of E_{in}. It indicated that the specimen C is a more sensitive structure to the delamination damage as compared with specimen D, and the slope of specimen C was $111 mm^2/J$ but it was $69.5 mm^2/J$ for specimen D.

At the test above this experiment range in the incident impact energy, additional matrix cracking and fiber breaking were produced at both side surfaces of specimen. From these results, the damage induced in CFRP laminates due to low energy impact could be characterized as the delamination occurred at the interface among each layers (Grag 1988, Shin et al 1994).

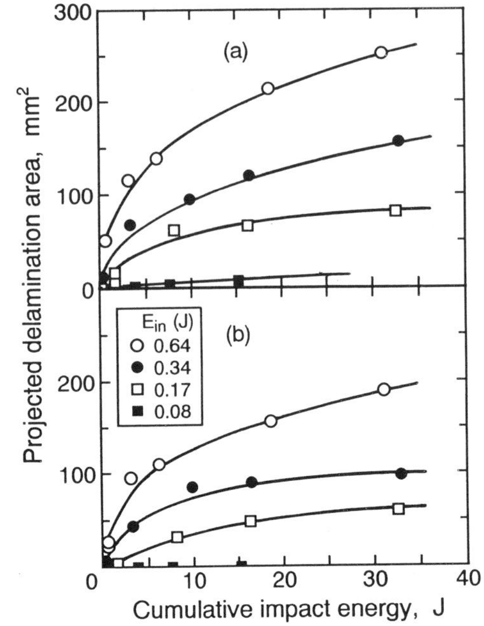

Fig. 4 Relation between delamination area and cumulative impact energy after repeated impacts in (a) specimen C and (b) specimen D

3.2 *Damage behavior due to repeated impacts*

In repeated impact experiments, four levels of E_{in} were selected; two subcritical impact energy levels($E_{in} < E_c$) and two supercritical ones($E_{in} > E_c$), compared with the critical impact energy level corresponding to the initiation of delamination. The values are corresponding to 0.08J, 0.17J, 0.34J and 0.64J, respectively. In the repeated impacts, the cumulative impact energy added to the specimen was determined from multiplying repeated number N by the incident impact energy, E_{in}.

The projected images of delamination developed with increasing N were shown in Fig. 4 (b), at a supercritical incident energy level of 0.34J. As N increased, the delamination extended, taking a similar damage pattern to the case of single impact shown in Fig. 4(a) and having a characteristic pattern according to stacking sequence. On the other hand, at subcritical energy levels, many numbers of impact were needed to initiate the delamination in the specimen, depending on the E_{in} level, but the behavior of delamination after that was very similar to the cases of supercritical energy impacts.

In cases of the repeated impact, the relationship between the projected area of delamination and the cumulative impact energy was shown in Fig. 4 (a) and (b). Where (a) and (b) represent for the cases of specimens C and D, respectively. In both specimens, although there existed some differences depending upon the E_{in} level, the projected area of delamination increased rapidly at first with increase of the cumulative impact energy, but it soon became slow. Especially, in the case of a subcritical energy impact, E_{in}=0.17J, the delamination area showed a nearly saturated value when the cumulative impact energy exceeded 20J. At the same value of cumulative impact energy, the higher the E_{in} level, the larger the delamination area produced. And the specimen C showed a somehow larger delamination area than the specimen D.

The influences of E_{in} levels upon the behavior of the delamination resulted from the difference in the damage mechanism during the subcritical and supercritical energy impacts. Firstly, at the supercritical energy impacts, the occurrence of delamination were directly resulted from the shearing stress generated at the interface between layers having different fiber directions due to bending deformation during repeated impacts (Grag 1988). Therefore, as impacts were repeated, the delamination extended in plane at first, then propagated across the thickness. More increase in N produced matrix cracking at the surface of specimen, therefore the increase of tensile or compressive stresses at the surface layers due to these additional damage finally resulted in the fiber breaking.

On the other hand, in the case of subcritical energy impact, there occurred no detectable damage during a single impact, but microcracking occurred at the surface matrix, and it was accumulated during repeated impacts, and eventually initiated delamination in the interior of specimen. Thus, there was a slight increase in the delamination with N, but the energy consumed to extend the delamination would still be relatively small. At this point, the number of repeated impacts to the initiation of delamination was determined and plotted in Fig. 5 against the incident impact energy level. When the E_{in} level became small, more numbers of impact were needed to initiate the delamination. There also existed the effect of stacking sequence, and more numbers of impact were needed in specimen D as compared with the specimen C. This might be related with the fact that the flexural stiffness of specimen C is larger

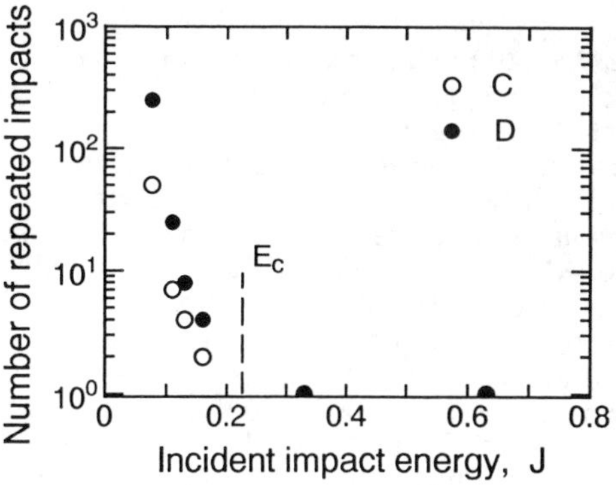

Fig. 5 The numbers of impact to delamination initiation for repeated impacts under subcritical energy levels.

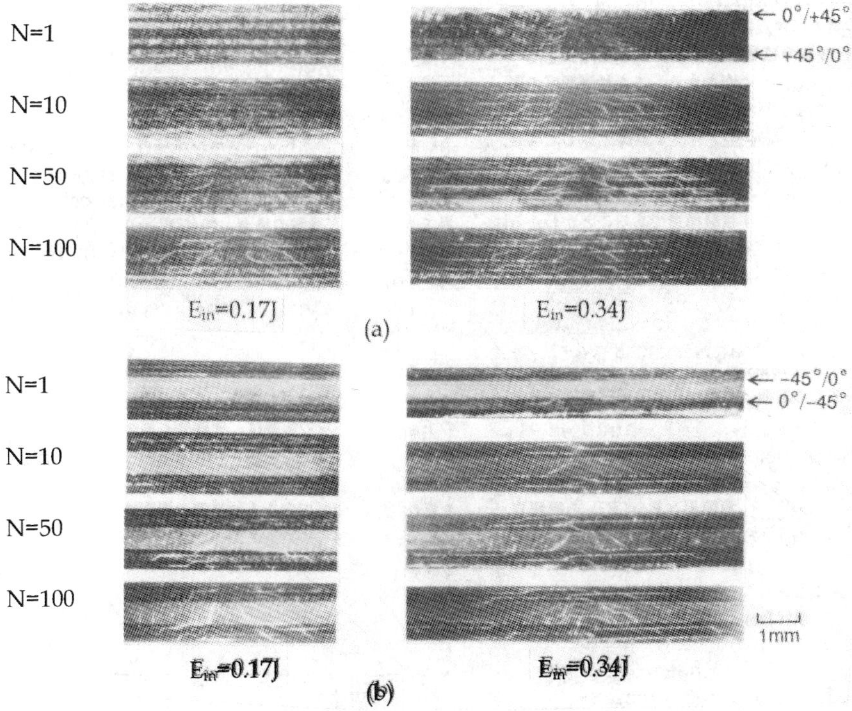

Fig. 6 Cross-sectional views of internal damage after repeated impact in (a) specimen C and (b) specimen D.

than D, so that a higher contacting pressure generated at the matrix part of specimen C made it easy to initiate microcracks during repeated impacts.

In order to examine the morphology of damage growth, especially in the interior of specimen during repeated impacts, it was sectioned passing through the center of the impact site along the longitudinal direction of specimen. Representative cross-sectional views at two different energy levels were shown in Fig. 6 (a) and (b). In both specimens of C and D, the delamination occurred mainly at the interfaces between $+45^{\circ}/-45^{\circ}$ layers or $45^{\circ}/0^{\circ}$ layers. In specimen C of Fig. 6 (a), a significant delamination can be seen at the interface between $+45^{\circ}/0^{\circ}$ layers located on the back side of impact, this fact corresponds well to the shape of projected delamination along the 0° direction, shown in Fig. 2 (a).

In the subcritical energy impact of E_{in}= 0.17J, no damage occurred at N=1. At N=10, the delamination occurred at the interfaces between $+45^{\circ}/-45^{\circ}$ layers located closely to the back side of impact in both specimens. Increasing N in specimen C, the delaminated layers multiplied within $\pm 45^{\circ}$ layers of the central region of the specimen And in specimen D, it could be classified by two types of damage: the delamination at interfaces of $+45^{\circ}/-45^{\circ}$ layers located on both sides of the specimen and the transversely generated matrix cracks through the thickness into the central region of specimen.

In the supercritical energy impact of E_{in}= 0.34J, the damage existed already at N=1. With increasing N, the delamination extended significantly in plane and also in thickness. The transverse matrix cracks were connected to the delamination generated at the upper and lower interfaces of the layer. Especially, in specimen D, multiple transverse matrix cracks were developed having a cone shape within the 0° layers of the central region. When N exceeded 50, many matrix cracks in the layers close to the impact side and a significant delamination in the layers close to the back side could be seen. From these results, it could be thought that the repeated impacts of low energy in CFRP laminates, at first produced the expansion of delamination in plane and then multiplied it through the thickness direction and the transverse matrix cracks connected to the delaminated parts. Therefore, it would be expected to degrade the load bearing ability of CFRP laminates, since matrix cracks were developed along the thickness direction by repeated impacts.

3.3 *Residual strength after repeated impacts*

The delamination, which was a characteristic damage generated in CFRP laminates by repeated impacts of low energy, will produce a significant strength degradation under the compressive loading(Prichard & Hogg 1990). After repeated impacts, a compressive test for specimen C and a four point bending test for specimen D, respectively, were performed to evaluate the residual strength.

Figure 7 shows an example of a load-displacement relation obtained from the compressive test, after repeated impacts at a subcritical energy (E_{in}=0.17J) in specimen C. The load increased linearly at first, and at point S there existed a change on the load curve which represents that a local buckling fracture was initiated at a layer from the delamination. Then, the load continued to increase representing a tooth shape as the buckling fracture was extended to the other layers delaminated already. Finally, when the load reached to point F, a sudden drop of load occurred because the buckling fracture spreads over all of the layers in thickness direction and the specimen did not sustain the compressive load. The slope of load curve within the early region to point S was nearly similar to the case of the smooth specimen without damage, although there existed the delamination damage in the interior of specimen. This fact indicated that there occurred little

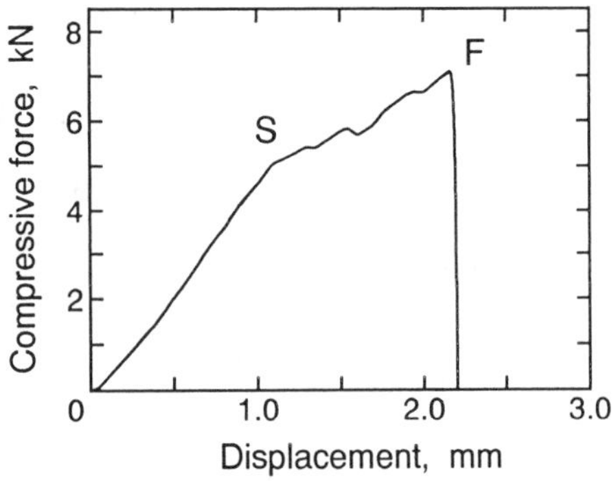

Fig. 7 An example of compressive load-displacement curve after repeated impacts(Ein=0.17J, N=100cycles).

degradation in the flexural stiffness of specimen, even if the delamination existed in the interior of specimen. During repeated impacts of low energy, the existence of delamination, which would be supposed to produce the degradation of the stiffness, did not influence the delamination growth.

In specimen C, the relationship between residual compressive strength and the cumulative impact energy was shown in Fig. 8. In the case of a subcritical energy impact, E_{in}=0.08J, it showed no degradation in the residual strength until the cumulative impact energy approached to 20J. As E_{in} level became higher, the residual strength started to degrade when the value of the cumulative impact energy exceeded 1.5J. At each level of E_{in}, the residual strength degraded gradually with an increase of the cumulative impact energy. At the same value of cumulative energy, the degradation became greater with increase of the E_{in} level and appeared significant at the super-critical energy impacts. These behaviors of strength degradation in CFRP laminates are well corresponding to the ones of delamination area against the cumulative impact energy, as already discussed previously, indicating that both are closely relating with each other in specimen C.

On the other hand, in specimen D, the influence of the delamination damage on the residual strength couldn't be investigated through the compressive test above. Specimen D collapsed completely under a small compressive load, even though it had a significant delamination in the interior, and the application of the compressive load could not extend the delamination through the specimen because its stiffness was low as compared with specimen C. Therefore, the residual strength of specimen D was determined by a four point bending test. The results are shown in Fig. 9. In this figure, although it is impossible to compare the value directly with specimen C, the degradation behavior was similar to the case of specimen C, shown in Fig. 8. And the residual flexural strength degraded gradually with increase in the cumulative impact energy excepting for the case of E_{in}=0.08J, similarly to specimen C. At the same cumulative impact energy, the degradation became significant with increase in the level of E_{in}, although the obtained values were slightly dispersed.

From a viewpoint of damage tolerance design, it would be meaningful to investigate the relation between residual strength and the damage size developed in the interior of specimens. At each value of the cumulative impact energy, the obtained residual strength was plotted against the corresponding projected delamination area and shown in Fig. 10. In both specimens of C and D, the residual strength began to degrade when the delamination area exceeded to a certain critical value. The critical value was $50mm^2$ in specimen C and $30mm^2$ in specimen D, respectively. There existed a nearly linear degradation with increasing the delamination area regardless of E_{in} levels. This relation is very important in the point of the damage tolerance design of composite laminates, because the damage size of specimen can be well predicted through a nondestructive technique. The existence of a critical delamination size might result from the fact that the delamination was isolated or disconnected

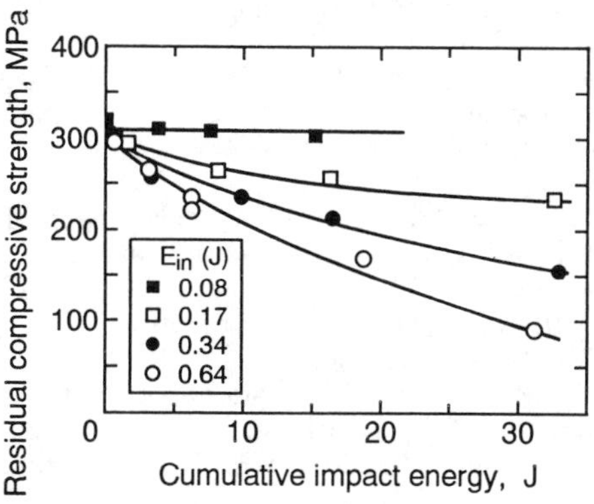

Fig. 8 Degradation of residual compressive strength due to repeated impacts for specimen C.

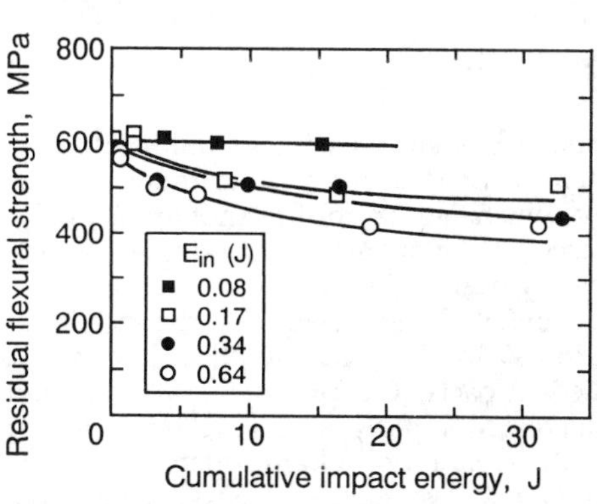

Fig. 9 Degradation of residual flexural strength due to repeated impacts for specimen D.

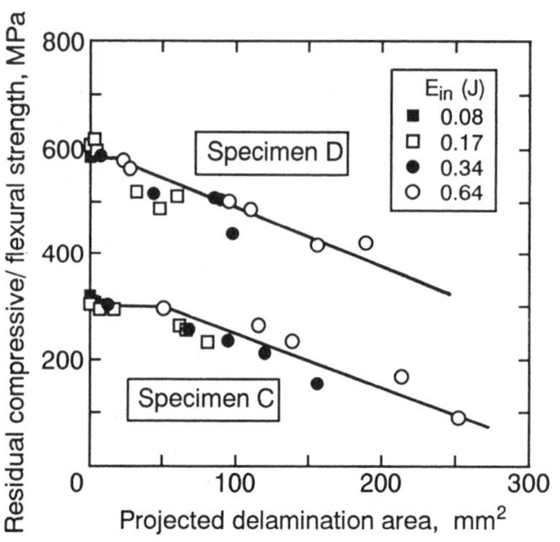

Fig. 10 Relation between residual strength and projected delamination area.

when its size was small, therefore it did not lower the load bearing ability. It might be corresponding to the delamination area when the matrix cracks generated and passed transversely through the thickness of specimen, shown in Fig. 6

From these results, it could be found that in the evaluation of residual strength of laminated specimens after an impact test, the compressive test was suitable for the stiffer specimens like C, but the flexural test for flexible specimens like D. Through the repeated impacts of low energy, it could be also found that the evaluations of the impact damage behavior and the residual strength in CFRP laminates, was possible, even though a small size of specimen was used.

4 CONCLUSIONS

The obtained results were as follows:

There existed a critical incident impact energy which led to the delamination at the interfaces of laminates at a single impact, and a critical size of delamination area which initiated to degrade the residual strength. With increase of the number of repeated impacts, the projected delamination area increased rapidly at first, after which it showed a relatively slight increase, and dense transverse matrix cracks across the thickness direction also occurred. The delamination behavior in the laminated specimens depended largely upon both the level of incident impact energy and the stiffness of specimen. The stiffer the specimen, the larger the projected delamination area generated in the interior of specimen.

In each laminated specimen, there existed a linear relationship between the residual strength and the projected delamination area, regardless of the incident impact energy levels. Through the repeated impacts of low energy, it was possible to evaluate the impact damage behavior in laminated specimens, even though a small size specimen was used.

REFERENCES

Abrate, S. 1991. Impact on laminated composite materials, *Appl. Mech. Rev.*, 44:155-190.

Grag, A. C., 1988 Delamination-A damage mode in composite structures, *Eng. Fracture Mech.*, Vol. 29, pp. 557-584.

Gweon, S. Y. and W. D. Bascom 1992. Damage in carbon fibre composites due to repetitive low-velocity impact loads, *J. Mater. Sci.*, 27: 2035-2047.

Hong, S. and D. Liu, 1989, On the relationship between impact energy and delamination area, *Exp. Mech.* 29: 115-120.

Ishikawa, H., T. Koimai and T. Natsumura 1992 Studies on impact loading damage in CFRP laminates by the three dimentional dynamic finite element analysis, *Proc. JSME/ MMD*, 920-72: 359-360.

Prayogo, G., H. Homma and Y. Kanto 1996. Repeated rain-drop impact damage in glass-fiber reinforced plastics, *Proc. 2nd ISIE'96*, Beijing, 350-364.

Prichard, J. C. and P. J. Hogg 1990, The role of impact damage in post-impact compression testing, *Composites* 21: 503-511.

Shin, H. S., I. Maekawa et al 1994 Effect of pre-bending on the properties of impact damage in CFRP laminates, *Trans. of KSME*, 18: 1144-1149 (in Korean).

Shin, H. S., I. Maekawa et al, 1991 Damage induced by the repetition of particle impact in silicon carbide, Trans. Jap. Soc. Mech. Eng., 57 A:3057-3062 (in Japanese).

Damage and Failure of Interfaces, Rossmanith (ed.) © *1997 Balkema, Rotterdam, ISBN 90 5410 899 1*

Critical impact energy density for welding aluminium plate onto surface of different metal by gas-gun method

T. Suhara
Graduate School, University of East Asia, Fukuoka, Japan

J. Omotani
Department of Mechanical Engineering, University of East Asia, Japan (Presently: Hiroshima Daihatsu Co., Ltd, Japan)

T. Matsubara
Research Institute for Applied Mechanics, Kyushu University, Japan

ABSTRACT: The lower limits of impact energy density for welding an aluminum plate onto another metal surface are evaluated by experiments using the gas-gan method. An aluminum thin plate stuck on the front surface of a polyacetal projectile is shot at high velocity (150~300 m/s) onto a target metal, such as mild steel, stainless steel or titanium and is thus welded. To confirm the welding of the two metal plates, the shear strength of joints is measured and microstructure of the interface is observed by optical microscope and analysis is made by an electron probe microanalyzer. The critical impact energy density is evaluated by subtracting the plastic work of the polyacetal portion from the total kinetic energy of the projectile.

1 INTRODUCTION

There have been many investigations on the instantaneous bonding of two metal plates by high velocity impact and these have been applied to the industrial processes such as explosive welding. Recently, an aluminum cylindrical projectile was shock welded to another metal target using the gas-gun method, and the mechanical and metallurgical properties of the interface were studied. The bonding mechanism is depend upon the impact conditions and is very complicated. For example, the interface of two metals bonded by explosive welding shows a ripple pattern, but in the case of an aluminum projectile shot on a metal surface, no ripple pattern is observed in the central part of the interface.

In this report, the experimental values of the lower limit of impact energy density for welding are evaluated using the gas-gun method by which two metal plates are welded without a ripple pattern in the interface. An aluminum plate is stuck on the flat front of a 20 mmϕ polyacetal(POM) projectile whose front half is tapered like the frustum of a cone, and is shot at high velocity (150~300 m/s) onto the surface of another metal plate by a compressed air gun. The POM portion of projectile separates from the aluminum plate at the instant of shooting.

To verify the weld of two metals, the shear strength of the welded joints is measured by a specially devised apparatus and the microstructure of the interface is observed by an optical microscope and analyzed by an electron probe microanalyzer.

The critical impact energy density is estimated as follows. We first assume that the dynamic deformation process is adiabatic and all plastic work is converted to heat energy, and further, that the kinetic energy is converted completely into plastic work in uniaxial compression. The plastic work of aluminum plate, is obtained by subtracting the plastic work of the POM portion from the total kinetic energy of the projectile. To evaluate the plastic work of the POM portion, an analysis is used which employs an energy equilibrium equation derived by J.B.Hawkyard resulting in a solution for the strain distribution. This analysis is extended to calculate the strain distribution of a uniaxially compressed projectile of variable cross section. From the results of measurements of impact velocity and final deformed shape of the projectile, it follows that the plastic work of the POM portion should be evaluated. Although the effects of temperature rise caused by plastic work on the stress-strain relation of POM are not included in the preceding discussion, they are minimal in the present case.

2 EXPERIMENTAL PROCEDURE

2.1 *Material*

The target materials are mild steel (C:0.16%, Mn: 0.57%), stainless steel (Ni:8~11%, Cr:18~20%, Mn:2%,Si:1%) and titanium (Ti:99.5%). The size

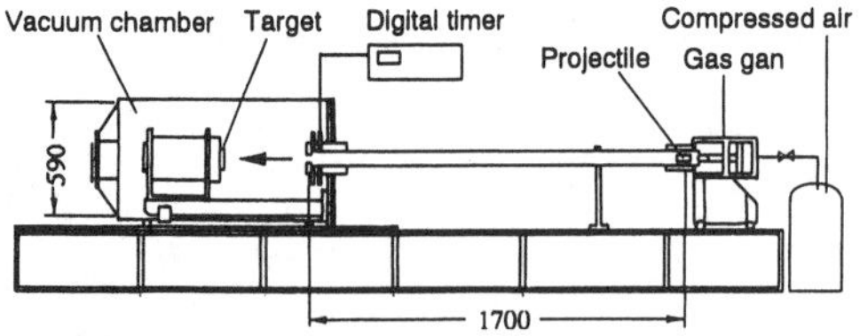

Figure 1. Schematic representation of impact welding by gas-gun method

of the target plate is 40×40×5 mm. Shooting alumum (Al:99.2%) is a circular plate of 12 mm diameter and 1 mm thickness. The contact surfaces are polished using CC 1500-CW waterproof paper.

2.2 *Gas-gun and projectile*

Figure 1 shows the schematic system of impact welding by gas-gun method. In the present experiments, the highest pressure of compressed air is 1 MPa and the maximum impact velocity is 300 m/s. The pressure in the vacuum chamber is 1KPa.

As shown in Fig.2, an alumum plate is stuck on the flat front of a polyacetal (POM) projectile which is shot on the surface of target metal plate using a gas-gun. The diameter of a aluminum plate is decided experimentally to be 12 mm which makes the mass of the projectile as small as possible, makes the energy flow into the aluminum plate as large as possible, and the temperature of POM portion of projectile remaines as low as possible.

2.3 *Measurements of the shear strength of welded joints*

The testing apparatus shown in Fig.3 is used to evaluate the shear strength of impact welded joints. The specimens are made by cutting off the welded aluminum plate leaving a central rectangular parallelepiped as shown in the figure.

2.4 *Observation and analysis of micro-structure of interface of welded metals*

The micro-structure of the interface of welded metals is observed by an optical microscope and is analized by EPMA microanalyzer.

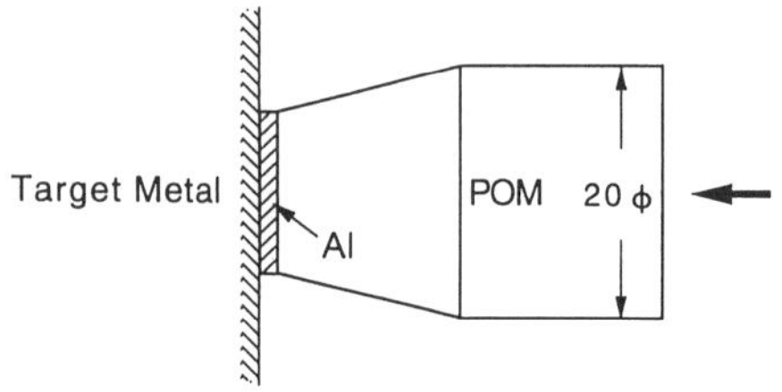

Figure 2. Projectile of impact welding

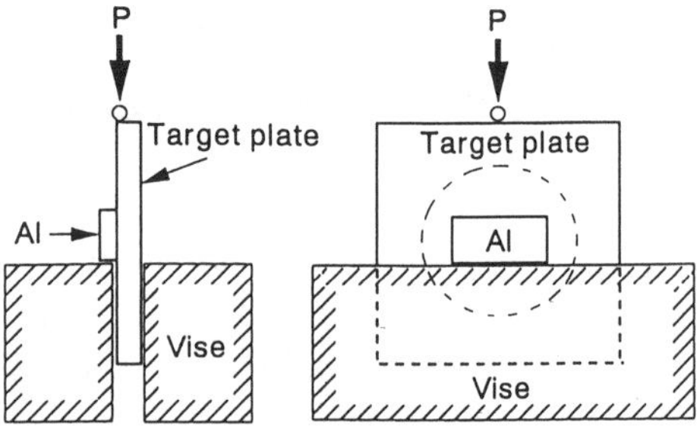

Figure 3. Testing apparatus of shear strength of welded joint

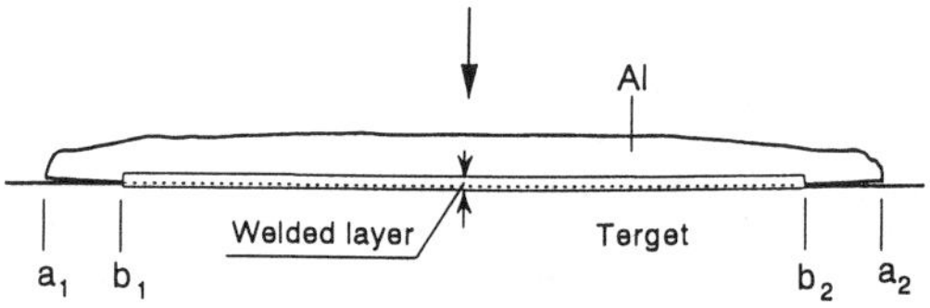

Figure 4. Schematic drawing of an aluminum plate welded to a target metal surface

2.5 *Impact experiments of polyacetal projectile*

Impact experiments of the POM projectile are carried out to obtain the relation between the impact velocity and the deformation after impact. Two types of projectiles A and B shown in Fig.11 are used.

3 EXPERIMENTAL RESULTS AND DISCUSSION

3.1 *Optical microphotograph of interface of welded metals*

A schematic drawing of an aluminum plate welded to the target metal is shown in Fig.4. The welded layer has, as illustrated later, fairly uniform throughout the welded region, though a width of about 1 mm at the edges a_1b_1 and a_2b_2 is not welded as shown. Figure 5 shows optical microphotographs of the welded layers. The upper dark parts are the aluminum and the lower parts are the target metals. Figure 5 (a) (b) and (c) are photographs of the central region of the welded layer and (d), (e), (f) are those of a region a few mm apart from the edge of the welded layer.

3.2 *X-ray analysis of interface by EPMA*

Figure 6(a), (b) and (c) are X-ray image analysis of

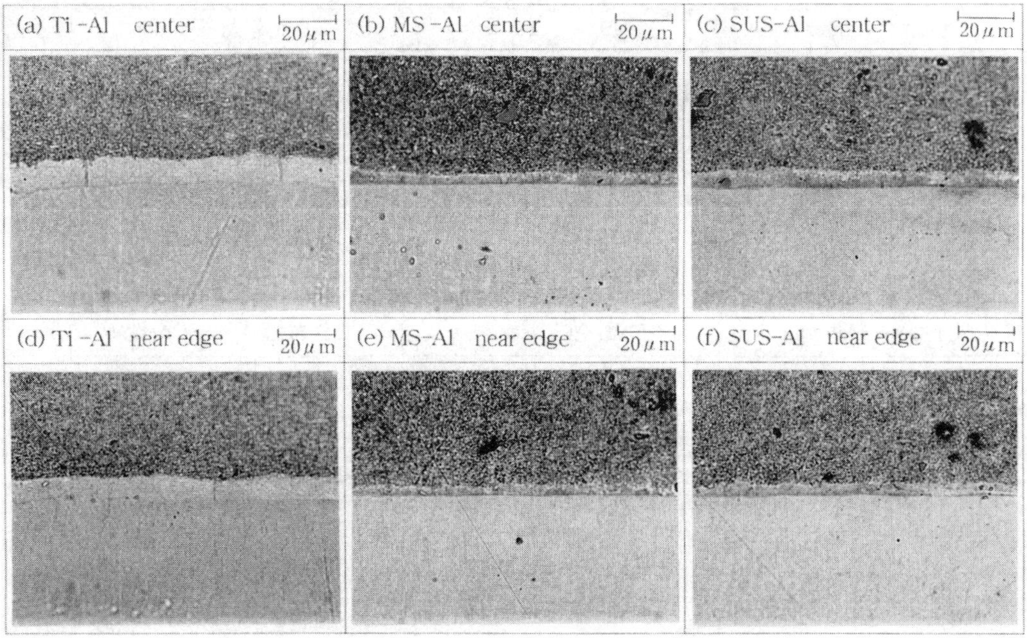

Figure 5. Optical microphotograph of welded joint

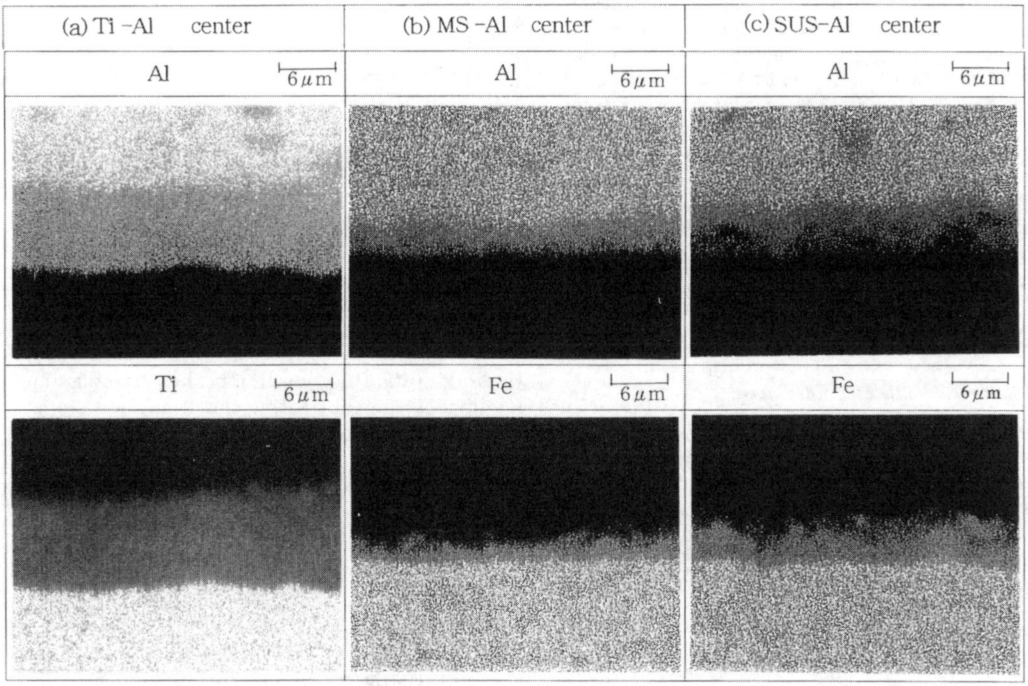

Figure 6. X-ray image analysis of interface by EPMA.

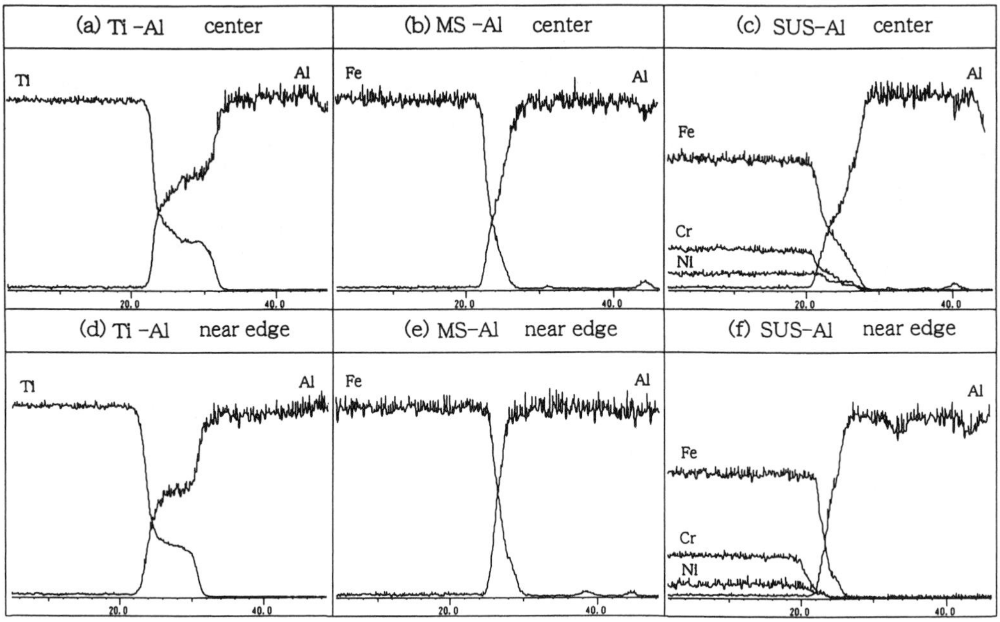

Figure 7. X-ray line analysis of interface by EPMA.

the central region of the welded layer of aluminum and target metals. Generally the layer thicknesses is dependent on the impact velocity: the higher the velocity, the greater the thickness. The impact velocity of Al to Ti, Al to Ms and Al to SUS in Fig.6 are 291 m/s, 234 m/s and 258 m/s respectively, and the corresponding mean thickness is 7.4 μm, 3.6 μm and 5.2 μm respectively. The x-ray line analysis of the central region and the region near the welding edge are shown in Fig.7. As shown, there is no difference in layer thickness of the two regions. Based on the observations in Fig.5 and Fig.7, it is concluded that the thickness of the welding layer is nearly equable confirming the uniformity.

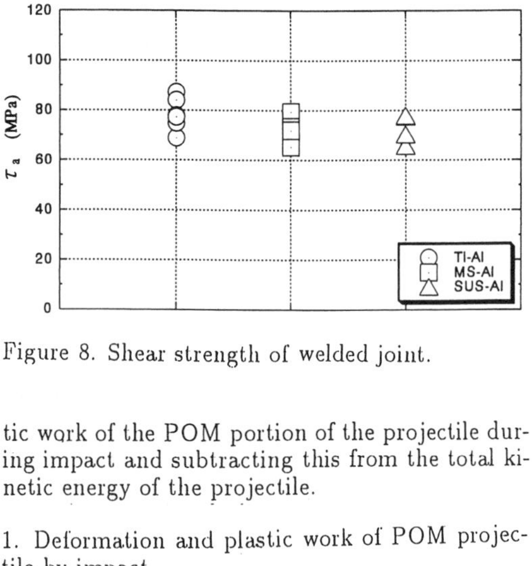

Figure 8. Shear strength of welded joint.

3.3 *Shear strength of welded joint*

The shear strength of the welded joint is evaluated using the apparatas of Fig.3. Figure 8 shows the results of measurement which indicate that the mean shear strengths of the welded joints of aluminum to three kind of metals are nearly equal. These results and observations of the fracture surface reveal that fracture of the interface occurs due to the shear fracture of aluminum.

3.4 *Evaluation of impact energy density for welding*

Evaluation of the amount of energy flowing into the aluminum plate requires estimation of the plastic work of the POM portion of the projectile during impact and subtracting this from the total kinetic energy of the projectile.

1. Deformation and plastic work of POM projectile by impact.

Strain ϵ of projectile is expressed by

$$\epsilon = 1 - (d/d_p)^2, \qquad (1)$$

where d and d_p are diameter of arbitrary section of the projectile before and after the impact. The relation between impact velocity v_0 and maximum strain ϵ_0 is given by

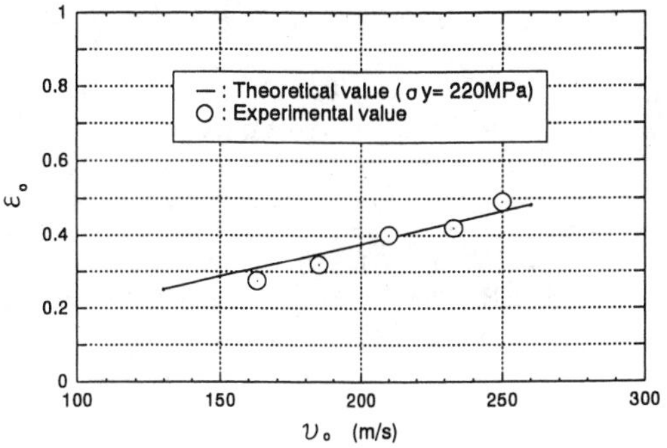

Figure 9. Relation between maximum strain ϵ_0 and impact velocity v_0.

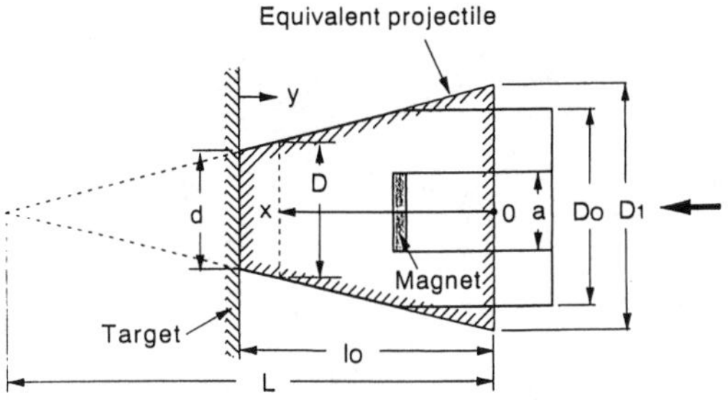

Figure 10. Equivalent frustum projectile.

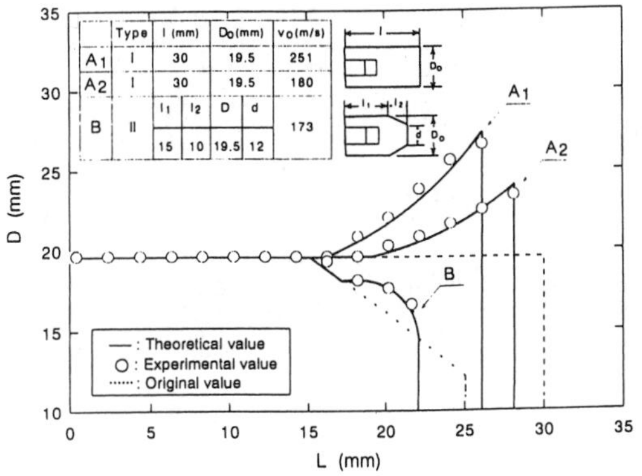

Figure 11. Change in projectile diameter by impact.

$$\rho_p v_0^2/2\sigma_{yp} = \ln\{1/(1-\epsilon_0)\} - \epsilon_0\,, \tag{2}$$

where ρ_p is the density and σ_{yp} is the dynamic yield stress of polyacetal. Since the difference of impact velocities in the present experiments is small, we can approximate from the experimental results of ϵ_0 and using Eq.(2) that σ_{yp} is constant and is equal to 22 kgf/mm^2. Figure 9 shows the relation between the impact velocity v_0 and maximum strain ϵ_0, where the line is the calculated value under this assumption which agrees well with the results of experiments. Deformation of the cylindrical projectile is calculated using Hawkyard's theory. The calculated values of changes in diameter by deformation of projectile A_1 and A_2 agree well with the results of experiments (Fig.11).

To calculate the deformation of a projectile whose front half is tapered like a frustum, we approximate it as a equivalent frustum (Fig.10), where the latter half of the original projectile is a hollow cylinder to decrease the mass. The mass of the magnet attached to measure velocity is small and is included in the total mass of the equivalent projectile.

The diameter of the equivalent projectile after the impact can be calculated using Eqs.(3) to (6)

$$D_p = D_1\{1-(x/L)\}\sqrt{1/(1-\epsilon)}\,, \tag{3}$$

$$y = -Lf(l_0/L)(1-\epsilon_0)\int_{\epsilon_0}^{\epsilon}[\{1-(x/L)\}^2(1-\epsilon)\}^{-1}d\epsilon\,, \tag{4}$$

$$1-\epsilon = (1-\epsilon_0)f(l_0/L)/f(x/L)\,, \tag{5}$$

$$f(x/L) = (x/L)\{1-(x/L)+(x/L)^3/3\}\,, \tag{6}$$

The value of ϵ_0 is given as a function of the impact velocity v_0 in Fig.9. The result of the calculated value of the diameter after impact is shown in Fig.11 and agrees well with the results of the experiment.

Effect of the stuck aluminum plate on deformation of the projectile

Since the 1 mm thick aluminum plate is stuck on the front of projectile in the present experiment (Fig.2), the impact velocity of the front plane of the POM part of the projectile is decreased. To calculate this velocity, we consider the impact of a 12 mm diameter aluminum cylinder having the same mass as the projectile and calculate the impact velocity of its section 1 mm behind the front surface. Assuming that this velocity is equal to the impact velocity of the POM portion of the projectile, the strain ϵ_0' of front of the POM portion can be calculated by Eq.2. Using the above method of calculation, the plastic work W of the POM part of projectile is given as follows

$$W = \sigma_{yp}\pi(D_1/2)^2[\{(\ln(x_f/L)-1)f(x_f/L) -(\ln f(l_0/L)-1)f(l_0/L)\} -\{f(X_f/L)-f(l_0/L)\}\ln\{(1-\epsilon_0')f(l_0/L)\}]\,, \tag{7}$$

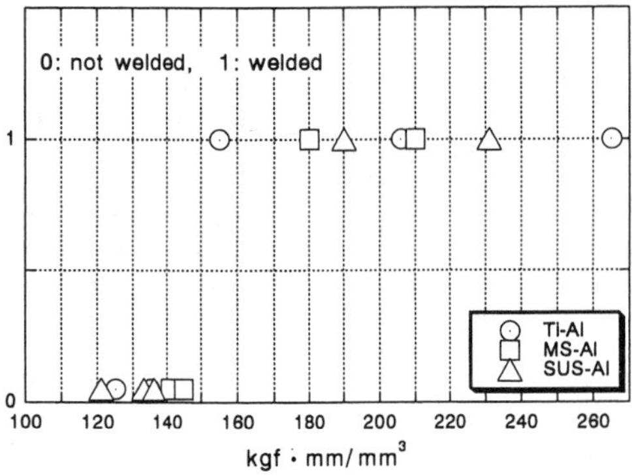

Figure 12. Critical impact energy density for welding.

where x_f is the value of x when $\epsilon = 0$ in Eq.5.

Since the kinetic energy of the projectile is expressed by $Mv_0^2/2$, where M is the mass of the projectile, and v_0 is the impact velocity. then the energy E which flows into the aluminum plate is expressed by

$$E = (Mv_0^2/2) - W, \tag{8}$$

The energy density is given by E/V, where V is the volume of the welded part of aluminum between b_1 and b_2 in Fig.4. The critical energy density for welding a aluminum plate to the surface of mild steel, stainless steel or titanium is given by Fig.12. Although the critical values differ slightly with the target metal, the corresponding temperaturs of aluminum plate are little more than its melting point approximately.

4 CONCLUSION

The method of impact welding of thin aluminum plate onto the surface of a different metal using a gas-gun has been described and experiments of welding aluminum to mild steel, stainless steel and titanium were carried out. The critical impact energy densities of welding these metals have been discussed.

ACKNOWLEDGMENTS

The authors of University of East Asia are grateful for the support by the Grant-in-Aid for the scientific research of private university by the Ministry of Education, Science and Culture, Japan.

REFERENCES

Date, H. Y.Sato, T.Abe & M.Naka 1993. Impact welding of aluminum to steel by gas-gun method. *Jour. High Temperature Society,* 19-2, 69-76.
Date, H. & M.Naka 1995. Impact welding of aluminum to copper by gas-gun method. *Jour. High Temperature Society,* 21-6, 312-319
Hawkyard, J.B. 1969. Theory of the mushrooming of flat-ended projectiles impinging on a flat rigid anvil using energy consideration. *Int. Jour. mechanical Society,* 11, 313-333

Damage and Failure of Interfaces, Rossmanith (ed.) © 1997 Balkema, Rotterdam, ISBN 90 5410 899 1

Damage assessment of two-layer targets under nominal ordnance impact

C.K.Chao
Department of Mechanical Engineering, National Taiwan Institute of Technology, Taipei, Taiwan

ABSTRACT: Perforation and penetration of two-layer targets composed of aluminum and steel by a taper-nosed projectile is investigated. The proposed method in analysing the present study is based upon the energy density damage theory. Both aluminum and steel are modeled as elastic/plastic materials which were used as the based materials at the beginning of penetration process. After the target is damaged, the behavior of material elements in the target varies from one location to another depending on local strain rates which were derived in the energy desity damage theory. An initial velocity of 1300 m/sec of the projectile made of tungsten is considered to thrust through the targets with aluminum/steel and steel/aluminum layered systems. The complete time history of damage until the targets were broken in two has been obtained and provided in graphic form. The reults presented in this work will be helpful in deep understanding of penetration process.

1 INTRODUCTION

Numerical simulations of perforation and penetration of targets by projectiles have become the important subjects for many years. The work of Wilkins (1978), Zukas et al. (1982) and Anderson and Bodner (1988) provide comprehensive reviews to the subject matters. A series of articles (Chen, 1987; Johnson and Stryk, 1987) has appeared recently in the literature using the finite element method. Since it involves a very complex process due to the large strains, strain rates, the one-dimensional model of Alekeseevskii (1966) and Tata (1967, 1969) has been regarded for over 25 years in spite of simplistic assumptions. This simple model was used to analyse the interaction of long rods with thick ceramic tiles. In general, penetrators with length to diameter ratio greater than or equal 10 are considered as long rods. The one-dimensional model predicts that the curve of normalized penetration versus impact velocity has an S shape, starting at a certain critical velocity and approaching the hydrodynamic limit asymptomatically. Recently, two-dimensional computer simulations (Anderson, 1991; Sorensen, 1992) predict that, despite variations during the initial and final phases of penetration, the penetrator-target interaction as described by the analytical model is similar to that observed with the 2-D simulations.

All the aforementioned investigations only concentrate on the global response, studies on the local response of damage process is very rare to appear in the literature. The interaction between the projectile and target material under nominal ordnance and hypervelocity impact was recently studied by Chao (1994) by application of the energy density damage theory (Sih, 1985). As a continuation, in the present work, we aim to extend the previous study to consider the two-layer targets impacted by the projectile with conical-nosed shape. The response of the materials for change in local strain rates and strain history is successfully predicted by applying the energy density damage theory without making further assumptions. Only the uniaxial stress and strain response at low strain rates is needed at the very begining of impact procedure. Changes in local strain rates and strain rate history is included in the analysis such that the dynamic stress and stress behavior of each material elment is different and derived independently. Damage by permanent deformation and/or fragmentation is assumed to occur as the corresponding surface and/or volume energy density reach their critical values. These elements are thus removed from the analysis and the grid pattern must be modified accordingly. This process should be repeated for each time increment until the target is completely fractured. Damage of the projectile/target system can be observed by the fluctuation of the dW/dV and dW/dA on the damage plane. Initially, the surface energy density is dominant as the projectile first strikes the target where the energy is transferred across the interface between the target and projectile. The primary failure mode dominates and occurs before the onset of the secondary failure mode. Suppose that (dW/dA) first becomes critical, then failure by the creation of new surfaces is the primary mode. On the other hand, failure tends to spread over the volume if dW/dV first reaches the threshold. This leads to void creation as the material is fragmented into small pieces.

2 DAMAGE THEORY OF ENERGY DENSITY

The objective of this study is to predict different failure modes of the layered target impacted by a projectile. In the study of material and structure failure, say by creep, fatigue, fracture and/or collapse, there involves several pertinent parameters, each of which may be dominant for a given situation. Generally speaking, loading rate, which should be distinguished from strain rate being a local effect which is defined and assumed to be the same as the loading rate in the uniaxial test that can be measured globally, component geometry and material inhomogeneity interact each other and influence the outcome. The major aim is to express their combined effects in a form that a design engineer can follow.

2.1 Concept of energy density

The original concept of the strain energy density theory was based upon the view that a continuum may be regarded as an assembly of small elments, each of which contain a unit volume of material and can store a finite amount of energy at a given instant of time. This quantity generally referred to as the strain energy density function, dW/dV, that can fluctuate from one location to another and its stationary values can be associated with failure by deformation and/or fracture. The strain energy density function is defined as

$$\frac{dW}{dV} = \int \sigma_{ij} d\varepsilon_{ij} + f(\Delta T, \Delta C) \tag{1}$$

Equation (1) shows that energy can be stored in a material even when the stresses are zero. Hence, any failure criteria based on stress quantities alone are definitely limited in application. The assumption is that progressive damage material can be associated with the rate at which energy is dissipated in a unit volume of material. In equation (1), ΔT and ΔC stand for the changes in temperature and moisture concentration. While dW/dV serves a useful failure criterion, it is not sufficient for the determination of the stress and/or strain fields and must be applied in conjunction with the theory of plasticity that frequently leads to incompatibility of stress (or strain) and failure analysis. Moreover, dW/dV considers only volume energy although surface energy can be easily incorporated into the energy density theory via the quantity dV/dA that represents the rate change of volume with respect to surface area.

2.2 Corollary of the strain energy density theory

As a departure from the conventional approach of analyzing fracture and yielding, a nonlinear damage theory follows immediately as a corollary of the strain energy density theory (Sih,1972). Assuming that progressive material damage can be uniquely identified with the rate at which energy is dissipated across a unit area of material, the failure of an element is hypothesized to occur on the ith plane by matching

$$\left(\frac{dW}{dA}\right)_i = \left(\frac{dV}{dA}\right)_i \left(\frac{dW}{dV}\right) \tag{2}$$

with known uniaxial data. Since equation (2) applies uniquely to a given element and dW/dV is a scalar, the condition

$$\left(\frac{dV}{dA}\right)_i = \text{constant}, \quad i = \xi, \eta, \zeta \tag{3}$$

determines the damage plane.

A distinct feature of equation (2) is that $(dV/dA)_i$ may be related to $(dV/dA)_0$ obtained from a uniaxial test which may be regarded as a length of homogeneity "*l*". A regoin may thus be defined within which the stress or energy state is uniform or homogeneous. As the uniaxial stress and strain relation becomes nonlinear, dV/dA changes with the slope of the stress-strain curves. A data bank consisting of uniaxial stress and strain curves can thus be developed to cover the range of local strain rates experienced by the material elements that behave differently from one location to another. Only a single parameter $(dV/dA)_0$ is sufficient for describing the mechanical properties of the material instead of using Poisson's ratio, Young's modulus, yield strength, etc.

2.3 Determination of damage plane

Equation (3) gives the condition for determining the orientation of the damage plane and enables the consistent transformation of uniaxial data to multiaxial stress or strain states. If the material is isotropic and homogeneous, equation (3) implies that

$$\left(\frac{dV}{dA}\right)_\xi = \left(\frac{dV}{dA}\right)_\eta = \left(\frac{dV}{dA}\right)_\zeta = \left(\frac{dV}{dA}\right)_0 \qquad (4)$$

The quantity $(dV/dA)_0$ in (4) can be determined from the uniaxial test. The homogeneity of the multiaxial stress or energy state is described by $[(dV/dA)_\xi, \varepsilon_\xi]$ or $[(dV/dA)_\eta, \varepsilon_\eta]$ which are related to the uniaxial data $[(dV/dA)_0, \varepsilon_0]$. A theory of material damage based upon the energy density concept is thus formulated. As the uniaxial data bank provides known values of dV/dA and dW/dV for each energy state, the actual stress and strain path of each element can be derived as the system is loaded incrementally.

3 RESULTS

The present problem of a taper-nosed projectile impacting a target plate will be solved by the computer software program referred to as the Plane Energy Density Damage Analysis (PEDDA) code. This program has a clear advantage that the sequence of a damage pattern as the projectile penetrates through the target in an incremental manner is analysed without making any assumptions on the failure modes and the local strain rates. The material data bank has the capacity of covering strain rates from 10^{-4} to 10^{6} sec^{-1} which is more than adequate for the following examples considered in this study. A tuugsten projectile with

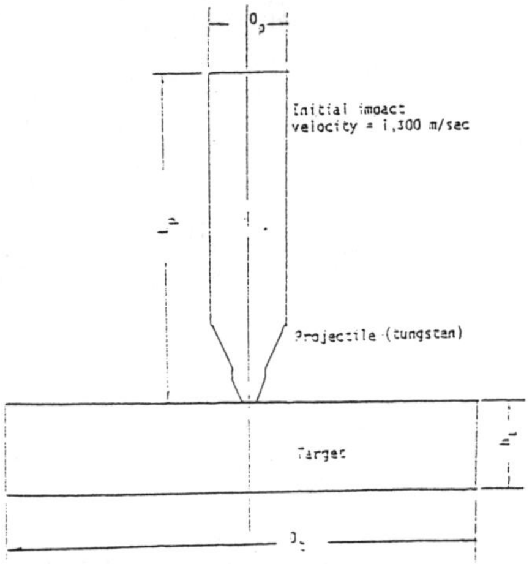

Fig. 1 Schematic of projectile/target system

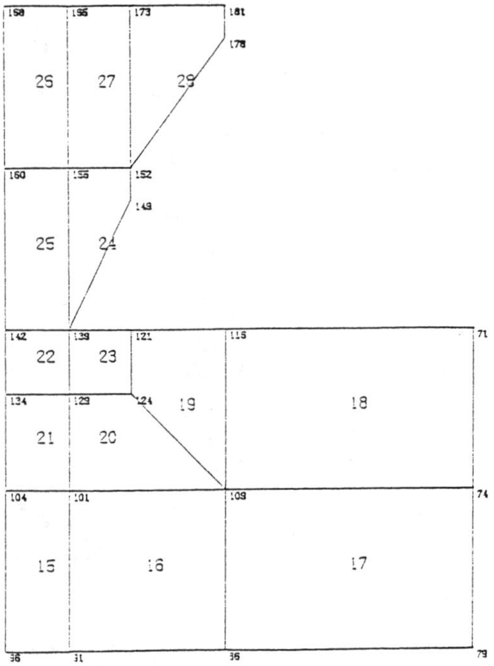
Fig.2 Numbering of nodes and elements near contact area

dimensions D_p=14 mm and L_p=47 mm is traveling at a velocity of 1300 m/sec and comes into contact with a plate with dimensions h_t = 20 mm and D_t = 100 mm as illustrated in Fig. 1. Because of symmetry, only one-half of the problem needs to be discreted by application of quadrilateral elements. A total of 34 elements and 217 nodes are addressed

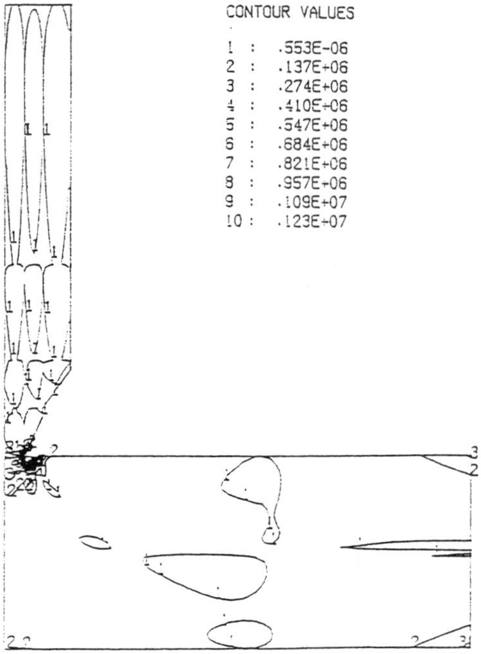

Fig. 3 Contours of dW/dA at t=0.077 μsec

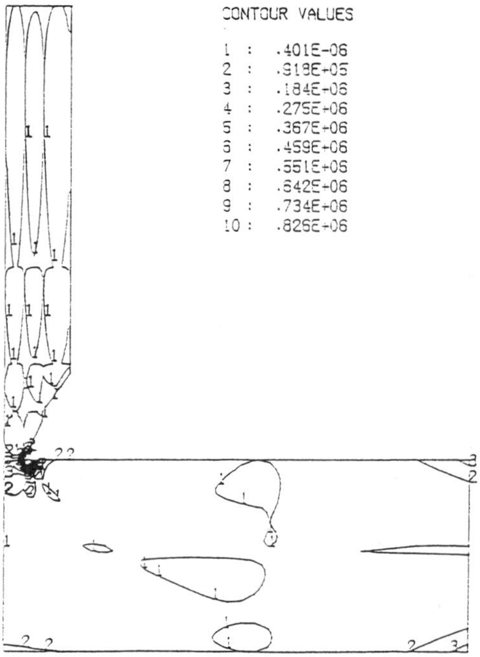

Fig. 4 Contours of dW/dA at t=0.077 μsec

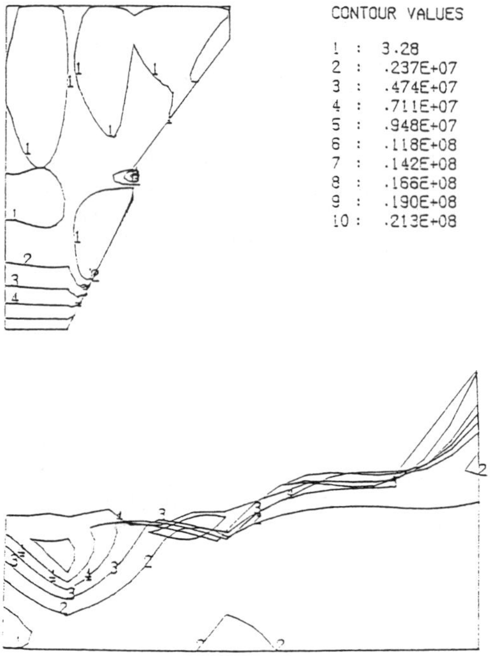

Fig. 5 Enlarged view of contours of dW/dA at t=1.54 μsec

in the initial grid pattern as shown in Fig. 2. At the early onset of impact, all 34 elements and 217 nodal points in Fig. 2 are assumed to respond in the same way. The contours of dW/dA on the damage plane and dW/dV are first obtained for t = 0.077 μsec as displayed in Figs. 3 and 4.. As the dynamic disturbances have now been further propagated into the system at t = 0.769μsec, both the surface energy density dW/dA and volume energy density dW/dV increased and further disintegration of the fragments occured as the volume energy density intensified substantially beyond $(dW/dV)_c$ = 3.0×10^6 N/m^2 as indicated in Fig. 5 and Fig. 6. After this time interval, all fragmentation is completed and the remaining energy from the projectile is concentrated along the path that outlines a plug.

4 CONCLUSION

The complete time history of damage process for a tungsten projectile impacting on a steel/aluminum target or aluminum/steel target at an initial velocity of 1300 m/sec has been obtained by application of the energy density damage theory. Impact damage

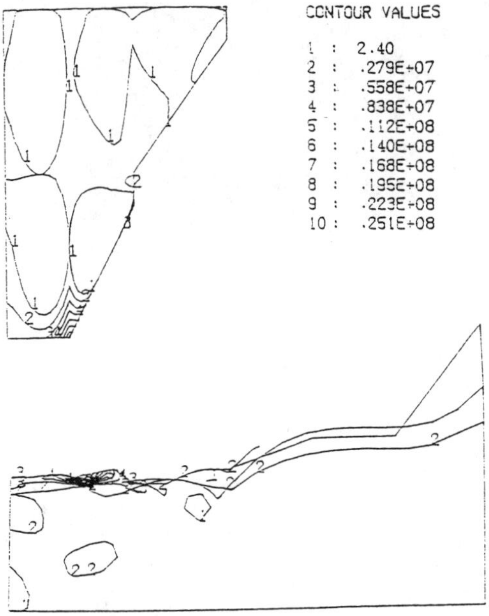

Fig. 6 Enlarged view of contours of dW/dA at $t=1.54\,\mu$sec

is a highly nonequilibrium process because the states traversed by the system can not be described in terms of constitutive parameters that present the system as a whole. Not a priori assumptions were made on the constitutive relations of the materials that are not only highly stressed but their strain rates change with location and time. Changes in local strain rates and strain rate history are derived rather than preassumed. This enables a realistic evaluation of intense deformation, fracture and fragmentation.

REFERENCES

Alekseevskii, V. P., 1966. Combusion Explosions and Shork Waves, Vol. 2 pp. 63-71.

Anderson, C. E. and Bodner, S. R., 1988, International Journal of Impact Engineering, Vol. 7, pp. 9-35.

Anderson, C. E. and Walker, J. D., 1991. International Journal of Impact Engineering, Vol. 9, pp. 247-261.

Chao, C. K., 1994. National Science Council Report, NSC 82-0401-D011-003, Republic of China.

Chen, E. P., 1987. Journal of Theoretical and Applied Fracture Mechanics, Vol. 8, pp. 125-135.

Johnson, G. R. and Stryk, R., 1987. The Post-Smirt Seminar on Impact, Laussane, Switzerland.

Sih, G. C., 1972. International Journal of Engineering Fracture Mechanics, Vol. 5, pp. 365-377.

Sih, G. C., 1985. Journal of Theoretical and Applied Fracture Mechanics, Vol. 4, pp. 157-173.

Sorensen, B. R., Kimsey, K. D., Silsby, G. F., Scheffler, R. D., Sherrick, T. M., and Derosset, W. S., 1991. International Journal of Impact Engineering, Vol. 11, pp. 107-121.

Tate, A., 1967. Journal of the Mechanics and Physics of Solids, Vol. 15, pp. 387-393.

Tate, A. 1969. Journal of the Mechanics and physics of Solids, Vol. 17, pp. 141-158.

Wilkins, M. I., 1978. International Journal of Impact Engineering, Vol. 16, pp. 793-807.

Zukas, J. A., Nichols, T., Swift, H. F., Greszezuk, L. B., and Curran, C. R., 1982. Impact Dynamics, Wiley, New York.

Debonding and pull-out

Damage and Failure of Interfaces, Rossmanith (ed.) © 1997 Balkema, Rotterdam, ISBN 90 5410 899 1

Analysis of singularities and interfacial crack propagation in micromechanical tests

C. Marotzke
Federal Institute for Materials Research and Testing (BAM), Department VI.2, Berlin, Germany

ABSTRACT: Micromechanical tests such as single fiber pull-out, fragmentation and indentation tests are used for the characterisation of the interface strength between fiber and matrix. At kinks in the interface, e. g. at fiber ends, severe stress concentrations are found, which result from singularities within the linear theory of elasticity. The dependency of the singularities on the elastic constants and on the phase angles is studied in this paper. Furthermore, the failure process taking place in pull-out and fragmentation tests is analysed by means of the finite element method using a contact algorithm. The mixed mode energy release rate is calculated taking into account interfacial friction. Furthermore, the influence of thermal residual stresses due to the cooling process of thermoplastic matrices is studied.

1 INTRODUCTION

The failure of fiber reinforced polymers is a process usually starting at microdefects, e.g. local imperfections in fiber, matrix or interface. During increase of loading, the microscopical failure processes accumulate until the overall failure occurs. The way the failure develops is strongly influenced by the capability of the composite material to redistribute stresses around the microdefects, e.g. to transfer stresses from a broken fiber into the surrounding matrix and neighbouring fibers. This capability is governed by the interface strength. For the characterisation of the interface strength, several micromechanical test methods have been developed. These tests either are based on single filaments, e.g. pull-out and fragmentation test, or on a multifiber composite, e.g. indentation test. This paper is confined to single filament tests.

In the pull-out test, a single fiber is embedded into a polymer drop (fig 1a). The fiber is axially loaded in a micro-testing machine until the total interface is broken. Subsequently, the fiber is pulled out against interfacial frictional forces. The fragmentation test specimen consists of a single fiber embedded in a small polymer bar (fig. 1b). For this test, the matrix has to be transparent with a strain to failure

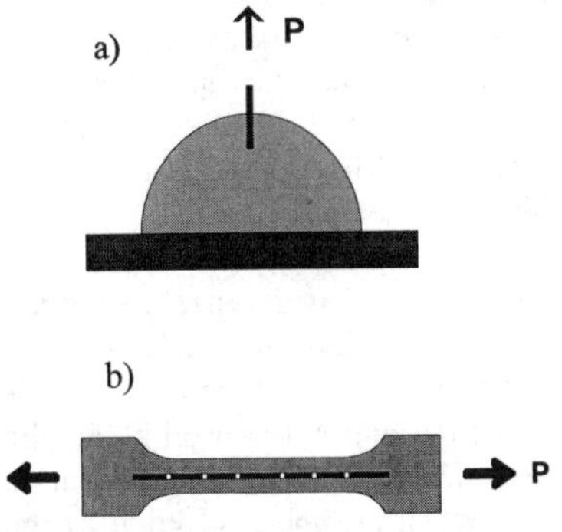

Figure 1. Sketch of pull-out test (a) and fragmentation test (b).

significantly higher than that of the fiber. The bar is drawn and the fiber is fragmented until saturation is reached, this is, no further fiber breaks occur during further loading.

Due to the small scale of the specimens of either test, a direct measurement of the stress field around the interface is nearly impossible. The only method which gives reasonable results for specific fibers and matrices is the Laser Raman spectroscopy (Young et al. 1995). With this method, the axial strains in the fiber can be measured and the interfacial shear stresses can be calculated by derivation, however, resulting in a loss of accuracy.

The evaluation of the test data commonly is performed by using so called 'data reduction schemes', which are based on strong simplifications of the stress field around the fiber. These procedures are of limited value, since a complex, threedimensional stress state is actually encountered around the interface, as shown by finite element calculations (Atkinson et al. 1982, Desarmot & Favre 1991, Marotzke 1994). Severe stress concentrations arise at the fiber break in the fragmentation test as well as at the fiber entry and at the fiber end in the pull-out test. These are related to singularities within the linear theory of elasticity, which are discussed in the next section.

The simple 'data reduction schemes' characterise the bond strength merely by shear stresses, neglecting the radial stresses, which, however, can be tensile. Two kinds of data reduction schemes are found in the literature, one is based on elasticity, the other one on plasticity. The first model simplifies the elastic stress field using the shear lag hypothesis (Piggott 1980, Hsueh 1988). This model is inconsistent in that it assumes constant axial stresses in the matrix, violating the boundary conditions in the interface and, furthermore, underestimates the stress concentrations at the critical points by far (Marotzke 1993). Still it is widely used, mainly for the pull-out test.

The second model uses an averaged shear stress, implying the assumption of unbounded yielding of the matrix near the interface (Kelly & Tyson 1965). This model is applied for the pull-out as well as for the fragmentation test. The two main inconsistencies inherent in this approach are the neglection of the other stress components and the unrealistic strains resulting from the unbounded yielding of the matrix.

Due to the nature of the stress field, it is not reasonable to conduct a stress analysis in order to characterise the bond strength between fiber and matrix, because the stresses at the critical points depend on the local geometry and on the local material properties, which may differ significantly from that of the bulk material. The alternative to a stress analysis is a fracture mechanical analysis of the micromechanical tests, characterising a fracture toughness or an energy release rate. This is outlined in section 3. First, however, the singularities arising in the interface at specific locations are studied.

2 STRESS SINGULARITIES

For a stress field or fracture mechanical analysis of the pull-out and fragmentation test, simplified models have to be used. The geometry commonly is idealised by regarding the fiber as a cylinder. Furthermore, the coupling conditions in the interface and the material behaviour of the two phases have to be idealised. A perfect bond with no microdefects usually is assumed except of the specified cracked regions. In this paper, both phases are assumed to be homogeneous and to behave linear elastically.

Obviously, with these idealisations, singularities arise in the models of the pull-out and fragmentation test at kinks in the interface. In the pull-out test, singularities arise at the circumference of the cylinder, one at the fiber entry, i.e. at the intersection with the matrix surface and a second one at the embedded fiber end. Another type of singularity is found at a broken fiber end in the fragmentation test (penny shaped crack) or at a debonded fiber end in the pull-out test.

2.1 *Wedge problem*

The problem of singularities at kinks of surfaces in one-phase materials, known as 'wedge problems', was solved by Williams (1952), using the stress function:

$$F(r,\theta) = r^{(2-\lambda)}\,[a \sin(2-\lambda)\theta + b\cos(2-\lambda)\theta + c\sin\lambda\theta + d\cos\lambda\theta\,]$$

The stresses then read as

$$\sigma_{ij} = r^{-\lambda}\, h_{ij}\,(\lambda,\theta)$$

with $h_{ij}(\lambda,\theta)$ being shape functions. The eigenvalue problem leads to the solution of a 4x4 determinant corresponding to the four boundary conditions. Zak (1964) developed the solution for axisymmetrical problems. He showed that the plane strain conditions are asymptotically fulfilled with $r \to 0$. In the bimaterial case, a 8x8 determinant has to be solved as a result of four additional coupling conditions at the interface. Solutions for this problem were given by Dempsey and Sinclair (1979, 1981) for several boundary conditions. For the investigation of micromechanical tests, two cases are of interest, the 'open bimaterial wedge' and the 'closed bimaterial wedge' problem (fig.2). In case of the open wedge, boundary conditions for the normal and shear stresses are prescribed at L_0 and L_2 while coupling conditions for

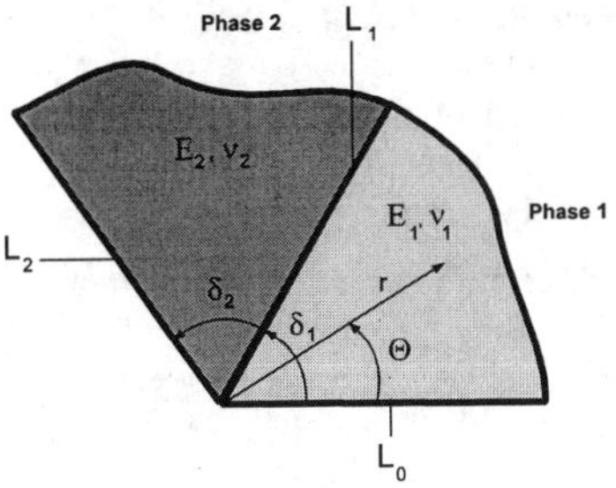

Figure 2. Sketch of bimaterial wedge.

the stresses and displacements are given at L_0. In case of a closed wedge, coupling conditions also are given at $L_0=L_2$.

The eigenvalue equation for the open bimaterial wedge, which corresponds to the situation at the fiber entry or at the debonded fiber end, reads as

$$\begin{aligned}0 = 4\,&((1+\beta)\sin^2\varrho\delta_1 - \varrho^2(\beta-\alpha)\sin^2\delta_1)\\ &((1-\beta)\sin^2\varrho\delta_2 + \varrho^2(\beta-\alpha)\sin^2\delta_2)\\ &-(1-\alpha)^2(\sin^2\varrho\delta_1 - \varrho^2\sin^2\delta_1)\\ &-(1+\alpha)^2(\sin^2\varrho\delta_2 - \varrho^2\sin^2\delta_2)\\ &-2(1-\alpha^2)(\sin\varrho\delta_1\sin\varrho\delta_2\cos\varrho(\delta_2-\delta_1)\\ &-\varrho^2\sin\delta_1\sin\delta_2\cos(\delta_2+\delta_1))\,.\end{aligned}$$

$$\begin{aligned}0 = &((1+\beta)^2\sin^2\varrho\delta_1 - \delta^2(\beta-\alpha)^2\sin^2\delta_1)\\ &((1-\beta)^2\sin^2\varrho\delta' - \varrho^2(\beta-\alpha)^2\sin^2\delta')\\ + &(1-\alpha^2)\sin^2\varrho(\pi-\delta_1)(2\varrho^2(\beta-\alpha)^2\sin^2\delta_1\\ + &2(1-\beta^2)\sin\varrho\delta_1\sin\varrho\delta' + (1-\alpha^2)\sin^2\varrho(\pi-\delta_1))\end{aligned}$$

With $\rho = 1-\lambda$, δ_i being the angles of phases with $\delta' = 2\pi - \delta_1$ and

$$\alpha = \frac{\mu_2(\kappa_1+1) - \mu_1(\kappa_2+1)}{\mu_2(\kappa_1+1) + \mu_1(\kappa_2+1)},$$

$$\beta = \frac{\mu_2(\kappa_1-1) - \mu_1(\kappa_2-1)}{\mu_2(\kappa_1+1) + \mu_2(\kappa_2+1)}$$

are the Dundurs constants with μ_i = shear moduli and $\kappa_i = 3 - 4\nu_i$ for plane strain.

In order to ensure a finite strain energy, the range of the real part of λ is confined to $0 \le Re\ \lambda < 1$. Within this range, the maximal λ is searched, i. e. the highest order of singularity, both, from the real as well as from the complex solution. Since the eigenvalue found by a numerical solution of the nonlinear

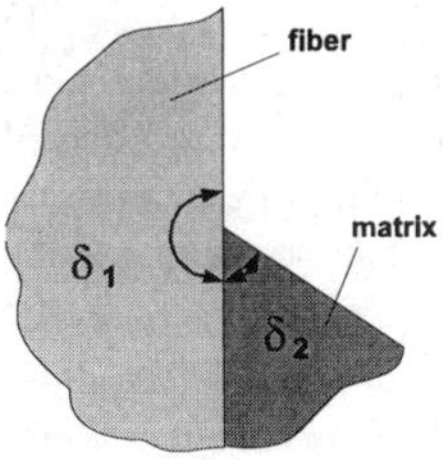

Figure 3 Sketch of fiber entry.

equation depends on the initial value, several solutions with different start values are calculated.

2.2 *Singularity at fiber entry*

First, the singularity arising at the fiber entry in the pull-out test is studied. The angle of the first phase, say the fiber, is taken as 180°. However, the angle between fiber and matrix can vary over a wide range in an actual specimen, since the matrix often forms a meniscus as a result of shrinking, either due to cooling from melt in case of a thermoplastic matrix or due to curing in case of a thermosetting matrix. The shape of the menisci is very different, since it strongly depends on the manufacturing process. The matrix angle is varied between 0° and 180° in order to take into account all kinds of matrix shapes (fig. 3).

The eigenvalues are plotted in figure 4 for various elastic moduli of fiber and matrix, presuming zero Poisson ratios for both phases. The low stiffness ratios correspond to ceramic composites. The calculations show that all traces reach the value of 0.5 at 180°, this is, the classical crack tip singularity. This means that the order of singularity of an interface crack is equal to that of a crack in a homogeneous material,

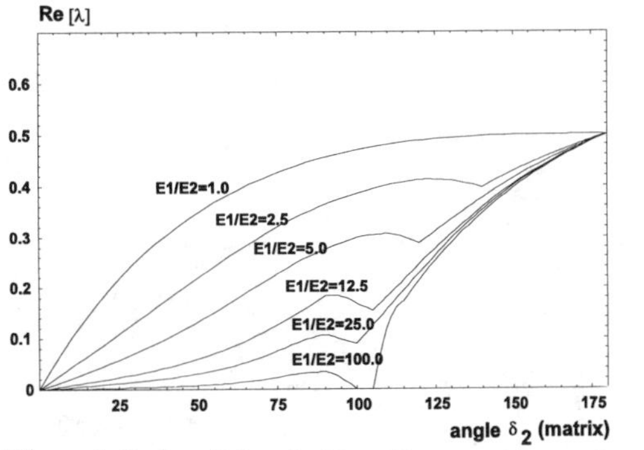

Figure 4. Order of singularity at fiber entry for various stiffness ratios.

irrespective of the elastic moduli of the two phases. In case of identical stiffnesses, i.e. a single phase, the maximal eigenvalue is determined by the solution of the real eigenvalue problem, leading to a continuous course. With increasing fiber modulus, the eigenvalues are diminished, especially within the domain of the real solution (small angles). After reaching a maximum, the real eigenvalue decrease and fall below the complex one. The kinks in the traces, accordingly, mark the transition from the real to the complex solution. Within the complex domain, the dependency on the elastic moduli is much less than within the real domain.

2.3 *Singularity at bonded fiber end*

Concerning the fiber end, two configurations are of interest, this is, a bonded and a debonded fiber end (bottom surface). In actual specimens, the fiber end usually does not meet the idealised shape, i.e. a flat bottom surface under 90° inclined to the cylinder wall, but is more ore less rugged. This is due to the fact that the fibers are cut with a razor blade or broken by bending before embedding into the matrix. The angle between the bottom surface and the wall may therefore vary over a wide range. To take this effect into account, the angel of the first phase (fiber) is varied between 0° and 180° (fig. 5).

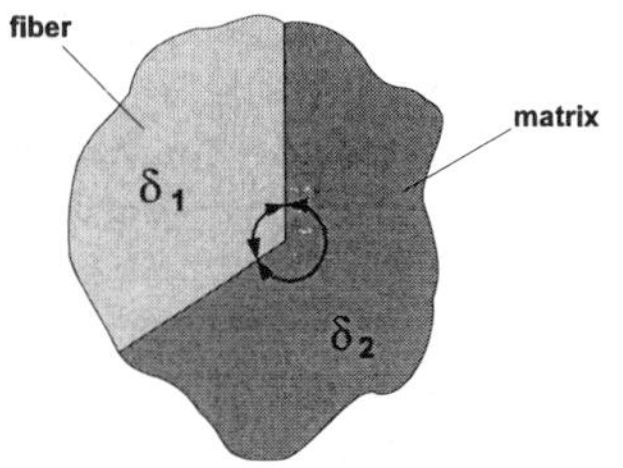

Figure 5. Sketch of bonded fiber end.

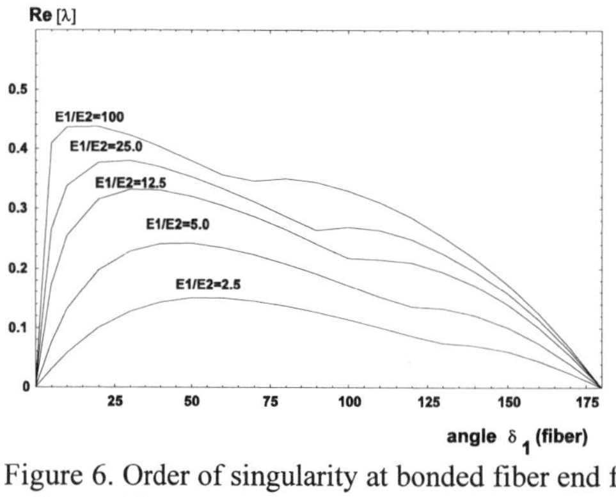

Figure 6. Order of singularity at bonded fiber end for various stiffness ratios.

As expected (fig.6), the singularity vanishes at a fiber angle of 0° (no fiber) as well as at 180° (continuous fiber). In case of a high fiber stiffness, the order of singularity increases very steeply reaching a maximum at small angles. This means that in case of very sharp edges and very stiff fibers nearly the interface crack conditions are met. With decreasing stiffness ratio, the maximum decreases and moves to greater angles. The kinks in the traces again mark the change from the real to the complex eigenvalue problem.

2.4 *Singularity at debonded fiber end*

In practical pull-out experiments, a debonding of the bottom surface is often observed long before an interface crack forms. A similar situation is found in the fragmentation test after fiber breakage. While the fiber ends in the pull-out specimen often are rugged, as mentioned above, the fiber ends created by fiber breakage in the fragmentation test are almost perfect, i.e., they have flat bottom surfaces. This configuration is studied by prescribing the angle of the fiber as 90° and varying the matrix between 90° and 270° (fig. 7). At matrix angles less or equal to 90°, the singularity vanishes. With increasing matrix angle, the order of the singularity grows (fig. 8). It reaches 0.5 at 270° in case of identical stiffnesses (single phase), corresponding to the classical crack problem. With increasing fiber stiffness, the order of singularity exceeds 0.5, this is, the order of singularity at the tip of the penny shaped crack meeting the cylindrical fiber surface is higher compared with an interface crack along the fiber.

While the order of singularities arising at kinks of surfaces in homogeneous materials only depends on the angle between the two parts, it additionally depends on the elastic constants of the respective

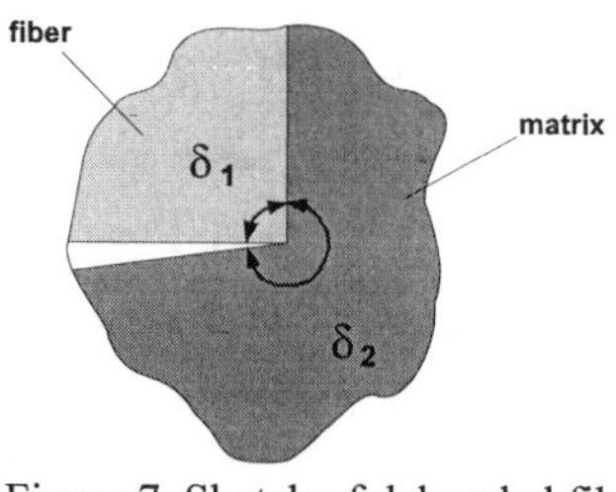

Figure 7. Sketch of debonded fiber end.

phases in case of bimaterial joints. Due to the dependency on the elastic constants, it is not possible to utilise the stress intensities for a characterisation of the interface strength. Accordingly, a fracture mechanical analysis of the failure process arising in the respective tests has to be performed (Atkinson et al. 1982, Gao et al. 1988, Liu et al. 1994)

3 PROPAGATION OF INTERFACE CRACKS IN PULL-OUT AND FRAGMENTATION TESTS

The failure processes taking place in pull-out and fragmentation tests are different. In the pull-out test, an interface crack starting at the fiber entry propagates stably towards the fiber end. Some fiber diameters before the fiber end is reached, the crack becomes unstable, indicated by a sharp decay of the load (Hampe et al. 1995, Piggott et al. 1995). In the fragmentation test, the fiber break is accompanied by a short interface crack or by short matrix cracks, perpendicular or inclined to the fiber axis (Pegoretti et al. 1996). During further loading, either the interface crack propagates or, depending on the fracture toughness of interface and matrix, the matrix cracks grow. In order to gain some insight into the process of interfacial failure, the crack propagation is simulated by means of the finite element method.

3.1 *Finite element model*

The pull-out test model consists of a fiber of ten microns in diameter embedded perpendicular into a halfsphere of matrix up to a length of 150 microns. The fragmentation test model of 400 microns in length corresponds to one half of a fragment, applying

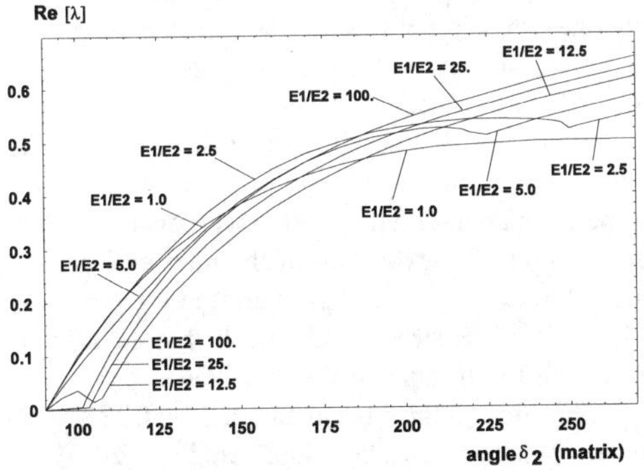

Figure 8. Order of singularity at debonded fiber end for various stiffness ratios.

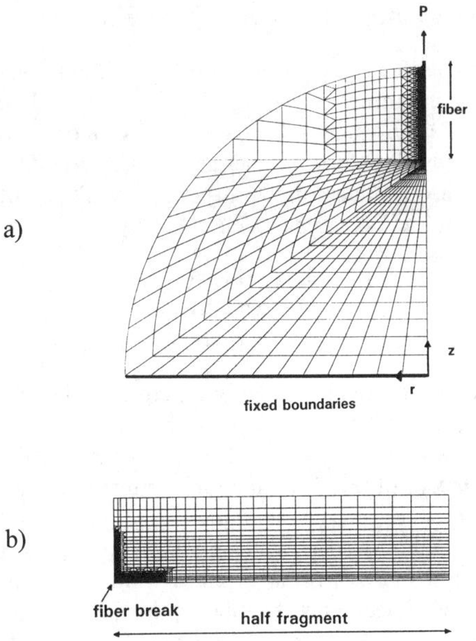

Figure 9. Finite Element meshes; pull-out test (a), fragmentation test (b)

symmetry conditions at the edges. For both specimens, axisymmetrical finite elements are used (fig. 9a, 9b). In the pull-out test, 8-node isoparametrical elements are used while in the fragmentation test 4-node elements applied. The meshes are strongly refined around the interface with respect to the high stress gradients arising there.

A crack increment of 1/10 of the fiber diameter is prescribed in the pull-out test, while in the fragmentation test, it is even 1/60 of the fiber diameter, i.e. 0.167 μm, in order to achieve a stable numerical solution. In the interface, contact elements are arranged together with links between the anticipated crack faces, which are removed to simulate crack propagation.

The analysis of interface cracks with contact and friction by finite elements is not a trivial task. The solution is very sensitive against the crack increment and parameters like the sliding tolerance and the penetration tolerance. Small changes of the numerical parameters or of the geometry or the material data can result in a failure of convergence or in oscillating stresses along the cracked interface. Hopefully, these problems will be overcome with a new release of the program.

3.2 *Energy release rate during pull-out test*

In the pull-out test, fixed load conditions are prescribed, corresponding to the level at failure initiation as observed in the experiments. In order to estimate the influence of thermal stresses resulting from the manufacturing process as well as of interfacial friction, three models are analysed:

- no thermal stresses, no interfacial friction
- thermal stresses, no interfacial friction
- thermal stresses, interfacial friction

In case of a frictionless interface excluding thermal stresses, the energy release rate decreases during the first phase of the crack (fig. 10). This is due to the redistribution of the shear and radial stresses around the matrix surface. The radial tensile stresses, which are very high in a small zone around the interface at the matrix surface before crack initiation, are significantly decreased by the formation of the interface crack. At the same time, the shear stresses are dramatically increased, resulting in a decrease of the energy release rate. With the further extension of the crack, the energy release rate starts to grow again very rapidly within a crack length of two fiber diameters, followed by an almost constant phase. If the crack approaches the fiber end, the energy release rate again increases very steeply, resulting in an unstable crack propagation as observed in practical experiments.

In case of a thermoplastic matrix, a large portion of energy is stored in the specimen as a result of the cooling process. The fiber is compressed while the matrix around the fiber is in tension. If the interface crack propagates, a part of the stored energy is released. Furthermore, the external work done by the pull-out force acting at the fiber tip is increased by the additional displacement of the fiber tip as a result of the initial stresses. The thermal stresses are taken into account approximately by simulating the cooling process from glass transition to room temperature, which corresponds to a temperature difference of 130°. Due to the thermal stresses, the energy release rate is enhanced dramatically. It increases very rapidly during the first phase of the crack, followed by a plateau. If the crack approaches the fiber end, the energy release rate is growing again. These results show that stored energy due to thermal stresses has to be taken into account for the evaluation of the test data, which, however, is often neglected in practice.

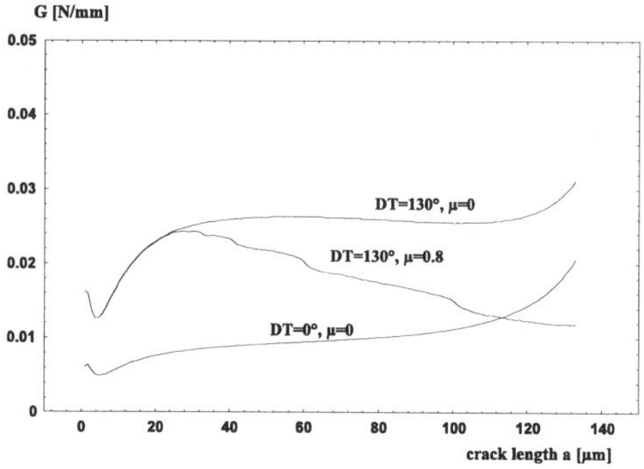

Figure 10: Energy release rate arising in pull-out test with and without thermal stresses and friction.

In pull-out experiments, frictional sliding can be observed after total failure of the interface. The level of the frictional forces varies over a wide range, depending on the specific material combination. In order to get an estimation of the upper limit of the influence of interfacial friction, a friction coefficient of 0.8 is taken. Within the first phase of crack propagation, no remarkable friction develops in the debonded interface. At a crack length of about three fiber diameters, frictional shear stresses develop resulting in an almost linearly increasing work of friction, which diminishes the energy release rate significantly. Due to the frictional forces, a stable crack propagation is found in the experiments, while the force is raised.

3.3 *Energy release rate during fragmentation test*

In the fragmentation test, fixed grips conditions are prescribed. The average axial strain of the specimen corresponds to the strain to failure of the fiber. The thermal residual stresses are taken into account. While the extension of the interface crack results in an observable increase of the compliance in the pull-out test, the compliance of the fragmentation test specimen is nearly insensitive against fiber fractures or interface cracks as a result of the large matrix volume compared with the fiber. The redistribution of stresses due to the extension of the interface crack is confined to a very small region around the crack tip. An analysis of the stress field reveals that the force in the fiber in the middle of the fragment decreases by only 0.7 % due to the fiber break and by 0.5% by an interface crack of two fiber diameters in length.

The course of the energy release rate of a

fragmentation test is totally different from that of a pull-out test. With the breakage of the fiber, a large amount of stored energy is released, which is much higher than that dissipated by the fiber break. The unused energy causes an unstable interface or matrix crack. In general, both cracks may develop simultaneously. The process occurring directly after fiber break is very complex and requires an analysis including dynamical effects. However, no reliable local material data are given at present for the respective matrices under high speed conditions. The present analysis, therefore, is confined to static conditions, which are valid for the second phase of stable crack propagation.

The energy release rate of an interface crack is plotted in fig. 11 with and without interfacial friction. The abscissa corresponds to the pure interface crack length, not including the fiber break. The first point (a=0) represents the average of the energy release rate of the fiber break. The maximum, on the other hand, represents the energy release rate of the first increment of the interface crack. In the first phase of the crack, a large amount of energy is released. However, the energy release rate drops to about 20% of the maximum value within a crack length of only 20% of the fiber diameter. Beyond this region, it slightly decreases, leading to a stable crack propagation. In order to get an estimation of the upper limit of the influence of friction, the same system is analysed with a friction coefficient of 1.0. As a result of friction, the energy release rate is significantly diminished to about one quarter of the value without friction. Even though the effect of friction is overestimated (μ=1.0), the analysis shows its essential influence on the released energy. It comes out that the evaluation of fragmentation tests without consideration of the work of friction, which sometimes is found in the literature, leads to unreasonable results.

The question arises, where the released energy comes from. To this end, the contributions of fiber, matrix as well as the energy dissipated by friction are calculated. By the fracture of the fiber (a=0), the fiber is unloaded, releasing a large amount of strain energy (fig.12). At the same time, the matrix around the fiber break is stressed dramatically, accordingly consuming energy. If the crack has run a few microns, the strain energy released by the fiber is growing almost linearly, while the energy consumed by the matrix remains constant. Due to the increasing frictional work, the total amount of released energy increases

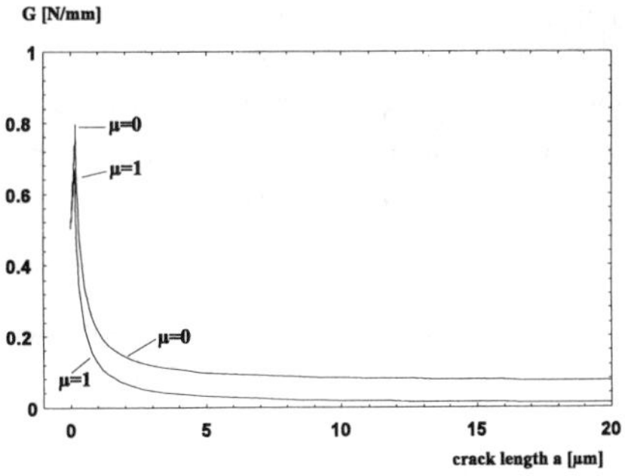

Figure 11. Energy release rate arising in fragmentation test with and without friction.

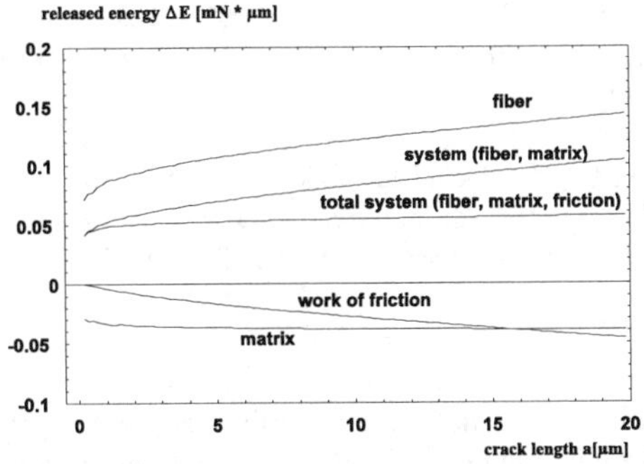

Figure 12. Released energy by fiber and matrix and frictional work in fragmentation test

very slowly resulting in a low energy release rate (fig. 11). Obviously, the energy released without frictional work is growing much faster leading to a significantly enlarged energy release rate (fig.11).

4 CONCLUSIONS

The analysis of the singularities arising at discontinuities of interfaces in micromechanical tests shows that they strongly depend on the material parameters as well as on the local geometry. As a result, it is not possible to characterise the bond strength in micromechanical tests by the stress intensities before the interface fails. Instead, a fracture mechanical analysis should be performed. The simulation of the interfacial failure arising in pull-out and fragmentation tests reveals the strong influence of the thermal stresses as well as of interfacial friction. Both effects, accordingly, have to be taken into account in the evaluation of the experimental data.

REFERENCES

Atkinson, C., J. Avila, E. Betz & R. E. Smelser 1982. The rod pull out problem, theory and experiment, *Journal of the Mechanics and Physics of Solids* 30: 97-120.

Dempsey, J. P. & G. B. Sinclair 1979. On the singular behaviour at the vertex of a bi-material wedge, *Journal of elasticity* 9: 373-391.

Dempsey, J. P. & G. B. Sinclair 1981. On the stress singularities in the plane elasticity of the composite wedge, *Journal of elasticity* 11: 317-321.

Desarmot, G., & J. P. Favre 1991. Advances in pull-out Testing, *Composites Science and Technology* 42: 151-187.

Gao, Y. C., Y. W. Mai & B. Cotterell 1988. Fracture of fiber reinforced materials, *Journal of Applied Mathematics and Physics* (ZAMP) 39: 550-572.

Hampe, A. & C. Marotzke 1995. The fracture toughness of glass fibre/polymer matrix interfaces: measurement and theoretical analysis, *Proceedings of the third International Conference on Deformation and Fracture of Composites*, Surrey.

Hsueh, C. H. 1988. Elastic load transfer from partially embedded axially loaded fibre to matrix, *Journal of Materials Science Letters* 7: 487-500.

Kelly, A., & W. R. Tyson 1965. Tensile properties of fibre reinforced metals: copper/tungsten and copper/molybdenium, *Journal of the Mechanics and Physics of Solids* 13: 329-350.

Liu, H.-Y., Y. W. Mai, L. M. Zhou & L. Ye. 1994. Simulation of the fibre fragmentation process by a fracture mechanics analysis, *Composites Science and Technology* 52: 253-260.

Marotzke, C. 1993. Influence of the fiber length on the stress transfer from glass and carbon fibers into a thermoplastic matrix in the pull-out test, *Composite Interfaces* 1: 153-166.

Marotzke, C. 1994. The elastic stress field arising in the single fiber pull-out test, *Composite Science and Technology* 50: 393-405.

Piggott, M. R. 1980. *Load bearing fibre composites*, Oxford, Pergamon Press.

Piggott, M. R. & Y. J. Xiong 1995. Visualisation of debonding of fully and partially embedded glass fibres in epoxy resins, *Composite Science and Technology* 52: 535-540.

Young, R. J., Y.-L.Huang, X. Gu & R. J. Day 1995. Analysis of composite test methods using Raman spectroscopy, *Plastics, Rubber and composites processing and applications* 23: 11-19.

Williams, M. L. 1952. Stress singularities resulting from various boundary conditions in angular corners of plates in extension, *Journal of applied mechanics* 9: 526-528.

Zak, A. R. 1964. Stresses in the vicinity of boundary discontinuities in bodies of revolution, *Journal of applied mechanics* 31:150-152.

Damage and Failure of Interfaces, Rossmanith (ed.) © 1997 Balkema, Rotterdam, ISBN 90 5410 899 1

Characterization of damage and failure during the pull-out of a reinforcement bar by acoustic emission and numerical simulation

S.Weihe, F.Ohmenhäuser & B.-H.Kröplin
Institute for Statics and Dynamics of Aerospace Structures, University of Stuttgart, Germany

C.U.Grosse & H.-W.Reinhardt
Institute of Construction Materials, University of Stuttgart, Germany

ABSTRACT: The presented initiative is an attempt to combine both, numerical simulation and experimental techniques, to further understand the phenomena which characterize the pull-out of a reinforcement bar from the surrounding concrete 'matrix'. Based on the Finite-Element Method, non-linear constitutive relations are introduced to reflect the sliding of the reinforcement bar inside the concrete and to account for fracture in the concrete mass. The experimental setup is equipped with quantitative acoustic emission. The spatial and temporal location of the sliding and fracture events is recorded during the test and the underlying fracture mode for the acoustic events is evaluated via a relative moment tensor inversion technique. Thus, characteristic values, such as the relative amount of the occurring fracture modes, the specific energy release rates separated for the generic fracture modes, and the specific location of the degradation events are available from both approaches. The first results of this ongoing research effort are very encouraging and strongly support the combined experimental and numerical investigation of such complex challenges.

1 INTRODUCTION

Fibre reinforced materials are todays and tomorrows materials due to their versatile potential: Fibre reinforcement provides in the field of polymers and metals increased strength and stiffness especially in the high temperature range, fibre reinforced ceramics become damage tolerant. In concrete technology, fibre reinforcement consisting of steel or corrosion-resistant polymer is well established and steel fibre reinforced shot concrete is one of the most challenging new developments. The prototype for all these materials is nature: Fibre reinforcement is one of its key structural paradigms: Human bone, grain stalks or wood are impressive examples.

The quality of these materials is decisively influenced by the interface between fibre and the embedding matrix. Although being of negligible volume ratio, this mechanism is of paramount importance in providing the desired material characteristics. In ambivalence, it is a dominant factor in the initiation and propagation of damage and failure. In this qualitative sense, most fibre reinforced materials bear a common ground which is the scientific foundation behind the Collaborative Research Centre SFB 381 'Characterization of Failure Propagation in Fibre Reinforced Materials Using Non-Destructive Methods' which is being granted to the University of Stuttgart by the German Research Foundation (DFG). Ten institutes of two universities, and regional as well as

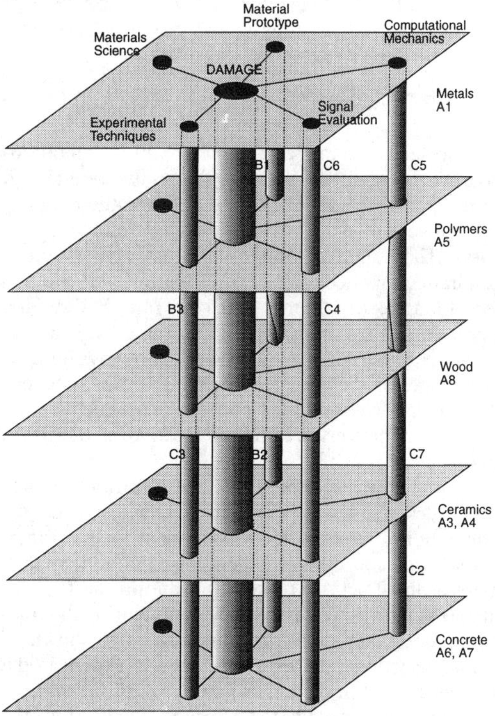

Figure 1: The structure of the SFB 381 (SFB 381, 1997).

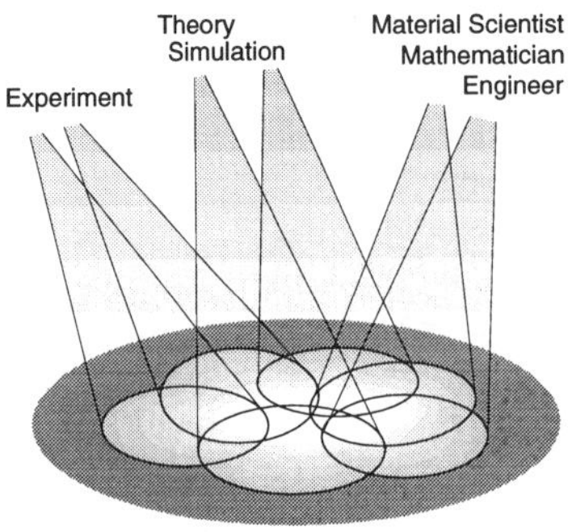

Figure 2: Various exploring techniques reveal different aspects of an object.

national research organizations cooperate within thirteen research projects A1–C7 in aiming for an optimal symbiosis of the scientific expertise, which is available for the various material classes as shown in Figure 1 (SFB 381, 1997). The individual exploring techniques, based on the specific hypothesis as they are stated by material scientists, physicists, mathematicians, and engineers, reveal different aspects of the failure process in fibre reinforced materials. The combination of all these views may provide a better understanding of the damage accumulation and propagation in these mechanically very complex materials.

2 THE COMMON TASK

The presented example of a successful bi-lateral cooperation within the SFB combines theoretical and experimental techniques to characterize the debonding between concrete and embedded reinforcement bars. The concrete samples are rectangular with a quadratic cross-section of 150x150 mm^2 and two different lengths of 150 mm and 1000 mm, both with a single reinforcement along the central axis. The long specimens are used to study the effect of environmental changes on the pre-stressed configuration, whereas the cubical ones serve for pull-out experiments. The reinforcement consist either of Aramid or of ribbed steel bars.

From a theoretical point of view, the fracture process is described using Fracture Mechanics. Thus, the orientation of the cracks with respect to the principal stresses and the corresponding relative displacement at the crack tip (both representing the fracture mode) as well as the dissipated energy during the fracture process are of primary interest. As shown in the next sections, these parameters are present in the simulation and in the experiment. Additionally, both methodologies provide their specific insight into the process: The simulation yields the stress and strain state in the specimen and the process can be simulated

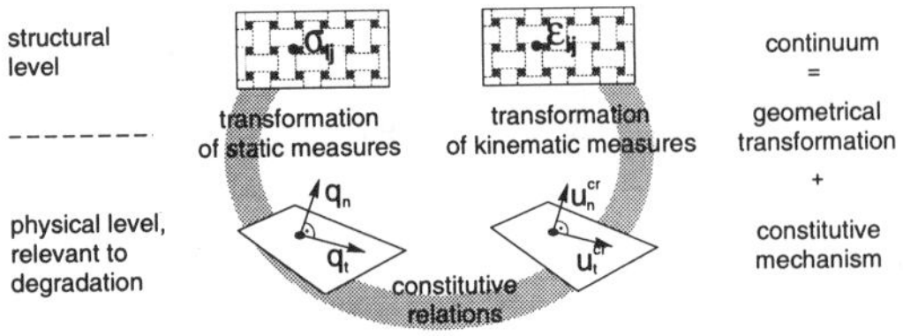

Figure 3: Transformation of kinematic and static measures between physical and structural level.

with any desired spatial and temporal resolution. The experiment, however, provides the ultimate reference to the simulation as being the 'true solution'.

3 SIMULATION

3.1 *Hypothesis*

Fracture will be introduced as a phenomena which is of physical as well as of geometrical nature. The physical degradation of the material is characterized by a softening mechanism and describes the degradation of the material strength and/or the material stiffness. The geometrical aspect is reflected by the usually well defined failure pattern, which defines an orientational preference for the material degradation. Thus, the initiation and evolution of the failure mechanism is controlled in the so called plane of degradation (POD), which constitutes a micromechanical perception of the fracture plane. The physical deterioration processes in the two-dimensional POD are homogenized by a purely geometrical transformation in order to obtain the behaviour of the effective three-dimensional continuum. This geometrical transformation can be uniquely defined by the orientation of the emerging displacement discontinuities.

The presented hypotheses are implemented in the framework of the Finite Element Method. Thus, it is possible to predict the anisotropic evolution of the residual strength capacity, which is initiated through the individual fracture processes. Due to the microphysical motivation of the numerical simulation, the predicted results can be correlated directly to data obtained from non-destructive testing methods, i.e. the quantitative acoustic emission technique. Thus, the micromechanically based approach of the POD presents a valuable alternative to phenomenological constitutive models, which in general are not tailored to provide an in depth understanding of the individual failure processes but concentrate on the accurate prediction of the overall behaviour.

3.2 *Material Degradation*

The basic hypothesis for fracture is based on the characterization of the failure initiation by a strength criterion, whereas the subsequent failure propagation is

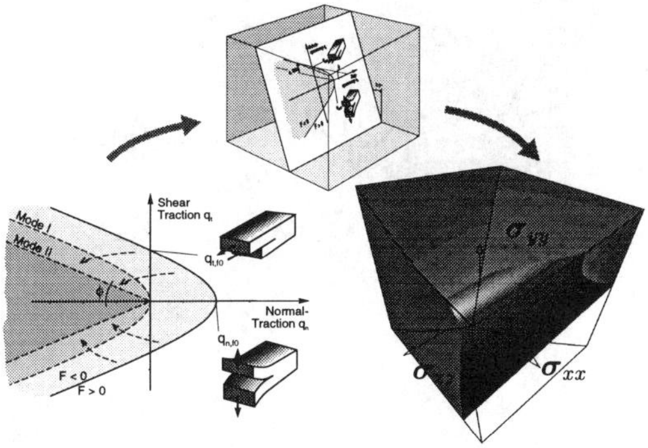

Figure 4: 2D-fracture embedded in a 3D-continuum: A geometrical transformation.

controlled by the dissipated energy. The model is explained in further detail in Kröplin and Weihe (1997): The strength capacity of the material is defined in terms of the fracture criterion F, which is a function of the tractions on the critical POD. It discriminates between stress states that the material is able to sustain ($F \leq 0$), and inadmissible stress states ($F > 0$) where failure mechanisms are activated until an admissible stress state ($F = 0$) is achieved. The fracture criterion is calibrated by the current tensile strength $q_{n,f}$, the current shear strength $q_{t,f}$, and the coefficient of internal friction ϕ between the two faces of the emerging crack (Figure 4).

It is noted that the term 'material strength' has been introduced with respect to the strength of the POD, i.e. the physical limit state condition which characterizes the maximum tractions that a specific (fictitious) plane in the material can withstand without (further) degradation. Clearly, the initial strength of the POD and the conventional strength of the (bulk) material need not coincide, e.g. $q_{n,f0} \neq f_t'$. However, a unique relationship between both has been established (Kröplin and Weihe, 1997). It is used to calibrate the micromechanical strength parameters from the macroscopic strength of the bulk material, the latter being usually available from conventional test data.

An adequate (non-associated) hypothesis for the slip and frictional behaviour of established cracks provides an elegant way of including the contact behaviour due to crack closure.

The fracture criterion is augmented by a energy controlled degradation process. Thus, the potential influence of stress singularities, e.g. due to the geometrical discretization of loading patterns, remains bounded and the overall response is dominated by the dissipated energy due to fracture. The fracture toughness G_f^I and G_f^{II} for Mode I and Mode II, respectively, are used as basic calibration parameters.

3.3 *Interaction between Cracks and Continuum: Fracture Induced Anisotropy*

The transformation between the POD, being the level relevant to the physical degradation mechanism, and the continuum as the structural level is formulated within the smeared crack approach: The evolving discontinuity at the POD with the resulting displacement jump is introduced into the continuous displacement field via 'smearing' out over a characteristic distance h. The transformation between the global structural and the physical level of the POD thus can be defined for the static and the kinematic quantities in a strong sense by application of the (Gaussian) transformation (Figure 3).

Figure 4 depicts the result of the geometrical transformation for the case that a crack will develop under Mode II conditions, 30° inclined with respect to the x-axis and perpendicular to the z-axis. The material strength of this particular POD (only!) is reduced as shown in the left hand figure. The reduced strength is then geometrically embedded in the 3D-continuum (Weihe, 1995; Weihe et al., 1997) as depicted in the central insert. The resulting residual strength of the effective 3D-continuum with the embedded crack is shown in the right hand figure: The initial strength is still present in most directions, however, the stress capacity with respect to the x-direction is significantly reduced. It is obvious that the presented approach yields to a highly anisotropic representation of the residual strength due to the oriented fracture process.

3.4 *Fracture Mode*

For a given stress state, the tractions in a fictitious POD depend on its orientation with respect to the axes of principal stress. The most critical orientation, in which fracture eventually will occur, is defined by the maximum value of the fracture criterion F. Since the fracture criterion is a function of normal and tangential (shear) tractions, the crack is not necessarily introduced under Mode I conditions. Depending on the relative shear strength $\gamma_0 = q_{t,f0}/q_{n,f0}$ of the material

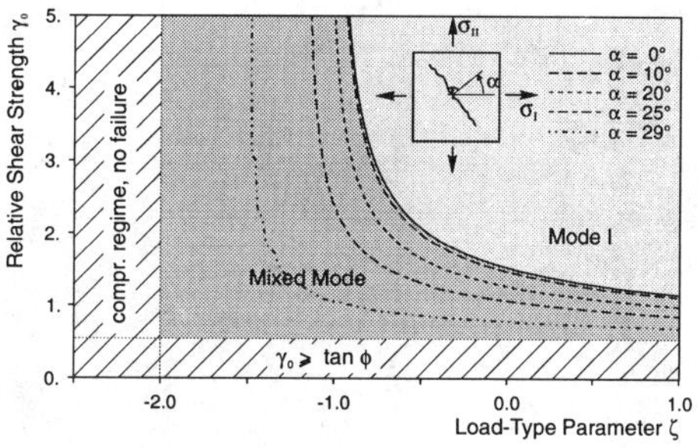

Figure 5: Correlation between fracture mode, stress state and material parameters.

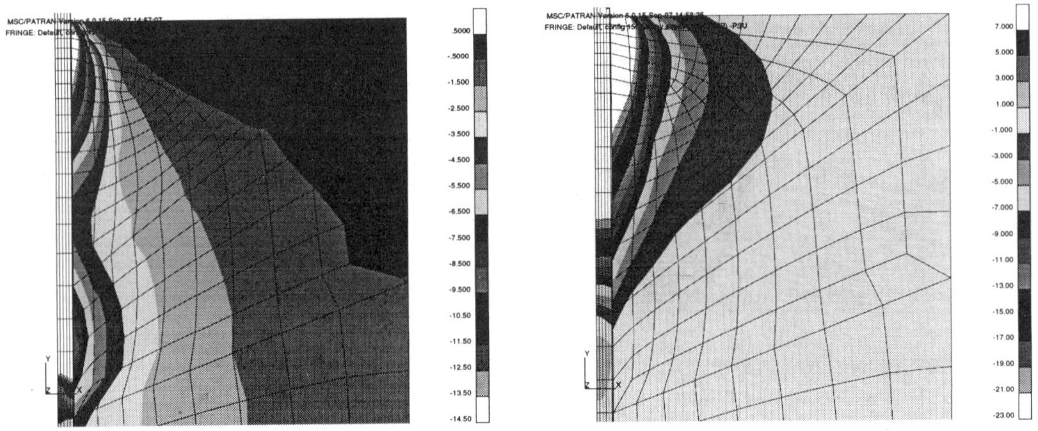

Figure 6: Response due to relaxation of the pre-stress: Axial stress σ_{zz} (left) and radial stress σ_{rr} (right).

and the local stress state $\zeta = (\sigma_1 + \sigma_3)/(\sigma_1 - \sigma_3)$, the corresponding fracture mode can be predicted analytically as shown in Figure 5.

It is seen that typical brittle materials as concrete ($\gamma_0 \approx 2$) will exhibit predominantly Mode I failure, since significant compressive stresses have to be present to initiate fracture under mixed mode. With decreasing relative shear strength – which is usually accompanied by higher ductility of the material – mixed mode (shear) failure becomes much more likely also under tensile loading situations.

In conclusion, the theoretical model is capable of representing mixed mode fracture in dependence of the current loading conditions and the specific material parameters, which makes it superior to many of the classic yield criteria of the Rankine- or Mohr-type. The crack propagation induces a very dominant anisotropic evolution of the residual material strength and makes the presented approach especially valuable to predict the behaviour of structures that undergo complex fracture and eventually develop characteristic failure patterns.

3.5 *Results of the Simulation*

The finite element discretization using elements with quadratic approximation of the displacement field is shown in Figure 7, considering the twofold symmetry of the specimen under axial symmetric conditions. In a first analysis, the interfacial behaviour due to the relaxation of the pre-stress of the Aramid during the casting process is simulated. As shown in Figure 6, the stress transfer between the reinforcement and the concrete concentrates at two different locations: the end of the specimen and near the current location of the crack tip. The latter being anticipated, the former results from secondary effects: (1) Due to the Poissons effect, the bar expands radially with decreasing axial stress. The sand-coating of the Aramid surface is pulled off the interface and (2) acts like a wedge

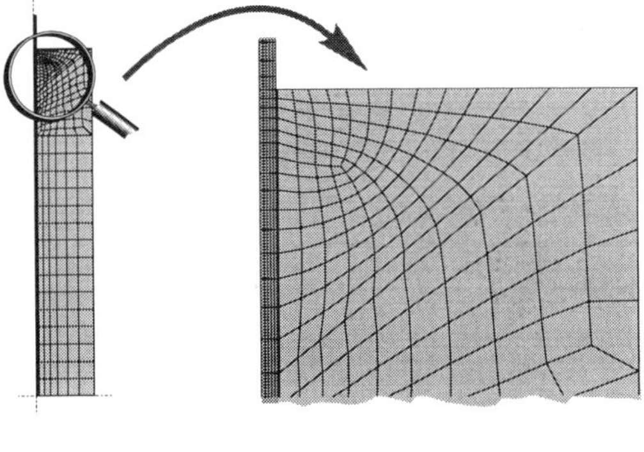

Figure 7: Discretization for the FE-analysis.

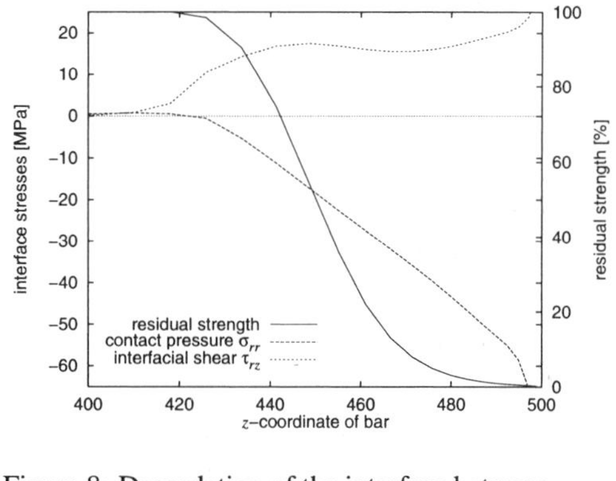

Figure 8: Degradation of the interface between concrete and reinforcement.

between the bar and the concrete which also increases the compressive radial stress, i.e. the contact pressure (Figure 8). Thus, although the interface is completely deteriorated at the end of the specimen due to debonding, the induced contact pressure provides significant potential for frictional contact forces. This effect is finally responsible for arresting the interfacial debonding. The concrete exhibits radial cracking in the vicinity of the reinforcement bar. These cracks relax the tangential (hoop) stresses and affect the stress state in the interface as well.

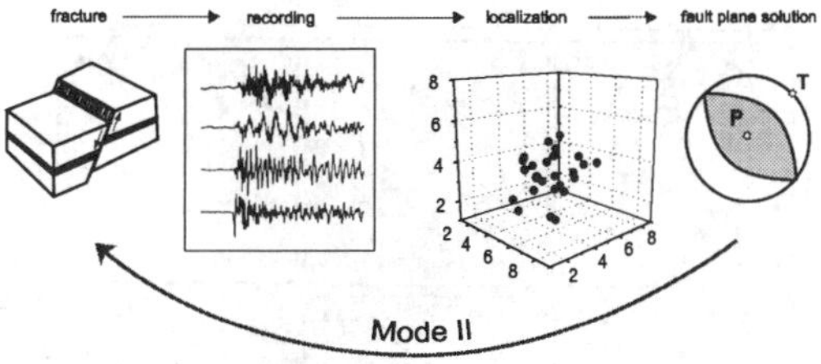

Figure 9: From failure to analysis – stages of quantitative acoustic emission analysis.

4 EXPERIMENT

4.1 *Introduction*

To compare the model with experimental results, quantitative acoustic emission (AE) techniques have been applied. On the basis of a calculation of the AE source coordinates, this enables numerous analysis techniques in the time and frequency domain including waveform comparisons and fault plane solutions. The radiation pattern of acoustic emission sources and the seismic moment (as an equivalent for the emitted energy) as well as the type and orientation of the cracks can be determined using moment tensor inversion methods.

There are several different approaches to the evaluation of acoustic emission data for material characterization, depending on the way of recording, the equipment used, and the analysis technique applied. In general, it has to be distinguished between the classical statistical analysis technique and a more quantitative approach to the problem including waveform recording and interpretation. Both methods offer the advantage to investigate a specimen or a structural part under loading conditions in an integral way. Very often the acoustic emissions are directly correlated with the failure of structural integrity. The difference between these two methods are the extent of data acquisition and data analysis as well as the degree of reliability of the interpretation. Obviously, concrete is a very suitable material for the application of quantitative acoustic emission technique. A modern quantitative acoustic emission technique is capable of giving detailed information on the defect formation and the failure process in materials. It provides a deep insight into material behaviour under load and can largely influence the optimization of material design.

4.2 *Quantitative Acoustic Emission Technique and Moment Tensor Inversion*

The broadband recording of acoustic emissions during the loading of a specimen with a reasonable number of sensors and the 3D localization of the sources are the fundamentals of these techniques. The measuring device used for the experiments is based on a multi-channel transient recorder. Between eight and twelve channels with sampling frequencies of 1 or 10 MHz at 12 bit resolution have been used for the experiments as described earlier in more detail (Grosse, 1996; Grosse et al., 1997). A software called Hypo^{AE} to localize the sources of acoustic emissions was developed in cooperation with the Institute of Geophysics at the University of Karlsruhe in Germany (Oncescu and Grosse, 1996). Assuming a time resolution of at least 1 s, a localization accuracy of 8 to 10 mm has to be accepted. To improve the P-wave-onset picking, an automatic picker based on the approach by (Mikhailov and Grosse, 1995) is currently being developed. To determine the fracture type and orientation of a crack as well as the seismic moment (which describes the released energy) the waveforms of the recorded acoustic emission events have to be interpreted by the inversion of the moment tensor. As illustrated by Figure 9, the failure of a brittle specimen is accompanied by a sudden release of energy in form of acoustic waves. Using a moment tensor inversion in combination with the three-dimensional localization, a fault plane solution can be applied which enables the analysis of the fracture process in the material.

There are several ways to determine the crack type and orientation of AE sources. One concept is the determination of the polarity of the initial P-wave pulses. The distribution of the two senses of the wave polarity around the focus is determined by the radiation pattern of the source. This way, it is possible to estimate the orientation of the nodal planes and thus the mechanism of the source. Unfortunately, it is not possible to quantify the deviation from a pure shear dislocation source (Double Couple or DC) with this method. The analysis of the moment tensor is a different approach to the problem. The symmetric moment tensor with six independent components mathematically defines the strength and the 3D radiation pattern of a general seismic point source. The diagonal and the off-diagonal elements of the moment tensor represent force dipoles without or with moment, respectively. With this method, deviations of DC mechanisms can be extracted as well as the radiation pattern of the whole damage process. A determination of the DC, compensated linear vector

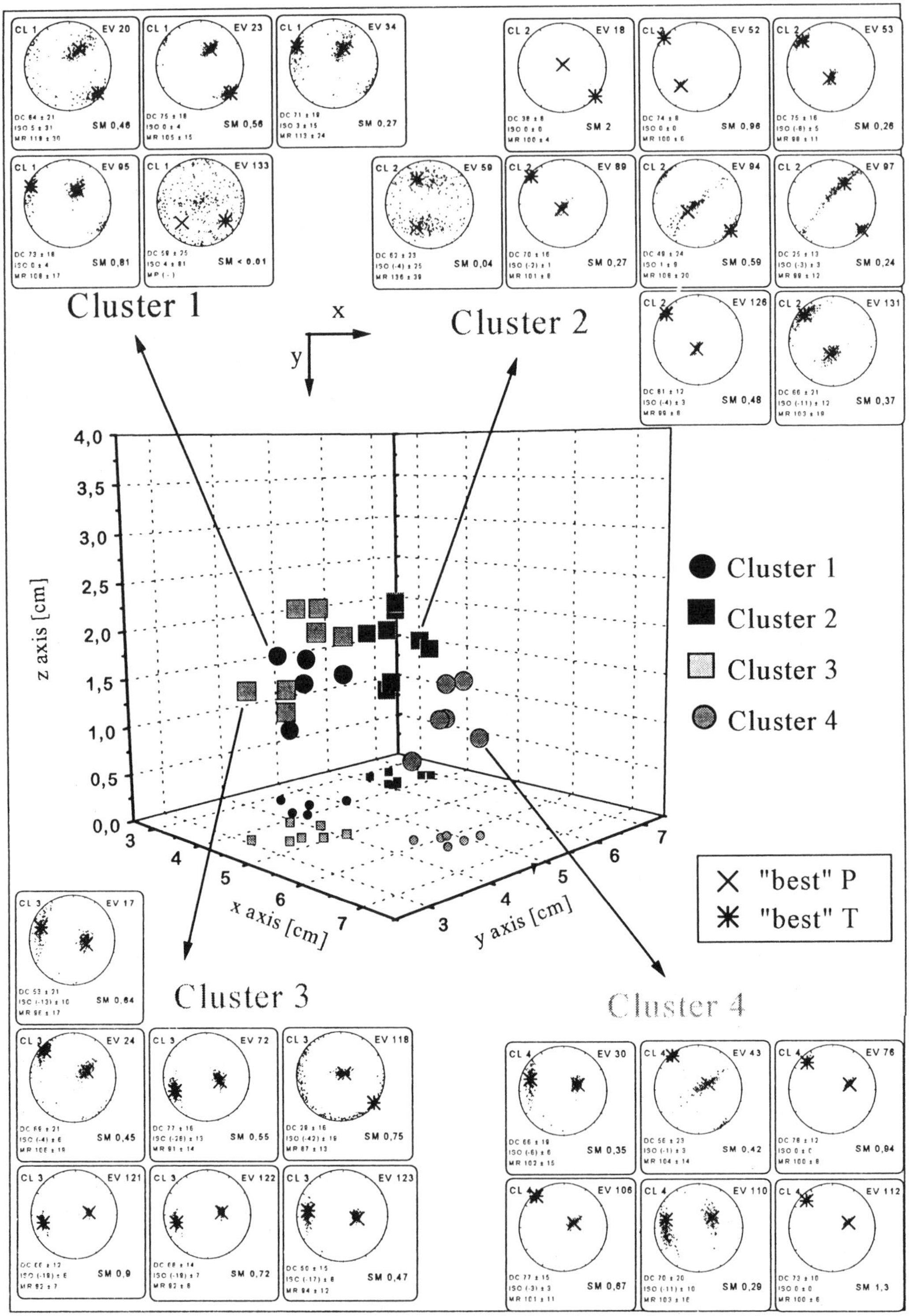

Figure 10: Results of the relative moment tensor inversion for four clusters of acoustic emissions: Detail of the concrete cube projection of the AE's to the x/y plane and P/T plots (Grosse et al., 1997).

dipole (CLVD), and the isotropic tensile components are the basis for the fracture mechanics analysis. To estimate the six moment tensor components, the amplitude informations of at least six AE recordings have to be used - a reasonable number in acoustic emission experiments. Solving the problem with this method, the Green's functions of the specimen, describing the wave propagation in a medium, and the transfer function of the recording system have to be known. Considering the wave propagation in concrete, a unique Green's function is hardly to be found for a specimen because of the heterogeneity of the medium. Consequently, a moment tensor inversion based on P-wave amplitudes was employed in a relative way to eliminate the influences of inhomogeneity and anisotropy. The method was developed for the determination of the radiation pattern using cluster analysis (Dahm, 1996). This relative approach is suitable for the requirements in AE experiments, as will be demonstrated in the next section. Up to hundreds of acoustic emissions are recorded which occur commonly in certain regions. This is called clustering. The travel path from different events of a cluster to a certain sensor is approximately the same, and thus the dynamic part of the Green's functions can be eliminated. A description of the mathematical procedure can be found in Grosse et al. (1997).

4.3 *Example of an Application*

Investigating the steel-concrete interaction, a series of acoustic emission tests have been carried out. The specimens were concrete cubes of 100 mm side length and centrically reinforced with a single deformed bar of 16 mm diameter. The bond length was limited to twice the rib spacing (20 mm) in order to minimize the number of sources producing local damage. In these experiments, eight sensors were coupled to the sides of the cube. During the pull-out tests, the actual tensile force together with the relative displacements and the AE signals were recorded simultaneously. The results of the force and slip measurements for tests with different load histories (monotonically increasing displacements, cyclic loads and long term loads) as well as the details of the test setup are summarized in Balázs et al. (1996).

Because of the load history, cracks with fault planes parallel to the pull-out direction are expected as the dominant failure mechanisms. On the basis of the 3D localization, the AE signals were separated into clusters of up to ten events. A moment tensor inversion of every single event of the clusters resulted in numerous P/T diagrams as shown in Figure 10.

Considering the polarity constraints, it seems that shear faults in downward direction are predominant. The P-axes, pointing to the principal stress axes, are generally vertical with a shift of 10° to 25°. Regarding the T-axes, east-west directions are dominant but are not as well-fixed as the P-axes. Obviously, fractures of the Mode-II-type (normal shear faults) are the sources of these acoustic emissions. This is illustrated by Figure 11, showing the results of a moment tensor inversion for a single event of a cluster with nine signals. The best solution for the P and T axis is indicated with a cross and a star, respectively. The estimated P and T axis are the eigenvectors corresponding to the minimal and maximal eigenvalue of the moment tensor. For pure shear dislocation sources, the fault plane normal is at 45° between the P and T axis, which are also the principal stress axes if the fault plane is a plane of maximum shear. The errors of the decomposed source components are estimated with a bootstrap analysis and are shown as small circles. The values below the graphs indicate the relative values (expressed as a percentage) for the shear component (DC), the extensional component (ISO) and the relative strength (MR). In addition, the relative seismic moment (SM) was calculated, which is proportional to the energy emitted by the source. The data quality and hence the scattering of the error points is closely related to the seismic moment. The scattering of the error points is smaller with higher SM-values.

In the middle of Figure 11, the fault plane solution resulting from the inversion of the moment tensor is shown, indicating the regions of pressure and dilatation. On the right, the fracture mechanical interpretation is represented summarizing the results of the fault plane solution and taking into account the polarity informations of the AE recordings. In this example, the normal fault is inclined by 20°.

Some events of the examined clusters show variations of normal faulting. As represented in Figure 10, very few events had to be interpreted as Double-Couple mechanisms with a strong strike slip component. This could be explained as a turn around of a matrix crack formation caused by the aggregates in concrete. Even though these interpretations seem plausible, it is important to know that this conclusions are based on an inherent data base (disfavourable hypocenter distance to wave- length relation). A summary of the results of the relative moment tensor inversion for the events and clusters of this experiment is reported in Grosse et al. (1997). Experiments investigating the fracture mechanisms in more detail as well as enhancing and improving the data basis are conducted nowadays.

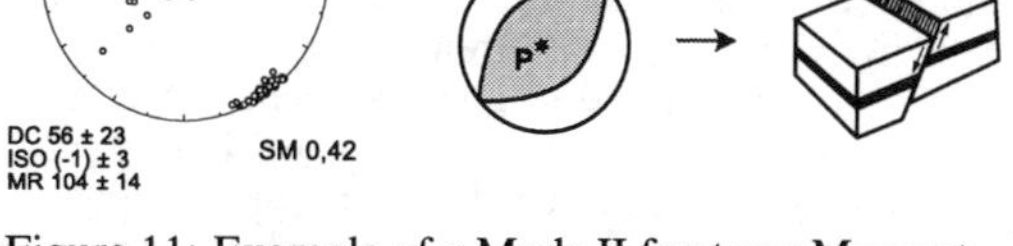

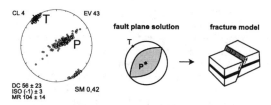

Figure 11: Example of a Mode II fracture: Moment tensor inversion and interpretation.

5 CONCLUSIONS

Numerical simulations and enhanced experimental techniques have been combined to further understand the pull-out of reinforcement from the surrounding concrete. This approach is very promising, since the experimental and the theoretical approach share in this special setting the same set of characteristic parameters for the problem, such as the occurring fracture modes, the specific energy release rates separated for the generic fracture modes, and the specific location of the degradation events.

The presented theoretical model adopts a two-level constitutive model, based on the decomposition of the complex behaviour of anisotropic degradation into a generic physical process which is embedded in a geometric transformation. Thus, the calibration effort is substantially reduced when compared to phenomenological anisotropic models since the geometric transformation needs not be calibrated. Crack shielding is inherently contained in the approach. The model is predicting very well phenomena of fracture induced material anisotropy as the result of a complex interaction between multiple fracture planes. In the framework of the Finite Element Method, the interface failure between the reinforcement and the concrete has been studied in detail. Secondary effects, e.g. the Poisson and hygrothermal effects of the Aramid as well as the dilatation due to the sanded surface of the bar, are shown to play an important role in the crack arrest of the interfacial debonding.

The AE technique was used to study the fracture processes within steel-reinforced concrete. A more quantitative analysis of the signals including a 3D localization based on the recordings of a sensor network is particularly valuable for making inferences on microscopic damage. The moment-tensor method can be used to investigate the fracture process in detail, extracting the fracture type (Mode I, Mode II or mixed mode), the crack orientation and the released energy. The present method, called relative moment-tensor inversion, has some advantages compared to others. Procedures of cluster analysis can be used to avoid uncertainties in Green's functions associated with the inhomogeneities of the material and in the sensor characteristics. In common with an efficient localization technique, the relative moment-tensor method is a valuable tool for materials research.

With this knowledge of basic fracture properties using on one hand a variety of AE parameters and on the other hand appropriate numerical simulation techniques, a comprehensive characterization of the fracture propagation in fiber-reinforced, cement-based materials can be developed. A combined theoretical and experimental approach as presented in this paper can be used to understand the failure process in concrete and to optimize the bond between matrix and reinforcement.

ACKNOWLEDGMENTS

The second and fifth author gratefully acknowledge the help of Dr. T. Dahm from the Institute of Meteorology and Geophysics of the University of Frankfurt and Dr. L. Oncescu from the Geophysical Institute, University of Karlsruhe. The presented work is partly funded by the German Research Foundation DFG within the Collaborative Research Centre SFB 381 'Characterization of the Failure Propagation in Fibre Reinforced Materials using Non-Destructive Methods'.

REFERENCES

Balázs, G.L., Grosse, C.U., Koch, R. and Reinhardt, H.W. 1996. Damage Accumulation on Deformed Steel Bar to Concrete Interaction Detected by Acoustic Emission Technique. *Mag. of Concrete Res.*, **48**(177), 311–320.

Dahm, T. 1996. Relative Moment Tensor Inversion Based on Ray Theory: Theory and Synthetic Tests. *Geophys. J. Int.*, **124**, 245–257.

Grosse, C.U. 1996. *Quantitative zerstörungsfreie Prüfung von Baustoffen mittels Schallemissionsanalyse und Ultraschall.* Ph.D. thesis, Universität Stuttgart (D).

Grosse, C.U., Reinhardt, H.W. and Dahm, T. 1997. Localization and Classification of Fracture Types in Concrete with Quantitative Acoustic Emission Measurement Techniques. *NDT&E Intern.*, **30**(4), 223–230.

Kröplin, B. and Weihe, S. 1997. Constitutive and Geometrical Aspects of Fracture Induced Anisotropy. *Pages 255–279 of:* Owen et al. (ed), *Int. Conf. on Computational Plasticity, COMPLAS 5.*

Mikhailov, N. and Grosse, C.U. 1995. An Automatic Picker of the Onset Time of Acoustic Emission Signals. *Otto-Graf-Journal*, **6**, 168–187.

Oncescu, L. and Grosse, C.U. 1996. *$HYPO^{AE}$ - A Program for the Localization of Hypocenters of Acoustic Emissions.* Software Documentation for PC and Workstations, Rev. 2.0.

SFB 381. 1997. *Ergebnisbericht 1994–1997.* Sonderforschungsbereich 381 'Charakterisierung des Schädigungsverlaufes in Faserverbundwerkstoffen mittels zerstörungsfreier Prüfung. Universität Stuttgart (D).

Weihe, S. 1995. *Modelle der fiktiven Rißbildung zur Berechnung der Initiierung und Ausbreitung von Rissen: Ein Ansatz zur Klassifizierung.* Ph.D. thesis, Universität Stuttgart (D).

Weihe, S. 1997. Failure Induced Anisotropy in the Framework of Multi-Surface Plasticity. *Pages 1049–1056 of:* Owen et al. (ed), *Int. Conf. on Computational Plasticity, COMPLAS 5.*

Weihe, S., Kröplin, B. and de Borst, R. 1997. Classification of Smeared Crack Models Based on Material and Structural Properties. *Int. J. Solids Structures.* (acc. for publ.).

Damage and Failure of Interfaces, Rossmanith (ed.) © *1997 Balkema, Rotterdam, ISBN 90 5410 899 1*

Stable crack growth and damage tolerance evaluation of Ti-24Al-11Nb matrix composite unidirectionally reinforced with SiC fibers

M.Okazaki & N.Tokiya
Department of Mechanical Engineering, Nagaoka University of Technology, Japan

H.Nakatani
Gifu Technical Institute, Kawasaki Heavy Industries Ltd, Kakamigahara, Japan

ABSTRACT: Stable fracture in a Ti-24Al-11Nb matrix composite reinforced unidirectionally with continuous SiC fibers, SCS-6, was studied by using the notched specimens with different notch length, notch radius and specimen width. During the study, special attention is paid to understand what are the mechanisms and mechanics of stable crack growth, what are the role of reaction zone near the fiber/matrix interface and that of residual stress, and how the damage tolerance level should be assessed and evaluated simply, on the basis of non-linear fracture mechanics parameter; J-integral.

1. INTRODUCTION

Metal matrix composites have received considerable interests as the materials under severe service conditions, because they generally possess many attractive characteristics: higher specific strength at high temperature and higher specific stiffness, compared with traditional monolithic materials. Especially, silicon-carbide fiber reinforced titanium alloy composites provide great potential for aerospace structural applications. On this kind of composites many studies, such as on fracture under static loading (Chiu & Yang 1995), fatigue (Davidson 1991) and creep (Eggleston &Ritter 1995) have been carried out. On applying these composites to actual components and structural materials in engineeering field, on the other hand, many people must face on a very simple, but very indispensable question to be clarified: how high is the fracture toughness of the material and how should the damage tolerance be assessed ? This is because local damage does not always mean the catastrophic failure in composite materials. In addition, the fracture mechanism is very complicated, since many phenomena, such as fiber breakage, matrix cracking and/or fiber bridging, contribute to the fracture in an complex manner (Marshall et. al. 1985). Moreover, the situation is not always unique, or dependent on the property of interface relating to the fabricating process of the composite, as is often the case with advanced materials. In ordet to answer the above question, therefore, it would be a first way to catch the information on the fracture process in detail, in relation to the interface property.

In this work the stable crack growth behavior of a Ti-24Al-11Nb matrix composite reinforced unidirectionally with continuous SCS-6 fibres is studied under quasi-static loading. During the study, special attention is paid to understand what is the mechanism of stable crack growth, what are the role of reaction zone near the fiber/matrix interface and that of residual stress, and how the damage tolerance level should be assessed and evaluated simply, on the basis of non-linear fracture mechanics parameter; J-integral.

2. EXPERIMENTAL PROCEDURE

2.1 Material

The material tested in this work is a Ti-24Al-11Nb matrix composite reinforced unidirectionally with continuous SiC fiber, SCS-6, where the SCS-6 is a β-SiC fiber of 140 μm diameter with a carbon rich graded silicon carbide coating on the surface. The 7-plies composite sheet of 180x40x1.5 mm was fabricated at Kawasaki Heavy Industries Co. Ltd. (Kakamigahara, Gifu, Japan), by hot pressing alter-

Table 1. Mechanical properties of the matrix alloy and the SCS-6 fiber.

	Youngs mudulus (GPa)	Poisson ratio	Thermal expansion coefficient (/°C)	Tensile strength (MPa)
SCS-6 Fiber	400	0.17	4.5×10^{-6}	3800
Ti-24Al-11Nb	94	0.32	9.6×10^{-6}	685

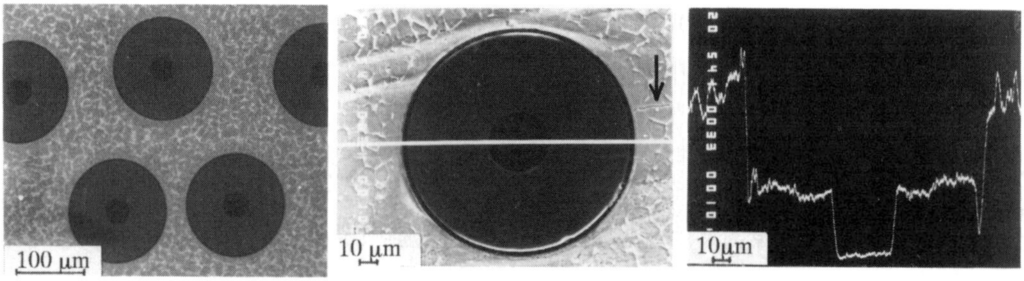

(a) Overview (b) Metallograph near the interface (c) Distribution of Nb near the interface.

Figure 1 Metallograph of the composite tested.

nate layers of thin-foils of Ti-24Al-11Nb alloy and green tapes of the SCS-6 fibers at 1050℃ for 2 hrs. in vacuum. The mechanical properties of the matrix alloy and the fibers are summarized in Table 1. The volume fraction of the fibers in the composite is about 33 %. The fibers orientated to 0 deg. from the longitudinal direction of the pannnel. The metallograph of the cross section of the composite is given in Figure 1. The microstructure of the matrix consist of equiaxied β (light) phase in α_2 matrix (dark) phase. It is worth noting that the reaction zone of about 5 μm in thickness is formed at the SCS-6/matrix interface, where the brittle hexagonal close-packed α_2 phase is dominant relatively to the β phase (Figure 1(a) and (b), resulting from the depletion of Nb (Baumann et al 1990); an element to stabilize the ductile β phase (Figure 1(c)). Some cracks, which must be originated from the residual stress in tension produced during the cooling process after the hot pressing, has already initiated at the β phase depleted zone neighbouring to interfaces (see arrow mark in Figure 1 (b)), although they are not found so often.

2.2 Stable crack growth test

From the composite sheet, the plate specimens as shown in Figure 2 were cut, and then an initial notch was introduced by electro-discharge machining. In order to study the effect of the initial crack length, a_i and the relative ratio, a_i/W to the specimen width, W, on the stable crack growth behavior, several combinations of these values are designed in the specimens. The radius of the notch was about 75 μm in almost specimens, and 35 μm in a few specimens which was prepared to investigate the influence of initial notch radius. For the purpose to explore the stable crack growth process some specimens was mechanically polished by emery paper so that some fibres near the initial notch root appeared on the specimen surface. On conducting the test, two couples of steel end tabs were affixed to the end of the composite specimen to grip, as illustrated in Figure 2.

Almost according to the ASTM method for J_{Ic} determination, ASTM E-813 (ASTM E813-81 1981), the stable crack growth process was observed and measured as follows (see Figure 3): after applying a load upto a certain level under deflection-controlled condition of 0.1 mm/min. in rate, the deflection was hold to take the replica of the specimen surface morphology, followed by a partial unloading of about 30 % in load to measure the specimen unloading compliance. Then a series of these processes were repeated in a given specimen until the specimen finally ruptured. During this process the relation between the applied load, P and the crack opening displacement, δ (i.e., P-δ curve), was continued to monitor. A clip gauge of which interval was about 0.2 mm and was attached to the mouth of the initial notch, was used to measure the crack

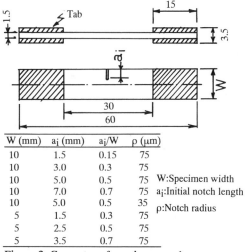

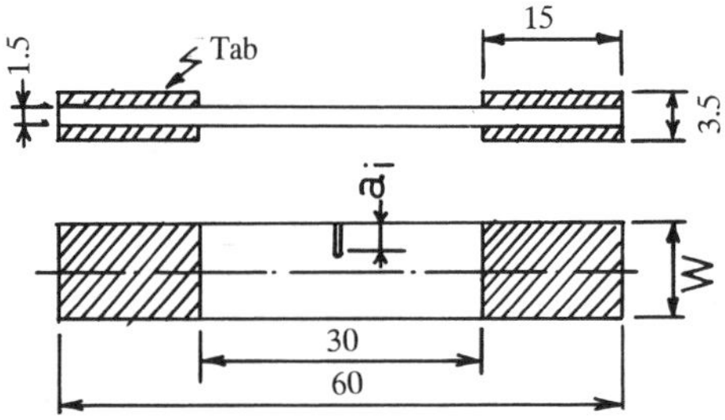

W (mm)	a_i (mm)	a_i/W	ρ (μm)
10	1.5	0.15	75
10	3.0	0.3	75
10	5.0	0.5	75
10	7.0	0.7	75
10	5.0	0.5	35
5	1.5	0.3	75
5	2.5	0.5	75
5	3.5	0.7	75

W:Specimen width
a_i:Initial notch length
ρ:Notch radius

Figure 2. Geometry of specimen used.

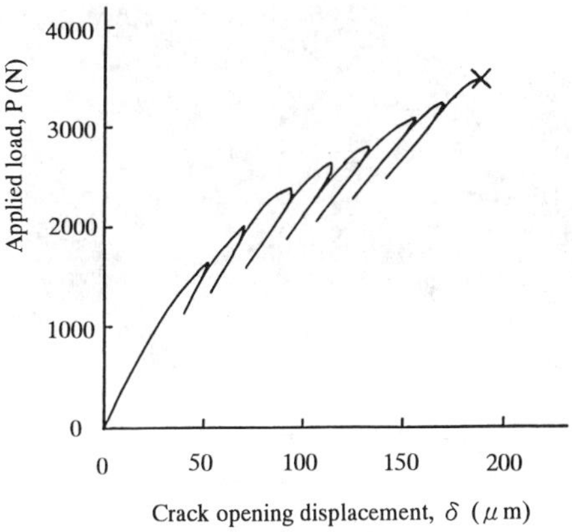

Figure 3. Procedure to observe the stable crack growth and an actual P-δ curve monitorred.

opening displacement. Figure 3 presents an actual P-δ curve obtained, where the mark of x indicates the unstable fracture point. The amount of stable crack extension, Δa, was measured by the medium of the calibration curve between the specimen unloading compliance and the crack length, which had been previously obtained by employing the specimens with several initial notch lengths. Δa thus obtained was correlated with elastic-platic fracture mechanics parameter, J-integral, to evaluate the stable crack growth curve, or R-curve. For simplicity, the value of J-integral was calculated by applying the conventional method for monolithic materials:

$$J=2A/(bB) \qquad (1)$$

where A, b and B are the area under P-δ curve, ligament length and specimen thickness, respectively. Through the study, particular attentions are paid to understand what is the mechanism of stable crack growth in the composite material, what are the role of reaction zone near the fiber/matrix interface and that of residual stress, what are the effects of initial notch length and specimen width on the R-curve andon the critical J-integral value at which the unstable fracture was reached, and how the damage tolerance level should be assessed and evaluated simply.

3. RESULTS.

3.1 Stable crack growth process

An example of the stable crack growth process is given in Figure 4, where these photpgraphs cover the same area and the history advances from Figs. (a) to (d). In order to easily understand the fracture process the tentative numbers are given to the fibers in order from the initial notch root; 1, 2 and so on. It is found that the quasi-static fracture evolves according to the following procedure in Figure4: at the beginning of the loading some small cracks initiate

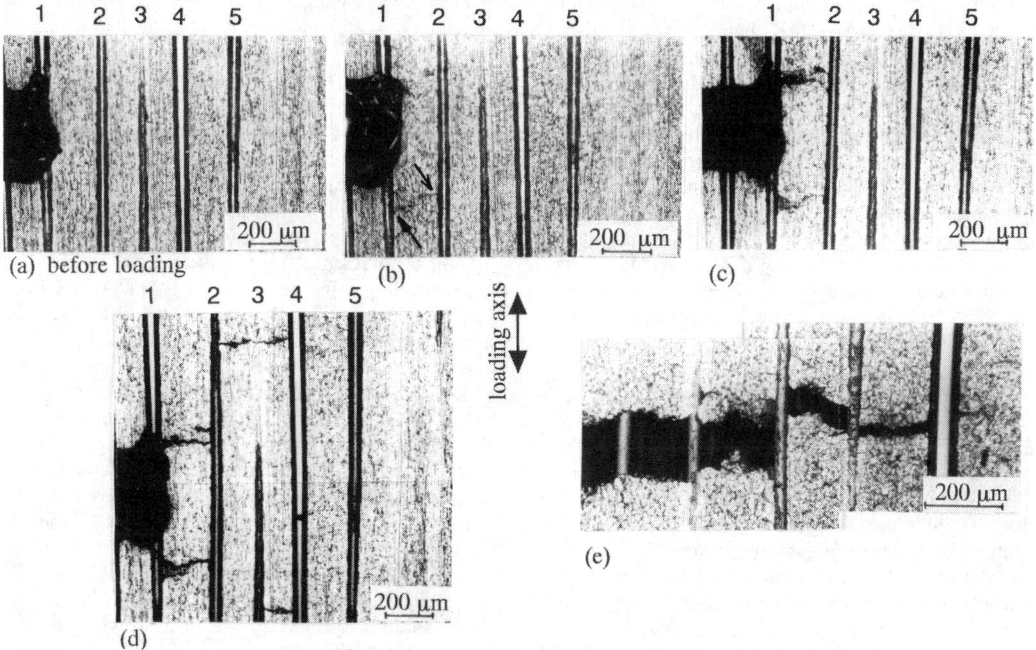

Figure 4. Stable crack growth process.

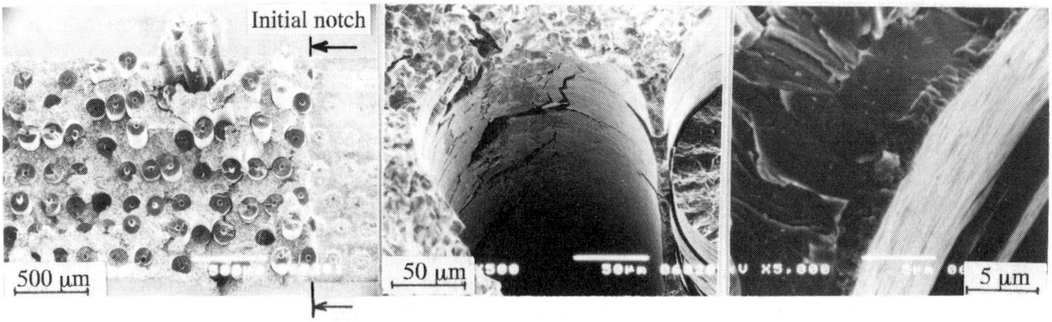

(a) overview (b) fiber pull-out. (c) matrix cracking near the interface.

Figure 5. Fracture surface.

from the interface which neighbours to the fiber-1 and fiber-2 near the notch root (compare Figs. 4(a) with (b), and see the arrow marks in Figure 4(b)). It is worth noting that these small crack predominantly originate from the β phase depleted zone near the interface (see Sec. 2). It is also important to remind that the thermal expansion coefficient of the matrix relatively larger than that of the fiber induces the residual stress in tension in the matrix during the cooling process after hot pressing to produce the material (see Table 1). As the external load increases, these cracks open so that the profile are identified clearly (Figure 4(c)), but the growth of these cracks are inhibitted by the fiber 2. When the load increases more, the new cracks nucleate in the matrix between the fibers 3 and 4 (Figure 4 (d)). What is interesting in this stage is that the fibers 2 and 3 are still intact even after the fiber 4 far apart from the notch-root was broken: the fibers 2 and 3 bridge the crack plane. The statistical scatter in the tensile strength of the fibers would be attributed to the above phenomenon that the nearest fiber from notch root-did not always break at first. The matrix cracking, the fiber breakage and their coalescence became significant after Figure 4(d), resulting in the stable crack growth. However, these did not introduce the unstable fracture immediately. In the other words, a sigfinicant stable crack growth supported by the crack face bridging by the intact fibers was found, as given in Figure 4 (e).

The fracture surface of the specimen is presented in Figure 5. The fracture surface is very three-dimensionally torturous (Figure 5 (a)). The noticeable matrix cracking in the transverse and longitudinal direction can be also seen at the prior fiber/matrix interface from which the fibers have been pull-out (Figure 5(b)). Peeping into the crack plane, the quasi-cleavage fracture surface which started from the reaction zone, or the β-phase depleted zone are observed (Figure 5(c)).

Summarizing these obserevations it is reasonably concluded that the stable crack growth and the subsequent unstable fracture occurred according to the following process: (i) first matirix cracking starting from the reaction zone (i.e., the brittle β-phase depleted zone), with possible interfacial debonding near the initial notch tip; (ii) the increase of the matrix cracking density, resulting in the crack face bridging by the intact fibers; (iii) the breakage of some bridging fibers; and finally (iv) unstable fracture, accompanied with the coalescence of cracks and with the fiber pull-out. The residual stress in tension in the matrix played the role to promote the processes (i) and (ii).

3.2 Stable crack growth curve.

The relationship between the J-integral value and the stable crack growth extension, or J-R curve, is given in Figure 6 as the functions of initial crack length ratio, a_i/W and the specimen width, W, where the arrow marks depict the points at which the unstable fracture was reached. The abscissa in Figure 6 expresses the amount of stable crack growth extension, Δa, normalized by the initial crack length, a_i. This normalization results from the characteristics in stress shielding effect near crack tip due to the fiber bridging, which will come up in detail in the next section. Some important values obtained in the test are also summarized in Table 2. From these figures and table the following features should be noted:

First of all, the critical J-integral value, J_c, at which the unstable fracture began, does not exhibit unique value; i.e., varies with the initial crack length, a_i and with the relative ratio, a_i/W. In this work the J_c reveals maximum when the a_i/W was 0.5. This is the case on the basis of linear elastic fracture mechanics parameter, critical stress intensity factor, K_c (see Table 2). Fortunately, on the other hand, the critical values, J_c and K_c, are insensitive to the ini-

Table 2. Sumary of test results.

W (mm)	a_i (mm)	a_i/W	ρ (mm)	K_c(MPa$\sqrt{m}$)	J_c(kJ/m^2)	dJ/d(Δa) (MJ/m^3)
10	1.5	0.15	75	73	110	110
10	3.0	0.3	75	90	144	150
10	5.0	0.5	75	141	185	145
10	7.0	0.7	75	131	136	172
10	5.0	0.5	35	145	192	159
5	1.5	0.3	75	78	110	98
5	2.5	0.5	75	141	184	162
5	3.5	0.7	75	105	103	171

W:specimen width a_i:initial notch length ρ:notch radius

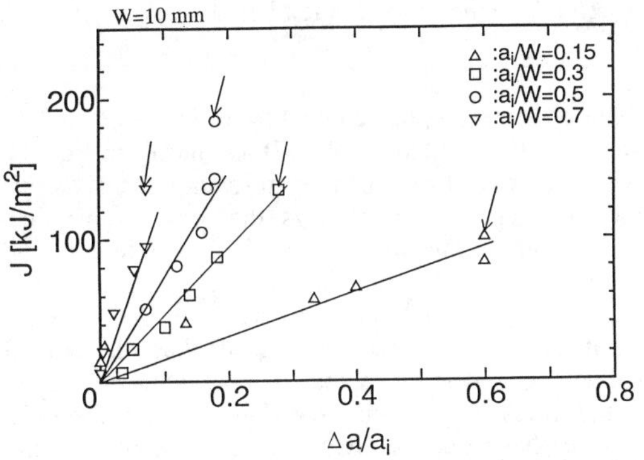

(a) W= 10 mm

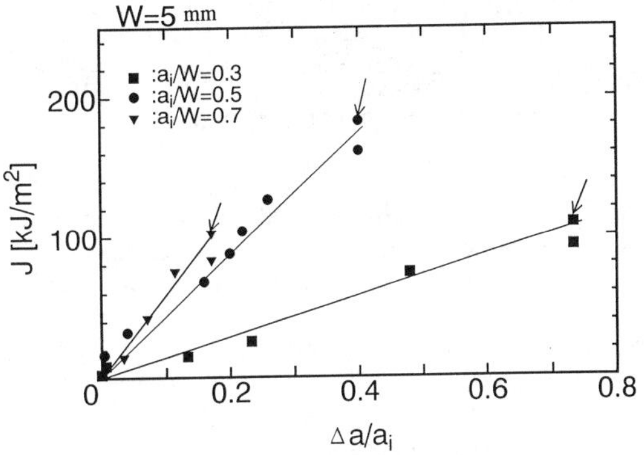

(b) W= 5 mm

Figure 6. Resistance curve for stable crack growth.

tial notch radius (Table 2). The fracture toughness of the Ti-24Al-11Nb matrix alloy has been confirmed about 20MPa by other reseracher (Chan K.S. 1990). Since the values of K_c in all specimens are large enough compared with the Ti-24Al-11Nb matrix alloy, the primary purpose of reinforcement is achieved. However, the above characteristics does not appprove the validity of the traditional treatment for monolithic materials in this work; hence some refined parameters should be developed for this kind of composite material.

Secondly, although the initial resistance to the stable crack growth is not high, the resistance significantly and almost proportionallly increases with the crack extension in the respective specimens. Although the slopes in the J-R curves are also dependent on the a_i and a_i/W, they monotonically varies with the a_i/W.

4. DISCUSSION

It must be interesting to discuss the J-R curve obtained in Sec. 3.2 in connection with the stress shielding due to the crack face bridging by fibers, because the fiber bridging was found to play an important role in the stable crack growth process as shown in Sec. 3.1. At first, let's consider phenomenologically the features of the bridging effect, employing the model schematically illustrated in Figure 7, in which a cohesive force, $\alpha\sigma_B V_f$, is acting between the distance, r, along the crack wake behind the crack tip, where σ_B, V_f and α are the fiber

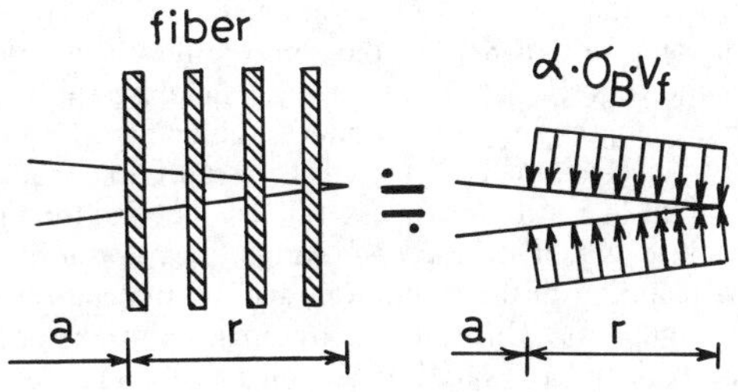

Figure 7. Cohesive model for fiber bridging.

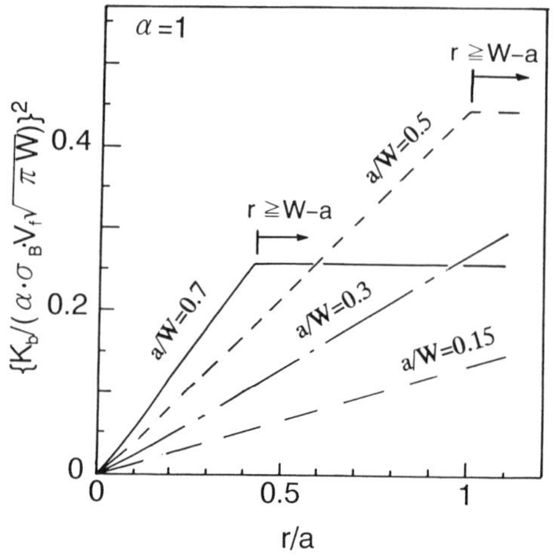

Figure 8. Numerical calculation of fiber bridging on the basis of the cohesive model in Fig. 7.

strength, the fiber volume fraction, and the material parameter representing the level of bridging, respectively. Provided the uniform and continuous (hence, not discrete) distribution of the cohesive force, the shielding amount in stress intensity, K_b, is given by (Duddale 1960):

$$K_b = 2\alpha \cdot \sigma_B \cdot V_f \times \{\sqrt{\pi}\ (a+r)/\pi\} \cos^{-1}\{(a/(a+r)\} \quad (2)$$

where a is crack length. Then the actual crack tip stress intensity factor, K, is

$$K = K_{app} - K_b \quad (3)$$

where K_{app} is the stress intensity by the external load.

Figure 8 depicts the numerical calculation of K_b, where in order to keep a pace with the treatment in Figure 6 on the basis of energy concept, the vertical axis is given in the form of square of K_b normalized by $\alpha \cdot \sigma_B \cdot V_f\{\sqrt{\pi W}$ (where W is the specimen width). Of particular interests in Figure 8 are as followings: at first, K_b^2 almost proportionally increases with the evolution of bridging region, r. The developing rate of K_b^2, or the slope in the linear part of the curve, also monotonically increases with the crack length ratio, a/W, under a given r/a. Secondary, because of the geometrical restriction that the amount of r can not be essentially beyond the ligament length, W-a, the K_b^2 can not develop infinitely, resulting in the saturation (see the plateau region in Figure 8). Once the above restriction appears, the K_b^2 under small a/W can excel that under large a/W in some cases, in spite of the above first

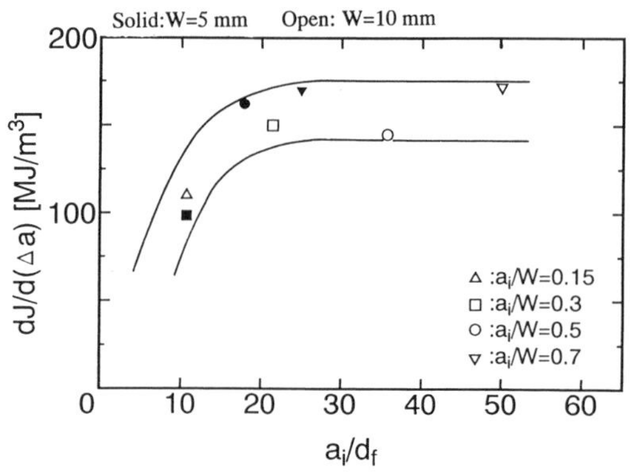

Figure 9. Relationship between the slope of the J-R curve and the initial crack length.

feature (for example, compare the two plateaus between a/W=0.5 and a/W=0.7). Note from Eq. (1), when the fiber bridging is significant, the K_c value measured in the experiment is the apparent one corresponding to the sum of an intrinsic toughness and the K_b. Hence, the R-curve given in Figure 6. can be directly related to the preceding first characteristics. The experimental result that the critical values, J_c and K_c, revealed maximum at a given initial crack length ratio (see Figure 6 and Table 2), can be also naturally attributed to the above second features resulting from the limitation of ligament. From these, it can be reasonably concluded that the fiber bridging plays an intrinsic role in the stable crack growth in this work.

The above is not beyond the qualitative discussion. In order to quantitative assessment the most frontal is to quantitatively know how the cohesive force distributes along the crack wake and what is the size of bridging region, as many researchers have been trying (Marshall et. al. 1985, Davidson 1991). Is this type of frontier attack appropriate from engineering purpose ? It would be a present situation that these have been inversely estimated from the experimental results, employing some approximation, because there are many unknown factors controlled by statistical phenomena, such as friction force at fiber/matrix interface. It is another aspect that the level of crack tip stress shielding has been confirmed to be insensitive to the distribution of bridging force (Sakai 1991).

In order to find a simple and useful fracture criterion independent on the material, special focus is given to the slope in the R curve in this work, because it must represent the developing rate of the fiber bridging with the stable crack extension. The slope, dJ/d(Δa), obtained from Figure 6 is summarized in Table 2, and is correlated with the relative initial crack length, a_i/d_f, in Figure 9, where d_f is the fiber diameter. The slope seems to exhibit an unique value independent of the initial crack length ratio and the specimen width, when the a_i/d_f is large enough: $a_i/d_f>20$ in this work. In the other words, although the K_c and J_c themselves are not always adequate as the material parameter, the slope in the R-curve is a noteworthy parameter which provides the level of damage tolerance in composite materials.

As described in Sec. 2, the amount of stable crack extension, Δa, was evaluated from the calibration curve between the specimen compliance and the crack length which had been previously obtained employing the specimen with a single crack. From such a evaluation method, the Δa in this work may represent the amount of "equivalent" crack extension corresponding to the decrease of elastic stifness of the composite. However, the actual stable crack grows accompanied with multiple small cracks and their coalescence, as shown in Figure 4. Such a difference of crack advance morphologies asks us other questions: how multiple cracks should be related to damage and how the crack length should be defined on the stable crack growth tests ? Further investigaion is recommended to clarify these points.

5. CONCLUSIONS

The stable fracture behavior in a Ti-24Al-11Nb matrix composite reinforced unidirectionally with continuous SiC fibers, SCS-6, was studied by using the notched specimens with different notch length and specimen width. Through the work special attention was paid to understand the stable crack growth process, and to assess the relationship between the crack and the damage tolerance of the composite.

The main conclusions reached are summarized as follows:

(1) The stable crack in the notched specimen grew according to the following process: (i) first matrix cracking emanating from the reaction zone (i.e., Nb-depleted α_2 phase) at the fiber/matrix interface near the initial notch tip, and/or the interfacial debonding; (ii) the increase of small crack density in the matrix, resulting in the crack face bridging by the intact fibers; (iii) the breakage of some bridging fibers; and finally (iv) unstable fracture, accompanied with the coalescence of cracks and with the fiber pull-out. The residual stress in tension in the matrix played the role to promote the processes (i) and (ii).

(2) The material exhibited the stable crack growth curve, or J-R curve, almost linearly increasing with the crack extension, Δa. The observation of the fracture process as well as the numerical calculation based on the cohesive model indicated that the fiber bridging would be the most dominant mechanism in the above R-curve.

(3) Not only the critical stress intensity factor but also the critical J-integral value at which the unstable fracture was reached, were identified to be inadequate as fracture criteria: they did not exhibit an unique value, and varied depending on the initial crack length, a_i and on the relative ratio, a_i/W. But instead, the slope in the R-curve, which must reflect the developing rate of the fiber bridging with the stable crack growth, was found to be an useful material parameter which provides the level of damage tolerance in composite materials. In this work this slope exhibited an almost constant value independent on the specimen configuration, when the following condition was satisfied: the relative ratio, $a_i/d_f \geqq 20$, where d_f is fiber diameter.

ACKNOWLEDGEMENT

Authors express their great gratitude to the financial support by the Grant-in-Aid for sceintific reserach by Japan Ministry of Education (No.08555023) .

REFERENCES

ASTM E813-81 1981. Standard test for J_{IC}, A measure of fracture toughness, *Annual book of ASTM standard*, Philadelphia, ASTM.

Baumann S.F., Pk. Brindley & S.D. Smith 1990. Recation zone microstructure in a Ti_3Al+Nb/SiC composite. *Metall. Trans.-A*, 21-A: 1559-690.

Chan K.S. 1990. Fracture and Toughnening mecahnisms in an α_2 titanium aluminide alloy. *Metall. Trans.-A* , 21-A: 2687-2699.

Chiu H.P & J.M. Yang 1995. Effect of fiber coating on the fracture and fatigue resistance of SCS-6/Ti_3Al composite. *Acta Met.*, 43: 2581-2587.

Davidson D.L. 1991. The micromechanics of fatigue crack growth at 25 C in Ti-6Al-4V reinforced with SCS-6 fibers. *Metall. Trans.-A* , 23-A: 865-879.

Dugadale D.S. 1960. Yielding in steel sheets containing slits. *J. Mech. & Phy. Solids*, 8:100-104.

Eggleston M.R. & A.M. Ritter 1995, The transverse creep deformation and failure characteristics of SCS-6/Ti-6Al-4V metal matrix composite at 482 C, *Metall. Trans.-A*, 26-A: 2733-2744.

Marshall D.B., B.N.Cox & A.G.Evans 1985. The mechanics of matrix cracking in brittle matrix fiber composites, *Acta Metall* , 33: 2013-2021.

Sasai M. 1991. Fracture mechanics and mechanisms of fiber reinforced britle matrix composite, *The Journal of The Ceramic Society of Japan* 99: 983-992.

Damage and Failure of Interfaces, Rossmanith (ed.) © *1997 Balkema, Rotterdam, ISBN 90 5410 899 1*

Identification of interfacial parameters with analytical and numerical models from Pull-out tests on a micro-composite

P. Levasseur & J. L. Chaboche
Structural Department, ONERA, Chatillon, France

J.C. Sangleboeuf
Materials Department (LMS), Ecole Polytechnique, Palaiseau, France

O. Sudre
Materials Department, ONERA, Chatillon, France

ABSTRACT: This paper deals with the identification of interfacial parameters of SiC/Pyrex composite from Pull-out tests. The response load versus crack opening displacement of a monofilament bridging a matrix crack is measured. It is analysed using a classical model developed by Marshall which takes roughness effects into account. We focus attention on both the unloading/reloading cycles and monotonic loading. A Finite Element analysis is also presented and compared to analytical solutions. The two approaches are in agreement but are unable to model a whole test with a single set of parameters.

1. INTRODUCTION

Brittle Ceramic Matrix Composites put up with global strains thanks to a matrix multi-cracking held by fibre-reinforcement. Their toughness relies on a crack-bridging mechanism (Evans & Zok 1994). It is well-known that interfaces play a significant role as sites for dissipation mecanisms such as friction, debonding, all delaying fibre rupture.

Their mechanical characteristics depend on the bonded fibre/matrix, the friction behaviour of materials, the residual stress state and also its rough topography (Evans & Marshall 1989).

The direct way to evaluate interfacial properties is to get the response of the applied force against the fibre/matrix relative displacement on a monofilament composite. Several micromechanical tests: push-out, push-in, pull-out are appropriate but their intepretation is still difficult (Liang & Hutchinson 1992, Marshall & Oliver 1990, Tsuda 1992). These techniques require the resolution of structural composite problems; under some assumptions, Marshall has proposed analytical solutions which take debonding and friction into account (Marshall 1992). Recent analyses include the interface roughness in the models. For a sliding displacement greater than the dominant half-wavelength of the nonperiodic roughness, a simple method consists of introducing effective interfacial clamping stress which contains thermal misfit strain and a constant additional roughness misfit strain (Kerans & Parthsarathy 1993).

The present study deals with the identification of interfacial parameters from a modified pull-out test similar to the one performed by Mumm & Faber (1995). This technique develops a stable progressively debonding and sliding interfacial crack and measures an accurate response of a crack-bridging monofilament. We focus attention on both the loading curves and unloading/reloading cycle curves in order to determine quantitative parameters. The materials used are a SiC fibre: SCS-6, surrounded by a transparent glass: a pyrex, which allows a visual observation of the interface during the test. Analysis of experimental measurements are made by using a classical model developed by Marshall (1991). Finite Element calculations are also presented. They simulate the whole test and give specific information on the analytical model and experimental procedure.

2. SPECIMEN PREPARATION

2.1 Materials

A SiC SCS-6 fibre from Textron was sandwiched between two Pyrex glass plates. The SCS-6 outer layer, rich in carbon, allows the bifurcation of the matrix crack and smoothes out processing flaws. The surface roughness presents a nonperiodic trace with an average roughness amplitude of about 50 nm and a peak-to-valley value of about 150 nm (Jero & Parthasarathy 1992). Mechanical properties are listed in Table 1

2.2 Elaboration

Glass plates and fibres are placed in a graphite mould. The composite is uniaxially hot pressed in a

Table 1 : fibre and matrix properties

	SCS-6	Pyrex glass
E (Gpa)	400	70
μ	0.15	0.2
α (10^{-6} C^{-1})	4.0	3.3
R_f (μm)	140	

graphite resistors furnace in vacuum. The temperature increases upto 750°C at 10°C/min, then it is maintained for 20 min. At this temperature level, a pressure of about 3 Mpa is applied for 10 min. Room temperature is reached without any control of the temperature cooling ramp.

The composite is then cut along the fibre axis to obtain a parallelepipedic specimen. The polishing of each face to one micron surface finish allows to center the fibre accurately and to observe interface surface during the test. A notch is made at the middle of the specimen with a 200 μm diamond blade, through which a crack is propagated leaving a single fibre bridging the crack. The final specimen has a cross section of 3.5 mm x 5.8 mm and a length of 70 mm (see fig1).

Residual thermal stresses take place during cooling when the glass viscosity is high enough, only the range between the softening temperature and the room temperature were taken into account. As the thermal expansion coefficient difference is small, a correct calculation requires solving a viscoelasticity problem which has not yet been done. An elasticity calculation puts the interface in tension due to $\alpha_f > \alpha_m$.

Optical micrographs of composite cross sections reveal that the fibre outer layer is unaffected during elaboration, no reaction product being detected. Interface adhesion is expected to be low. Other post-mortem observations of the matrix show an inprint shape of the fibre roughness surface. The interlocking asperities of the rough interface caused the seating-drop phenomenon in pushout experiments (Jero & Kerans 1990).

3. EXPERIMENTAL PROCEDURE

3.1 Experimental set-up

The test is performed in traction as illustrated in fig1. The crosshead displacement is controlled at a ramp of 1mm/min. Applied loads do not exceed 100 N and are measured using a load cell mounted on the crosshead. Specimens are bonded on grips with a

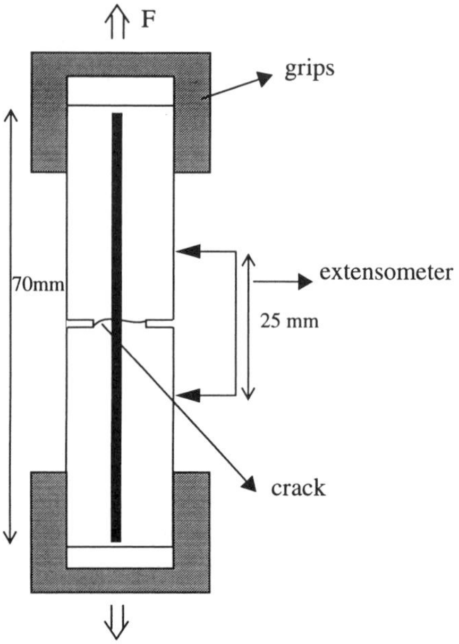

Fig.1. Schematic representation of pull-out test.

cyano-based adhesive. The whole system is fixed to the support by ball joints to avoid any flexion.

A crack was propagated through the specimen before loading. Through the transparent matrix, a bright zone locally extending along the interface from the crack plane was observed. This change of contrast results from a sliding fibre/matrix. This initial condition, being uncontrolled, varies from one test to another. The matrix cracking, initiating interfacial debonding and friction, leaves an unknown mechanical state.

The crack opening displacement (COD) is measured with an extensometer located at 12.5 mm on both sides of the crack plane. Numerical computations confirmed that the displacement induced by the matrix axial strains on a free side are insignificant relative to the COD measurement (see fig 10).

3.2 Results from load(COD) curves

A representative load versus COD curve is shown in fig2. A monotonic loading and several unloading/reloading cycles were performed until the fibre broke. The fibre rupture seldom occured at the crack plane where the axial stress level is maximum. This is a classical statistical effects of the fibre strength. However, it suggests that axial fibre stress varies slowly along the interface.

The test is divided into three parts:

-1st : the initial response fluctuates from one test to another. This sensitivity is explained by expe-

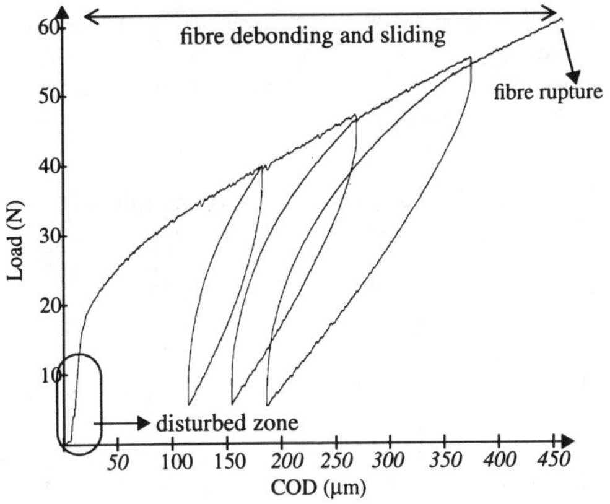

Fig. 2. Representative Load / COD curve

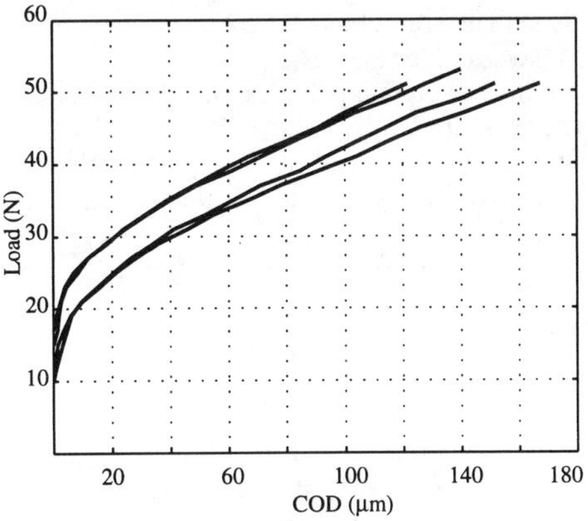

Fig.3. Load versus COD during initial loading for 4 representative tests

rimental initial conditions: defective alignment and initial stress state induced by matrix cracking.

- 2nd : The non-linearity of the curve results from the development of debonding and sliding zone during which the compliance increases. The transition between the two parts is not clearly defined.

- 3rd : After the fibre rupture, the fibre is extracted from the matrix. This response is not considered in the present work.

Loading curves :

The reproducibility is illustrated in fig. 3 where four experiments are shown together. They exhibit the same shape but they are shifted on the load axis. The matrix cracking has generated different initial conditions without changing interfacial behaviour.

In all tests, small amplitude load fluctuations (about 0.25 N) have been recorded. These instabilities were directly connected to the rough sliding behavior (Mumm & Faber 1995). On the contrary of Mumm & Faber's tests, their amplitudes do not increase with the load and are rather low. Two experimental conditions explain this difference. The higher crosshead speed displacement (1mm/min instead of 10 µm/min) which smoothes friction instabilities and the opening stress at interface, created by thermal misfit, which reduces roughness sliding effects.

Unloading/reloading curves (see fig. 2):

During a cycle, the debonding zone is assumed to no longer progress. Friction dissipates energy by reversing the shear stress on a portion of the sliding zone. The affected region increases with the reverse load.

The cycle shape is clearly asymmetrical. The curvature is more pronounced at the beginning of reloading than at the beginning of unloading.

The initial slopes remain vertical before reversing the displacement. The stress gap, needed to reverse sliding, may be due to a roughness effect or to slack in the loading train.

Two other characteristics are worth noting. First, the cycles are not closed for high loads. A degradation at the interface by friction may occur. Second, load fluctuations are visible at the end of each sequence. The amplitudes of these instabilities increased as the sliding zone got longer (Mumm & Faber 1995).

4. ANALYTICAL MODELLING

4.1 Analytical expressions

A Marshall's model (1992) based on the solutions of Lamé developed by Hutchinson & Jensen (1990) is used to analyse the load/COD response. Its application is straightfoward since we consider an axisymetric geometry for the specimen.

Friction is modelled using the Coulomb's law. The constant friction model was not considered as it did not allow to describe the asymmetrical shape of experimental cycles. The Poisson's contraction effect is not negligible relative to the residual radial strains at the interface (Marshall & Shaw 1995).

Debonding is viewed as a mode II interface fracture. The half COD, δ, is explicitly expressed as a function of the load F through the normalized applied fibre stress S_a :

$$S_a = \frac{F}{\pi R_f^2 \cdot \sigma_p} \quad (1)$$

where R_f is the fibre radius and σ_p is the maximum applied fibre stress for the tests.

The original expressions are expressed in another form to allow a straightfoward identification of μ and S_{R0} when the stress does not reach zero after a complete unloading. For monotonic loading, the relation is :

$$\delta = \frac{\delta'}{\mu}\left[-AS_{R0}\ln\left(\frac{S_{R0}-S_a}{S_{R0}-\Gamma'}\right)+\Gamma'-S_a\right] \qquad (2)$$

δ' and A are a function of the materials properties and the geometry of the problem (elastic and thermoelastic coefficients, fibre radius and volume fraction of fibre). Three parameters are unknown : Γ', μ and S_{R0}.

Γ' characterizes the interfacial fracture energy; the stress condition causing initial debonding is $S_a = \Gamma'$. A distinctive feature of equation (2) is that different values of Γ' translate the function $\delta(S_a)$ along the stress axis.

μ is the friction coefficient in the Coulomb law. The curvature of the representation $\delta(S_a)$ is proportionnal to μ.

S_{R0} represents the effective radial clamping stress by the relation:

$$S_{R0} = \frac{-\sigma_{r0}}{b_1\sigma_p} \qquad (3)$$

σ_{r0} is a component of the radial stress at the sliding interface due exclusively to roughness and thermal misfit effects. S_{R0} is an upper bound; when S_a reaches S_{R0} the contact between fibre and matrix is lost at the crack plane.

For unloading/reloading sequences, the debonding zone does not extend; only friction governs the $\delta(S_a)$ relations.

For unloading:

$$\delta_u - \delta = \frac{\delta'}{\mu}(S_{R0}-S_a)\left[1-\sqrt{\frac{S_{R0}-S_u}{S_{R0}-S_a}}\right]^2 \qquad (4)$$

where S_u and δ_u are respectively the normalized stress and half COD at which an unloading sequence starts (see fig. 7).

For reloading:

$$\delta - \delta_{re} = \frac{\delta'}{\mu}(S_{R0}-S_a)\left[1-\sqrt{\frac{S_{R0}-S_{re}}{S_{R0}-S_a}}\right]^2 \qquad (5)$$

where S_{re} and δ_{re} are respectively the normalized stress and half COD at which a reloading sequence starts (see fig. 7).

Only the two parameters μ and S_{R0} are unknown in equations (4) and (5).

4.2 Discussion on the identification of monotonic loading and cyclic loading

Equations (2), (4) and (5) were fitted to experimental data by using an optimization least square algorithm to find μ, S_{R0} and Γ'. The results are illustrated in fig. 4, fig. 5 and fig. 6 from the same representative tests.

For unloading and reloading, equations (4) and (5) of the model are well suited (see fig. 4 and fig. 5). The two fitting parameters μ and S_{R0} are very close in each case: less than a 2% difference. The reloading curves are remarkably superposed for different unloading stress levels. This feature is predicted by the model because equation (5) only depends on the reloading stress level.

For initial loading, the extra parameter Γ' should facilitate the fitting procedure. However, the curvature of the model does not respect those of experimental curves (see fig. 6).

Furthermore, the physical meaning of the parameter Γ' needs to be clarified. The initial conditions of the tests do not correspond to the initial loading condition of the model. Matrix cracking has created an unknown mechanical state which cannot be taken into account. As Γ' is a constant in equation (2), the fitting parameter Γ' represents the normalized stress at which the tests begin (see fig. 6).

Therefore, the characteristic debonding energy value cannot be quantified by these tests; only the two identified parameters μ and S_{R0} are physically meaningful.

The parameter S_{R0} identified on initial loading curve is greater than those identified on the cycles: 1.68 and 1.30 respectively. The effective clamping stress appears greater when the debonding zone extends.

One consequence is the incapacity to fit a whole test as is shown in fig. 7. Using parameters identified on the cycle loops, the initial loading equation (2) predicts a greater compliance of the crack-specimen whatever the choice of Γ'. This difference may be explained by the modelling of the rough friction. During the initial loading, the work required to dislocate the interlocking asperities when the debonding zone is extending, is not taken into account.

4.3 Determination of interfacial parameters

The parameters identified on unloading/reloading loops appear more reliable: the cracking condition does not take place and only 2 parameters are used

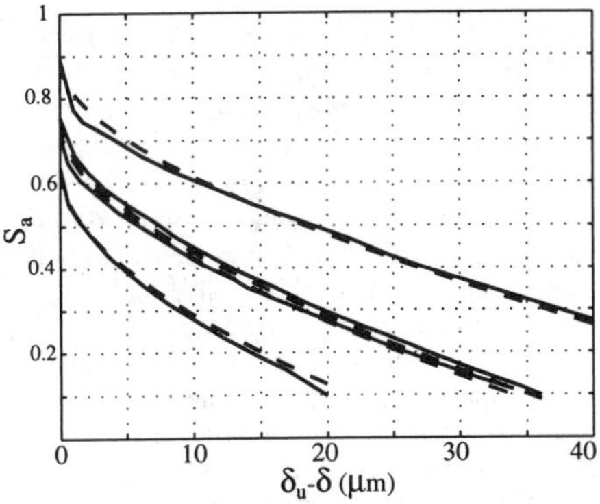

Fig.4. Experimental unloading data (solid lines) fitted with model equation (dashed lines).

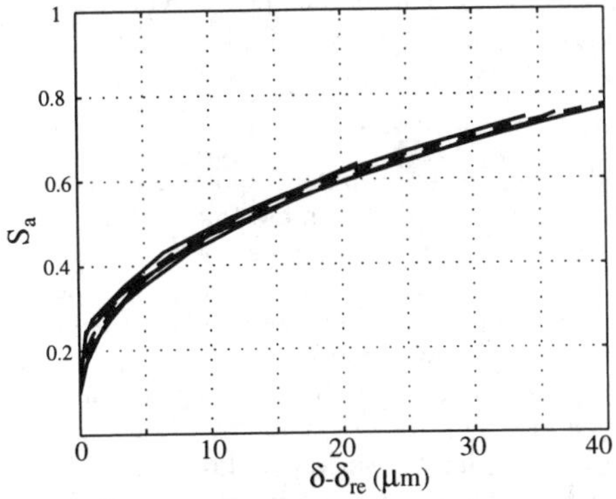

Fig.5. Experimental reloading data (solid lines) fitted with model equation (dashed lines).

for the fit. These values, presented in table 2, characterize interfacial behaviour only when sliding occurs in a fixed debonding zone.

The effective clamping stress σ_{r0} is calculated using equation (3) and roughness amplitude A' using an equation proposed by Kerans and Parthasarathy (1993):

$$\sigma_{r0} = -(a_4 - a_2 b_1)\, E_m\left((\alpha_m - \alpha_f)\, \Delta T + \frac{A'}{R_f}\right) \quad (6)$$

where ΔT is the range between the softening temperature and room temperature. a_i and b_i are from Hutchison and Jensen's works (1990).

Results are gathered in Table 2.

The friction coefficient is lower than those determined by Mumm & Faber (1995) for a similar material system. This would be explained by the experimental conditions: processing under vacuum instead of argon atmosphere decreases μ as it has

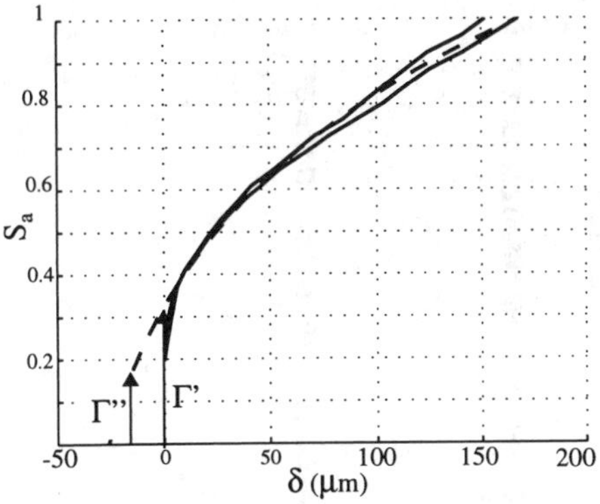

Fig.6. Experimental initial loading data (solid lines) fitted with model equation (dashed line). Γ' is normalized stress at which the experimental data are fitted. The analytical equation (2) is represented with an arbitrary debonding energy parameter Γ''.

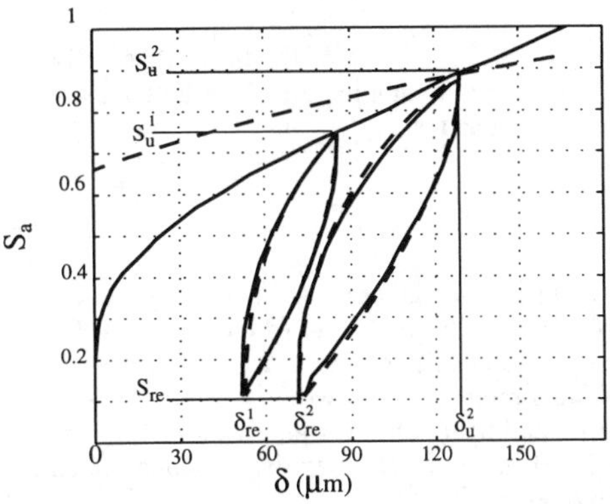

Fig.7. A whole test data fitted with the identified parameters by unloading/reloading sequences (dashed lines).

Table 2 : interfacial parameters

S_{R0}	μ	σ_{r0} (Mpa)	A' (nm)
1.27	0.04	-115.0	172

been observed by Tsuda & Enoki (1996). Moreover, the greater loading velocity should modify interfacial sliding. Recently performed tests at a lower loading velocity show a lower compliance of the crack-specimen.

σ_{r0} is in agreement with Mumm & Faber's value. As processing leaves an interface in tension, roughness causes important clamping stresses during interfacial sliding. Calculated roughness amplitude is close to the peak-to-valley value determined by Jero & Kerans (1990). An accurate determination of ther-

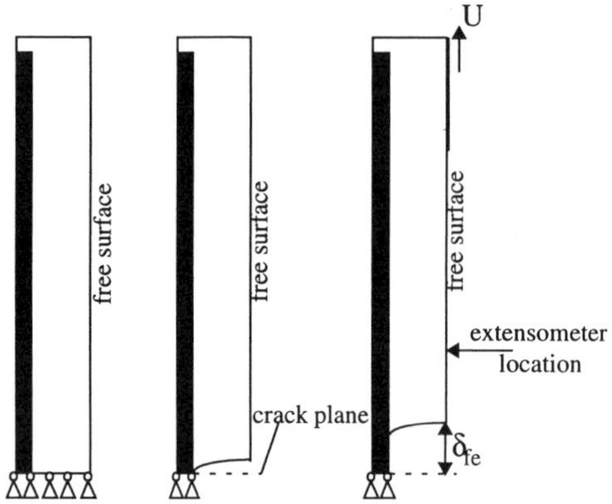

Fig.8. Boundary conditions and caculations steps to model the pull-out test

mal expansion coefficients of the fibre, especially in the radial direction, should provide a better evaluation of this parameter.

5. FINITE ELMENT ANALYSIS

Finite element calculations are another approach to analyse micromechanical tests which require solving structural problems. For an axisymmetric cell, calculations are simple and of low cost. A numerical tool to verify analytical solutions and to develop further studies, such as to model an unidirectional composite was developed.

5.1 Methodology

The finite elements modelling is based on the same assumptions as those of Marshall's models. An axisymmetric cell represents one half of the specimen. Friction is taken into account by the Coulomb law. Debonding is not considered, this is consistent with a low interfacial fracture energy as evaluated by Mumm & Faber (1995). Moreover, as Γ' is a constant in the relation between Load and COD equation (2), the curve shape is independent of the value Γ'.

Effective clamping stresses which contain roughness effects are introduced by using an equivalent thermal coefficient mismatch; this is also the case in the analytical solutions shown in fig. 9,10.

Experimental conditions have been respected as far as possible: applied force zone, length of the specimen and the volume fraction of fibre. Calculation follows three steps represented in fig. 8:

- a thermal calculation which creates clamping stresses all along the interface.
- the relaxing of matrix crack plane nodes
- the mechanical loading under displacement control. It 's imposed on the side face to represent the applied force by the adhesive and grips system.

5.2 Results and discussion

The analytical response is in total agreement with FE solution (see fig. 9 and 10). In the frame of the problem : the Coulomb friction with a low coefficient μ and without taking debonding into account, Marshall's equations (2), (4) & (5) are an excellent solution to the response of a microcomposite.

The opening due to matrix cracking, taken into account in the calculation of δ_{fe} but not in analytical solutions, is negligible. Since μ is low, the COD is only caused by the relative sliding between the fibre and the matrix. Furthermore, the displacement of the location extensometer node is identical to δ_{fe} (see fig. 10). Experimental tests provide a correct measure of the COD.

The inversion of the sliding zones observed in FE calculations are perfectly described by the model (see fig. 10). For each sequence, shear stress changes sign at the crack plane. At the end of a reloading sequence, the mechanical state reached is identical to that of before the cycle.

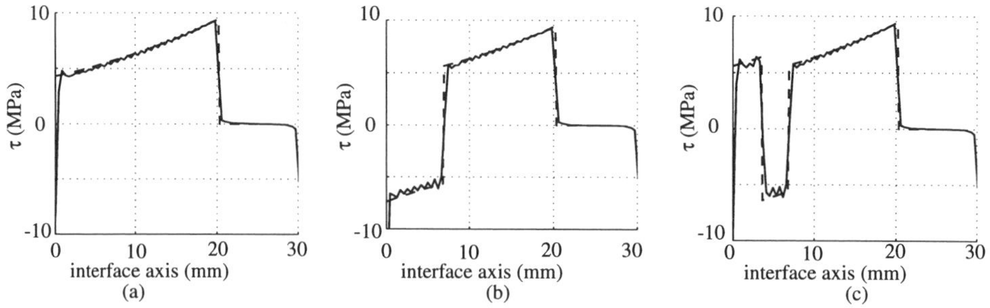

Fig.9. Shear stress repartition along interface for Marshall's model (dashed lines) and FE calculations (solid lines).
(a) : Initial loading, (b) : unloading sequence, (c) reloading sequence.

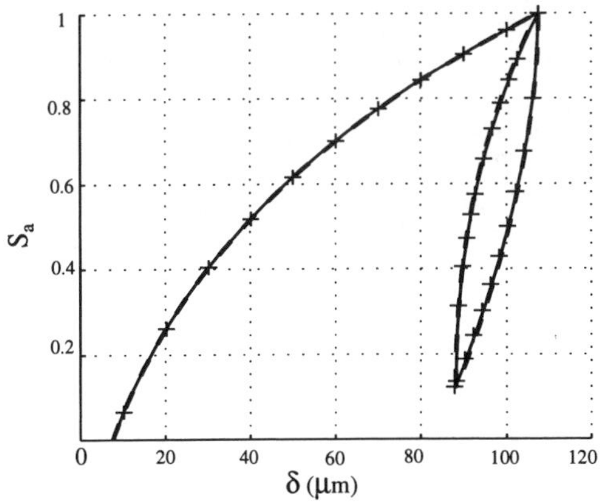

Fig. 10. response of analytical model equations (dashed lines) and FE calculations (solid lines). The "+" marks represent numerical displacements of the extensometer location node.

The agreement between the two approaches, demonstrates that the solution of the problem, even exact, cannot succeed in simulating a whole test (see fig. 7). More sophisticated interfacial laws should be considered.

6 CONCLUSION

The performed pull-out tests accurately measure the load versus COD response of a microcomposite. A stable debonding develops along the interface where rough friction occurs.

The use of Marshall's model allows us to determine relevant interfacial parameters in unloading/ reloading sequences. These values are in agreement with other measurements reported in the literature.

The interfacial fracture energy is not characterized by this test due to the initial cracking of the matrix. This unknown condition cannot be taken into account by the model equations.

The model gives an excellent solution with a simple Coulomb interfacial behaviour as is verified by FE calculations. However it is not able to fit the whole test. Interfacial friction is different when the debonding zone is extending during the monotonic loading and when it stays fixed during unloading/ reloading. The model seems insufficient by considering rough friction through a single parameter.

Interface laws should be further developed, Stupkiewicz (1996) has introduced dilatency in interface behaviour. We also propose a numerical development from FE solutions (Chaboche & Girard 1997).

REFERENCES

J.L Chaboche, R. Girard, A. Schaff, "Numerical analysis of composite systems by using interphase/ interface models.", to be published in Computational Mechanics.

A.G. Evans, D.B. Marshall, "The mechanical behavior of Ceramic Matrix Composites", Acta metall. Vol 37, No 10, 2567-2583, (1989).

A.G. Evans, F.W. Zok, "The physics and mechanics of fibre-reinforced brittle matrix composites", Journal of Materials Science, 29, (1994), 3857-3896.

J. W. Hutchinson, M. Jensen, "Models of fiber debonding and pullout in brtittle composites with friction", Mechanics of Materials 9 (1990), 139-163.

P.D. Jero, R.J. Kerans, "The contribution of interfacial roughness to sliding friction of ceramic fibers in a glass matrix", Scipta Metall. Mater., 24, (1990), 2315-2318.

P.D. Jero, T.A. Parthasarathy, R.J. Kerans, "Interfacial Roughness in Ceramic Matrix Composites", Ceram. Engig. Proc. 13, 64 (1992).

R.J. Kerans, T.A. Parthasarathy, P.D. Jero, A. Chatterjee, N.J. Pagano, "fracture and sliding in the fibre/matrix interface and failure processes in Ceramic composites", Br. Ceram. Trans., 92 (1992), No5.

C. Liang, J.W. Hutchinson, "Mechanics of the fiber pushout test", Mechanics of Materials 14 (1993) 207-221.

D.B. Marshall, "Analysis of fiber debonding and sliding experiments in brittle matrix composites", Acta. metall. mater., Vol 40, No 3, (1992), 427-441.

D.B. Marshall, W.C. Oliver, "An indentation method for measuring residualstresses in fiber-reinforced ceramics", Materials Science and Engineering,A126 (1990) 95-103.

D.B. Marshall, M.C. Shaw, W.L. Morris "The determination of interfacial properties from fiber sliding experiments: the roles of misfit anisotropy and interfacial roughness", Acta. metall. mater., Vol 43, No 5, (1995), 2041-2051.

D.R. Mumm, K.T. Faber, "Interfacial debonding and sliding in brittle-matrix composites measured using an improved fiber pullout technique", Acta. metall. mater., vol 43, no3 (1995), 1259-1270.

S. Stupkiewicz, "Fiber sliding model accounting for interfacialmicro-dilatency", Mechanics of Materials 22 (1996) 65-84.

H. Tsuda, M. Enoki, T. Kishi, "Interfacial mechanical properties of fiber-reinforced ceramic composites", J. Am. Ceram. Soc. 100 [4] 530-555 (1992).

H. Tsuda, M. Enoki, T. Kishi, "Influence of loading rate atmosphere onpull-out of sic fibers from pyrex glass", J. Am. Ceram. Soc. 79 [5] 1319-23 (1996).

Damage and Failure of Interfaces, Rossmanith (ed.) © 1997 Balkema, Rotterdam, ISBN 90 5410 899 1

Interlaminar fracture analysis of GFRP influenced by fibre surface treatment

F.-G. Buchholz & H. Wang
Institute of Applied Mechanics, University of Paderborn, Germany

R. Rikards & A. Korjakin
Institute of Computer Analysis of Structures, Riga Technical University, Latvia

A. K. Bledzki
Institute of Material Science, University of Kassel, Germany

ABSTRACT: In this paper the interlaminar fracture behaviour of unidirectionally glass fibre reinforced composites with two different fibre surface treatments has been investigated. DCB specimens and ENF specimens are used to determine the mode I and mode II interlaminar fracture toughness of this composite respectively. The data obtained from these tests have been analysed by using different analytical and finite element approaches in order to evaluate the critical energy release rates. The results show that the interlaminar fracture toughness for both types of fibre surface treatments depends on the loading mode. For the DCB test under mode I loading the interlaminar fracture toughness for crack initiation is the same for both surface treatments but it differs for the ENF test under mode II loading. For both fibre surface treatments the interlaminar fracture toughness increases considerably with increasing crack length. The geometric nonlinearity for the DCB test and the influence of friction for the ENF test are also considered in the finite element calculations.

1 INTRODUCTION

One of the major disadvantages of laminated composites is their tendency to delaminate. In composites with brittle matrices initiation and growth of delamination can reduce the stiffness and compressive strength of the composites considerably.

The quality of the fiber/matrix interface will influence the initiation and propagation of delamination cracks. The resistance to delamination growth has been characterized by the fracture toughness or the critical energy release rates under mode I, mode II or mode III or mixed mode loading conditions. In composite structures, for example, mixed mode fracture is observed under impact loading. Therefore, the interlaminar fracture toughness has been identified as a prime factor in controlling the growth of delamination in laminated composites.

However, as may be seen from the data reported in the literature, the values of the critical energy release rates are frequently dependent on the method of analysis employed for the data reduction. For example, it has been observed that the area and the compliance calibration method, which are two direct approaches for the calculation of critical energy release rates from experimental data, may give different values compared to those obtained by using the linear beam theory. In papers (Hashemi et al 1990) corrections in the formulae based on the linear beam theory were proposed by taking into consideration such effects as the end rotation and deflection of the crack tip, the effective shortening of the beam due to large displacements and the stiffening of the beam due to the presence of the end blocks bonded to the specimens.

One way to improve the delamination fracture toughness of composites is focused on improving the fracture toughness of the matrix by using toughened resin systems or thermoplastic matrices (Sela & Ishai 1989). However, an increase in resin toughness is not completely realized in the interlaminar fracture toughness of composites having poor adhesion between fibers and matrix. Interlaminar fracture toughness is highly dependent on the fiber/matrix adhesion (Chai 1986). Therefore, another way to increase the interlaminar fracture toughness of a composite is to improve the fiber/matrix adhesion by fiber surface treatments, which changes the fiber surface morphology and chemistry (Albertsen et al 1995).

There are different methods available for fiber surface treatments of carbon (Drzal et al 1982) and aramid (Gutowski et al 1993) fibers. Increasing the level of adhesion in the fiber/matrix interface yields significant improvements in the interlaminar fracture toughness (Madhukar & Drzal 1992). In the present paper composites reinforced by glass fibers with surface treatments are investigated. Although glass fibers with surface treatments have been used for a long time the fiber/matrix interface morphology and chemistry are not completely understood (Drzal 1986). The surface of the glass fibers used in the present investigation were improved by a technology, which was outlined previously (Bledzki 1993).

The goal of this investigation is to study the influence of the fiber surface treatment on the interlaminar fracture toughness. Delamination of unidirectional glass/epoxy laminates under mode I and mode II loading conditions is investigated. In the present investigation for the mode I tests DCB specimens and for the mode II tests ENF specimens are used. The data reduction from experiments for the DCB test have been carried out by using six different approaches: the area method, the Berry method, the modified beam analysis, the compliance method and the linear or geometric non-linear finite element analysis. For the ENF tests only the data reduction by using the beam analysis and the linear finite element analysis were considered to be sufficient. For the ENF test contact and friction along the crack surfaces have been taken into account. The critical energy release rates G_{IC} and G_{IIC} are calculated by using the virtual crack closure integral (VCCI) method (Rybicki & Kanninen 1977, Buchholz 1984).

2 MATERIAL

For the production of the composite material to be considered in this investigation borosilicate glass fibers (E-glass) with small parts of alkali and a brittle epoxy matrix (Araldite LY 556 with curing agent Hy 917, Ciba Geigy Ltd) were used. For the fibers embedded in the brittle epoxy matrix two different types of fiber surface treatments were chosen. The first type of sizing (surface treatment SI/EP) is based on using a silane coupling agent (Aminosilane 1100) in a mixture with an epoxy modifying agent (Wacker 1996). The second type of sizing (surface treatment PE) is based on a modification of the fiber surface with polyethylene.

Unidirectionally reinforced glass fiber plates were produced through a winding technology. The plates contained a starter crack, which was introduced by a Teflon film (thickness 40 μm) placed between the central plies. The total number of plies was 12 in order to get a plate thickness of approximately 3 *mm*. The laminates were produced by following the standard cure cycle recommended by Ciba Geigy Ltd. In order to improve the quality of the plates and to reduce the void content, the plates were placed in a vacuum before curing. The specimens were cut with a diamond wheel and kept at room temperature until testing (23° C and 50 % of relative humidity). All tests were performed at the same conditions. The number of specimens tested for the composite with fiber surface treatment SI/EP was 9, and for the composite with fiber surface treatment PE it was 10. For both composites (SI/EP and PE) the fiber volume content of the specimens was 50% (±1.2%). The elastic properties of the composites investigated in this paper are given in Table 1.

3 TEST SPECIMENS AND EXPERIMENTAL PROCEDURE

The double cantilever beam (DCB) was used for those tests in which mode I loading conditions at the crack front had to be achieved. The DCB specimen is shown in Figure 1 for which the dimensions have been chosen as recommended in the literature (Hojo et al 1995).

Table 1. Elastic properties of the composites.

SI/EP	PE
E_1=40.7 GPa	E_1=39.8 GPa
E_2, E_3=10.7 GPa	E_2, E_3=7.45 GPa
G_{12}, G_{13}=4.2 GPa	G_{12}, G_{13}=4.2 GPa
G_{23}=3.98 GPa	G_{23}=3.98 GPa
ν_{12}, ν_{13}=0.27	ν_{12}, ν_{13}=0.27
ν_{23}=0.29	ν_{23}=0.29

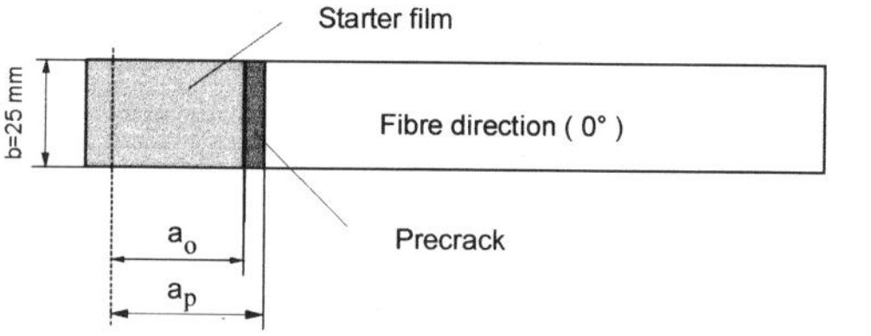

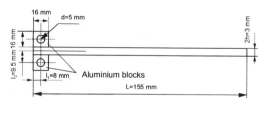

Figure 1. Double cantilever beam specimen (DCB) for the mode I tests and dimensions of the specimen and the aluminium blocks (in *mm*).

Precracks of length $a_p - a_0 = 2.5mm$ were introduced into the specimens by a starter film of length $a_0 = 31\,mm$. The loads were applied to the specimen via pins through universal joints and aluminium blocks that were bonded on the specimen.

For the experiments the testing machine Zwick 1446 was used. The loading was realized through a given displacement rate 1.0 *mm/min*$^{-1}$. The specimen was loaded until a quasi-static crack extension of approximately 10 *mm* was observed and then it was unloaded. This procedure was repeated until the crack had reached a final length of about 100 to 120 *mm*. The crack opening displacement (COD) δ on the load line was measured from the displacement of the testing machine and the crack length a was measured from the distance between crack tip and load line. A typical load crack opening displacement diagram as function of crack length is shown in Figure 2. It can be seen that UD laminates show stable crack growth under mode I loading conditions of the DCB test even for a brittle epoxy matrix.

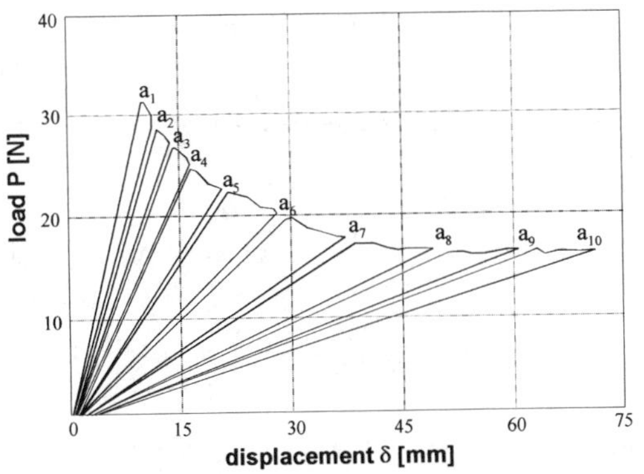

Figure 2. Typical load P against displacement δ diagram for the DCB test with composite SI/EP. Crack lengths a (*mm*) are presented for the critical values of the load.

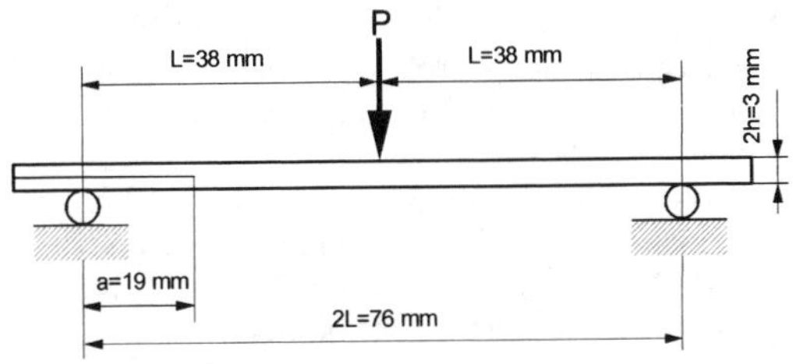

Figure 3. End notched flexure specimen (ENF) for the pure mode II tests.

The mode II tests were performed by the aid of end notched flexure (ENF) specimens (see Figure 3), and a three point bending fixture with a span distance of $2L$=76 *mm* and a width of the specimen of b=25 mm, i.e. the same as for the DCB specimen. The thickness of the specimen is $2h$=3 *mm* and it contains a symmetric split. Precracks of length from 4 *mm* to 10 *mm* were introduced before testing by loading with the crack opening mode, so that the initial position of the crack tip was at one fourth of the span of the beam. The loading during the test was carried out again by a given displacement rate of 1.0 *mm/min*$^{-1}$. When the limit load was reached, unstable crack growth occurred until crack arrest in the mid-span below the compression load. The limit load P and the critical displacement δ (deflection) at the mid-span of the beam were measured at the beginning of the crack extension.

4 CALCULATION OF THE INTERLAMINAR FRACTURE ENERGIES

4.1 *Analytical approaches*

A direct method for the evaluation of the total energy release rate is the area method. According to this method the released energy due to crack extension can be calculated by using Irwin's formula for the critical energy release rate

$$G_{TC} = \frac{\Delta U}{b\Delta a}, \qquad \Delta U = \frac{1}{2}(P_1\delta_2 - P_2\delta_1) \tag{1}$$

where P_1 and δ_1 are the load and displacement at crack length a and P_2 and δ_2 are the respective values at crack length $a+\Delta a$. Formula (1) can be used to calculate the critical energy release rates for mode I, mode II or mixed mode I/II loading conditions. However, only the total energy release rate can be calculated in the case of mixed mode conditions by eq. (1). In this case the separate energy release rates can only be calculated if the mode I and mode II values can be determined by another approach.

Since the strain energy for linear response is $U = P\delta/2$ and $\Pi = -U$, the energy release rate in eq. (1) can be expressed through the compliance (Broek 1986).

$$G_{TC} = \frac{P_c^2}{2b}\frac{\partial C}{\partial a} = \frac{P_c\delta}{2bC}\frac{\partial C}{\partial a} \tag{2}$$

where P_c is the critical load at which crack growth is observed, $C=\delta/P$ is the compliance, δ is the displacement and a is the crack length. In order to evaluate G_{TC} the compliance must be calculated by using experimental data, i.e. the measurements of loads and displacements at different crack lengths. The curve fit for the compliance must be performed with great accuracy, since in formula (2) the first derivative of the compliance function determines about the precision of the calculated critical energy release rate.

For the DCB specimen the critical mode I energy release rate calculated by the linear beam theory is given by (Broek 1986)

$$G_{IC} = \frac{3P_c\delta}{2ba} = \frac{3P_c^2 C}{2ba} \tag{3}$$

If the data reduction is applied together with the least square technique (Madhukar & Drzal 1992) formula (3) can be used for the determination of the mean value of G_{IC}. However, the coefficient 3, obtained by the classical beam theory can be corrected by using the data of experiments (Berry 1963). In this case eq. (3) is rewritten by introducing an unknown quantity n

$$G_{IC} = \frac{nP_c\delta}{2ba} = \frac{nP_c^2 C}{2ba} \tag{4}$$

which can be determined by the least square method from the experimental data.

For the mode II ENF test the critical energy release rate calculated by linear beam theory is given through (Hashemi 1990)

$$G_{IIC} = \frac{9P_c a^2\delta}{2b(3a^3 + 2L^3)} \tag{5}$$

It was observed that different data reduction methods result in different values of the critical energy rates. This is because in linear beam theory some effects are not taken into account. These effects are the rotation and deflection at the crack tip, the large displacements occurring in some of the test specimen and stiffening effects due to the presence of the bonded end blocks. In order to take these effects into account some correction coefficients were introduced (Hashemi 1990), with which the data reduction formulae of the linear beam theory (3) and (5) can be corrected by

$$G_{IC} = \frac{12FP_c^2(a + \chi h)^2}{b^2 h^3 E_1} \tag{6}$$

$$G_{IIC} = F\frac{9P_c(a + \chi h)^2\delta}{2b[3(a + \chi h)^3 + 2L^3]} \tag{7}$$

where χ, F are correction coefficients for the rotation effects at the crack tip and for the large displacement effect in the test specimen respectively.

4.2 *Finite element analysis*

The finite element analyses of the experiments are performed by using the program ABAQUS for plane strain conditions and with standard four-node quadrilateral elements. It was shown (Rybicki & Kanninen 1977) that by using the virtual crack closure integral (VCCI) method in combination with standard elements also good results for the energy release rates can be obtained in comparison with reference solutions or compared to approaches using higher order elements (Buchholz 1984). Here $\Delta a = 0.2\ mm$ is taken for the values of successive crack extensions, which is coinciding with the finite element length along the crack.

For the calculation of the energy release rates G_I and G_{II} the VCCI method is used here. Then the separated strain energy release rates are obtained by two calculations (VCCI or 2C method) for the actual crack length a and for the extended crack of length $a+\Delta a$

$$\begin{aligned} G_I^{2C}(a + \frac{\Delta a}{2}) &= \frac{1}{b\,\Delta a}\frac{1}{2}\left[F_y^i(a)\Delta u_y^{j-1}(a + \Delta a)\right] \\ G_{II}^{2C}(a + \frac{\Delta a}{2}) &= \frac{1}{b\,\Delta a}\frac{1}{2}\left[F_x^i(a)\Delta u_x^{j-1}(a + \Delta a)\right] \end{aligned} \tag{8}$$

where $F_x^i(a)$ and $F_y^i(a)$ are the nodal point forces at the crack tip node i in x and y directions, respectively, while $\Delta u_x^{j-1}(a + \Delta a)$ and $\Delta u_y^{j-1}(a + \Delta a)$ are the corresponding relative nodal point displacements at node j-1 of the opposite crack faces respectively. Therefore $\Delta u_x^{j-1}(a + \Delta a)$ is the crack sliding displacement, while $\Delta u_y^{j-1}(a + \Delta a)$ is the crack opening displacement at a distance Δa behind the crack tip of the extended crack of length $a+\Delta a$.

In the case of the ENF test contact and friction along the crack surfaces has to be considered in the region of the support (see Figure 3). In this case the global potential energy of the beam is

$$\Pi = U - W + W^D \tag{9}$$

with U for the elastic strain energy of the beam, $W^D \equiv W^F$ for the energy dissipated due to friction along the crack faces and W for the work of the external forces. The dissipated energy is equal to the frictional work W^F performed along the crack faces, where Coulomb's friction law is used

$$C_x = \mu C_y \tag{10}$$

Here μ is the coefficient of friction, C_x is the frictional force and C_y is the contact force at correlated nodal points along the crack faces (Rikards et al 1995). They correspond to the nodal point forces F_x and F_y at the crack tip in formula (8) but a different notation is used for all nodes behind the crack tip, that means along the crack faces which may open or may be in sliding contact. In the case of contact and friction along the crack surfaces the frictional work is calculated by

$$W_{ij}^F = \frac{1}{2} C_x^{j-1}(a+\Delta a)\Delta u_x^{j-1}(a+\Delta a) + \sum_{k=1}^{j-1} \frac{1}{2}\left[C_x^{i-k}(a) + C_x^{j-(1+k)}(a+\Delta a)\right] \left[\Delta u_x^{j-(1+k)}(a+\Delta a) - \Delta u_x^{i-k}(a)\right] \qquad k=1,2,\dots,i \quad j=i+1 \tag{11}$$

Therefore, the total energy release rate $G_T^{2C} = G_I^{2C} + G_{II}^{2C}$ of the specimen can also be calculated from the separated energy release rates (eq. (8)).

5 RESULTS FOR CRITICAL ENERGY RELEASE RATES

The critical energy release rates G_{IC} and G_{IIC} are calculated from the data of the experiments by the aid of different data reduction formulae, which were presented in the previous section. The data reduction for the mode I test by using the DCB specimen is performed by 6 approaches: the area method, the compliance method, the Berry method, the modified beam analysis and in addition by the linear or geometric non-linear finite element analysis. The data reduction for the mode II ENF test are carried out by the linear beam analysis and by the finite element analysis taking into account contact and friction along the crack surfaces.

In the mode I DCB test 10 steps of crack extension were measured. For the composite SI/EP the experiments on the delamination crack propagation were performed for 9 specimens, whereas for the composite PE 10 specimens have been available. For these tests the crack growth was carefully controlled so that at each step the values of crack length a were about the same for all specimens. Therefore, the critical loads and the corresponding crack opening displacements can be averaged for each step from the data obtained from all specimens. The corresponding mean values of the load P, the crack length a and the displacement δ for the composite SI/EP and the composite PE are presented in Table 2.

Table 2. Experimental results for the DCB tests with composites SI/EP and PE

No.	P (N)		a (mm)		δ (mm)	
	SI/EP	PE	SI/EP	PE	SI/EP	PE
1	38.93	38.41	33.45	33.43	3.536	4.280
2	35.63	36.98	41.30	40.88	5.474	6.173
3	31.63	32.56	50.80	50.75	8.932	10.09
4	28.21	28.43	60.59	60.57	14.08	14.39
5	25.11	24.83	70.33	70.28	19.42	20.09
6	23.11	22.06	79.87	79.87	26.87	26.88
7	21.60	20.30	89.29	89.31	35.03	34.85
8	20.31	19.14	98.56	98.53	44.04	44.31
9	19.16	18.13	107.8	107.7	53.14	53.90
10	18.33	17.40	116.6	116.6	64.28	64.48

The critical energy release rates obtained from the DCB test with the different data reduction methods are presented in Table 3 for the composite SI/EP and in Table 4 for the composite PE. From a detailed comparison of the results it can be concluded that for crack lengths between 40 and 70 *mm* the critical energy release rates are in good agreement with the different data reduction methods. For longer cracks (a>70 *mm*) only the results obtained through the

Table 3. Mode I energy release rates $G_{IC}(a)$ (kJ/m^2) for the composite SI/EP calculated by six different methods

No.	eq. (1)	eq. (2)	eq. (4)	eq. (6)	FEM linear analy.	FEM non-linear analy.
1	0.186	0.229	0.244	0.234	0.258	0.242
2	0.269	0.292	0.280	0.299	0.323	0.299
3	0.311	0.348	0.332	0.355	0.381	0.348
4	0.344	0.392	0.392	0.400	0.427	0.385
5	0.410	0.418	0.413	0.424	0.453	0.404
6	0.423	0.456	0.464	0.459	0.490	0.429
7	0.441	0.498	0.505	0.495	0.532	0.456
8	0.448	0.536	0.541	0.524	0.573	0.478
9	0.476	0.569	0.563	0.548	0.607	0.493
10	0.481	0.609	0.603	0.570	0.649	0.508

nonlinear finite element analysis and the area method are in good agreement, whereas all other data reduction methods deliver critical energy release rates that are by 15 to 30% too high. Results obtained by the modified beam analysis show a trend to decrease in comparison with the results obtained by the conventional beam analysis. Also some differences for the crack initiation values in the first step of crack propagation are observed. In particular the crack initiation values obtained by the area method are lower than the values obtained by the methods based on the beam analysis and its modifications. Crack initiation values obtained by the finite element solution (linear or nonlinear) are in good agreement with those obtained by the compliance method or the conventional and the modified beam analysis.

Table 4. Mode I energy release rates $G_{IC}(a)$ (kJ/m^2) for the composite PE calculated by six different methods

No.	eq. (1)	eq. (2)	eq. (4)	eq. (6)	FEM linear analy.	FEM non-linear analy.
1	0.157	0.246	0.285	0.245	0.259	0.243
2	0.294	0.334	0.323	0.340	0.365	0.336
3	0.323	0.390	0.374	0.405	0.415	0.378
4	0.400	0.415	0.390	0.438	0.446	0.401
5	0.414	0.420	0.410	0.447	0.458	0.407
6	0.428	0.422	0.429	0.451	0.459	0.404
7	0.454	0.441	0.458	0.472	0.482	0.417
8	0.453	0.472	0.498	0.501	0.524	0.441
9	0.453	0.501	0.525	0.527	0.560	0.459
10	0.489	0.532	0.556	0.553	0.602	0.477

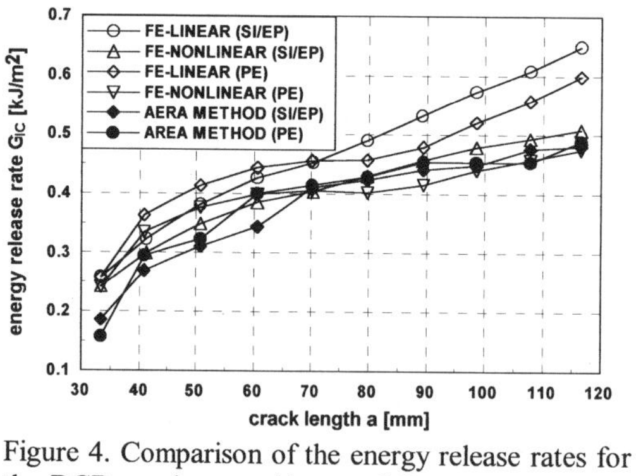

Figure 4. Comparison of the energy release rates for the DCB specimens with two different fibre surface treatments

Furthermore it is of interest to analyse how the critical energy release rates depend on crack length. In Figure 4 it is seen that with crack growth a considerable increase of interlaminar fracture toughness was found and that stabilisation of the crack resistance curves obtained by the nonlinear FE analysis and also by the area method is observed only for rather long cracks. For the composite PE the considerable increase of the critical energy release rates at crack propagation is due to extensive fiber bridging, but for the composite SI/EP practically no fiber bridging was observed in the DCB tests.

From the detailed analysis of the results presented in Tables 3 and 4, and also in Fig 4 it is seen that the critical energy release rates G_{IC}^{init} at crack initiation are much lower than the critical energy release rates G_{IC}^{prop} at further crack propagation. For comparison the mean values of G_{IC}^{init}, calculated as average of all values obtained by the 6 methods of data reduction applied in the present investigation, have been calculated. In the same way the mean values of the energy release rates G_{IC}^{prop} for steady state crack propagation can be obtained by averaging of the G_{IC} values for the last 2 steps of crack propagation (the 2 last rows in Tables 3 and 4). For the composite SI/EP the averaging gives the following results: G_{IC}^{init}=0.232 kJ/m^2 and G_{IC}^{prop}=0.557kJ/m^2. The corresponding values for the composite PE are as follows: G_{IC}^{init}=0.238kJ/m^2 and G_{IC}^{prop}=0.518kJ/m^2.

Thus it is found that for both composites the mode I critical energy release rates are of about the same value. However, by transverse tension tests it was obtained, that the transverse interlaminar strength in tension for the composite PE is about 3 times lower than for the composite SI/EP. The higher values of G_{IC} that are found here for the composite PE can be explained by extensive fiber bridging and pull-out in the DCB test, whereas for the composite SI/EP practically no fiber bridging and pull-out was observed in the DCB tests. Therefore, in the case of fiber bridging (composite PE) the crack surface on the micro level is much greater than the crack macro surface $A=b \times a$, which has been considered for the calculations of G_{IC}.

Table 5. Mode II energy release rates $G_{IIC}(a)$ (kJ/m^2) for the composites SI/EP and PE

Comps.	Experiment P (N)	δ (mm)	G_{IIC}^{init} eq. (5)	FEM μ=0.0	FEM μ=0.7
SI/EP	870	4.87	2.070	1.873	1.741
PE	540	3.18	0.839	0.744	0.690

For the mode II ENF test (see Figure 3) 4 specimens for each type of the composite SI/EP and PE were used. Before the tests mode I precracks of length from 4 to 10 *mm* were introduced and since in the ENF test the crack growth is unstable, only crack initiation values G_{IIC}^{init} can be obtained. In Table 5 the mean values of the experimental data for the critical load and displacement are presented.

In Table 5 the critical energy release rates G_{IIC}^{init} for crack initiation are presented. From the comparison of the results for the two different surface treatments it seen that the mode II critical energy release rate G_{IIC}^{init} for the composite SI/EP is about 2.5 times higher than for the composite PE. This result is in good agreement with the value of the transverse interlaminar strength in tension, which is about three times higher for the composite SI/PE (Wacker 1996). For both composites mode II cracks propagate without fiber bridging and, therefore, the value G_{IIC}^{init} indirectly characterises the fiber/matrix adhesion properties. It should be mentioned that for the ENF test the area and the compliance method can not be employed, since the crack growth in the experiments was unstable so that only the first critical load and the corresponding displacements can be measured.

In Table 5 it can also be seen that the mode II critical energy release rates obtained by the FE analysis taking into account contact and friction along the crack surfaces are about 15% lower than those obtained on the basis of the conventional beam analysis. For the frictional coefficient μ = 0.7 the frictional work in the ENF test is about 6% of the total potential energy of the specimen. Since the actual coefficient of friction is an unknown quantity it

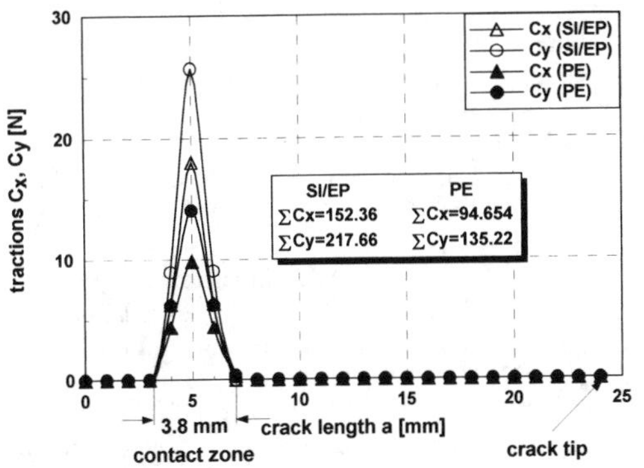

Figure 5. Distribution of tractions along the crack faces for the mode II ENF specimen (coeff. of friction $\mu = 0.7$).

can be supposed that in the present experiment, where the pre-crack was made by inserting one layer of Teflon film of thickness 40 μm, the actual coefficient of friction in the contact area is to be in the limits $0 \le \mu \le 0.7$.

It is also of interest to analyse the length of the contact zone and the contact forces along the crack surfaces. In Figure 5 the distribution of the normal surface tractions C_y at the contact zone, where sliding friction will occur, is shown. The resultant of the contact forces for both composites SI/EP and PE is found to be $\Sigma C_y \approx P/4$ (*P* is the total load applied at the midspan of the beam), i. e. of the same value as in the literature (Carlsson et al 1989). The contact area is found to be for both composites SI/EP and PE about 2.5*h* in length, where h is the thickness of the beam in the delamination region of the ENF specimen. This result is slightly different from the result obtained in the paper (Carlsson et al 1989), where a contact zone length 4*h* has been found.

6 CONCLUSIONS

Mode I and mode II delamination crack propagation experiments have been performed. Two types of fiber surface treatments have been investigated. The fiber surface treatment with the silane coupling agent in combination with an epoxy modifying agent (composite SI/EP) results in good fiber/matrix adhesion properties, whereas the fiber sizing with polyethylene (composite PE) results in comparably poor fiber/matrix adhesion properties. However, due to extensive fiber bridging in the mode I DCB test a significant difference of the critical energy release rates between both composites have only been found for the mode II ENF tests.

The results obtained for the critical energy release rates show that in mode I delamination crack propagation the interlaminar fracture toughness for both composites are approximately of the same value due to extensive fiber bridging in the case of the composite PE. However, also the mode I critical energy release rates for crack initiation are about of the same value for both composites, but the mode II critical energy release rate for crack initiation of the composite SI/EP is about 2.5 times higher than for the composite PE. This result is in good agreement with the interlaminar transverse strength in tension, which for the composite SI/EP is about 3 times higher.

From comparison of the critical energy release rates obtained by using different methods of data reduction it can be concluded that for the longer cracks in the mode I DCB tests the area method or the nonlinear finite element analysis must be used. The critical energy release rates obtained by the compliance method, by the Berry method and by the linear finite element analysis are overestimated for the longer cracks. The critical energy release rates for crack initiation for the mode I DCB tests can be calculated by all methods considered here. In the ENF tests the mode II critical energy release rates obtained by the FE analysis taking into account contact and friction along the crack surfaces are about 15% lower than those obtained on the basis of conventional beam analysis. For the frictional coefficient μ=0.7 the frictional work in the ENF test is about 6% of the total potential energy of the specimen.

ACKNOWLEDGEMENTS

This study was performed in the framework of a project on design of interfaces of glass fiber reinforced composites. The project is performed in cooperation between the universities of Riga, Kassel and Paderborn. The authors gratefully acknowledge the financial support by VW Foundation (Hannover, Germany) through grant I/70 665. The authors also would like to thank Dr. H. Frenzel at the Institute of Polymer Research (Dresden, Germany) for preparing and providing the glass fibers with different surface treatments.

REFERENCES

Albertsen, H., J. Ivens, P. Peters, M. Wevers & I. Verpoest. 1995. Interlaminar Fracture Toughness of CFRP Influenced by Fibre Surface Treatment, Part 1: Experimental Results. *Composites Science and Technology*. 54: 133-145.

Berry, J.P. 1963. Determination of fracture surface energies by the cleavage technique. *J. Appl. Physics*. 34: 62-68.

Bledzki, A. K., G. Wacker, & H. Frenzel 1993. Effect of surface treated glass fibres on the dynamic behaviour of fibre-reinforced composites. *Mechanics of Composite Materials*, 29: 585-591.

Broek, D. 1986. *Elementary Engineering Fracture Mechanics*. 4th, ed., Dordrecht, The Netherlands: Martinus Nijhoff Publishers.

Buchholz, F.-G. 1984. Improved Formulae for the Finite Element Calculation of the Strain Energy Release Rate by Modified Crack Closure Integral Method. In J. Robinson (ed.), *Accuracy, Reliability and Training in FEM Technology, Proc. of the 4th World Congr. and Exib. on Finite Element Methods, Interlaken, Switzerland, September 1984*: 650-659. Dorset: Robinson and Associates.

Chai, H. 1986. On the Correlation between the Mode I Failure of Adhesion Joints and Laminated Composites. *Engineering Fracture Mechanics*. 24:413-431.

Carlsson, L. A., Gillespie, J. W. & Jr. 1989. Mode-II Interlaminar Fracture of Composites. In K. Friedrich (ed.), *Application of Fracture Mechanics to Composite Materials, Vol. 6, Composite Materials series*: 113-157. Amsterdam: Elsevier.

Drzal, L. T. 1986. The interphase in epoxy composites. *Advances in Polymer Science*. 75:1-31.

Drzal, L. T., M. J. Rich & P. F. Lloyd. 1982. Adhesion of graphite fibres to epoxy matrices: 1. The role of fibre surface treatment. *J. Adhesion*. 16: 1-30.

Gutowski, W. S., E. R. Pankevicius & D. Y. Wu. 1993. Controlled interfaces of ultra-high modulus of polyethylene and aramid fibres for advanced composites. In *Proc. Int. Conf. Interfaces II*, CAMT, Ballart, Australia.

Hashemi, S., A. J. Kinloch and G. Williams. 1990. Mechanics and mechanisms of delamination in a poly(ether sulphone) - fibre composite. *Comp. Sci. Technol.*. 37: 429-426.

Hojo, M., K. Kageyama & K. Tanaka. 1995. Prestandartization study on mode I interlaminar fracture toughness test for CFRP in Japan. *Composites*. 26: 243-255.

Madhukar, M. S. & L. T. Drzal. 1992. Fibre-matrix adhesion and its effect on composite mechanical properties, IV. Mode I and mode II fracture toughness of graphite/epoxy composites. *J. Comp. Mater.*. 26: 936-968.

Rikards, R., F.-G. Buchholz, & H. Wang. 1995. Finite Element Analysis of Delamination Cracks in Bending of Cross-Ply Laminates. *Mech. Comp. Mater. and Struct.*. 2: 281-294.

Rybicki, E. F. & M. F. Kanninen. 1977. A Finite Element Calculation of Stress Intensity Factors by Modified Crack Closure Integral. *Engng. Fracture Mech.*. 9: 931-938.

Sela, N. & O. Ishai. 1989. Interlaminar fracture toughness and toughening of laminated composite materials: A review. *Composites*. 20: 423-435.

Wacker, G. 1996, Experimentell gestützte Identifikation ausgewählter Eigenschaften glasfaserverstärkter Epoxidharze unter Berücksichtigung der Grenzschicht. PhD Dissertation, University of Kassel, Kassel, Germany, (in German).

Damage and Failure of Interfaces, Rossmanith (ed.) © 1997 Balkema, Rotterdam, ISBN 90 5410 899 1

Trapping mechanism of interface crack at concrete/engineered cementitious composite bimaterial interface

V.C. Li
Advanced Civil Engineering Material Research Laboratory, Department of Civil and Environmental Engineering, The University of Michigan, Ann Arbor, Mich., USA

Y.M. Lim
Applied Mechanics Laboratory, Department of Civil Engineering, College of Engineering, Yonsei University, Seoul, Korea

ABSTRACT: This paper reports on an interfacial crack trapping mechanism experimentally observed in a concrete/Engineered Cementitious Composite system. The mechanism involves cycles of interfacial crack extension, kinking of this interfacial crack into the ECC, and kinked crack arrest. During this process, the macroscopic load capacity continues to increase, allowing for high strength and ductile characteristics of this bi-material system.

1 INTRODUCTION

Bimaterial systems can be found in many engineered devices and structures, including thin film/substrate systems for electronic device, coated systems for mechanical devices, and layered composites for many structural applications (Hsueh and Evans 1985; Evans et al. 1990). In these bimaterial systems, the interface between the two different materials is usually the weakest part and fracture failure initiated from an interfacial defect is often observed. Thus extensive research on the mechanics of interfacial failure has been conducted (e.g. Suo and Hutchinson 1989; Cao and Evans 1989; Charalambides et al. 1989; Wang and Suo 1990). Two distinct fracture behavior, interfacial fracture and kink out fracture are typically identified. Analytic tools for evaluation of whether an interfacial defect will advance straight ahead or kink out of the interface have been developed.

In a recent study, intermittent interface crack propagation and kink-out was observed in a cementitious bi-material system in connection with investigations on durability of concrete repair (Lim 1996). In this paper, we report on this *interface crack trapping mechanism*. The concept of the interface crack trapping mechanism is discussed below. This is followed by a presentation of the experimental observations in a bimaterial system containing a normal concrete and an Engineered Cementitious Composite (ECC) (Li, 1996).

2 CONCEPT OF TRAPPING MECHANISM

When a bimaterial system is subject to applied load, the relative driving force in terms of energy release rates $\frac{G}{G'_{max}}$ for in-plane extension to out-of-plane kinking of an interfacial crack can be analytically determined (Suo and Hutchinson 1989), assuming small scale yielding condition. A crack at the interface between two cementitious materials may satisfy this small scale yielding condition. There is no aggregate interlocking nor fiber bridging across the interface so that a small process zone relative to other body geometry, associated with break down of cement paste material, can be expected. The relative toughness $\frac{\Gamma(\hat{\psi})}{\Gamma_c}$ between the interface and the material into which the kink crack propagates (labeled as material #2 in this paper for convenience) should be evaluated experimentally. The condition for kinking has been expressed in terms of these relative driving force and relative toughness.

$$\frac{G}{G'_{max}} < \frac{\Gamma(\hat{\psi})}{\Gamma_c} \qquad (1)$$

Figure 1 plots these relative quantities schematically. As expected from Eqn. (1), low phase angle (low Mode II to Mode I loading) leads to low values of $\Gamma(\hat{\psi})$, which promotes in-plane interfacial crack extension. Conversely, high phase angle promotes kinking.

Consider a bimaterial system which possesses varying levels of Γ_c, and therefore $\frac{\Gamma(\hat{\psi})}{\Gamma_c}$. This is

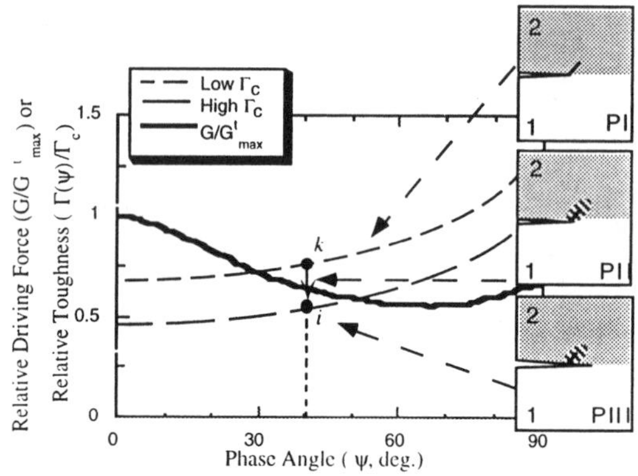

Figure 1: Possible patterns of interface cracking and kinking behavior

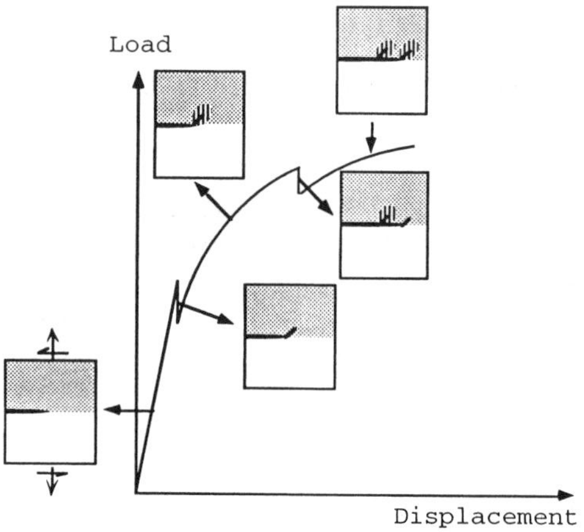

Figure 2: Trapping mechanism in a bimaterial interface system

possible if, e.g., Material 2 has an R-curve behavior, with low initial toughness which rises with the extension of the kinked crack. This scenario together with the corresponding crack propagation patterns are illustrated in Figure 1, with two levels of $\frac{\Gamma(\hat{\psi})}{\Gamma_c}$ indicated (labeled as Low and High Γ_c curves).

The initial low toughness in Material 2 (shaded in Figure 1) attracts the interfacial crack to kink (phase angle around 40 degrees, condition k, Pattern PI). Subsequent rise of the toughness associated with development of a process zone in Material 2 then arrests the kink-out crack (Pattern PII). As the $\frac{\Gamma(\hat{\psi})}{\Gamma_c}$ curve shifts down because of the rising Γ_c, the kinking condition (Eqn. (1)) is no longer satisfied, and interfacial in-plane crack extension then prevails (condition i, Pattern PIII).

In Figure 2, the conceptual trapping mechanism described above is illustrated together with the load-displacement relation. After the kinked crack is arrested and propagation is forced back into the interface, the relative toughness curve (Figure 1) can move back upward again as the interface crack moves out of the regime of the first damage process zone. Under this new condition, the propagated interface crack can kink out from the interface again. Thus, the sequence of kinking, damaging, trapping, and interface propagation will be repeated under continued increase loading until the full interface is exhausted or other failure modes take over.

The conceptual interface crack trapping behavior is motivated by the consideration of interface fracture mechanics. Such a failure process is desirable as it is expected to involve a large amount of energy absorption associated with sub-interface damage in a bimaterial system. The ECC material is a fiber reinforced mortar micromechanically tailored to exhibit high damage tolerance (Li and Hashida, 1993; Li, 1997).

3 EXPERIMENTAL PROGRAM

Besides the concrete/ECC bimaterial system, two additional systems were tested as control. All involve concrete as Material 1, while the two control systems have concrete and regular fiber reinforced concrete as Material 2.

The material compositions are tabulated in Table 1. The concrete (material #1) was five weeks old when the other material (material #2) was cast on it. The bimaterial system had two weeks curing before testing. The material properties and Dundar's elastic mismatch parameters are reported in Table 2. The difference of elastic modulus between base

Table 1: Material composition

Material	Cement	Water	FA	CA	SF	SP	Fiber†
Concrete	1.0	0.5	2.27	1.8	-	-	-
FRC	1.0	0.5	2.27	1.8	-	-	0.01 (steel)
ECC	1.0	0.35	0.5	-	0.1	0.01	0.02 (poly ethylene)

(FA: Fine Aggregate, CA: Coarse Aggregate (maximum size < 9.5 mm), SF: Silica Fume, SP: Superplasticizer, †: Volume fraction)

Table 2: Elastic Modulus and Dundar's elastic mismatch parameter α

Material	Elastic Modulus (GPa)	α
Base Concrete	25.8	-
Concrete	24.9	0.018
FRC	26.1	-0.005
ECC	18.0	0.178

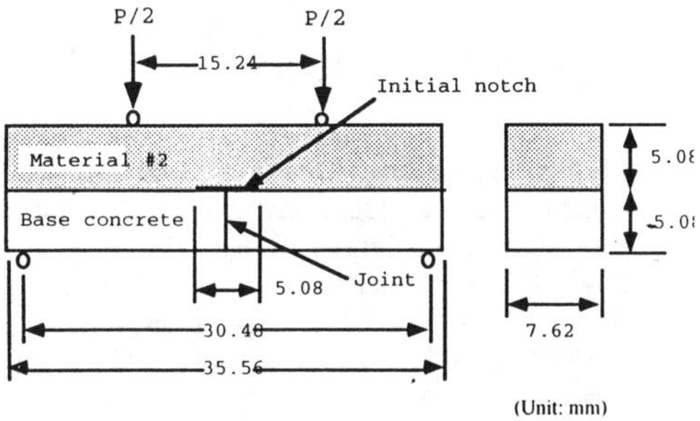

Figure 3: Loading configuration and dimensions of designed specimen

concrete (material 1) and concrete for material #2 is due to differences in their age of curing.

Figure 3 shows the loading configuration and the dimensions of the designed specimen. This specimen include an initial defect in the form of an interfacial crack between the base material and the material #2 as well as a joint in the base material. These specimens can provide stable interface crack propagation condition under this loading configuration (Charalambides et al. 1989). The phase angle of this specimens is about 41°~45°.

4 EXPERIMENTAL RESULTS

As expected, the concrete/ECC bimaterial system shows the distinctive trapping mechanism under this experimental condition. The failure process with trapping mechanism shows tremendous differences compared with those of the control specimens (Figure 4). In the concrete/ECC system, the initial interface crack propagated along the interface about 5 mm, followed by kinking out from the interface. Subsequently, the kinked crack appeared trapped inside the ECC and the mother crack (interface crack) propagated along the interface again (about 27 mm). Then, the interface

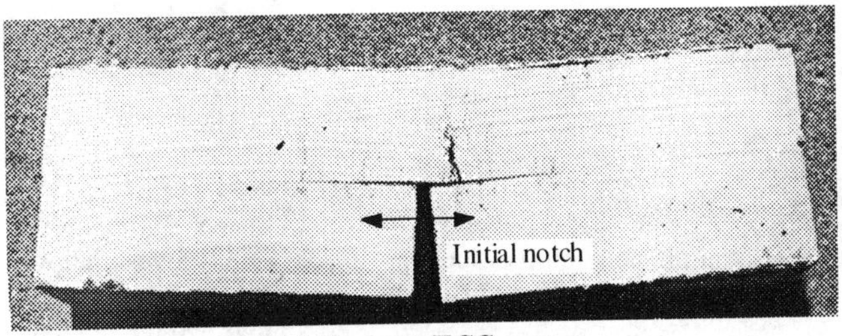

(a) Concrete/ECC system

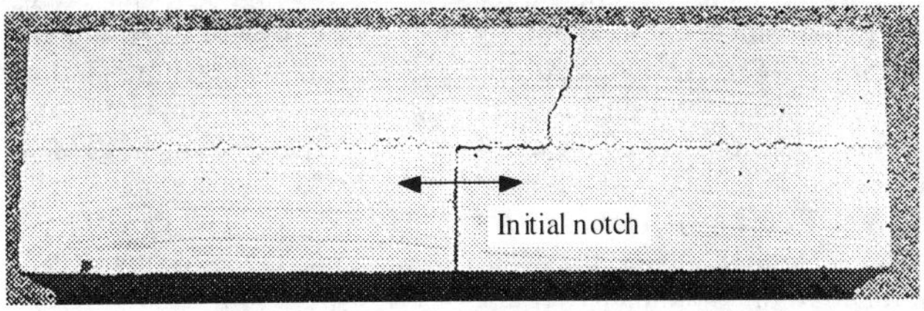

(b) Concrete/Concrete system (reassembled after test)

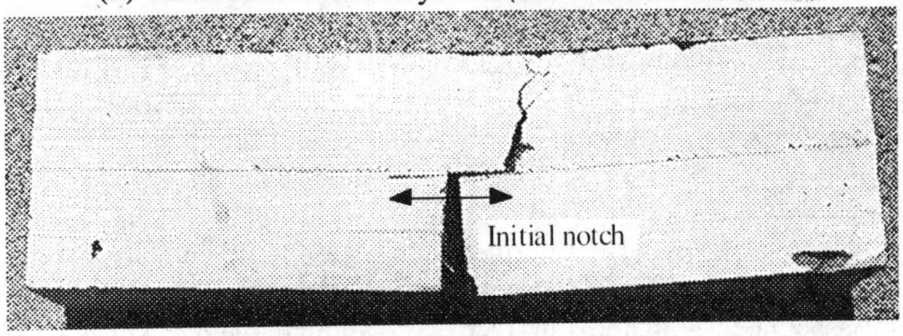

(c) Concrete/FRC system

Figure 4: Cracking in specimens

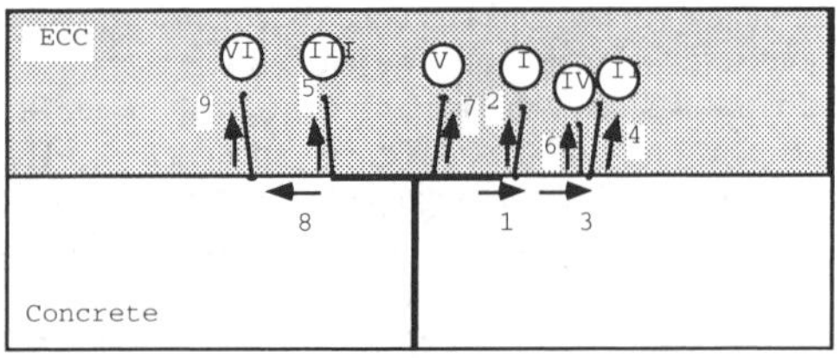

(a) Sequence of kinked crack development

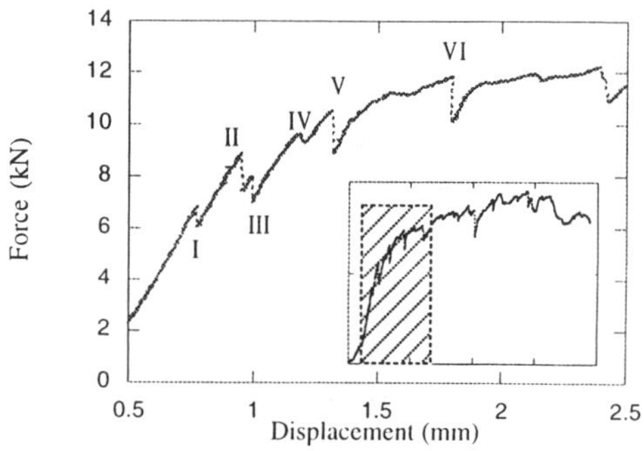

(b) Load-deflection with kinked sequence

Figure 5: Kinked crack development in concrete/ECC bimaterial system

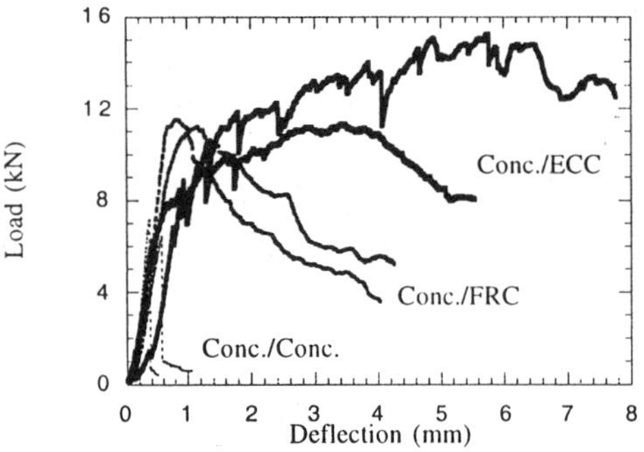

Figure 6: Load-deflection behavior in different bimaterial system

crack kinked and was trapped again. After several cracking and kinking, the final failure occurred due to the large opening of a flexural crack in the middle of the specimen.

The sequence of cracking behavior for one of the two tested concrete/ECC bimaterial system is illustrated in Figure 5(a). The arrows and the number beside the arrow indicate the direction of crack propagation and the time sequence of cracking. The load drop associated with the kinked cracks (numbered by Roman numerals) are illustrated in Figure 5(b), which shows the load-deflection curve of one of the specimens in the concrete/ECC bimaterial system. The kinked cracks occurred immediately following interface crack propagation. Thus, it is difficult to distinguish between interface cracking and kink cracking in terms of load drops. After the sixth kinked crack (VI) occurred, the bimaterial system failed with flexural crack development in the ECC. All kink cracks emanated from the tip of the current interfacial crack. In contrast, the flexural cracks developed at sites not associated with the interface crack tip.

Both control specimens showed only one macro-crack opening due to the kinked crack at the initial interface crack tip (Figures 4b and 4c). In the concrete/concrete bimaterial system, brittle behavior was observed. Quasi-brittle behavior was observed in the concrete/FRC bimaterial system. This quasi-brittle behavior comes from the fiber bridging, but the bridging effect in the FRC is not enough to create the trapping behavior. The concrete/ECC system shows macroscopic strain-hardening response. These behaviors can clearly be found in the load-deflection curves in Figure 6.

5 CONCLUSION

The concept of interface crack trapping mechanism is introduced, and its existence is confirmed in experimental investigations. Among the three bimaterial systems tested, only the concrete/ECC system revealed the trapping behavior. This system shows high ultimate strength and large deflection capacity with large amount of energy absorption. The ultimate failure mode has been shifted from one associated with interface crack extension to one associated with the flexural strength of the ECC. This dramatic improvement in terms of strength, deflection, energy absorption capacity and ultimate failure mode is not feasible without the trapping mechanism.

ECC is characterized by strong fiber bridging property combined with a low cement matrix toughness. The low matrix toughness promotes the satisfaction of the kink condition represented by Eqn. (1), circumventing brittle delamination of the interface. The strong fiber bridging property leads to a rapidly rising R-curve as the kinked crack extends, eventually forcing the cracking behavior back into the interface under continued rising applied load. On a more macroscopic level, the observed crack pattern may also be alternatively interpreted as a result of the strain-hardening characteristics and damage tolerance of the ECC material. The high tensile stress near the interface crack tip causes the ECC to go into strain-hardening, and to accommodate the local strain with microcrack inelastic deformation. In this interpretation, the interface crack never (macroscopically) kinks out, but trapped inside the interface due to an effectively toughened interface. This interpretation implies an interface with R-curve behavior. The damage in the ECC becomes part of the interfacial fracture characteristics. For either interpretation of trapping in interface or in the ECC, improvement in mechanical performance is unequivocally demonstrated to exist in the concrete/ECC bimaterial system. Implications of these findings on the durability of repaired concrete structure using ECC as a repair material are discussed in Lim and Li (1996) and Lim (1996).

REFERENCES

Cao, H. C., and Evans, A. G. (1989). "An experimental study of the fracture resistance of bimaterial interface." *Mechanics of Materials*, 7, 295-304.

Charalambides, P. G., Lund, J., Evans, A. G., and McMeeking, R. M. (1989). "A test specimen for determining the fracture resistance of bimaterial interfaces." *J. of Applied Mechanics*, 56, 77-82.

Evans, A. G., Rühle, M., Dalgleish, B. J., and Charalambides, P. G. (1990). "The fracture energy of bimaterial interfaces." *Mater. Sci. Engng.*, A 126, 53-64.

Hsueh, C. H., and Evans, A. G. (1985). "Residual Stresses and Cracking in Metal/Ceramic Systems for Microelectronics Package." *Journal of American Ceramic Society*, 68(3), 120-127.

Li, V.C. (1997). "Tailored Composites Through Micromechanical Modeling," to appear in Fiber Reinforced Concrete: Present and the Future, Eds: N. Banthia, A. Bentur, and A. Mufti, Canadian Society of Civil Engineers.

Li, V.C. and T. Hashida (1993). "Engineering Ductile Fracture in Brittle Matrix Composites" *J. of Materials Science Letters*, 12, 898-901.

Li, V.C. "Damage Tolerance Of Engineered Cementitious Composites," in Advances in Fracture Research, Proc. 9th ICF Conference on Fracture, Sydney, Australia, Ed. B.L.Karihaloo, Y.W. Mai, M.I. Ripley and R.O. Ritchie, Pub. Pergamon, UK, pp. 619-630, 1997.

Li, V. C., Maalej, M., and Lim, Y. M. (1995). "Fracture and flexural behavior in strain-hardening cementitious composites." Fracture of Brittle Disordered Materials: Concrete, Rock and Ceramics, G. Baker and B. L. Karihaloo, eds., E & FN Spon, London, 101-115.

Lim, Y. M. (1996) "Interface Fracture Behavior of Rehabilitated Concrete Infrastructures using Engineered Cementitious Composites," Doctoral Thesis, The University of Michigan, Ann Arbor, USA.

Lim, Y. M., and Li, V. C. (1996) "Durable Repair of Aged Infrastructures Using Trapping Mechanism of Engineered Cementitious Composites," accepted, the *Journal of Cement and Concrete Composites*.

Suo, Z., and Hutchinson, J. W. (1989). "Sandwich Test Specimens for Measuring Interface Crack Toughness." *Material Science and Engineering*, A107, 135-143.

Wang, J. S., and Suo, Z. (1990). "Experimental determination of interfacial toughness using Brazil-nut-sandwich." *Acta Met.*, 38, 1279-1290.

Damage and Failure of Interfaces, Rossmanith (ed.) © *1997 Balkema, Rotterdam, ISBN 90 5410 899 1*

Interface roughness and sliding in ceramic matrix composites

H.Cherouali, P.Reynaud & D.Rouby
GEMPPM UMR CNRS, INSA de Lyon, Villeurbanne, France

ABSTRACT: Results obtained by push-out and push-back tests with a instrumentated indentation device on single-filament model composites are presented and discussed. The composites are made with Pyrex matrix containing two kinds of SiC filaments, SCS-6 from Textron and Sigma from BP. Because of the hot-pressing processing in viscous stage, the matrix surface mirrors exactly the rough fibre surface. The sliding is analysed by taking into account the interfacial roughness and the device stiffness which leads to stick-slip or steady sliding. The interface degradation phenomena due to an abrasion of the roughness or to matrix microcracking are described. The seating drop, which appears when the filament returns in its initial position and the rough surfaces come back in coincidence, is explained by a stress transfer model where the interfacial shear stress depends on the displacement of the fibre surface relatively to that of the matrix.

1 INTRODUCTION

Ceramic matrix composites exhibit good resistance against damage extension, mainly because the matrix cracks are bridged by the fibres. The crack bridging efficiency, which can be described by the so-called P(u) curve (Brenet et al. 1996), is critically dependent upon the characteristics of the fibre/matrix interface (residual thermal stresses, debonding conditions, frictional behaviour, interfacial wear), where the roughness plays an important role (Martin et al. 1993) (Kerans & Parthasarathy 1991). Concerning cyclic fatigue, it is yet known that the fatigue life is related with interface degradation due to the see-saw slip under cyclic load and this was first pointed out by Rouby & Reynaud (1993) Therefore, a better knowledge of the friction mechanisms, including roughness and degradation effects, is needed.

Some progress have been made shortly in the approach of this problem from experimental and theoretical viewpoints. Kerans & Parthasarathy 1991 have been improved existing models by taking into account the contribution of interface topography on the sliding friction. Experimentally, Jero & Kerans (1990) have found an interesting phenomenon, the so-called seating drop when they pushed back a fibre beyond its original position in the matrix. This phenomenon shows the importance of the "mirrored" nature of the fibre and matrix surfaces and of the relative position of these two surfaces. The interfacial shear stress (ISS) should be low near the debond tip and increasing up to a steady value as the relative fibre/matrix displacement increases (Parthasarathy et al. 1994, Marshall et al. 1995, Stupkiewicz 1996).

Mumm & Faber (1995) have addressed a improved pull-out test on monofilament composite samples that describes well the behaviour of a bridged matrix crack. The load-crack opening displacement curves showed load fluctuations during fibre sliding in the debonded zone, during loading and unloading as well. These fluctuations, whose amplitude scales with debond length, can be readily explained by the fact that the fibre slippage results from a succession of micro-instabilities, leading to small drops of the effective ISS.

The aim of this work is to study, in monofilament SiC/Pyrex model composite, the friction and degradation mechanisms as the fibre slides into the matrix. For the analysis, distinction is made between long and short distance sliding as regards to roughness periodicity. The stiffness of the testing device controls stable or unstable slip conditions and this feature is exploited in order to obtain more information about the operating mechanisms.

2 EXPERIMENTAL PROCEDURE

The monofilament composites used in this study are made with a borosilicate glass matrix (Pyrex 732-01, from J. Bibby Science Products) and two types of SiC filaments: SiC SCS-6 from Textron Specialty Materials and SiC Sigma from British Petroleum. These filaments have been described by Mumm & Faber (1995).

The composites are processed by hot pressing, the filament being placed between two Pyrex plates (pressure: 1 MPa, temperature: 810°C, duration: 30

mn). After cooling, the composite is carefully cut perpendicularly to the fibre axis, by means of a diamond wire saw, into slabs with thickness, noted H, from 0.6 to more than 3 mm.

The tests are performed on an instrumentated indentation device, described in more detail by Cherouali et al. (1997a). The indenter is a WC/Co cylinder with a conical tip (base diameter ≈ 1 mm, angle ≈ 30°), machined at the tip in order to get a flat pushing surface of 50 or 80 μm diameter. The displacement is measured by means of a capacitive sensor fixed very closely to the indenter.

The special feature of this device is that the overall stiffness can be changed. Coil springs or flexural beams of different sizes are used for this purpose, in series with the load cell. By taking into account the other stiffness (also in series) of the indenter tip and of the specimen, the effective device stiffness, noted K, can be changed from about 50 to 1200 N/mm.

The testing procedure consists of submitting the specimen to a succession of push-out (PO) and push-back (PB) cycles, as shown in Figure 1: PO1 (debonding and first push-out), PB1 (first push-back) after specimen reversal, PO2 after turn of specimen again, PB2, etc. The drive speed is 1 μm/s.

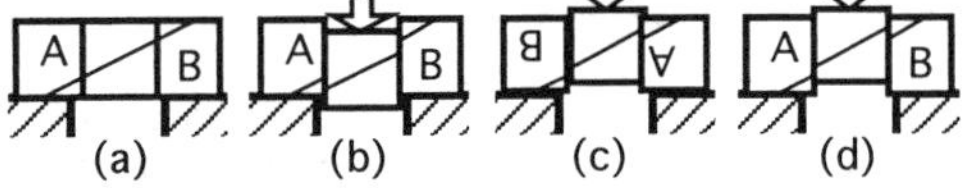

Figure 1. Push-out/push-back cycles.
(a): initial position ; (b): PO1; (c): PB1; (d):PO2; etc.

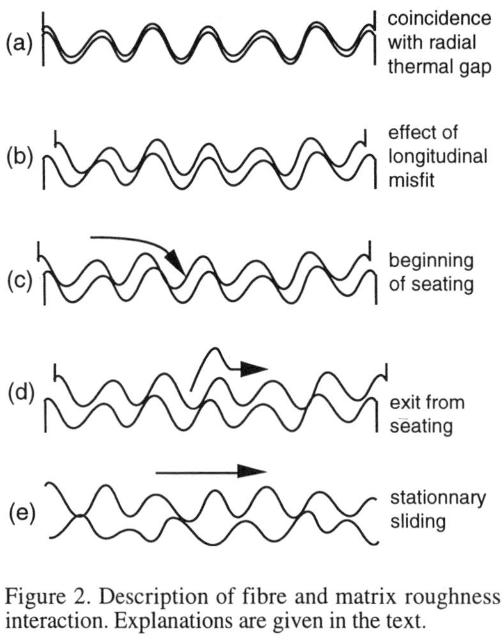

Figure 2. Description of fibre and matrix roughness interaction. Explanations are given in the text.

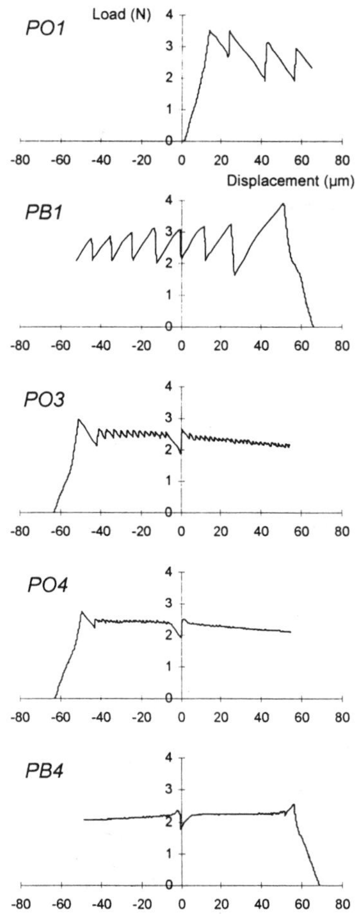

Figure 3. Examples of push-out/push-back cycles (all the runs are not represented)
SiC SCS-6/Pyrex system.
Device stiffness: K = 94 N/mm.

3 LONG RANGE SLIDING

This Section concerns the most part of the PO/PB cycles. The effect of roughness, schematically depicted in Figure 2, is discussed, as well as the role of the stiffness, the mechanisms and the results of interface degradation.

3.1 *Results of push-out/push-back cycles*

The SiC SCS-6/Pyrex composite exhibits stick-slip under low device stiffness (Figure 3) and steady sliding under high stiffness (Figure 4). The serrated slip disappears after a certain number of PO/PB cycles. The effect of stiffness will be discussed in Section 3.3. In the case of the SiC Sigma/Pyrex system, steady sliding is observed, even at low stiffness (Figure 5).

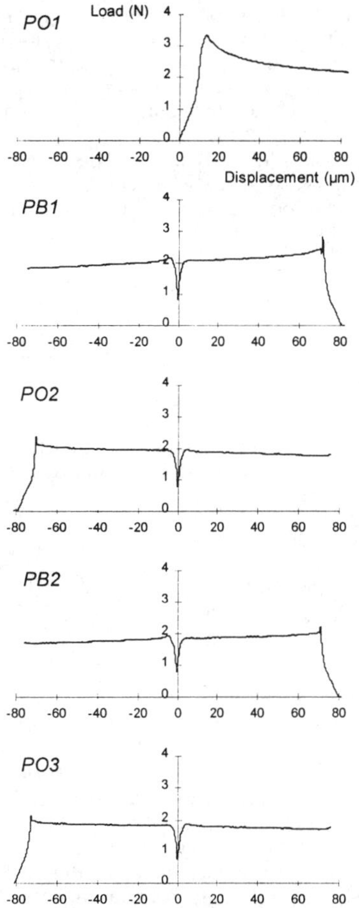

Figure 4. Examples of push-out/push-back cycles (only the first cycles are represented).
SiC SCS-6/Pyrex system. K = 1150 N/mm.

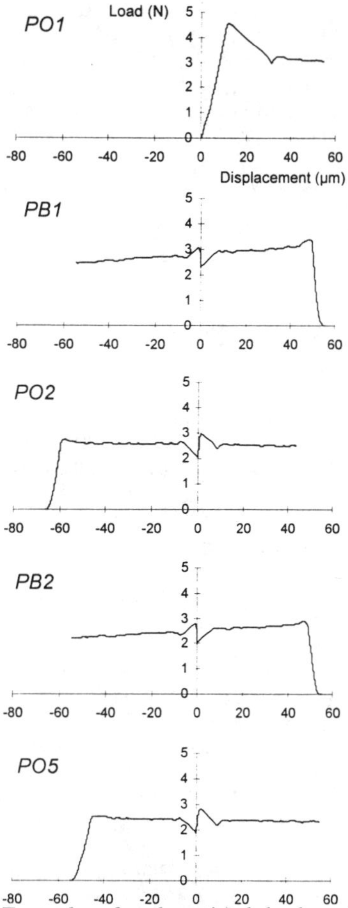

Figure 5. Examples of push-out/push-back cycles (all the runs are not represented).
SiC Sigma/Pyrex system. K = 94 N/mm.

In all cases, the sliding force decreases, as the number of cycles increases, indicating an interface degradation phenomenon (see Section 3.4).

Before testing, the rough surfaces of fibre and matrix are more or less in coincidence, depending on the thermal misfit (see Figure 2a and 2b). During the first run, long range sliding, the surfaces are in stationary interaction, giving an overall constant sliding force (see Figure 2e and Section 3.2). During the subsequent cycles, a seating drop is observed whenever the fibre reaches again its initial position (see Figure 2c and 2d). The seating drop is also visible in the serrated state, the valley load being markedly lower than that of the stick-slip serration's. Despite the large reduction in sliding force due to interfacial degradation, the seating drop is few changed, indicating that the degradation does not sharply affect the main roughness characteristics. The seating drop is the manifestation of the mechanisms operating during short range sliding. This feature is described and modelled in Section 4.

3.2 *Effect of roughness on sliding conditions*

Because of the hot-pressing at viscous state, the matrix surface mirrors the roughness of the fibre surface, leading to a very interesting system:

1. The surfaces in contact are almost identical (the stiffness of its asperities is however different), so two cases arise: the surfaces are more or less in coincidence (short range displacement) and the surfaces are in statistical stationary interaction by few randomly contacts located (long range displacement).

2. The distance between the first bodies (fibre cylinder and surrounding matrix) is fixed, so that any change in roughness topography or change in the physical or chemical nature of the surfaces, leads to a well identified mechanical response.

The main characteristics of the filaments, as reviewed by Mumm & Faber (1995) are as follows:

- Amplitude of roughness: a = 200 nm for SCS-6
  a = 400 nm for Sigma
- Width of the asperities: 2b = 8 µm for SCS-6

$2b = 10$ μm for Sigma

- Fibre radius: $R = 140$ μm for SCS-6

$R = 100$ μm for Sigma

During push-out test, the sliding force, F, is related to the interfacial shear stress (ISS), τ, assumed here constant along fibre axis:

$$F = 2 \pi R (H - U) \tau \quad (1)$$

where H = specimen thickness, U = displacement of the filament, relatively to the matrix.

Assuming Coulombic friction, the ISS is related to an effective coefficient of friction, μ, by a radial (compressive) stress σ_{rad}:

$$\tau = - \sigma_{rad} \mu \quad (2)$$

This clamping stress finds its origin in the thermal misfit and in a certain radial dilatancy due to the fact that sliding needs to overcome the height of the asperities:

$$\sigma_{rad} = E_{rad} \left(\Delta\alpha \ \Delta T - \frac{D}{R} \right) \quad (3)$$

where $\Delta\alpha \ \Delta T$ = radial thermal misfit ($\Delta\alpha = \alpha_m - \alpha_f$ and $\Delta T = T - T_0$, T_0 = stress free temperature), E_{rad} = effective radial modulus (obtained from Lamé's problem, (Hutchinson & Jensen 1990). In fact, the bodies are in contact through rough surfaces, so E_{rad} is not a constant parameter (Kuntz et al. 1994). D = roughness induced dilatancy. D is mainly taken equal to the roughness amplitude (D = a).

It should be noted here that the above equations are very simplified, because the Poisson's effect due to the axial stress is neglected (see Kerans & Parthasarathy (1991) for a more complete analysis).

SCS-6 and Sigma filaments cannot be directly compared by considering the roughness and the measured sliding force, because they have a carbon coating of different nature and thickness. So, the parameters such as D, E_{rad}, σ_{rad} and μ should be different and almost unknown.

3.3 *Effect of device stiffness*

Unstable displacement occurs when the effective device stiffness becomes low. This stiffness includes all the parts belonging to the mechanical chain of the testing device (load cell, specimen, indenter, etc.).

Stable/unstable bifurcation arise from the mismatch between the resistance to slide of the system (characterised by the intrinsic constitutive law: $F_{intrinsic}(U)$, the sliding behaviour under infinite stiffness), and the drive stiffness (given by K).

When a given event leads to a drop of and an corresponding increment in displacement, the system becomes unstable if the following condition is fulfilled (Rabinowicz 1965):

$$\text{Unstable if:} \quad \frac{d(F_{intrinsic})}{dU} < - K \quad (4)$$

This condition can be fulfilled for geometrical reasons. For example, the entry in seating (Figure 2c) gives a load drop, the so-called seating drop.

It is well known that, in most case, the coefficient of friction becomes lower as the sliding velocity increases. Mostly, the relationship between μ and the logarithm of velocity is linear. This is because as velocity increases, the current contacts have not the time to develop higher strength, for chemical or mechanical reasons (Persson & Tossati 1966). Recently, Heslot et al. (1994) proposed a model based on local creep phenomena: under the applied load at the end of "stick phase", creep takes place at the level of the contacts, leading to accelerating displacement and decreasing resistance, down to instability. The unstable phase ("slip phase") stops when there is no more excess of strain energy, and the "stick phase" is reached again.

According to that model, a critical stiffness giving the bifurcation between stick-slip and steady sliding, can be expressed:

$$\text{Unstable if:} \quad K < K^* \quad (5)$$

where:

$$K^* = - \frac{Mg}{U_0} \frac{d(\mu)}{d(\ln V)} = - \frac{2 \pi R H |\sigma_{rad}|}{U_0} \frac{d(\mu)}{d(\ln V)} \quad (6)$$

With $d(\mu)/d(\ln V)$: sensitivity of coefficient of friction to sliding velocity (negative); Mg: normal load applied to the sliding body which, in the case of a cylindrical fibre, comes down to $2 \pi R (H - U) |\sigma_{rad}|$; U_0: displacement needed to break a contact.

Equations 5 and 6 account for the observed behaviour: occurrence of stick-slip under low stiffness. The test corresponding to Figure 3 begins with $K < K^*$. As the displacement induces abrasion of the asperities, the clamping stress σ_{rad} slopes progressively down. So K^* decreases and becomes at least lower than K. At this level of interface degradation, stick-slip phenomenon disappears. Equation 6 predicts also that thicker specimens (larger H) have, under same stiffness K, more tendency to exhibit stick-slip (larger K^*). This trend is experienced in practice as well (Rausch et al. 1992).

3.4 *Interface degradation*

Figure 6 shows the reduction of sliding force as cumulative displacement increases (this displacement is cumulated over PO sense and PB sense). The load is measured just after the seating drop and post-seating peak, so its value corresponds always to the same filament position.

For the SCS-6 system (Figure 6a), the degradation extends more slowly (as a function of displacement) in the stick-slip situation (the stick-slip disappears here at 2.3 load. This confirms the weaker interaction between the surfaces as they move more rapidly (in the slip phase, the typical velocity is about 200 μm/s, instead of less than the drive velocity (1 μm/s) in the steady sliding state.

In fact, the degradation, due to the breaking of one (or more in a sequence) contact, occurs mainly at the end of the stick stage and initiates the instability.

Some evidence of this mechanism has been reported by Cherouali et al. (1997b). After a first run with stick-slip periodic serration's, by using the proper surfaces (sliding in return steadily) as a probe, the periodic damage induced by the first former periodic stick phases, has been indicated by small drops in the F(U) curve.

For example (Figure 6a), a given amount of degradation, corresponding to F = 2.5 N, is reached after 160 μm cumulative displacement in the steady sliding case, and after 600 μm or nearly 50 serration's in the stick-slip case. So, we can deduce an estimate of the displacement, U_0, needed for breaking the contacts at the end of the stick phase, by writing: $U_0 = 160/50 = 3.2$ μm. Notice that this value is close to the third of the asperity width. More work should be done in order to understand the real meaning of this value.

The law of interface degradation can be modelled as follows: the asperity height (proportional to τ, Equations 2 and 3) is worn down with a rate hardly increasing with the interfacial pressure (also proportional to τ, Equation 2):

$$\frac{d\tau}{dU} = - A \tau^n \qquad (7)$$

where A is scaling factor and the exponent, n, account for the acuteness of the pressure on the degradation kinetics. This expression can be easily integrated and the best fit with experiments leads to an exponent of 4 to 5. The cause of that high value is not presently understood.

In the SiC Sigma/Pyrex system, the load drops very quickly during the first few cycles and remains then unchanged. In-situ observations of the interface through the matrix during testing, showed initiation and propagation of radial cracks in the matrix, in correlation with the load reduction. The cracking configuration remains then stable as well as the sliding force. In that system, interface degradation is not related to wear phenomena as observed with the SCS-6 filament. The difference comes probably from a higher roughness amplitude on the SiC Sigma filament, and a more abrasive coating on the SiC SCS-6 filament (turbostratic carbon containing SiC nanoparticles).

4 SHORT RANGE SLIDING

In the case of long distance sliding, the rough surfaces are in statistical stationary interaction through a limited number of contacts. In that situation, the roughness induced additional dilatancy has the maximum value, D. The long range sliding is reached when the sliding distance exceeds a certain value, L (u > L, origin of displacement at coincidence). The L and D parameters are related to

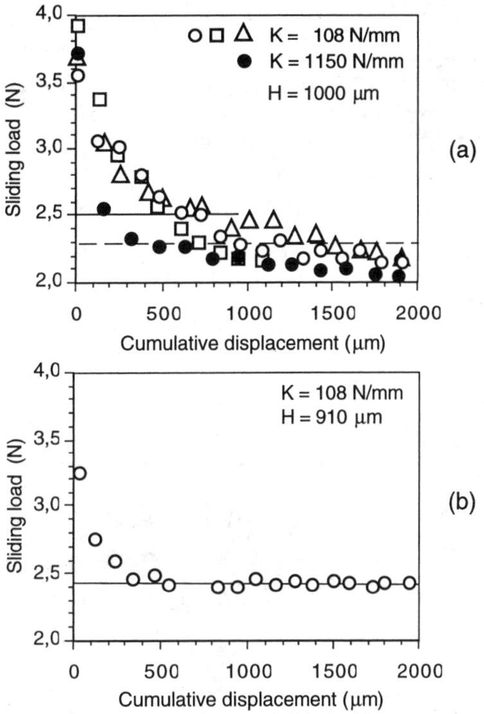

Figure 6. Evolution of sliding force as a function of cumulative displacement over PO and PB runs.
(a): SiC SCS-6 system; (b) SiC Sigma system.

the roughness topography: D is a portion of roughness amplitude, L is linked, probably equal, to the asperity width.

The case of short distance sliding corresponds to a sliding distance smaller than L (u < L). The corresponding additional dilatancy is therefore lower than D, because the rough surfaces are not far from coincidence and remain engaged.

Actually, there is a progressive increase in dilatancy and consequently in sliding resistance, from perfect coincidence (u = 0) to stationary sliding, when u > L or u < -L. This defines the short range sliding region.

4.1 *Description of the seating drop*

Whether push-out or push-back is considered, the entry in seating corresponds to negative displacement (u < 0, increasing) and the exit to positive one (u > 0, increasing). As shown in Figures 4 and 5, the seating is unstable and the post-seating displacement is stable. A so-called post-seating peak occurs before the sliding force reaches its stationary value.

Figure 7 shows detail of the seating drop in the case of SiC Sigma/Pyrex system, particularly the effect of stiffness. In this experiment, during the PO-PB cycling, the stiffness has been increased at each specimen turn. Except the first cycles (where the

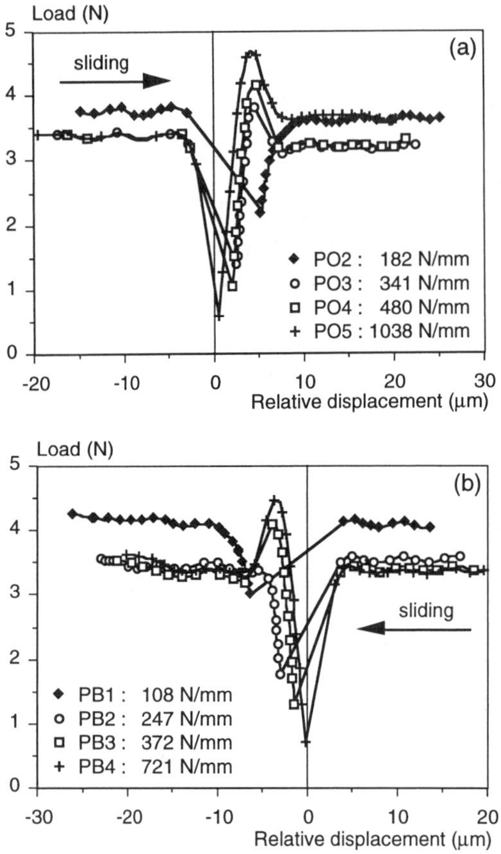

Figure 7. Evolution of the sliding force as a function of displacement on both sides of the initial position. SiC Sigma/Pyrex system The stiffness is specified for each run. (a): push-out runs (sliding from left to right); (b): push-back runs (sliding from right to left).

matrix cracking occurs, see Section 3.4), the stationary load remains constant (even the irregularities of the curves can be well superimposed), indicating a lack of interfacial wear. So, the seating drops recorded can be easily compared with respect to the device stiffness.

At low stiffness (PB1, PO2, PB2), the load increases monotonously, from the valley value reached by instability, up to the stationary value again. In this case, at instability, a large amount of strain energy is available so that the system runs, during the "slip phase", largely over the seating position. At the end of "slip phase" the surfaces are again in stationary interaction, but beyond the seating position. An increase in load is only needed to make the system in stationary sliding state again.

At higher stiffness (PO3, PB4, etc.), a well defined post-seating peak appears, the peak height being increasing with increasing stiffness. In this case, despite the instability, the system remains

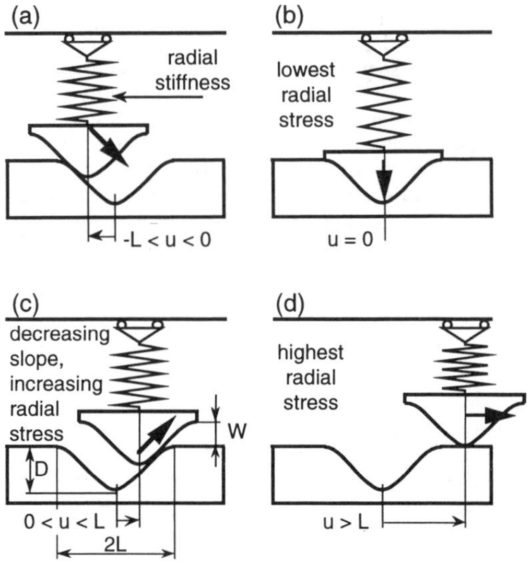

Figure 8. Modelling of the interaction between two identical rough surfaces. (a): entrance in seating; (b): coincidence, initial position; (c): exit of seating; (d): stationary sliding.

located inside the seating zone. A large increases in load is needed for reaching again the stationary state, not only for starting the sliding but also for "climbing" the exit of seating. The "climbing" phenomenon, explaining the post-seating peak, is modelled in the next Section.

4.2 *Modelling of short distance sliding*

Several authors (Parthasarathy et al. 1994, Marshall et al. 1995, Stupkiewicz 1996) have proposed that the ISS increases linearly from u = 0 up to the stationary value operating when $u \geq L$. These models are reversible, in the sense that the same ISS profile is taken wether the system moves away from coincidence or draws nearer. These features are not sufficient to explain the experimental observations concerning the seating drop.

The present model is based on the interaction (friction at the contact point) of two surfaces, one simulating the effective asperity of the filament and the other the corresponding matrix trough.

The relative displacement, u, of these arches gives rise to a radial dilatancy, W, being symmetric as regards to the coincidence position:

$$\frac{W}{D} = \sin\left(\frac{\pi\,|u|}{2\,L}\right) \qquad \text{if } |u| < L \qquad (8a)$$

$$\frac{W}{D} = 1 \qquad \text{if } |u| \geq L \qquad (8b)$$

This dilatancy generates a compressive radial stress, given by Equation 3, with W instead of D. This radial compression is maximum ($\sigma_{rad} = \sigma_{rad}^{max}$) in the stationary state (Figure 8d, W = D, |u| > L). In the present model, to simplify, the radial thermal misfit is neglected ($\Delta\alpha\ \Delta T = 0$ in Equation 3) as well as the situation, depicted in Figure 2b, due to the axial thermal misfit.

At the contact point, the combination of the radial compression, the coefficient of friction and the slope of sliding leads to an ISS, depending on displacement as follows:

$$\tau(u) = -\sigma_{rad}^{max} \frac{W}{D} \frac{\mu + P}{1 - \mu P} \quad \text{if } 0 < u < L \qquad (9)$$

$$\tau(u) = -\sigma_{rad}^{max} \frac{W}{D} \frac{\mu - P}{1 + \mu P} \quad \text{if } -L < u < 0 \qquad (10)$$

$$\tau(u) = \tau_{stationary} \quad \text{if } u > L \text{ or } u < -L \qquad (11)$$

where P is the slope of the plane of contact, given by:

$$P = \frac{\pi D}{2 L} \cos\left(\frac{\pi u}{2 L}\right) \qquad (12)$$

and σ_{rad}^{max} is the maximum radial compressive stress, given from Equation 2 by:

$$\sigma_{rad}^{max} = \frac{-\tau_{stationary}}{\mu} \qquad (13)$$

To simplify, the same coefficient of friction is taken whether the surfaces are in seating zone or in stationary state. The ISS operating in stationary state, $\tau_{stationary}$, can be evaluated by using Equation 1.

The evolution of sliding force F(U) is then obtained by solving together the following expressions along filament axis:

axial filament stress: $$\frac{d\sigma_f}{dx} = \frac{2\,\tau(u)}{R} \qquad (14)$$

displacement: $$\frac{du}{dx} = \frac{\sigma_f(u)}{E_f} \qquad (15)$$

where E_f = filament Young's modulus.

The load applied at one end of the filament (x = 0) is: $F = \pi R^2 \sigma_f(x = 0)$ and the recorded displacement of the fibre end is: U = u(x = 0). At the other end, the fibre stress vanishes and the filament can protrude from the matrix.

Figure 9a shows some theoretical results obtained with a specimen of 1 mm thickness. The geometrical parameters first chosen are directly linked to the roughness characteristics of the SiC Sigma filament (see Section 3.2): L = b= 4 μm and D = a = 0.4 μm. Concerning the friction coefficient, a value μ = 0.15 seems typical for glass/carbon or carbon/carbon frictional systems (Mumm & Faber 1995). With this set of parameters, the load falls down very sharply by seating and a well pronounced post-seating peak appears. Only the combination of μ, P and σ_{rad} can give rise to that peak. This load evolution is well consistent with experimental observations, however the chosen set of parameters must be discussed.

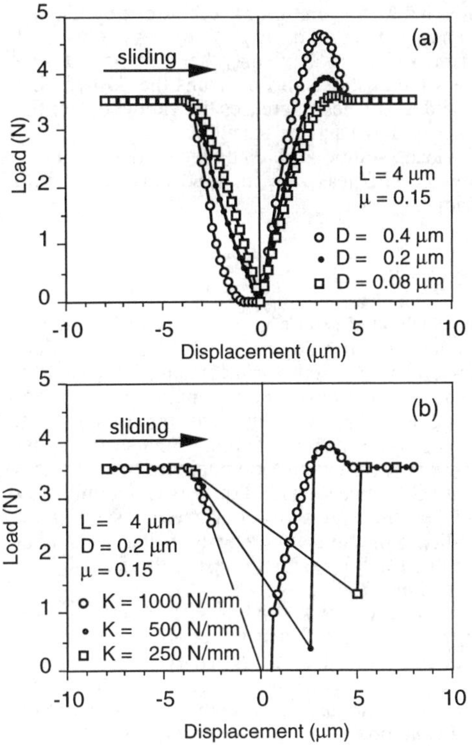

Figure 9. Theoretical evolution of sliding force as a function of displacement. Specimen thickness: 1 mm. (a): case of infinite stiffness; (b): effect of stiffness on the unstable drop.

As a matter of fact, the experimental stationary sliding force corresponds to an effective ISS of about 11 MPa and from Equation 13 this value leads to a clamping stress 5 times lower than the one obtained from Equation 3 which involves the roughness induced dilatancy and the radial stiffness.

This means that the radial stiffness, simply deduced from Lamé's problem (E_{rad} = 56 MPa for the SiC/glass system with smooth cylindrical surfaces), or the roughness induced dilatancy (taken to be equal to the amplitude of the asperities) are overestimated.

By taking D = 0.08 μm (lower dilatancy and nominal radial stiffness), one can see that the post-seating peak becomes very small and the load evolution tends to be symmetrical. A median dilatancy (D = 0.2 μm) combined with a median radial stiffness (approximately the half of the above value) is probably closer to the real situation. Notice that the coefficient of friction plays the same role than the inverse of dilatancy. More work must be done, at the scale of the asperities in contact, for a better evaluation of these parameters.

Figure 9b shows the effect of device stiffness in the median situation. The end of "slip phase" is reached when the excess of strain energy (given by the area below the slip line) is consumed by friction (the area above the slip line). As observed experimentally, it is clearly shown that at low stiffness the unstable slip overruns the post-seating peak, and as stiffness increases the valley load of the drop slopes down and the instability starts later.

The load evolutions given by the model, including the device stiffness, are in good agreement with experiment.

5 CONCLUSIONS

Fibre/matrix sliding in a microcomposite produce a very interesting tribological system because sliding occurs under imposed displacement of the bodies, whereas classical devices commonly operate under imposed normal force. In addition, this system consists obviously in two facing rough surfaces, but the more or less random topography of both surfaces is identical. Hence, this kind of tests is not only very useful for the study of interface stress transfer capability in the field of composite materials and but also powerful for basic studies in the field of the physics of dry friction.

In some cases, stick-slip has been observed and it is important to note that it is not an artefact of the experiments. On the contrary, analysis of stick-slip gives access to detailed information about the sliding mechanisms and probably about the degradation phenomena too. In fact, the serration's play the role of amplification of the micro-phenomena occurring during sliding.

The stiffness of the measuring device is a particularly important parameter, as for all mechanical tests. For testing single filament model composites (push-out and pull-out), the device stiffness should be controlled or, at least, mentioned together with the results. In real composites, where the architecture is more complicated, the effective stiffness comes from a combination of elastic elements in parallel (neighbouring fibres, neighbouring plies, etc.) and in series (the material, the structure and the loading device). In that case, the occurrence or not of stick-slip is more difficult to control.

During interface degradation, many intricate mechanisms are operating and more experimental and theoretical work must be done in this field. Emphasis must be put on this feature because interface degradation controls the larger part of damage kinetics in brittle matrix composites, particularly under cyclic fatigue.

The seating drop shows the importance of roughness upon the sliding conditions. It remains quite visible, even after a large number of push-out/push-back cycles. This clearly signifies that the interfacial degradation does not destroy the main characteristics of the initial roughness. Actually, the seating drop represents the constitutive law of relative displacement when the facing surfaces leave and return in coincidence. This kind of information is particularly useful for studying the debonding conditions and more precisely what happens just behind the debond crack tip.

REFERENCES

Brenet Ph., Conchin F., Fantozzi G., Reynaud P., Rouby D. & Tallaron C. 1996. Direct measurement of the crack bridging traction's, a new approach of the fracture behaviour of ceramic-ceramic composites. *Composite Science and Technology* 56:817-823.

Cherouali H., Fantozzi G., Reynaud P. & Rouby D. 1997a. Analysis of interfacial sliding in brittle-matrix composites during push-out and push-back tests. *Mater. Sci. Engng.*, to be published.

Cherouali H., Reynaud P., Berthier Y. & Rouby D. 1997b. Influence of serrated sliding on subsequent steady slip in single filament model composite SiC/Pyrex. *Scripta Mater.*, to be published.

Heslot F., Baumberger T., Perrin B., Caroli B. & Caroli C. 1994. Creep, stick-slip, and dry-friction dynamics: Experiment and a heuristic model. *Phys. Rev. E* 49:4973-4988.

Hutchinson J.W. & Jensen H.M. 1990. Models of fiber debonding and pullout in brittle composites with friction. *Mech. Mater.* 9:139-163.

Jero P.D. & Kerans. R.J. 1990. The contribution of interfacial to sliding friction of ceramic fibres in a glass matrix. *Scripta Metall. Mater.* 24:2315-2318.

Kerans R.J. & Parthasarathy T.A. 1991. Theoretical analysis of the fiber pullout and pushout tests. *J. Amer. Ceram. Soc.* 74(7):1585-1596.

Kuntz M., Schlapschi K.H., Meier B. & Grathwohl G. 1994. Evaluation of interface parameters in push-out and pull-out tests. *Composites* 25:476-481.

Marshall D.B., Shaw M.C. & Morris W.L. 1995. Measurements of interfacial properties from fiber sliding experiments: the role of misfit anisotropy and interfacial roughness. *Acta Matall. Mater.* 43(5):2041-2051.

Martin B., Benoit M. & Rouby D. 1993. Interfacial sliding strength in fibre reinforced ceramic matrix composites involving positive radial thermal misfit strain. *Scripta Metall. Mater.* 28:1429-1433.

Mumm D.R. & Faber K.T. 1995. Interfacial debonding and sliding in brittle-matrix composites measured using an improved fiber pullout technique. *Acta Metall. Mater.* 43(3)1259-1270.

Parthasarathy T.A., Marshall D.B. & Kerans R.J. 1994. Analysis of the effects of interfacial roughness of fiber debonding and sliding in brittle matrix composites. *Acta Metal. Mater.* 42(11)3773-3784.

Persson B.N.J. & Tossati E. (editors) 1996. *Physics of sliding friction*, NATO ASI Series E. 311, Dortrecht:Kluwer Academic Pub.

Rabinowicz E., *Friction and wear of materials* 1965. New York:Wiley & Sons.

Rausch G., Meier B. & Grathwohl G. 1992. A push-out technique for the evaluation of interfacial properties of fiber-reinforced materials. *J. Eur. Ceram. Soc.* 10:229-235.

Rouby D. & Reynaud P. 1993. Fatigue behaviour related to interface modifications during load cycling in ceramic-ceramic matrix fibre composites. *Composite Science and Technology* 48:109-118.

Stupkiewicz S. 1996. Fiber sliding model accounting for interfacial micro-dilatancy. *Mech. Mater.* 22:65-84.

Damage and Failure of Interfaces, Rossmanith (ed.) © 1997 Balkema, Rotterdam, ISBN 90 5410 899 1

On the role of the steel-concrete interface in dowel action: Some experimental and numerical results

M.di Prisco & L.Ferrara
Dipartimento di Ingegneria Strutturale, Politecnico di Milano, Italy

ABSTRACT: When concrete structural members are cast, the steel reinforcing bars behave like a sort of obstacle to the normal flow of the fluid mixture, expecially to a hc:mogeneous positioning of the aggregate grains. This phenomenon, together with the well known "wall-effect", makes the quality of the concrete in the immediate surroundings of the rebars significantly lower than elsewhere in the structure. The results of some experimental tests, performed on "dowel-plate" specimens, and of the related numerical analyses, carried out using a non-local damage model to describe concrete and interface behaviour are shown here. The final aim is to better understand the role of the steel-concrete interface in the whole development of the structural resistant mechanism as well as to attempt to identify the geometrical and mechanical characteristics of the interface itself.

1. INTRODUCTION

In the last three decades dowel action has been recognized as one of the most important shear resistant mechanisms in R/C structures. Many authors have devoted their efforts to study it, both from an experimental and a theoretical point of view (see Johnston and Zia 1971, Walraven and Reinhardt 1981, Paschen and Schönhoff 1983, Qureshi 1993). Several significant contributions have been made, namely in quantifying the maximum bearing strength of the mechanism - in a strong as well as in a weak condition, i.e. when the steel bar respectively acts against the concrete core or against the cover - or the subgrade stiffness of the concrete for one or several dowel bars embedded in it (see Soroushian et al. 1986, 1987a, 1987b, 1988). Quite appreciable results have furthermore been achieved in the field of mathematical modelization of this mechanism: thus it has been possible both to comprehend and describe its physical nature and to quantify its bearing strength under static and cyclic loading as a function of the displacement parallel to the discontinuity (Figure 1), for different types of concrete (normal, high-strength, fiber-reinforced etc. - see Vintzeleou and Tassios 1986, Dei Poli et al. 1992, 1993, di Prisco et al. 1994).

The resistant features of dowel action essentially spring from the interaction between the reinforcing bars and the surrounding concrete. The evolution of the mechanism, especially in its initial stages, is characterized by the onset - in the material close to the dowel bar - of high strain gradients, both compressive and tensile. These respectively take the form of concrete crushing and of microcraks, growing and coalescing until the bar is completely detached from the stretched concrete.

As a matter of fact, and as one can reasonably argue by considering the technological problems arising

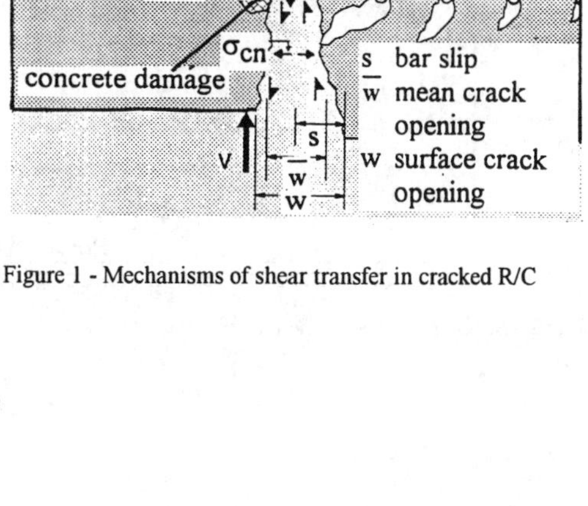

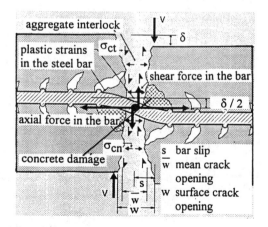

Figure 1 - Mechanisms of shear transfer in cracked R/C

when structural members are cast, the mechanical properties of the material, in the "halo-shaped" zone, just around the bars, cannot be equal to the ones elsewhere in the whole structure, but must be considered to be genuinely poorer. The mechanical and geometrical characteristics of such a "local unhomogeneity" of the material field properties can be hypothesized as having a quite significant influence on the global behaviour of the mechanism, mainly on the stiffness in the early stages of its evolution.

To the authors' knowledge, few researchers have considered such an important feature in investigating the dowel action mechanism: some information about the experimental detection of the material behaviour in the proximity of the bar, and about the relative mathematical modelization, can be found in [di Prisco and Gambarova 1992 - see also di Prisco et al. 1992]. The existence of such an interface zone, justified by means of intuitive and practical arguments, was hypothesized by di Prisco and Mazars (1992) in a numerical analysis of some experimental tests, previously performed by the first author, in the field of concrete damage mechanics. Attention to the local phenomena occurring around a steel dowel bar embedded in concrete or in a stone masonry wall has been paid respectively by Biolzi and Giuriani (1990) and by Felicetti et al. (1997).

In this paper the results obtained from some experimental tests carried out on dowel plate specimens, pushing the bar against the core, are first shown. Moirè interferometry technique was employed in addition to displacement measures carried out by means of inductive transducers. Then, on the basis of those results and by means of further numerical analyses, performed using a non-local damage model, the influence of the mechanical and geometrical characteristics of the interface zone on the global behaviour of the structure is more carefully checked. At the same time, an attempt to evaluate the above mentioned properties (both mechanical and geometrical) is made, once again by means of numerical analyses.

2. EXPERIMENTAL RESULTS

As already pointed out in the previous chapter, the experimental tests were performed on "dowel plate specimens", investigating the dowel action mechanism in its strong condition, i.e. pushing the bar against the concrete core. The geometrical characteristics of the specimens are shown in Figure 2. The drawing also shows the loading frame, with all the measuring instruments in place. Four specimens were cast: in two of them a ribbed bar was used as the dowel, while, in the others, a smooth bar was chosen. In both cases 18 mm diameter (d_b) FeB44k steel bars were employed. Concrete was made with 350 kg/m^3 Portland cement, using a water/cement ratio equal to 0.6 and aggregates, whose grading curve was calculated according to the Bolomey formula, with a maximum size d_a of 10 mm (one third of the specimen thickness). Once cast, the specimens were cured for 28 days (90% RH, 20°C) and then left at room conditions (70% RH, 20 °C) until testing. They were tested 47 days after being cast. The mean cubic compressive strength, at the test date, as measured on the companion cube specimens (10 cm side), was 42 MPa.

Each specimen was supplied with two Moirè grids (density 40 lines per mm), one on each face, in order

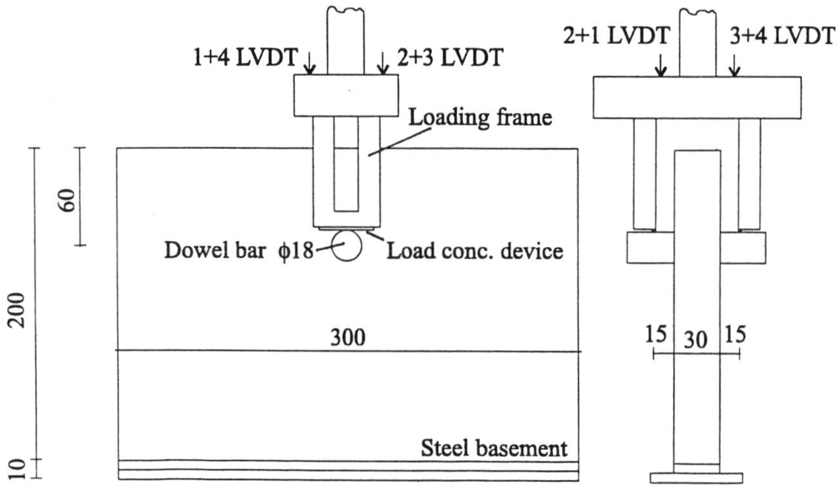

Figure 2 - Schema of the specimen

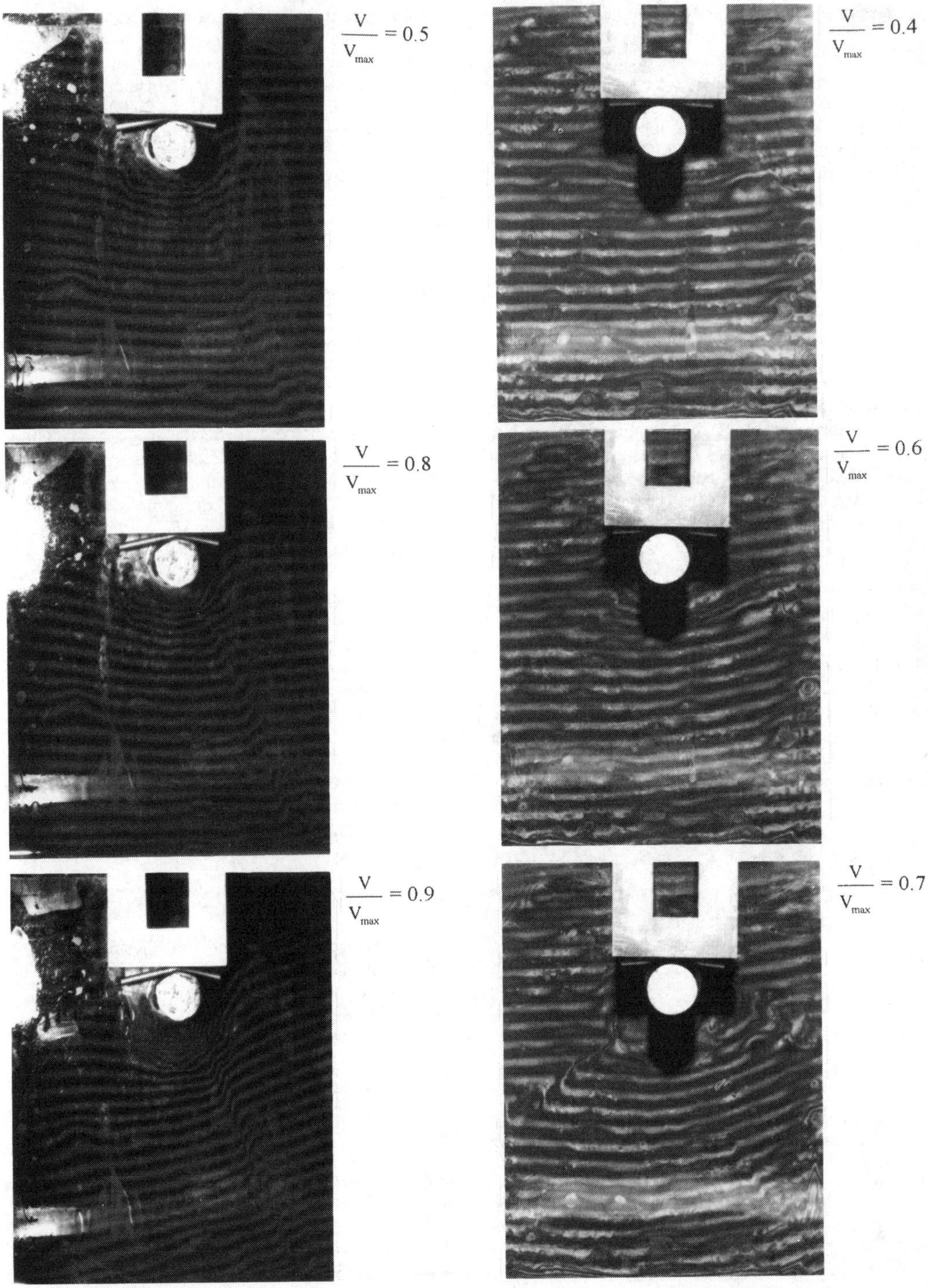

Figure 3 - Moirè fringes, for both ribbed (left column) and smooth dowel bar specimens (right column), at different stages of the loading path.

to detect both the vertical and horizontal components of the displacement field in the specimen. Some Moirè photographs are shown in Figure 3. The failure mechanism of the specimen can be clearly detected in these photographs: two diagonal macrocracks, starting from the bar and propagating with an inclination of about 45°. They highlight the shear stress role in the development of the whole resistant mechanism. It is worth noting that a large deformation zone is formed just around the bar, for an extension of about 1÷1.5 times the bar diameter. These results seem to agree with those obtained by other authors (see Biolzi and Giuriani 1990, di Prisco et al. 1992).

The experimental results, in terms of load-displacement curves, are shown in Figure 4; the displacement value plotted on the x-axis corresponds to the mean value of the displacements recorded by the four transducers (n°1 - n°4; Figure 2). This experimental mean value has obviously been cleansed of the deformation introduced by the load concentrating device, and of the flexural behaviour of the dowel bar, owing to the eccentricity of the applied load. The former deformation was experimentally quantified while the latter has been estimated according to the model of a beam on an elastic foundation. Totally, their contibution is roughly a 20% of the deformation measured by means of LVDTs.

While the global trend of the different load-displacement curves is similar for both the ribbed and smooth dowel bars, significant differences exist among the various specimens tested. First of all a difference in the maximum load attainable between the smooth dowel specimens and the ribbed ones can be detected. In the first case, the mean bearing strength of the mechanism is lower than in the other case. This can be explained by considering that the ribs, placed alongside the bar surface, enlarge the contact surface between concrete and steel, enhancing the bar capability of transferring the applied load to the surrounding material. This is also confirmed by the failure mechanism which can be seen in the Moirè photographs. The differences that can be detected between two nominally identical curves (i.e. curves a and b, and curves c and d; Figure 3), especially in what concerns the secant stiffness of the resistant mechanism, appear even more significant. The main reason for such differences seems to be the interface zone; still looking at the Moirè photographs (which refer to different steps of the loading program) and at the graphs developed from them, it can be easily noted that the most significant part of the deformation and of the energy dissipation phenomena is concentrated in this zone. The observed differences can thus be attributed to the randomness of the field properties in the interface halo, whose existence is essentially due to non exactly repeatable causes, e.g. the wall effect and the fact that the dowel is a sort of obstacle to the normal flow of the fluid concrete mixture and to the homogeneous positioning of the aggregate grains when the specimen is cast. The need to further study this problem follows from such experimental evidences. In particular, the characteristics of the interface zone around the dowel (also by means of numerical aids) must be investigated, aiming first of all at confirming the hypothesized role of the zone in issue, and, secondly, at the correct modelization of its geometry and constitutive behaviour.

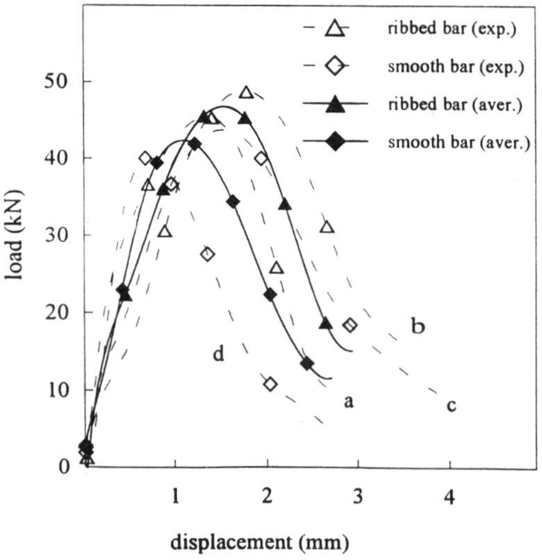

Figure 4 - Experimental load - displacement curves

3. NUMERICAL ANALYSES

In order to better understand the physical problems pointed out in the previous chapter, by analizing the experimental results, the tests performed were numerically simulated by the Finite Element code CASTEM 2000. The "crush-crack" damage model, conceived by M. di Prisco and J. Mazars (1996), was adopted to describe the concrete constitutive behaviour in the interface zone as well. The model couples elastic stiffness degradation with cumulation of irreversible strains, both in tension and compression. It furthermore allows, by introducing a second internal variable δ, to take into account the growth of the Poisson coefficient ν, thus trying to model the reversible dilatancy phenomenon, which plays, as is known, a key-role in shear transfer of cracked plain and reinforced concrete. The model requires, as an input, three curves describing the evolution of the scalar damage variable D, with respect to a strain invariant (built with the positive part of the strain tensor components), in three basic load conditions: uniaxial compression and tension, and biaxial compression. The above mentioned curves can be simply built, point-by-point, starting from the constitutive curves referring to the same three basic load situations.

The main purpose of the whole set of analyses is to investigate on the role played by the steel-concrete interface in the specimen behaviour, especially in its early stages. To this end, it was decided to carry out a set of numerical simulations thus structured:

- first of all the case in which no interface exists was considered, as a reference;
- secondly the case in which an interface layer 3 mm thick was present, and with its stiffness set respectively at 0.75, 0.50, 0.25, 0.10 times the concrete stiffness, was dealt with;
- finally the presence of an interface layer of different thicknesses (namely 6 mm = $d_b/3$ and 10 mm = d_a) was also hypothesized, in which the material had a stiffness equal to 0.5 times the material stiffness elsewhere in the specimen.

In all the analyses, steel was modeled as an elastic-perfectly plastic material, with f_y = 440 MPa, E_s = 206 GPa and $\nu = 0.3$.

Concrete and mortar were modeled as elastic-damageable materials with irreversible strains growing in tension and compression. For concrete, starting from the experimentally measured mean cubic compressive strength - equal, as quoted above, to 42 MPa - the mean cylindrical compressive strength was determined as

$$f_c = 0.83\ f_{cc} = 35 \text{ MPa} \tag{1}$$

The concrete stiffness and tensile strength were calculated by means of the EC2 formulae:

$$E_c = 9500\ f_c^{\ 1/3} = 31100 \text{ MPa} \tag{2}$$

$$f_t = 0.27\ f_c^{\ 2/3} = 2.90 \text{ MPa} \tag{3}$$

(from such values, assuming a linear behaviour up to the peak, a damage threshold $\varepsilon_{t0} = f_t/E_c = 9.325\text{E-}5$ was obtained). The biaxial compressive strength was set equal to 1.25 times the uniaxial one. Uniaxial and biaxial compressive constitutive behaviour were described by means of some empirically derived rational formulae (see di Prisco and Ferrara 1996). The uniaxial tensile behaviour was modeled by means of the formula proposed by Hordijk (1993):

$$\sigma=f_t\{[1 + (c_1\eta)^3] \exp(-c_2\eta) + (1+c_1)^3\ \eta \exp(c_2)\} \tag{4}$$

where c_1 = 3, c_2 = 6.93, $\eta = \varepsilon/\varepsilon_u$, $\varepsilon_u = 5.14\ G_f/hf_t$, G_f is the material tensile fracture energy and h the characteristic length. For concrete G_f was estimated, from the results available in literature, equal to 110 N/m, while h was set equal to d_a = 10 mm.

The mechanical characteristics of the mortar in the interface halo were modeled according to the following assumptions:

- the Young modulus E^{in} was set equal, as above stated, respectively to 0.75 E_c, 0.5 E_c, 0.25 E_c and 0.10 E_c;
- the same ratios were adopted for the mortar compressive and tensile strength, with respect to the concrete ones; no difference was therefore introduced in relation to the damage threshold nor in relation to the crushing threshold ε_{c0}, which was set at 1.E-3, for both concrete and mortar;
- with reference to the first two cases (E^{in} = 0.75 E_c or 0.5 E_c) the same ratios were adopted also for the tensile fracture energy of the mortar, G_f^{in}, with respect to that of concrete G_f. In the other two cases, i.e. for E^{in} = 0.25 E_c or 0.10 E_c, it was set at G_f^{in} = 0.5 G_f, in order not to limit the uniaxial tensile curve to very small values, that would have caused some convergency problems in the analysis.

In all the analyses performed, the concrete was modeled as a non-local material, and its representative volume characterized by the parameter l_c, which controls the weighting function introduced in the non-local approach chosen (see di Prisco and Mazars 1996). The parameter l_c was set equal to $d_a/2$ = 5mm (see di Prisco and Ferrara 1996, di Prisco et al. 1997 for a motivation of such a ratio between l_c and d_a); mortar interface around the dowel bar was treated as local, except in the case in which its thickness was set at 10 mm. Due to such a relevant dimension of the interface, in the analysis performed treating the interface as a local zone, convergence was lost for very small values of the loading parameter (the displacement applied at the bar center), probably due to a snap-back, that could be captured only by means of an arc-length control procedure. For this reason a non-local modelization was proposed and pursued. The value of the parameter l_c was set equal to the one adopted for concrete, as the procedure that allows to describe l_c as a field property in the specimen is not yer available. In any case the minimum element dimension for the interface layer was always set at 3 mm.

It is worth noting that, in dealing with an interface between a non-local and a local zone, the F.E. code does not provide a modification of the weigth function, as it could be reasonable and physically justified, but simply cuts out part of the averaging domain for the usual symmetrical bell-shaped weight function (see Bazant and Chang, 1984).

The meshes adopted for numerical analyses are shown in Figure 5. Only half of the specimen was discretized, due to its symmetry; all the points along the side A-B were considered fixed both along the vertical and the orizontal directions. The load was applied, in displacement control, at the bar center. The convergency requirements of the adopted meshes were checked by means of simple elastic analyses. The energy stored in the specimen with reference to three different degrees of mesh accuracy was calculated, for a unit displacement imposed at the bar center. Medium mesh was adopted in every case for the analyses. The whole set of results on elastic convergence analyses is shown in Figure 6. With reference to an elastic situation, the dependence of the structural stiffness on the interface thickness and on the E^{in} / E_c ratio was checked. The results are shown in Figure 7. While an almost linear trend can be observed for the structural stiffness K_{st} versus the interface thickness th^{in} curve, a strong non-linear dependence on the Young moduli ratio was found. Furthermore, the structural stiffness descrease is much more significant if one considers the variability of the mortar interface stiffness (almost 33%), than if, keeping the E^{in} / E_c ratio constant , one varies only the interface thickness (about 10 ÷ 12%). It follows that to attempt a correct identification of the mechanical parameters for the material in the steel concrete interface layer is much more interesting and profitable, in order to well fit the experimental results, than focusing one's attention on the geometrical characteristics af the layer itself. These can be reasonably estimated on the basis of intuitive and phenomenological arguments, starting from the dimensions of the specimen and from the maximum aggregate size.

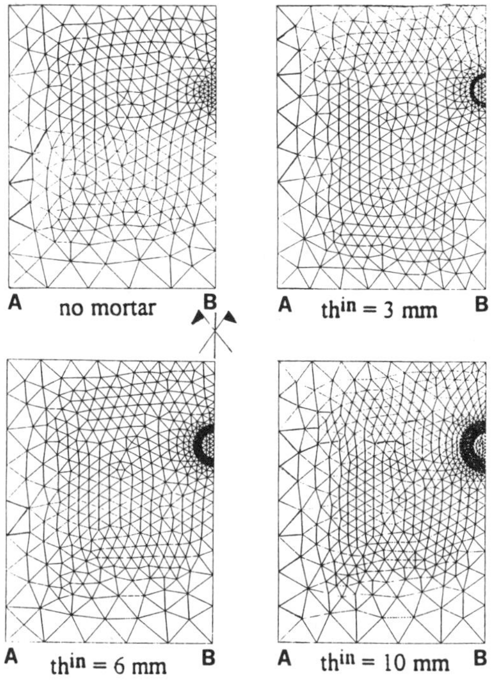

Figure 5 - Meshes adopted for numerical analyes (the interface is grey shaded).

The elastic analyses in issue were performed, as can be seen still in Figure 7, both in plane strain and in plane stress conditions. The difference is really insignificant (about 5%, as one could reasonably expect by a rough estimation). In a nonlinear analysis with the crush-crack damage model, where phenomena such as dilatancy and confinement are taken into account, such a difference could become more and more notable as the calculation proceeds along the load-path steps. The plane stress algorithms for the crush-crack model have not yet been implemented in the F.E. code CASTEM 2000. For this reason, having decided to focus on the role played by the steel concrete interface in the early, almost elastic, stages of the mechanism development, the above mentioned nonlinear analyses were performed in plane strain condition up to a value of the bar center displacement equal to 0.05 mm.

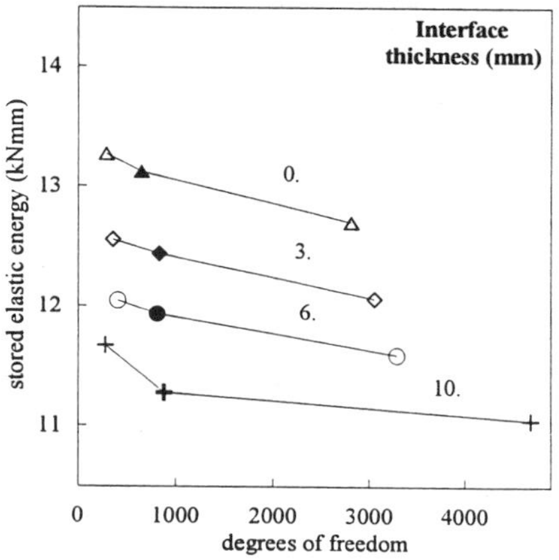

Figure 6 - Results of the elastic convergence analyses.

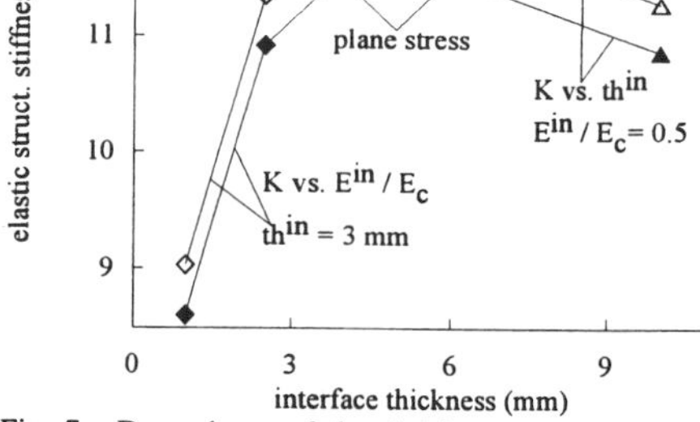

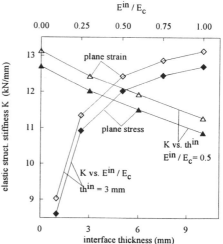

Fig. 7 - Dependence of the elastic structural stiffness on the interface thickness and on the E^{in} / E_c ratio

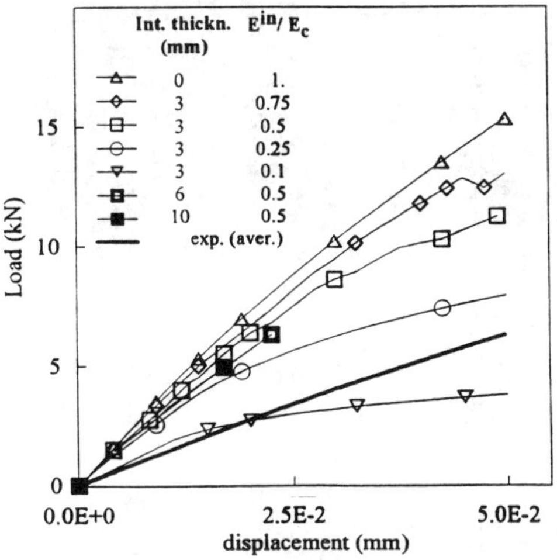

Fig. 8 - Numerical-experimental comparison - Load-displacement curves

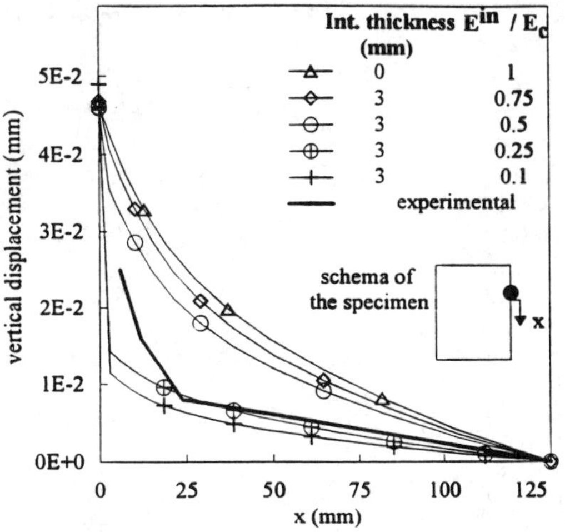

Figure 9 - Numerical-experimental comparison - Vertical displacements along the symmetry axis

The results of the numerical analyses, in terms of load-displacement curves, are shown and compared with the experimental ones, in Figure 8. It can be seen that the best results were obtained with an interface layer 3 mm thick, in which the stiffness of the material was set at about 0.10 ÷ 0.25 times the concrete stiffness. The same remarks can be made by looking at the curves in Figure 9, showing the numerical-experimental comparison referring to the vertical displacements along the symmetry axis. It must also be noted that both the cases of interface thickness equal to 6 and to 10 mm showed some convergency problems, for a value of the bar center displacement value equal to about 0.02 mm. In the first case this was probably caused by a structural snap-back, while, in the other case, by the low number of elements in the interface representative volume . It is also worth pointing out that almost no difference exists, in terms of load-displacement curves, between the two cases at issue It seems reasonable that, as in the case with an interface 10 mm thick the layer was treated as non-local while in the other case it was treated as local, the averaging operation practically cancelled the differences that would have occurred.

In Figure 10 the damage patterns for all the analysed cases are shown, with reference to the last step of the calculation. In the authors' opinion the assumptions which best fit the experimental tests refer to the case with an interface layer 3 mm thick and an interface stiffness equal to a quarter of the concrete stiffness. In this case also the damage evolution, the stress field and the vertical irreversible strain patterns at the end of the calculation are shown in Figure 11. It can be seen that a correct calibration of interface thickness and stiffness, allows both the local phenomena occurring just around the bar, and the trend of the global structure behaviour to be captured. It is also evident, in Figure 10, the onset of a crack that will propagate in a radial direction inclined at almost 45°, starting from the upper semiperimeter of the bar. The excessive spreading of the damaged zone, that can be observed for the case where no interface exists, but also for a 10 mm thick interface layer, is thus avoided. An equally undesirable excessive concentration of the damage itself is also avoided, as in the case of an over soft interface.

4. CONCLUDING REMARKS AND FURTHER DEVELOPMENTS

The existence of a steel-concrete interface, in all r/c structures, can be justified by means of some intuitive physics arguments. By looking, in the experimental P-δ curves, at the relevant scattering, especially in terms of stiffness, it can be hypothesized that such an interface plays quite a significant role in the development of the dowel action shear resistant mechanism of a "plate" specimen, especially in its early stages. That role was more carefully checked by a numerical simulation, in the field of concrete damage mechanics, of the experimental tests.

The introduction, around the bar, of an halo-shaped interface layer of mortar can be a decent approximation of a continuous variation of the material field properties in the immediate surroundings of the reinforcing bar. A rough identification of the geometrical and mechanical characteristics of the material in the interface was also attempted. The best results, in terms of fitting of the experimental curves, have been obtained with an

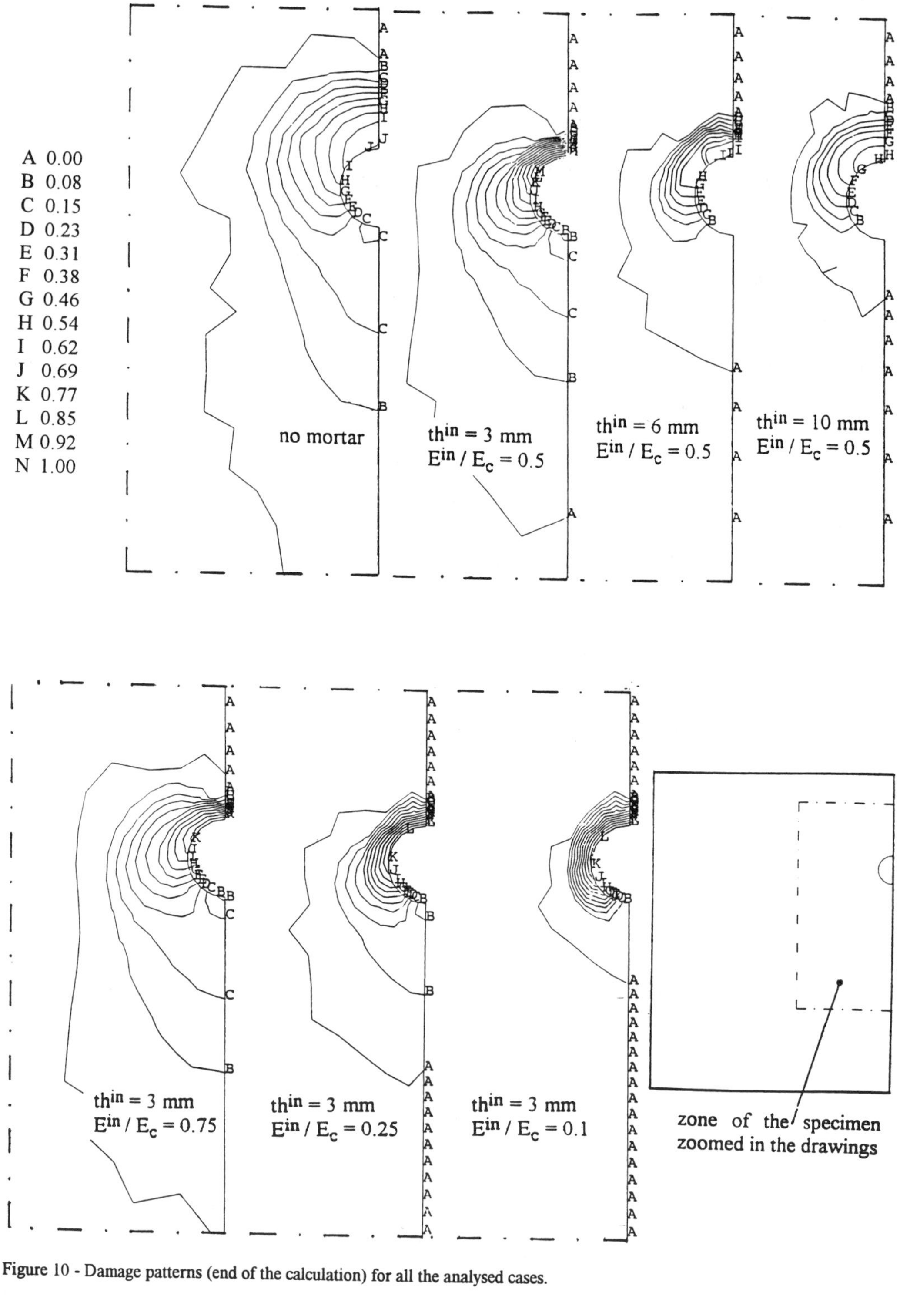

Figure 10 - Damage patterns (end of the calculation) for all the analysed cases.

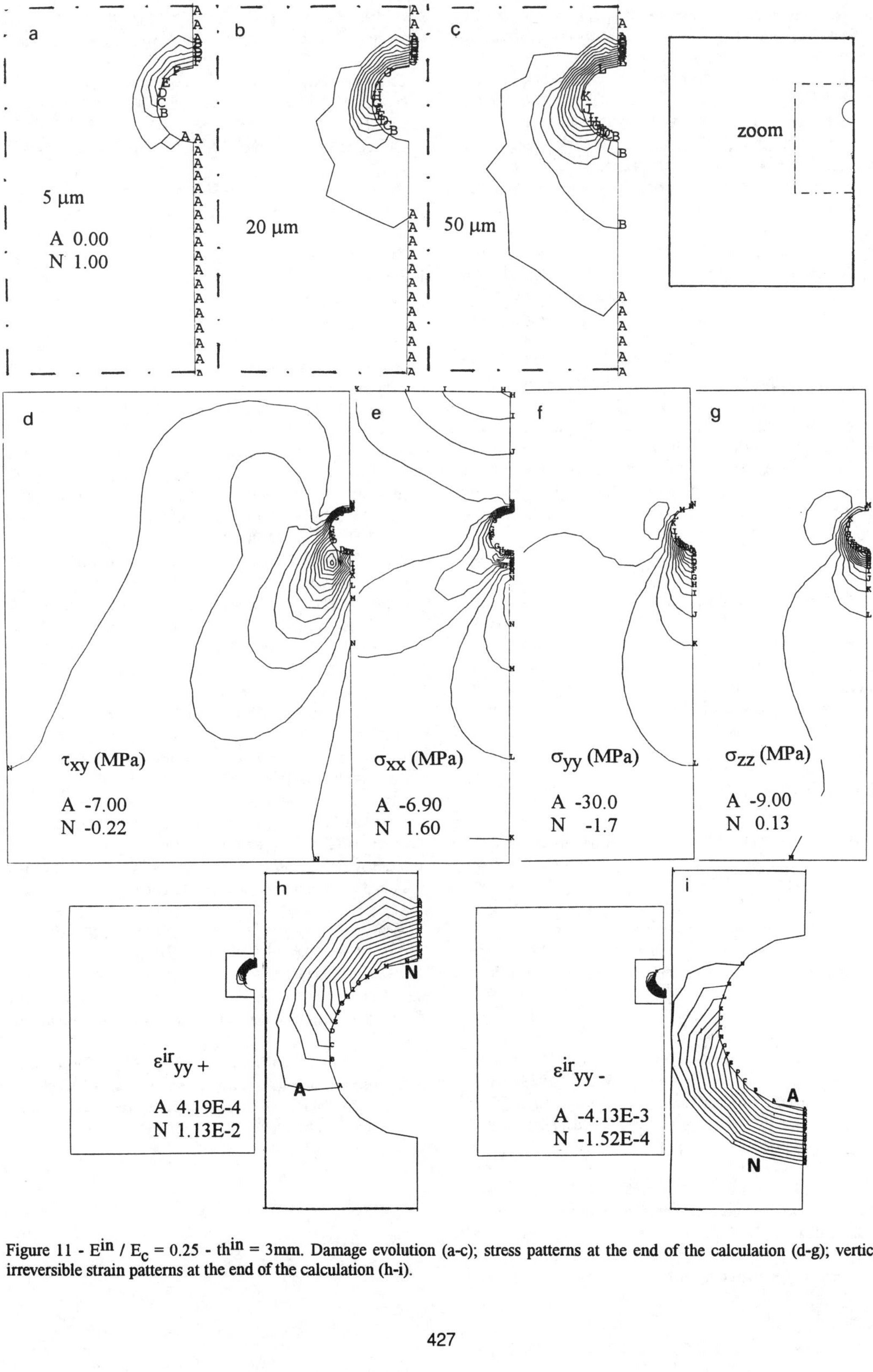

Figure 11 - $E^{in} / E_c = 0.25$ - $th^{in} = 3mm$. Damage evolution (a-c); stress patterns at the end of the calculation (d-g); vertical irreversible strain patterns at the end of the calculation (h-i).

interface layer 3 mm thick ($\approx d_a / 3$), whose elastic stiffness was equal to 0.10÷0.25 times the stiffness in the material elsewhere in the structure.
As a further development of the work herein presented, the influence of other mechanical characteristics of the mortar in the interface can be checked (crushing threshold, damage threshold, fracture energy, absence or presence of dilatancy). Furthermore a more enriched modelization of the interface itself could be attempted, for example by treating it as a non-local zone, but with a characteristic length different from the one adopted for concrete. Regarding the characteristic length as a field property of the structure seems to be more realistic and useful in dealing with all problems where geometrical and mechanical discontinuities have to be taken into account. A very similar approach should consist in regarding the characteristic length as a structural parameter whose value decreases with increasing damage, thus providing a field modelization with a lower l_c where the strain gradient is greater (see again di Prisco and Ferrara 1996). Work is in progress on this subject and the authors will publish the results in a forthcoming paper.

ACKNOWLEDGEMENTS

The financial support of the European Community to **A**lliance of **L**aboratories in **E**urope for **R**esearch and **T**echnology - Geomaterials of **C**.**E**.**C**. in the framework of program *Human Capital and Mobility* is gratefully acknowledged.

REFERENCES

Bazant, Z.P., Chang, T.S.M. 1984. Instability of nonlocal continuum and strain averaging. *ASCE J. of Engrg. Mech.*, 110:1441-1450.

Biolzi, L., Giuriani, E. 1990. Bearing capacity of a bar under transversal loads. *Materials and Structures*, 23: 449-456.

Dei Poli, S., di Prisco, M., Gambarova, P.G. 1992. Shear response, deformation and subgrade stiffness of a dowel bar embedded in concrete. *ACI Struct. J*,. 89 (6): 665-675.

Dei Poli, S., di Prisco, M., Gambarova, P.G. 1993. Cover and stirrup effects on the shear response of dowel bar embedded in concrete. *ACI Struct. J.*, 90 (4): 441-450.

di Prisco, M. Gambarova, P.G. 1992. Dowel action: experimental analysis of local damage. *Proc. 1st Int. Conf. on Fracture Mechanics of Concrete Structures FraMCoS 1*, Breckenridge, Colorado, USA, Z.P. Bazant Ed., Elsev. Appl. Sc.:542-549.

di Prisco, M., Cerrini, P., Gambarova, P.G. 1992. Dowel action in r/c structural elements: experimental analysis of local damage. *Studi & Ricerche* 13: 319-358. School for the Design of R/C Structures, Milan University of Technology, Italy (in Italian).

di Prisco, M., Mazars, J. 1992. On damage modelling of reinforced concrete subject to high strain gradients. *Studi & Ricerche* 13: 33-82. School for the Design of R/C Structures, Milan University of Technology, Italy.

di Prisco, M., Caruso, M.L., Piatti, S. 1994. On fiber role in dowel action. *Studi & Ricerche* 15: 151-194. School for the Design of R/C Structures, Milan University of Technology, Italy

di Prisco, M., Ferrara, L. 1996. A trial method to follow the progressive localization of fracture processes in concrete. *Studi & Ricerche* 17: 1-35. School for the Design of R/C Structures, Milan University of Technology, Italy.

di Prisco, M., Mazars, J. 1996. Crush-crack: a non-local damage model for concrete. *Mechs. of Cohesive and Frictional Materials,* 1: 321-347.

di Prisco, M., Felicetti, R., Gambarova, P.G. 1997. On the evaluation of the characteristic length in high strength concrete. Accepted for presentation to Int. Conf. on HSC, 13-18 July 1997, Kona, Hawaii (USA), Engrg. Found. ASCE, 10017 New York, N.Y.

Felicetti, R., Gattesco, N., Giuriani, E. 1997. Local phenomena around a steel dowel embedded in a stone masonry wall. *Material and Structures,* 30 (5): 238-246.

Hordijk, D. A. 1991. Local approach to fatigue of concrete. Ph. D Thesis T.U. Delft, The Netherlands.

Johnston, D.W., Zia, P. 1971. Analysis of dowel action. *ASCE J. of the Struct. Div.* 97 (ST5): 1611-1630.

Qureshi, J. 1993. Modeling of stress transfer across r/c interfaces. *Ph. D. Thesis* Dept. of Civ. Engng. Graduate School of the Univ. of Tokyo, Japan.

Paschen, H., Schönhoff, T. 1983. Investigation on steel dowels subjected to shear and embedded in a concrete mass. *DAfSt, Report 346*: 105-149, Berlin (in German).

Soroushian, P., Obaseki, K., Rojas, M.C., Sim, J. 1986. Analysis of dowel bars acting against concrete core. *ACI J.*, 83 (4): 642-649.

Soroushian, P., Obaseki, K., Rojas, M.C., Najm, H.C. 1987. Behaviour of bars in dowel action acting against concrete cover. *ACI Struct. J.*, 84 (2): 170-176.

Soroushian, P., Obaseki, K., Rojas, M.C. 1987. Bearing strength and stiffness of concrete under reinforcing bars. *ACI Mat. J.*, 84 (3): 179-184.

Soroushian, P., Obaseki, K., Baiyasi, M.I., El-Sweidan, B., Choi, K.B. 1988. Inelastic behaviour of dowel bars. *ACI Struct. J.*, 85 (1): 23-29.

Vintzeleou, E.N., Tassios, T.P. 1986. Mathematical models for dowel action under monotonic and cyclic conditions. *Mag. Concr. Res.*, 38 (134): pp. 13-28.

Walraven, J.C., Reinhardt, H. W. 1981. Theory and experiments on the mechanical behaviour of cracks in plain and reinforced concrete sujected to shear loading. *HERON* 26 (1A), Delft, The Netherlands.

Miscellaneous problems

Damage and Failure of Interfaces, Rossmanith (ed.) © 1997 Balkema, Rotterdam, ISBN 90 5410 899 1

Microstructural analysis of cancellous bone using interface elements

M. Pini
University of Trento, Italy

C. M. López & I. Carol
ETSECCPB-UPC, Barcelona, Spain

R. Contro
Dipartimento di Ingegneria Strutturale, Politecnico di Milano, Italy

ABSTRACT: The mechanical properties of trabecular bone are studied experimentally and numerically. Experimental studies include compression tests of 1 x 1 x l cm cubes of bovine femoral bone. The distribution of voids is measured on digitalized images and statistical parameters are obtained. In the FE analysis, the microstructure is discretized explicitly. The continuum elements are assumed elastic and the possibility of cracking is introduced by inserting interface elements along all potential crack paths. Interface behavior is given by a fracture energy-based work-softening plastic model with a coupled normal-shear failure surface. Numerical results of two different microstructural arrangements subjected to pure tension are presented and discussed.

1 INTRODUCTION

Cancellous bone tissue is an inhomogeneous porous structure. The cancellous structure is a lattice of narrow rods and plates of calcified tissue called trabeculae, surrounded by vascular marrow which provides nutrients and waste disposal for the bone cells. Due to a progressive increase in the use of implanted bone devices, a clinical need is developing to understand the mechanical and remodelling behaviour of bone tissue. This is the motivation to develop a quantitative stereology technique to study the morphological pattern of the trabecular structure and how it aligns itself with principal stress trajectories (Cowin 1985). The underlying idea is that it is possible to compare the stress trajectories predicted by the elastic stress analysis with the actual trabecular orientation and consequently to predict the response of bone at the implant process of the artificial prosthetic devices (Huiskes 1987, Turner et al. 1990). Detailed understanding of complex aspects of fracture of quasi-brittle materials may be improved with explicit consideration of their microstructure. Some studies of this kind, using the FEM, can be found in literature for concrete (Stankowski 1990, Vonk 1992, López et al. 1996) and for cancellous bone (Huiskes and Hollister 1993, Prendergast 1994, Van Rietbergen 1995). In this paper, on-going experimental and numerical research along this line being carried out at ETSECCPB-UPC and Technical University of Milan, is summarized, and some preliminar results are presented.

2 EXPERIMENTAL WORK

Trabecular bone distributes load from articular surfaces to cortical bone through its bone matrix. In this matrix the trabeculae constitute the actual loadcarrying skeleton, and therefore the material properties of the trabeculae together with their geometry determine the strength and stiffness of the trabecular bone. A series of experimental studies are currently under way at the Department of Structural Engineering of the Technical University of Milan to characterize the overall mechanical properties of this material and to try to relate them to the microstructural properties and configuration. The experiments have been carried out on samples of bovine femoral bones. The bones were excised from the same fresh sample with no history of metabolic bone diseases. Cubical specimens of approximately l x l x 1 cm were sectioned from the proximal end on a low-speed Isomet diamond saw under constant water irrigation (Fig. 1).

Surface roughness remaining from the cutting process was eliminated by gently polishing with 2000-grit sand paper. Prior to testing, each specimen

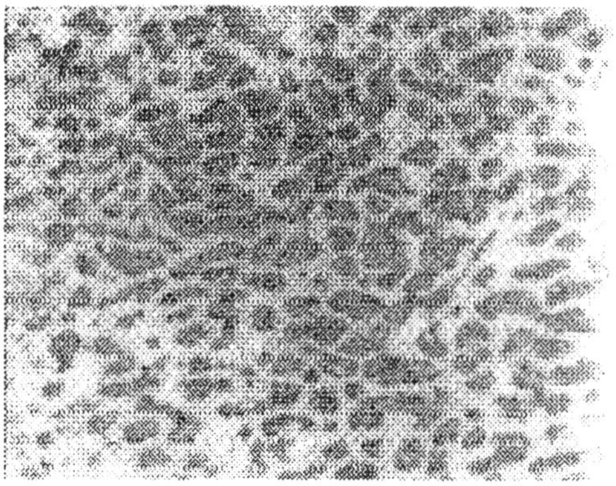

Fig. 1 Side view of a trabecular bone specimen.

was lightly blotted dry with a paper towel. In this way the soft tissue had been finally removed. The inhomogeneity of cancellous bone, and the fact that its properties are less documented, required a great number of specimens.

In order to determine the mechanical properties, a biaxial loading MTS Bionix system was used. It consists of an actuator, a Bionix servohydraulic load frame, and a digital control and data-acquisition system. The device is designed for axial and torsional loading of up to ±100kN and ±1100Nm respectively. Although trabecular bone is subjected primarily to compressive stresses *in vivo*, knowledge of both compressive and tensile failure properties is useful for various reasons. First, because in this way a comparison of the two failure properties at the continuum level provide information on the underlying failure mechanics of the individual trabeculae; and also for a better understanding of a number of pathological effects. For tensile testing, typical experimental results in terms of stress-strain curves are given in Keaveny (1993)(Fig. 2a). In compression, traditional platen-to-platen tests of bone material have been carried out in the context of this study. A typical curve obtained is shown in Fig. 2b. For both compression and tension cases, the behaviour was initially elastic followed by yield regions that appeared to start earlier for the tensile specimens. In the compression test, a locking behaviour due to the compaction of the bony porous microstructure was observed in accordance with similar effects reported in the literature (Keaveny 1993, Turner 1989). Experimentally, the differences between the mean tensile and compressive values of moduli, strengths or ultimate strains are significant. Thin layers of teflon were used to reduce end effects such as friction and surface roughness due to the cutting process. In spite of that, the results still exhibit significant dispersion. This dispersion can also be caused by the different precedent locations (from the epiphysis to the diaphysis in the proximal and distal femur).

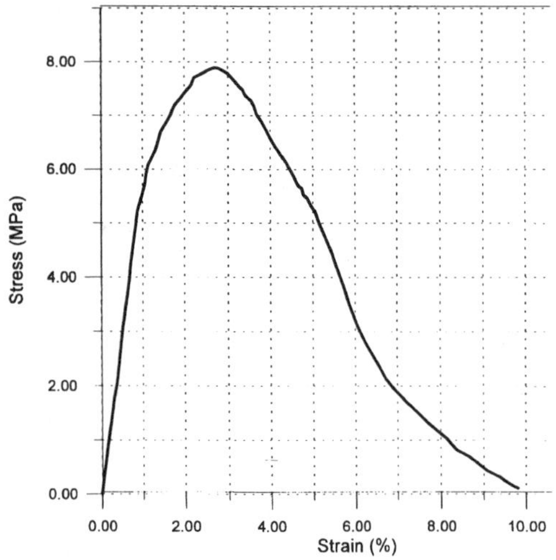

(a)

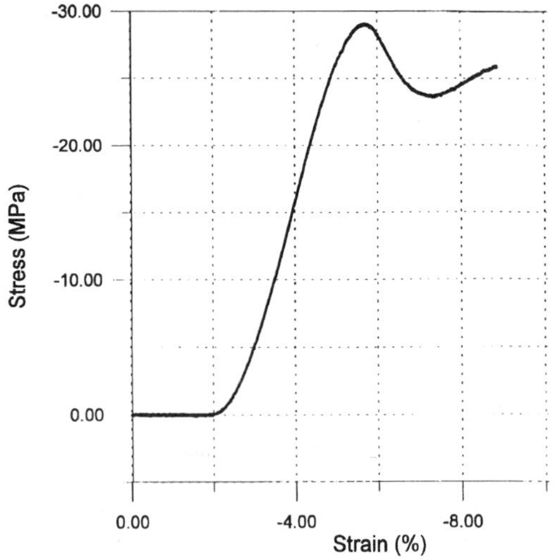

(b)

Fig. 2 Experimental stress-strain curves for: (a) tensile loading test; (b) compressive loading test.

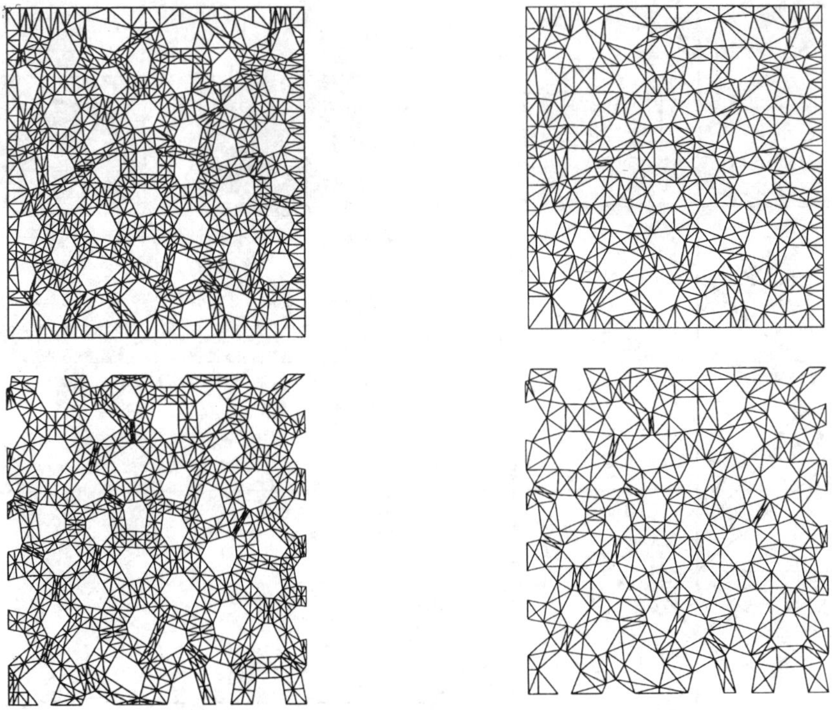

Fig. 3 FE Discretization of trabecular architecture for mesh 1 (up) and mesh 2 (down) with all continuous elements (left) and with only interface elements (right).

3 NUMERICAL ANALYSIS

3.1 Microstructural discretization and meshes

The textural characteristics of the trabecular bone can be measured by means of the mean intercept length, L, the average distance between two bone marrow interfaces measured along a line as a function of its direction on polished plane sections of bone (Whitehouse 1974). When the mean intercept lenghts measured in cancellous bone were plotted in a polar diagram as a function of the orientation, the polar diagram produced ellipses. The concepts can be used to simulate the trabecular architecture with 2-D meshes describing the geometrical anisotropy in terms of the density and the microstructural disorder (Turner 1987).

In the work presented, the computer analyses have been run on meshes generated numerically according to the average geometric characteristics measured in real bone. A mesh of regular hexagons has been defined as a function of some stochastic parameters, such as the number of hexagons in the two directions *x* and *y*, and the semi-length of the hexagon edge. Subsequently, a distortion and shrinking of the mesh is applied to reproduce the distribution and density of pores within the matrix. The trabecular matrix is discretized with triangular finite elements with linear elastic behavior. The FE model mesh includes a number of zero-thickness interface elements in between the continuum triangles. The constitutive behaviour of the interfaces is described in the following section. A 10 x 10 mm specimen is represented in 2-D by two alternative arrangements which differ in the discretization of the boundary: in one case the specimen boundary goes through the matrix, in the other the boundary crosses voids. Fig. 3 shows the two meshes with all the continuum elements (left) and with only interface elements (right). The interface elements have been included along selected boundaries to provide non-tortuous failure paths. Mesh 1 contains 1360 triangles, 1061 interface elements and 2736 nodes, while mesh 2 has 1336 triangles, 974 interface elements and 2500 nodes.

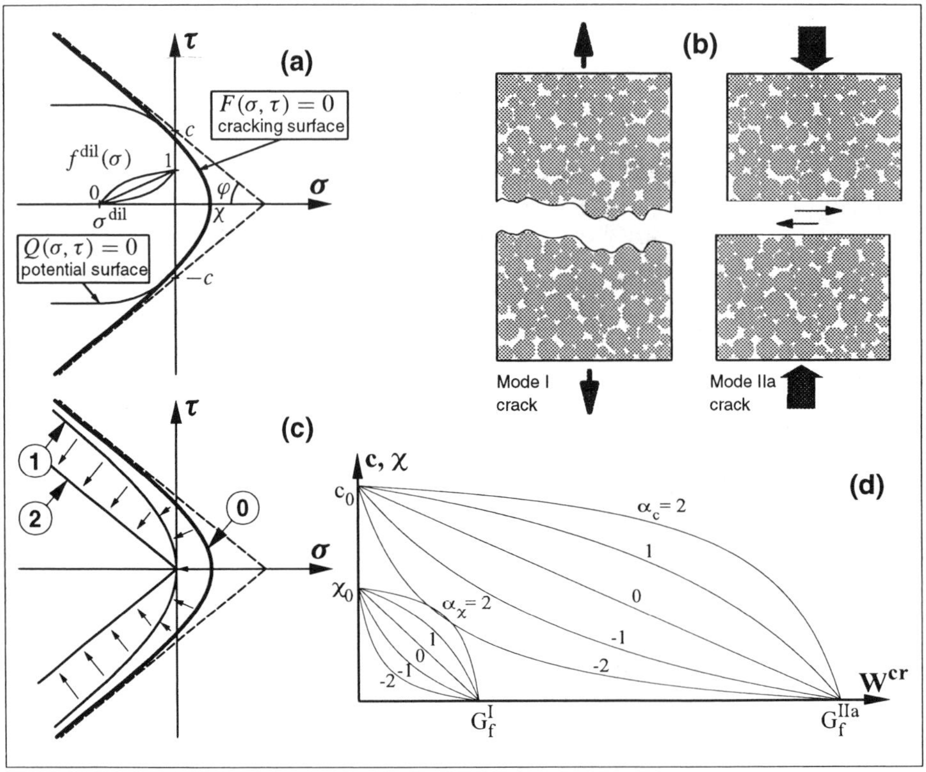

Fig. 4 Interface model: (a) failure surface and plastic potential, (b) basic modes of fracture, (c) softening laws, and (d) evolution of the failure surface.

3.2 Interface constitutive model

Interface behavior is formulated in terms of the normal and shear components of stresses (tractions) on the interface plane, $\sigma = [\sigma_N, \sigma_T]^t$, and corresponding relative displacements $u = [u_N, u_T]^t$ (^{t}=transposed).

The model is analogous to that used for individual discrete cracks (Carol et al., 1997) and for each potential crack plane in a continuum-type multicrack model (Carol and Prat, 1990, Carol and Prat, 1995) for concrete. It conforms to work-softening elasto-plasticity, where plastic relative displacements can be identified with crack openings. The main features of the plastic model are represented in Fig. 4. The initial loading (failure) surface $F = 0$ is given as a three-parameter hyperbola (tensile strength χ, c and $\tan\phi$; Fig. 4a). The model is associated in tension ($Q = F$), but not in compression, where dilatancy vanishes progressively for $\sigma_N \rightarrow \sigma^{dil}$. Classic Mode I fracture occurs in pure tension. A second Mode IIa is defined under shear and high compression, with no dilatancy allowed (Fig. 4b). The fracture energies G^I_f and G^{IIa}_f are two model parameters. After initial cracking, c and χ decrease (Fig. 4d), and the loading surface shrinks, degenerating in the limit case into a pair of straight lines representing pure friction (Fig. 4c). The process is driven by the energy spent in fracture process, W^{cr}, the increments of which are taken equal to the increments of plastic work, less frictional work in compression. Total exhaustion of tensile strength ($\chi = 0$) is reached for $W^{cr} = G^I_f$, and residual friction ($c = 0$) is reached for $W^{cr} = G^{IIa}_f$. Additional parameters α_χ and α_c allow for different shapes of the softening laws (linear decay for $\alpha_\chi = \alpha_c = 0$). The elastic stiffness matrix is diagonal with constant K_N and K_T, that can be regarded simply as penalty coefficients. A more detailed description can be found in (Carol et al., 1997).

3.3 Numerical results

The results presented correspond to the two meshes of figure 4, subject to uniaxial tension along y-axis.

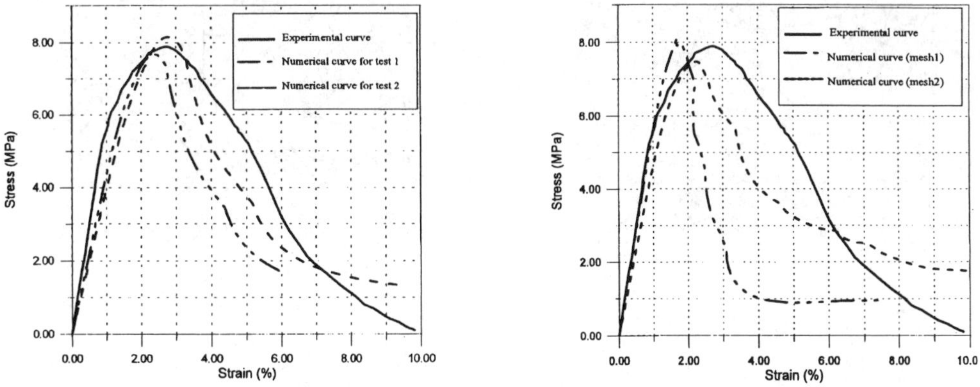

Fig. 5 Average stress-average strain curves

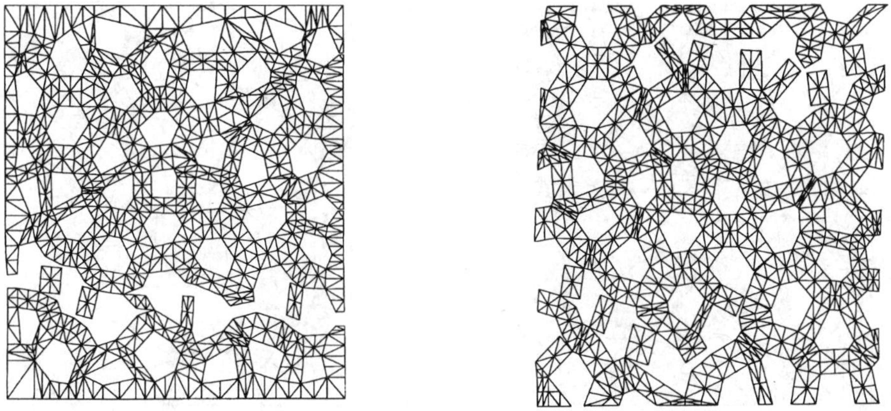

Fig. 6 Deformed mesh for mesh 1 (left) and mesh 2 (right)

In each case, uniform displacements are prescribed to all nodes of the corresponding specimen edges, while transverse displacements are left free. Average stresses are obtained by summing nodal reactions and dividing by specimen size. For the mesh with no voids intercepting the boundaries (mesh 1) two cases with different parameters calibrating the experimental stress-strain curve are presented (Fig. 2a). For the first case (test 1), the material parameters for the continuum elements are :E = 900MPa, ν = 0.18, while for the interface: $K_N = K_T = 10^9$MPa/m, tensile strength χ_0 = 30MPa, c_0 = 100MPa, $\tan\phi$ = 0.8, G^I_f = 0.0018MPaxm, $G^{IIa}_f = 10G^I_f$. For the second case (test 2) the parameters are the same except for χ_0 = 32MPa, G^I_f = 0.0022MPaxm. Elastic stiffnesses for interfaces are assigned high values compatible with not causing numerical difficulties. The iterative strategy used is an arc length-type procedure (García Álvarez et al. 1994), which seems necessary to obtain convergence near and after the peak load. Resulting average stress-average strain curves for the two tests are represented in Fig. 5 (left). In this figure the numerical results appear to be qualitative very similar to the experimental ones. Fig. 5 (right) shows the stress-strain curves for the previous mesh and the one characterized by boundaries crossing voids (mesh 2) obtained with the same material parameters, in order to underline the influence of boundary conditions. It can be observed that mesh 1 presents a brittler behaviour. More insight into the results may be gained with the detailed representation of the crack patterns through the deformed mesh (Fig. 6) and the evolution of the cracking process in terms of the fracture energy spent W^{cr} (Fig. 7).

In the latter figure, each vertical sequence corresponds to three stages (near the peak load, at an intermediate and an advanced stage of the softening branch) for mesh 1 (left diagrams) and for mesh 2

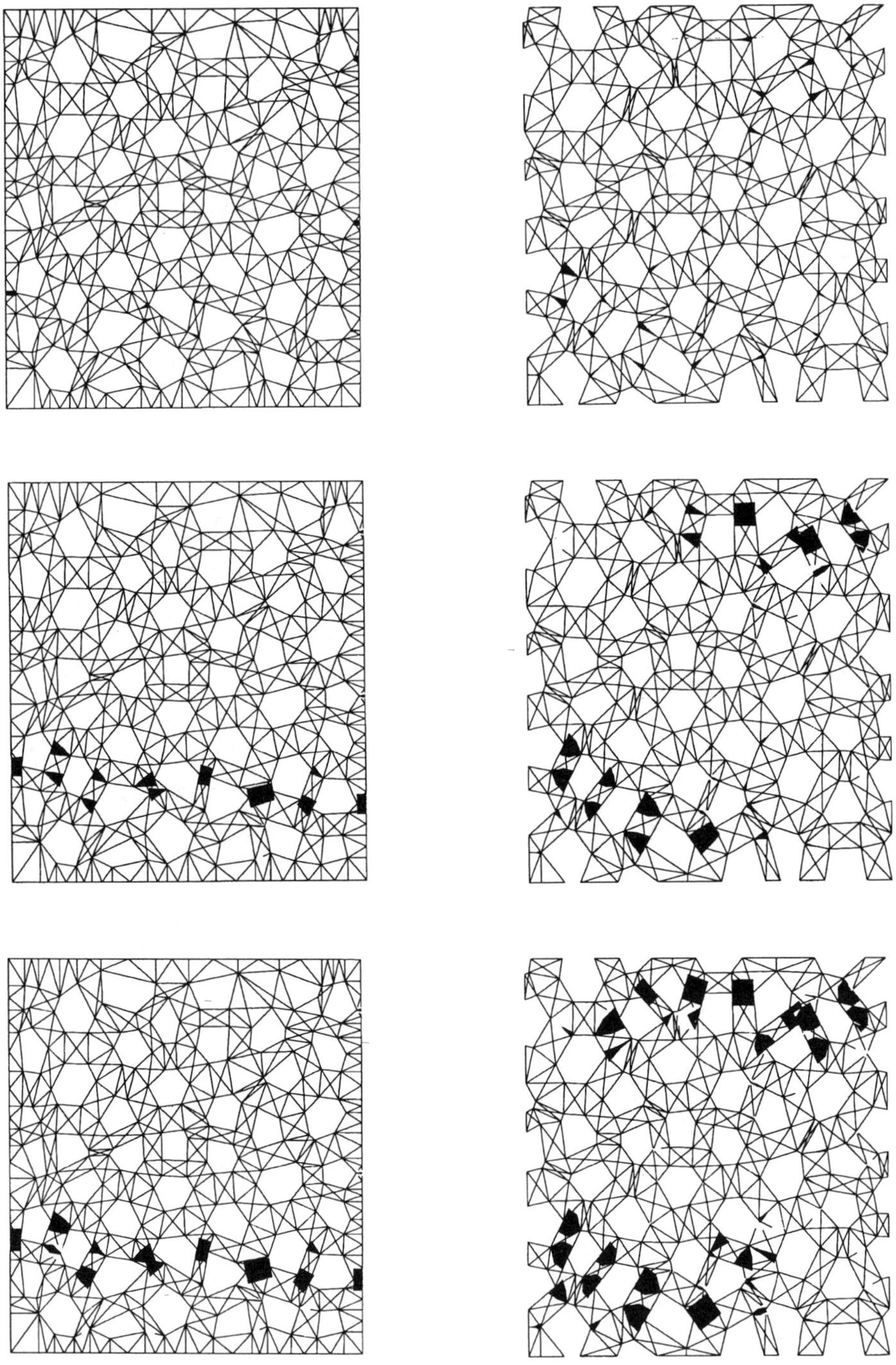

Fig. 7 Progressive cracking for mesh 1 (left diagrams) and for mesh 2 (right diagrams).

(right diagrams). From the figures it is apparent that an initially distributed crack pattern turns, at some point near the peak load, into a highly localized state, with a single crack developing through the specimen, and all other existing cracks unload. It is also apparent that the fracture process is more localized for mesh 1, explaining its brittler behaviour.

4 CONCLUDING REMARKS

The results obtained seemed to be reasonable and to fit well the experimental data in tension. The research described continues to consolidate and improve the initial results obtained in tension and further efforts are currently aimed at simulating the compressive behaviour.

ACKNOWLEDGEMENTS

Partial financial support from DGICYT (Madrid, Spain) under research grant PB93-0955 is gratefully acknoledged. The first author wish to thank the Italian Ministry for the Scientific and Technological Research (MURST) and the School of Civil Engineering at Barcelona (ETSECCPB) for the support received under the agreements with Departament de Política Territorial i Obras Públiques and Colegio de Ingenieros de Caminos, Canales y Puertos, as well as UPC. The second author thanks the Direcció General de Recerca of the Generalitat de Catalunya for the grant received under the program Grups de Recerca Consolidats.

REFERENCES

Carol, I., Prat, P. C., López, C. M., 1997. A normal/shear cracking model. Application to discrete crack analysis. *J. of Engineering Mechanics*.Vol. 123,No 8

Carol, I. and Prat, P.C., 1990. A statically constrained microplane for the smeared analysis of concrete cracking. In Bicanic and Mang, editors, *Computer aided analysis and design of concrete structures,* 919-930.

Carol, I., Prat, P. C., 1995. A multicrack model based on the theory of multisurface plasticity and two fracture energies. In Owen, D.R.J., Oñate, E., and Hinton, E., editors, *Computational plasticity 4*, 1583-1594, Pineridge Press (Swansea, UK).

Cowin, S. C. 1985. The relationship between the elasticity tensor and the fabric tensor. *Mech. Mater.*, A-137-147

García Alvarez, V. O., Carol, I. and Gettu, R., 1994. Numerical simulation of fracture in concrete using joint elements. *Anales de Mecánica de la Fractura*, 11:75-80.

Huiskes, R., 1987. Adaptative bone - remodelling theory applied to prosthetic-design-analysis. *J. Biomechanics*, 20:1135-1150.

Huiskes, R. and Hollister, S. J., 1993. From structure to process organ to cell: recent developments of FE-analysis in orthopaedic biomechanics. *Transactions of the ASME*, 115:520-527.

Keaveny, T. M. and Hayes, W. C., 1993. A 20 Year Perspective on the mechanical properties of trabecular bone. *Trans. of ASME*, 115:534-542.

López, C. M., Carol, I. and Aguado, A., 1996.New results in fracture analysis of concrete microstructure using interface elements. *Anales de Mecánica de la Fractura*, 13:92-97.

Prat, P.C., Gens, A., Carol, I., Ledesma, A. and Gili, J.A., 1993. DRAC: A computer software for the analysis of rock mechanics problems. In H. Liu, editor, *Application of computer methods in rock mechanics, 2*: 1361 - 1368, Xian, China. Shaanxi Science and Technology Press.

Prendergast, P. J., 1994, Prediction of Bone Adaptation using damage accumulation. *J. Biomechanics, 27*: 1067-1076.

Stankowski, T., 1990. Numerical simulation of progressive failure in particle composites. *PhD thesis, Dept. CEAE, University of Colorado, Boulder, CO 80309-0428, USA.*

Turner, C. H, 1987. Dependence of elastic constants of an anisotropic porous material upon porosity and fabric. *J. of Materials Science, 22*:3178-3184.

Turner, C. H, 1989. Yield Behaviour of Bovine Cancellous Bone. *Transactions of ASME, 111* :256-260.

Turner, C. H, Cowin, S. C., Ashman, R. B. and Rice, J. C., 1990, The Fabric Dependence of Orthotropic Elastic Constants of Cancellous Bone. *J. Biomechanics 23*:549-561.

Van Rietbergen, B., Weinans, H., Huiskes, R. and Odgaard, A., 1995. A new method to determine trabecular bone elastic properties and loading using micromechanical Finite-Element method. *J. Biomechanics, 28(1)*:69-81.

Vonk, R., 1992. Softening of concrete loaded in compression. *PhD thesis, Technische Universiteit Eindhoven, Postbus 513, 5600 MB Eindhoven, Netherlands.*

Whitehouse W. J., 1974, The quantitative morphology of anisotropic trabecular bone. *J. Microscopy*, 101:153.

Damage and Failure of Interfaces, Rossmanith (ed.) © *1997 Balkema, Rotterdam, ISBN 90 5410 899 1*

The bone-prosthesis interface: An overview and preliminary tests

R.Contro & V.Quaglini
Dipartimento di Ingegneria Strutturale, Politecnico di Milano, Italy

R.Pietrabissa
Dipartimento di Bioingegneria, Politecnico di Milano, Italy

S.Rizzo, R.Rodriguez y Baena
Istituto di Odontoiatria, Università degli Studi di Pavia, Italy

ABSTRACT: The mechanical stability of the bone-prosthesis interface is crucial for the long term reliability of implanted devices. The chance of bone to self-remodelling involves the bone-prosthesis interface which evolves during time according to the transmitted stresses; the design of the prosthesis, including shape, materials and surface finishing, should take into consideration this phenomenon in order to avoid the loosening of the mechanical interface. The numerical analysis allows to calculate the stresses at the interface due to the loads applied to the prosthesis but, being a suitable model of bone remodelling not yet available, the prevision of the interface evolution is not reliable. The experimental approaches, like the experimental procedure that we have set up to evaluate the mechanical interface between dental implants and bone, evaluate the mechanical stability of the interface by measuring the load necessary to break it.

1 INTRODUCTION

One of the main problems in orthopaedics and dentistry deals with the mechanical stability of the bone-prosthesis interface, which is crucial for the long term reliability of an implanted device: the most frequent cause of failure is the loosening of the interface and the consequent sliding between the bone and the surface of the prosthesis, which shall be removed.

The stability of an implanted prosthesis is due firstly after implantation to the geometrical fit of the device into the prepared site (through drilling, milling, tapping, etc.) in living bone, that is referred to as the primary stability, while in the long time a secondary stability can be attained through the integration of the implant surface by the bone growth. The last phenomenon is referred to as "osteointegration", that means "a direct connection between living bone and a load-carrying endosseous implant at the light microscopical level" (Brånemark, 1985), able to transmit load from the implant to and within the bone tissue.

For the proper evaluation of the interface mechanical stability it is therefore fundamental to take into consideration the properties of bone to change itself in time as a consequence of a "functional adaptation" mechanism which is known as bone remodelling: the morphology of bone is firstly established by genetic factors, and afterward the bone goes through dynamic shape and density optimisation (self remodelling) to adapt its mechanical properties and its structural behaviour to the applied loads (Lanyon et al. 1992).

As a consequence, it follows that the bone-prosthesis interface is a dynamic phenomenon, showing evolution in time according to the response of the tissue to external stimuli of mechanical and chemical nature. In the short time the problem concerns the creation of the interface itself, which depends upon the absence of relative motion between prosthesis and bone and the existence of mechanical or chemical sites for bone anchorage on the implant surface. After the osteointegration of the prosthesis has been achieved, the problem becomes the maintenance of the interface, which can be lost not only under high peaks of load, but even for the resorption of bone due to the functional adaptation to low mechanical stresses (Jacob & Huggler, 1980, Van Rietbergen et al. 1993).

To achieve and maintain an adequate osteointegration of the prosthesis, key factors are:

- the design of the implant, which the initial implant fixation depends upon: if the space between the surface of the implant and the bone is too large, a soft tissue proliferation could occur preventing a direct bone interface; the design plays a fundamental role even after osteointegration is attained, for a high concentration of stresses can result in the rupture of the interface;
- the "stress-shielding" phenomenon: the prosthesis takes, according to its major stiffness, the load which is normally taken by the bone alone, and the consequent anomalous reduction of stresses in the

surrounding bone can lead to tissue resorption instead of growth (Engh et al. 1987);

- the surface finishing and the chemical nature of the materials of the prosthesis at the interface, which can aid the bone integration (Rohlmann et al. 1988);
- the "healing period", that is the post-operative time without loading: the premature application of load to the implant during healing can destroy primary fixation causing the relative motion of the implant itself which opposes to osteointegration.

2 NUMERICAL ANALYSES

The approaches to the problem of this particular interface include both numerical analyses and experimental tests. The numerical analysis with the Finite Element Method allows to calculate the stresses at the interface due to the loads applied to the prosthesis and to predict the interface evolution consequent to bone remodelling (Huiskes et al. 1987, Cowin 1987). The numerical model requires specific elements able to modify their dimensions and elastic characteristics according to the changing of some mechanical parameters (Cowin & Hegedus 1976).

At present the prediction of the evolution of the bone-prosthesis interface based on numerical simulations is affected by some major limits, consisting in the unavailability of the following:

- a suitable model of bone, for the lack of information about the distribution of mechanical properties and constitutive laws of the tissue, and the internal and external geometry of bone segments;
- the mechanical characteristics of bone-prosthesis bounding;
- the definition of a "mechanical control variable" for the bone remodelling process and the identification of the transducers of the mechanical stimuli.

Another problem which results from the complex three-dimensional models is the great amount of time required: in qualitative analyses it's therefore common the use of bi-dimensional models, linear constitutive laws, hypothesis of isotropy for the bone tissue and simplified loads.

Two approaches are used in numerical analysis: the first one is restricted to the search of the final "condition of equilibrium", while the second, that has been only lately developed, follows the temporal evolution of the process of bone remodelling (Beauprè et al. 1990, Carter et al. 1987, Hart et al. 1984).

3 EXPERIMENTAL PROCEDURE

The experimental approach consists in the evaluation of the mechanical stability of the interface by measuring its strength in a load test: the osteointegrated prosthesis is subjected to an increasing force or torque till the rupture of the interface (Carr et al. 1995, Philips et al. 1991), allowing to estimate the effectiveness of the design of a prosthesis in promoting bone integration of the implant (Clemow et al. 1981, Hoshaw et al. 1994).

We have set up a procedure which allows to evaluate the mechanical interface between dental implants and bone. The two major aims of the research are:

a) the identification of the resistance to masticatory load through characterisation of the mechanical properties of different samples;

b) the evaluation of the influence of healing time, surgical techniques and implant surface coating with regenerative materials on bone growth around the implant, and consequently on the interface stability.

3.1 Methods

The experimental procedure consists in:

a) surgical application of dental implants - shaped as screws, cylinders or blades - in the shinbones of sheep;

b) stalling of the walking sheep and following animal sacrifices: the sacrifices of different sheep shall occur at different times from implantation when the influence of healing time on the interface stability is investigated too;

c) preparation of the test sample, which consists of a portion of the sheep bone that integrates the implanted device;

d) mounting of the sample in a special apparatus on a MTS Bionix 319.10S (MTS Minneapolis, USA) testing machine; to obtain the alignment of the sample to the load axis, a tapered abutment is screwed on the implant and mates to a taper seat on the hydraulic actuator of the machine;

e) breaking of the implant-bone interface through compression of the implant within the surrounding bone at a constant rate of 0,25 mm/min and recording of the force-displacement curve.

Preliminary tests were carried out on six samples obtained from prostheses implanted in two sheep. The prostheses differ for surface coating and technique of implantation.

3.2 Preliminary tests

In the proximal end of the shinbones of two sheep, six Astra-Tech titanium screw fixtures for dental prosthesis (Astra-Tech, Goteborg, Sweden) were inserted:

- four Tio-blast fixtures (surface blasted with particles of titanium bi-oxide) were coupled to surgically created peri-implantar circular defects of about 3 mm diameter (keeping in any case enough amount of bone to allow the primary stability of the implant). After the insertion of the fixture within the

Table 1. Samples and results of compressive tests

	Days	Notes	Load (N)
P3 A	24	standard, no defect	**2200**
P3 B	24	Tio-blast with defect	**370**
P3 C	24	Tio-blast with defect	**420**
P4 A	45	Tio-blast with Vycril	**2400**
P4 B	45	Tio-blast with Vycril	**950**
P4 C	45	Tio-blast with Vycril	**2900**

surgical defect, the gap between the outer surface of the implant and the surrounding healthy bone was filled up with deantigenate bovine bone (Bio-Oss, Geisthich) and covered with an absorbable membrane in Poliglactin (Vycril, Johnson & Johnson);

- a Tio-blast fixture and a standard fixture in titanium with machined surface (for comparison), were both inserted without creation of defects.

The aim of the experiment was to compare the natural and the artificially driven bone regeneration, investigating in detail the influence on the interface strength of the following parameters:

a) healing time;
b) surface coating of the fixture (Tio-blasting);
c) presence of surgical bone defects filled in with Bio-Oss.

The first sheep was sacrificed 24 days after implantation, and the second one 45 days after.

For testing, six samples were prepared (Table 1):

a) from sheep P3 sacrificed at 24 days:
P3 A: standard fixture without defect
P3 B: Tio-blast fixture with defect
P3 C: Tio-blast fixture without defect

b) from sheep P4 sacrificed at 45 days:
P4 A: Tio-blast fixture with Vycril
P4 B: Tio-blast fixture with Vycril
P4 C: Tio-blast fixture with Vycril

3.3 Results

Compression tests as described in Methods were accomplished.

The maximum load measured for each sample, that correspond to the failure of its interface, is reported in Table 1.

Data in Table 1 show that by increasing healing time from 24 to 45 days, the mechanical strength of the interface raises from 10 to 30% (the different strength of samples P4 C and P4 A may be ascribed to the difference in bone density of the sites of insertion of the respective fixtures), for it has allowed the calcification of a larger amount of bone (Figure 1); unfortunately, it has not been possible to evaluate the effects of the surface coating of the implant because the threads of the Tio-blasted fixtures of samples P3 B and P3 C were partially bared during sample preparation.

The fixture of P4 B was implanted in a region of

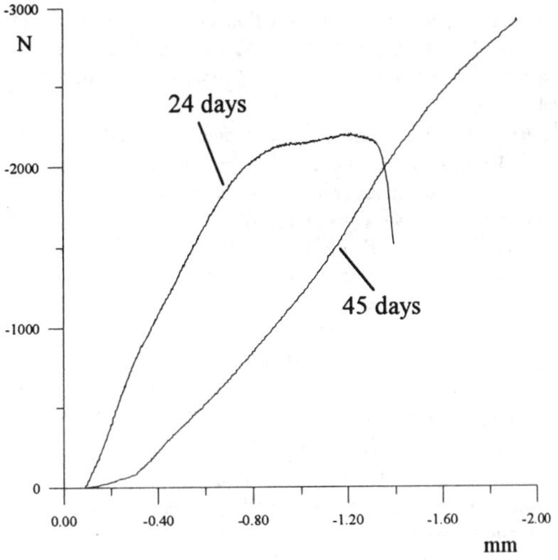

Figure 1. Load-displacement curves versus healing time and bone density of the site of insertion

thin bone thickness and in consequence its anterior threads did not grip cancellous bone, but emerged in the medullary channel of the shinbone, reducing in consequence the extension of the interface.

4 DISCUSSION

1. The experimental procedure allows to compare the influence of several parameters, such as healing time, surface coating and presence of surgical defects, on the bone integration of implanted fixtures; a microradiographic analysis of the interface between the implants and bone has been accomplished with corroboration of the results of mechanical tests.

2. It is possible to evaluate the masticatory load resistance of the interface versus healing time: this information allows to estimate the necessary time before loading the implant.

3. Two crucial steps in the procedure are the implantation of the fixture and the sample preparation: the site for the fixture must be properly chosen both for density and thickness of bone, which shall be uniform for all the tested fixtures, and particular attention must be paid, during the preparation of the samples, in avoiding to uncover the device.

REFERENCES

Beauprè, G.S., T.E. Orr, D.R. Garter & D.J. Churman 199. Computer predictions of bone remodelling around porous-coated implants. *J. Arthropl.* 5-3:191-200.

Brånemark, P.I. 1985. Introduction to osseointegration. In Brånemark, P.I., G. Zarb & T. Albrektsson (eds.), *Tissue-integrated prostheses*:11-76. Chicago: Quintessence Publishing.

Carr, A.B., P.E. Larsen, E. Papazoglou & E. McGlumphy 1995. Reverse torque failure of screw shaped implants in baboons: baseline data for abutment torque application. *Int. J. Oral Maxillofac. Implants* 10(2):167-174.

Carter, D.R., D.P. Fyhrie & R.T. Whalen 1987 Trabecular bone density and loading history: regulation of connective tissue biology by mechanical energy. *Journal of Biomechanics* 20: 1111-1120.

Clemow, A.J.T., A.M. Weinstein, J.J. Clawitter, J. Coeneman & J. Anderson 1981. Interface mechanics of porous titanium implants. *Journal of Biomed. Mat. Res.* 15:77-82.

Cowin, S.C. & D.H. Hegedus 1976. Bone remodelling I: Theory of adaptive elasticity. *J. Elasticity* 6:613-626.

Cowin, S.C. 1987. Bone remodelling of diaphysial surfaces by torsional loads: theoretical predictions. *J. Biomechanics* 20:1111-1120.

Engh, C.A., .J.D. Bobyn, & A.M. Glassman 1987. Porous coated hip replacement: the factors governing bone ingrowth, stress shielding and clinical results. *J. Bone Jt. Surgery* 69B:45-51.

Hart R.T., D.T. Davy & K.G. Heiple 1984. Mathematical modelling and numerical solutions for functionally dependent bone remodelling. *Calcif. Tissue Int.* 36: S104-S109.

Hoshan, S.T. J.B. Brunsky & G.V.B. Cochran 1994. Mechanical loading of Brånemark implants affects interfacial bone modelling and remodelling. *Int. J. Oral Maxillofac. Implants* 9:345-360.

Huiskes, R., H. Weinans, H.J. Grootenboer, M. Dalstra, B. Fudala & T.J. Sloff 1987. Adaptive bone-remodelling theory applied to a prosthetic-design analysis. *J. Biomechanics* 20:1135-1150.

Jacob, H.A.C. &. A.H. Huggler 1980. Biomechanical causes of prosthesis stem loosening. *J. Biomechanics* 20:159-173.

Lanyon, L.E., A.E. Goodship, C.J. Pye & J.H. McFie 1982 Mechanically adaptive bone remodelling. *J. Biomechanics* 15:141-154.

Philips, T.W., L.T. Nguyen & S.D. Munro 1991. Loosening of cementless femoral stems. A Biomechanical analysis of immediate fixation with loading vertical, femur horizontal. *J. Biomechanics* 24:37-48.

Rohlmann, A., E.J. Cheal W.C., Hayes & G. Bergmann 1988. A nonlinear finite element analysis of interface conditions in porous coated hip endoprostheses. *J. Biomechanics* 21:605-611.

Van Rietbergen, B., R. Huiskes, H. Weinans, D.R. Sumner, T.H. Turner & O.J. Galante 1993. The mechanism of bone remodelling and resorption around press-fitted THA stems. *JJ. Biomechanics* 26:369-382.

Damage and Failure of Interfaces, Rossmanith (ed.) © *1997 Balkema, Rotterdam, ISBN 90 5410 899 1*

A mechanical constitutive model for rock and concrete joints under cyclic loading

P. Divoux
Laboratoire 3S, Université Joseph Fourier, Grenoble & Electricité de France/National Hydro Engineering Center, Savoie Technolac, Le Bourget du Lac, France

M. Boulon
Laboratoire 3S, Université Joseph Fourier, Grenoble, France

E. Bourdarot
Electricité de France/National Hydro Engineering Center, Savoie Technolac, Le Bourget du Lac, France

ABSTRACT: A constitutive model for the description of the behaviour of rock joints taking into account observations made on experimental cyclic tests is presented. Two types of tests have been studied in particular: *cyclic shear tests under constant normal load* and *cyclic normal compression tests*, which are the most frequently performed on rock joints. The model relates the normal behaviour of a joint to its tangential and dilatant behaviour. In order to remain efficient in finite element computations, the model has to give an accurate response to all loading processes, either monotonous or cyclic, normal and tangential. The possibility of taking into account contact surface degradation or joint damage makes the model more accurate under cyclic or seismic loading. Damage is a function of the tangential work and of the normal compression work. Their mathematical definition is mentioned in the paper. These two state variables are sensitive to all kinds of loading and seem very accurate for evaluating the contact surface degradation. Experimental and numerical tests have been carried out on different types of rock, different types of joint and following a large range of loading paths.

1. INTRODUCTION

The constitutive model presented in this paper has been developed in order to model the behaviour of rock joints from the analysis of experimental test results.

The main purpose was to make a model able to reproduce the essential characteristics of joints observed on experimental test results taking into account the changes in their behaviour under cyclic or dynamic loading. In order to be used in finite element codes, an accurate prediction must be obtainable for any type of loading path. All the parameters have to be independent from each other and be easily determined from experimental results.

The main characteristics of the behaviour of rock joints have been studied by analysing the experimental investigations of Barton (1973 and 1976), Bandis & al (1981), Huston & al (1990, Leichnitz (1985), Jing & al (1992 and 1995), Benjelloun (1991).

A large amount of joint modelling work is available in the literature. Goodman & al (1968 and 1972), Ghaboussi & al (1973), Ladanyi & al (1970), Plesha (1987 and 1995), Barton & al (1985), Saeb & al (1992), Souley & al (1995), and Boulon (1988 and 1995) are some of the investigators who have derived the basic equations describing joint behaviour. Their numerical investigations have been the basis of our research, which has lead to the conception of the model proposed here.

2. RELATIVE NORMAL DISPLACEMENT DECOMPOSITION

Figure 1 shows the decomposition into two parts of the relative normal displacement δ_n.

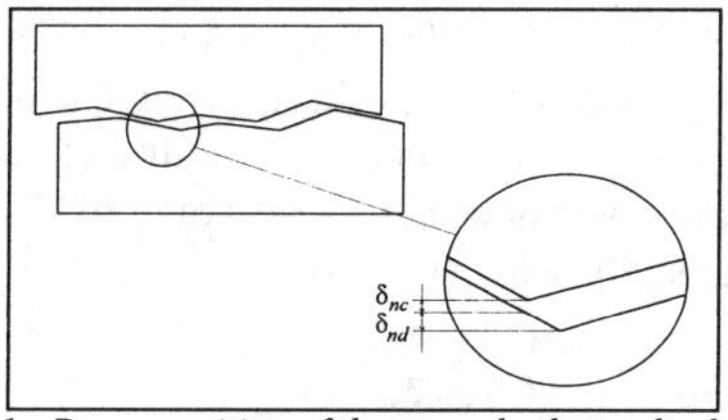

Figure 1 - Decomposition of the normal relative displacement

δ_n is considered to be the sum of the joint dilatancy δ_{nd} and the joint compression δ_{nc}.

$$\delta_n = \delta_{nc} + \delta_{nd} \qquad (1)$$

- when performing a normal compression test, δ_{nd} remains constant,
- when performing a cyclic shear test under a constant normal stress, δ_{nc} remains constant.

3. STATE VARIABLES

Cyclic experimental tests on rock joints show some damage when the joint is mechanically perturbed: the peak shear strength, the dilatancy angle, and the joint thickness usually decrease. To quantify this damage, the two state variables W_s and W_n have been introduced into the model. The tangential work W_s and the normal compression work W_n are defined in the following.

3.1. Definition of the tangential work W_s

W_s, represented on figure 2 is defined incrementally as,

$$dW_s = \sigma_s . d\delta_s \qquad (2)$$

The state variable W_s is very similar to the plastic tangential work proposed by Plesha (1987) to model the degradation of the tangential properties of the joint.

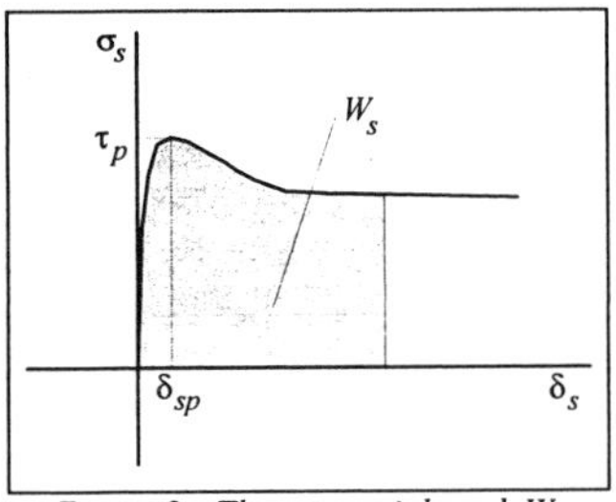

Figure 2 - The tangential work W_s

However, the model presented here is incrementally non-linear, and the tangential work cannot be divided into the two parts, elastic and plastic.

3.2. Definition of the normal compression work W_n

W_n is the sum of the positive increment of normal work. It is the energy that has been used to compress the joint. The mathematical definition of W_n is given incrementally as:

$$\begin{cases} dW_n = \sigma_n . d\delta_{nc} & \text{if } \sigma_n . d\delta_{nc} > 0 \\ dW_n = 0 & \text{if } \sigma_n . d\delta_{nc} \leq 0 \end{cases} \qquad (3)$$

Figure 3 shows W_n as a cyclic compression test is performed.

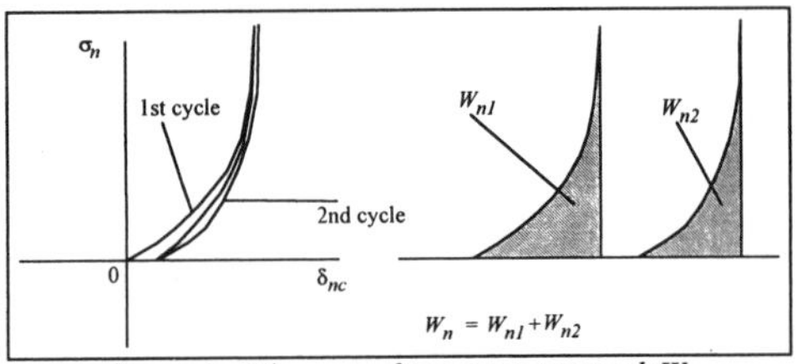

Figure 3 - The normal compression work W_n

4. TANGENTIAL BEHAVIOUR

In the model, the tangential behaviour of the joint is determined from the results of cyclic shear tests under constant normal stress by analysing the curves $\sigma_s = f(\delta_s)$ obtained.

Figure 4 shows a typical result. Experimental results of such tests are given by Huston (1990).

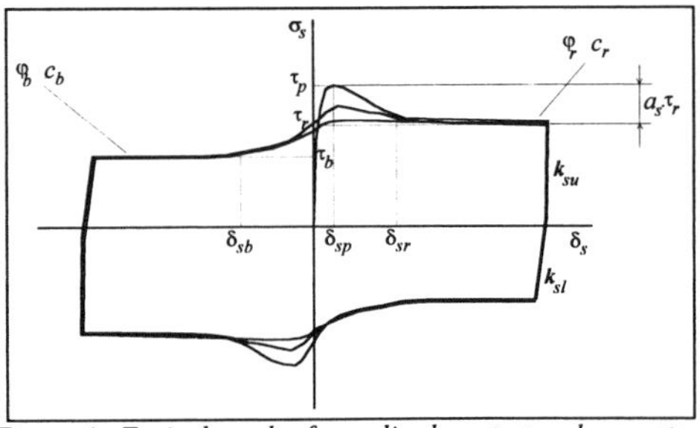

Figure 4 - Typical result of a cyclic shear test under constant normal stress

4.1. Main characteristics

- The stress increment $d\sigma_s$ is different when loading or unloading. It is necessary to define a loading parameter and to use an incrementally non-linear model to predict the tangential behaviour of the joints.
- There is a non-linear failure criterion (shear stress peak, residual stresses, etc.),
- The residual stress τ_b measured before the peak is usually different from the post-peak residual stress τ_r. The joint is usually contractant before the peak and dilatant after the peak.
- The residual stresses τ_b and τ_r depend on σ_n and remain constant when cycling.
- The peak shear stress τ_p decreases when cycling.
- The tangential stiffness k_{sl} when loading is usually different from the tangential stiffness k_{su} when unloading.

4.2. Definition of a loading parameter R_l

A loading parameter R_l has been chosen to model the change in the tangential stress increment $d\sigma_s$

observed in the results of cyclic shear tests when loading or unloading.

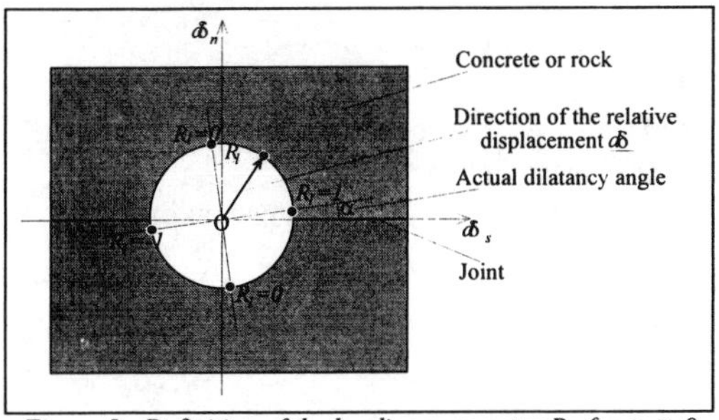

Figure 5 - Definition of the loading parameter R_l for $\sigma_s > 0$

We propose the following mathematical formulation of R_l:

$$\begin{cases} R_l = +\dfrac{d\delta_s}{\|d\delta\|}.\cos\alpha + \dfrac{d\delta_n}{\|d\delta\|}.\sin\alpha & \text{si } \sigma_s \geq 0 \\ R_l = -\dfrac{d\delta_s}{\|d\delta\|}.\cos\alpha - \dfrac{d\delta_n}{\|d\delta\|}.\sin\alpha & \text{si } \sigma_s < 0 \\ R_l = 0 & \text{si } \sigma_s = 0 \end{cases} \qquad (*)$$

where $\tan\alpha = \dfrac{\partial\delta_{nd}}{\partial\delta_s}$

When performing a shear test under a constant normal stress :

- $R_l = 1$ when loading,
- $R_l = -1$ when unloading.

Thus, the experimental results of a shear test under constant normal stress give the behaviour of the joint for $R_l = 1$ (loading phases) and $R_l = -1$ (unloading phases).

4.3. The tangential behaviour when loading *($R_l = 1$)*

When loading, the shear stress changes as a function of a failure shear stress σ_{sr} and a « pseudo-elastic » behaviour:

$$d\sigma_{sl} = \left(1 - \frac{\sigma_s}{\sigma_{sr}}\right)^{\gamma} d\sigma_{se} + \left(\frac{\sigma_s}{\sigma_{sr}}\right)^{\gamma} d\sigma_{sr} \qquad (7)$$

where γ controls the failure of the joint, $d\sigma_{se}$ is a « pseudo elastic » stress increment; σ_{sr} and its increment $d\sigma_{sr}$ are determined from the failure criterion equation.

For a high value of γ, the failure is fragile. We propose γ=5 as a usual value for rock and concrete joints. For other type of joints, γ can be reduced to model a more ductile failure.

<u>The « pseudo-elastic » stress increment $d\sigma_{se}$:</u>

$$d\sigma_{se} = k_{sl}.d\delta_s \qquad (4)$$

where k_{sl} is the loading tangential stiffness.

<u>Equation of the failure criterion, σ_{sr} and determination of the stress increment $d\sigma_{sr}$:</u>

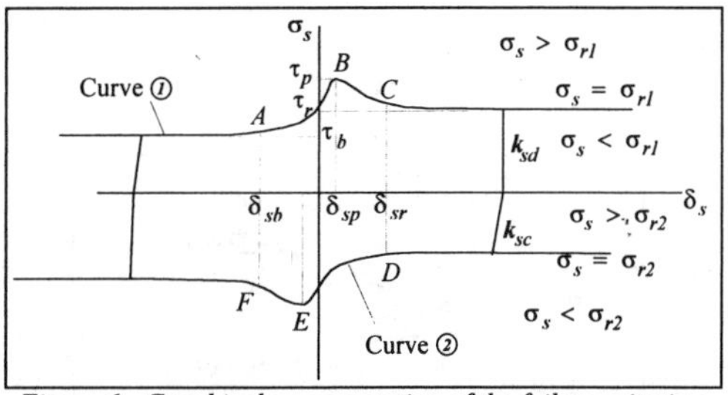

Figure 6 - Graphical representation of the failure criterion.

The failure criterion is represented by curves ① and ② on figure 6.

$$\begin{cases} \sigma_{sr} = \sigma_{r1} & \text{if } \sigma_s \geq 0 \\ \sigma_{sr} = \sigma_{r2} & \text{if } \sigma_s < 0 \end{cases} \qquad (5)$$

Equation of ①:

$$\begin{cases} \sigma_{r1} = +\tau_b - (\tau_b - \tau_p).10^{-\frac{(\delta_{sp}-\delta_s)^{\alpha}}{(\delta_{sp}-\delta_{sb})^{\alpha}}} & \text{si } \delta_s \leq \delta_{sp} \\ \sigma_{r1} = +\tau_r - (\tau_r - \tau_p).10^{-\frac{(\delta_s-\delta_{sp})^{\alpha}}{(\delta_{sr}-\delta_{sp})^{\alpha}}} & \text{si } \delta_s > \delta_{sp} \end{cases} \qquad (6)$$

Equation of ②:

$$\begin{cases} \sigma_{r2} = -\tau_b + (\tau_b - \tau_p).10^{-\frac{(\delta_{sp}+\delta_s)^{\alpha}}{(\delta_{sp}-\delta_{sb})^{\alpha}}} & \text{si } \delta_s > -\delta_{sp} \\ \sigma_{r2} = -\tau_r + (\tau_r - \tau_p).10^{-\frac{(-\delta_s-\delta_{sp})^{\alpha}}{(\delta_{sr}-\delta_{sp})^{\alpha}}} & \text{si } \delta_s \leq -\delta_{sp} \end{cases} \qquad (7)$$

The shape of the curve is a function of α. Figure 7 shows curve ① for different values of α.

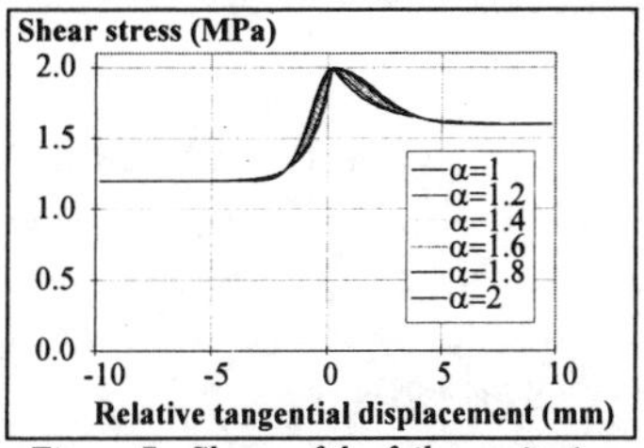

Figure 7 - Shape of the failure criterion

δ_{sb} and δ_{sr} are two parameters of the law, the changes in τ_b, τ_r, τ_p and δ_{sp} with the loading is given in the following. We propose α=1.4 as a typical value for rock and concrete joints.

Residual shear stresses τ_b, τ_r

A hyperbolic criterion shown on figure 8 is used to determine the dependency of the residual stresses with σ_n

$$\tau_r = \sqrt{\sigma_n^2.\tan^2\varphi_r - 2.\frac{c_r}{\tan\varphi_r}.\sigma_n}$$
$$\tau_b = \sqrt{\sigma_n^2.\tan^2\varphi_b - 2.\frac{c_b}{\tan\varphi_b}.\sigma_n} \quad (8)$$

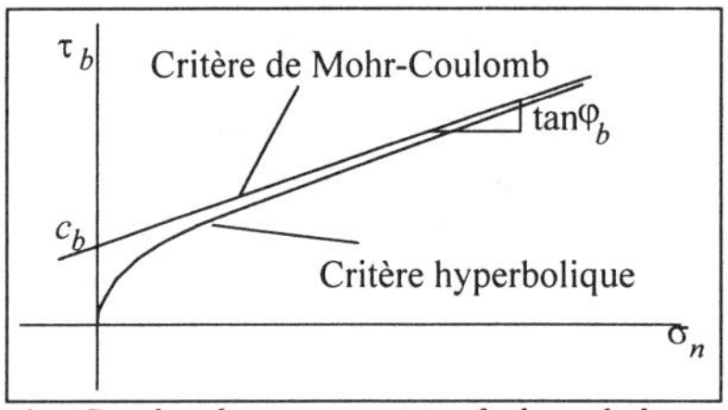

Figure 8 - Graphical representation of a hyperbolic criterion.

The friction angles φ_r and φ_b and the cohesions c_r and c_b are the four parameters used to model the changes in τ_b and τ_r.

Peak shear stress τ_p and associated relative tangential displacement δ_{sp}

τ_p and δ_{sp} are determined by the following set of equations:

$$\begin{cases} \tau_p = (1+a_s).\tau_r \text{ with } a_s = a_{s0}e^{-c_s.W_s} \\ \delta_{sp} = 2.\dfrac{\tau_p}{k_{sl}} \end{cases} \quad (9)$$

where $1+a_{s0}$ is the initial ratio between τ_p and τ_r, and c_s is a coefficient which controls the decrease of the shear stress peak W_s.

4.4. Tangential behaviour when unloading *(R_l = -1)*

When unloading, the tangential behaviour of the joint is considered to be « pseudo elastic ».

$$d\sigma_{su} = k_{su}.d\delta_s \quad (10)$$

4.5. Tangential behaviour for any value of R_l

The tangential stress increment is determined from the value of the unloading stress increment $d\sigma_{su}$ (if R_l=-1), the loading stress increment $d\sigma_{sl}$ (if R_l=1) and the loading parameter R_l.

$$d\sigma_s = \frac{1+R_l}{2}.d\sigma_{sl} + \frac{1-R_l}{2}.d\sigma_{su} \quad (11)$$

5. DILATANCY

The dilatancy of a joint can be evaluated by analysing the curve $\delta_{nd} = f(\delta_s)$ of cyclic shear tests under constant normal stress.

A typical curve $\delta_{nd} = f(\delta_s)$ is shown on figure 9.

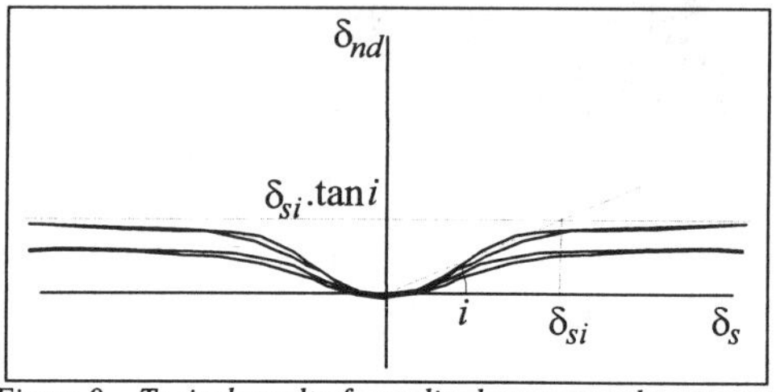

Figure 9 - Typical result of a cyclic shear test under constant normal stress - Dilatancy

5.1. Main characteristics

- The roughness of rock joints usually makes them dilatant.
- The relative normal displacement δ_{nd} reaches a maximum for large relative tangential displacements.
- When cycling, the average dilatancy angle *i* decreases.
- For a zero relative displacement δ_{nd}, the relative normal displacement vanishes.

5.2. Modelling

Equation (12) is the mathematical formulation of the dilatancy curve typically shown on figure 9.

$$\delta_{nd} = \delta_{si}.\tan i.\left(1 - 10^{-(\delta_s/\delta_{si})^\beta}\right) \quad (12)$$

where i is the average dilatancy angle, δ_{si} is the relative tangential displacement such that the maximum normal relative displacement is equal to $\delta_{si}.\tan i$.

Figure 10 shows the shape of the curve for different values of β.

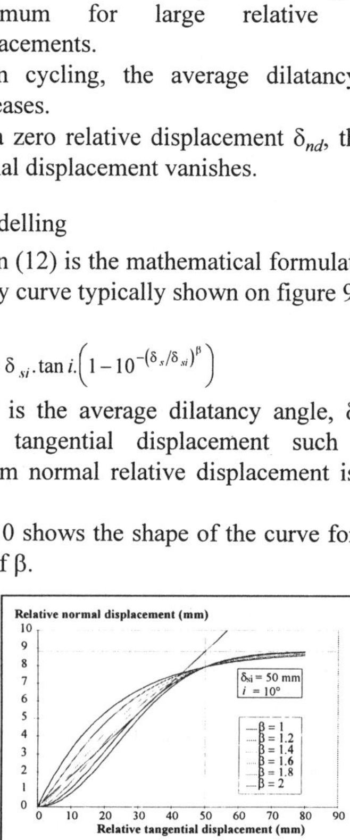

Figure 10 - Influence of β on the dilatancy

The value of $\beta = 1.3$ has been retained for rock and concrete joints.
To model the damage, the average dilatancy angle i is considered to be a function of σ_n and W_s:

$$i = i_0 . \frac{\sigma_{n0}}{\sigma_n + \sigma_{n0}} . 2^{-W_s / W_{s0}} \qquad (13)$$

where i_0 is the initial average dilatancy angle of the joint, σ_{n0} is the normal stress at the level at which the average dilatancy angle is halved, W_{s0} is the tangential work at the level at which the average dilatancy angle is halved.
This formulation of the damage modelled from the relation between the dilatancy angle and the normal stress is different from the one presented by Plesha (1987). The formulation proposed is a result of the analysis of the experimental shear tests under constant normal stiffness performed by Benjelloun (1991) on a granite.

6. NORMAL BEHAVIOUR

The normal behaviour of the joint is determined from the analysis of the results of experimental compression tests.
Results of normal compression tests are given in the graph $\delta_{nc} = f(\sigma_n)$.

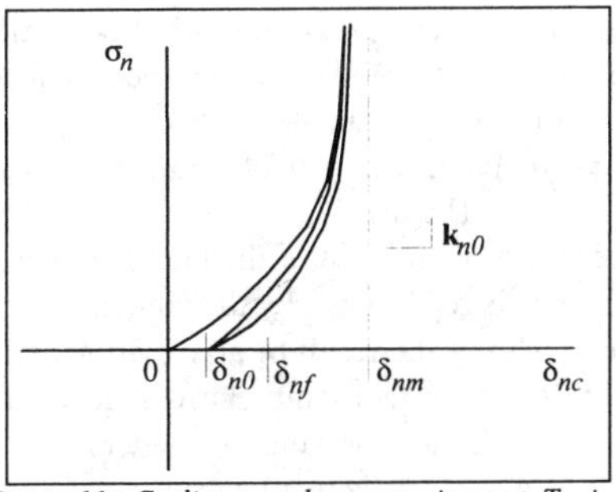

Figure 11 - Cyclic normal compression test. Typical result.

6.1. Main characteristics of normal behaviour

- The joint is supposed to have no tensile strength.
- Under compressive normal stress, the Bandis & al (1981) hyperbolic model is proposed to describe the behaviour of the joint.
- When cycling, the joint thickness decreases.

6.2. Mathematical formulation

The normal behaviour is modelled with the following set of equations.

$$\begin{cases} \sigma_n = k_{n0} \dfrac{(\delta_{nc} - \delta_{n0})(\delta_{nm} + \delta_{n0})}{(\delta_{nm} + \delta_{nc})} \\ \text{and } \delta_{n0} = \delta_{nf} . \left(1 - e^{-c_n W_n}\right) \end{cases} \qquad (14)$$

δ_{nm} is the minimum relative normal displacement, δ_{nf} is the relative normal displacement at which the normal stress vanishes, k_{n0} is the initial normal stiffness, c_n controls the decrease of the joint thickness δ_{nf} - δ_{nm} when the normal compression work W_n increases.

7. DETERMINATION OF PARAMETERS, SIMPLIFICATION OF THE MODEL

The incremental formulation of the model can be determined from the constitutive equations of the model given above. The law has been developed in order to be introduced into the finite element code GEFDYN. A law driver has also been developed to model experimental shear and compression tests on joints and to determine the parameters of the joint. It has also been used to analyse the incremental responses.
All the parameters are evaluated from the results of cyclic compression tests and cyclic shear tests under constant normal stress. These are the two tests most

Table 1 - Parameters of the model

Tangential behaviour		
k_{sl}	Pa/m	Loading normal stiffness
k_{su}	Pa/m	Unloading normal stiffness
φ_r	°	Post-peak residual friction angle
c_r	Pa	Post-peak residual cohesion
δ_{sr}	m	Post-peak failure relative tangential displacement
φ_b	°	Post-peak residual friction angle
c_b	Pa	Pre-peak residual cohesion
δ_{sb}	m	Pre-peak failure relative tangential displacement
a_{s0}	None	Initial ratio between the peak shear stress and the residual shear stress
c_s	$(Pa.m)^{-1}$	Damage parameter which controls the decrease in the peak shear stress
Dilatancy		
i_0	°	Average dilatancy angle under low normal stress
δ_{si}	m	Relative tangential displacement which controls the diminution of the dilatancy angle
σ_{n0}	N/m^2	Normal stress under which the average dilatancy angle is half i_0
W_{s0}	Pa.m	Damage parameter Tangential work after which the dilatancy angle is half i_0
Normal behaviour		
k_{n0}	Pa/m	Initial normal stiffness
δ_{nm}	m	Max. relative normal displacement
δ_{nf}	m	Max. decrease of the joint thickness
c_n	$(Pa.m)^{-1}$	Damage parameter which control the decrease in the joint thickness

frequently performed on joints. They are all independent of one another and they control all the phenomena described above.
In its more complete form, the model is controlled by 18 parameters (table 1) : 10 to model the tangential behaviour, 4 to model the dilatancy and 4 to model the normal behaviour of the joint.

The model can also easily be simplified. For example, only 10 parameters are used to model monotonic behaviours. In its simplest form, and considering the formulation given in table 2, we obtained a six-parameter model.

Table 2 - Simplification of the model

	Parameters
Tangential behaviour	$k_{sc}= k_{sd}$
	$\varphi_r= \varphi_b$
	$c_r= c_b$
	$a_{s0}= 0$
	no influence of δ_{sr} , δ_{sb} and c_s
Dilatancy	i_0
	δ_{si}
	$\sigma_{n0} \rightarrow\infty$
	$W_{s0} \rightarrow\infty$
Normal behaviour	k_{n0}
	$\delta_{nm} \rightarrow\infty$
	$\delta_{nf}= 0$
	no influence of c_n

The six parameters are k_{sc}, φ_r, c_r, i_0, δ_{si} and k_{ni}. The main characteristics of such a model are:

- No tensile strength.
- A constant normal stiffness in compression.
- The failure criterion is defined with a friction angle and a cohesion.
- No tangential stress peak.
- No damage (decrease in the dilation angle, decrease in the peak shear stress or decrease in the thickness).

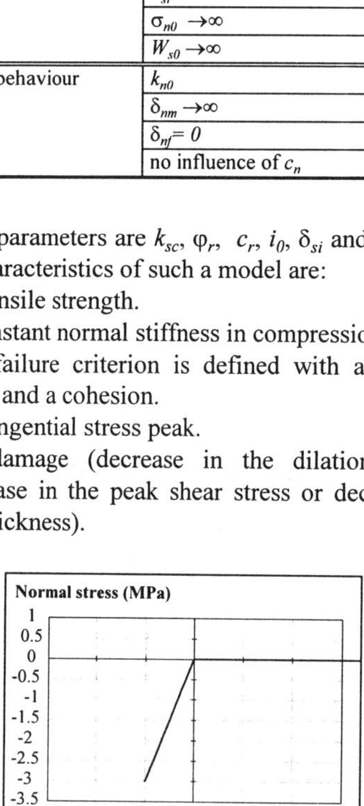

Figure 12 - Results of a cyclic normal compression with a maximum compression stress equal to 3 MPa - Simplified model

Figure 12 and 13 show the results of a cyclic normal compression test and a cyclic shear test under constant normal stress with the set of the six following parameters: $k_{sl} = 10^{10}$ Pa.m, $\varphi_r = 37°$, $c_r = 10^5$ Pa, $i_0 = 10°$, $\delta_{si} = 20$ mm, $k_{n0} = 1.5\text{x}10^{10}$ Pa.m.

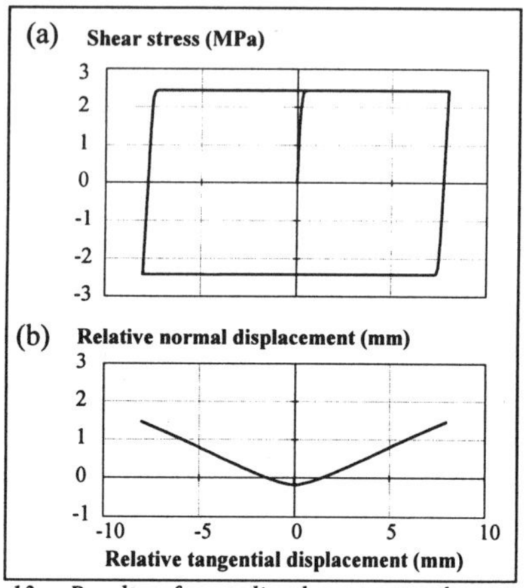

Figure 13 - Results of a cyclic shear test under a constant normal stress equal to 3 MPa - Simplified model
(a) - Tangential behaviour, curve $\sigma_s=f(\delta_s)$
(b) - Dilatancy, curve $\delta_n=f(\delta_s)$

8. VALIDATION

A cyclic normal compression test has been modelled in figure 14. The modelling has been carried out with the following parameters:
$k_{n0} = 6\text{x}10^9$ Pa.m, $\delta_{nm} = 0.245$ mm, $\delta_{nf} = 0.12$ mm and $c_n= 3\text{x}10^{-3}$ $(\text{Pa.m})^{-1}$.
The decrease of the joint's thickness shown by the experiments is reproduced by the model.
Figure 15 shows the result of an experimental cyclic shear test under constant normal load and its modelling with the following parameters:
$k_{sl} = 10^{10}$ Pa.m, $k_{su} = 2\text{x}10^{10}$ Pa.m,
$\varphi_r = 38.7°$, $c_r = 0$ Pa, $\delta_{sr}= 0.02$ m, $\varphi_b= 33°$, $c_b = 0$ Pa, $\delta_{sb}= -2.2$ mm, $a_s = 0.5$, $c_s = 10^{-5}$ $(\text{Pa.m})^{-1}$, $i_0 = 10°$, $\delta_{si} = 12$ mm, $\sigma_{n0} = 10^9$ Pa, $W_{s0}= 2\text{x}10^5$ Pa.m.
In this example, one can visualise the capability of the model to describe a peak shear stress and how it changes, the post-peak and pre-peak behaviours and the decrease of the average dilatancy angle.
Figure 16 shows the stress responses of small relative displacement loadings. These computations have been done at certain steps of the first cycle of the shear test shown on figure 15.
The continuity of the responses can be observed and the ability of the model to follow any displacement path can be demonstrated. When loading near the

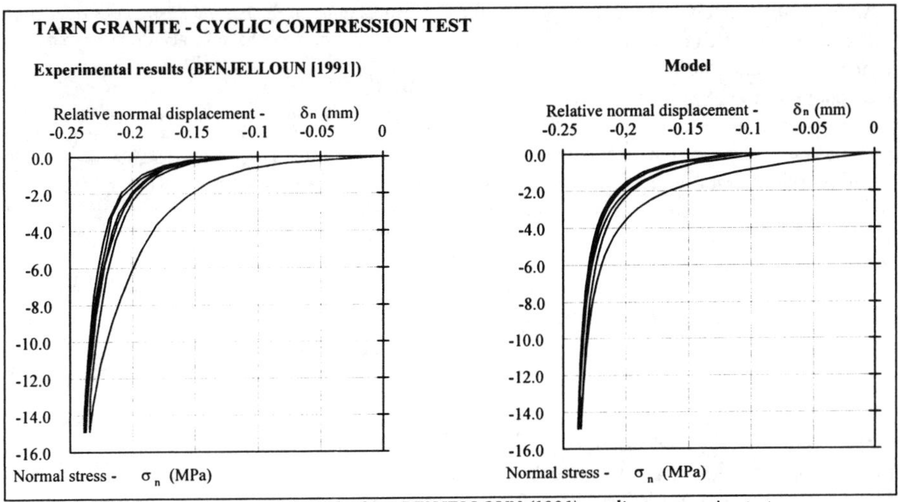

Figure 14 - Experimental results from BENJELLOUN (1991), cyclic compression test on a granite specimen.

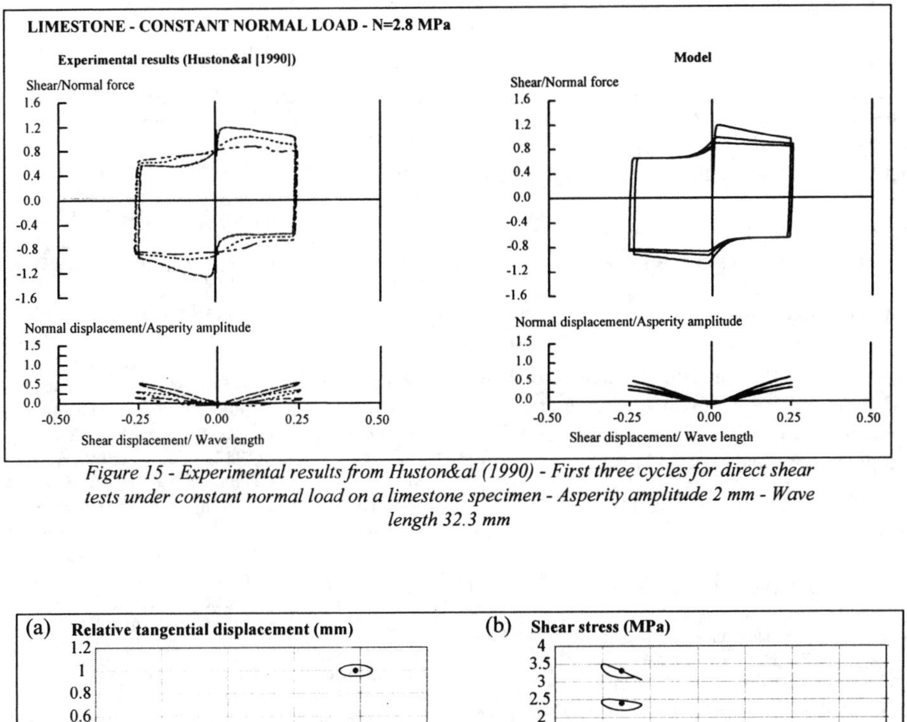

Figure 15 - Experimental results from Huston&al (1990) - First three cycles for direct shear tests under constant normal load on a limestone specimen - Asperity amplitude 2 mm - Wave length 32.3 mm

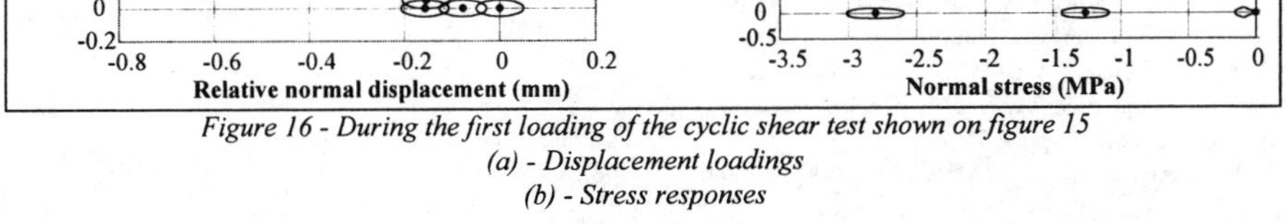

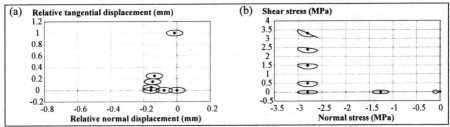

Figure 16 - During the first loading of the cyclic shear test shown on figure 15
(a) - Displacement loadings
(b) - Stress responses

failure surface, the stress response follows the hyperbolic criterion but still remains continuous.

Some other validation studies of the model have been done using Benjelloun's (1991) experimental results on Gueret granite specimens. The set of parameters has been determined from the results of normal compression tests and shear tests under constant normal stress. Then, shear tests under constant normal stiffness have been modelled in order to demonstrate the ability of the model to relate the normal behaviour to the tangential behaviour with different types of loading paths.

9. CONCLUSION

The model presented in this paper has been developed using the results of experimental investigations performed on rock joints. The two main characteristics of the proposed incrementally non-linear model are:

- The decomposition of the relative normal displacement into two parts in order to relate the tangential behaviour of the joint to its normal behaviour.
- The modelling of the cyclic or dynamic behaviour of joints taking into account damage and how it changes with the loading. The damage is a function of the energy that has been used during the loading.

The parameters are determined from experimental cyclic shear and compression tests. A particular set of parameters can be used to invoke a simpler model and to perform preliminary studies.

Numerical tests have been carried out on different types of rock (granite, limestone, gneiss), different types of joints and following a wide range of loading displacement paths. A good match was found between predictions and experiments.

REFERENCES

BANDIS S., LUMSDEN A. C., BARTON N., 1981, *" Experimental studies of scale effects on the shear behaviour of rock joints. "*, Int. jour. of Rock Mechanics and Mining Science & Geomech. Abstr., Vol. 18, N°1, pp. 1-21

BANDIS S., LUMSDEN A. C., BARTON N., 1983, *" Fundamentals of rock joint deformation. "*, Int. jour. of Rock Mechanics and Mining Science & Geomech. Abstr., Vol. 20, N°6, pp. 249-268

BARTON N., 1973, *" Review of a new shear-strength criterion for rock joints. "*, Engineering Geology, Vol. 7, pp. 287-332

BARTON N., CHOUBEY V., 1976, *" The shear strength of rock joints in theory and practice. "*, Rock Mechanics and Rock Engineering, Vol. 10, pp 1-54

BARTON N., BANDIS S., K. BAKHTAR, 1985, *" Strength, deformation and conductivity coupling of rock joints. "*, Int. jour. of Rock Mechanics and Mining Science & Geomech. Abstr., Vol. 22, N° 3, pp. 121-140

BENJELLOUN Z. H., 1991, *"Etude expérimentale et modélisation du comportement hydromécanique des joints rocheux. "*, Thèse de l'Université de Grenoble, France

BOULON M., 1988, *"Contribution à la mécanique des interfaces sols-structures. Application au frottement lattéral des pieux. "*, Mémoire d'habilitation, Université de Grenoble, France

BOULON M., 1995, *"Soil-structure interaction: FEM computations. "*, Studies in applied mechanics 42, Mechanics of geomaterial interfaces, ELSEVIER, pp. 147-171

GHABOUSSI J., WILSON E. L., ISENBERG J., 1973, *" Finite element for rock joints and interfaces. "*, Journal of the soil mechanics and foundations division, ASCE, Vol. 99, SM 10, pp. 833-848

GOODMAN R. E., TAYLOR R. L., BREKKE T. L., 1968, *" A model for the mechanics of jointed rock. "*, Journal of the soil mechanics and foundations division, ASCE, Vol. 94, SM 3, pp. 637-659

GOODMAN R. E., DUBOIS J., 1972, *" Duplication of dilatancy in analysis of jointed rocks. "*, Journal of the soil mechanics and foundations division, ASCE, Vol. 98, SM 4, pp. 399-422

HUSTON R. W., DOWDING C. H., 1990, *"Joint asperity degradation during cycling shear. "*, Int. jour. of Rock Mechanics and Mining Science & Geomech. Abstr., Vol. 27, N°2, pp. 109-119

JING L., NORDLUND E., STEPHANSSON O., 1992, *" An experimental study on the anisotropy and stress-dependency of the strength and deformability of rock joints. "*, Int. Jour. Rock Mech. Min. Sci. & Geomech. Abstr., Vol 29, N° 6, pp. 535-542

JING L., STEPHANSSON O., 1995, *"Mechanics of rock joints: Experimental Aspects "*, Studies in applied mechanics 42, Mechanics of geomaterial interfaces, ELSEVIER, pp. 317-342

LADANYI B., ARCHAMBAULT G., 1970, *"Simulation of the shear behavior of a jointed rock mass. "*, Proc. 11th Symp. on Rock Mechanics, pp. 105-125, Berkeley

LEICHNITZ W., 1985, *" Mechanical properties of rock joints. "*, Int. jour. of Rock Mechanics and Mining Science & Geomech. Abstr., Vol. 22, N°5, pp. 313-321

PLESHA M. E., 1987, *" Constitutive models for rock discontinuities with dilatancy and surface degradation. "*, Int. Jour. for Num. and Anal. Methods in Geomech., Vol. 11, pp. 345-362

PLESHA M. E., 1995, *" Rock joints: Theory, constitutive equations. "*, Studies in applied mechanics 42, Mechanics of geomaterial interfaces, ELSEVIER, pp. 375-394

SAEB S., AMADEI B., 1992, *" Modelling rock joints under shear and normal loading. "*, Int. jour. of Rock Mechanics and Mining Science & Geomech. Abstr., Vol. 29, N° 3, pp. 267-278

SOULEY M., HOMAND F., AMADEI B., 1995, *"An extension to the Saeb and Amadei constitutive model for rock joints to include cyclic loading paths. "*, Int. jour. of Rock Mechanics and Mining Science & Geomech. Abstr., Vol. 32, N° 2, pp. 101-109

Damage and Failure of Interfaces, Rossmanith (ed.) © 1997 Balkema, Rotterdam, ISBN 90 5410 899 1

Modelling of the influence of interfaces on the fracture pattern during sequential mining

W.B.Tomlin & M.U.Ozbay
University of the Witwatersrand, Johannesburg, South Africa
E.J.Sellers
CSIR Division of Mining Technology, Johannesburg, South Africa

ABSTRACT: Stress fracturing invariably occurs around deep gold mine stopes in South Africa. Where possible, vertical fracturing is preferred to intense shallow dipping fracturing since the latter is liable to cause rock falls either through gravity or mining induced seismic activity. Currently, a boundary element based computer code DIGS is being developed to be used as a tool for predicting fracture patterns resulting from different stoping configurations. This paper presents the results from a series of numerical and physical modelling experiments conducted with the twin objectives of improving our understanding of the mechanisms of fracturing and secondly to calibrate and validate the numerical code DIGS. Brief descriptions of DIGS and the poly-axial cell developed for testing of physical models are given. The emphasis is on the effect of geological discontinuities on fracture development and real time observations of fracture onset and growth by means of acoustic emission monitoring. The results obtained so far appear to be encouraging in the sense that a good agreement was observed between the fracture patterns obtained from both the physical and numerical modelling experiments.

1 INTRODUCTION

In South African gold mines, the sedimentary origin of the gold bearing strata leads to shallow dipping tabular reefs, which are currently being mined at depths of about 500 to 3000 meters. These reefs form part of laminated quartzite and shale strata separated by bedding or parting planes. On occasions, the reefs are overlain by massive lava. The tectonic history of the Witwatersrand basin is known to have led to also the formation of extensive sets of secondary geological structures, such as faults, dykes, veins and joints.

Considerable fracturing occurs around the deep mine stopes because of the high overburden stresses encountered at these depths. The frequency and orientation of fracturing is largely affected by the presence of the interfaces mentioned above, e.g. the lithological boundaries and secondary geological structures, as well as the stope layout and mining configurations. Being able to accurately predict the effect of these interfaces and mining configurations would greatly assist in designing less dangerous stoping layouts and support systems which could in turn drastically reduce the hazards associated with rock falls and rock bursts.

A computer program based on the boundary element method is being developed with the objective of predicting the mining induced fracture patterns. The program, called the Discontinuity Iteration and Growth Simulation code (DIGS), obviously requires extensive verification to validate that the numerical simulations of fracture growth can represent real situations (Napier 1990). The philosophy of the validation process has been that the program should be able to model fracture growth in idealised, laboratory, tests where the loading is known and the fracture patterns can be extensively studied before embarking on the analysis of fracturing around underground excavations where qualitative data relating the boundary conditions to the fracture pattern is difficult to obtain. The paper compares the results of physical experiments in cubic blocks of rock containing tabular slots with numerical simulations to determine whether the program can correctly predict how the presence of interfaces will influence the final fracture pattern.

2. EXPERIMENTAL SET-UP

The samples being tested are typically cubic blocks with a side length of 80 mm and all sides are machined perpendicular to within 0.01°. The block is cut in half and a slot machined to the required depth half way across the lower half of the block. Figure 1 shows a schematic of a typical sample.

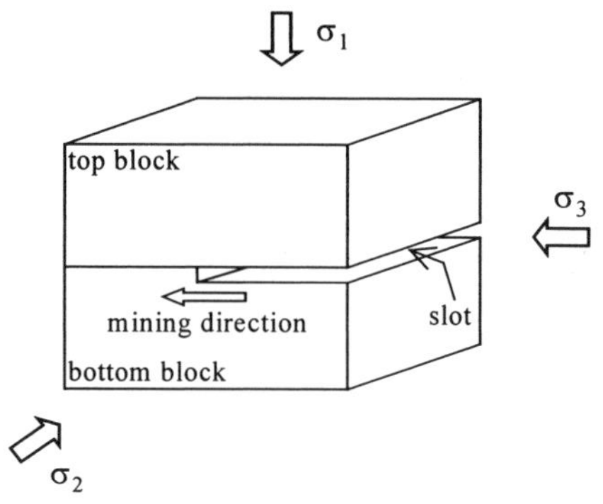

Figure 1: Schematic of model used for testing

The samples are made from Elsburg Quartzite for the discontinuity experiment and Black Reef Quartzite for the parting plane tests. The Elsburg Quartzite has a uniaxial compressive strength (UCS) of 180 MPa, Young's Modulus of 69.6 GPa and a Poisson's ratio of 0.2. The Black Reef Quartzite has a UCS of 203 MPa, Young's Modulus of 78 GPa and Poisson's ratio of 0.17. Friction between the sample and platens is minimised by the application of stearic acid to the sample before testing.

Once the sample has been prepared, the testing takes place in a standard sequence. A 2.5 MN MTS loading frame applies the principal stress for the experiments. Two independent confining stresses are applied via pistons embedded in the polyaxial cell. The pistons are driven by a centrifugal pump and the pressure is controlled manually using two valves. The sample is confined triaxially to a pressure of 27 MPa. Once this is achieved, the valves on the cell are closed which means that the confinement will increase as the sample deforms in the horizontal directions. Vertical loading continues at a rate of 0.005mm/sec until a vertical pressure of 80 MPa is reached. This state of stress is held for approximately 5 minutes after which the vertical stress is reduced slowly down to 27 MPa, then both vertical and horizontal stresses are reduced to zero. The slot is extended in length by 1mm and the process is repeated.

The MTS data acquisition system calculates the vertical displacement and load from the loading piston. An external computer is used to record the vertical load separately, confining pressures, and vertical and horizontal displacements. The formation of the fractures within the sample is monitored by locating the source of acoustic emission (AE) events which are associated with the energy release during fracturing (Lockner, 1993). The arrival times of the AE events are recorded using 1 MHz piezoelectric sensors. The AE system operates at a data acquisition rate of 4 MHz per channel. Three dimensional location is achieved using 9 sensors positioned around the sample in slots machined into the confining platens. A 1 mm thickness of steel separates the sensors from the rock. The measured p-wave velocities of 5200 m/s for the Elsburg Quartzite and 5500 m/s for the Black Reef Quartzite were used in the least squares location algorithm.

3. NUMERICAL MODELLING

The numerical code DIGS is based on the displacement discontinuity formulation of the boundary element method in which a set of fictitious cracks or dislocations in an elastic body, are used to represent the jump in the displacement field across each fracture (Napier, 1990). The displacement discontinuity method can be extended to the modelling of fracture growth as a pseudo-static process if each crack tip is advanced incrementally in a direction that maximises a specified growth criterion (Napier, 1990). Several criteria have been proposed, including local strain-energy density, angular tensile stress, energy release rate, or combinations of Mode I and Mode II stress intensity factors (Napier and Hildyard, 1992). In this study, two basic modes of fracture in brittle rock are distinguished, namely extension, or cleavage, fracturing and shear fracturing. To simulate fracture patterns in the numerical model, it is necessary to define seed positions which define the origin of fractures and the fracture growth rule. Fracturing is assumed to continue from a seed point or an existing crack tip if the specified failure criterion is met at a designated point ahead of the crack tip. The failure criterion is formulated in terms of the components of the local stress tensor at the designated point.

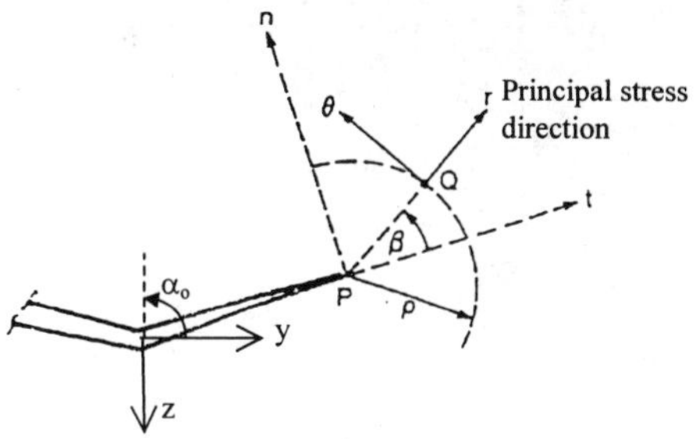

Figure 2: Diagram illustrating the fracture growth rule (Napier, 1990)

The point at which the specified failure criterion is to be evaluated is determined by searching around the seed point or crack tip at a fixed distance ρ with a given angular increment. For extension fractures, the growth angle, β, is chosen to be the direction in which the stress component normal to the crack is a maximum, see Figure 2. The fracture directions in the shear growth mode are based on shear band theory, with no dilation, and there are two potential angles at $\beta = \pm(\pi/4 + \varphi/4)$ to the directions of the two principal stresses. The growth direction is selected as the angle closest to a specified base angle α_0. Growth occurs when the stress state is such that the energy release rate G_s exceeds the critical energy release rate determined from the Mohr-Coulomb failure criterion. The energy release rate for an element of length l, incremental energy release ΔW, and slip and opening displacements related by a Mohr-Coulomb slip law, can be calculated as

$$G_s = \frac{\Delta W}{l} \approx [\overline{T}_y^2 - \tan^2 \varphi \overline{T}_z^2]/2k, \qquad (1)$$

where $\overline{T}_y$ and $\overline{T}_z$ are the external tractions, including the effects of the primitive stresses and all mobilized elements, acting parallel and perpendicular to the potential crack element, respectively (Napier et al., 1995). The friction angle is φ, and k is the self-effect influence which arises in the displacement discontinuity formulation when the point receiving the influences of other elements coincides with the collocation point of the proposed element during the calculation of the stress. For an element of half length b and collocation point position $c = b/\sqrt{2}$, the self effect influence is given by

$$k = \frac{E}{8c\pi(1-\nu^2)}\log[(\frac{1+c}{1-c})^2 + \frac{8}{1-c^2}] \qquad (2)$$

4. INVESTIGATING THE INFLUENCE OF AN EXISTING DISCONTINUITY ON THE FINAL FRACTURE PATTERN

4.1 *Physical Models*

Two samples of Elsburg Quartzite were tested to investigate the influence of a pre-existing discontinuity on the fracture pattern around the slot. This situation can be thought of as a small scale representation of mining towards a fault. The one sample was free of visible discontinuities and was machined according to the standard procedure shown in Figure 1. Another sample was machined with an existing discontinuity plane at an angle to the slot. The plane consisted of a welded joint with quartz infilling and was angled so that it passed through the centre point of one side of the sample. The test involved sequential machining of the slot towards the discontinuity. Figure 3 shows tracings of the final fracture pattern in the blocks. These patterns will be subsequently analysed and compared with the numerical simulations

4.2 Numerical Simulations

In numerical simulations, a constant confining pressure of 27 MPa is applied to the sides of the sample. The sides and base contacts are modelled to be frictionless. The vertical loading is generated from uniform vertical displacement. The sample without discontinuities can be represented by a single plane strain approximation. Two analyses were required to investigate the effect of discontinuities because of the different distances between the slot and the angled discontinuity on either side of the sample.

The slot was modelled by linear displacement discontinuity elements having no normal or shear stress and a limited amount of closure, corresponding to the slot width of 0.5mm. Material properties for this simulation were a Poisson's ratio of 0.17 and a Young's Modulus of 68 GPa. The fracture seed points are positioned close to the slot as this is where the high stresses are expected to occur and hence failure is likely to take place. The shear growth rule is specified on each crack with an initial cohesion of 35.1 MPa, initial friction angle φ of 35°, dilation angle of 5°, no residual cohesion, a residual friction angle of 32°, and a tensile cut-off of 5 MPa. A termination rule is specified so that the growth of a fracture would be terminated on intersection with another fracture. The plots of the

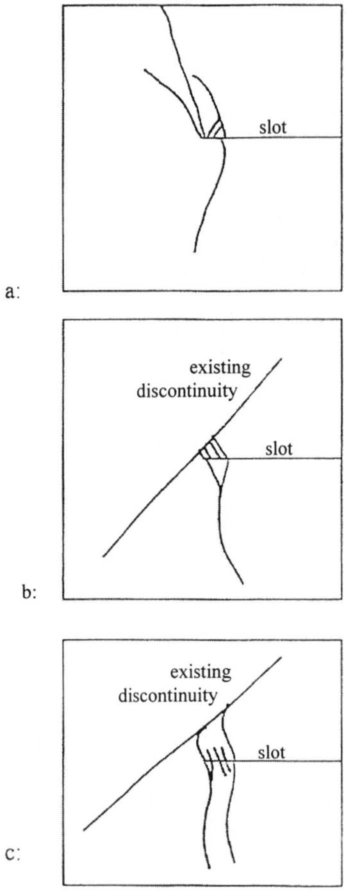

Figure 3: Tracing of fracture patterns observed on sides of physical models. a: block without discontinuity, b: side with discontinuity close to slot, c: opposite side with discontinuity further from slot

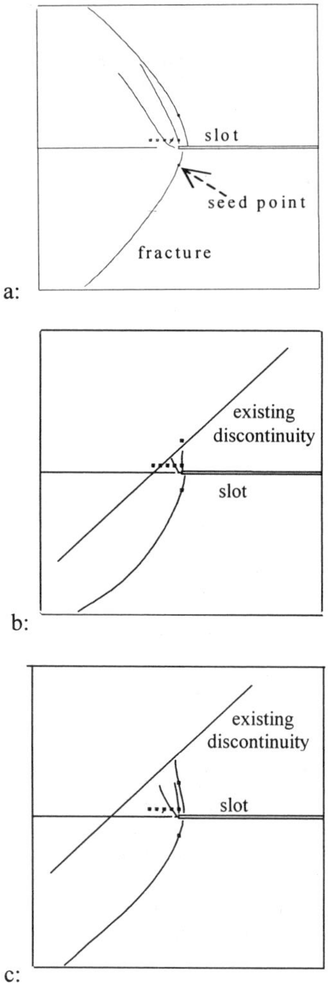

Figure 4: Fracture patterns predicted in numerical simulations. a: block without discontinuity, b: discontinuity close to slot, c: discontinuity further from slot

fractures grown in each of the three analyses are shown in Figure 4.

4.3 Comparison of physical and numerical models

In the experiment (Figure 3a) and the numerical model (Figure 4b), the fractures which formed in the first step extend ahead of the face, both in the footwall and the hangingwall, until almost intersecting with the outer boundary. Subsequent steps in both the experiment and the simulation resulted in shorter fractures, in much the same direction as the initial fracture. In the experiment, the major fractures were observed to consist of sets of small, en-echelon, fractures. These en-echelon fractures would indicate that the mode of failure was shear fracture (Kranz, 1983; Lockner et al. 1992). This would provide confirmation that the shear criterion applied in the numerical solution, shown in Figure 4a, can be applied for modelling shear fractures in the quartzite rock.

The influence of the existing discontinuity can be seen by comparing Figures 3b, 3c, 4b and 4c. The physical and numerical models show similar fracture patterns. In all cases, a footwall fracture extends ahead of the face in the first step, as was observed in the case without a discontinuity. A difference in the fracture pattern is apparent in the hangingwall block. The initial fracture was formed in much the same way as the homogeneous sample and terminated at the discontinuity. The fracture terminating at the point of intersection between the fracture and the discontinuity suggests that either the quartz infilling of the joint was too tough to fracture, or that the deformation was accommodated by activation of the discontinuity. Observation of the tested samples showed evidence that the discontinuity has been activated.

On the side of the block where the discontinuity intersects the slot, subsequent steps in the experiment showed little fracturing occurring in either the footwall or the hangingwall (Figure 3b). In the numerical solution, see Figure 4b, a single fracture extended to the discontinuity, and then continued mining induced no more fracturing. On the other side of the block, smaller fractures extended in the same direction as the initial fracture, increasing in length with each mining step. Then, the eighth step of the experiment produced an extended fracture in the footwall, and no fracturing in the hangingwall. Finally, in the tenth step, one fracture extended from the corner of the slot to the discontinuity in the hangingwall, and another grew in the footwall towards the large fracture formed in the previous step. These results suggest that the deformation will occur preferentially on the largest existing weakness (e.g. fracture or discontinuity), until the stress state has been altered sufficiently to require the formation of a new fracture.

5. INVESTIGATING THE INFLUENCE OF PARTING PLANES ON THE FINAL FRACTURE PATTERN

5.1 Physical Models

Two samples were prepared to compare the influence of parting planes on the final fracture pattern, each of Black Reef Quartzite. The first was an ordinary sample as shown in Figure 1 and the other sample consisted of 4 plates each 20mm thick. This sample represents an approximation of a rockmass containing a series of horizontal bedding planes. The friction angle on each interface was measured as 26 degrees. In both cases, the thickness of the slot was 0.5mm which was too large to induce closure of the slot in the stiffer Black Reef Quartzite.

a:

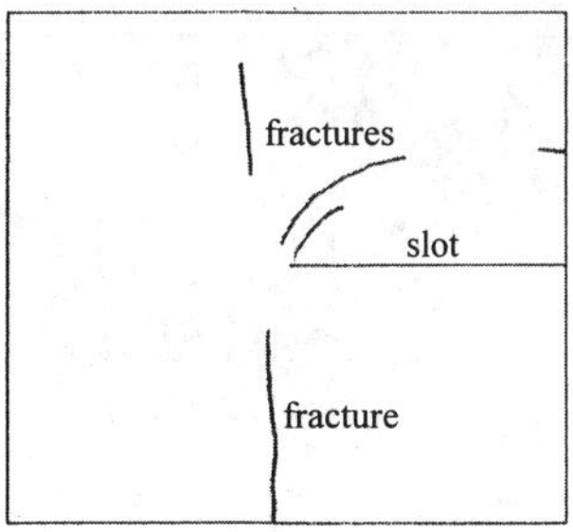

b:

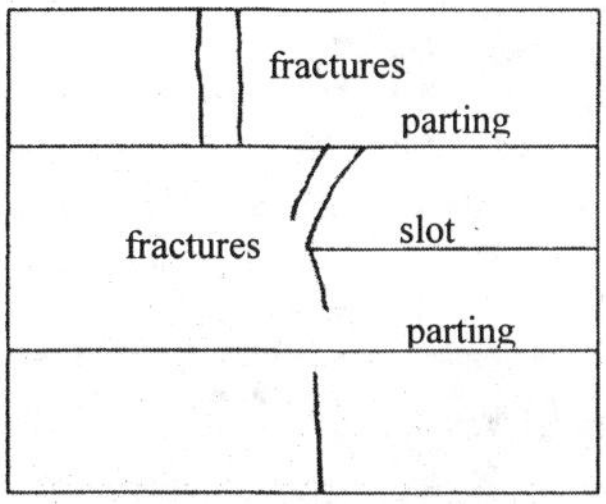

Figure 5: Schematic of fracture patterns observed in physical models using Black Reef Quartzite. a: block without discontinuity, b: block with parting planes

5.2 Acoustic emission

Figure 6a shows the located AE events in the sample without parting planes. The events are clustered in the same regions as the observed fractures. The events in the centre of the sample correspond to the fractures which curve into the hangingwall from the edge of the slot. The vertical fractures ahead of the slot, in the hangingwall and footwall are also evident. A cluster of events at the edge of the sample corresponds to the small horizontal fracture in the hangingwall.

Two steps were undertaken with the sample containing parting planes and, in each step, the positions of the AE events (shown in Figure 6b) correspond closely to the cracks observed (Figure 5b). In the first step, the vertical cracking is evident in the two footwall blocks, ahead of the stope. There is less AE activity in the immediate hangingwall block, and the located events correspond to the angled crack extending from the slot tip, and the small fractures which are initiated at the upper parting planes. In the top block, the events relate to the two vertical fractures ahead of the face. In the second step, the vertical fractures ahead of the slot extended through the sample and the angled cracks in the hangingwall connected the slot and the parting

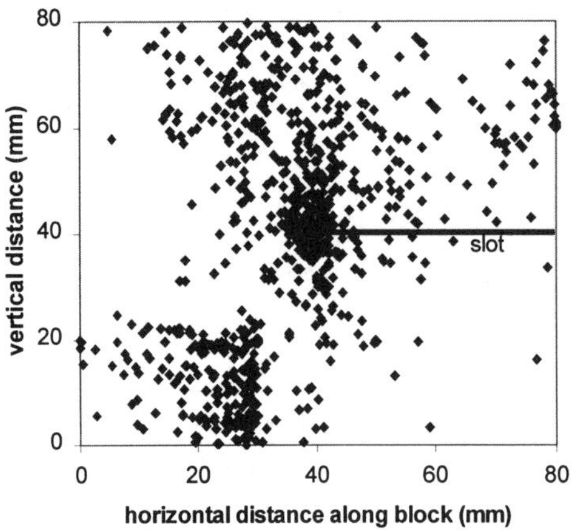

a:

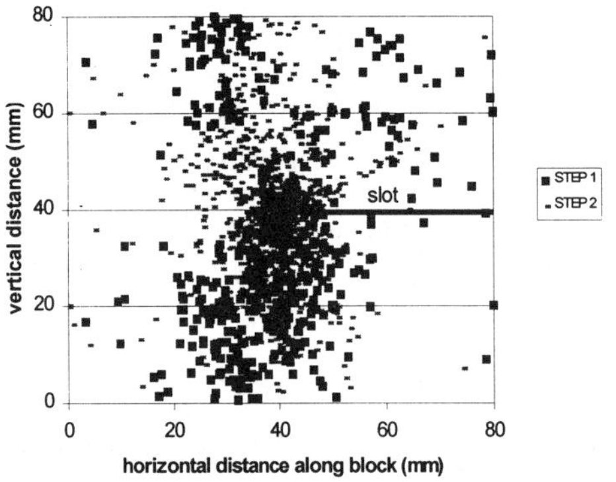

b:

Figure 6: Locations of acoustic emissions determined in physical models using Black Reef Quartzite. a: block without discontinuity, b: block with parting planes

plane above, as shown by the AE events above the slot.

5.3 Numerical Simulations

For all the numerical simulations, plane strain conditions were considered. Two simulations were carried out, one with parting planes and one without. The material properties used in the simulations were a Young's Modulus of 78 GPa and a Poisson's ratio of 0.2. The parting planes had a friction angle of 26°. A tension mode of failure was selected for the cracks with zero tensile strength and a residual friction of 20°.

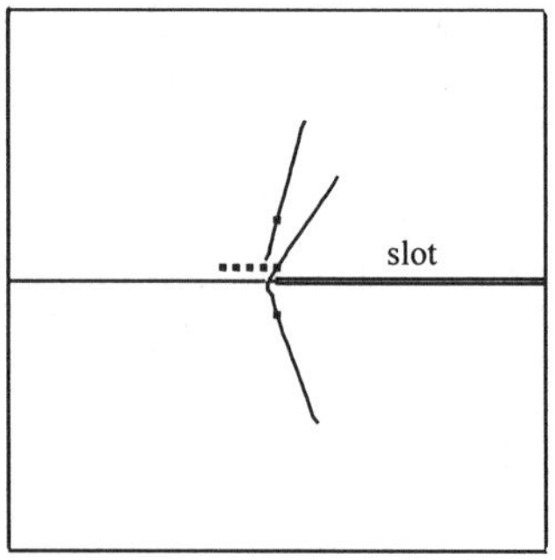

a:

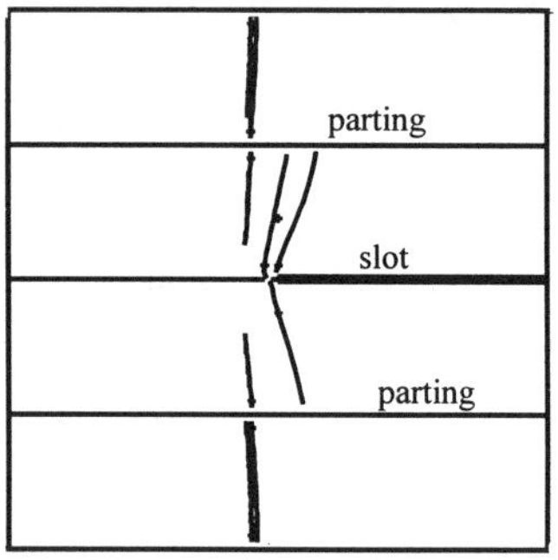

b:

Figure 7: Fracture patterns predicted in numerical simulations of Black Reef Quartzite blocks. a: block without discontinuity, b: block with parting planes

5.4 Comparison of physical and numerical models

In Figures 5a and 7a, the cases without parting planes, no closure was observed. The resulting fracture patterns show fractures extending behind the slot in both the hangingwall and footwall. In the hangingwall of the experiment (Figure 5a) two fractures are evident, apparently initiating near the stope and extending out in two different directions. Similar fractures are observed in the numerical model. The vertical fractures ahead of the slot are not observed in the model as no seeds were placed at the top surface.

In Figure 5b, the fractures extend behind the advancing slot. Two fractures occur in the hangingwall at angles of 70° and 79° to the slot. In the footwall, a single fracture is angled at 81° to the slot. These crack angles are very close to those predicted in the model which are 71° and 80° in the hangingwall and 74° in the footwall, as shown in Figure 7b. On the parting plane in the hangingwall of the test sample, there appears to be a slight shift of the fracture from left to right of about 2mm

across the parting plane, and another fracture extends almost vertically to intersect with the boundary.
The direction of fracture growth is not clear from the physical model, but comparison of the AE results (Figure 6b) with the numerical modelling (Figure 7b) suggests that that each crack has grown downwards, one from the top boundary and the other from the parting plane. This supports the numerical modelling of Napier and Hildyard (1992), who suggest that it is the presence of parting planes that causes extension fractures to initiate above and ahead of a stope and not in the high stress region at the stope face.

6. CONCLUSION

The results of experiments on blocks of rock containing a tabular slot under triaxial loading show that the presence of interfaces significantly alters the position and inclination of the resulting fractures. In the sample without discontinuities, the shear fractures which are observed to extend ahead of the stope in the experiments. The shear fracture criterion correctly predicts corresponding fractures in the numerical models. In the models with a pre-existing discontinuity, fractures extending from the tip of the slot are observed to terminate on the discontinuity. No fractures are evident once the slot passes through the discontinuity suggesting that the deformation is occurring on the discontinuity. Similar trends are evident in the numerical simulations.

Fractures were apparently initiated at horizontal parting planes in both the physical and numerical models, indicating that the discontinuity can alter the stress distribution within the sample, depending on the interface properties. The analysis with a tensile fracture criterion correctly predicted the final fracture directions. The three dimensional acoustic emission location provides useful information regarding the sequence of fracture growth and the growth directions, which correlate well with the numerical simulations. The displacement discontinuity program DIGS can therefore predict the fracture patterns in these samples, and the influence of the various interfaces. The requirement to pre-select the seed point may however bias the final fracture pattern. The physical experiments demonstrate that the fracturing may occur in shear or tension depending on whether the slot is closed, or remain open. The logic to decide on the appropriate fracture criterion needs further consideration.

REFERENCES

Kranz, R.L., (1983) Microcracks in rocks: a review. *Tectonophysics,* 100:449-480.

Lockner D., (1993) The role of acoustic emission in the study of rock fracture. *Int. J. Rock Mech. Min Sci. & Geomech Abstr.* 30:883-889.

Lockner, D.A., Moore D.E., Reches, Z., (1992) Microcrack interaction leading to shear fracture. *Rock Mechanics*. Tillerson and Wawersik (eds). 807-816. Balkema, Rotterdam.

Napier, J.A.L., (1990) Modelling of fracturing near deep level gold mine excavations using a displacement discontinuity approach. *Proceedings of Conference on Mechanics of Jointed and Faulted Rock.* 709-715. Vienna, Balkema, Rotterdam.

Napier, J.A.L., Hildyard, M.W., (1992) Simulation of fracture growth around openings in highly stressed , brittle rock. *Journal of the South African Institute of Mining and Metallurgy.* 92:159-168.

Napier, J.A.L., Hildyard, M.W., Kuijpers, J.S., Daehnke, A., Sellers, E.J., Malan, D.F., Siebrits, E., Ozbay, M.U., Dede, T., Turner, P.A., (1995) Develop a quantitative understanding of rockmass behaviour near excavations in deep mines. *Final report to Safety in Mines Research Advisory Committee*. CSIR Division of Mining Technology.

ACKNOWLEDGEMENT

The support of the Safety in Mines Research Advisory Committee (SIMRAC) under project GAP332 is gratefully acknowledged.

Damage and Failure of Interfaces, Rossmanith (ed.) © 1997 Balkema, Rotterdam, ISBN 90 5410 899 1

Modelization of the damage and failure of a finger joint

O. Dinkel, Ph. Jodin & G. Pluvinage
Laboratory of Mechanical Reliability, University of Metz, France

ABSTRACT : Several experiments on a model of a finger joint in a timber beam have given data about the mechanical behaviour of such a joint. This behaviour is analyzed in terms of damage and the damage versus applied load is drawn. A model of this damage curve is proposed to determine the behaviour of glued finger joints. The role of the quality of the bonding process is evidenced and the behaviour of the interface between wood and adhesive is of primary importance. The results are compared with the recommandations of the recently published EUROCODE 5 (Timber construction).

INTRODUCTION

As a building technique, gleud laminated timber (glulam) is nowadays largely used for large buildings as well as for individual houses and also for bridges. The possibilities for use are quite infinite, due to the very long span length available and the very easy shaping and joining processes.

In this paper we are dealing with the technique of finger joints. To obtain the desired length of glulam beams, it is necessary to join the lamellae together. This is obtained with the technique of finger joints. These joints are designed to have a strength closed to that of the material joined. In some particular cases this technique is also used to join beams together, either directly, either by the mean of a corner piece in case of angled beams.

The quality of the finger joints is particularly important for the safety of the construction. In factories where the fabrication process is very well controlled, the localizaton of finger joints in the most loaded parts of the beams is avoided, if possible. If not, the reliability of the beams depends highly on the quality of the finger joints.

It has been shown that this quality depends on several factors, which are :

- the degree of sharpness of the tool which has machined the fingers
- the moisture content of the wood
- the gluing pressure
- the viscosity of th applied adhesive.

In order to simplify the problem we have studied a model of the finger joint which is the so-called "scarf joint" (Figure. 1).

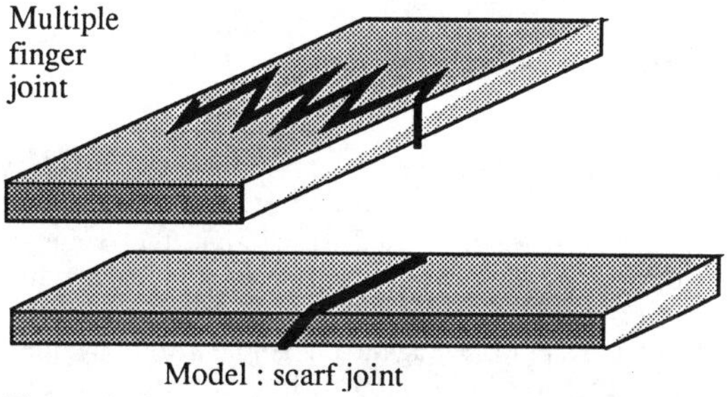

Figure 1. A multiple finger joint and the model used in this study.

These specimens are loaded in tension and the load-displacement curve is recorded. As it exhibits a brittle behaviour, a step-by-step procedure has been used to load the specimen progressively until rupture occurs.

From these curves, a damage is caculated, using the amplitude of strain as an indicator. The calculated damage is plotted versus the corresponding mean level of load. A S-shaped curve is observed. An empirical law of evolution of damage is deduced and a tentative explanation of physical significance of parameters is given.

From these data, it is possible to calculate with the help of finite elements calculation a stress-strain static behaviour curve which shows that the joint begins to fail when the stress reaches about 80% of the ultimate stress. This value is compared with the safety factors deduced from the EUROCODE 5 rules.

DAMAGE DEFINITIONS

Damage and degradation.

It is essential in this work to distinguish damage and degradation of the material.

Damage is usually defined as a variable representing the evolution of a given property of the material with progressive loading. The relationship between the damage D and the property P is chosen so that when there has been no loading D=0 and when fracture occurs D=1. It is clear, from this definition, that the evolution of damage is highly dependant of the chosen property. Different shapes of evolution can be obtained, following the observed property (Raguet, 1981).

Conversely the degradation of the material induces variation of mechanical properties. For instance, in some cases, it is possible to observe the occurence of micro-cracks. Some authors have tried to establish a correlation between the occurence of these degradations and the evolution of a mechanical property. A good correlation is not always observed.

For all these reasons, a definition of the damage is always arbitrary.

Definition of damage.

In our case we first tried to estimate the damage from the evolution of the slope of the load-strain curve, when the specimen is loaded progressively, unloaded and loaded again at a slightly higher value. But, due to the very brittle character of the behaviour of the joint, this method did not give coherent results.

We then chose to load the specimen step-by-step. Around each load level, the load oscillates during 20 cycles and the amplitude of strain is determined. The damage is defined with the following relationship :

$$D_{level\,i} = \frac{(\varepsilon_{\max} - \varepsilon_{\min})_{level\,i}}{(\varepsilon_{\max} - \varepsilon_{\min})_{fracture\,level}} \tag{1}$$

The standardized load is defined as :

$$N_{level\,i} = \frac{Mean\,load_{level\,i}}{Mean\,load_{fracture\,level}} \tag{2}$$

EXPERIMENTAL PROCEDURE

Test specimen

The test specimens are cut from *Spruce (Picea abies)* logs. They have the shape of usual tensile specimens (Figure 2).

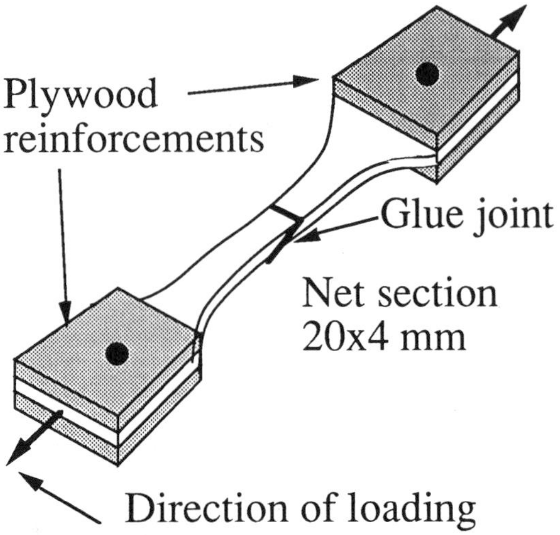

Figure 2. General view of the test specimen.

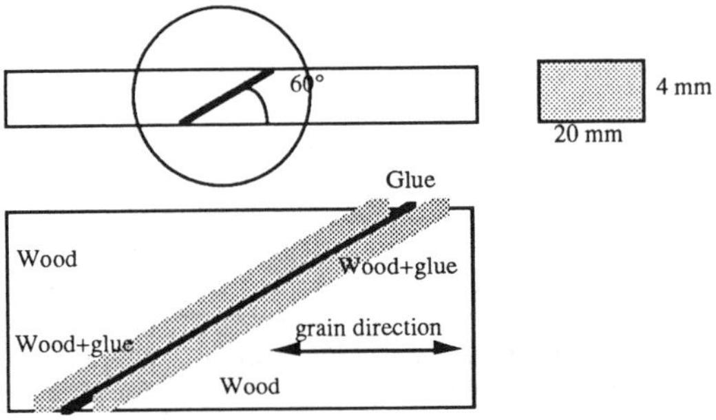

Figure 3. Details about the model of scarf joint. The three layers (wood, wood and glue, glue) are shown.

The details of the glue joint are given on figure 3.

The adhesive used in these experiments is a phenol-formol-resorcinol adhesive which is widely used in the gluelam industry. This is a two-components adhesive which should be applied with a pressure of 1.2 N/mm^2.

The whole specimen is first machined, then cut into two parts along the future bond line, then reconstituted by gluing in a special device which allows to apply the convenient pressure.

A preliminary study has shown that the mode of failure of the joint depends on the angle shown on figure 3. If this angle is higher than 45°, the mode of failure is mainly a brittle tensile mode. For higher angles, shear progressively appears and the fracture becomes a little more ductile. The angle of 60° which has been chosen here is a compromise between a ductile failure and the load capacities of the testing device.

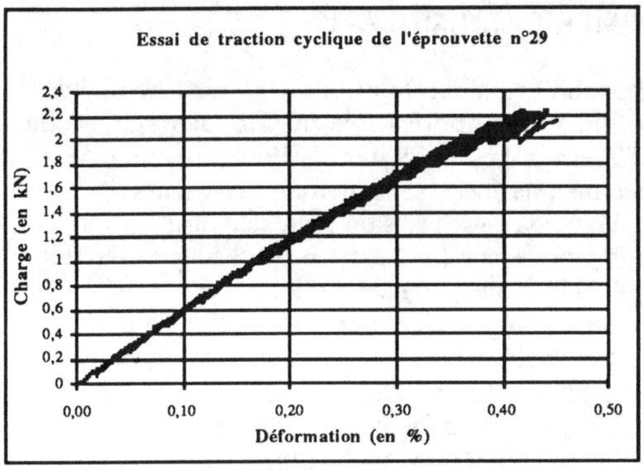

Figure 4. Load - strain curve of a whole experiment.

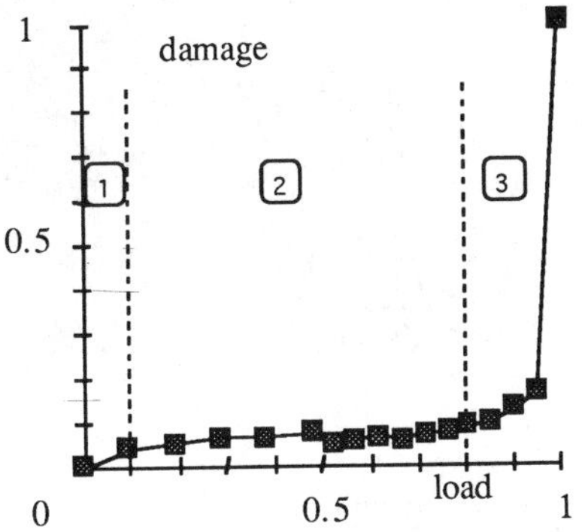

Figure 5. Typical damage vs load curve.

It should be noticed that for a multiple finger joint, optimization of the strenfgth of such a joint leads to an angle of the fingers of 6° (say total angle of 12°).

Loading scheme.

Some of the specimens are uniformely loaded in tension until fracture. These are called the static monotonic tests.

The other specimens are submitted to progressive damaging and loading following this scheme :

1. Static loading from 0 to 0.2 kN
2. The load is oscillated during 20 cycles around this level with an amplitude of 0.2 kN.
3. The mean level is increased to 0.4 kN.
4. Step 2. is applied again.
5. And so on…

When fracture occurs, the number of cycles is recorded.

The specimen is equiped with an extensometer which allows to record the strain of the specimen over a gage length of 25 mm. The load cell of the tensile machine allows recording of applied load. All the test and data acquisition is computer controlled.

In these conditions it is possible to draw the load-strain curve (Figure 4).

Data processing.

For each mean load level, the mean amplitude of strain is determined from the experimental data. The mean load level is reported and this procedure is achieved for each level.

Then, relationships (1) and (2) can be computed for each load level and a damage - load curve can be drawn (Figure 5).

RESULTS

Monotonic loading tests

Figure 6 shows the different load - strain curves obtained for a monotonc loading.

The appearance of the surfaces of fracture is visually analyzed. It seems that, most of fractures occur both in the intermediate zone (wood+adhesive) and in the adhesive itself. They are called "mixed fractures". It must be noticed that the presence of a sharp angle makes this part of the joint more brittle. A finer surface analysis using fractal concepts (Dekiouk, 1996) is under performance.

Cyclic tests.

All damage vs load curves are gathered in figure 7. They all show three parts :

1. An initial fast and small increase
2. A stable level
3. A final fast increase preceding fracture.

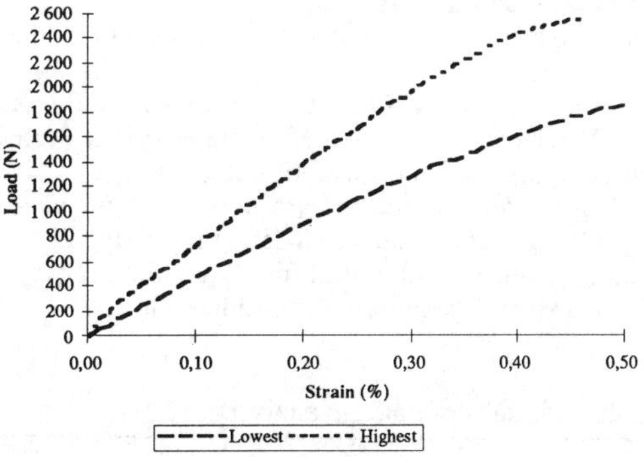

Figure 6. Results of the monotonic loading tests.

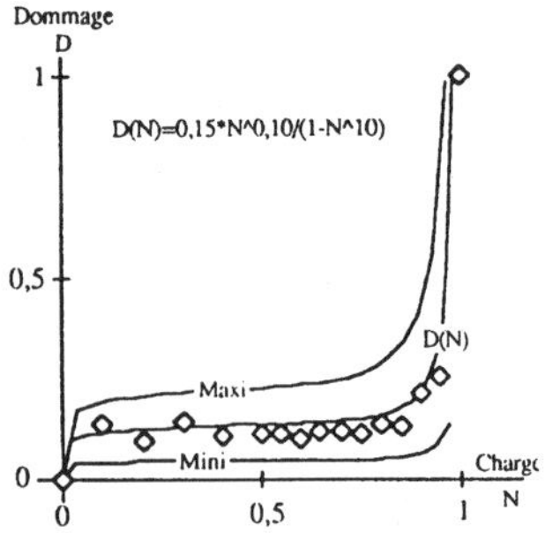

Figure 7. Damage vs load curves.

It can be seen that there is a significant discrepancy on the second level.

ANALYSIS OF RESULTS

Such S-shaped damage curve have been already reported in the litterature (Raguet, 1981). A mathematical model is defined for such a curve :

$$D(N) = a.\frac{N^b}{1-N^c} \qquad (3)$$

where D(N) is the damage,

N is the normalized load,

a, b and c are experimental parameters which respectively control the level of the second part of the damage curve, the slope of the first part and the initial point of the third part.

Parameters a b and c can be considered as indicators of the quality of the interface as :

a translate the initial damage caused by the very first loading of the joint. It could be related to the number of small defects.

b translates the speed at which this initial damage occurs. It could be related to the nature of the preceding small defects.

c translate the brittleness (or the ductility) of the joint.

A very good quality of interface would result in a zero level of second part of damage curve (and, consequently, in a null speed of initial damage).

These three parameters are computed for each damage vs load curve with the help of a computerized bi-logarithmic method (Mankowsky, 1982). The results are summarized in table 1.

Table 1. Results of damage analysis

	a	b	c
mean	0.13	0.07	10
standard déviation	0.03	0.02	2.20

NUMERICAL SIMULATION

The numerical simulation has been realized to simulate the application of the damage to the different zones of material. For convenience, the specimen has been separated into five parts :

the wood (two parts on each side of the joint)

th wood+glue (two parts on each side of the joint)

the glue (one part) i.e. the joint itself

The damage experimentally measured is applied following six ways :

1.application only on the wood
2.application only on the wood+glue
3.application only on the glue
4.application on wood and wood+glue
5.application on wood and glue
6.application on wood, wood+glue and glue

The numerical simulation has also been realized to test the effects of the variation of parameters a, b and c on the behaviour of the specimen. At least, a simulated load-strain curve is computed using the damage law hereup established and is compared with the experimental results.

Finite elements model.

The specimen has been modelized using the finite elements method. The programme used is the CASTEM 2000® software distributed by the French Atomic Energy Agency (CEA).

Only the calibrated part of the specimen is modelized, assuming that the load is uniformly applied at the ends of this part. Quadrangular specimen with four nodes and four integration points have been used, as the computation is done assuming an elastic behaviour.

Five zones have been defined, i.e. three different materials, say wood, wood and glue, pure glue.

The following material properties are entered into the programme :

Wood :

Orthotropic material
$E_L=E_y=13100$ MPa
$E_T=E_x=800$ MPa
$E_R=E_z=1000$ Mpa
$\nu_{TL}=0.02$
$\nu_{TR}=0.4$
$\nu_{LR}=0.4$
$G_{LT}=800$ Mpa
$G_{TR}=100$ Mpa
$G_{LR}=800$ MPa

Glue :

Isotropic material
E=2700 Mpa
$\nu=0.3$

G=1000 MPa

Wood and glue :

Orthotropic material. The mechanical properties are calculated using the mixing law, assuming that the proportions of wood and glue in this small zone are equal in volume.

E_L=7900 Mpa
E_T=1750 Mpa
E_R=1850 Mpa
ν_{TL}=0.16
ν_{TR}=0.35
ν_{LR}=0.35
G_{LT}=900 Mpa
G_{TR}=550 Mpa
G_{LR}=900 MPa

The boundaries conditions are reported on figure 8 : on the upper end, the load is applied uniformly on the nodes of the line and displacement in the x direction is fixed ; on the lower end, the displacement of the nodes in the y direction are fixed and the displacement in the x direction of the middle node is also fixed. In addition to these conditions, the to nodes are defined as near as possible to the position of the knives of the extensometer which has been used to measure the strain. These points are referenced as D1 and D2.

The load is applied step by step, and the damage D is computed for each level. The the simple law :

$$E_i = E_0(1 - D_i) \tag{4}$$

is applied to each material property, and relationship (3) is used to compute D_i.

Numerical determinaton of the damaged zone.

The results of the computation have shown that the best fit with experimental results is obtained in case 3, i.e. pure glue. This means that, in case of a glue joint, the most sensitive part to the load is the adhesive itself.

Influence of the parameters a, b and c on the load-strain curve.

The sensitivity of the load-strain curve to the variation of the parameters of the damage law is studied by simulating such a curve with the finite elements calculation described hereup.

Different simulation results are obtained fixing two parameters to their mean value and varying thethird between given limits. The computation protocol is given in Table 2.

The results of these computations are given in figures 8 to 10.

Table 2. Set of parameters for simulation.

	Mean value	Lower limit	Upper limit	Step
a	0.14	0.06	0.22	0.04
b	0.1	0.0	0.2	0.05
c	7	3	11	2

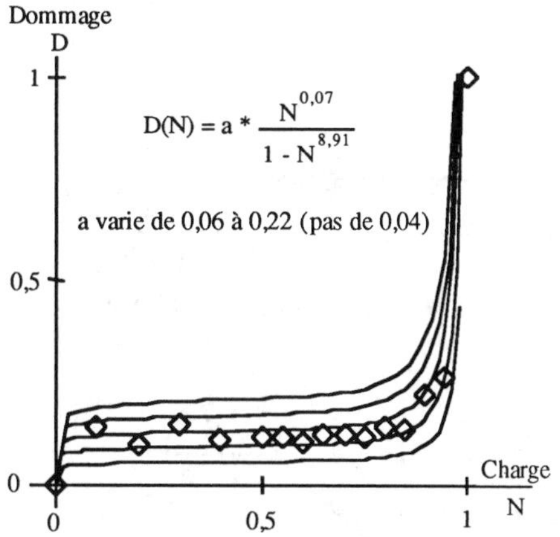

Figure 8. Effect of the variation of parameter a.

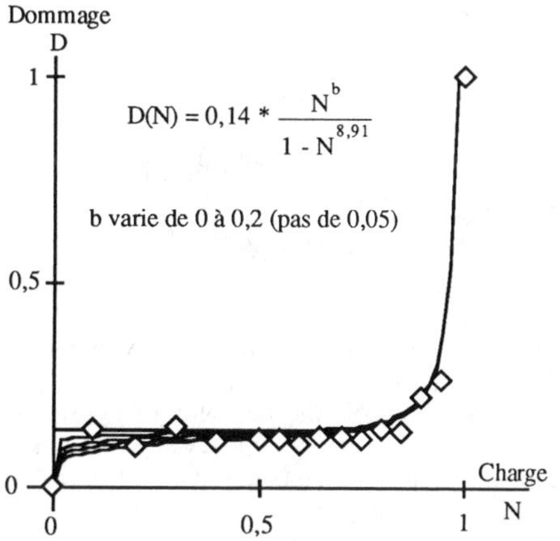

Figure 9. Effect of the variation of parameter b.

It can be seen that the effect of the variation of the parameters is sensitive on the behaviour curve. Parameter a influences the level of fracture, parameter b has very little effect and parameter c influences the beginning level of the non linear part of the curve.

This can be interpreted as follows :

parameter a controls the level of initial damage, i.e. thenecessary initial adjustment of the joint to the load ;

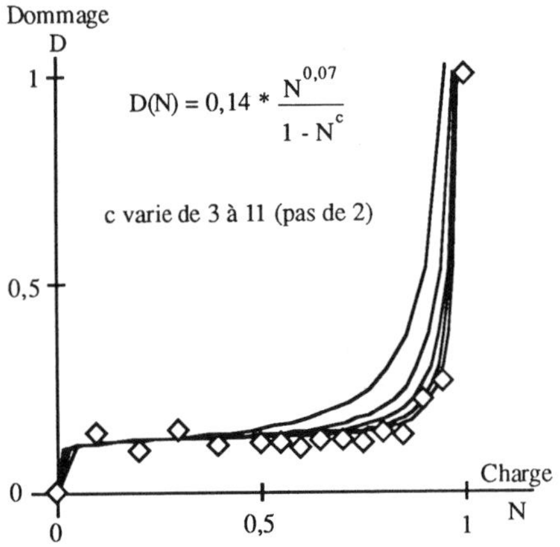

Figure 10. Effect of the variation of parameter c.

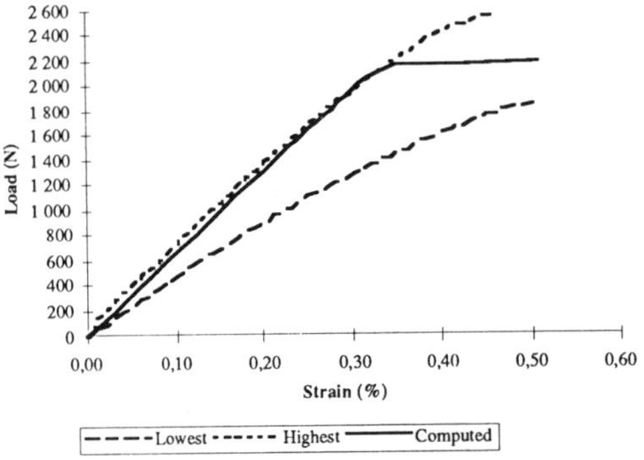

Figure 11. Comparison between numerical simulation and experimental results.

parameter b controls the speed at which this initial level is reached ;

parameters c controls the beginning of fracture, i.e. the occurrence of irremediable damage.

Parameter a could be considered as an indicator of the quality of the joint and parameter c as an indicator of its strength.

Numerical simulation of the load-strain curve.

The numerical simulation of the load-strain curve is obtained by achieving the same calculation as before using the mean values experimentally determined in the previous experimental part. This numerical simulation is compared to the experimental results. This is shown on figure 11.

It can be seen on this figure that the numerical simulation passes in the range defined by the highest and the lowest experimental curves. This result validates the model of damage and the simulation procedure. However it should be noticed that the numerical simulation uses data from cyclic damage experiments, when the experiments are monotonic loading. It can also be seen on this figure that the non-linearity begins at about 80% of the maximum load.

Table 3. Results of tensile tests on glue joints.

Mean	Stand. dev.	coeff. variat.	size
19.28	2.61	0.1352	20

COMPARISON WITH EUROCODE 5 RECOMMENDATIONS

EUROCODE 5 gives no particular recommendations on this kind of glued joint. It can only be established a reduction factor from the experimental results of the fracture tests. They are given in Table 3.

The applicable relationship is :

$$k = \exp\left[-\left(2.645 + \frac{1}{\sqrt{n}}\right).v(x) + 0.15\right] \qquad (5)$$

where k is the reduction factor, n the sample size and v(x) the coefficient of variation.

In these conditions, k=0.78, which is to be compared with the coefficient 0.8 approximately determined with the preceeding numerica simulation. Thus, in this particular case, EUROCODE 5 recommendations fit quite good with the experimental results.

CONCLUSIONS

The load-strain behavior of a inclined glued joint has been experimentally established. The investigation of the progressive damage has shown that, with respect to load, the representative curve is S-shaped. This shape is represented by a three-parameters relationship, each parameter representing respectively :

- the size and number of small defects taken into account by the first load
- the nature of these small defects
- the ductility of the joint.

A numerical simulation has permitted to determine the zone in which the damage mainly operates, which is the adhesive itself.

Last, a comparison has been established with the recommendations of EUROCODE 5 which, in this case gives similar prediction for the allowed stress of a glued joint.

REFERENCES

DEKIOUK, R., F. BOUYGE, Z. AZARI & G. PLUVINAGE "Relation entre ténacité et dimension fractale au point d'amorçage du polycarbonate" Récents progrès en Génie des Procédés : les Applications des Théories du Chaos en Sciences pour l'Ingénieur, *Groupe Français de Génie des Procédés,* vol. 11, n.50, pp 209-214, 1997.

EUROCODE 5 DAN ENV Timber construction. (1995)

MANKOWSKY, V. A. "Non-linear parametric creep function " Mechanics of composite materials n°4 pp. 579-584 (1982) (In Russian)

RAGUET, M. " Endommagement en fatigue oligocyclique à haute température, application aux esais à deux niveaux de déformations dans le cas d'un acier " Doctor Thesis. University of Metz (1981)

Damage and Failure of Interfaces, Rossmanith (ed.) © 1997 Balkema, Rotterdam, ISBN 90 5410 899 1

An acoustic emission study on tensile behavior of welded steel joints

C.S.Lee & J.H.Huh
Center for Advanced Aerospace Materials, Pohang University of Science and Technology, Korea
K.S.Kim
Department of Mechanical Engineering, Pohang University of Science and Technology, Korea

ABSTRACT: The tensile behavior of the welded joints of a low-carbon low-alloy steel has been studied by means of acoustic emission (AE) technique. The results indicate that the AE characteristics of the base metal is distinctly different from those of the heat affected zone (HAZ) and the weld metal. For the base metal having a ferrite-pearlite microstructure, most of the AE events occur around the yield point, chiefly due to the dislocation activities associated with the tensile flow. For the HAZ and the weld metal, in addition to the AE peak occurring around the yield point, another AE peak with higher energy occurs post yielding. This second AE peak is attributed to the existence of martensite and, its amplitude is proportional to the volume fraction of the martensite. The ferrite-martensite interfacial debonding and the martensitic plate cracking are responsible for the occurrence of the second AE peak. This is further confirmed by the microstructural and the frequency spectrum analyses.

1 INTRODUCTION

As modern buildings are becoming increasingly higher, improvements on the performance of the structural steels are necessitated in order to resist earthquakes and strong winds. Therefore a comprehensive requirement on strength, ductility and weldability is issued for the steels served for these purposes. In recent years the steels treated via the thermo-mechanically controlled process (TMCP) are especially attractive for their high strength and high toughness due to their fine grain sizes and low carbon contents (DeArdo 1995). During welding three zones are usually generated in the steel joints, namely, the base metal, the heat affected zone (HAZ) and the weld metal. Mechanical properties, especially the toughness (Haze 1986), are seriously deteriorated in those regions with evidently coarsened grains (typically in HAZ and the weld metal) because of the dynamic growth of grains and transformed microstructures during cooling.

As one of non-destructive evaluation techniques, acoustic emission (AE) method has been widely utilized in recent decades to inspect the safety of atomic power plants, bridges and fuel tanks owing to its capability of real time monitoring (Non-Destructive Testing Handbook, 1981). Since AE is very sensitive to some specific micro-events in local areas, *e.g.* dislocation movement and multiplication, microcrack initiation and growth, and interfacial cracking, the AE method is useful in identifying the origin, magnitude and distribution of the source that causes the characteristic AE events. Studies using the AE method have been conducted on the deformation behavior of steels (Wadley 1981, Webbor 1981) and Al alloys (Hong 1995, Lee 1996) and on the evaluation of welds (Bentley 1988, Scruby 1988, Whittaker 1987). However, information is still lacking for a better understanding of the correlation between the AE behavior and microstructure in steel welds. Consequently the present study is aimed to investigate the AE behavior during tensile deformation of welded steel joints and to correlate the behavior with the specific microstructures in the base metal, the HAZ and the weld metal, as well as the transformed microstructures reproduced by heat treatment.

2 EXPERIMENTAL PROCEDURES

The material used in the present study was a TMCP steel PILAC-BT33 with a chemical composition of (in weight percent) 0.14 C, 0.23 Si, 1.23 Mn, 0.0117

P, 0.003 S, 0.033 Cr, 0.015 Ti and balance Fe. Submerged arc welding (SAW) method is used to weld the steel joints, with an arc voltage of 36 V, a welding current of 980 A and a heat input of 92 kJ/cm. No heat-treatment was conducted post welding. To simulate the microstructural changes in the HAZ and the weld metal regions, another batch of samples were heated to 760 °C, 800 °C, 830 °C and 860 °C, respectively, for 1.5 hours and then quenched in water, to produce different amount of transformed martensite. They are hereafter referred to as HT_1, HT_2, HT_3 and HT_4, respectively. The volume fraction of the martensite was measured using the Mossbauer analysis to be 26%, 43%, 61% and 82%, respectively, corresponding to the increase in heating temperatures given above. The microstructures were observed under a scanning electron microscope (SEM) and a transmission electron microscope (TEM).

The tensile sheet specimens were prepared according to the relevant ASTM code and they had a width of 6.25 mm, a thickness of 3 mm and a gauge length of 25 mm. The specimens of the welded joints were cut from the base metal, the HAZ and the weld metal regions, respectively, with the tensile axis in the rolling direction. The tensile tests were conducted at room temperature using an Instron 1361 testing machine with a cross-head speed of 0.27 mm/min. The AE analysis was performed using the AE detector AEDSP-32/16 made by PAC Co., with the sensor R15. The data processing was done under the condition of a pre-amplification of 40 dB plus a main amplification of 40 dB, giving a total amplification of 80 dB. The sensor was coupled to the polished jig using a grease under a constant pressure. To remove the noise from actuator, *etc*., a threshold amplitude is specified to be 30 dB, which was determined in the preliminary tests by using an oscilloscope. The jig pin is coated with a layer of grease to minimize the friction between the specimen and the jig. The AE data acquisition was realized by means of the MISTRAS 2001 program operated during the tensile tests.

3 RESULTS AND DISCUSSION

3.1 *Microstructure and tensile properties*

The microstructures of the weld joints in different regions are shown in Figure 1. The base metal is composed of ferrite and pearlite (Figure 1*a*). The microstructure in the HAZ is characteristic of very coarse grains (Figure 1*b*). Especially, embrittlement of the steel is usually attributed to the martensite-austenite (M-A) constituent, upper bainite and martensite formed during the heating-cooling cycle of welding (Chen 1984). In the region of the weld metal, the microstructural constituents similar to those in the HAZ may also exist, but typically an area of coarse columnar microstructure is observed (Figure 1*c*), which is chiefly due to the temperature gradient formed during the cooling process while being welded.

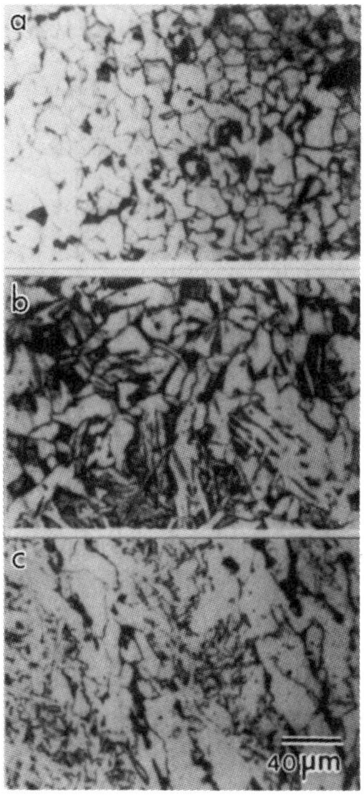

Figure 1. Optical micrographs showing microstructures in (*a*) base metal, (*b*) HAZ and (*c*) weld metal.

Figure 2 shows the tensile properties of specimens representing the base metal, the HAZ and the weld metal, respectively. For the base metal, the strength and elongation are slightly higher in longitudinal (L) direction than in transverse (T) direction. For the specimens in the HAZ, the yield strength (Y.S.) is about 9% lower and the elongation is about 15% lower than the base metal, presumably due to the grain coarsening and the transformed constituents-induced embrittlement. In the weld metal, the elongation is about 10% lower than the

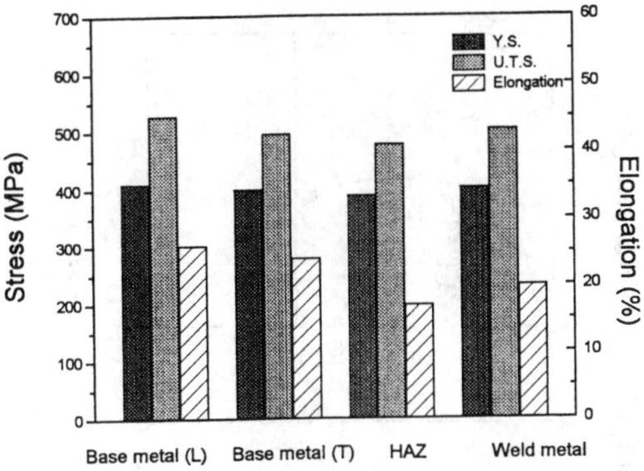

Figure 2. Tensile properties of the welds in different regions.

base metal, which is probably caused by the structural inhomogeneity occurring in the coarse columnar solidification area or in the segregated central area.

3.2 *The AE characteristics*

As one of main AE parameters, the AE energy rate measured during the tensile tests is shown in Figure 3 together with the corresponding tensile curve, for the specimens representative of different regions in the welded joint. In the base metal (Figure 3*a*), most of the AE events occur around the yielding stage and the energy rate peak is detected around the yield point. By contrast, the AE behavior in the HAZ and the weld metal is quite different from that in the base metal. As is seen in Figure 3 *b* - *c*, the AE events are evidenced not only around the yield point, but also in the post-yielding stage. A second and higher peak of the AE energy rate is detected in the post-yielding stage.

Figure 4 shows the AE amplitude distribution for the tensile specimens from different regions of the welds. Only amplitudes above 30 dB are plotted due the threshold set at this value. It is found that for all the three regions most AE events occur around the yield point but with relatively low amplitudes (30-35 dB). For the specimens in the HAZ and the weld metal, however, the AE events with higher amplitudes (typically around 54 dB) are also evidenced (Figure 4, *b-c*), in contrast to the case in the base metal in which the high-amplitude AE events are basically not detectable. Therefore the specimens in the HAZ and the weld metal are characteristic of the appearance of the second AE peak with high amplitudes occurring in the post-yielding stage.

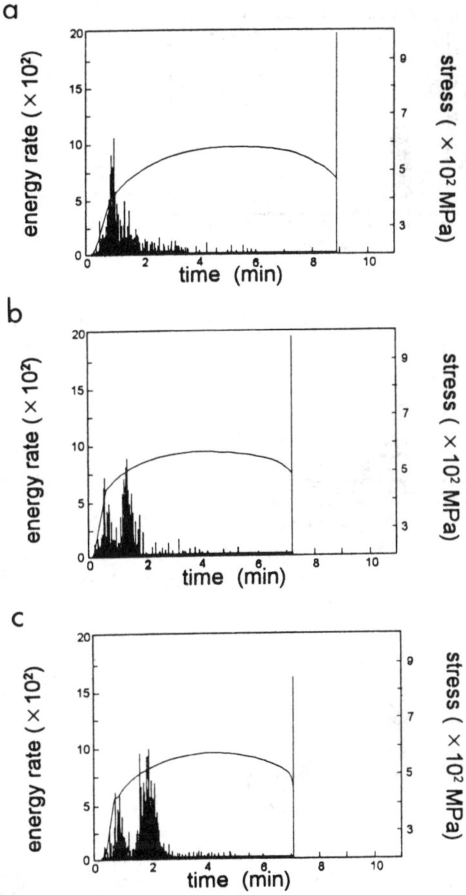

Figure 3. AE energy rate during tensile tests for specimens in (*a*) base metal, (*b*) HAZ and (*c*) weld metal.

The occurrence of low-amplitude AE events and the high amplitude AE events have also been reported for some Al alloys (Hong 1995, Lee 1996), and the former is attributed to the dislocation movement while the latter, to the intergranular delamination cracking. Scruby (1981) has also documented that the AE signals due to the dislocation movement get to the maximum near the yield point and then gradually decrease. Referring to these observations, the AE peak around the yield point for the present cases is also assumed to be related to the onset of large amount of dislocation movements. On the other hand, it is highly probable that the second AE peak appearing in the HAZ and the weld metal might be attributed to the existence of martensite. To examine this assumption, the AE investigations have further been conducted on the

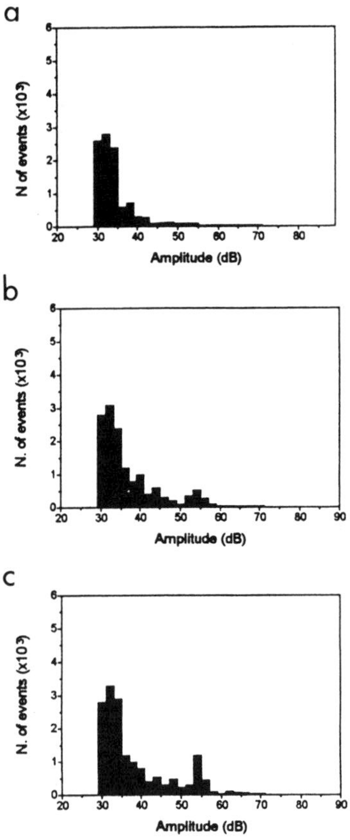

Figure 4. AE amplitude distribution measured for specimens in (*a*) base metal, (*b*) HAZ and (*c*) weld metal.

TMCP steel heat-treated to result in different amount of martensite.

The AE energy rate variation during the tensile tests is shown Figure 5 for the specimens with different martensite volume fractions.

It is found that all the heat-treated specimens exhibit a second peak of AE energy rate in the post-yielding stage, similar to the cases in the HAZ and the weld metal of the welded joints (Figure 3 *b-c*). The AE amplitude distribution has been analyzed for the heat-treated specimens and the results are given in Figure 6. As is indicated in this figure, in addition to the low-amplitude heavy AE activities occurring around the yield point, there are quite a lot of AE events occurring at amplitudes higher than 50 dB, characteristic of the second AE peak. The general trend is that the amplitude and the number of events both increase with increasing the volume fraction of

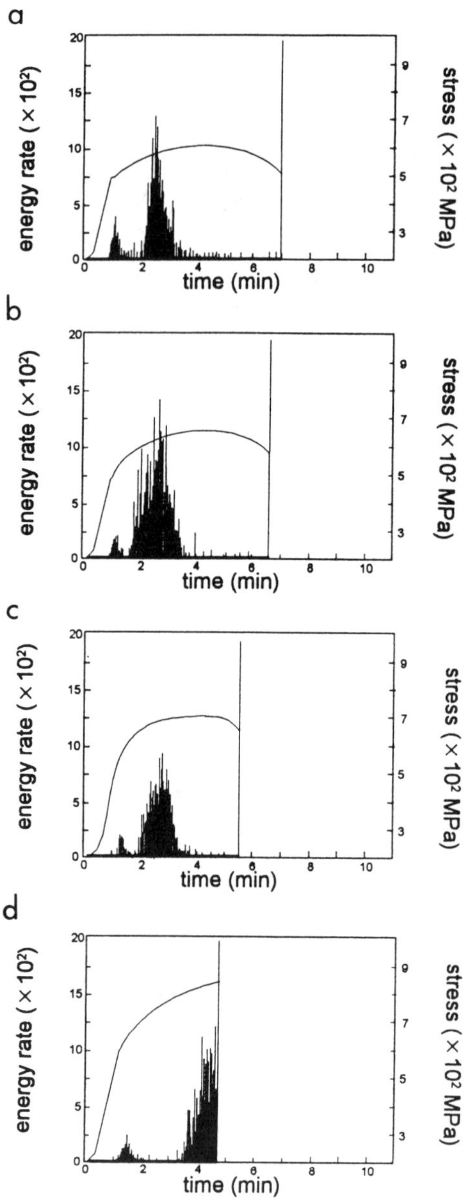

Figure 5. AE energy rate during tensile tests for specimens with different heat-treatments of (*a*) HT_1, (*b*) HT_2, (*c*) HT_3 and (*d*) HT_4.

the martensite for the second peak of the heat-treated specimens.

As an example, Figure 7 shows the SEM micrographs taken from the tensile specimens of the heat-treated steel. It is evidenced that microcracks

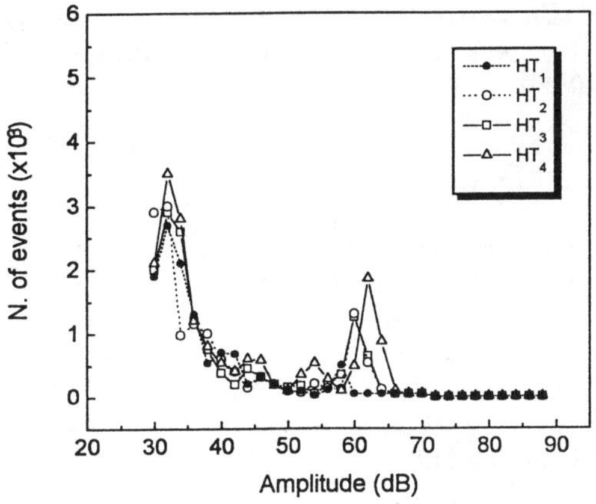

Figure 6. AE amplitude distribution for specimens with different martensite contents.

can be found both at the ferrite-martensite interface (Figure 7*a*) and in the interior of martensite (Figure 7*b*). Combining with the AE characteristics shown above, it is justified that the appearance of the second AE peak is attributed to the existence of the

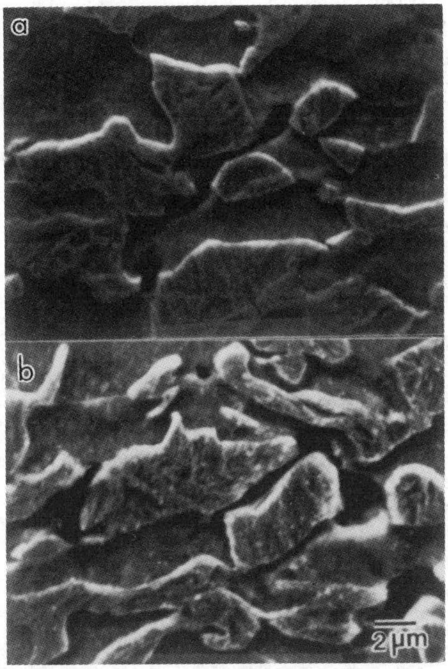

Figure 7. SEM micrographs showing cracks (*a*) at the ferrite-martensite interface and (*b*) inside the martensite for heat-treated specimens in tensile tests.

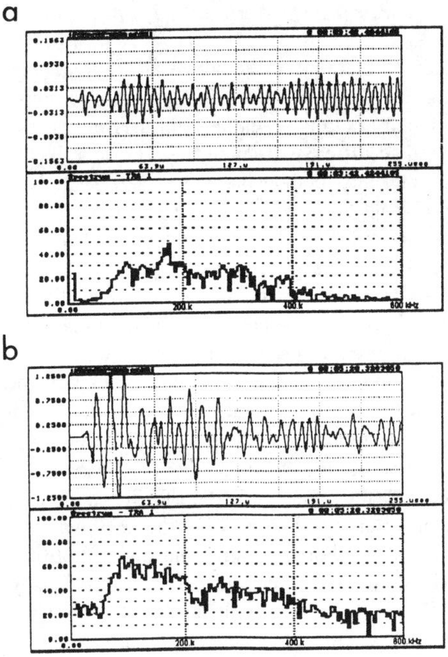

Figure 8. AE Waveform (upper part) and frequency spectrum (lower part) for (*a*) ferrite deformation and (*b*) ferrite-martensite interface cracking.

martensite in which the micro-cracking events occur due to the ferrite-martensite interfacial debonding and/or the martensite plate cracking itself.

The difference in the strength level between ferrite and martensite is assumed to be mainly responsible for the cracking *via* the ferrite-martensite interfacial debonding. Moreover, the residual stresses resulted from the inhomogeneous contraction during cooling and expansion during martensitic transformation may also contribute to the interfacial cracking. Therefore the interface is highly potential for the initiation of microcracks and hence, the AE from these regions possesses an evidently higher energy than in the case when just plastic deformation occurs. For the heat-treated specimens with various martensite contents, the first AE peak around the yield point is obviously lower than the second peak (Figure 5). This is probably due to the Kaiser effect (Non-destructive Testing Handbook, 1981) since, during the heat-treatment, the inhomogeneous contraction causes local plastic deformation with which the Kaiser effect may happen.

3.3 *Fast Fourier transformation (FFT) analysis*

Using FFT method to analyze the frequency spectrum for the AE waves is important because different deformation or damage processes usually correspond to different frequency spectrums. In the present study this method has been used to analyze the AE signals detected in the cases of the ferrite deformation and the ferrite-martensite interfacial cracking. The results are given in Figure 8 as typical examples. In the case of ferrite deformation (Figure 8*a*), the AE wave is characterized by long rise time (the time required to reach the maximum amplitude) and low amplitude. This is in accordance with the present fractographic observation that the base metal fractures in a typical ductile mode. The frequency corresponding to the peak is about 175 kHz. By contrast, in the case of the ferrite-martensite interfacial cracking (Figure 8*b*), the AE wave is characterized by short rise time and high amplitude (approximately 5 - 10 times that of the former case). This is also in agreement with the fact that a brittle fracture occurs with a high energy releasing rate in the steel with the martensitic microstructure. The frequency for the peak is about 110 kHz, obviously lower than in the case of ferrite deformation. This frequency shifting behavior is consistent with the well recognized fact that high-amplitude signals move to low frequency region.

4 CONCLUSIONS

1. For the base metal, most AE events are associated with the dislocation movement during the yielding and only one AE peak is found around the yield point. For the HAZ and the weld metal, in addition to the first AE peak around the yield point, a second peak appears in the post-yielding stage and is related to the microcracking events at the ferrite-martensite interface and within the martensite.

2. The AE behavior of the heat-treated steel with different martensite contents is also characterized by the existence of the second AE peak. The AE amplitude and the number of events both increase as the volume fraction of martensite increases. Cracks are found both at the ferrite-martensite interface and in the interior of martensite.

3. When the ferrite matrix deforms, the AE wave is characteristic of long rise time, low amplitude and the peak occurs at a relatively high frequency of 170 kHz. However, when the ferrite-martensite interfacial cracking occurs, the AE wave is characteristic of short rise time and high amplitude and the peak occurs at a relatively low frequency of 110 kHz.

5 ACKNOWLEDGEMENT

The present work is financially supported by Pohang Iron & Steel Co. Ltd. (POSCO). The authors are also grateful to Dr. H. D. Jeong (RIST) for his assistance in the welding.

6 REFERENCES

Bentley, P. G. and M. J. Beesley 1988. Acoustic emission measurements on PWR weld material with inserted defects using advanced instrumentation. *J. of Acoustic Emission* 7 (2): 59-79.

Chen, J. H., Y. Kikuta, T. Araki, M. Yoneda and Y. Matsuda 1984. Micro-fracture behavior induced by M-A constituent (island martensite) in simulated welding heat affected zone of HT80 high strength low alloy steel. *Acta Metall.* 32: 1779-1788.

DeArdo, A. J. 1995. *Proc. of Intl. Conf. on Microalloying*: 15-23. Pittsburgh: Iron and Steel Society.

Haze, T., S. Aihara and H. Mabuchi 1986. *Accelerated Cooling of Rolled Steel.* Warrendale: TMS.

Hong, K. J., H. D. Jeong and C. S. Lee 1995. Acoustic emission behavior during plastic deformation of Al-Li 8090 alloy. *Bull. of the Korean Inst. of Metals* 33 (6): 806-813.

Lee, K. A., H. D. Jeong and C. S. Lee 1996. The effects of microstructure and plate orientation the acoustic emission behavior in Al-Li 8090 alloy. *J. of the Korean Soc. for Non-destructive Testing* 16 (4): 215.

Leslie, W. C. 1981. *The Physical Metallurgy of Steels*. New York: McGraw-Hill.

Non-Destructive Testing Handbook (2nd Ed.) 1981. Vol. 5. Columbus: American Society for Non-Destructive Testing.

Scruby, C., H. Wadley and J. E. Sinclair 1981. The origin of acoustic emission during deformation of aluminum and aluminum-magnesium alloy. *Phil. Mag.* 44A: 249.

Scruby, C. B. and K. A. Stacey 1988. Acoustic emission measurements on PWR welded material with inserted defects. *J. of Acoustic Emission* 7 (2): 81-93.

Wadley, H. N. G., C. B. Scruby, P. Lane and J. A. Hudson. 1981. Influence of microstructure on acoustic emission during deformation and fracture of Fe-3.5Ni-0.21C steel. *Met. Sci.* 15: 514-524.

Webbor, T. J. C. and R. D. Rawlings 1981. Acoustic emission from structural steels and Fe-C alloy. *Met. Sci.* 15: 533-540.

Whittaker, J. M. and M. W. Richey 1987. The detection of hydrogen-assisted-crack formation in V-0.8%Ti alloy electron-beam weldments. *J. of Acoustic Emission* 6 (4): 257-260.

Damage and Failure of Interfaces, Rossmanith (ed.) © 1997 Balkema, Rotterdam, ISBN 90 5410 899 1

Local crack-tip strain behavior during fatigue crack initiation under mixed mode loading

A.Shimamoto
Department of Mechanical Engineering, Saitama Institute of Technology, Japan

E.Umezaki
Department of Mechanical Engineering, Nippon Institute of Technology, Japan

F.Nogata
Department of Mechanical Engineering, Himeji Institute of Technology, Japan

ABSTRACT: Generally fatigue cracks initiated in machine parts and structural members such as automobiles, aircrafts and ships have been investigated under opening-mode loadings. However fatigue crack initiation must be evaluated under mixed-mode loadings because actual fatigue crack are caused by combinations of tension, compression, in-plane shear and out-plane shear. In this study, local strain at the notch root during crack-initiation process and local strain at the crack tip during crack-propagation process are measured in real time in polycarbonate specimens with crack orientations of 0,15,30,45 and 60deg subjected to low-cycle fatigue tests under mixed-mode loadings by the fine-grid method. As a result, the linear damage accumulation law based on local-strain range is found to be effective for estimating fatigue-crack-initiation life under mixed-mode loadings.

1 INTRODUCTION

Polymers are widely used as machine parts and structural members such as automobiles, aircrafts and ships. As they are used under severer conditions, complex fracture phenomena, which are caused by interaction among circumstances, stresses and materials used, are found in them. The phenomena have been unsolved quantitatively until now. Therefore understanding of fracture phenomena characteristic of polymers is important for improving the safety and reliability of machines and structures as well as design, maintenance and life evaluation of those.

Although initiation and propagation of fatigue cracks have a very important significance for the strength to fatigue fracture which accounts for most of fracture in polymers, the process of fatigue fracture is not well known yet because it occurs in the small zone of crack tip. In addition, actual fatigue fractures are caused by combinations of tensile stress (mode I), in-plane shear stress (mode II) and out-plane shear stress (mode III) rather than uniaxial stress (mode I) only.

In order to understand the process of fatigue fractures under mixed-mode loadings, as the first step, local-strain behavior at the root of notch and the tip of crack in real time has to be known. Iida and Kobayashi (1969) have reported the crack-propagation rate in 7075-T6 plates under cyclic tensile and transverse shear loadings. Otsuka and Tohgo (1987) have investigated fatigue crack initiation and growth under mixed-mode loading in 2017-T3 and 7075-T6 plates. Shimamoto et al. (1992) have studied strain behavior near the fatigue crack tip in polycarbonate under mode-I loadings and its life evaluation. However, there have been not many studies made on the fatigue fracture of polymers under mixed-mode loadings.

In this study, local strain at the notch root during the process of crack initiation and local strain at the crack tip during the process of crack propagation were measured in real time in a polycarbonate subjected to low-cycle fatigue tests under mixed-mode loadings by the fine-grid method, and were discussed.

2 SPECIMENS AND EXPERIMENTAL METHOD

Specimens used were polycarbonate (Lexan 9030) commercially available. Dimensions of the specimens of 2mm thickness are given in Figure 1. A round notch with a depth of 25mm and a radius of 0.4mm was machined on one side. In order to continuously measure local strain at the notch root and the crack tip, fine-dot grids with 25.4μm pitch, 4μm diameter and 2μm depth were printed on the surface of the specimen, as shown

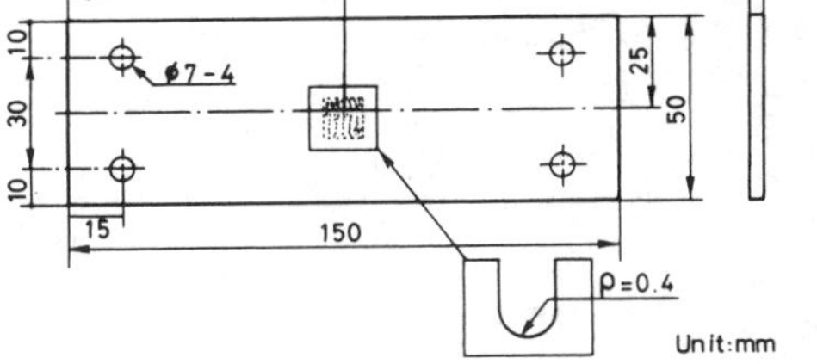

Figure 1. Shape and dimensions of specimen.

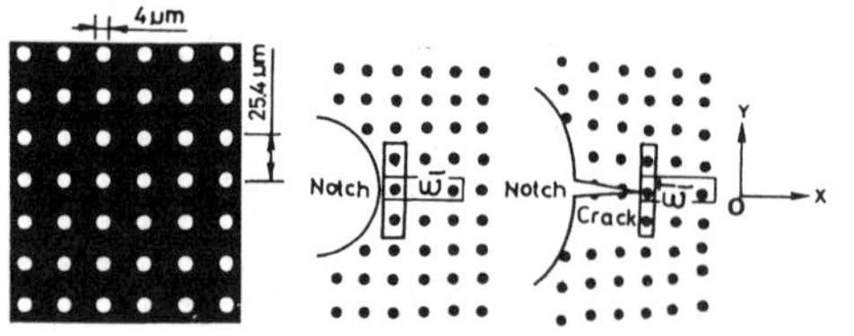

Figure 2. Fine dot grid and schematic view of local strain measurement area.

in Figure 2. The specimens were naturally dried keeping them in a thermostatic room with temperature of 20°C and humidity around 65% for a sufficiently long period of time before use. The fatigue tests were conducted in the same room.

A loading jig was used for fatigue tests under mixed-mode loadings, which is composed two sets of a crescent-shaped loading device with seven holes made at 15-degree intervals and bolts, as shown in Figure 3. The loading devices were made according to the concepts of loading jigs created by Richard and Benitz (1983), Otsuka and Tohgo (1987) and Arcan et al. (1978). It facilitated the attainment of a step by step transition from pure tensile stress to pure shear stress in cross section AB of such a specimen, adjusting the angle between the load axis of a fatigue test machine and loading jig. Five crack orientations of α=0,15,30,45 and 60deg were used.

Fatigue tests were carried out using a hydraulic servo controlled uniaxial-tension-compression-type fatigue machine of 9807N capacity with a digital servo controller under the conditions of constant load-amplitudes of σ_{max}=5.88 and 11.76MPa at a stress ratio $R=\sigma_{min}/\sigma_{max}$ of 0. Cyclic frequency was 0.02Hz with a sine waveform. In these fatigue tests, K-increasing method was used, which implies increasing ΔK with crack propagation. The length of fatigue crack was measured on one side of specimen by using a X-Y stage.

Local strain at the notch root during fatigue crack initiation, ε_l, and local strain at the propagating crack tip, ε_l^T, were measured in the X and Y directions which are parallel and perpendicular to the notch-depth direction, respectively, by the fine-grid method as shown in Figure 2. Photographs were taken of grids in the area of the notch root and the local zone of the crack tip using a camera equipped with automatic exposure and automatic-film-advancing functions, a relay lens, and an optical microscope of 100 magnification. Local strain, ε_l and ε_l^T, was obtained by directly measuring the deformation of grids on the negative with reference to the grids in the pretested state through an enlarging profile projector of 10 magnification. For strain calculation, the gauge length used was 76.2μm (three grid long). Local regions chosen for measuring the notch-root and crack-tip strain are shown in Figure 2.

The crack-initiation cycle, Nc, was determined by two methods. One was the confirmation of initiated small crack propagation (crack length=5-10μm) from the notch root by microscope observation, and the other used the inflection point on the curve of maximum local strain, $\varepsilon_{l(max)}$, at the first-line grid ahead of the crack initiated from the notch root versus number of cycles.

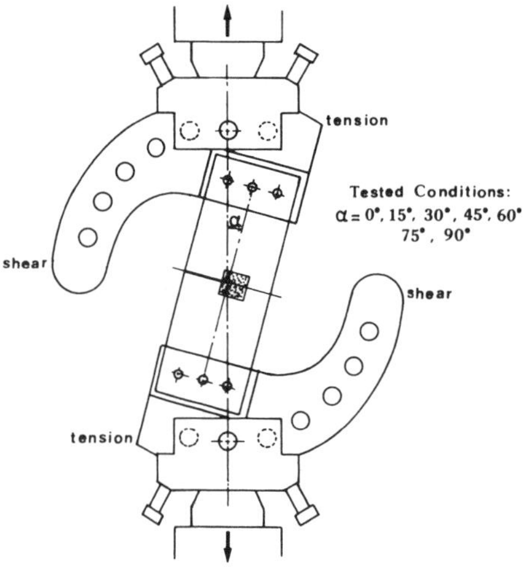

Figure 3. Constitution of loading jig.

3 RESULTS AND DISCUSSIONS

3.1 *Crack Initiation from Notch Root*

Figures 4 and 5 present the relations between crack length, a, including an initial notch depth and number of load cycles, N, and maximum local strain, ε_l, at the notch root and number of load cycles, N, during the process of crack initiation at σ_{max}=5.88MPa,

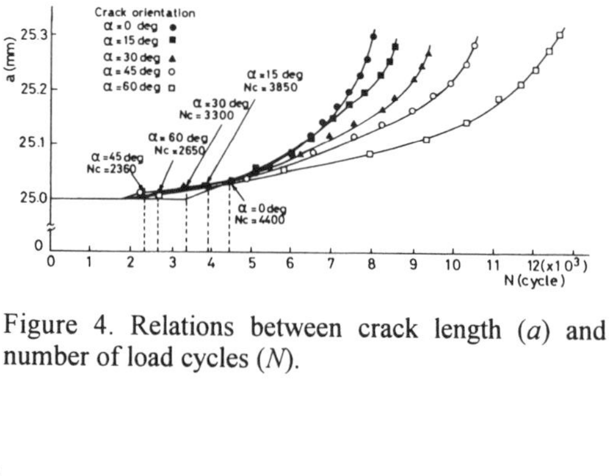

Figure 4. Relations between crack length (a) and number of load cycles (N).

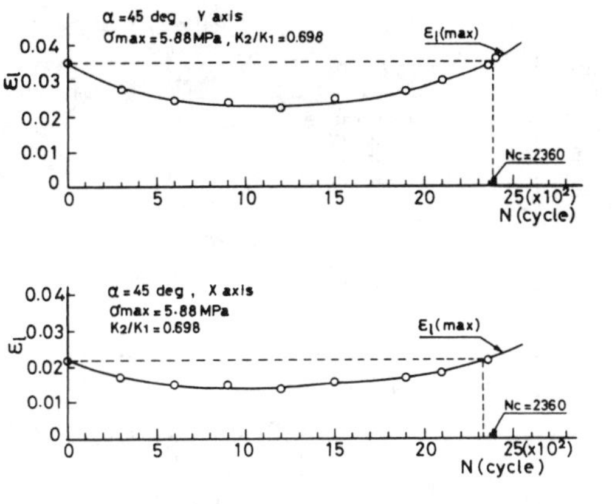

Figure 5. Relations between local strain at the notch root (ε_l) and number of load cycles *(N)*.

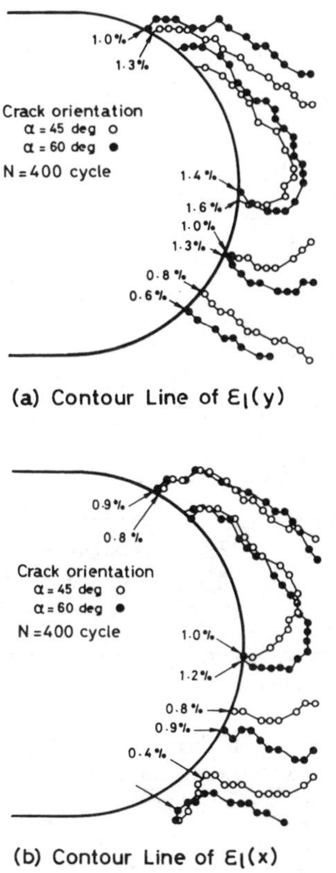

Figure 6. Examples of $\varepsilon_l(x)$, $\varepsilon_l(y)$ distribution near the notch tip.

respectively. ε_l was measured in the X and Y directions which are parallel and perpendicular to the notch, respectively. The value of fatigue-crack-initiation cycle, N_c, at both σ_{max}=5.88 and 11.76MPa was smallest at a crack orientation of α=45deg, and larger in the order, 60, 30, 15, 0deg. Increasing α resulted in decrease in the crack-propagation rate, *da/dN*. This implies decrease in mode I and increase in mode II with increasing α.

As shown in Figure 5, ε_l decreased at once after starting the fatigue test, became a constant after the decrease, and increased gradually with an increase in the number of load cycles. When ε_l regained its initial value, crack initiation occurred. It is estimated that the phenomena is caused by a softening or hardening of the material at the plastic zone near the notch root subjected to cyclic loading. ε_l in the X direction was lower than that in the Y direction, as shown in Figure 5. Increasing α caused a little increase in ε_l in the X direction and decrease in that in the Y direction.

Figure 6 shows examples of the contour maps of maximum local stain at the notch root in the X and Y directions, $\varepsilon_{l(x)}$ and $\varepsilon_{l(y)}$, at N=400 cycles for α=45 and 60deg. Increasing α greatly affects the distributions of ε_l at the notch root, and the value of ε_l in the direction of the line drawn perpendicular to the load axis and through the center of a half circular which partly shapes the round notch is higher than those in the other directions. Consequently the plastic zone is found to growth in the direction where a higher value of ε_l exists.

3.2 *Relations between Mean Local-Strain Range and Crack-Initiation Cycle, and Crack Orientation*

Fatigue crack initiation has been found to be controlled by mean local-strain range in the Y direction, $\Delta\bar{\varepsilon}_l$, proposed as follows (Shimada & Furuya 1987).

$$\Delta\bar{\varepsilon}_l = \frac{1}{N_c}\int_0^{N_c} \Delta\varepsilon_l \, dN \qquad (1)$$

The relations between mean local-strain range in the Y direction, $\Delta\bar{\varepsilon}_l$, and crack-initiation cycle, N_c, as a function of α are shown in Figures 7(a) and (b). Each data point falls on a line of slope -0.38 in a log-log coordinate graph, regardless of the values of α. Therefore, the linear cumulative damage law based on local-strain range (Shimada & Furuya 1987) is confirmed for fatigue crack initiation in polycarbonate under mixed-mode loading. This suggests that fatigue-crack-initiation life is estimated from local-strain ranges measured at several cyclic numbers.

Figure 8 shows the relation between mean local-strain range in the Y direction, $\Delta\bar{\varepsilon}_l$, and crack orientation, α at σ_{max}=5.88MPa. $\Delta\bar{\varepsilon}_l$ linearly increases with increase in α to α=45deg and decreases at α=60deg. This fact shows that $\Delta\bar{\varepsilon}_l$ is affected by α.

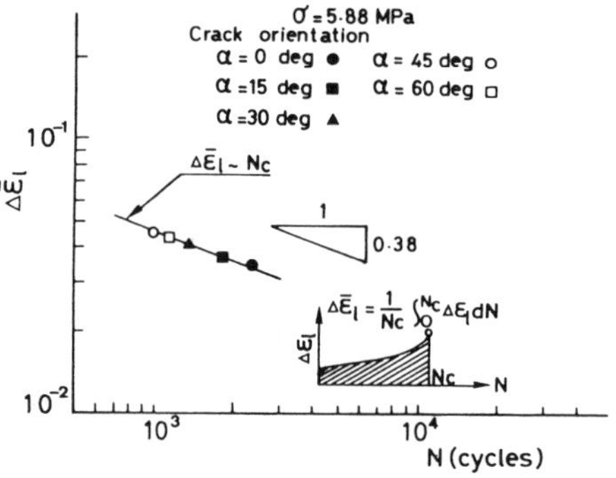

(a) σ_{max}=5.88MPa

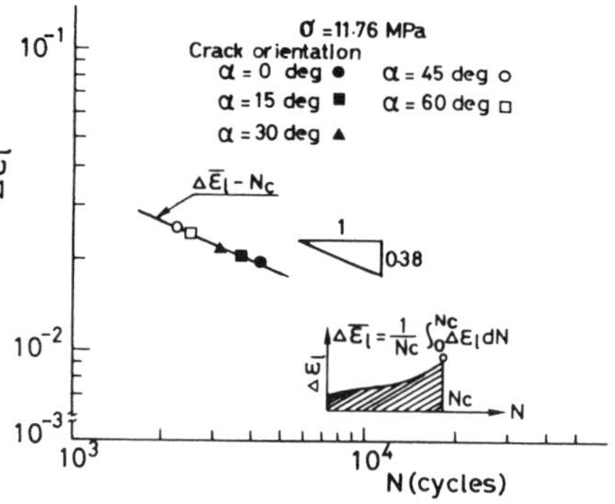

(b) σ_{max}=11.76MPa

Figure 7. Relations between mean local strain range ($\Delta \bar{\varepsilon}_l$) and number of crack initiation.

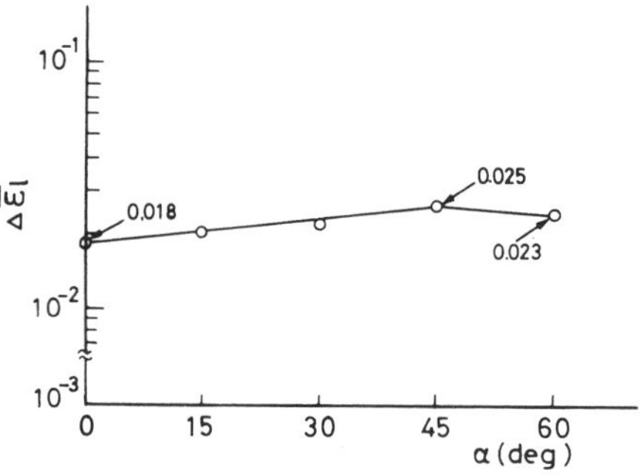

Figure 8. Relation between mean local strain range in the Y direction ($\Delta \bar{\varepsilon}_l$) and crack orientation (α).

3.3 *Relation of Crack-Propagation Rate to Stress-Intensity Factor and Local-Strain Behavior at the Crack Tip*

Figures 9(a) and (b) show the relations between crack-propagation rate, *da/dN*, and stress-intensity factor, K_I.

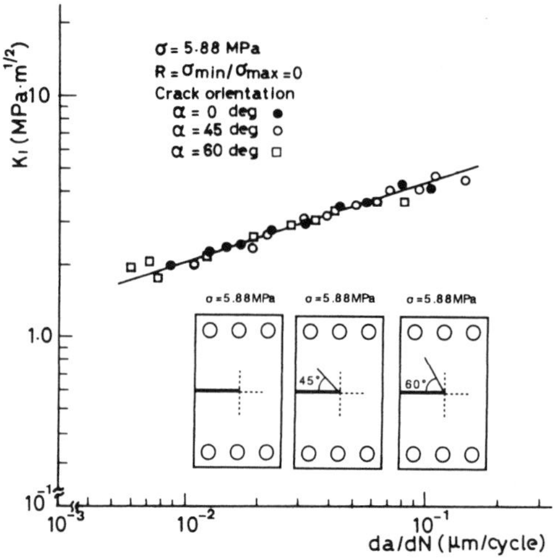

(a) σ_{max}=5.88MPa

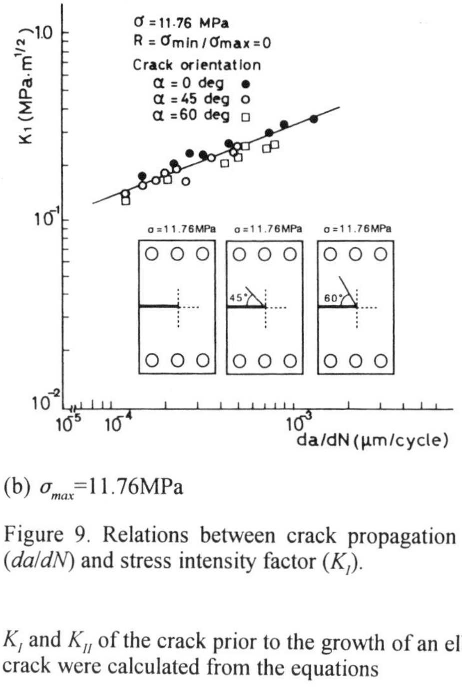

(b) σ_{max}=11.76MPa

Figure 9. Relations between crack propagation rate (*da/dN*) and stress intensity factor (K_I).

K_I and K_{II} of the crack prior to the growth of an elbow crack were calculated from the equations

$$K_I = \sqrt{\pi}\, \sigma_{max} \sqrt{a \sec\alpha}\, \cos^2\alpha$$
$$K_{II} = \sqrt{\pi}\, \sigma_{max} \sqrt{a \sec\alpha}\, \sin\alpha \cos\alpha \qquad (2)$$

From Figure 9, K_I is found to increase with increase in *da/dN*. These tendencies were consistent with those obtained by Iida and Kobayashi (1969). *da/dN* increases in the presence of K_{II} as shown in Tables 1 and 2.

Table 1. da/dN at K_I=3.147MPa·m$^{1/2}$.

Crack orientation α(deg)	K_{II}/K_I	da/dN (μm/cycle)
45	0.698	0.0391
60	0.888	0.0558

Table 2. da/dN at K_I=0.208MPa·m$^{1/2}$.

Crack orientation α(deg)	K_{II}/K_I	da/dN (μm/cycle)
45	0.647	0.233
60	0.841	0.691

Figures 10(a) and (b) show the relations between crack-propagation rate, da/dN, and local-strain range at the crack tip, $\Delta\varepsilon_l^T(=\varepsilon_{l\,(max)}^T-\varepsilon_{l\,(min)}^T)$. As shown in Figure 10, each data point falls near a line whose slope is 3.9 in a log-log coordinate graph. The data do not show a dependence on α. This fact shows that da/dN is controlled by $\Delta\varepsilon_l^T$ over the wide range of da/dN, and effectiveness of $\Delta\varepsilon_l^T$ in the evaluation of crack-propagation rate.

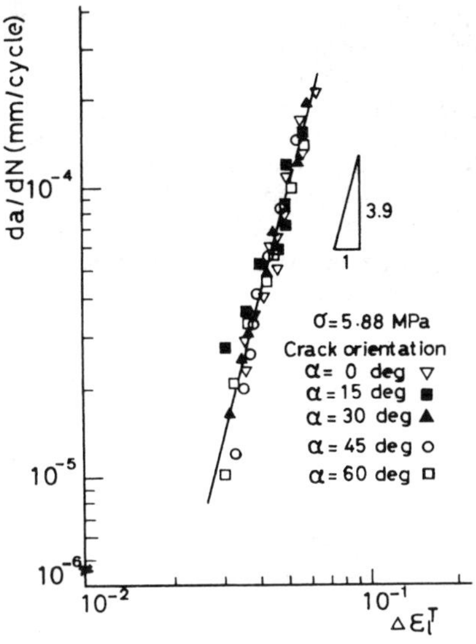

(a) σ_{max}=5.88MPa

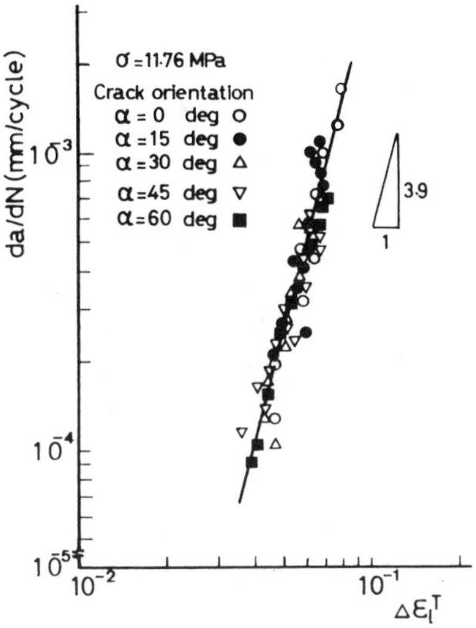

(b) σ_{max}=11.76MPa

Figure 10. Relations between crack propagation rate (da/dN) and strain range at the crack tip ($\Delta\varepsilon_l^T$).

4 CONCLUSIONS

Fatigue crack initiation from the notch root and fatigue crack propagation in polycarbonate, a polymer, have been investigated under mixed-mode loading using the real-time fine-grid method. The findings are as follows:
(1) The fatigue crack-initiation cycle, N_c, decreased with increasing crack orientation, α to α=45deg and inversely increased more than α=45deg.
(2) The relation between mean local-strain range, $\Delta\varepsilon_l$, and crack-initiation cycle number, N_c, is expressed by a line whose slope is -0.38 in a logarithmic graph. In addition, local-strain damage accumulation law in the fatigue crack-initiation stage of polymers under mixed-mode loading is effective, too.
(3) The crack-propagation rate, da/dN, increases in the presence of K_{II} whose value is smaller than that of K_I.
(4) The relation between local-strain range at the crack tip, $\Delta\varepsilon_l^T$ and crack-propagation rate, da/dN, is expressed by a line whose slope is 3.9 in a logarithmic graph without a dependence on crack orientation, α.

REFERENCES

Arcan, M., Z. Hashin & A. Voloshin 1978. A method to produce uniform plane stress state with applications to fiber-reinforced materials. *Exp. Mech.* 18(4):141-146.

Iida, S. & A. S. Kobayashi 1969. Crack-propagation rate in 7075-T6 plates under cyclic tensile and transverse shear loadings. *Trans. ASME*, Ser.D 91: 764-769.

Otsuka, A. & K. Tohgo 1987. Fatigue crack initiation and growth under mixed mode loading in aluminum alloys 2017-T3 and 7075-T6. *Eng. Fract. Mech.* 28(5/6):721-732.

Richard, H. A. & K. Benitz 1983. A loading device for the creation of mixed mode in fracture mechanics. *Int. J. Fract.* 22:R55-R58.
Shimada, H. & Y. Furuya 1987. Local crack-tip strain concept for fatigue crack initiation and propagation. *J. Eng. Mater. Technol.* 109(4):101-105.
Shimamoto, A., E. Umezaki, F. Nogata & S. Takahashi 1992. Strain behavior near fatigue crack tip in polycarbonate and its life evaluation (in Japanese). *Trans. Jpn. Soc. Mech. Eng.* 58(555),A :2023-2027.

List of participants

Dr.C. Adam
Department of Civil Engineering
Vienna University of Technology
Wiedner Hauptstr. 8-10/E201
A-1040 Vienna
Austria
E-mail: ca@allmech9.tuwien.ac.at
Fax: 43 1 5876093

Dipl.-Ing.H. Assler
German Aerospace Research Establishment (DLR)
Institute of Materials Research
Linder Höhe
D-51147 Köln
Germany
E-mail: herwig.assler@dlr.de
Fax: 49 02203 696480

Prof.J. Ballmann
Lehr- und Forschungsgebiet für Mechanik
der RWTH Aachen
Templergraben 64
D-52062 Aachen
Germany
E-mail: ballmann@lufmech.rwth-aachen.de
Fax: 49 241 8888 126

Dr.M. Benzeggagh
Univ. de Technologie de Compiegne LG2mS
UPRES 6066 associee au CNRS
BP 20 549
60206 Compiegne cedex
France
E-mail: Malk.Benzeggagh@utc.fr
Fax: 33 44 20 48 13

Prof.J. Botsis
Department of Mechanical Engineering
Swiss Federal Inst. of Technology, LMAF-EPFL
CH-1015 Lausanne
Switzerland
E-mail: john.botsis@epfl.ch
Fax: 41 21 693 2999

Prof.F.-G. Buchholz
Institute of Applied Mechanics
University of Paderborn
Pohlweg 47-49
D-33098 Paderborn
Germany
E-mail: buchholz@fam.uni-paderborn.de
Fax: 49 5251 603719

Prof.C.K. Chao
Department of Mechanical Engineering
National Taiwan Institute of Technology
43, Keelung Road, Section 4
Taipei, Taiwan 106
Republic of China
E-mail: ckchao@mail.ntit.edu.tw

Dr.C.M. Chimani
Institute of Light Weight Structures (E317)
Vienna University of Technology
Gusshausstrasse 27-29
A-1040 Wien
Austria
Fax: 43 1 505 44 68

Dr.T.J. Chotard
Univ. de Technologie de Compiegne LG2mS
UPRES 6066 associee au CNRS
BP 20 549
60206 Compiegne cedex
France
E-mail: Thiery.chotard@mx.univ-compiegne.fr
Fax: 33 44 20 48 13

Dr.T.-J. Chuang
Ceramics Division
National Institute of Standards & Technology
Gaithersburg, Maryland 20899
USA
E-mail: chuang@nist.gov
Fax: 1 301 990 8729

Prof.A. Corigliano
Department of Structural Engineering
Politecnico di Milano
Piazza L. da Vinci 32
20133 Milano
Italy
E-mail: coriglia@hp710.stru.polimi.it
Fax: 39 22 23994220

Prof.A. Crocker
Department of Physics
University of Surrey
Guildford
Surrey GU2 5XH
UK
E-mail: A.Crocker@surrey.ac.uk
Fax: 44 014 83 259 501

Prof.S.K. Datta
Department of Mechanical Engineering & CIRES
University of Colorado at Boulder
Engineering Center, Campus Box 427
Boulder, CO 80309-0427
USA
E-mail: dattas@spot.Colorado.edu
Fax: 1 303 492 3498

Mr.P. Divoux
Laboratoire 3S
Universite J.Fourier
BP 53
38041 Grenoble Cedex 9
France
E-mail: patrick@divoux@cnh.de.edf.fr
Fax: 33 79 60 62 98

Prof.F. Erdogan
Department of Mechanical Engineering and Mechanics
Lehigh University
19 Memorial Drive West
Bethlehem, PA 18015-3085
USA
E-mail: fe00@lehigh.edu
Fax: 1 610 758 4099

Mr.L. Ferrara
Department of Structural Engineering
Politecnico di Milano
Piazza L. da Vinci 32
20133 Milano
Italy
E-mail: ferrara@ester.stru.polimi.it
Fax: 39-2-23994220

Dr.P.E.J. Flewitt
Department of Physics
University of Surrey
Guildford
Surrey GU2 5XH
UK
Fax: 44 014 83 259 501

Dr.L. Gornet
Lab. Mec. et Mat., LMS
Division Mecanique des Structures
Ecole Centrale de Nantes
1 rue de la Noe, BP 92101
44321 Nantes cedex 03
France
E-mail: Laurent.gornet@ec-nantes..fr
Fax: 33 2 40372573

Prof.N. Hasebe
Department of Civil Engineering
Nagoya Institute of Technology
Gakisocho, Showaku
Nagoya 466
Japan
E-mail: hasebe@kozo4.ace.nitech.ac.jp
Fax: 81 052 8735 5482

Prof.C. Hwu
Inst. of Aeronautics & Astronautics
National Cheng Kung University
Tainan
Taiwan 70101
Republic of China
E-mail: chwu@mail.ncku.edu.tw
Fax: 06 238 99 40

Dr.S.G. Ivanov
Mechanics of Composite Mat. & Struct. Dept.
Perm State Technical University
Komsomolsky Ave. 29a
614600 Perm
Russia
E-mail: ivmel@pi.ccl.ru
Fax: +7 3422-33-11-47

Dr.Ph. Jodin
Laboratoire de Fiabilite Mecanique
Universite de Metz - ENIM
Ile du Saulcy
F-57045 Metz Cedex 01
France
E-mail: jodin@lfm.univ-metz.fr
Fax: 33 03 87 31 53 03

Dr.A. Kaczynski
Institute of Mathematics
Warsaw University of Technology
Plac Politechniki 1
00-661 Warsaw
Poland

Mr.K. Keller
Institut für Statik und Dynamik der Luft- und Raumfahrtkonstruktionen
Universität Stuttgart
Pfaffenwaldring 27
D-70550 Stuttgart
Germany
E-mail: keller@isd.uni-stuttgart.de
Fax: 49 711 685 3706

Prof.K.S. Kim
Department of Mechanical Engineering
Pohang University of Science and Technology
San 31, Hyoja-dong
Pohang 790-784
Korea
E-mail: illini@vision.postech.ac.kr
Fax: 82-562-279-5899

Prof.C.S. Lee
Department of Materials Science and Engineering
Pohang University of Science and Technology
San 31, Hyoja-dong
Pohang 790-784
Korea
E-mail: cslee@postech.ac.kr
Fax: 82-562-279-2399

Prof.D. Leguillon
Lab. de Modelisation en Mecanique, CNRS-URA 229
Universite P. et M.Curie, t. 66, case 162
4 place Jussieu
75252 Paris Cedex 05
France
E-mail: dol@ccr.jussieu.fr
Fax: 33 44 27 52 59

Dr.P. Levasseur
ONERA
Structural Department
29 Ave. de la Division Leclerc
92322 Chatillon
France
E-mail: levas@onera.fr
Fax: 33 46 73 41 58

Dr.D. Leveque
Lab. de Mec. et Tech., ENS, CNRS
Universite Pierre et Marie Curie
61 ave du President Wilson
94235 Cachan Cedex
France
Fax: 33 01 47 40 27 85

Dr.G.R. Leverant
Southwest Research Institute
6220 Culebra Road
PO Drawer 28510
San Antonio, TX 78238
USA
E-mail: GLeverant@swri.edu
Fax: 1 210 522 6965

Prof.T. Lewinski
Institute of Structural Mechanics
Warsaw University of Technology
Al. Armii Ludowej 16
PL-00637 Warsaw
Poland
E-mail: TOLEW@omk.il.pw.edu.pl
Fax: 48 22 256 985

Dr.S. Li
Department of Mechanical Engineering
University of Manchester-UMIST
Sackville St., P.O.Box 88
Manchester M60 1QD
UK
E-mail: Shuguang.Li@umist.ac.uk
Fax: 44 161 200 3849

Dr.Y.L. Li
IMF II
Forschungszentrum Karlsruhe GmbH
Postfach 36 40
D-76021 Karlsruhe
Germany
E-mail: yulanli@hdiriscg.fzk.de
Fax: 49 7247 822347

Prof.V.C. Li
Dept. of Civil and Environmental Engineering
University of Michigan
2326 G.G.Brown Building
Ann Arbor, MI 48109-2125
USA
E-mail: vcli@engin.umich.edu
Fax: 1 313 764 4292

Dr.C. Marotzke
BAM
Fiber Reinforced Composites, Lab. YI.22
Unter den Eichen 87
D-12205 Berlin
Germany
E-mail: christian.marotzke@bam-berlin.de
Fax: 49 30 8104 1627

Prof.S.J. Matysiak
Inst. of Hydrogeology & Eng. Geology
University of Warsaw
Al.Zwirki i Wigury 93
02-089 Warsaw
Poland
E-mail: matysiak@albit.geo.uw.edu.pl
Fax: 48 22 22 02 48

Prof.H. Mihashi
Department of Architecture and Building Engineering
Tohoku University
Sendai 980-77
Japan
E-mail: mihashi@timos.str.archi.tohoku.ac.jp
Fax: 81 22 217 7886

Dr.G. Mishuris
Department of Mathematics
Technical University of Rzeszow
W.Pola 2
35-959 Rzeszow
Poland
E-mail: miszuris@prz.rzeszow.pl

Prof.L. Myer
Lawrence Berkeley National Laboratory
Earth Sciences Division, 90-1116
One Cyclotron Road
Berkeley, CA 94720
USA
E-mail: Larry_Myer@macmail.lbl.gov
Fax: 1 510 4865686

Dr.K.-F. Nilsson
Aeronautical Research Institute of Sweden
Structures Department
P.O.Box 11021
S-16111 Bromma
Sweden
E-mail: nnk@ffa.se
Fax: 46 8 258919

Prof.T. Nishioka
Department of Ocean Mechanical Engineering
Kobe University of Mercantile Marine
5-1-1 Fukae Minamimachi
Higashinada-ku 658
Japan
E-mail: nishioka@cc.kshosen.ac.jp
Fax: 81-78-431-6365

Mr.F. Ohmenhäuser
Institut für Statik und Dynamik der Luft- und Raumfahrtkonstruktionen
Universität Stuttgart
Pfaffenwaldring 27
D-70550 Stuttgart
Germany
E-mail: ohmenhaeuser@isd.uni-stuttgart.de
Fax: 49 711 685 3706

Prof.M. Okazaki
Department of Mechanical Engineering
Nagaoka University of Technology
Tomioka
Nagaoka 940-21
Japan
E-mail: okazaki@mech.nagaokaut.ac.jo
Fax: 81 258 47 9770

Prof.S.L. Phoenix
Department of Theoretical & Applied Mechanics
Cornell University
321 Thurston Hall
Ithaca, NY 14853
USA
E-mail: slp6@cornell.edu
Fax: 1 607 255 2011

Mrs.M.Pini
Viale Monte Ceneri 78
20155 Milan
Italy
E-mail: pini@rachele.stru.polimi.it
Fax: 39 2 23 99 43 69

Dr.R.M. Pradieilles-Duval
Laboratory of Tribology
CREA/PS
16 Bis Avenue Prieur de la Cote d'Or
94114 Arcueil Cedex
France
E-mail: rachel@athena.polytechnique.fr

Dr.V. Quaglini
Dipartimento di Ing. Strutturale
Politecnico di Milano
Piazza Leonardo da Vinci 32
20133 Milano
Italy
Fax: 39 2 23994243

Prof.D. Rouby
Groupe d'Etudes de Metallurgie Physique
GEMPPM URA CNRS 341, Batiment 502
Institut National des Sci. Appliquees de Lyon
69621 Villeurbanne Cedex
France
E-mail: rouby@gemppm.insa-lyon.fr
Fax: 33 472 43 85 28

Dr.M. Scherzer
TU Chemnitz-Zwickau
Fakultät für Mathematik
SFB 393
D-09107 Chemnitz
Germany
E-mail: scherzer@mathematik.tu-chemnitz.de
Fax: 49 0371 5312657

Prof.A. Shimamoto
Saitama Institute of Technology
1690 Fusiji, Okabe
Ohsato
Saitama 369-02
Japan
Fax: 81 485 85 6717

Prof.H.S. Shin
Department of Mechanical Engineering
Andong National University
388 Songchun-dong, Andong
Kyungbuk 760-740
Korea
E-mail: hsshin@anu.andong.ac.kr
Fax: 82 571 841 1630

Dr.R.P. Singh
Graduate Aeronautical Laboratories
California Institute of Technology
M.C.105-50
Pasadena, CA 91125
USA
E-mail: raman@athens.caltech.edu
Fax: 1 818 449 2677

Mrs.G. Smith
Department of Physics
University of Surrey
Guildford
Surrey GU2 5XH
UK
E-mail: phs1gs@surrey.ac.uk
Fax: 44 014 83 259 501

Dr.C. Stolz
C.N.R.S.-U.R.A. 317
Laboratoire de Mecanique des Solides
Ecole Polytechnique
91128 Palaiseau Cedex
France
E-mail: stolz@athena.polytechnique.fr
Fax: 33 1 69 33 30 26

Dr.T. Suhara
Najima 4-52-25
Higashiku
Fukuoka 813
Japan
Fax: 81 92 681 1768

Prof.T.C.T. Ting
Department of Civil & Materials Engineering
University of Illinois at Chicago
842 West Taylor Str. (M/C246)
Chicago, IL 60607-7023
USA
E-mail: tting@uic.edu
Fax: 1 312 996 2426

Mr.W.B. Tomlin
22 Venter St., Parkdene
Boksburg, 1459
Gauteng
South Africa
E-mail: tomlin@egoli.min.wits.ac.za

Prof.E. Umezaki
Department of Mechanical Engineering
Nippon Institute of Technology
4-1 Gakuendai, Miyashiro
Saitama 345
Japan
E-mail: umezaki@nit.ac.jp
Fax: 81 480 33 7645

Prof.I. Verpoest
Department of Metallurgy & Materials Engineering
Katholieke Universiteit Leueven
De Croylaan 2
3001 Leuven
Belgium
E-mail: Ignaas.Verpoest@mtm.kuleuven.ac.be
Fax: 32 16 32 19 90

Mr.H. Wang
Institute of Applied Mechanics
University of Paderborn
Pohlweg 47-49
D-33098 Paderborn
Germany
E-mail: wang@fam.uni-paderborn.de
Fax: 49 5251 603719

Dr.S. Weihe
Institute for Statics and Dynamics of Aerospace Structures (ISD)
University of Stuttgart
Pfaffenwaldring 27
D-70550 Stuttgart
Germany
E-mail: weihe@isd.uni-stuttgart.de
Fax: 49 711 685 3706

Prof.K.W. White
College of Engineering
University of Houston
Houston, TX 77204-4792
USA
E-mail: KWWhite@UH.EDU
Fax: 1 713 743-4527

Dr.J. Zhang
Laboratorium für Technische Mechanik
Universität Paderborn
Pohlweg 47-49
33098 Paderborn
Germany
E-mail: jzhan1@ltm.uni-paderborn.de
Fax: 49-5251-603483

Prof.S.S. Lee
Yoo-sung, P.O.Box 148
Reactor Mechanical Engineering Group
Korea Power Engineering Company, Inc.
Taejon, 305-600
South Korea
Fax: 82 42 863 4862

Author index